STUDENT STUDY GUIDE
FOR
BIOLOGY

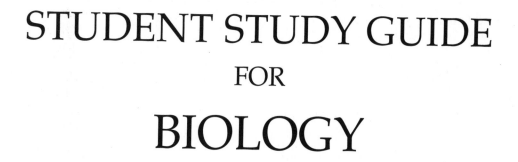

CAMPBELL • REECE

MARTHA R. TAYLOR
Cornell University

SEVENTH EDITION

PEARSON

Benjamin
Cummings

San Francisco Boston New York
Capetown Hong Kong London Madrid Mexico City
Montreal Munich Paris Singapore Sidney Tokyo Toronto

Editor-in-Chief: Beth Wilbur
Editorial Project Manager: Ginnie Simione Jutson
Marketing Manager: Jeff Hester
Publishing Assistant: Julia Khait
Production Supervisor: Vivian McDougal
Production Service: The GTS Companies: Griselda Gonzales, Sandie Sigrist
Text Designer: Brad Greene
Cover Designer: Mark Ong, Stacy Wong

ISBN 0-8053-7155-9

PEARSON
Benjamin Cummings

7 8 9 10 – BRR – 09 08 07
www.aw-bc.com

Contents

Preface

Another name for the *Student Study Guide* for Campbell/Reece, *Biology*, Seventh Edition, could be "A Student Structuring Guide." The purpose of this guide is to help you to structure and organize your developing knowledge of biology and to create your own personal understanding of the topics covered in the text.

Biology is a visual science, and this *Student Study Guide* includes many diagrams taken from *Biology*, Seventh Edition, to help you see and understand the relationships among the parts of structures and processes. Interactive Questions, which are interspersed throughout each chapter's narrative, are designed to help you develop an active learning approach to your study of biology. The numerous multiple choice questions in each chapter will help you test your growing understanding of biological facts and concepts. (See the end of this Preface for tips on taking objective tests.) The Word Roots section helps you learn the large new vocabulary that is a necessary part of a biology course.

The seven divisions of each study guide chapter are as follows:

- The listing of the *Key Concepts* from the text keeps you focused on the major themes of the chapter.
- The *Framework* identifies the overall picture; it provides a conceptual framework into which the chapter information fits.
- The *Chapter Review* is a condensation of each textbook chapter, with page references given for each concept heading within the chapter. All text bold terms are also shown in boldface in this section. Interspersed in this summary are **Interactive Questions** that help you stop and synthesize the material just covered. You are asked to complete tables, label diagrams, respond to short answer or essay questions, and complete or construct concept maps.
- The *Word Roots* section presents the derivation of key biological prefixes, suffixes, and word roots. Examples of bold-faced terms from the chapter are then defined. Breaking a complicated term down to identifiable components will help you to recognize and learn many new biological terms.

- The *Structure Your Knowledge* section directs you to organize and relate the main concepts of the chapter. It helps you to piece together the key ideas into a bigger picture.
- In the *Test Your Knowledge* section, you are provided with objective questions to test your understanding. The multiple choice questions presented in each chapter ask you to choose the best answer. Some answers may be partly correct; almost all choices have been written to test your ability to think and discriminate among alternatives. Make sure you understand why the other choices are incorrect as well as why the correct answer is correct.
- Suggested answers to the Interactive Questions, the Structure Your Knowledge sections, and the Test Your Knowledge questions are provided in the *Answer Section* at the end of the book.

Using Concept Maps: What are the **concept maps** that appear throughout this study guide? A concept map is a diagram that shows how ideas are organized and related. The *structure* of a concept map is a hierarchically organized cluster of concepts, enclosed in boxes and connected with labeled lines, that explicitly shows the relationships among the concepts. The *function* of a concept map is to help you structure your understanding of a topic and create meaning. The *value* of a concept map is in the thinking and organizing required to create a map.

Developing a concept map for a group of concepts requires you to evaluate the relative importance of the concepts (Which are most inclusive and important? Which are less important and subordinate to other concepts?), arrange the concepts in a meaningful cluster, and draw connections between them that help you make their meanings explicit.

This book uses concept maps in several ways. A map of a chapter may be presented in the Framework section to show the organization of the key concepts in that chapter. An Interactive Question may provide a skeleton concept map, with some concepts provided and boxes for you to complete. This technique is intended to help you become more familiar with concept maps and to illustrate one possible approach to organizing the concepts of a particular section.

You will also be asked to develop your own concept maps on certain subsets of ideas. In these cases, the Answer Section will present a suggested concept map. A concept map is an individual picture of your understanding at the time you make the map. As your understanding of an area develops, your concept map will evolve—sometimes becoming more complex and interrelated, sometimes becoming simplified and streamlined. Do not look to the Answer Section for the "right" concept map. After you have organized your own thoughts, look at the answer map to make sure you have included the key concepts (although you may have added more), to check that the connections you have made are reasonable, and perhaps to see another way to organize the information.

Tips for Using this Study Guide: This guide certainly is not a replacement for your textbook or biology class. But it should help support and even streamline your learning process. Some students read the assigned text chapters before lecture and then the study guide chapters after class to reinforce and review. Others reverse that order, skimming the study guide chapter before lecture and then carefully dissecting the text after listening to the professor's presentation. This study guide is especially helpful in preparation for tests. With the text open beside it so you can refer to the essential textbook diagrams, reread the relevant study guide chapters for a quick review. Return to the text for a more complete description of any sections that you don't fully understand.

Because this book is intended to help you learn, most incorrect choices given for multiple choice questions are written to be educational: to review other concepts covered in the chapter, to help you distinguish between closely related ideas, to point out common misconceptions of a particular concept. Get your money's worth out of this book. Even when you can quickly identify the correct answer to a question, take the time to read the other choices to see what they can teach you. Page back through the Chapter Summary section and the textbook to review material that you have not yet fully understood. Take your time with the Test Your Knowledge section—don't rush through it right before an exam. Treat it as an important learning opportunity. And use it to practice good test-taking skills.

Tips for Taking Multiple Choice Tests: Don't do all of your thinking in your head. Write in the margins and blank spaces of your test. Interact with each question. Read the stem of the question carefully, underlining or even using a highlighter to identify the key concept. Read each answer slowly. Cross out the ones you know are wrong. Circle the key idea that you think identifies the correct answer. Now read the question and the answer you chose together, making sure your choice really does answer what is asked. If you aren't sure about a question, try rereading the question and each choice individually. Draw yourself diagrams and pictures. Write down what you do know, and it may jog your memory. If you still are not sure, mark the question to come back to later. As you work through related questions, you may find information that helps you figure out that question. And remember, there is no substitute for good preparation, proper rest and nutrition, and a positive attitude.

Biology is a fascinating, broad, and exciting subject. Campbell/Reece, *Biology* is filled with terminology and facts organized in a manner that will help you build a conceptual framework of the major themes of modern biology. This *Student Study Guide* is intended to help you learn and recall information and, most importantly, to encourage and guide you as you develop your own understanding of and appreciation for biology.

Acknowledgement: A special word of thanks to Scott T. Meissner, PhD, for sharing his ideas, and for his overall assistance in preparing this Seventh Edition of the *Student Study Guide.*

This Student Study Guide is dedicated to Neil A. Campbell, my friend and colleague through many years, and many editions of this study guide.

Martha R. Taylor
Cornell University

Chapter 1
Exploring Life

Framework

This chapter outlines the broad scope of biology, describes themes that unify the study of life, and examines the scientific construction of biological knowledge. A course in biology is neither a vocabulary course nor a classification exercise for the diverse forms of life. Biology is a collection of facts and concepts structured within theories and organizing principles. Recognizing the common themes within biology will help you to structure your knowledge of this fascinating and challenging study of life.

Chapter Review

Biology is the scientific study of life. The properties and processes of life include highly ordered structure, evolutionary adaptation, response to the environment, regulation, energy processing, growth and development, and reproduction.

1.1 Biologists explore life from the microscopic to the global scale

A Hierarchy of Organization   The levels of biological structure extend from the biosphere to molecules.

■ **INTERACTIVE QUESTION   1.1**

Write a brief description of each of these levels of biological organization.

a. biosphere

b. ecosystem

c. community

d. population

e. organism

f. organs and organ systems

g. tissues

h. cells

i. organelles

j. molecules

A Closer Look at Ecosystems   An ecosystem includes all the organisms and the nonliving factors in an area. Both organisms and the environment are affected by the interactions between them. The dynamics of an ecosystem include the cycling of nutrients and the flow of energy from sunlight to **producers** (photosynthesizers) to **consumers.** In each energy transformation, some energy is converted to thermal energy, which is dissipated to the surroundings as heat. Thus energy flows through ecosystems, entering as solar energy and exiting as heat.

A Closer Look at Cells   The cell is the lowest structural level capable of performing all the activities of life. The heritable information of a cell is coded in **DNA, deoxyribonucleic acid,** the substance of genes. **Genes** are the units of inheritance, which transmit information from parents to offspring. Genes are located on chromosomes, long DNA molecules that replicate before cell division and provide identical copies to daughter cells.

The biological instructions for the development and functioning of organisms are coded in the arrangement of the four kinds of nucleotides in DNA molecules. Most genes program the cell's production of proteins, and almost all cellular actions involve one or more proteins. Enzymes are proteins that catalyze a cell's chemical reactions.

All forms of life use essentially the same genetic code of nucleotides. This universal genetic code permits the engineering of cells to produce proteins of other organisms. All the genetic instructions an organism inherits are called its **genome.** The human genome is about 3 billion nucleotides long, and it codes for the production of more than 70,000 proteins, each with a specific function.

Every cell uses DNA as its genetic information and is enclosed by a membrane. The simpler and smaller **prokaryotic cell,** unique to bacteria and archaea, lacks both a nucleus to enclose its DNA and most cytoplasmic organelles. The **eukaryotic cell,** with a nucleus containing its DNA, and numerous membrane-bound organelles, is typical of all other living organisms.

■ INTERACTIVE QUESTION   1.2

How do DNA nucleotides relate to proteins?

1.2   Biological systems are much more than the sum of their parts

Biology attempts to understand the behavior of **systems,** the complex organization resulting from the integration of component parts.

The Emergent Properties of Systems   Interactions among components at each level of biological organization lead to the emergence of novel properties at the next level. The structural arrangement and interactions of parts lead to these **emergent properties.**

The Power and Limitations of Reductionism   Biology combines the powerful and pragmatic strategy of **reductionism,** which breaks down complex systems to simpler components, with the study of the highly complex organizational levels of life.

Systems Biology   Many researchers are seeking to understand the emergent properties of life by looking at the functional integration of the parts of a system. **Systems biology** seeks to model biological systems and predict their responses as variables change. Systems biology is carrying this approach, already useful for analyzing ecosystem dynamics and physiological interactions in organisms, to the study of cellular and molecular interactons.

The first step of the systems strategy is to inventory all the known parts of a system. The second step is to explore how each part behaves in relation to others in the working system. Then data from many research teams are combined using computers and software to model a system network. Three research developments are contributing: high-throughput technology or mega-data-collection methods such as the automatic DNA-sequencing machines; **bioinformatics,** which provides the computing power, software, and mathematical models to process and integrate data from enormous data sets; and interdisciplinary research teams with diverse specialists from many scientific fields.

■ INTERACTIVE QUESTION   1.3

a. What types of diverse specialists might be involved in a systems biology team?

b. Give examples of how systems biology may impact medical practice or environmental policy making.

Feedback Regulation in Biological Systems   Many biological systems self-regulate by a mechanism called feedback, in which the output or product of a process regulates the process. In **negative feedback,** an end-product slows down the process. In **positive feedback,** less common in biological processes, an end-product speeds up its own production. Regulatory mechanisms ensure a dynamic balance in living systems.

1.3   Biologists explore life across its great diversity of species

About 1.8 million species, out of an estimated total of 10–200 million, have been identified and named.

Grouping Species: The Basic Idea   Taxonomy is the branch of biology that names organisms and groups species into ever broader categories from genera to family, order, class, phylum, kingdom, and domain.

The Three Domains of Life   Traditionally, life-forms have been organized into five kingdoms, although schemes ranging from six to dozens of kingdoms have recently been proposed. These varying numbers of kingdoms are grouped, based primarily on molecular data, into three domains. The prokaryotes are divided into **domain Archaea** and **domain Bacteria,** and each of these includes multiple kingdoms. The eukaryotes are placed in **domain Eukarya.** Within the Eukarya, the traditional kingdom Protista contains mostly unicellular or simple multicellular forms. Biologists are currently debating how to split the protists into several kingdoms that would better represent evolution and diversity. The other three are plants, fungi, and animals.

Unity in the Diversity of Life   Within this diversity, living forms share a universal genetic language of DNA and similarities in cell structure.

■ INTERACTIVE QUESTION   1.4

What are the main criteria for separating plants, fungi, and animals into kingdoms?

1.4   Evolution accounts for life's unity and diversity

Evolution connects all of the diverse forms of life by common ancestry. In *The Origin of Species*, published in 1859, Charles Darwin presented his case for evolution, or "descent with modification," that present forms evolved from a succession of ancestral forms.

Natural Selection   Darwin synthesized the theory of natural selection as the mechanism of evolution by drawing an inference from two observations: Individuals vary in many heritable traits, and the overproduction of offspring sets up a competition. Individuals with traits best suited for an environment leave a larger proportion of offspring than do less fit individuals. This natural selection, or differential reproductive success within a population, results in the gradual accumulation of favorable adaptations to the challenges of an environment.

The Tree of Life   The underlying unity seen in the structures of related species, both living and in the fossil record, reflects the inheritance of that structure from a common ancestor. Their diversity results from natural selection acting over millions of generations in different environments. According to Darwin, natural selection could produce new species from ancestral ones when isolated populations diversify over time in response to different sets of environmental factors. The tree-like diagrams of evolutionary relationships reflect the branching genealogy extending from ancestral species. Similar species, such as the Galápagos finches, share a common ancestor at a more recent branch point on the tree of life. Birds and mammals share a more ancient common ancestor. And all of life traces back to the earliest ancestral forms.

■ INTERACTIVE QUESTION   1.5

Describe in your own words Darwin's theory of natural selection as the mechanism of evolution.

1.5   Biologists use various forms of inquiry to explore life

Science is a way of knowing that involves **inquiry,** searching for information by asking and endeavoring to answer questions about nature.

Discovery Science   Careful and verifiable observation and analysis of data are the basis of **discovery science.** Observations involve our senses and tools that extend our senses; **data,** both quantitative and qualitative, are recorded observations. Using **inductive reasoning,** a generalized conclusion can often be drawn from collections of observations.

Hypothesis-Based Science   Observations and inductions often lead to the search for explanations. Hypothesis-based science is the search for explanations. A **hypothesis** is a tentative answer to a question. Using "if . . . then" logic, **deductive reasoning** proceeds from the general to the specific, from a general hypothesis to specific predictions of results if the general premise is true. A hypothesis is usually tested by performing experiments or making observations to see whether predicted results occur.

A hypothesis must be testable and falsifiable—there must be some observation or experiment that could reveal if the hypothesis is actually not true. In hypothesis-based science, the ideal is to frame two or more alternative hypotheses and design experiments to falsify each candidate explanation. Hypotheses cannot be proven, they can simply be not eliminated through falsification. The more attempts to falsify it that fail, however, the more a hypothesis gains credibility.

The scientific method, as outlined by a structured series of steps, is rarely adhered to rigidly in a scientific inquiry. Scientists often backtrack to make more observations, or make progress on answering a question only after other research projects provide a new context. In some cases, research must be redirected and refocused by asking different or more productive questions.

A Case Study in Scientific Inquiry: Investigating Mimicry in Snake Populations   In 2001, D. and K. Pfennig designed field experiments to test the Batesian mimicry hypothesis that mimics benefit because predators avoid them, confusing them with the harmful species they resemble. The scarlet king snake mimics the ringed coloration of the highly poisonous coral snake. Predators rarely attack coral snakes, apparently as a result of an inherited instinctive ability to recognize the warning coloration. The Pfennigs tested the hypothesis that predators will attack king snakes more frequently in non–coral snake areas than in areas where predators have adapted to the warning coloration of coral snakes. They placed equal numbers of plain brown and ringed colored artificial king snakes in regions with and without coral snakes. Compared to the brown "control" snakes, the ringed "experimental" snakes were attacked less frequently only in field sites within the range of the poisonous coral snakes.

The Pfennigs' experimental design illustrates a **controlled experiment** in which subjects are divided into an experimental group and a control group. Both groups are treated alike except for the one variable that the experiment is trying to test.

■ **INTERACTIVE QUESTION   1.6**

a. How did predators "learn" to avoid coral snakes?

b. Why were the results of the mimicry study presented as the percent of attacks on king snakes in each area rather than the total number of attacks?

Limitations of Science   Scientific inquiry is limited by the requirements that hypotheses be testable and falsifiable and that observations and experimental results be repeatable. Science seeks natural causes for natural phenomena.

Theories in Science   A **theory** is broader in scope than a hypothesis and is supported by a large body of evidence. Still, theories can be modified or even rejected through the testing of the specific, falsifiable hypotheses they generate.

Model Building in Science   Scientific **models,** such as diagrams, graphs, computer programs, or mathematical equations, help explain ideas and processes. The test of a good model is that it fits available data, incorporates new observations, and makes accurate predictions of new experiments.

The Culture of Science   Most scientists work in teams and share their results with a broader research community in seminars, publications and websites. The political and cultural environment influences the ways in which scientists approach their work. But the adherence to the criteria of verifiable observations and hypotheses that are testable and falsifiable sets science apart from other ways of "knowing nature."

Science, Technology, and Society   Science and technology are interdependent as the information generated by science is applied by **technology** in the development of goods and services, and as technological advances are used to extend scientific knowledge. The uses of scientific knowledge and technologies are influenced by and will influence politics, economics, and cultural values.

1.6   A set of themes connects the concepts of biology

Biology is a demanding science—partly because living systems are so complex and partly because biology incorporates concepts from chemistry, physics, and math. This book presents a wealth of information. The basic themes of biology will help you understand, appreciate, and structure your growing knowledge of biology.

Word Roots

bio- = life (*biology:* the scientific study of life; *biosphere:* all the environments on Earth that are inhabited by life; *bioinformatics:* using information technology to extract useful information from large sets of biological data)

eu- = true (*eukaryotic cell:* a cell that has a true nucleus)

-ell = small (*organelle:* a small formed body with a specialized function found in the cytoplasm of eukaryotic cells)

pro- = before; **karyo-** = nucleus (*prokaryotic cell:* a cell that has no nucleus)

Structure Your Knowledge

1. This chapter presents eleven unifying themes of biology. Briefly describe each of these in your own words:
   a. the cell
   b. heritable information
   c. emergent properties of biological systems
   d. regulation
   e. interaction with the environment
   f. energy and life
   g. unity and diversity
   h. evolution
   i. structure and function
   j. scientific inquiry
   k. science, technology, and society

Test Your Knowledge

MULTIPLE CHOICE: *Choose the one best answer.*

1. The core idea that makes sense of the unity and all the diversity of life is
   a. the scientific method.
   b. inductive reasoning.
   c. deductive reasoning.
   d. evolution.
   e. systems biology.

2. In an experiment similar to the mimicry experiment performed by the Pfennigs, a researcher found that there were more total predator attacks on model king snakes in areas with coral snakes than in areas outside the range of coral snakes. From this the researcher concluded that
   a. the mimicry hypothesis is false.
   b. there were more predators in the areas with coral snakes.
   c. king snakes do not resemble coral snakes enough to protect them from attack.
   d. the data that should be compared to draw a conclusion must include a control—a comparison with the number of attacks on model brown snakes.
   e. more data must be collected before a conclusion can be drawn.

3. Why can a hypothesis never be "proven" to be true?
   a. One can never collect enough data to be 100% sure.
   b. There may always be alternative hypotheses that might account for the results and that were not tested.
   c. Science is limited by our senses.
   d. Experimental error is involved in every research project.
   e. Science "evolves;" hypotheses and even theories are always changing.

4. Which of the following is an example of positive feedback regulation?
   a. The hormones insulin and glucagon regulate blood-sugar levels.
   b. In the birth of a baby, uterine contractions stimulate release of chemicals that stimulate more uterine contractions.
   c. A rise in temperature when you exercise stimulates sweating and increased blood flow to the skin.
   d. When cells have sufficient energy available, the pathways that break down sugars are turned off.
   e. A rise in CO_2 in the atmosphere correlates with increasing global temperature.

5. Which of the following areas is mismatched with its description?
   a. discovery science—data collection and analysis; deductive reasoning
   b. hypothesis-based science—hypothesis generation; predictions; experiments or observations
   c. systems biology—high throughput technology, bioinformatics, interdisciplinary teams
   d. taxonomy—identify and name organisms; place in hierarchical categories
   e. technology—inventing practical uses of scientific knowledge

6. In a pond sample you find a unicellular organism that has numerous chloroplasts and a whiplike flagella. In which of the following groups do you think it should be classified?
   a. plant
   b. animal
   c. domain Archaea
   d. one of the proposed kingdoms of protists
   e. You cannot tell unless you see if it has a nucleus or not.

7. What is DNA?
   a. the substance of heredity
   b. a double helix made of four types of nucleotides
   c. a code for protein synthesis
   d. a component of chromosomes
   e. all of the above

8. Which of the following represents the correct sequence in the life's hierarchical levels, proceeding upward?
   a. organ, tissue, organ system, organism, population
   b. organism, community, population, ecosystem, biosphere
   c. molecule, organelle, cell, tissue, organ, organism
   d. tissue, cell, organ, organism, community
   e. Both b and c are correct sequences.

UNIT 1

The Chemistry of Life

Chapter 2
The Chemical Context of Life

▶ Framework

This chapter considers the basic principles of chemistry that explain the behavior of atoms and molecules and that form the basis for our modern understanding of biology. You will learn how the subatomic particles—protons, neutrons, and electrons—are organized into atoms and atoms are combined by covalent or ionic bonds into molecules. Weak chemical bonds help to create the shapes and functions of molecules. Emergent properties are associated with each new level of structural organization in the hierarchy from atoms to life.

▶ Chapter Review

2.1 Matter consists of chemical elements in pure form and in combinations called compounds

Elements and Compounds **Matter** is anything that takes up space and has mass. (Although sometimes used interchangeably, *mass* reflects the amount of matter in an object, whereas *weight* reflects gravity's pull on that mass.) The basic forms of matter are **elements,** substances that cannot be chemically broken down to other types of matter. A **compound** is made up of two or more elements combined in a fixed ratio. A compound usually has characteristics quite different from its constituent elements, an example of the emergence of novel properties in higher levels of organization.

Essential Elements of Life Carbon (C), oxygen (O), hydrogen (H), and nitrogen (N) make up 96% of living matter. The seven elements listed in Interactive Question 2.1 make up most of the remaining 4%. Some elements, like iron (Fe) and iodine (I), may be required in very minute quantities and are called **trace elements.**

■ INTERACTIVE QUESTION 2.1

Fill in the names beside the symbols of the following elements commonly found in living matter.

Symbol	Element
Ca	
P	
K	
S	
Na	
Cl	
Mg	

2.2 An element's properties depend on the structure of its atoms

An **atom** is the smallest unit of an element retaining the physical and chemical properties of that element.

Subatomic Particles Three stable subatomic particles are important to our understanding of atoms. Uncharged **neutrons** and positively charged **protons** are packed tightly together to form the **atomic nucleus** of an atom. Negatively charged **electrons** orbit rapidly about the nucleus.

Protons and neutrons have a similar mass of about 1.7×10^{-24} g or 1 **dalton** each. A dalton is the measurement unit for atomic mass. Electrons have negligible mass.

Atomic Number and Atomic Mass Each element has a characteristic **atomic number**, or number of protons in the nucleus of its atom. Unless indicated otherwise, an atom has a neutral electrical charge, and thus the number of protons is equal to the number of electrons. A subscript to the left of the symbol for an element indicates its atomic number; a superscript indicates mass number. The **mass number** is equal to the number of protons and neutrons in the nucleus and approximates the mass of an atom of that element in daltons. The term **atomic mass** refers to the total mass of an atom.

■ INTERACTIVE QUESTION 2.2

The difference between the mass number and the atomic number of an atom is equal to the number of _____ . An atom of phosphorus, $_{15}^{31}P$, contains _____ protons, _____ electrons, and _____ neutrons. The atomic mass of phosphorus is approximately _____ .

Isotopes Although the number of protons is constant, the number of neutrons can vary among the atoms of an element, creating different **isotopes** that have slightly different masses but the same chemical behavior. Some isotopes are unstable, or **radioactive**; their nuclei spontaneously decay, giving off particles and energy.

Radioactive isotopes are important tools in biological research and medicine. Chemical processes can be located and monitored within an organism using radioactive tracers and PET (positron-emission tomography). Too great an exposure to radiation from decaying isotopes poses a significant health hazard.

The Energy Levels of Electrons **Energy** is defined as the ability to cause change. **Potential energy** is energy stored in matter as a consequence of the relative position of masses. Matter naturally tends to move toward a more stable lower level of potential energy and requires the input of energy to return to a higher potential energy.

The potential energy of electrons increases as their distance from the positively charged nucleus increases. Electrons can orbit in several different potential energy states, called **energy levels** or **electron shells**, surrounding the nucleus.

■ INTERACTIVE QUESTION 2.3

To move to a shell farther from the nucleus, an electron must _____ energy; energy is _____ when an electron moves to a closer shell.

Electron Configuration and Chemical Properties The chemical behavior of an atom is a function of its electron configuration—in particular, the number of **valence electrons** in its outermost electron shell, or **valence shell**. A valence shell of eight electrons is complete, resulting in an unreactive or inert atom. (The first shell holds only two electrons; thus $_2$He is inert.) Atoms with incomplete valence shells are chemically reactive because of their unpaired electrons. The *periodic table of the elements* is arranged in order of the sequential addition of electrons to orbitals in the electron shells.

■ INTERACTIVE QUESTION 2.4

Draw the electron shell diagram for these atoms.

a. $_7$N **c.** $_{12}$Mg

b. $_8$O **d.** $_6$C

Electron Orbitals　An **orbital** is the three-dimensional space or volume within which an electron is most likely to be found. No more than two electrons can occupy the same orbital. The first electron shell can contain two electrons in a single spherical orbital, called the 1*s* orbital. The second electron shell can hold a maximum of eight electrons in its four orbitals, which are a 2*s* spherical orbital and three dumbbell-shaped *p* orbitals located along the *x, y,* and *z* axes.

■ INTERACTIVE QUESTION　2.5

Look again at the electron shell diagram you drew for carbon (d.) in Interactive Question 2.4. Did you show the outer shell electrons unpaired? Why?

■ INTERACTIVE QUESTION　2.6

Fill in the blanks in the following concept map to help you review the atomic structure of atoms.

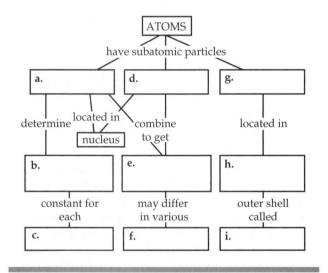

2.3　The formation and function of molecules depend on chemical bonding between atoms

Atoms with incomplete valence shells can either share electrons with or completely transfer electrons to or from other atoms such that each atom is able to complete its valence shell. These interactions usually result in attractions, called **chemical bonds,** that hold the atoms together.

Covalent Bonds　When two atoms share a pair of valence electrons, a **covalent bond** is formed. A **molecule**

consists of two or more atoms held together by covalent bonds. A **structural formula,** such as H—H, indicates both the number and type of atoms and also the bonding within a molecule. The dash indicates a single covalent bond, or just a **single bond.** A **molecular formula,** such as O_2, indicates only the kinds and numbers of atoms in a molecule. In an oxygen molecule, two pairs of valence electrons are shared between oxygen atoms, forming a double covalent bond, or simply a **double bond.**

The **valence,** or bonding capacity, of an atom equals the number of unpaired electrons in its valence shell. (Even though phosphorus has three unpaired electrons and a valence of three, it has a valence of five in some important biological molecules.)

■ INTERACTIVE QUESTION　2.7

What are the valences of the four most common elements of living matter?

a. hydrogen　　　　　**c.** nitrogen

b. oxygen　　　　　　**d.** carbon

Electronegativity is the attraction of a particular type of atom for shared electrons. If the atoms in a molecule have similar electronegativities, the electrons remain equally shared between the two nuclei, and the covalent bond is said to be a **nonpolar covalent bold.** If one element is more electronegative, it pulls the shared electrons closer to itself, creating a **polar covalent bond.** This unequal sharing of electrons results in a slight negative charge ($\delta-$) associated with the more electronegative atom and a slight positive charge ($\delta+$) associated with the atom from which the electrons are pulled.

■ INTERACTIVE QUESTION　2.8

Explain whether the following molecules contain nonpolar or polar covalent bonds. (Hint: N and O both have high electronegativities.)

a. nitrogen molecule　N≡N

c. methane

$$\text{H}-\overset{\displaystyle \overset{\text{H}}{|}}{\underset{\displaystyle \underset{\text{H}}{|}}{\text{C}}}-\text{H}$$

b. ammonia

d. formaldehyde

Ionic Bonds   If two atoms are very different in their attraction for the shared electrons, the more electronegative atom may completely transfer an electron from another atom, resulting in the formation of charged atoms called **ions.** The atom that lost the electron is a positively charged **cation.** The negatively charged atom that gained the electron is called an **anion.** An **ionic bond** may hold these ions together because of the attraction of their opposite charges.

Ionic compounds, called **salts,** often exist as three-dimensional crystalline lattice arrangements held together by electrical attractions. The number of ions present in a salt crystal is not fixed, but the atoms are present in specific ratios. Salts have strong ionic bonds when dry, but the crystal dissolves in water.

Ion also refers to entire covalent molecules that are electrically charged. Ammonium (NH_4^+) is a cation; this covalently bonded molecule is missing one electron.

■ **INTERACTIVE QUESTION   2.9**

Calcium ($_{20}$Ca) and chlorine ($_{17}$Cl) can combine to form the salt calcium chloride. Based on the number of electrons in their valence shells and their bonding capacities, what would the molecular formula for this salt be? a. _____
Which atom becomes the cation? b. _____

Weak Chemical Bonds   Weak bonds, such as ionic bonds in water, form temporary interactions between molecules and are involved in many biological signals and processes. Weak bonds within large molecules such as proteins help to create the three-dimensional shape and resulting activity of these molecules.

When a hydrogen atom is covalently bonded with an electronegative atom, and thus has a partial positive charge, it can be attracted to another electronegative atom and form a **hydrogen bond.**

All atoms and molecules are attracted to each other when in close contact by **van der Waals interactions.** Momentary uneven electron distributions produce changing positive and negative regions that create these weak attractions.

■ **INTERACTIVE QUESTION   2.10**

Sketch a water molecule, showing oxygen's electron shells and the covalently shared electrons. Indicate the areas with slight negative and positive charges that enable a water molecule to form hydrogen bonds with other polar molecules.

Molecular Shape and Function   A molecule's characteristic size and shape affect how it interacts with other molecules. When atoms form covalent bonds, their *s* and three *p* orbitals hybridize to form four teardrop-shaped orbitals in a tetrahedral arrangement. These hybrid orbitals dictate the specific shapes of different molecules.

■ **INTERACTIVE QUESTION   2.11**

Look at your diagram of a water molecule in Interactive Question 2.10. Why should its shape be roughly like a V?

2.4   Chemical reactions make and break chemical bonds

Chemical reactions involve the making or breaking of chemical bonds in the transformation of matter into different forms. Matter is conserved in chemical reactions; the same number and kinds of atoms are present in both **reactants** and **products,** although the rearrangement of electrons and atoms causes the properties of these molecules to be different.

■ **INTERACTIVE QUESTION   2.12**

Fill in the missing coefficients for respiration, the conversion of glucose and oxygen to carbon dioxide and water, so that all atoms are conserved in the chemical reaction.

$$C_6H_{12}O_6 + \underline{\quad} O_2 \longrightarrow \underline{\quad} CO_2 + \underline{\quad} H_2O$$

Most reactions are reversible—the products of the forward reaction can become reactants in the reverse reaction. Increasing the concentrations of reactants can speed up the rate of a reaction. **Chemical equilibrium** is reached when the forward and reverse reactions proceed at the same rate, and the relative concentrations of reactants and products no longer change.

Word Roots

an- = not (*anion:* a negatively charged ion)

co- = together; **-valent** = strength (*covalent bond:* an attraction between atoms that share one or more pairs of outer-shell electrons)

electro- = electricity (*electronegativity:* the tendency for an atom to pull electrons toward itself)

iso- = equal (*isotope:* an element having the same number of protons and electrons but a different number of neutrons)

neutr- = neither (*neutron:* a subatomic particle with a neutral electrical charge)

pro- = before (*proton:* a subatomic particle with a single positive electrical charge)

Structure Your Knowledge

Take the time to write out or discuss your answers to the following questions. Then refer to the suggested answers at the end of the book.

1. Fill in the following chart for the major subatomic particles of an atom.

Particle	Charge	Mass	Location

2. Atoms can have various numbers associated with them.
   a. Define the following and show where each of them is placed relative to the symbol of an element such as C: atomic number, mass number, atomic mass.
   b. Define valence.
   c. Which of these four numbers is most related to the chemical behavior of an atom? Explain.
3. Explain what is meant by saying that the sharing of electrons between atoms falls on a continuum from nonpolar covalent bonds to ionic bonds.

Test Your Knowledge

MULTIPLE CHOICE: *Choose the one best answer*

1. Each element has its own characteristic atom in which
   a. the atomic mass is constant.
   b. the atomic number is constant.
   c. the mass number is constant.
   d. two of the above are correct.
   e. all of the above are correct.

2. Radioactive isotopes can be used in studies of metabolic pathways because
   a. their half-life allows a researcher to time an experiment.
   b. they are more reactive.
   c. the cell does not recognize the extra protons in the nucleus, so isotopes are readily used in metabolism.
   d. their location or quantity can be experimentally determined because of their radioactivity.
   e. their extra neutrons produce different colors that can be traced through the body.

3. In a reaction in chemical equilibrium,
   a. the forward and reverse reactions are occurring at the same rate.
   b. the reactants and products are in equal concentration.
   c. the forward reaction has gone further than the reverse reaction.
   d. there are equal numbers of atoms on both sides of the equation.
   e. a, b, and d are correct.

4. Oxygen has eight electrons. You would expect the arrangement of these electrons to be:
   a. eight in the second energy shell, creating an inert element.
   b. two in the first energy shell and six in the second, creating a valence of six.
   c. two in the 1*s* orbital and two each in the three 2*p* orbitals, creating a valence of zero.
   d. two in the 1*s* orbital, one each in the 2*s* and three 2*p* orbitals, and two in the 3*s* orbital, creating a valence of two.
   e. two in the 1*s* orbital, two in both the 2*s* and 2*px* orbitals, and one each in the 2*py* and 2*pz* orbitals, creating a valence of two.

5. A covalent bond between two atoms is likely to be polar if
   a. one of the atoms is much more electronegative than the other.
   b. the two atoms are equally electronegative.
   c. the two atoms are of the same element.
   d. the bond is part of a tetrahedrally shaped molecule.
   e. one atom is an anion.

6. A triple covalent bond would
   a. be very polar.
   b. involve the bonding of three atoms.
   c. involve the bonding of six atoms.
   d. produce a triangularly shaped molecule.
   e. involve the sharing of six electrons.

7. A cation
   a. has gained an electron.
   b. can easily form hydrogen bonds.
   c. is more likely to form in an atom with seven electrons in its valence shell.
   d. has a positive charge.
   e. Both c and d are correct.

8. What types of bonds are identified in the following illustration of a water molecule interacting with an ammonia molecule?

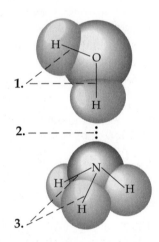

   a. Bonds 1 are polar covalent bonds, bond 2 is a hydrogen bond, and bonds 3 are nonpolar covalent bonds.
   b. Bonds 1 and 3 are polar covalent bonds, and bond 2 is a hydrogen bond.
   c. Bonds 1 and 3 are polar covalent bonds, and bond 2 is an ionic bond.
   d. Bonds 1 and 3 are nonpolar covalent bonds, and bond 2 is a hydrogen bond.
   e. Bonds 1 and 3 are polar covalent bonds, and bond 2 is a nonpolar covalent bond.

9. Which of the following weak bonds may form between any closely aligned molecules?
   a. nonpolar covalent
   b. polar covalent
   c. ionic
   d. hydrogen
   e. van der Waals interactions

10. The ability of morphine to mimic the effects of the body's endorphins is due to
   a. a chemical equilibrium developing between morphine and endorphins.
   b. the one-way conversion of morphine into endorphin.
   c. molecular shape similarities that allow morphine to bind to endorphin receptors.
   d. the similarities between morphine and heroin.
   e. hydrogen bonding and other weak bonds forming between morphine and endorphins.

Use this information to answer questions 11 through 16.

The six elements most common in living organisms are:

$$^{12}_{6}C \quad ^{16}_{8}O \quad ^{1}_{1}H \quad ^{14}_{7}N \quad ^{32}_{16}S \quad ^{31}_{15}P$$

11. How many electrons does phosphorus have in its valence shell?
   a. 3
   b. 5
   c. 7
   d. 15
   e. 16

12. What is the atomic mass of phosphorus?
   a. 15
   b. 16
   c. 31
   d. 46
   e. 62

13. A radioactive isotope of carbon has the mass number 14. How many neutrons does this isotope have?
   a. 2
   b. 6
   c. 8
   d. 12
   e. 14

14. How many covalent bonds is a sulfur atom most likely to form?
   a. 1
   b. 2
   c. 3
   d. 4
   e. 5

15. Based on electron configuration, which of these elements would have chemical behavior most like that of oxygen?

    a. C

    b. H

    c. N

    d. P

    e. S

16. How many of these elements are found next to each other (side by side) on the periodic table?

    a. one group of two

    b. two groups of two

    c. one group of two and one group of three

    d. one group of three

    e. all of them

17. Taking into account the bonding capacities or valences of carbon (C) and oxygen (O), how many hydrogen (H) must be added to complete the structural diagram of this molecule?

    $$O\!\!\diagdown\!\!\diagup\!\!O \;\; C-C-C-C=C-C-C$$

    a. 9

    b. 10

    c. 11

    d. 12

    e. 13

18. A sodium ion (Na^+) contains 10 electrons, 11 protons, and 12 neutrons. What is the atomic number of sodium?

    a. 10

    b. 11

    c. 12

    d. 23

    e. 33

19. What type of bond would you expect potassium ($^{39}_{19}K$) to form?

    a. ionic; it would donate one electron and carry a positive charge

    b. ionic; it would donate one electron and carry a negative charge

    c. covalent; it would share one electron and make one covalent bond

    d. covalent; it would share two electrons and form two bonds

    e. none; potassium is an inert element

20. What is the molecular shape of methane (CH_4)?

    a. planar or flat, with the H arranged around the C

    b. pentagonal, or a flat five-sided arrangement

    c. tetrahedral, due to the hybridization of the s and three p orbits of the C

    d. circular, with the four H attached in a ring around the C

    e. linear, since all the bonds are nonpolar covalent

21. Which of the following is a molecule capable of forming hydrogen bonds?

    a. CH_4

    b. H_2O

    c. NaCl

    d. H_2

    e. a, b, and d can form hydrogen bonds.

22. Chlorine has an atomic number of 17 and a mass number of 35. How many electrons would a chloride ion have?

    a. 16

    b. 17

    c. 18

    d. 33

    e. 34

23. What is the difference between a molecule and a compound?

    a. There is no difference; the terms are interchangeable.

    b. Molecules contain atoms of a single element, whereas compounds contain two or more elements.

    c. A molecule consists of two or more covalently bonded atoms; a compound contains two or more atoms held by ionic bonds.

    d. A compound consists of two or more elements in a fixed ratio; a molecule has two or more covalently bonded atoms of the same or different elements.

    e. Compounds always consist of molecules, but molecules are not always compounds.

24. Which of the following atomic numbers would describe the element that is least reactive?

    a. 1

    b. 8

    c. 12

    d. 16

    e. 18

25. What coefficients must be placed in the blanks to balance this chemical reaction?

    $$C_5H_{12} + \underline{\ 8\ }\,O_2 \longrightarrow \underline{\ 5\ }\,CO_2 + \underline{\ 6\ }\,H_2O$$

    a. 5; 5; 5

    b. 6; 5; 6

    c. 6; 6; 6

    d. 8; 4; 6

    e. 8; 5; 6

Chapter 3
Water and the Fitness of the Environment

Framework

Water makes up 70% to 95% of the cell content of living organisms and covers 75% of the Earth's surface. Its unique properties make the external environment fit for living organisms and the internal environments of organisms fit for the chemical and physical processes of life.

Hydrogen bonding between polar water molecules creates a cohesive liquid with a high specific heat and high heat of vaporization, both of which help to regulate environmental temperature. Ice floats and protects oceans and lakes from freezing. The polarity of water makes it a versatile solvent. An organism's pH may be regulated by buffers. Acid precipitation poses a serious environmental threat.

Chapter Review

3.1 The polarity of water molecules results in hydrogen bonding

A water molecule consists of two hydrogen atoms each covalently bonded to a more electronegative oxygen atom. This **polar molecule** has a shape like a wide V with a slight positive charge on each hydrogen atom ($\delta+$) and a slight negative charge ($\delta-$) associated with the oxygen. Hydrogen bonds, electrical attractions between the hydrogen atom of one water molecule and the oxygen atom of a nearby water molecule, create a

higher level of structural organization and lead to the emergent properties of water.

3.2 Four emergent properties of water contribute to Earth's fitness for life

Cohesion Liquid water is unusually cohesive due to the constant forming and reforming of hydrogen bonds that hold the molecules together. This **cohesion** creates a more structurally organized liquid and helps water to be pulled upward in plants. The **adhesion** of water molecules to the walls of plant vessels also contributes to water transport. Hydrogen bonding between water molecules produces a high **surface tension** at the interface between water and air.

■ INTERACTIVE QUESTION 3.1

Draw the four water molecules that can hydrogen-bond to this water molecule. Show the bonds and the slight negative and positive charges that account for the formation of these hydrogen bonds.

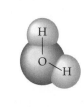

Moderation of Temperature In a body of matter, **heat** is a measure of the total quantity of **kinetic energy,** the energy associated with the movement of atoms and molecules. **Temperature** measures the average kinetic energy of the molecules in a substance.

Temperature is measured using a **Celsius scale.** Water at sea level freezes at 0°C and boils at 100°C. A **calorie (cal)** is the amount of heat energy it takes to raise 1 g of water 1°C. A **kilocalorie (kcal)** is 1,000 calories, the amount of heat required to raise 1 kg of water 1°C. A **joule (J)** equals 0.239 cal; a calorie is 4.184 J.

Specific heat is the amount of heat absorbed or lost when 1 g of a substance changes its temperature by 1°C. Water's specific heat of 1 cal/g/°C is unusually high compared with that of other common substances; water must absorb or release a relatively large quantity of heat in order for its temperature to change. Heat must be absorbed to break hydrogen bonds before water molecules can move faster and the temperature can rise, and conversely, heat is released when hydrogen bonds form as the temperature of water drops. The ability of large bodies of water to stabilize air temperature is due to the high specific heat of water. The high proportion of water in the environment and within organisms keeps temperature fluctuations within limits that permit life.

The transformation from a liquid to a gas is called vaporization or evaporation and happens when molecules with sufficient kinetic energy overcome their attraction to other molecules and escape into the air as gas. The **heat of vaporization** is the quantity of heat that must be absorbed for 1 g of a liquid to be converted to a gas. Water has a high heat of vaporization (580 cal/g at 25°C) because a large amount of heat is needed to break the hydrogen bonds holding water molecules together. This property of water helps moderate the climate on Earth as solar heat is dissipated from tropical seas during evaporation and heat is released when moist tropical air condenses to form rain.

As a substance vaporizes, the liquid left behind loses the kinetic energy of the escaping molecules and cools down. **Evaporative cooling** helps to protect terrestrial organisms from overheating and contributes to the stability of temperatures in lakes and ponds.

Insulation of Water Bodies by Floating Ice   As water cools below 4°C, it expands. By 0°C, each water molecule becomes hydrogen-bonded to four other molecules, creating a crystalline lattice that spaces the molecules apart. Ice is less dense than liquid water, and therefore, it floats. The floating ice insulates the liquid water below.

The Solvent of Life   A **solution** is a liquid homogeneous mixture of two or more substances; the dissolving agent is called the **solvent** and the substance that is dissolved is the **solute.** Water is the solvent in an **aqueous solution.** The positive and negative regions of water molecules are attracted to oppositely charged ions

or partially charged regions of polar molecules. Thus, solute molecules become surrounded by water molecules (a **hydration shell**) and dissolve into solution.

■ INTERACTIVE QUESTION   3.2

The following concept map is one way to show how the breaking and forming of hydrogen bonds is related to temperature moderation. Fill in the blanks and compare your choice of concepts to those given in the answer section. Or, better still, create your own map to help you understand how water stabilizes temperature.

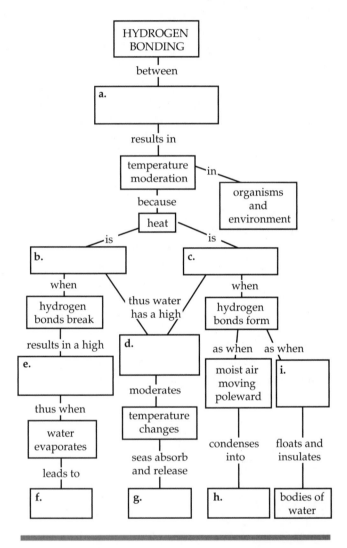

Ionic and polar substances are **hydrophilic;** they have an affinity for water due to electrical attractions and hydrogen bonding. Large hydrophilic substances may not dissolve but become suspended in an aqueous solution, forming a mixture called a **colloid.** Nonionic and nonpolar molecules are **hydrophobic;** they will not easily mix with or dissolve in water.

■ INTERACTIVE QUESTION 3.3

Indicate whether the following are hydrophilic or hydrophobic. Do these substances contain ionic, polar covalent bonds, or nonpolar covalent bonds?

a. olive oil

c. salt

b. sugar

d. candle wax

Most of the chemical reactions of life take place in water. A **mole (mol)** is the amount of a substance that has a mass in grams numerically equivalent to its **molecular mass** (sum of the mass of all atoms in the molecule) in daltons. A mole of any substance has exactly the same number of molecules—6.02×10^{23}, called Avogadro's number. The **molarity** of a solution (abbreviated M) refers to the number of moles of a solute dissolved in 1 liter of solution.

■ INTERACTIVE QUESTION 3.4

a. How many grams of lactic acid ($C_3H_6O_3$) are in a 0.5 M solution of lactic acid? (^{12}C, ^1H, ^{16}O)

b. How many grams of salt (NaCl) must be dissolved in water to make 2 liters of a 2 M salt solution? (^{23}Na, ^{34}Cl)

3.3 Dissociation of water molecules leads to acidic and basic conditions that affect living organisms

A water molecule can dissociate into a **hydrogen ion**, H^+ (which binds to another water molecule to form a hydronium ion, H_3O^+) and a **hydroxide ion**, OH^-. Although reversible and statistically rare, this dissociation into the highly reactive hydrogen and hydroxide ions has important biological consequences. In pure water at 25°C, the concentrations of H^+ and OH^- ions are the same; both are equal to 10^{-7} M.

Effects of Changes in pH   When acids or bases dissolve in water, the H^+ and OH^- balance shifts. An **acid** adds H^+ to a solution, whereas a **base** reduces H^+ in a solution by accepting hydrogen ions or by adding hydroxide ions (which then combine with H^+ and thus remove hydrogen ions). A strong acid or strong base may dissociate completely when mixed with water. A weak acid or base reversibly dissociates, releasing or binding H^+.

In an aqueous solution, the product of the $[H^+]$ and $[OH^-]$ is constant at 10^{-14}. Brackets, [ ], indicate molar concentration. If the $[H^+]$ is higher, then the $[OH^-]$ is lower, due to the tendency of excess hydrogen ions to combine with the hydroxide ions in solution and form water. Likewise, an increase in $[OH^-]$ causes an equivalent decrease in $[H^+]$. If $[OH^-]$ is equal to 10^{-10} M, then $[H^+]$ will equal 10^{-4} M.

The logarithmic pH scale compresses the range of hydrogen and hydroxide ion concentrations, which can vary in different solutions by many orders of magnitude. The **pH** of a solution is defined as the negative log (base 10) of the $[H^+]$: pH $= -\log [H^+]$. For a neutral aqueous solution, $[H^+]$ is 10^{-7} M, and the pH equals 7. As the $[H^+]$ increases in an acidic solution, the pH value decreases. The difference between each unit of the pH scale represents a tenfold difference in the concentration of $[H^+]$ and $[OH^-]$.

■ INTERACTIVE QUESTION 3.5

Complete the following table to review your understanding of pH.

$[H^+]$	$[OH^-]$	pH	Acidic, Basic, or Neutral?
	10^{-11}	3	acidic
10^{-8}			
	10^{-7}		
		1	

Most cells have an internal pH close to 7. **Buffers** within the cell maintain a constant pH by accepting excess H^+ ions or donating H^+ ions when H^+ concentration decreases. Weak acid-base pairs that reversibly bind hydrogen ions are typical of most buffering systems.

■ INTERACTIVE QUESTION 3.6

The carbonic acid/bicarbonate system is an important biological buffer. Label the molecules and ions in this equation and indicate which is the H^+ donor and which is the acceptor.

$$H_2CO_3 \rightleftharpoons HCO_3^- + H^+$$

In which direction will this reaction proceed

a. when the pH of a solution begins to fall?

b. when the pH rises above normal level?

The Threat of Acid Precipitation   **Acid precipitation,** rain, snow, or fog with a pH lower than normal (pH 5.6), is due to the reaction of water in the atmosphere with the sulfur oxides and nitrogen oxides released by the combustion of fossil fuels. Aquatic life is damaged by acid precipitation, and lowering the pH of the soil solution affects the solubility of minerals needed by plants.

Word Roots

kilo- = a thousand (*kilocalorie:* a thousand calories)
hydro- = water; **-philos** = loving; **-phobos** = fearing
    (*hydrophilic:* having an affinity for water; *hydrophobic:* having an aversion to water)

Structure Your Knowledge

1. Fill in the table below that summarizes the properties of water that contribute to the fitness of the environment for life.

2. To become proficient in the use of the concepts relating to pH, develop a concept map to organize your understanding of the following terms: pH, $[H^+]$, $[OH^-]$, acidic, basic, neutral, buffer, 1–14, acid-base pair. Remember to label connecting lines and add additional concepts as you need them. *A suggested concept map is given in the answer section, but remember that your concept map should represent your own understanding. The value of this exercise is in organizing these concepts for yourself.*

Property	Explanation of Property	Example of Benefit to Life
a.	Hydrogen bonds hold molecules together and adhere them to hydrophilic surface.	**b.**
High specific heat	**c.**	Temperature changes in environment and organisms are moderated.
d.	Hydrogen bonds must be broken for water to evaporate.	**e.**
f.	Water molecules with high kinetic energy evaporate; remaining molecules are cooler.	**g.**
Ice floats	**h.**	**i.**
j.	**k.**	Most chemical reactions in life involve solutes dissolved in water.

Test Your Knowledge

MULTIPLE CHOICE: *Choose the one best answer.*

1. Each water molecule is capable of forming
    a. one hydrogen bond.
    b. three hydrogen bonds.
    c. four hydrogen bonds.
    d. two covalent bonds and two hydrogen bonds.
    e. two covalent bonds and four hydrogen bonds.

2. The polarity of water molecules
    a. promotes the formation of hydrogen bonds.
    b. helps water to dissolve nonpolar solutes.
    c. lowers the heat of vaporization and leads to evaporative cooling.
    d. creates a crystalline structure in liquid water.
    e. does all of the above.

3. What accounts for the movement of water up the vessels of a tall tree?
    a. cohesion
    b. hydrogen bonding
    c. adhesion
    d. hydrophilic vessel walls
    e. all of the above

4. Climates tend to be moderate near large bodies of water because
    a. a large amount of solar heat is absorbed during the gradual rise in temperature of the water.
    b. water releases heat to the environment as it cools.
    c. the high specific heat of water helps to moderate air temperatures.
    d. a great deal of heat is absorbed and released by the breaking and forming of hydrogen bonds.
    e. of all of the above.

5. Temperature is a measure of
   a. specific heat.
   b. average kinetic energy of molecules.
   c. total kinetic energy of molecules.
   d. Celsius degrees.
   e. joules.

6. Evaporative cooling is a result of
   a. a low heat of vaporization.
   b. a high specific heat.
   c. absorption of heat as hydrogen bonds break.
   d. a reduction in the average kinetic energy of a liquid after energetic water molecules enter the gaseous state.
   e. release of heat caused by the breaking of hydrogen bonds when water molecules escape.

7. Ice floats because
   a. air is trapped in the crystalline lattice.
   b. the formation of hydrogen bonds releases heat; warmer objects float.
   c. it has a smaller surface area than liquid water.
   d. it insulates bodies of water so they do not freeze from the bottom up.
   e. hydrogen bonding spaces the molecules farther apart, creating a less dense structure.

8. The molarity of a solution is equal to
   a. Avogadro's number of molecules in 1 liter of solvent.
   b. the number of moles of a solute in 1 liter of solution.
   c. the molecular mass of a solute in 1 liter of solution.
   d. the number of solute particles in 1 liter of solvent.
   e. 342 g if the solute is sucrose.

9. Some archaea are able to live in lakes with pH values of 11. How does pH 11 compare with the pH 7 typical of your body cells?
   a. It is four times more acidic than pH 7.
   b. It is four times more basic than pH 7.
   c. It is a thousand times more acidic than pH 7.
   d. It is a thousand times more basic than pH 7.
   e. It is ten thousand times more basic than pH 7.

10. A buffer
    a. changes pH by a magnitude of 10.
    b. releases excess OH^-.
    c. releases excess H^+.
    d. is often a weak acid-base pair.
    e. always maintains a neutral pH.

11. Which of the following is least soluble in water?
    a. polar molecules
    b. nonpolar molecules
    c. ionic compounds
    d. hydrophilic molecules
    e. anions

12. Which would be the best method for reducing acid precipitation?
    a. Raise the height of smokestacks so that exhaust enters the upper atmosphere.
    b. Add buffers and bases to bodies of water whose pH has dropped.
    c. Use coal-burning generators rather than nuclear power to produce electricity.
    d. Tighten emission control standards for factories and automobiles.
    e. Reduce the concentration of heavy metals in industrial exhaust.

13. What bonds must be broken for water to vaporize?
    a. polar covalent bonds
    b. nonpolar covalent bonds
    c. hydrogen bonds
    d. ionic bonds
    e. polar covalent and hydrogen bonds

14. How would you make a 0.1 *M* solution of glucose ($C_6H_{12}O_6$)? The mass numbers for these elements are approximately: C = 12, O = 16, H = 1.
    a. Mix 6 g C, 12 g H, and 6 g O in 1 liter of water.
    b. Mix 72 g C, 12 g H, and 96 g O in 1 liter of water.
    c. Mix 18 g of glucose with enough water to yield 1 liter of solution.
    d. Mix 29 g of glucose with enough water to yield 1 liter of solution.
    e. Mix 180 g of glucose with enough water to yield 1 liter of solution.

15. How many molecules of glucose would be in the 1 liter solution made in question 14?
    a. 0.1
    b. 6
    c. 60
    d. 6×10^{23}
    e. 6×10^{22}

16. Why is water such an excellent solvent?
    a. As a polar molecule, it can surround and dissolve ionic and polar molecules.
    b. It forms ionic bonds with ions, hydrogen bonds with polar molecules, and hydrophobic interactions with nonpolar molecules.
    c. It forms hydrogen bonds with itself.
    d. It has a high specific heat and a high heat of vaporization.
    e. It is wet and has a great deal of surface tension.

17. Which of the following when mixed with water would form a colloid?
    a. a large hydrophobic protein
    b. a large hydrophilic protein
    c. sugar
    d. cotton
    e. NaCl

18. Adding a base to a solution would
    a. raise the pH.
    b. lower the pH.
    c. decrease $[H^+]$.
    d. do both a and c.
    e. do both b and c.

19. A hydration shell is most likely to form around
    a. an ion.
    b. a fat.
    c. a sugar.

d. both a and c.
e. both b and c.

20. The following are the pH values for each item: cola–2; orange juice–3; beer–4; coffee–5; human blood–7.4. Which of these liquids has the *highest* molar concentration of OH^-?
    a. cola
    b. orange juice
    c. beer
    d. coffee
    e. human blood

21. Comparing the $[H^+]$ of orange juice and coffee, the $[H^+]$ of
    a. orange juice is 10 times higher.
    b. orange juice is 100 times higher.
    c. orange juice is 1,000 times higher.
    d. coffee is two times higher.
    e. coffee is 100 times higher.

22. The ability of water molecules to form hydrogen bonds accounts for water's
    a. high specific heat.
    b. evaporative cooling.
    c. high heat of vaporization.
    d. cohesiveness and surface tension.
    e. All of the above result from water's hydrogen-bonding capacity.

Chapter 4
Carbon and the Molecular Diversity of Life

Key Concepts

4.1 **Organic chemistry is the study of carbon compounds**

4.2 **Carbon atoms can form diverse molecules by bonding to four other atoms**

4.3 **Functional groups are the parts of molecules involved in chemical reactions**

Framework

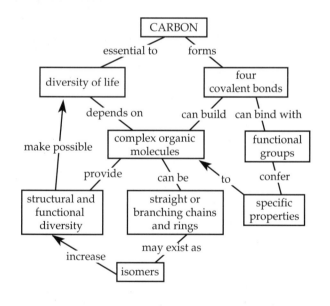

Chapter Review

4.1 Organic chemistry is the study of carbon compounds

Organic chemistry is the study of carbon-containing molecules. Early organic chemists could not synthesize the complex molecules found in living organisms and, therefore, attributed the existence of life and the for-

mation of these molecules to a life force independent of physical and chemical laws, a belief known as vitalism. Mechanism, the philosophy underlying modern organic chemistry, holds that physical and chemical laws and explanations are sufficient to account for all natural phenomena, even the origin of life.

4.2 Carbon atoms can form diverse molecules by bonding to four other atoms

Formation of Bonds with Carbon Carbon has six electrons. To complete its valence shell, carbon forms four covalent bonds with other atoms. This *tetravalence* is at the center of carbon's ability to form large and complex molecules with characteristic three-dimensional shapes and properties. When carbon forms four single covalent bonds, its hybrid orbitals create a tetrahedral shape. When two carbons are joined by a double bond, the other carbon bonds are in the same plane, forming a flat molecule.

Molecular Diversity Arising from Carbon Skeleton Variation Carbon atoms readily bond with each other, producing chains or rings of carbon atoms. These molecular backbones can vary in length, branching, placement of double bonds, and location of atoms of other elements. The simplest organic molecules are **hydrocarbons,** consisting of only carbon and hydrogen. The nonpolar C—H bonds in hydrocarbon chains account for their hydrophobic properties.

Isomers are compounds with the same molecular formula but different structural arrangements and, thus, different properties. **Structural isomers** differ in the covalent arrangement of atoms and often in the location of double bonds. **Geometric isomers** have the same sequence of covalently bonded atoms but differ in spatial arrangement due to the inflexibility of double bonds. A *cis* isomer has non-hydrogen atoms attached to double-bonded carbons on the same side of the double bond; a *trans* isomer has these atoms on opposite sides of the double bond. **Enantiomers** are left- and right-handed versions of each other and can differ greatly in their biological activity. An asymmetric carbon is one that is covalently bonded to four different kinds of atoms or groups of atoms. Due to the tetrahedral shape of the asymmetric carbon, the four groups can be attached in spatial arrangements that are not superimposable on each other.

■ INTERACTIVE QUESTION   4.1

Identify the structural isomers, geometric isomers, and enantiomers from the following compounds. Which of the geometric isomers is the *cis* isomer?

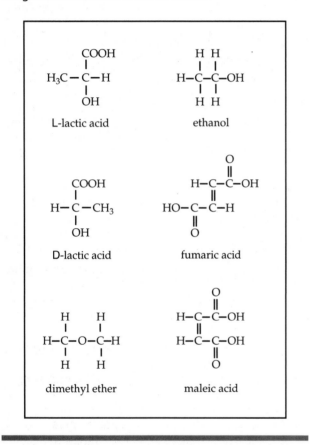

Carbonyl groups consist of a carbon double-bonded to an oxygen ($\gt$CO). If the carbonyl group is at the end of the carbon skeleton, the compound is called an **aldehyde.** Otherwise, the compound is called a **ketone.**

A **carboxyl group** consists of a carbon double-bonded to an oxygen and also attached to a hydroxyl group (—COOH). Compounds with a carboxyl group are called **carboxylic acids** or organic acids because they tend to dissociate to release H^+, becoming —COO$^-$.

An **amino group** consists of a nitrogen atom bonded to two hydrogens (—NH$_2$) and to the carbon skeleton. Compounds with an amino group, called **amines,** can act as bases. The nitrogen, with its pair of unshared electrons, can attract a hydrogen ion, becoming —NH$_3^+$.

The **sulfhydryl group** consists of a sulfur atom bonded to a hydrogen (—SH). **Thiols** are compounds containing sulfhydryl groups.

A **phosphate group** is bonded to the carbon skeleton by an oxygen attached to PO$_3^{2-}$, a phosphorus atom that is bonded to three other oxygen atoms (—OPO$_3^{2-}$). It is an ionized form of a phosphoric acid group (—PO$_3$H$_2$). Compounds containing phosphate groups are called **organic phosphates.** The group is an anion due to the dissociation of hydrogen ions.

ATP: An Important Source of Energy for Cellular Processes   Adenosinetriphosphate (ATP) is the cell's primary energy-transferring molecule. The splitting off of the third phosphate group releases energy as ATP is converted to ADP.

The chemical elements of life: a review

Carbon, oxygen, hydrogen, nitrogen, and smaller quantities of sulfur and phosphorus, all capable of forming strong covalent bonds, are combined into the complex organic molecules of living matter. The versatility of carbon in forming four covalent bonds, linking readily with itself to produce chains and rings, and binding with other elements and functional groups makes possible the incredible diversity of organic molecules.

4.3   Functional groups are the parts of molecules involved in chemical reactions

The Functional Groups Most Important in the Chemistry of Life   The properties of organic molecules are largely determined by groups of atoms, known as **functional groups,** that bond to the carbon skeleton and behave consistently from one carbon-based molecule to another. Because the functional groups considered here are hydrophilic, they increase the solubility of organic compounds in water.

The **hydroxyl group** consists of an oxygen and hydrogen (—OH) covalently bonded to the carbon skeleton. Organic molecules with hydroxyl groups are called **alcohols,** and their names often end in *-ol.*

Word Roots

hydro- = water (*hydrocarbon:* an organic molecule consisting only of carbon and hydrogen)

iso- = equal (*isomer:* one of several organic compounds with the same molecular formula but different structures and, therefore, different properties)

enanti- = opposite (*enantiomer:* molecules that are mirror images of each other)

carb- = coal (*carboxyl group:* a functional group present in organic acids, consisting of a carbon atom double-bonded to an oxygen atom and a hydroxyl group)

sulf- = sulfur (*sulfhydryl group:* a functional group that consists of a sulfur atom bonded to an atom of hydrogen)

thio- = sulfur (*thiol:* organic compounds containing sulfhydryl groups)

Structure Your Knowledge

1. Construct a concept map that illustrates your understanding of the characteristics and significance of the three types of isomers. *A suggested map is in the answer section. Comparing and discussing your map with that of a study partner would be most helpful.*

2. Fill in the following table on the functional groups.

Functional Group	Molecular Formula	Names and Characteristics of Organic Compounds Containing Functional Group
	—OH	
		Aldehyde or ketone; polar group
Carboxyl		
	—NH$_2$	
		Thiols; cross-links stabilize protein structure
Phosphate		

Test Your Knowledge

MULTIPLE CHOICE: *Choose the one best answer.*

1. The tetravalence of carbon most directly results from
   a. its tetrahedral shape.
   b. its very slight electronegativity.
   c. its four electrons in the valence shell that can form four covalent bonds.
   d. its ability to form single, double, and triple bonds.
   e. its ability to form chains and rings of carbon atoms.

2. Hydrocarbons are not soluble in water because
   a. they are hydrophilic.
   b. the C—H bond is nonpolar.
   c. they do not ionize.
   d. they store energy in the many C—H bonds along the carbon backbone.
   e. they are lighter than water.

3. Which of the following is *not* true of an asymmetric carbon atom?
   a. It is attached to four different atoms or groups.
   b. It results in right- and left-handed versions of a molecule.
   c. It can be a part of enantiomers.
   d. Its configuration is in the shape of a tetrahedron.
   e. It can be a part of geometric isomers.

4. A reductionist approach to considering the structure and function of organic molecules would be based on
   a. mechanism.
   b. holism.
   c. determinism.
   d. vitalism.
   e. evolution.

5. The functional group that can cause an organic molecule to act as a base is
   a. —COOH.   c. —SH.   e. —OPO$_3^{2-}$.
   b. —OH.   d. —NH$_2$.

6. The functional group that confers acidic properties to organic molecules is
   a. —COOH.   c. —SH.   e. $>$C $=$ O.
   b. —OH.   d. —NH$_2$.

7. Which is *not* true about structural isomers?
   a. They have different chemical properties.
   b. They have the same molecular formula.
   c. Their atoms and bonds are arranged in different sequences.
   d. They are a result of restricted movement around a carbon double bond.
   e. Their possible numbers increase as carbon skeletons increase in size.

8. The fats stored in your body consist mostly of
   a. amino acids.
   b. alcohols.
   c. carboxylic acids.
   d. hydrocarbons.
   e. organic phosphates.

9. How many asymmetric carbons are there in the sugar ribose?

   a. 1   **c.** 3   **e.** 5

   b. 2   **d.** 4

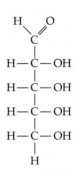

MATCHING: *Match the formulas (**a–f**) to the terms at the right. Choices may be used more than once; more than one right choice may be available.*

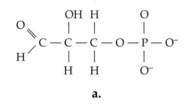

a.

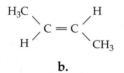

b.

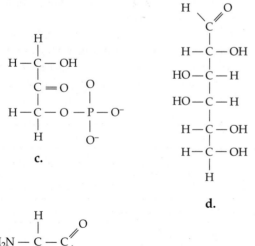

c.

d.

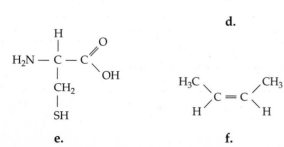

e.   **f.**

_____ **1.** structural isomers

_____ **2.** geometric isomers

_____ **3.** can have enantiomers

_____ **4.** carboxylic acid

_____ **5.** can make cross-link in protein

_____ **6.** hydrophilic

_____ **7.** hydrocarbon

_____ **8.** amino acid

_____ **9.** organic phosphate

_____ **10.** aldehyde

_____ **11.** amine

_____ **12.** ketone

Chapter 5

The Structure and Function of Macromolecules

▶ Framework

The central ideas of this chapter are that molecular function relates to molecular structure and that the diversity of molecular structure is the basis for the diversity of life. Combining a small number of monomers or subunits into unique sequences and three-dimensional structures creates a huge variety of macromolecules. The table below briefly summarizes the major characteristics of the four classes of macromolecules.

▶ Chapter Review

Smaller organic molecules are joined together to form carbohydrates, lipids, proteins, and nucleic acids. These molecules, many of which are giant **macromolecules,** represent another level in the hierarchy of biological organization, and their functions derive from their complex and unique architectures.

5.1 Most macromolecules are polymers, built from monomers

Polymers are chainlike molecules formed from the linking together of many similar or identical small molecules, called **monomers.**

Synthesis and Breakdown of Polymers Monomers are joined by **condensation reactions** (or **dehydration reactions**), in which one monomer provides a hydroxyl group ($-OH$) and the other contributes a hydrogen ($-H$) to release a water molecule. With an input of energy and the help of enzymes, a covalent bond between the monomers is formed.

Hydrolysis is the breaking of bonds between monomers through the addition of water molecules. A hydroxyl group is joined to one monomer while a hydrogen is bonded with the other. Enzymes also control hydrolysis.

Class	Monomers	Functions
Carbohydrates	Monosaccharides	Energy, raw materials, energy storage, structural compounds
Proteins	Amino acids	Enzymes, transport, movement, receptors, defense, structure
Nucleic acids	Nucleotides	Heredity, code for amino acid sequence
Lipids	Glycerol and fatty acids → fats; phospholipids; steroids (do not form polymers)	Energy storage, membranes, hormones

Diversity of Polymers   Macromolecules are constructed from about 40 to 50 common monomers and a few rarer molecules. The seemingly endless variety of polymers arises from the essentially infinite number of possibilities in the sequencing and arrangement of these basic building blocks.

5.2   Carbohydrates serve as fuel and building material

Carbohydrates include sugars and their polymers.

Sugars   **Monosaccharides** have the general formula of $(CH_2O)_n$. The number of these units forming a sugar varies from three to seven, with hexoses $(C_6H_{12}O_6)$, trioses, and pentoses found most commonly. Sugar molecules may be enantiomers due to the spatial arrangement of parts around asymmetric carbons.

Glucose is broken down to yield energy in cellular respiration. Monosaccharides serve also as the raw materials for synthesis of other organic molecules and as monomers that are synthesized into disaccharides or polysaccharides.

■ **INTERACTIVE QUESTION   5.1**

Fill in the blanks to review the structure of monosaccharides.

You can recognize a monosaccharide by its multiple **(a)** _____ groups and its one **(b)** _____ group, whose location determines whether the sugar is an **(c)** _____ or a **(d)** _____ . In aqueous solutions, most monosaccharides form **(e)** _____ .

Sucrose, or table sugar, is a **disaccharide** consisting of a glucose and a fructose molecule. A **glycosidic linkage** is a covalent bond formed by a dehydration reaction between two monosaccharides.

Polysaccharides   **Polysaccharides** are storage or structural macromolecules made from a few hundred to a few thousand monosaccharides. **Starch,** a storage molecule in plants, is a polymer made of glucose molecules joined by 1–4 linkages that give starch a helical shape. Most animals have enzymes to hydrolyze plant starch into

glucose. Animals produce **glycogen,** a highly branched polymer of glucose, as their energy storage form.

Cellulose, the major component of plant cell walls, is the most abundant organic compound on Earth. It differs from starch by the configuration of the ring form of glucose and the resulting geometry of the glycosidic bonds. In a plant cell wall, hydrogen bonds between hydroxyl groups hold parallel cellulose molecules together to form strong microfibrils.

Enzymes that digest the α linkages of starch are unable to hydrolyze the β linkages of cellulose. Only a few organisms (some bacteria, microorganisms, and fungi) have enzymes that can digest cellulose.

Chitin is a structural polysaccharide formed from glucose monomers with a nitrogen-containing group and found in the exoskeleton of arthropods and the cell walls of many fungi.

■ **INTERACTIVE QUESTION   5.2**

Circle the atoms of these two glucose molecules that will be removed by a dehydration reaction. Then draw the resulting maltose molecule with its 1–4 glycosidic linkage (between the number 1 carbon of the first glucose and the number 4 carbon of the second).

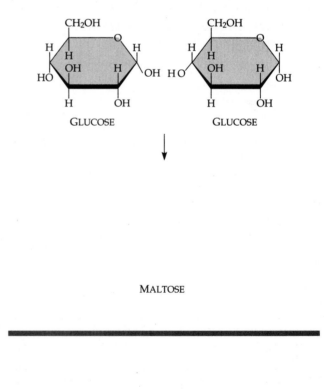

MALTOSE

■ INTERACTIVE QUESTION 5.3

Fill in the following concept map that summarizes this section on carbohydrates.

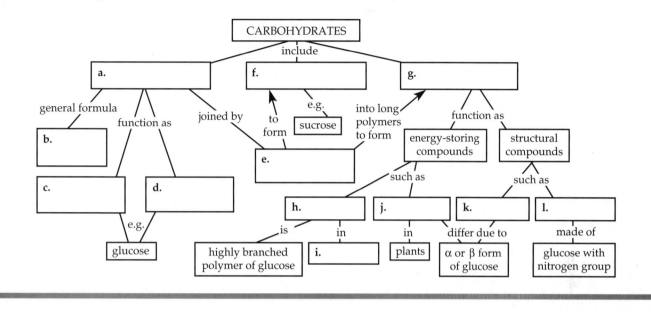

5.3 Lipids are a diverse group of hydrophobic molecules

Fats, phospholipids, and steroids are a diverse assemblage of macromolecules that are classed together as **lipids** based on their hydrophobic behavior. Lipids do not form polymers.

Fats　**Fats** are composed of fatty acids attached to the three-carbon alcohol, glycerol. A **fatty acid** consists of a long hydrocarbon chain with a carboxyl group at one end. The nonpolar hydrocarbons make a fat hydrophobic.

A **triacylglycerol,** or fat, consists of three fatty acid molecules, each linked to glycerol by an ester linkage, a bond that forms between a hydroxyl and a carboxyl group. Triglyceride is another name for fats.

Fatty acids with double bonds in their carbon skeletons are called **unsaturated fatty acids.** The *cis* double bonds create a kink in the hydrocarbon chain and prevent fat molecules that contain unsaturated fatty acids from packing closely together and becoming solidified at room temperature. **Saturated fatty acids** have no double bonds in their carbon skeletons. Most animal fats are saturated and solid at room temperature. The fats of plants and fish are generally unsaturated and are called oils. Diets rich in saturated fats and in *"trans* fats" made in the process of hydrogenating vegetable oils have been linked to cardiovascular disease.

Fats are excellent energy storage molecules, containing twice the energy reserves of carbohydrates such as starch. Adipose tissue, made of fat storage cells, also cushions organs and insulates the body.

Phospholipids　**Phospholipids** consist of a glycerol linked to two fatty acids and a negatively charged phosphate group, to which other small molecules may be attached. The phosphate head of this molecule is hydrophilic and water soluble, whereas the two fatty acid chains are hydrophobic.

The unique structure of phospholipids makes them ideal constituents of cell membranes. Arranged in a bilayer, the hydrophilic heads face toward the aqueous solutions inside and outside the cell, and the hydrophobic tails mingle in the center of the membrane.

■ INTERACTIVE QUESTION 5.4

Sketch a section of a phospholipid bilayer of a membrane, and label the hydrophilic head and hydrophobic tail of one of the phospholipids.

Steroids　**Steroids** are a class of lipids distinguished by four connected carbon rings with various functional groups attached. **Cholesterol** is an important steroid that is a common component of animal cell membranes and a precursor for other steroids, including many hormones.

■ **INTERACTIVE QUESTION   5.5**

Fill in this concept map to help you organize your understanding of lipids

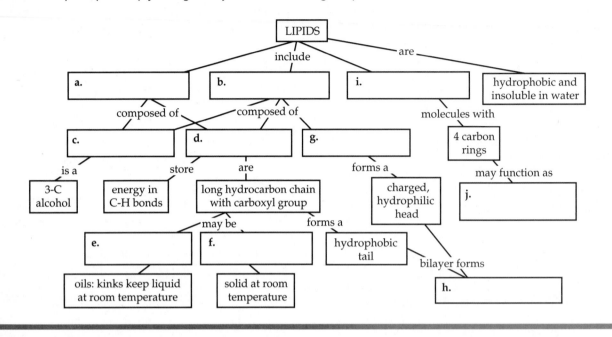

5.4   Proteins have many structures, resulting in a wide range of functions

Proteins are central to almost every function of life. As **catalysts,** protein **enzymes** selectively speed up the chemical reactions of a cell.

Polypeptides   A **polypeptide** is a polymer of amino acids. A **protein** consists of one or more polypeptide chains folded into a specific three-dimensional shape or conformation.

Amino acids are composed of an asymmetric carbon (called the α carbon) bonded to a hydrogen, a carboxyl group, an amino group, and a variable side chain called the R group. At the pH in a cell, the amino and carboxyl groups are usually ionized. The R group confers the unique physical and chemical properties of each amino acid. Side chains may be either nonpolar and hydrophobic, or polar or charged (acidic or basic) and thus hydrophilic.

A **peptide bond** links the amino group of one amino acid with the carboxyl group of another. A polypeptide chain has a free amino group at one end and a free carboxyl group at the other. Polypeptides vary in length from a few to a thousand or more amino acids.

In the late 1940s and early '50s, Sanger determined the primary structure of insulin through the laborious process of hydrolyzing the protein into small peptide chains, determining their amino acid sequences, and then overlapping the sequences of small fragments created with different agents to reconstruct the whole polypeptide. Most of these steps are now automated.

■ **INTERACTIVE QUESTION   5.6**

a. Draw the amino acids alanine (R group—CH_3) and serine (R group—CH_2OH) and then show how a dehydration reaction will form a peptide bond between them.

b. Which of these amino acids has a polar R group? _____ a nonpolar R group? _____

c. What does this molecule segment represent? Note the N-C-C-N-C-C sequence.

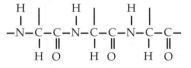

Protein Conformation and Function   Proteins have unique three-dimensional shapes created by the twisting or folding of one or more polypeptide chains. Protein conformation is dependent upon the interactions among the amino acids making up the polypeptide chain and usually arises spontaneously as soon as the protein is synthesized in the cell. The unique conformation of a protein, which results from its sequence of amino acids, enables it to recognize and bind to other molecules.

Primary structure is the unique, genetically coded sequence of amino acids within a protein.

Secondary structure involves the coiling or folding of the polypeptide backbone, stabilized by hydrogen bonds between the electronegative oxygen of one peptide bond and the weakly positive hydrogen attached to the nitrogen of another peptide bond. An **α helix** is a coil produced by hydrogen bonding between every fourth amino acid. A **β pleated sheet** is also held by repeated hydrogen bonds along the polypeptide backbone. This secondary structure forms when regions of the polypeptide chain lie parallel to each other.

Interactions between the various side chains (R groups) of the constituent amino acids produce a protein's **tertiary structure. Hydrophobic interactions** between nonpolar side groups clumped in the center of the molecule due to their repulsion by water, van der Waals interactions among those nonpolar side chains, hydrogen bonds between polar side chains, and ionic bonds between negatively and positively charged side chains produce the stable and unique shape of the protein. Strong covalent bonds, called **disulfide bridges,** may occur between the sulfhydryl side groups of cysteine monomers that have been brought close together by the folding of the polypeptide.

Quaternary structure occurs in proteins that are composed of more than one polypeptide chain. The individual polypeptide subunits are held together in a precise structural arrangement.

Even a slight deviation from the sequence of amino acids can severely affect a protein's function by altering the protein's conformation. A substitution of only one of the 146 amino acids in the primary structure of hemoglobin causes sickle-cell disease.

The interactions that create and maintain secondary and tertiary structure can be disrupted by changes in pH, salt concentration, temperature, or other aspects of the environment, and the protein may **denature,** losing its native conformation and thus its function.

The amino acid sequences of more than 100,000 proteins have been determined. Using the technique of **X-ray crystallography,** coupled with computer modeling and graphics, biochemists have established the three-dimensional shape of thousands of these molecules. Researchers have developed methods for following a protein through its intermediate states on the way to its final form and have discovered **chaperonins,** chaperone proteins that assist other proteins during the folding process, perhaps by providing a sheltered environment.

■ **INTERACTIVE QUESTION 5.7**

In the following diagram of a portion of a polypeptide, label the types of interactions that are shown. What level of structure are these interactions producing?

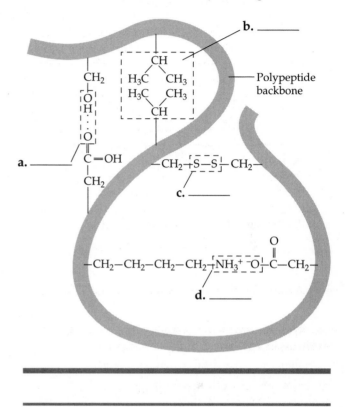

■ **INTERACTIVE QUESTION 5.8**

a. Why would a change in pH cause a protein to denature?

b. Why would transfer to a nonpolar organic solvent (such as ether) cause denaturation?

c. A denatured protein may re-form to its functional shape when returned to its normal environment. What does that indicate about a protein's conformation?

■ **INTERACTIVE QUESTION 5.9**

Now that you have gained more experience with concept maps, create your own map to help you organize the key concepts you have learned about proteins. Try to include the concepts of structure and function and look for cross-links on your map. One version of a protein concept map is included in the answer section, but remember that the real value is in the thinking process you must go through to create your own map.

5.5   Nucleic acids store and transmit hereditary information

Genes are the units of inheritance that determine the primary structure of proteins. **Nucleic acids** are polymers that carry and transmit this code.

The Roles of Nucleic Acids   DNA, **deoxyribonucleic acid,** is the genetic material that is inherited from one generation to the next and is reproduced in each cell of an organism. The instructions coded in DNA are transcribed to **RNA, ribonucleic acid,** which directs the synthesis of proteins, the ultimate enactors of the genetic program. In a eukaryotic cell, DNA resides in the nucleus and messenger RNA carries the instructions for protein synthesis to ribosomes located in the cytoplasm.

■ INTERACTIVE QUESTION   5.10

Show the flow of genetic information in a cell.

_____ → _____ → _____

The Structure of Nucleic Acids   Nucleic acids, also called **polynucleotides,** are polymers of **nucleotides**—monomers that consist of a pentose (five-carbon sugar) covalently bonded to a phosphate group and a nitrogenous base. A monomer without the phosphate group is called a nucleoside. There are two families of nitrogenous bases. **Pyrimidines,** including cytosine (C), thymine (T), and uracil (U), are characterized by six-membered rings of carbon and nitrogen atoms. **Purines,** adenine (A) and guanine (G), add a five-membered ring to the pyrimidine ring. Thymine is only in DNA; uracil is only in RNA. In DNA, the pentose is **deoxyribose;** in RNA it is **ribose.**

Nucleotides are linked together into a polynucleotide by phosphodiester linkages, which join the phosphate of one nucleotide with the sugar of the next. The polymer has two distinct ends: a 5′ end with a phosphate attached to a 5′ carbon and a 3′ end with a hydroxyl group on a 3′ carbon. The nitrogenous bases extend from this backbone of repeating sugar-phosphate units. The unique sequence of bases in a gene codes for the specific amino acid sequence of a protein.

■ INTERACTIVE QUESTION   5.11

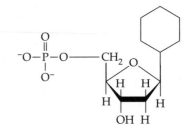

a. Label the three parts of this nucleotide. Indicate with an arrow where the phosphate group of the next nucleotide would attach to build a polynucleotide. Number the carbons of the pentose sugar.

b. Is this a purine or a pyrimidine?

c. Is this a DNA or RNA nucleotide?

The DNA Double Helix   DNA molecules consist of two chains of polynucleotides spiraling around an imaginary axis in a **double helix.** The two chains run in opposite 5′ to 3′ directions, an arrangement called **antiparallel.** In 1953 Watson and Crick first proposed this double-helix arrangement, which consists of two sugar-phosphate backbones on the outside of the helix with their nitrogenous bases pairing and hydrogen-bonding together in the inside. Adenine pairs only with thymine; guanine always pairs with cytosine. Thus, the sequences of nitrogenous bases on the two strands of DNA are complementary. Because of this specific base-pairing property, DNA can replicate itself and precisely copy the genes of inheritance.

DNA and Proteins as Tape Measures of Evolution   Genes form the hereditary link between generations. Closely related members of the same species share many common DNA sequences and proteins. More closely related species have a larger proportion of their DNA and proteins in common. This "molecular genealogy" provides evidence of evolutionary relationships.

■ INTERACTIVE QUESTION   5.12

Take the time to create a concept map that summarizes what you have just reviewed about nucleic acids. Compare your map with that of a study partner or explain it to a friend. One version of a map on nucleic acids is included in the answer section. Refer to Figures 5.26 and 5.27 in your textbook to help you visualize polynucleotides and the double helix of DNA.

**The Theme of Emergent Properties
in the Chemistry of Life: A Review**

At each stage in the hierarchy of levels from atoms through macromolecules, we have seen that novel properties arise with increasing structural organization.

Word Roots

con- = together (*condensation reaction:* a reaction in which two molecules become covalently bonded to each other through the loss of a small molecule, usually water)

di- = two (*disaccharide:* two monosaccharides joined together)

glyco- = sweet (*glycogen:* a polysaccharide sugar used to store energy in animals)

hydro- = water; **-lyse** = break (*hydrolysis:* breaking chemical bonds by adding water)

macro- = large (*macromolecule:* a large molecule)

meros- = part (*polymer:* a chain made from smaller organic molecules)

mono- = single; **-sacchar** = sugar (*monosaccharide:* simplest type of sugar)

poly- = many (*polysaccharide:* many monosaccharides joined together)

tri- = three (*triacylglycerol:* three fatty acids linked to one glycerol molecule)

Structure Your Knowledge

1. Describe the four structural levels in the conformation of a protein.

2. Identify the type of monomer or group shown by the formulas shown on the right. Then match the chemical formulae with their description. Answers may be used more than once.

_____ **1.** molecules that would combine to form a fat

_____ **2.** molecule that would be attached to other monomers by a peptide bond

_____ **3.** molecules or groups that would combine to form a nucleotide

_____ **4.** molecules that are carbohydrates

_____ **5.** molecule that is a purine

_____ **6.** monomer of a protein

_____ **7.** groups that would be joined by phosphodiester bonds

_____ **8.** most nonpolar (hydrophobic) molecule

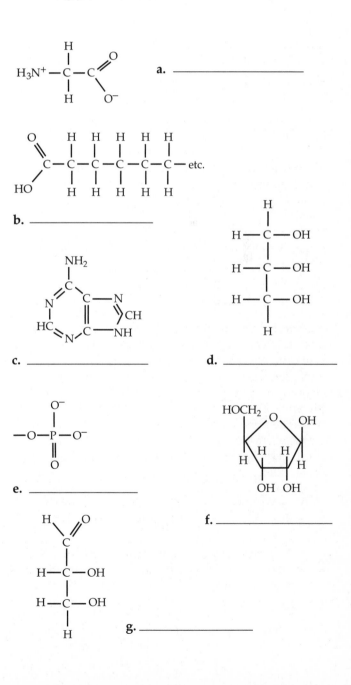

a. _____

b. _____

c. _____

d. _____

e. _____

f. _____

g. _____

Test Your Knowledge

MATCHING: *Match the molecule with its type of molecule.*

_____ **1.** glycogen   **A.** carbohydrate

_____ **2.** cholesterol   **B.** lipid

_____ **3.** RNA   **C.** protein

_____ **4.** collagen   **D.** nucleic acid

_____ **5.** hemoglobin

_____ **6.** a gene

_____ **7.** triacylglycerol

_____ **8.** enzyme

_____ **9.** cellulose

_____ **10.** chitin

MULTIPLE CHOICE: *Choose the one best answer.*

1. Polymerization is a process that
   a. creates bonds between amino acids in the formation of a peptide chain.
   b. involves the removal of a water molecule.
   c. links the sugar of one nucleotide with the phosphate of the next.
   d. requires a condensation or dehydration reaction.
   e. may involve all of the above.

2. Which of the following is *not* true of a pentose?
   a. It can be found in nucleic acids.
   b. It can occur in a ring structure.
   c. It has the formula $C_5H_{12}O_5$.
   d. It has one carbonyl and four hydroxyl groups.
   e. It may be an aldose or a ketose.

3. Disaccharides can differ from each other in all of the following ways *except*
   a. in the number of their monosaccharides.
   b. as enantiomers.
   c. in the monomers involved.
   d. in the location of their glycosidic linkage.
   e. in their structural formulas.

4. Which of the following is *not* true of cellulose?
   a. It is the most abundant organic compound on Earth.
   b. It differs from starch because of the configuration of glucose and the geometry of the glycosidic linkage.
   c. It may be hydrogen-bonded to neighboring cellulose molecules to form microfibrils.
   d. Few organisms have enzymes that hydrolyze its glycosidic linkages.
   e. Its monomers are glucose with nitrogen-containing appendages.

5. Plants store most of their energy as
   a. unsaturated fats.
   b. glycogen.
   c. starch.
   d. sucrose.
   e. cellulose.

6. What happens when a protein denatures?
   a. It loses its primary structure.
   b. It loses its secondary and tertiary structures.
   c. It becomes irreversibly insoluble and precipitates.
   d. It hydrolyzes into component amino acids.
   e. Its hydrogen bonds, ionic bonds, hydrophobic interactions, disulfide bridges, and peptide bonds are disrupted.

7. The α helix of proteins is
   a. part of the tertiary structure and is stabilized by disulfide bridges.
   b. a double helix.
   c. stabilized by hydrogen bonds and commonly found in fibrous proteins.
   d. found in some regions of globular proteins and stabilized by hydrophobic interactions.
   e. a complementary sequence to messenger RNA.

8. A fatty acid that has the formula $C_{16}H_{32}O_2$ is
   a. saturated.
   b. unsaturated.
   c. branched.
   d. hydrophilic.
   e. part of a steroid molecule.

9. Three molecules of the fatty acid in question 8 are joined to a molecule of glycerol ($C_3H_8O_3$). The resulting molecule has the formula
   a. $C_{48}H_{96}O_6$.
   b. $C_{48}H_{98}O_9$.
   c. $C_{51}H_{102}O_8$.
   d. $C_{51}H_{98}O_6$.
   e. $C_{51}H_{104}O_9$.

10. β pleated sheets are characterized by
   a. disulfide bridges between cysteine amino acids.
   b. parallel regions of the polypeptide chain held together by hydrophobic interactions.
   c. folds stabilized by hydrogen bonds between segments of the polypeptide backbone.
   d. membrane sheets composed of phospholipids.
   e. hydrogen bonds between adjacent cellulose molecules.

11. Cows can derive nutrients from cellulose because
   a. they can produce the enzymes that break the β linkages between glucose molecules.
   b. they chew and rechew their cud so that cellulose fibers are finally broken down.
   c. one of their stomachs contains bacteria that can hydrolyze the bonds of cellulose.
   d. their intestinal tract contains termites, which produce enzymes to hydrolyze cellulose.
   e. they can convert cellulose to starch and then hydrolyze starch to glucose.

12. Which of these molecules would provide the most energy (kcal/g) when eaten?
   a. glucose
   b. starch
   c. glycogen
   d. fat
   e. protein

13. What *determines* the sequence of the amino acids in a particular protein?
   a. its primary structure
   b. the sequence of nucleotides in RNA, which was determined by the sequence of nucleotides in the gene for that protein
   c. the sequence of nucleotides in DNA, which was determined by the sequence of nucleotides in RNA
   d. the sequence of RNA nucleotides making up the ribosome
   e. the three-dimensional shape of the protein

14. Sucrose is made from joining a glucose and a fructose molecule in a dehydration reaction. What is the molecular formula for this disaccharide?
   a. $C_6H_{12}O_6$
   b. $C_{10}H_{20}O_{10}$
   c. $C_{12}H_{22}O_{11}$
   d. $C_{12}H_{24}O_{12}$
   e. $C_{12}H_{24}O_{13}$

15. How are the nucleotide monomers connected to form a polynucleotide?
   a. hydrogen bonds between complementary nitrogenous base pairs
   b. ionic attractions between phosphate groups
   c. disulfide bridges between cysteine amino acids
   d. covalent bonds between the sugar of one nucleotide and the phosphate of the next
   e. ester linkages between the carboxyl group of one nucleotide and the hydroxyl group on the ribose of the next

16. Which of the following would be the most hydrophobic molecule?
   a. cholesterol - bec it's embedded bot. hydrophobic & hydrophelic membrane.
   b. nucleotide
   c. amino acid
   d. chitin
   e. glucose

17. What is the best description of this molecule?

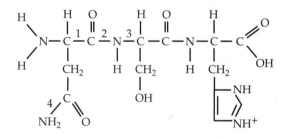

   a. chitin
   b. amino acid
   c. polypeptide (tripeptide)
   d. nucleotide
   e. protein

18. Which number(s) in the molecule in question 17 refer(s) to a peptide bond?
   a. 1   **c.** 3   **e.** both 2 and 4
   b. 2   **d.** 4

19. If the nucleotide sequence of one strand of a DNA helix is GCCTAA, what would be the sequence on the complementary strand?
   a. GCCTAA
   b. CGGAUU
   c. CGGATT
   d. ATTCGG
   e. TAAGCC

20. Monkeys and humans share many of the same DNA sequences and have similar proteins, indicating that
    a. the two groups belong to the same species.
    b. the two groups share a relatively recent common ancestor.
    c. humans evolved from monkeys.
    d. monkeys evolved from humans.
    e. the two groups first appeared on Earth at about the same time.

21. Which of the following would be the major component of the cell membrane of a fungus?
    a. cellulose
    b. chitin
    c. cholesterol
    d. phospholipids
    e. unsaturated fatty acids

22. Hydrophobic as well as hydrophillic interactions would be important for which of the following types of molecules?
    a. proteins
    b. unsaturated fats
    c. glycogen and cellulose
    d. polynucleotides
    e. all of the above

23. What are *trans* fats?
    a. hydrogenated vegetable oils that have been identified with health risks
    b. fats made from cholesterol that are components of plaques in the walls of blood vessels
    c. fats that are derived from animal sources and are associated with cardiovascular disease
    d. fats that contain *trans* double bonds and may contribute to atherosclerosis
    e. polyunsaturated fats produced by removing H from fatty acids and forming *cis* double bonds

24. Which of the following is *not* a function performed by proteins?
    a. transport of oxygen in blood
    b. catalyst for metabolic reactions
    c. protection against disease
    d. signals and receptors
    e. primary component of cell membranes

FILL IN THE BLANKS

1. The man who determined the amino acid sequence of insulin was _____.
2. Cytosine always pairs with _____.
3. Adenine and guanine are _____.
4. A pentose joined to a nitrogenous base and a phosphate group is called a _____.
5. The conformation of a protein is determined by its _____.
6. Proteins with more than one polypeptide chain have _____ structure.
7. The carbohydrate energy storage molecule of animals is _____.
8. Membranes are composed of a bilayer of _____.
9. The insoluble fiber listed on food packages consists primarily of _____.
10. Proteins that assist the proper folding of newly synthesized proteins are called _____.

UNIT 2

The Cell

Chapter 6

A Tour of the Cell

Framework

This chapter deals with the fundamental unit of life—the cell. The complexities in the processes of life are reflected in the complexities of the structure of the cell. It is easy to become overwhelmed by the number of new vocabulary terms for this array of cell organelles and membranes. The following concept map provides an organizational framework for the wealth of detail found in "a tour of the cell."

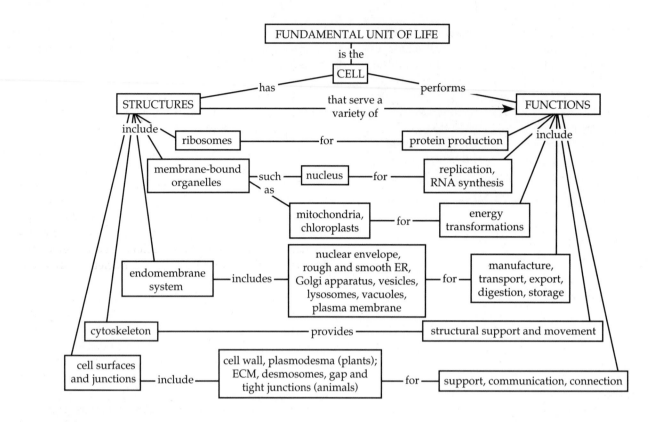

Chapter Review

The cell is the basic structural and functional unit of all living organisms. In the hierarchy of biological organization, the capacity for life emerges from the structural order of the cell. All cells are related through common descent, but evolution has shaped diverse adaptations.

6.1   To study cells, biologists use microscopes and the tools of biochemistry

Microscopy   The glass lenses of **light microscopes (LMs)** refract (bend) the visible light passing through a specimen such that the projected image is magnified. *Magnification* is the ratio of this projected image to the real size of the object. *Resolution* is a measure of how clear an image is, and is determined by the minimum distance two points must be separated to be distinguished. The resolution of the light microscope is limited by the shortest wavelength of visible light, so that details finer than 0.2 µm (micrometers = 10^{-3} mm) cannot be resolved. Staining of specimens and using techniques such as fluorescence, phase-contrast, and confocal microscopy improve visibility by increasing contrast between structures that are large enough to be resolved.

Most subcellular structures, or **organelles,** cannot be resolved by the light microscope. Cells were discovered by Robert Hooke in 1665, but their ultrastructure was largely unknown until the development of the **electron microscope (EM)** in the 1950s. The electron microscope focuses a beam of electrons through the specimen. The short wavelength of the electron beam allows a resolution of about 2 nm (nanometers = 10^{-3} µm), a hundred times greater than that of the light microscope.

In a **scanning electron microscope (SEM),** an electron beam scans the surface of a specimen usually coated with a thin gold film, exciting electrons from the specimen, which are detected and translated into an image on a video screen. This image appears three-dimensional.

In a **transmission electron microscope (TEM),** a beam of electrons is passed through a thin section of a specimen stained with atoms of heavy metals, and electromagnets, acting as lenses, focus the image onto a screen or film.

Modern cell biology integrates *cytology* with *biochemistry* to understand relationships between cellular structure and function.

■ INTERACTIVE QUESTION   6.1

a. Define cytology.

b. What do cell biologists use a TEM to study?

c. What does an SEM show best?

d. What advantages does light microscopy have over TEM and SEM?

Isolating Organelles by Cell Fractionation **Cell fractionation** is a technique that separates major organelles of a cell so that they can be identified and their functions can be studied. Cells are homogenized and the resulting cellular soup is separated into component fractions by centrifugation. The homogenate is first spun slowly, and dense structures settle to form a pellet. The supernatant is centrifuged at increasing speeds, each time isolating smaller and smaller cellular components in the pellet. **Ultracentrifuges** can spin at speeds up to 130,000 rpm.

6.2 Eukaryotic cells have internal membranes that compartmentalize their functions

Comparing Prokaryotic and Eukaryotic Cells All cells are bounded by a plasma membrane, which encloses a semifluid medium called **cytosol.** All cells contain chromosomes and ribosomes. Only members of the domains Bacteria and Archaea have **prokaryotic cells,** which are cells with no nucleus or membrane-enclosed organelles. The DNA of prokaryotic cells is concentrated in a region called the **nucleoid. Eukaryotic cells** have a true nucleus enclosed in a nuclear envelope and numerous organelles suspended in cytosol. **Cytoplasm** refers to the entire region between the nucleus and the plasma membrane, and also to the interior of a prokaryotic cell.

Most bacterial cells range from 1 to 10 μm in diameter, whereas eukaryotic cells are ten times larger, ranging from 10 to 100 μm.

The small size of cells is dictated by geometry and the requirements of metabolism. Area is proportional to the square of linear dimension, while volume is proportional to its cube. The **plasma membrane** surrounding every cell must provide sufficient surface area for exchange of oxygen, nutrients, and wastes relative to the volume of the cell.

■ INTERACTIVE QUESTION 6.2

a. If a eukaryotic cell has a diameter that is 10 times that of a bacterial cell, proportionally how much more surface area would the eukarotic cell have?

b. Proportionally how much more volume would it have?

A Panoramic View of the Eukaryotic Cell Membranes compartmentalize the eukaryotic cell, providing local environments for specific metabolic functions and participating in metabolism through membrane-

bound enzymes. Membranes are composed of a bilayer of phospholipid molecules associated with diverse proteins.

6.3 The eukaryotic cell's genetic instructions are housed in the nucleus and carried out by the ribosomes

The Nucleus: Genetic Library of the Cell The **nucleus** is surrounded by the **nuclear envelope,** a double membrane perforated by pores that regulate the movement of large macromolecules between the nucleus and the cytoplasm. The inner membrane is lined by the **nuclear lamina,** a layer of protein filaments that helps to maintain the shape of the nucleus. There is evidence of a framework of fibers, called a nuclear matrix, extending through the nucleus.

Most of the cell's DNA is located in the nucleus, where it is organized into units called **chromosomes,** which are made up of a complex of DNA and proteins called **chromatin.** Each eukaryotic species has a characteristic chromosomal number. Individual chromosomes are visible only when coiled and condensed in a dividing cell.

The **nucleolus,** a dense structure visible in the nondividing nucleus, synthesizes ribosomal RNA and combines it with protein to assemble ribosomal subunits that pass through nuclear pores to the cytoplasm.

■ INTERACTIVE QUESTION 6.3

How does the nucleus control protein synthesis in the cytoplasm?

Ribosomes: Protein Factories in the Cell Ribosomes are particles composed of protein and ribosomal RNA. Most of the proteins produced on *free ribosomes* are used within the cytosol. *Bound ribosomes,* attached to the endoplasmic reticulum or nuclear envelope, usually make proteins that will be included within membranes, packaged into organelles, or exported from the cell.

6.4 The endomembrane system regulates protein traffic and performs metabolic functions in the cell

The **endomembrane system** of a cell consists of the nuclear envelope, endoplasmic reticulum, Golgi apparatus, lysosomes, vacuoles, and the plasma membrane. These membranes are all related either through direct contact or by the transfer of membrane segments by membrane-bound sacs called **vesicles.**

The Endoplasmic Reticulum: Biosynthetic Factory

The **endoplasmic reticulum (ER)** is the most extensive portion of the endomembrane system. It is continuous with the nuclear envelope and encloses a network of interconnected tubules or compartments called cisternae. Ribosomes are attached to the cytoplasmic surface of **rough ER; smooth ER** lacks ribosomes.

Smooth ER serves diverse functions in different cells: Its enzymes are involved in phospholipid, steroid, and sex hormone synthesis; carbohydrate metabolism; and detoxification of drugs and poisons. Barbiturates, alcohol, and other drugs increase a liver cell's production of smooth ER, thus leading to an increased tolerance (and thus reduced effectiveness) for these and other drugs. Smooth ER also functions in storage and release of calcium ions during muscle contraction.

Proteins intended for secretion are manufactured by membrane-bound ribosomes and then threaded into the lumen of the rough ER, where they fold into their native conformation. Many are covalently bonded to small carbohydrates to form **glycoproteins.** Secretory proteins are transported from the rough ER in membrane-bound **transport vesicles.**

Rough ER manufactures membranes for the cell. Enzymes built into the membrane assemble phospholipids, and membrane proteins formed on bound ribosomes are inserted into the ER membrane. Transport vesicles transfer ER membrane to other parts of the endomembrane system.

The Golgi Apparatus: Shipping and Receiving Center

The **Golgi apparatus** consists of a stack of flattened membranous sacs. Vesicles that bud from the ER join to the *cis* face of a Golgi stack, adding to it their contents and membrane. Products that travel through the Golgi apparatus are usually modified or refined as they move from the *cis* region to the *trans* region, perhaps transferred in vesicles from one cisterna to the next. According to the *cisternal maturation model,* Golgi products are processed and tagged as the cisternae themselves progress from the *cis* to the *trans* face. Some polysaccharides, such as pectins, are manufactured by the Golgi of plant cells. Golgi products are sorted into vesicles, which pinch off from the *trans* face of the Golgi apparatus. These vesicles may have surface molecules that help direct them to the plasma membrane or to other organelles.

Lysosomes: Digestive Compartments

Lysosomes are membrane-enclosed sacs of hydrolytic enzymes used by animal cells to digest macromolecules. Lysosomes provide an acidic pH for these enzymes.

In some protists, lysosomes fuse with food vacuoles to digest material ingested by **phagocytosis.** Macrophages, a type of white blood cell, use lysosomes to destroy ingested bacteria. Lysosomes also recycle a cell's own macromolecules by engulfing damaged organelles or small bits of cytosol, a process known as *autophagy.* During development or metamorphosis, programmed cell death or apoptosis is mediated by lysosomes and helps to create a specific body form.

Vacuoles: Diverse Maintenance Compartments

Food vacuoles are formed as a result of phagocytosis. **Contractile vacuoles** pump excess water out of freshwater protists. A large **central vacuole** is found in mature plant cells, surrounded by a membrane called the **tonoplast.** This vacuole stores organic compounds and inorganic ions for the cell. Poisonous or unpalatable compounds, which may protect the plant from predators, and dangerous metabolic by-products may also be contained in the vacuole. Plant cells increase in size with a minimal addition of new cytoplasm as their vacuoles absorb water and expand.

The Endomembrane System: A Review

As membranes move from the ER to the Golgi and then to other organelles, their compositions, functions, and contents are modified.

■ INTERACTIVE QUESTION 6.4

Name the components of the endomembrane system shown in this diagram and review the functions of each of these membranes.

a.

b.

c.

d.

e.

f.

g.

6.5 Mitochondria and chloroplasts change energy from one form to another

Cellular respiration, the metabolic processing of fuels to produce ATP, occurs within the **mitochondria** of eukaryotic cells. Photosynthesis occurs in the **chloroplasts** of plants and algae, which produce organic compounds from carbon dioxide and water by absorbing solar energy. The membrane proteins of mitochondria and chloroplasts are made by ribosomes either free in the cytosol or contained within these organelles. They also contain a small amount of DNA that directs the synthesis of some of their proteins. **Peroxisomes** are oxidative organelles that are also not part of the endomembrane system.

Mitochondria: Chemical Energy Conversion Two membranes, each a phospholipid bilayer with unique embedded proteins, enclose a mitochondrion. A narrow intermembrane space exists between the smooth outer membrane and the convoluted inner membrane. The folds of the inner membrane, called **cristae,** create a large surface area and enclose the **mitochondrial matrix.** Many respiratory enzymes, mitochondrial DNA, and ribosomes are housed in this matrix. Other respiratory enzymes and proteins are built into the inner membrane.

Chloroplasts: Capture of Light Energy **Plastids** are plant organelles that include *amyloplasts,* which store starch; *chromoplasts,* which contain pigments; and *chloroplasts,* which contain the green pigment chlorophyll and function in photosynthesis.

Chloroplasts are bounded by at least two membranes separated by a thin intermembrane space. Inside the inner membrane is a fluid called the **stroma** surrounding a membranous system of flattened sacs called

thylakoids, inside of which is the thylakoid space. Thylakoids may be stacked together to form structures called **grana.**

■ INTERACTIVE QUESTION   6.5

Sketch a mitochondrion and a chloroplast and label their membranes and compartments.

Peroxisomes: Oxidation Peroxisomes are membrane-enclosed compartments filled with enzymes that function in a variety of metabolic pathways, such as breaking down fatty acids for energy or detoxifying alcohol and other poisons. An enzyme that converts hydrogen peroxide (H_2O_2), a toxic by-product of these pathways, to water is also packaged into peroxisomes.

■ INTERACTIVE QUESTION   6.6

Why are peroxisomes not considered part of the endomembrane system?

6.6   The cytoskeleton is a network of fibers that organizes structures and activities in the cell

Roles of the Cytoskeleton: Support, Motility, and Regulation The **cytoskeleton** is a network of fibers that give mechanical support, function in cell motility (of both internal structures and the cell as a whole), and transmit mechanical signals from the cell's surface to its interior. The cytoskeleton interacts with special proteins called **motor proteins** to produce cellular movements.

Components of the Cytoskeleton The three main types of fibers involved in the cytoskeleton are **microtubules, microfilaments,** and **intermediate filaments.**

All eukaryotic cells have microtubules, which are hollow rods constructed of columns of globular proteins called tubulins. In addition to providing the supporting framework of the cell, microtubules serve as tracks along which organelles move with the aid of motor molecules.

In many cells, microtubules radiate out from a region near the nucleus called a **centrosome.** In animal cells, a pair of **centrioles,** each composed of nine sets of triplet microtubules arranged in a ring, is associated with the centrosome and replicates before cell division.

Cilia and **flagella** are locomotor extensions of some eukaryotic cells. Cilia are numerous and short; flagella occur one or two to a cell and are longer. Many protists use cilia or flagella to move through aqueous media. Cilia or flagella attached to stationary cells of a tissue move fluid past the cell.

Both cilia and flagella are composed of two single microtubules surrounded by a ring of nine doublets of microtubules (a nearly universal "9 + 2" arrangement), all of which are enclosed in an extension of the plasma membrane. A **basal body,** structurally identical to a centriole, anchors the tubules in the cell. ATP drives the sliding of the microtubule doublets past each other as arms, composed of the motor protein **dynein,** alternately attach to adjacent doublets, pull down, release, and reattach. In conjunction with anchoring cross-linking proteins and radial spokes, this action causes the bending of the flagellum or cilium.

Microfilaments, probably present in all eukaryotic cells, are solid rods consisting of a twisted double chain of molecules of the globular protein **actin.** Also called actin filaments, microfilaments function in support, forming a network just inside the plasma membrane and the core of small cytoplasmic extensions called microvilli.

In muscle cells, thousands of actin filaments interdigitate with thicker filaments made of the protein **myosin.** The sliding of actin and myosin filaments past each other causes the contraction of muscles.

Actin and myosin also interact in localized contractions such as cleavage furrows in animal cell division and amoeboid movements. Actin subunits reversibly assemble into microfilaments and then networks, driving the conversion of cytoplasm from sol to gel during the extension and retraction of **pseudopodia.** Actin filaments interacting with myosin may propel cytoplasm forward into pseudopodia. **Cytoplasmic streaming** in plant cells appears to involve both actin–myosin interactions and sol–gel conversions.

Intermediate filaments are intermediate in size between microtubules and microfilaments and are more diverse in their composition. Intermediate fibers appear to be important in maintaining cell shape. The nucleus is securely held in a web of intermediate filaments, and the nuclear lamina lining the inside of the nuclear envelope is composed of intermediate filaments.

■ INTERACTIVE QUESTION 6.7

Fill in the following table to organize what you have learned about the components of the cytoskeleton. You may wish to refer to the textbook for additional details.

Cytoskeleton	Structure and Monomers	Functions
Microtubules	**a.**	**b.**
Microfilaments (actin filaments)	**c.**	**d.**
Intermediate filaments	**e.**	**f.**

6.7 Extracellular components and connections between cells help coordinate cellular activities

Cell Walls of Plants Plant **cell walls** are composed of microfibrils of cellulose embedded in a matrix of polysaccharides and protein.

The **primary cell wall** secreted by a young plant cell is relatively thin and flexible. Adjacent cells are glued together by the **middle lamella,** a thin layer of polysaccharides (called pectins). When they stop growing, some cells secrete a thicker and stronger **secondary cell wall** between the plasma membrane and the primary cell wall.

■ INTERACTIVE QUESTION 6.8

Sketch two adjacent plant cells, and show the location of the primary and secondary cell walls and the middle lamella.

The Extracellular Matrix (ECM) of Animal Cells Animal cells secrete an **extracellular matrix (ECM)** composed primarily of glycoproteins. **Collagen,** the most abundant glycoprotein of the ECM, forms strong fibers that are embedded in a network of proteoglycan complexes. **Proteoglycans** consist of a small core protein with many attached carbohydrate chains. Cells may be attached to the ECM by **fibronectins** and other glycoproteins that bind to **integrins,** receptor proteins that span the plasma membrane and bind to microfilaments of the cytoskeleton. Thus, information about changes inside and outside the cell can be exchanged through a mechanical signaling pathway involving fibronectins, integrins, and the microfilaments of the cytoskeleton. Signals from the ECM appear to influence the activity of genes in the nucleus.

Intercellular Junctions **Plasmodesmata** are channels in plant cell walls through which the plasma membranes of bordering cells connect, thus linking most cells of a plant into a living continuum. Water, small solutes, and even some proteins and RNA molecules can move through these channels.

There are three main types of intercellular junctions between animal cells. At **tight junctions,** proteins hold adjacent cell membranes tightly together, creating an impermeable seal across a layer of epithelial cells. **Desmosomes,** reinforced by intermediate filaments, are anchoring junctions between adjacent cells. **Gap junctions** (also called *communicating junctions*) are cytoplasmic connections that allow for the exchange of ions and small molecules between cells through protein-surrounded pores.

■ INTERACTIVE QUESTION   6.9

Return to your sketch of plant cells in Interactive Question 6.8 and draw in a plasmodesma.

The Cell: A Living Unit Greater Than the Sum of Its Parts   The compartmentalization and the many specialized organelles typical of cells exemplify the principle that structure correlates with function. The intricate functioning of a living cell emerges from the complex interactions of its multiple parts.

Word Roots

centro- = the center; **-soma** = a body (*centrosome:* material present in the cytoplasm of all eukaryotic cells and important during cell division)

chloro- = green (*chloroplast:* the site of photosynthesis in plants and eukaryotic algae)

cili- = hair (*cilium:* a short hair-like cellular appendage with a microtubule core)

cyto- = cell (*cytosol:* a semifluid medium in a cell in which are located organelles)

-ell = small (*organelle:* a small formed body with a specialized function found in the cytoplasm of eukaryotic cells)

endo- = inner (*endomembrane system:* the system of membranes within a cell that includes the nuclear envelope, endoplasmic reticulum, Golgi apparatus, lysosomes, vacuoles, and the plasma membrane)

eu- = true (*eukaryotic cell:* a cell that has a true nucleus)

extra- = outside (*extracellular matrix:* the substance in which animal tissue cells are embedded)

flagell- = whip (*flagellum:* a long whip-like cellular appendage that moves cells)

glyco- = sweet (*glycoprotein:* a protein covalently bonded to a carbohydrate)

lamin- = sheet/layer (*nuclear lamina:* a netlike array of protein filaments that maintains the shape of the nucleus)

lyso- = loosen (*lysosome:* a membrane-bounded sac of hydrolytic enzymes that a cell uses to digest macromolecules)

micro- = small; **-tubul** = a little pipe (*microtubule:* a hollow rod of tubulin protein in the cytoplasm of almost all eukaryotic cells)

nucle- = nucleus; **-oid** = like (*nucleoid:* the region where the genetic material is concentrated in prokaryotic cells)

phago- = to eat; **-kytos** = vessel (*phagocytosis:* a form of cell eating in which a cell engulfs a smaller organism or food particle)

plasm- = molded; **-desma** = a band or bond (*plasmodesmata:* an open channel in a plant cell wall)

pro- = before; **karyo-** = nucleus (*prokaryotic cell:* a cell that has no nucleus)

pseudo- = false; **-pod** = foot (*pseudopodium:* a cellular extension of amoeboid cells used in moving and feeding)

thylaco- = sac or pouch (*thylakoid:* a series of flattened sacs within chloroplasts)

tono- = stretched; **-plast** = molded (*tonoplast:* the membrane that encloses a large central vacuole in a mature plant cell)

trans- = across; **-port** = a harbor (*transport vesicle:* a membranous compartment used to enclose and transport materials from one part of a cell to another)

ultra- = beyond (*ultracentrifuge:* a machine that spins test tubes at the fastest speeds to separate liquids and particles of different densities)

vacu- = empty (*vacuole:* sac that buds from the ER, Golgi, or plasma membrane)

Structure Your Knowledge

1. The table below lists the general functions performed by an animal cell. List the cellular structures associated with each of these functions.

Functions	Associated Organelles and Structures
Cell division	a.
Information storage and transferal	b.
Energy conversions	c.
Manufacture of membranes and products	d.
Lipid synthesis, drug detoxification	e.
Digestion, recycling	f.
Conversion of H_2O_2 to water	g.
Structural integrity	h.
Movement	i.
Exchange with environment	j.
Cell to cell connections	k.

2. This table lists structures that are found in plant cells. Fill in the functions of these structures.

Plant Cell Structures	Functions
Cell wall	a.
Central vacuole	b.
Chloroplast	c.
Amyloplast	d.
Plasmodesmata	e.

3. Label the indicated structures in this diagram of an animal cell.

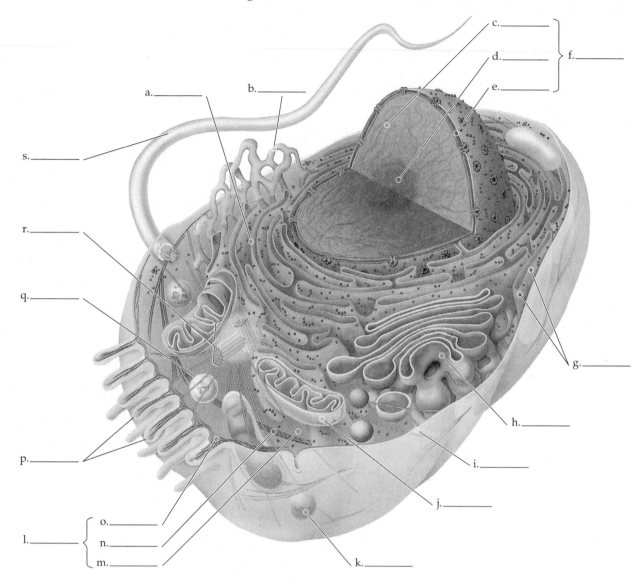

4. Create a diagram or flow chart in the space below to trace the development of a secretory product (such as a digestive enzyme) from the DNA code to its export from the cell.

Test Your Knowledge

MULTIPLE CHOICE: *Choose the one best answer.*

1. Which of the following is/are not found in a prokaryotic cell?
   a. ribosomes
   b. plasma membrane
   c. mitochondria
   d. a and c
   e. a, b, and c

2. Resolution of a microscope is
   a. the distance between two separate points.
   b. the sharpness or clarity of an image.
   c. the degree of magnification of an image.
   d. the depth of focus on a specimen's surface.
   e. the wavelength of light.

3. Which of the following is *not* a similarity among the nucleus, chloroplasts, and mitochondria?
   a. They contain DNA.
   b. They are bounded by two phospholipid bilayer membranes.
   c. They can divide to reproduce themselves.
   d. They are derived from the endoplasmic reticulum system.
   e. Their membranes are associated with specific proteins.

4. The pores in the nuclear envelope provide for the movement of
   a. proteins into the nucleus.
   b. ribosomal subunits out of the nucleus.
   c. mRNA out of the nucleus.
   d. signal molecules into the nucleus.
   e. all of the above.

5. The ultrastructure of a chloroplast could be seen with the best resolution using
   a. transmission electron microscopy.
   b. scanning electron microscopy.
   c. phase-contrast light microscopy.
   d. cell fractionation.
   e. fluorescence microscopy.

6. Which of the following is *incorrectly* paired with its function?
   a. peroxisome—contains enzymes that break down H_2O_2
   b. nucleolus—produces ribosomal RNA, assembles ribosome subunits
   c. Golgi apparatus—processes, tags, and ships cellular products
   d. lysosome—food sac formed by phagocytosis
   e. ECM (extracellular matrix)—supports and anchors cells, communicates information with inside of cell

7. The cytoskeleton is composed of which type of molecule?
   a. protein
   b. cellulose
   c. chitin
   d. phospholipid
   e. calcium phosphate

8. A growing plant cell elongates primarily by
   a. increasing the number of vacuoles.
   b. synthesizing more cytoplasm.
   c. taking up water into its central vacuole.
   d. synthesizing more cellulose.
   e. producing a secondary cell wall.

9. The innermost portion of the cell wall of a plant cell specialized for support is the
   a. primary cell wall.
   b. secondary cell wall.
   c. middle lamella.
   d. plasma membrane.
   e. plasmodesmata.

10. Contractile elements of muscle cells are
    a. intermediate filaments.
    b. centrioles.
    c. microtubules.
    d. actin filaments (microfilaments).
    e. fibronectins.

11. Microtubules are components of all of the following *except*
    a. centrioles.
    b. the spindle apparatus for separating chromosomes in cell division.
    c. tracks along which organelles can move using motor molecules.
    d. flagella and cilia.
    e. the pinching apart of the cytoplasm in animal cell division.

12. Of the following, which is probably the most common route for membrane flow in the endomembrane system?
    a. rough ER → Golgi → lysosomes → nuclear membrane → plasma membrane
    b. rough ER → transport vesicles → Golgi → vesicles → plasma membrane
    c. nuclear envelope → rough ER → Golgi → smooth ER → lysosomes
    d. rough ER → vesicles → Golgi → smooth ER → plasma membrane
    e. smooth ER → vesicles → Golgi → vesicles → peroxisomes

13. Proteins to be used within the cytosol are generally synthesized
    a. by ribosomes bound to rough ER.
    b. by free ribosomes.
    c. by the nucleolus.
    d. within the Golgi apparatus.
    e. by mitochondria and chloroplasts.

14. Plasmodesmata in plant cells are similar in function to
    a. desmosomes.
    b. tight junctions.
    c. gap junctions.
    d. the extracellular matrix.
    e. integrins.

15. In an animal cell fractionation procedure, the first pellet formed would most likely contain
    a. the extracellular matrix.
    b. ribosomes.
    c. mitochondria.
    e. nuclei.
    d. lysosomes.

Use the cells described as follows to answer questions 16–20.
    a. muscle cell in the thigh muscle of a long-distance runner
    b. pancreatic cell that manufactures digestive enzymes
    c. macrophage (white blood cell) that engulfs bacteria
    d. epithelial cell lining digestive tract
    e. ovarian cell that produces estrogen (a steroid hormone)

16. In which cell would you expect to find the most tight junctions?

17. In which cell would you expect to find the most lysosomes?

18. In which cell would you expect to find the most smooth endoplasmic reticulum?

19. In which cell would you expect to find the most bound ribosomes?

20. In which cell would you expect to find the most mitochondria?

FILL IN THE BLANKS *with the appropriate cellular organelle or structure.*

_____ 1. transports membranes and products to various locations

_____ 2. infoldings of the inner mitochondrial membrane with attached enzymes

_____ 3. consists of collagen, proteoglycans, and fibronectins

_____ 4. specialized metabolic compartments with enzymes that transfer hydrogen to oxygen, producing H_2O_2

_____ 5. stacks of flattened sacs inside chloroplasts

_____ 6. anchoring structure for cilia and flagella

_____ 7. semifluid medium between nucleus and plasma membrane

_____ 8. system of fibers that maintains cell shape, anchors organelles

_____ 9. connection between animal cells that creates impermeable layer

_____ 10. membrane surrounding central vacuole of plant cells

Chapter 7

Membrane Structure and Function

Framework

This chapter presents the fluid mosaic model of membrane structure, relating the molecular structure of biological membranes to their function of regulating the passage of substances into and out of the cell.

Chapter Review

The plasma membrane is the boundary of life; this **selectively permeable** membrane allows the cell to maintain a unique internal environment and to control the movement of materials into and out of the cell.

7.1 Cellular membranes are fluid mosaics of lipids and proteins

According to the currently accepted **fluid mosaic model,** the structure of biological membranes consists of various proteins that are attached to or embedded in a bilayer of **amphipathic** phospholipids.

Membrane Models: Scientific Inquiry   Early chemical analysis of membranes revealed a lipid and protein composition. In 1925 two Dutch scientists suggested that cell membranes must be a phospholipid bilayer, with the hydrophobic hydrocarbon tails in the center and the hydrophilic heads facing the aqueous solution on both sides of the membrane. In 1935 H. Davson and J. Danielli proposed a model in which a double layer of phospholipids is covered with a coat of hydrophilic proteins. This sandwich model was consistent with the first views of membranes seen with electron microscopy, and by the

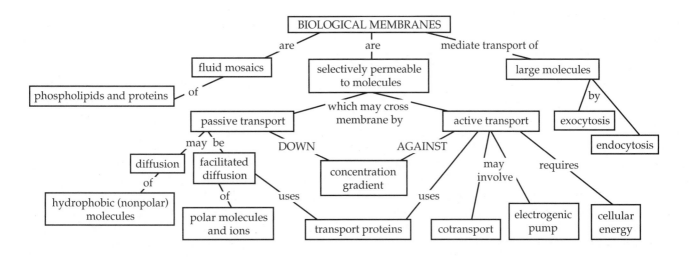

1960s, the Davson-Danielli model was widely accepted for all cellular membranes.

In 1972 S. J. Singer and G. Nicolson proposed their fluid mosaic model in which amphipathic membrane proteins are embedded in the phospholipid bilayer with their hydrophilic regions extending out into the aqueous environment. The phospholipid bilayer is envisioned as fluid, with a mosaic of protein molecules floating in it.

The fluid mosaic model is supported by evidence from freeze-fracture electron microscopy. This technique involves freezing a specimen, fracturing it with a cold knife, and examining the exposed interior of a membrane with the electron microscope.

The fluid mosaic model is currently the most accepted and useful model for organizing existing knowledge and extending further research on membrane structure.

■ INTERACTIVE QUESTION   7.1

Label the components in this diagram of the fluid mosaic model of membrane structure. Indicate the regions that are hydrophobic and those that are hydrophilic.

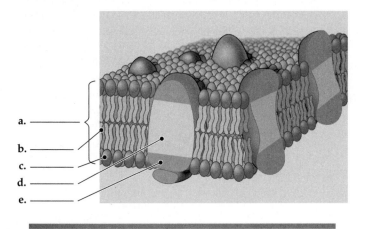

a. _____
b. _____
c. _____
d. _____
e. _____

The Fluidity of Membranes   Membranes are held together primarily by weak hydrophobic interactions that allow the lipids and some of the proteins to drift laterally. Some membrane proteins seem to be held rigid by attachments to the cytoskeleton; others appear to be directed in their movements by cytoskeletal fibers.

Phospholipids with unsaturated hydrocarbon tails maintain membrane fluidity at lower temperatures. The steroid cholesterol, common in plasma membranes of animals, restricts movement of phospholipids and thus reduces fluidity at warmer temperatures. Cholesterol also prevents the close packing of lipids and thus enhances fluidity at lower temperatures.

■ INTERACTIVE QUESTION   7.2

a. Cite some experimental evidence that shows that membrane proteins drift.

b. How might the plasma membrane of a plant cell change in response to the cold temperatures of winter?

Membrane Proteins and Their Functions   Each membrane has its own unique complement of membrane proteins, which determine most of the specific functions of that membrane. **Integral proteins** often extend through the membrane (transmembrane), with two hydrophilic ends and a hydrophobic midsection, usually consisting of one or more α-helical stretches of hydrophobic amino acids. **Peripheral proteins** are attached to the surface of the membrane, often to integral proteins. Attachments of membrane proteins to the cytoskeleton on the cytoplasmic side and fibers of the extracellular matrix on the exterior provide a supportive framework for the plasma membrane.

■ INTERACTIVE QUESTION   7.3

List the six major kinds of functions that membrane proteins may perform.

The Role of Membrane Carbohydrates in Cell-Cell Recognition   The ability of a cell to distinguish other cells is based on recognition of membrane carbohydrates. The **glycolipids** and **glycoproteins** attached to the outside of plasma membranes vary from species to species, from individual to individual, and even among cell types.

Synthesis and Sidedness of Membranes   Membranes have distinct inner and outer faces, related to the composition of the lipid layers, the directional orientation of their proteins, and the attachment of carbohydrates to the exterior surface. Carbohydrates are attached to membrane proteins as they are synthesized in the ER and modified in the Golgi, and attached to lipids in the Golgi. When transport vesicles fuse with the plasma membrane, these interior glycoproteins and glycolipids become located on the exterior face of the membrane.

7.2 Membrane structure results in selective permeability

The plasma membrane permits a regular exchange of nutrients, waste products, oxygen, and inorganic ions. Biological membranes are selectively permeable; the ease and rate at which small molecules pass through them differ.

The Permeability of the Lipid Bilayer Hydrophobic, nonpolar molecules, such as hydrocarbons, CO_2, and O_2, can dissolve in and cross through a membrane.

■ INTERACTIVE QUESTION 7.4

What types of molecules have difficulty crossing the plasma membrane? Why?

Transport Proteins Ions and polar molecules may move across the plasma membrane with the aid of **transport proteins.** Hydrophilic passageways through a membrane are provided for specific molecules by channel proteins, such as **aquaporins,** which facilitate water passage. Carrier proteins may physically bind and transport a specific molecule.

7.3 Passive transport is diffusion of a substance across a membrane with no energy investment

Diffusion Diffusion is the movement of a substance down its **concentration gradient** due to random thermal motion. The diffusion of one solute is unaffected by the concentration gradients of other solutes. The cell does not expend energy when substances diffuse across membranes down their concentration gradient; therefore, the process is called **passive transport.**

Effects of Osmosis on Water Balance Osmosis is the diffusion of water across a selectively permeable membrane. Water diffuses down its own concentration gradient, which is affected by the solute concentration. Binding of water molecules to solute particles lowers the proportion of unbound water that is free to cross the membrane.

■ INTERACTIVE QUESTION 7.5

A solution of 1 *M* glucose is separated by a selectively permeable membrane from a solution of 0.2 *M* fructose and 0.7 *M* sucrose. The membrane is not permeable to the sugar molecules. Indicate which side initially has more free water molecules and which has fewer. Show the direction of osmosis.

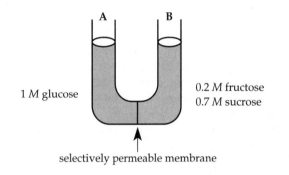

selectively permeable membrane

Tonicity, the tendency of a cell to gain or lose water in a given solution, is affected by the permeability of the plasma membrane and whether or not the cell has a wall. An animal cell will neither gain nor lose water in an **isotonic** environment. An animal cell placed in a **hypertonic** environment will lose water and shrivel. If placed in a **hypotonic** solution, the cell will gain water, swell, and possibly lyse (burst). Cells without rigid walls must either live in an isotonic environment, such as salt water or isotonic body fluids, or have adaptations for **osmoregulation,** the control of water balance.

■ INTERACTIVE QUESTION 7.6

a. What osmotic problems do freshwater protists face?

b. What adaptations may help them osmoregulate?

The cell walls of plants, fungi, prokaryotes, and some protists play a role in water balance within hypotonic environments. Water moving into the cell causes the cell to swell against its cell wall, creating a turgid cell. **Turgid** cells provide mechanical support for nonwoody plants. Plant cells in an isotonic surrounding are **flaccid.** In a hypertonic medium, a plant cell undergoes **plasmolysis,** the pulling away of the plasma membrane from the cell wall as water leaves and the cell shrivels.

■ **INTERACTIVE QUESTION   7.7**

a. The ideal osmotic environment for animal cells is _____.

b. The ideal osmotic environment for plant cells is _____.

Facilitated Diffusion: Passive Transport Aided by Proteins   **Facilitated diffusion** involves the diffusion of polar molecules and ions across a membrane with the aid of transport proteins, either channel proteins or carrier proteins. Aquaporins greatly speed up diffusion of water. Many **ion channels** are **gated channels,** which open or close in response to electrical or chemical stimuli.

The binding of the solute to a carrier protein may cause a conformational change that serves to translocate the binding site and attached solute across the membrane.

■ **INTERACTIVE QUESTION   7.8**

Why is facilitated diffusion considered passive transport?

7.4   Active transport uses energy to move solutes against their gradients

The Need for Energy in Active Transport   **Active transport,** requiring the expenditure of energy by the cell, is essential for a cell to maintain internal concentrations of small molecules that differ from environmental concentrations. The terminal phosphate group of ATP may be transferred to a carrier protein, inducing it to change its conformation and translocate the bound solute across the membrane. The **sodium-potassium pump** works this way to exchange Na^+ and K^+ across animal cell membranes, creating a greater concentration of potassium ions and a lesser concentration of sodium ions within the cell.

Maintenance of Membrane Potential by Ion Pumps   Cells have a **membrane potential,** a voltage across the plasma membrane due to the unequal distribution of ions. This electrical potential energy results from the separation of opposite charges: The cytoplasm of a cell is negatively charged compared to extracellular fluid.

The membrane potential favors the diffusion of cations into the cell and anions out of the cell. Both the membrane potential and the concentration gradient affect the diffusion of an ion; thus, an ion diffuses down its **electrochemical gradient.**

Electrogenic pumps are membrane proteins that generate voltage across a membrane by the active transport of ions. **A proton pump** that transports H^+ out of the cell generates membrane potentials in plants, fungi, and bacteria.

■ **INTERACTIVE QUESTION   7.9**

The Na^+-K^+ pump, the major electrogenic pump in animal cells, exchanges sodium ions for potassium ions, both of which are cations. How does this exchange generate a membrane potential?

Cotransport: Coupled Transport by a Membrane Protein   **Cotransport** is a mechanism through which the active transport of a solute is indirectly driven by an ATP-powered pump that transports another substance against its gradient. As that substance diffuses back down its concentration gradient through a specialized cotransporter, the solute is carried against its concentration gradient across the membrane.

7.5   Bulk transport across the plasma membrane occurs by exocytosis and endocytosis

Exocytosis   In **exocytosis,** the cell secretes macromolecules by the fusion of vesicles with the plasma membrane.

Endocytosis   In **endocytosis,** a region of the plasma membrane sinks inward and pinches off to form a vesicle containing material that had been outside the cell. **Phagocytosis** is a form of endocytosis in which pseudopodia wrap around a food particle, creating a vacuole that then fuses with a lysosome containing hydrolytic enzymes. In **pinocytosis,** droplets of extracellular fluid are taken into the cell in small vesicles. **Receptor-mediated endocytosis** allows a cell to acquire specific substances from extracellular fluid. **Ligands,** molecules that bind specifically to receptor sites, attach to proteins usually clustered in coated pits on the cell surface and are carried into the cell when a vesicle forms.

■ INTERACTIVE QUESTION 7.10

a. How is cholesterol, which is used for the synthesis of other steroids and membranes, transported into human cells?

b. Explain why cholesterol accumulates in the blood of individuals with the disease familial hypercholesteremia.

▶ Word Roots

amphi- = dual (*amphipathic molecule:* a molecule that has both a hydrophobic and a hydrophilic region)

aqua- = water; **-pori** = a small opening (*aquaporin:* a transport protein in the plasma membrane of a plant or animal cell that specifically facilitates the diffusion of water across the membrane)

co- = together; **trans-** = across (*cotransport:* the coupling of the "downhill" diffusion of one substance to the "uphill" transport of another against its own concentration gradient)

electro- = electricity; **-genic** = producing (*electrogenic pump:* an ion transport protein generating voltage across a membrane)

endo- = inner; **cyto-** = cell (*endocytosis:* the movement of materials into a cell; cell eating)

exo- = outer (*exocytosis:* the movement of materials out of a cell)

hyper- = exceeding; **-tonus** = tension (*hypertonic:* a solution with a higher concentration of solutes)

hypo- = lower (*hypotonic:* a solution with a lower concentration of solutes)

iso- = same; (*isotonic:* solutions with equal concentrations of solutes)

phago- = eat (*phagocytosis:* cell eating)

pino- = drink (*pinocytosis:* cell drinking)

plasm- = molded; **-lyso** = loosen (*plasmolysis:* a phenomenon in walled cells in which the cytoplasm shrivels and the plasma membrane pulls away from the cell wall when the cell loses water to a hypertonic environment)

▶ Structure Your Knowledge

1. Create a concept map to illustrate your understanding of osmosis. This exercise will help you practice using the words *hypotonic, isotonic,* and *hypertonic,* and it will help you focus on the effect of these osmotic environments on plant and animal cells. Explain your map to a friend.

2. The following diagram illustrates passive and active transport across a plasma membrane. Use it to answer questions a–d.

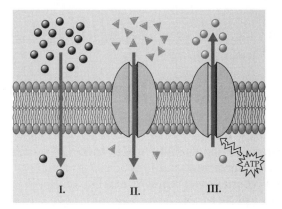

a. Which section represents facilitated diffusion? How can you tell?
   Does the cell expend energy in this transport? Why or why not?
   What types of solute molecules may be moved by this type of transport?

b. Which section shows active transport? List two ways that you can you tell?

c. Which section shows diffusion? What types of solute molecules may be moved by this type of transport?

d. Which of these sections are considered passive transport?

▶ Test Your Knowledge

MULTIPLE CHOICE: *Choose the one best answer.*

1. Glycoproteins and glycolipids are important for
   a. facilitated diffusion.
   b. active transport.
   c. cell-cell recognition.
   d. cotransport.
   e. signal-transduction pathways.

2. A single layer of phospholipid molecules coats the water in a beaker. Which part of the molecules will face the air?
   a. the phosphate groups
   b. the hydrocarbon tails
   c. both head and tail because the molecules are amphipathic and will lie sideways
   d. the phospholipids would dissolve in the water and not form a membrane coat
   e. the glycolipid regions

3. Which of the following is *not* true about osmosis?
   a. It is a passive process in cells without walls, but an active one in cells with walls.
   b. Water moves from a hypotonic to a hypertonic solution.
   c. Solute molecules bind to water and decrease the water available to move.
   d. It often occurs through channel proteins known as aquaporins.
   e. There is no net osmosis between isotonic solutions.

4. Support for the fluid mosaic model of membrane structure comes from
   a. the freeze-fracture technique of electron microscopy.
   b. the movement of proteins in hybrid cells.
   c. the amphipathic nature of many membrane proteins.
   d. both a and c.
   e. all of the above.

5. A freshwater *Paramecium* is placed into salt water. Which of the following events would occur?
   a. an increase in the action of its contractile vacuole
   b. swelling of the cell until it becomes turgid
   c. swelling of the cell until it lyses
   d. shriveling of the cell
   e. diffusion of salt ions out of the cell

6. Ions diffuse across membranes down their
   a. electrochemical gradient.
   b. electrogenic gradient.
   c. electrical gradient.
   d. concentration gradient.
   e. osmotic gradient.

7. The fluidity of membranes in a plant in cold weather may be maintained by
   a. increasing the number of phospholipids with saturated hydrocarbon tails.
   b. activating an H^+ pump.
   c. increasing the concentration of cholesterol in the membrane.
   d. increasing the proportion of peripheral proteins.
   e. increasing the number of phospholipids with unsaturated hydrocarbon tails.

8. A plant cell placed in a hypotonic environment will
   a. plasmolyze.
   b. shrivel.
   c. become turgid.
   d. become flaccid.
   e. lyse.

9. Which of the following is *not* true of the carrier molecules involved in facilitated diffusion?
   a. They increase the speed of transport across a membrane.
   b. They can concentrate solute molecules on one side of the membrane.
   c. They may have specific binding sites for the molecules they transport.
   d. They may undergo a conformational change upon binding of solute.
   e. They do not require an energy investment from the cell to operate.

10. The membrane potential of a cell favors the
    a. movement of cations into the cell.
    b. movement of anions into the cell.
    c. action of an electrogenic pump.
    d. movement of sodium out of the cell.
    e. action of a proton pump.

11. Cotransport may involve
    a. active transport of two solutes through a transport protein.
    b. passive transport of two solutes through a transport protein.
    c. ion diffusion against the electrochemical gradient created by an electrogenic pump.
    d. a pump such as the Na^+-K^+ pump that moves ions in two different directions.
    e. transport of one solute against its concentration gradient in tandem with another that is diffusing down its concentration gradient.

12. Exocytosis may involve all of the following *except*
    a. ligands and coated pits.
    b. the fusion of a vesicle with the plasma membrane.
    c. a mechanism to transport carbohydrates to the outside of plant cells during the formation of cell walls.
    d. a mechanism to rejuvenate the plasma membrane.
    e. a means of exporting large molecules.

13. The proton pump in plant cells is the functional equivalent of an animal cell's
    a. cotransport mechanism.
    b. sodium-potassium pump.
    c. contractile vacuole for osmoregulation.
    d. receptor-mediated endocytosis of cholesterol.
    e. ATP pump.

14. Pinocytosis involves
    a. the fusion of a newly formed food vacuole with a lysosome.
    b. receptor-mediated endocytosis and the formation of vesicles.
    c. the pinching in of the plasma membrane around small droplets of external fluid.
    d. pseudopod extension as vesicles move along the cytoskeleton and fuse with the plasma membrane.
    e. the accumulation of specific large molecules in a cell.

15. Watering a houseplant with too concentrated a solution of fertilizer can result in wilting because
    a. the uptake of ions into plant cells makes the cells hypertonic.
    b. the soil solution becomes hypertonic, causing the cells to lose water.
    c. the plant will grow faster than it can transport water and maintain proper water balance.
    d. diffusion down the electrochemical gradient will cause a disruption of membrane potential and accompanying loss of water.
    e. the plant will suffer fertilizer burn due to a caustic soil solution.

16. A cell is manufacturing receptor proteins for cholesterol. How would those proteins be oriented in the following membranes before they reach the plasma membrane?
    a. facing inside the ER lumen but outside the transport vesicle membrane
    b. facing inside the ER lumen and inside the transport vesicle
    c. attached outside the ER and outside the transport vesicle
    d. attached outside the ER but facing inside the transport vesicle
    e. completely embedded in the hydrophobic center of both the ER and transport vesicle membranes

17. Which of the following is the most probable description of an integral, transmembrane protein?
    a. amphipathic with a hydrophilic head and a hydrophobic tail region
    b. a globular protein with hydrophobic amino acids in the interior and hydrophilic amino acids arranged around the outside
    c. a fibrous protein coated with hydrophobic sugar residues

    d. a glycolipid attached to the portion of the protein facing the exterior of the cell and cytoskeletal elements attached to the portion facing inside the cell
    e. a middle region composed of α-helical stretches of hydrophobic amino acids, with hydrophilic regions at both ends of the protein

Use the U-tube setup to answer questions 18 through 20.

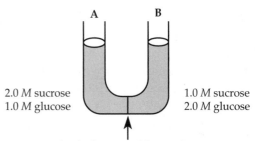

2.0 M sucrose
1.0 M glucose

1.0 M sucrose
2.0 M glucose

selectively permeable membrane

The solutions in the two arms of this U-tube are separated by a membrane that is permeable to water and glucose but not to sucrose. Side A is filled with a solution of 2.0 M sucrose and 1.0 M glucose. Side B is filled with 1.0 M sucrose and 2.0 M glucose.

18. *Initially,* the solution in side A, with respect to that in side B,
    a. has a lower solute concentration.
    b. has a higher solute concentration.
    c. has an equal solute concentration.
    d. is lower in the tube.
    e. is higher in the tube.

19. During the period *before* equilibrium is reached, which molecule(s) will show net movement through the membrane?
    a. water
    b. glucose
    c. sucrose
    d. water and sucrose
    e. water and glucose

20. *After* the system reaches equilibrium, what changes are observed?
    a. The water level is higher in side A than in side B.
    b. The water level is higher in side B than in side A.
    c. The molarity of glucose is higher in side A than in side B.
    d. The molarity of sucrose has increased in side A.
    e. Both a and c have occurred.

21. Facilitated diffusion across a cellular membrane requires _____ and moves a solute _____ its concentration gradient. (Assume solute is not an ion.)
    a. energy and transport proteins . . . down
    b. energy and transport proteins . . . up (against)
    c. energy . . . up
    d. transport proteins . . . down
    e. transport proteins . . . up

22. The extracellular fluids that surround the cells of a multicellular animal must be _____ to the cells.
    a. buffers
    b. isotonic
    c. hypotonic
    d. hypertonic
    e. homeostatic

23. LDLs (low-density lipoproteins) enter animal cells by
    a. diffusion through the lipid bilayer.
    b. pinocytosis.
    c. exocytosis.
    d. receptor-mediated endocytosis.
    e. diffusion through transport proteins

24. The fluid mosaic model describes biological membranes as consisting of
    a. a phospholipid bilayer with proteins sandwiched between the layers.
    b. a lipid bilayer with proteins coating the outside of this hydrophobic structure.
    c. a phospholipid bilayer with proteins embedded in and attached to it.
    d. a protein bilayer with phospholipids embedded in it.
    e. a cholesterol bilayer with proteins embedded in the hydrophobic center.

25. You observe plant cells under a microscope that have just been placed in an unknown solution. First the cells plasmolyze; after a few minutes, the plasmolysis reverses and the cells appear normal. What would you conclude about the unknown solution.
    a. It is hypertonic to the plant cells, and its solute cannot cross the plant cell membranes.
    b. It is hypotonic to the plant cells, and its solute cannot cross the plant cell membranes.
    c. It is isotonic to the plant cells, but its solute can cross the plant cell membranes.
    d. It is hypertonic to the plant cells, but its solute can cross the plant cell membranes.
    e. It is hypotonic to the plant cells, but its solute can cross the plant cell membranes.

Chapter 8
An Introduction to Metabolism

Framework

This chapter considers metabolism, the totality of the chemical reactions that take place in living organisms. Two key topics are emphasized: the energy transformations that underlie all chemical reactions and the role of enzymes in the "cold chemistry" of the cell. Enzymes are biological catalysts that lower the activation energy of a reaction and thus greatly speed up metabolic processes. An enzyme is a three-dimensional protein molecule with an active site specific for its substrate. Intricate control and feedback mechanisms produce the metabolic integration necessary for life.

The reactions in a cell either release or consume energy. The following illustration summarizes some of the components of the energy changes within a cell.

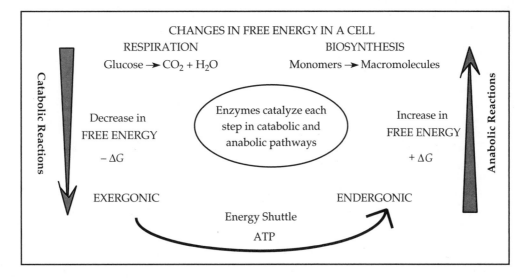

Chapter Review

8.1 An organism's metabolism transforms matter and energy, subject to the laws of thermodynamics

Organization of the Chemistry of Life into Metabolic Pathways   **Metabolism** includes all of the chemical reactions in an organism. These reactions are ordered into **metabolic pathways,** a sequence of steps, each controlled by an enzyme, that convert a specific molecule to a product. Through these pathways the cell transforms and creates the organic molecules that provide the energy and material for life.

Catabolic pathways release the energy stored in complex molecules through the breaking down of these molecules into simpler compounds. **Anabolic pathways,** sometimes called biosynthetic pathways, require energy to combine simpler molecules into more complicated ones. The energy released from catabolic pathways drives the anabolic pathways in a cell. The study of how organisms transform energy, called **bioenergetics,** is central to understanding metabolism.

Forms of Energy   **Energy** has been defined as the capacity to cause change. Some forms of energy can do work, such as moving matter against an opposing force. **Kinetic energy** is the energy of motion, of matter that is moving. This matter does its work by transferring its motion to other matter. The kinetic energy of randomly moving molecules is **heat,** or **thermal energy. Potential energy** is the capacity of matter to cause change as a consequence of its location or arrangement. **Chemical energy** is a form of potential energy stored in the arrangement of atoms in molecules and available for release in chemical reactions.

Energy can be converted from one form to another. Plants convert light energy to the chemical energy in sugar, and cells release this potential energy to drive cellular processes.

The Laws of Energy Transformation   **Thermodynamics** is the study of energy transformations. In an open

system, energy (and matter) may be exchanged between the system and the surroundings.

The **first law of thermodynamics** states that energy can be neither created nor destroyed. According to this *principle of conservation of energy,* energy can be transferred between matter and transformed from one kind to another, but the total energy of the universe is constant.

The **second law of thermodynamics** states that every energy transformation or transfer results in an increasing disorder within the universe. **Entropy** is used as a measure of disorder or randomness. In every energy transfer or transformation, some of the energy is converted to heat, the least ordered form of energy. For a process to occur spontaneously, without the input of external energy, it must result in an increase in entropy. A nonspontaneous process will occur only if energy is added to the system.

A cell may become more ordered, but it does so with an attendant increase in the entropy of its surroundings. An organism takes in and uses highly ordered organic molecules as a source of the energy needed to create and maintain its own organized structure, but it returns heat and the simple molecules of carbon dioxide and water to the environment.

Fill in the concept map in Interactive Question 8.1 to review the key concepts of energy.

■ INTERACTIVE QUESTION   8.1

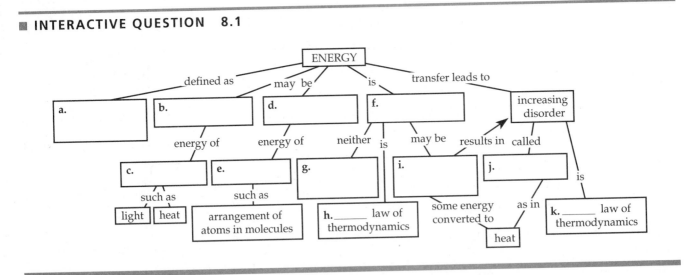

8.2   The free-energy change of a reaction tells us whether the reaction occurs spontaneously

Free-Energy Change, ΔG   **Free energy** can be defined as the portion of a system's energy available to perform work when the system's temperature and pressure are uniform. The change in free energy during a reaction is represented by $\Delta G = \Delta H - T\Delta S$. Enthalpy ($H$) is the

total energy of a system; entropy (S) is a measure of its disorder, and T is absolute temperature (°Kelvin, or °C + 273). For a reaction to be spontaneous, the free energy of the system must decrease ($-\Delta G$): the system must lose energy (H must decrease), become more disordered (S must increase), or both.

Free Energy, Stability, and Equilibrium   When ΔG is negative, the final state has less free energy than the initial state; thus the final state is less likely to change and is more stable. A system rich in free energy has a tendency to change spontaneously to a more stable state. At equilibrium in a chemical reaction, the forward and backward reactions are proceeding at the same rate and the relative concentration of products and reactants stays the same. Moving toward equilibrium is spontaneous; the ΔG of the reaction is negative. Once at equilibrium, a system is at a minimum of free energy and will not spontaneously change.

■ INTERACTIVE QUESTION 8.2

Complete the table to show how the free energy of a system or reaction relates to its stability, tendency for spontaneous change, equilibrium, and capacity to do work.

	System with high free energy	System with low free energy
Stability		
Spontaneous		
Equilibrium		
Work capacity		

Free Energy and Metabolism   An **exergonic reaction** ($-\Delta G$) proceeds with a net release of free energy and is spontaneous. The magnitude of ΔG indicates the maximum amount of work the reaction can do. **Endergonic reactions** ($+\Delta G$) are nonspontaneous; they must absorb free energy from the surroundings. The energy released by an exergonic reaction ($-\Delta G$) is equal to the energy required by the reverse reaction ($+\Delta G$).

Metabolic disequilibrium is essential to life. Metabolic reactions are reversible and could reach equilibrium if the cell did not maintain a steady supply of reactants and siphon off the products (as reactants for new processes or as waste products to be expelled).

■ INTERACTIVE QUESTION 8.3

Develop a concept map on free energy and ΔG. The value in this exercise is for you to wrestle with and organize these concepts for yourself. Do not turn to the suggested concept map until you have worked on your own understanding. Remember that the concept map in the answer section is only one way of structuring these ideas—have confidence in your own organization.

8.3 ATP powers cellular work by coupling exergonic reactions to endergonic reactions

Central to a cell's bioenergetics is **energy coupling**, using exergonic processes to power endergonic ones.

A cell usually uses ATP as the immediate source of energy for its mechanical, transport, and chemical work.

The Structure and Hydrolysis of ATP   **ATP (adenosine triphosphate)** consists of the nitrogenous base adenine bonded to the sugar ribose, which is connected to a chain of three phosphate groups. ATP can be hydrolyzed to ADP (adenosine diphosphate) and an inorganic phosphate molecule ($\textcircled{P}_i$), releasing 7.3 kcal (30.5 kJ) of energy per mole of ATP when measured under standard conditions. The ΔG of the reaction in the cell is estimated to be closer to -13 kcal/mol.

■ INTERACTIVE QUESTION 8.4

Label the three components (a through c) of the ATP molecule shown below.

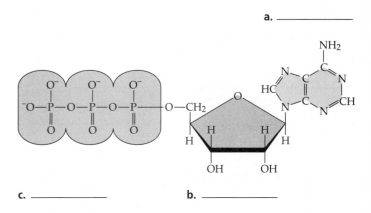

a. _____

c. _____   b. _____

d. Indicate which bond is likely to break. By what chemical mechanism is the bond broken?

e. Explain why this reaction releases so much energy?

How ATP Performs Work   In a cell, the free energy released from the hydrolysis of ATP is used to transfer the phosphate group to another molecule, producing a **phosphorylated** molecule that is more reactive (less stable). The phosphorylation of other molecules by ATP forms the basis for almost all cellular work.

The Regeneration of ATP   A cell regenerates ATP at a phenomenal rate. The formation of ATP from ADP and

Ⓟ$_i$ is endergonic, with a ΔG of $+7.3$ kcal/mol (standard conditions). Cellular respiration (the catabolic processing of glucose and other organic molecules) provides the energy for the regeneration of ATP. Plants can also produce ATP using light energy.

8.4   Enzymes speed up metabolic reactions by lowering energy barriers

Enzymes are biological **catalysts**—agents that speed the rate of a reaction but are unchanged by the reaction.

The Activation Energy Barrier   Chemical reactions involve both the breaking and forming of chemical bonds. Energy must be absorbed to make bonds unstable enough that they will break, and energy is released when bonds form and molecules return to stable, lower energy states. **Activation energy,** or the **free energy of activation** (E_A), is the energy that must be absorbed by reactants to reach the unstable *transition state*, in which bonds are likely to break, and from which the reaction can proceed.

How Enzymes Lower the E_A Barrier   The E_A barrier is essential to life because it prevents the energy-rich macromolecules of the cell from decomposing spontaneously. For metabolism to proceed in a cell, however, E_A must be reached. Heat, a possible source of activation energy in reactions, would be harmful to the cell and would also speed metabolic reactions indiscriminately. Enzymes are able to lower E_A for specific reactions so that metabolism can proceed at cellular temperatures. Enzymes do not change the ΔG for a reaction.

■ INTERACTIVE QUESTION   8.5

In this graph of an exergonic reaction with and without an enzyme catalyst, label the parts a through e.

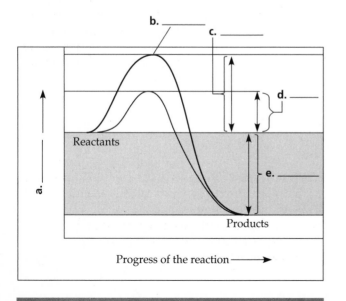

Substrate Specificity of Enzymes   Protein enzymes are macromolecules with characteristic three-dimensional shapes. The specificity of an enzyme for a particular **substrate** is determined by this shape. The substrate attaches at the enzyme's **active site**, a pocket or groove found on the surface of the enzyme that has a shape complementary to the substrate. The substrate is temporarily bound to its enzyme, forming an **enzyme-substrate complex.** Interactions between substrate and active site cause the enzyme to change shape slightly, creating what is called an **induced fit** that enhances the ability of the enzyme to catalyze the chemical reaction.

Catalysis in the Enzyme's Active Site   The substrate is often held in the active site by hydrogen or ionic bonds. The side chains (R groups) of some of the surrounding amino acids in the active site facilitate the conversion of substrate to product. The product then leaves the active site, and the catalytic cycle repeats, often at astonishing speed.

Whether an enzyme catalyzes the forward or backward reaction is influenced by the relative concentrations of reactants and products and the ΔG of the reactions. Enzymes catalyze reactions moving toward equilibrium.

Enzymes catalyze reactions involving the joining of two reactants by binding the substrates closely together and properly oriented. An induced fit can stretch critical bonds in the substrate molecule and make them easier to break. An active site may provide a microenvironment, such as a lower pH, that is necessary for a particular reaction. Enzymes may also actually participate in a reaction by forming brief covalent bonds with the substrate.

The rate of a reaction will increase with increasing substrate concentration up to the point at which all enzyme molecules are saturated with substrate molecules and working at full speed. Only adding more enzyme molecules will increase the rate of the reaction at that point.

Effects of Local Conditions on Enzyme Activity   The velocity of an enzyme-catalyzed reaction may increase with rising temperature up to the point at which increased thermal agitation begins to disrupt the hydrogen and ionic bonds and other interactions that stabilize protein conformation. A change in pH may also affect protein shape and enzyme function. Each enzyme has a temperature and pH optimum at which it is most active.

■ **INTERACTIVE QUESTION 8.6**

Outline a catalytic cycle using the diagrammatic enzyme below. Sketch two appropriate substrate molecules and two products, identify the enzyme-substrate complex, and describe the key steps of the cycle.

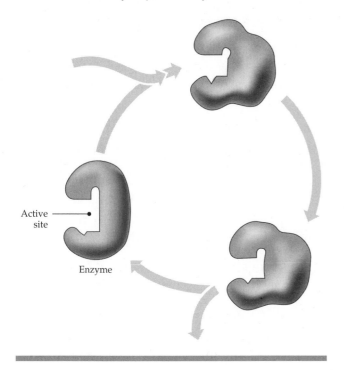

Active
site

Enzyme

Cofactors are small molecules that bind either permanently or reversibly with enzymes and are necessary for enzyme catalytic function. They may be inorganic, such as various metal ions, or organic molecules called **coenzymes.** Most vitamins are coenzymes or precursors of coenzymes.

Enzyme inhibitors selectively disrupt the action of enzymes, either reversibly by binding with the enzyme with weak bonds or irreversibly by attaching with covalent bonds. **Competitive inhibitors** compete with the substrate for the active site of the enzyme. Increasing the concentration of substrate molecules may overcome this type of inhibition. **Noncompetitive inhibitors** bind to a part of the enzyme separate from the active site and change the conformation of the enzyme, thus impeding enzyme action.

■ **INTERACTIVE QUESTION 8.7**

Return to your diagram in Interactive Question 8.6. Draw a competitive and a noncompetitive inhibitor, and indicate where each would bind to the enzyme molecule.

8.5 Regulation of enzyme activity helps control metabolism

Allosteric Regulation of Enzymes In **allosteric regulation** molecules may inhibit or activate enzyme activity when they bind to a site separate from the active site. Enzymes made of two or more polypeptides, each with its own active site, may have binding sites (sometimes called allosteric sites) located where subunits join. The entire unit may oscillate between two conformational states. The binding of an activator stabilizes the catalytically active conformation, whereas an inhibitor reinforces the inactive form of the enzyme. Allosteric enzymes may be critical regulators of metabolic pathways.

Through a phenomenon called **cooperativity,** the induced-fit binding of a substrate molecule to one subunit can change the conformation such that the active sites of all subunits are more active.

Metabolic pathways are commonly regulated by **feedback inhibition**, in which the product of a pathway acts as an allosteric inhibitor of an enzyme early in the pathway.

■ **INTERACTIVE QUESTION 8.8**

Both ATP and ADP serve as regulators of enzyme activity. In catabolic pathways, which of these molecules would you predict would act as an inhibitor?

Which molecule would you expect to act as an activator of anabolic pathways?

Specific Localization of Enzymes within the Cell Enzymes for several steps of a metabolic pathway may be associated in a multienzyme complex, facilitating the sequence of reactions. Specialized cellular compartments may contain high concentrations of the enzymes and substrates needed for a particular pathway. Enzymes are often incorporated into the membranes of cellular compartments. The complex internal structures of the cell facilitate metabolic order.

▶ Word Roots

cata- = down (*catabolic pathway:* a metabolic pathway that releases energy by breaking down complex molecules into simpler ones)

ana- = up (*anabolic pathway:* a metabolic pathway that consumes energy to build complex molecules from simpler ones)

bio- = life (*bioenergetics:* the study of how organisms manage their energy resources)

kinet- = movement (*kinetic energy:* the energy of motion)

therm- = heat (*thermodynamics:* the study of the energy transformations that occur in a collection of matter)

ex- = out (*exergonic reaction:* a reaction that proceeds with a net release of free energy)

endo- = within (*endergonic reaction:* a reaction that absorbs free energy from its surroundings)

allo- = different (*allosteric site:* a specific receptor site on some part of an enzyme molecule remote from the active site)

Structure Your Knowledge

This chapter introduced many new and complex concepts. See if you can step back from the details and answer the following general questions.

1. Relate the concept of free energy to metabolism.
2. What role do enzymes play in metabolism?

Test Your Knowledge

MULTIPLE CHOICE: *Choose the one best answer.*

1. Catabolic and anabolic pathways are often coupled in a cell because
   a. the intermediates of a catabolic pathway are used in the anabolic pathway.
   b. both pathways use the same enzymes.
   c. the free energy released from one pathway is used to drive the other.
   d. the activation energy of the catabolic pathway can be used in the anabolic pathway.
   e. their enzymes are controlled by the same activators and inhibitors.

2. When glucose and oxygen are converted to CO_2 and H_2O, changes in total energy, entropy, and free energy are as follows:
   a. $-\Delta H, -\Delta S, -\Delta G$
   b. $-\Delta H, +\Delta S, -\Delta G$
   c. $-\Delta H, +\Delta S, +\Delta G$
   d. $+\Delta H, +\Delta S, +\Delta G$
   e. $+\Delta H, -\Delta S, +\Delta G$

3. When a protein forms from amino acids, the following energy and entropy changes apply:
   a. $+\Delta H, -\Delta S, +\Delta G$
   b. $+\Delta H, +\Delta S, -\Delta G$
   c. $+\Delta H, +\Delta S, +\Delta G$
   d. $-\Delta H, +\Delta S, +\Delta G$
   e. $-\Delta H, -\Delta S, +\Delta G$

4. A negative ΔG means that
   a. the quantity G of energy is available to do work.
   b. the reaction is spontaneous.
   c. the reactants have more free energy than the products.
   d. the reaction is exergonic.
   e. all of the above are true.

5. According to the first law of thermodynamics,
   a. for every action there is an equal and opposite reaction.
   b. every energy transfer results in an increase in disorder or entropy.
   c. the total amount of energy in the universe is conserved or constant.
   d. energy can be transferred or transformed, but disorder always increases.
   e. potential energy is converted to kinetic energy, and kinetic energy is converted to heat.

6. What is meant by an induced fit?
   a. The binding of the substrate is an energy-requiring process.
   b. A competitive inhibitor can outcompete the substrate for the active site.
   c. The binding of the substrate changes the shape of the active site, which can stress or bend substrate bonds.
   d. The active site creates a microenvironment ideal for the reaction.
   e. Substrates are held in the active site by hydrogen and ionic bonds.

7. One way in which a cell maintains metabolic disequilibrium is to
   a. siphon products of a reaction off to the next step in a metabolic pathway.
   b. provide a constant supply of enzymes for critical reactions.
   c. use feedback inhibition to turn off pathways.
   d. use allosteric enzymes that can bind to activators or inhibitors.
   e. use the energy from anabolic pathways to drive catabolic pathways.

8. In an experiment, changing the pH from 7 to 6 resulted in an increase in product formation. From this we could conclude that
   a. the enzyme became saturated at pH 6.
   b. the enzyme's optimal pH is 6.
   c. this enzyme works best in a neutral pH.
   d. the temperature must have increased when the pH was changed to 6.
   e. the enzyme was in a more active conformation at pH 6.

9. When substance A was added to an enzyme reaction, product formation decreased. The addition of more substrate did not increase product formation. From this we conclude that substance A could be
   a. product molecules.
   b. a cofactor.
   c. an allosteric enzyme.
   d. a competitive inhibitor.
   e. a noncompetitive inhibitor.

10. The formation of ATP from ADP and inorganic phosphate
    a. is an exergonic process.
    b. transfers the phosphate to another intermediate that becomes more reactive.
    c. produces an unstable energy compound that can drive cellular work.
    d. has a ΔG of -7.3 kcal/mol under standard conditions.
    e. involves the hydrolysis of a phosphate bond.

11. At equilibrium,
    a. no enzymes are functioning.
    b. free energy is at a minimum.
    c. the forward and backward reactions have stopped.
    d. the products and reactants have equal values of H.
    e. a reaction has a $+\Delta G$.

12. In cooperativity,
    a. a cellular organelle contains all the enzymes needed for a metabolic pathway.
    b. a product of a pathway serves as a competitive inhibitor of an early enzyme in the pathway.
    c. a molecule bound to the active site of one subunit of an enzyme affects the active site of other subunits.
    d. the allosteric site is filled with an activator molecule.
    e. the product of one reaction serves as the substrate for the next in intricately ordered metabolic pathways.

13. When a cell breaks down glucose, only about 40% of the energy is captured in ATP molecules. The remaining 60% of the energy is
    a. used to increase the order necessary for life to exist.
    b. lost as heat because of the second law of thermodynamics.
    c. used to increase the entropy of the system by converting kinetic energy into potential energy.
    d. stored in starch or glycogen for use later by the cell.
    e. released when the ATP molecules are hydrolyzed.

14. An endergonic reaction could be described as one that will
    a. proceed spontaneously with the addition of activation energy.
    b. produce products with more free energy than the reactants.
    c. not be able to be catalyzed by enzymes.
    d. release energy.
    e. produce ATP for energy coupling.

15. What is most directly responsible for the specificity of a protein enzyme?
    a. its primary structure
    b. its secondary and tertiary structure
    c. the conformation of its allosteric site
    d. its cofactors
    e. the R groups of the amino acids in its active site

16. Which of the following parameters does an enzyme raise?
    a. ΔG
    b. ΔH
    c. equilibrium of a reaction
    d. speed of a reaction
    e. free energy of activation

17. Zinc, an essential trace element, may be found bound to the active site of some enzymes. What would be the most likely function of such zinc ions?
    a. coenzyme derived from a vitamin
    b. a cofactor necessary for catalysis
    c. a substrate of the enzyme
    d. a competitive inhibitor of the enzyme
    e. an allosteric activator of the enzyme

Use this diagram to answer questions 18 through 20.

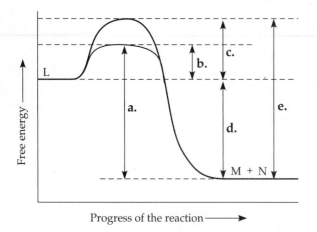

18. Which line in the diagram above indicates the ΔG of the enzyme-catalyzed reaction of L → M + N?

19. Which line in the diagram indicates the activation energy of the noncatalyzed reaction?

20. Which of the following terms would best describe this reaction?
    a. nonspontaneous
    b. −ΔG
    c. endergonic
    d. coupled reaction
    e. anabolic reaction

21. A reaction that is spontaneous
    a. has a +ΔG.
    b. occurs very rapidly.
    c. does not require enzyme catalysis in a cell.
    d. will decrease the entropy of a system.
    e. is exergonic.

22. In the metabolic pathway, A → B → C → D → E, what effect would molecule E likely have on the enzyme that catalyzes A → B?
    a. allosteric inhibitor
    b. allosteric activator
    c. competitive inhibitor
    d. feedback activator
    e. coenzyme

FILL IN THE BLANKS

_____ 1. the totality of an organism's chemical processes

_____ 2. pathways that require energy to combine molecules together

_____ 3. the energy of motion

_____ 4. enzymes that change between two conformations, depending on whether an activator or inhibitor is bound to them

_____ 5. term for the measure of disorder or randomness

_____ 6. the energy that must be absorbed by molecules to reach the transition state

_____ 7. inhibitors that decrease an enzyme's activity by binding to the active site

_____ 8. organic molecules that bind to enzymes and are necessary for their functioning

_____ 9. regulatory device in which the product of a pathway binds to an enzyme early in the pathway

_____ 10. more reactive molecules created by the transfer of a phosphate group from ATP

Chapter 9
Cellular Respiration: Harvesting Chemical Energy

Framework

The catabolic pathways of glycolysis and cellular respiration release the chemical energy in glucose and other fuels and store it in ATP. Glycolysis, occurring in the cytosol, produces ATP, pyruvate, and NADH; the latter two may then enter the mitochondria for respiration. A mitochondrion consists of a matrix in which the enzymes of the citric acid cycle are localized, a highly folded inner membrane in which enzymes and the molecules of the electron transport chain are embedded, and an intermembrane space between the two membranes to temporarily house H^+ that has been pumped across the inner membrane during the redox reactions of the electron transport chain. ATP is produced by oxidative phosphorylation, a chemiosmotic mechanism in which a proton motive force drives protons through ATP synthases located in the membrane.

Chapter Review

Photosynthesis stores the energy of sunlight in organic compounds. Through catabolic pathways, a cell breaks down these complex molecules to simpler ones, generating ATP for cellular work. Energy leaves an ecosystem as heat.

9.1 Catabolic pathways yield energy by oxidizing organic fuels

Catabolic Pathways and Production of ATP **Fermentation,** which occurs without oxygen, is the partial degradation of sugars to release energy. **Cellular respiration** uses oxygen in the breakdown of glucose (or other energy-rich organic compounds) to yield carbon dioxide and water and release energy as ATP and heat. This exergonic process has a free energy change of -686 kcal/mol of glucose.

■ INTERACTIVE QUESTION   9.1

Fill in this summary equation for cellular respiration.

$$\underline{\hspace{2cm}} + 6\,O_2 \rightarrow \underline{\hspace{2cm}} + 6\,H_2O + \underline{\hspace{2cm}}$$

Redox Reactions: Oxidation and Reduction Oxidation-reduction or **redox reactions** involve the partial or complete transfer of one or more electrons from one reactant to another. **Oxidation** is the loss of electrons from one substance; **reduction** is the addition of electrons to another substance. The substance that loses electrons becomes oxidized and acts as a **reducing agent** (electron donor) to the substance that gains electrons. By gaining electrons, a substance acts as an **oxidizing agent** (electron acceptor) and becomes reduced.

■ INTERACTIVE QUESTION 9.2

Fill in the appropriate terms in this equation.

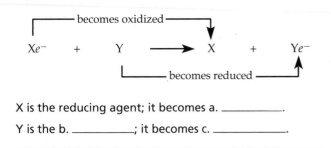

X is the reducing agent; it becomes a. _____.

Y is the b. _____; it becomes c. _____.

Oxygen strongly attracts electrons and is one of the most powerful oxidizing agents. As electrons shift toward a more electronegative atom, they give up potential energy. Thus, chemical energy is released in a redox reaction that relocates electrons closer to oxygen.

Organic molecules with an abundance of hydrogen are rich in "hilltop" electrons that release their potential energy when they "fall" closer to oxygen.

■ INTERACTIVE QUESTION 9.3

a. In the conversion of glucose and oxygen to carbon dioxide and water, which molecule becomes reduced?

b. Which molecule becomes oxidized?

c. What happens to the energy that is released in this redox reaction?

At certain steps in the oxidation of glucose, two hydrogen atoms are removed by enzymes called dehydrogenases, and the two electrons and one proton are passed to a coenzyme, **NAD$^+$** (nicotinamide adenine dinucleotide), reducing it to NADH.

Energy from respiration is slowly released in a series of small steps as electrons are passed from NADH down an **electron transport chain,** a group of carrier molecules located in the inner mitochondrial membrane, to a stable location close to a highly electronegative oxygen atom.

■ INTERACTIVE QUESTION 9.4

a. NAD$^+$ is called an _____.

b. Its reduced form is _____.

The Stages of Cellular Respiration: A Preview **Glycolysis,** occurring in the cytosol, breaks glucose into two molecules of pyruvate. The **citric acid cycle,** located in the mitochondrial matrix, converts a derivative of pyruvate into carbon dioxide. In some of the steps of glycolysis and the citric acid cycle, dehydrogenase enzymes transfer electrons to NAD$^+$. NADH passes electrons to the electron transport chain, from which they eventually combine with hydrogen ions and oxygen to form water. The energy released through this chain of redox reactions is used to synthesize ATP by **oxidative phosphorylation,** a process that includes electron transport and chemiosmosis.

Up to 38 molecules of ATP may be generated for each molecule of glucose oxidized to carbon dioxide. About 10% of this ATP is produced by **substrate-level phosphorylation,** in which an enzyme transfers a phosphate group from a substrate to ADP.

■ INTERACTIVE QUESTION 9.5

Fill in the three stages of respiration (a–c). Indicate whether ATP is produced by substrate-level or oxidative phosphorylation (d–f). Label the arrows indicating electrons carried by NADH.

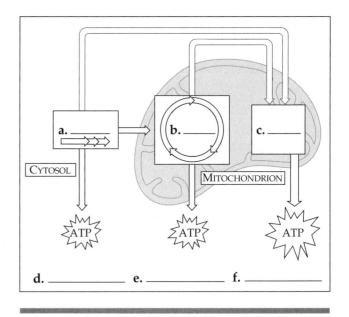

9.2 Glycolysis harvests chemical energy by oxidizing glucose to pyruvate

Glycolysis, a ten-step process occurring in the cytosol, has an energy-investment phase and an energy-payoff phase. Two molecules of ATP are consumed as glucose is split into two three-carbon sugars (glyceraldehyde-3-phosphate). The conversion of these molecules to pyruvate produces two NADH and four ATP by substrate-level phosphorylation. For each molecule of glucose, glycolysis yields a net gain of two ATP and two NADH.

Enzymes catalyze each step in glycolysis. Kinases transfer phosphate groups; a dehydrogenase oxidizes glyceraldehyde-3-phosphate and reduces NAD^+; and other enzymes rearrange atoms in substrate molecules.

■ INTERACTIVE QUESTION 9.6

Fill in the blanks in this summary diagram of glycolysis.

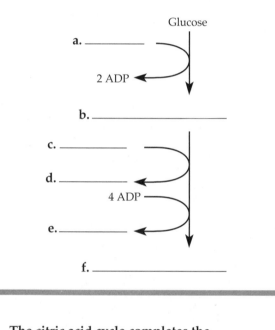

■ INTERACTIVE QUESTION 9.7

Fill in the blanks in this summary figure of the citric acid cycle. Balls represent carbon atoms.

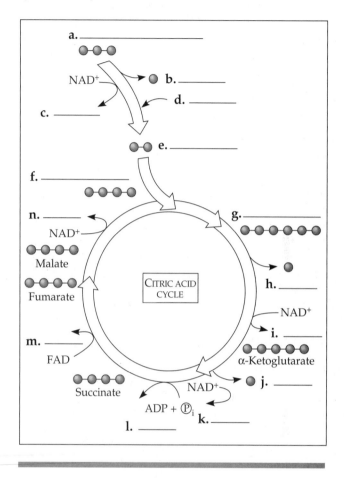

9.3 The citric acid cycle completes the energy-yielding oxidation of organic molecules

Pyruvate is actively transported into the mitochondrion. Before the citric acid cycle begins, a series of steps occurs within a multienzyme complex. A carboxyl group is removed from the three-carbon pyruvate and released as CO_2; the remaining two-carbon group is oxidized to form acetate with the accompanying reduction of NAD^+ to NADH; and coenzyme A is attached to the acetate by an unstable bond, forming **acetyl CoA.**

In the citric acid cycle (also called the Krebs cycle), the acetyl group of acetyl CoA is added to oxaloacetate to form citrate, which is progressively decomposed back to oxaloacetate. For each turn of the citric acid cycle, two carbons enter in the reduced form from acetyl CoA; two carbons exit completely oxidized as CO_2; three NADH and one $FADH_2$ are formed; and one ATP is made by substrate-level phosphorylation. There are two turns of the citric acid cycle for each glucose molecule oxidized.

9.4 During oxidative phosphorylation, chemiosmosis couples electron transport to ATP synthesis

Pathway of Electron Transport Thousands of electron transport chains are embedded in the cristae (infoldings) of the inner mitochondrial membrane. Most components of the electron transport chain are proteins, with tightly bound, nonprotein *prosthetic groups.* The electron carriers shift between reduced and oxidized states as they accept and donate electrons. The components are organized into four complexes. The transfer of electrons goes from NADH to a flavoprotein to an iron-sulfur protein (Complex I) to a mobile hydrophobic molecule, ubiquinone (Q). Next electrons pass down a series of molecules called **cytochromes (cyt),** proteins with an iron-containing heme group. The last cytochrome, cyt a_3, passes electrons to oxygen, which picks up a pair of H^+ and forms water.

66 *Unit Two: The Cell*

FADH$_2$ adds its electrons to the chain at a lower energy level (at Complex II); thus, one-third less energy is provided for ATP synthesis by FADH$_2$ as compared to NADH.

Chemiosmosis: The Energy-Coupling Mechanism
ATP synthase, a protein complex embedded in the inner membrane, uses the energy of a proton (H$^+$) gradient to make ATP, an example of the process called **chemiosmosis.** The electron transport chain creates the proton gradient. When some members of the chain pass electrons, they also accept and release protons, which are pumped into the intermembrane space at three points. The resulting proton gradient

stores potential energy, referred to as the **proton-motive force.**

The flow of H$^+$ ions down their gradient through the rotor and stator part of the ATP synthase complex causes the rotor and attached rod to rotate, activating catalytic sites on the knob portion where ADP and inorganic phosphate join to make ATP.

In mitochondria, exergonic redox reactions produce the H$^+$ gradient that drives the production of ATP. Chloroplasts use light energy to create the proton-motive force used to make ATP by chemiosmosis. Prokaryotes use H$^+$ gradients generated across their plasma membranes to transport molecules, make ATP, and rotate flagella.

■ INTERACTIVE QUESTION 9.8

Label this diagram of oxidative phosphorylation in a mitochondrial membrane.

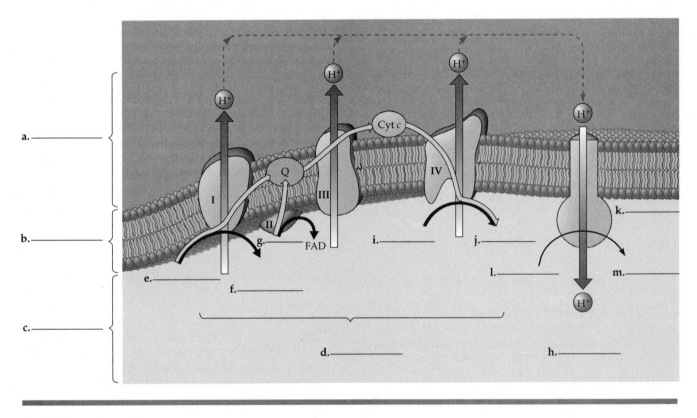

An Accounting of ATP Production by Cellular Respiration About 36 or 38 ATPs may be produced per glucose molecule oxidized. This number is only an estimate for three reasons: The 3 ATPs/NADH and 2 ATPs/FADH$_2$ are rounded off numbers; the electrons from the NADH produced by glycolysis may be passed across the inner mitochondrial membrane to

NAD$^+$ or FAD, depending on the type of shuttle used in the cell; and the proton-motive force generated by the electron transport chain is also used to power other work in the mitochondrion.

The efficiency of respiration in its energy conversions is approximately 40%. The rest of the energy is released as heat.

■ INTERACTIVE QUESTION 9.9

Fill in the tally for maximum ATP yield from the oxidation of one molecule of glucose to six molecules of carbon dioxide.

Process	# ATP
Initial phosphorylation of glucose:	a. _____
Substrate-level phosphorylation: in glycolysis	b. _____
in c. _____	2
Oxidative phosphorylation:*	d. _____
Maximum Total	e. _____

*approximately 3 ATP for each of the **f.** _____ NADH from pyruvate → acetyl CoA and the **g.** _____ NADH from citric acid cycle; **h.** _____ ATP for each of the **i.** _____ FADH$_2$ from citric acid cycle; 2 or 3 ATP of the **j.** _____ NADH from glycolysis, depending on whether the shuttle passes electrons across the membrane to NAD$^+$ or FAD.

9.5 Fermentation enables some cells to produce ATP without the use of oxygen

Types of Fermentation Glycolysis oxidizes glucose to produce a net of 2 ATP per glucose molecule, under both **aerobic** and **anaerobic** conditions. Without oxygen, glycolysis is part of fermentation—the anaerobic catabolism of organic nutrients to generate ATP by substrate-level phosphorylation and the reactions that regenerate NAD$^+$, the oxidizing agent for glycolysis.

In **alcohol fermentation,** pyruvate is converted into acetaldehyde, and CO$_2$ is released. Acetaldehyde is then reduced by NADH to form ethanol (ethyl alcohol), and NAD$^+$ is regenerated. In **lactic acid fermentation,** pyruvate is reduced directly by NADH to form lactate and recycle NAD$^+$. Muscle cells make ATP by lactic acid fermentation when energy demand is high and oxygen supply is low.

Fermentation and Cellular Respiration Compared Both fermentation and respiration use glycolysis with NAD$^+$ as the oxidizing agent to convert glucose and other organic fuels to pyruvate. To oxidize NADH back to NAD$^+$, fermentation uses pyruvate or acetaldehyde as the final electron acceptor, whereas respiration uses oxygen, via the electron transport chain.

Facultative anaerobes, such as yeasts and many bacteria, can make ATP by fermentation or respiration, depending upon the availability of oxygen.

Evolutionary Significance of Glycolysis Glycolysis is common to fermentation and respiration. This most widespread of all metabolic processes probably evolved in ancient prokaryotes before oxygen was available. The cytosolic location of glycolysis is also evidence of its antiquity.

■ INTERACTIVE QUESTION 9.10

How much more ATP can be generated by respiration than by fermentation? Explain why.

9.6 Glycolysis and the citric acid cycle connect to many other metabolic pathways

The Versatility of Catabolism Fats, proteins, and carbohydrates can all be used by cellular respiration to make ATP. Proteins are digested into amino acids, which are then deaminated (amino group removed) and can enter into respiration at several sites. The digestion of fats yields glycerol, which is converted to an intermediate of glycolysis, and fatty acids, which are broken down by **beta oxidation** to two-carbon fragments that enter the citric acid cycle as acetyl CoA.

Biosynthesis (Anabolic Pathways) The organic molecules of food also provide carbon skeletons for biosynthesis. Some monomers, such as amino acids, can be directly incorporated into the cell's macromolecules. Intermediates of glycolysis and the citric acid cycle serve as precursors for anabolic pathways. The molecules of carbohydrates, fats, and proteins can all be interconverted to provide for a cell's needs.

Regulation of Cellular Respiration Via Feedback Mechanisms Through feedback inhibition, the end product of a pathway inhibits an enzyme early in the pathway, thus preventing a cell from producing an excess of a particular substance. The supply of ATP in the cell regulates respiration. The allosteric enzyme that catalyzes the third step of glycolysis, phosphofructokinase, is inhibited by ATP and activated by AMP (derived from ADP). Phosphofructokinase is also inhibited by citrate released from the mitochondria into the cytosol, thus synchronizing the rates of glycolysis and the citric acid cycle. Other enzymes located at key intersections help to maintain metabolic balance.

Word Roots

aero- = air (*aerobic:* chemical reaction using oxygen)
an- = not (*anaerobic:* chemical reaction not using oxygen)
chemi- = chemical (*chemiosmosis:* the production of ATP using the energy of hydrogen ion gradients across membranes to phosphorylate ADP)
glyco- = sweet; **-lysis** = split (*glycolysis:* the splitting of glucose into pyruvate)

Structure Your Knowledge

1. This chapter describes how the catabolic pathways of glycolysis and respiration release chemical energy and store it in ATP. One of the best ways to learn the three main components of cellular respiration is to explain them to someone. Find two study partners and have each person be responsible for learning and explaining the important concepts and steps of either glycolysis, the citric acid cycle, or oxidative phosphorylation (electron transport and chemiosmosis). You may want to involve a fourth person to teach fermentation, or learn that together as a group. Use diagrams and sketches to help you explain your process to your partners.

2. Fill in the table below to summarize the major inputs and outputs of glycolysis, the citric acid cycle, oxidative phosphorylation, and fermentation. Base inputs and outputs on one glucose molecule.

3. Create a concept map to organize your understanding of oxidative phosphorylation.

Process	Main Function	Inputs	Outputs
Glycolysis			
Pyruvate to acetyl CoA			
Citric acid cycle			
Oxidative phosphorylation (Electron transport and chemiosmosis)			
Fermentation			

Test Your Knowledge

MULTIPLE CHOICE: *Choose the one best answer.*

1. When electrons move closer to a more electronegative atom,
   a. energy is released.
   b. energy is consumed.
   c. a proton gradient is established.
   d. water is produced.
   e. ATP is synthesized.

2. In the reaction $C_6H_{12}O_6 + 6 O_2 \rightarrow 6 CO_2 + 6 H_2O$,
   a. oxygen becomes reduced.
   b. glucose becomes reduced.
   c. oxygen becomes oxidized.
   d. water is a reducing agent.
   e. oxygen is a reducing agent.

3. A substrate that is phosphorylated
   a. has lost a phosphate group.
   b. has been formed by the reaction $ADP + \textcircled{P}_i \rightarrow ATP$.
   c. has an increased reactivity; it is primed to do work.
   d. has been oxidized.
   e. will pass its electrons to the electron transport chain.

4. Which of the following is *not* true of oxidative phosphorylation?
   a. It produces approximately three ATP for every NADH that is oxidized.
   b. It involves the redox reactions of the electron transport chain.
   c. It involves an ATP synthase located in the inner mitochondrial membrane.
   d. It uses oxygen as the initial electron donor.
   e. It is an example of chemiosmosis.

5. Substrate-level phosphorylation
   a. involves the shifting of a phosphate group from ATP to a substrate.
   b. can use NADH or $FADH_2$.
   c. takes place only in the cytosol.
   d. accounts for 10% of the ATP formed by fermentation.
   e. is the energy source for facultative anaerobes under anaerobic conditions.

6. The *major* reason that glycolysis is not as energy-productive as respiration is that
   a. NAD^+ is regenerated by alcohol or lactate production, without the high-energy electrons passing through the electron transport chain.
   b. it is the pathway common to fermentation and respiration.
   c. it does not take place in a specialized membrane-bound organelle.
   d. pyruvate is more reduced than CO_2; it still contains much of the energy from glucose.
   e. substrate-level phosphorylation is not as energy efficient as oxidative phosphorylation.

7. The net products of glycolysis are
   a. 2 ATP, 2 CO_2, 2 ethanol.
   b. 2 ATP, 2 NAD^+, 2 acetate.
   c. 2 ATP, 2 NADH, 2 pyruvate, 2 H_2O.
   d. 38 ATP, 6 CO_2, 6 H_2O.
   e. 4 ATP, 2 $FADH_2$, 2 pyruvate.

8. The electron carrier molecules Q and cytochrome *c*
   a. become reduced as they pass electrons on to the next molecule.
   b. contain heme prosthetic groups.
   c. shuttle protons to ATP synthase.
   d. transport H^+ into the mitochondrial matrix, establishing the proton-motive force.
   e. are mobile carriers that transfer electrons between the electron carrier complexes.

9. When pyruvate is converted to acetyl CoA,
   a. CO_2 and ATP are released.
   b. a multienzyme complex removes a carboxyl group, transfers electrons to NAD^+, and attaches a coenzyme.
   c. one turn of the citric acid cycle is completed.
   d. NAD^+ is regenerated so that glycolysis can continue to produce ATP by substrate-level phosphorylation.
   e. phosphofructokinase is activated and glycolysis continues.

10. How many molecules of CO_2 are generated for each molecule of acetyl CoA introduced into the citric acid cycle?
    a. 1   c. 3   e. 6
    b. 2   d. 4

11. In the chemiosmotic mechanism,
    a. ATP production is linked to the proton gradient established by the electron transport chain.
    b. the difference in pH between the intermembrane space and the cytosol drives the formation of ATP.
    c. the flow of H^+ through ATP synthases from the matrix to the intermembrane space drives the phosphorylation of ADP.
    d. the energy released by the reduction and subsequent oxidation of components of the electron transport chain is transferred as a phosphate to ADP.
    e. the production of water in the matrix by the reduction of oxygen leads to a net flow of water out of a mitochondrion.

12. Which of the following reactions is *incorrectly* paired with its location?
    a. ATP synthesis—inner membrane of the mitochondrion (produced in matrix) and cytosol
    b. fermentation—cell cytosol
    c. glycolysis—cell cytosol
    d. substrate-level phosphorylation—cytosol and matrix
    e. citric acid cycle—cristae of mitochondrion

13. When glucose is oxidized to CO_2 and water, approximately 40% of its energy is transferred to
    a. heat.
    b. ATP.
    c. acetyl CoA.
    d. water.
    e. the citric acid cycle.

14. From an energetic viewpoint, what do muscle cells in oxygen deprivation gain from the reduction of pyruvate?
    a. ATP and lactate
    b. ATP and recycled NAD^+
    c. CO_2 and lactate
    d. ATP, alcohol, and NAD^+
    e. ATP, lactate, and CO_2

15. Glucose, made from six radioactively labeled carbon atoms, is fed to yeast cells in the absence of oxygen. How many molecules of radioactive alcohol (C_2H_5OH) are formed from each molecule of glucose?

   a. 0   **c.** 2   **e.** 6

   b. 1   **d.** 3

16. Which of the following produces the most ATP per gram?

   a. glucose, because it is the starting place for glycolysis

   b. glycogen or starch, because they are polymers of glucose

   c. fats, because they are highly reduced compounds

   d. proteins, because of the energy stored in their tertiary structure

   e. amino acids, because they can be fed directly into the citric acid cycle

17. Fats and proteins can be used as fuel in the cell because they

   a. can be converted to glucose by enzymes.

   b. can be converted to intermediates of glycolysis or the citric acid cycle.

   c. can pass through the mitochondrial membrane to enter the citric acid cycle.

   d. contain unstable phosphate bonds.

   e. contain more energy than glucose.

18. Which is *not* true of the enzyme phosphofructokinase? It is

   a. an allosteric enzyme.

   b. inhibited by citrate.

   c. the pacemaker of glycolysis and respiration.

   d. inhibited by ADP.

   e. an early enzyme in the glycolytic pathway.

19. Substrate-level phosphorylation accounts for approximately what percentage of ATP formation when glucose is oxidized to CO_2 and water?

   a. 0%   **c.** 10%   **e.** 20%

   b. 4%   **d.** 15%

20. Cyanide is a poison that blocks the passage of electrons along the electron transport chain. Which of the following is a metabolic effect of this poison?

   a. The pH of the intermembrane space is much lower than normal.

   b. Electrons are passed directly to oxygen, causing cells to explode.

   c. Alcohol would build up in the cells.

   d. NADH supplies would be exhausted, and ATP synthesis would cease.

   e. No proton gradient would be produced, and ATP synthesis would cease.

21. Which enzyme would use NAD^+ as a coenzyme?

   a. phosphofructokinase

   b. phosphoglucoisomerase

   c. triose phosphate dehydrogenase

   d. hexokinase

   e. phosphoglyceromutase

22. List the order of the following compounds as you first encounter them during the process of cellular respiration.

   1. glyceraldehyde-3-phosphate

   2. pyruvate

   3. glucose

   4. acetyl CoA

   5. fructose bisphosphate

   6. CO_2

   a. 5, 3, 1, 2, 6, 4

   b. 3, 5, 1, 2, 6, 4

   c. 3, 5, 2, 1, 4, 6

   d. 6, 4, 1, 2, 3, 5

   e. 5, 3, 2, 1, 6, 4

23. Which compound has the highest free energy (will produce the most ATP when oxidized)?

   a. acetyl CoA

   b. glucose

   c. pyruvate

   d. fructose bisphosphate

   e. glyceraldehydes-3-phosphate

24. Why is glycolysis considered one of the first metabolic pathways to have evolved?
    a. It relies on fermentation, which is characteristic of the archaea and bacteria.
    b. It is found only in prokaryotes, whereas eukaryotes use their mitochondria to produce ATP.
    c. It produces much less ATP than does the electron transport chain and chemiosmosis.
    d. It relies totally on enzymes that are produced by free ribosomes, and bacteria have only free ribosomes and no bound ribosomes.
    e. It is nearly universal, is located in the cytosol, and does not involve O_2.

25. The metabolic function of fermentation is to
    a. oxidize NADH to NAD^+ so that glycolysis can continue in the absence of oxygen.
    b. reduce NADH so that more ATP can be produced by the electron transport chain.
    c. produce lactate during aerobic exercise.
    d. oxidize pyruvate in order to release more energy.
    e. make beer.

26. Which of the following conversions represents a reduction reaction?
    a. pyruvate $\rightarrow$ acetyl CoA + CO_2
    b. $C_6H_{12}O_6 \rightarrow 6\ CO_2$
    c. NADH + $H^+ \rightarrow NAD^+$ + 2 H
    d. glucose $\rightarrow$ pyruvate
    e. acetaldehyde (C_2H_4O) $\rightarrow$ ethanol (C_2H_6O)

27. The oxidation of a molecule of $FADH_2$ yields less ATP than a molecule of NADH yields because $FADH_2$
    a. carries fewer electrons.
    b. is formed in the cytosol and energy is lost when it shuttles its electrons across the mitochondrial membrane.
    c. passes its electrons to a transport molecule later in the chain and at a lower energy level.
    d. is the last molecule produced by the citric acid cycle, and little energy is left to be captured.
    e. has a much lower energy conformation than does NADH.

28. What is the role of oxygen in cellular respiration?
    a. It is reduced in glycolysis as glucose is oxidized.
    b. It provides electrons to the electron transport chain.
    c. It provides the activation energy needed for oxidation to occur.
    d. It is the final electron acceptor for the electron transport chain.
    e. It combines with the carbon removed during the citric acid cycle to form CO_2.

Chapter 10
Photosynthesis

Framework

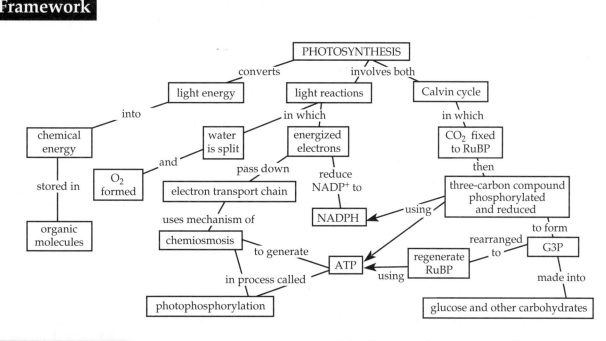

Chapter Review

In **photosynthesis,** the light energy of the sun is converted into chemical energy stored in organic molecules. Organisms obtain the organic molecules they require for energy and carbon skeletons by autotrophic or heterotrophic nutrition. **Autotrophs** "feed themselves" in the sense that they make their own organic molecules from inorganic raw materials. Plants, algae, some other protists, and some prokaryotes are photoautotrophs.

Heterotrophs are consumers. They may eat plants or animals or decompose organic litter, but almost all are ultimately dependent on photoautotrophs for food and oxygen.

10.1 Photosynthesis converts light energy to the chemical energy of food

Chloroplasts: The Sites of Photosynthesis in Plants Chloroplasts, found mainly in the **mesophyll** tissue of the leaf, contain **chlorophyll,** the green pigment that absorbs the light energy that drives photosynthesis. CO_2 enters and O_2 exits the leaf through **stomata.** Veins carry water from the roots to leaves and distribute sugar to nonphotosynthetic tissue.

A chloroplast consists of a double membrane surrounding a dense fluid called the **stroma** and an elaborate membrane system called **thylakoids,** enclosing the *thylakoid space.* Thylakoid sacs may be stacked to form grana. Chlorophyll is embedded in the thylakoid membrane.

■ INTERACTIVE QUESTION 10.1

Label the indicated parts in this diagram of a chloroplast.

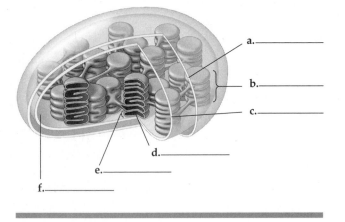

a._____

b._____

c._____

d._____

e._____

f._____

Tracking Atoms through Photosynthesis: Scientific Inquiry   If only the net consumption of water is considered, the equation for photosynthesis is the reverse of respiration:

$$6\,CO_2 + 6\,H_2O + \text{Light energy} \rightarrow C_6H_{12}O_6 + 6\,O_2$$

Using evidence from bacteria that utilize hydrogen sulfide (H_2S) for photosynthesis, C. B. van Niel hypothesized that all photosynthetic organisms need a hydrogen source and that plants split water as a source of electrons from hydrogen, releasing oxygen. Scientists confirmed this hypothesis by using a heavy isotope of oxygen (^{18}O). Labeled O_2 was produced in photosynthesis only when water, rather than carbon dioxide, contained the labeled oxygen.

Photosynthesis is a redox process like respiration but differs in the direction of electron flow. The electrons increase their potential energy when they travel from water to reduce CO_2 into sugar, and light provides this energy.

The Two Stages of Photosynthesis: A Preview   Solar energy is converted into chemical energy in the **light reactions.** Light energy absorbed by chlorophyll drives the transfer of electrons and hydrogen from water to the electron acceptor **$NADP^+$,** which is reduced to NADPH and temporarily stores energized electrons. Oxygen is released when water is split. ATP is formed during the light reactions, using chemiosmosis in a process called **photophosphorylation.**

In the **Calvin cycle,** carbon dioxide is incorporated into existing organic compounds by **carbon fixation,** and these compounds are then reduced to form carbohydrate. NADPH and ATP from the light reactions supply the reducing power and chemical energy needed for the Calvin cycle.

■ INTERACTIVE QUESTION 10.2

Fill in the blanks in this overview of photosynthesis in a chloroplast. Indicate the locations of the processes c. and h.

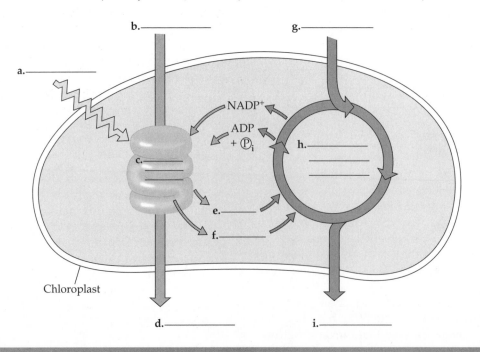

b._____   g._____

a._____

$NADP^+$

ADP + P_i

c._____

h._____

e._____

f._____

Chloroplast

d._____   i._____

10.2 The light reactions convert solar energy to the chemical energy of ATP and NADPH

The Nature of Sunlight Electromagnetic energy, also called electromagnetic radiation, travels as rhythmic wave disturbances of electrical and magnetic fields. The distance between the crests of the electromagnetic waves, their **wavelength,** ranges across the **electromagnetic spectrum,** from short gamma waves to long radio waves. The small band of radiation from about 380 to 750 nm is called **visible light.**

Light also behaves as if it consists of discrete particles called **photons,** which have a fixed quantity of energy. The amount of energy in a photon is inversely related to its wavelength.

■ INTERACTIVE QUESTION 10.3

A photon of which color of light would contain more energy: orange (620 nm) or blue (480 nm)? Why?

Photosynthetic Pigments: The Light Receptors A **spectrophotometer** measures the amounts of light of different wavelengths absorbed by a pigment. The **absorption spectrum** of **chlorophyll *a*,** the pigment that participates directly in the light reactions, shows that it absorbs violet-blue and red light best. Accessory pigments such as **chlorophyll *b*** and some **carotenoids** absorb light of different wavelengths and broaden the spectrum of colors useful in photosynthesis. Some carotenoids function in photoprotection by absorbing excessive light energy that might damage chlorophyll or interact with oxygen to form reactive molecules. (These pigments act as antioxidants.)

■ INTERACTIVE QUESTION 10.4

An **action spectrum** shows the relative rates of photosynthesis under different wavelengths of light. Label the absorption and action spectra on this graph. Why are these lines different?

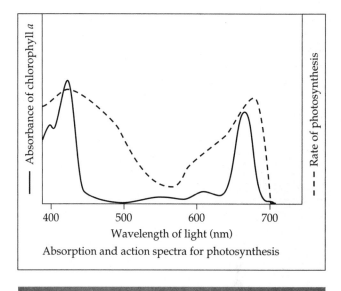

Absorption and action spectra for photosynthesis

Excitation of Chlorophyll by Light When a pigment molecule absorbs energy from a photon, one of the molecule's electrons is elevated to an orbital where it has more potential energy. Only photons whose energy is equal to the difference between the ground state and the excited state for that molecule are absorbed.

The excited state is unstable. Energy is released as heat as the electron drops back to its ground-state orbital. Isolated chlorophyll molecules also emit photons of light, called fluorescence, as their electrons return to ground state.

A Photosystem: A Reaction Center Associated with Light-Harvesting Complexes Photosystems, located in the thylakoid membrane, contain a number of **light-harvesting complexes** and a **reaction center,** a protein complex with two special chlorophyll *a* molecules and a **primary electron acceptor.** When a pigment molecule in a light-harvesting complex absorbs a photon, the energy is passed from pigment to pigment until it reaches the reaction center. In a redox reaction, an excited electron of a reaction-center chlorophyll *a* is trapped by the primary electron acceptor before it can return to the ground state.

There are two types of photosystems in the thylakoid membrane. The chlorophyll *a* molecule at the reaction center of **photosystem II (PSII)** is called P680, after the wavelength of light (680 nm) it absorbs best. At the reaction center of **photosystem I (PSI)** is a chlorophyll *a* molecule called P700.

■ INTERACTIVE QUESTION   10.5

Describe the components of a photosystem.

Noncyclic Electron Flow In **noncyclic electron flow,** electrons pass continuously from water to NADP⁺. An excited electron of P680 in photosystem II is trapped by the primary electron acceptor. P680 is a strong oxidizing agent, and its electron hole is filled when an enzyme removes electrons from water, splitting it into two electrons, two H⁺ and an oxygen atom that immediately combines with another oxygen to form O_2.

The primary electron acceptor passes the photoexcited electron to an electron transport chain made up of plastoquinone (Pq), a cytochrome complex, and plastocyanin (Pc). The energy released as electrons "fall" through the electron transport chain is used in the synthesis of ATP.

At the bottom of the electron transport chain, the electron passes to P700 in photosystem I to replace the photoexcited electron captured by its primary electron acceptor. This primary electron acceptor passes the electron down a second electron transport chain through ferredoxin (Fd), from which the enzyme NADP⁺ reductase transfers electrons to NADP⁺. (Review this electron path by completing Interactive Question 10.6.)

■ INTERACTIVE QUESTION   10.6

Fill in the steps of noncyclic electron flow in the diagram below. Circle the important products that will be used to provide chemical energy and reducing power to the Calvin cycle.

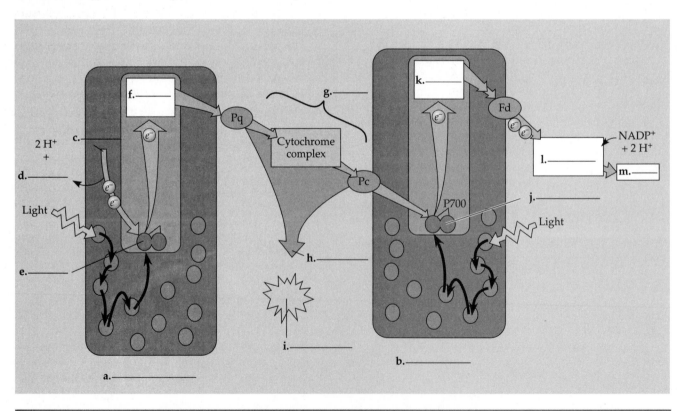

Cyclic Electron Flow   In **cyclic electron flow**, electrons excited from P700 in PSI are passed from Fd to the cytochrome complex and back to P700. Neither NADPH nor O_2 is generated. The Calvin cycle requires more ATP than NADPH, and the additional ATP may be supplied by the cyclic flow of electrons, perhaps in response to a buildup of NADPH.

■ **INTERACTIVE QUESTION   10.7**

a. On the diagram in Interactive Question 10.6 sketch the path that electrons from P700 take during cyclic electron flow.

b. Why is neither oxygen nor NADPH generated by cyclic electron flow?

c. How, then, is ATP produced by cyclic electron flow?

A Comparison of Chemiosmosis in Chloroplasts and Mitochondria   Chemiosmosis in mitochondria and in chloroplasts is very similar. The key difference is that in respiration, chemical energy from food is transferred to ATP, whereas in chloroplasts light energy is transformed to the chemical energy of ATP.

In chloroplasts, the electron transport chain pumps protons from the stroma into the thylakoid space. As H^+ diffuses back through ATP synthase, ATP is formed on the stroma side, where it is available for the Calvin cycle.

■ **INTERACTIVE QUESTION   10.8**

a. In the light, the proton gradient across the thylakoid membrane is as great as 3 pH units. On which side is the pH lowest?

b. What three factors contribute to the formation of this large difference in H^+ concentration between the thylakoid space and the stroma?

10.3   The Calvin cycle uses ATP and NADPH to convert CO_2 to sugar

The Calvin cycle turns three times to fix three molecules of CO_2 and produce one molecule of the three-carbon sugar **glyceraldehyde-3-phosphate (G3P).** The cycle can be divided into three phases:

1. *Carbon fixation:* CO_2 is added to a five-carbon sugar, ribulose bisphosphate (RuBP), in a reaction catalyzed by the enzyme RuBP carboxylase **(rubisco).** The unstable six-carbon intermediate that is formed splits into two molecules of 3-phosphoglycerate.

2. *Reduction:* Each molecule of 3-phosphoglycerate is then phosphorylated by ATP to form 1,3-bisphosphoglycerate. Two electrons from NADPH reduce this compound to create G3P. The cycle must turn three times to create a net gain of one molecule of G3P.

3. *Regeneration of CO_2 acceptor (RuBP):* The rearrangement of five molecules of G3P into the three molecules of RuBP requires three more ATP.

Nine molecules of ATP and six of NADPH are required to synthesize one G3P. (To review this process, complete the summary diagram of the Calvin cycle in Interactive Question 10.9.)

■ INTERACTIVE QUESTION   10.9

Label the three phases (a through c) and key molecules (d through o) in this diagram of the Calvin cycle.

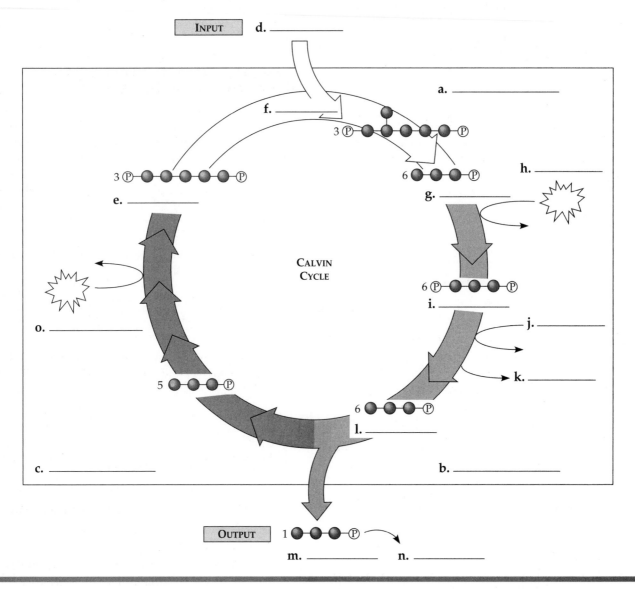

10.4   Alternative mechanisms of carbon fixation have evolved in hot, arid climates

Photorespiration: An Evolutionary Relic?   In most plants, CO_2 enters the Calvin cycle and the first product of carbon fixation is 3-phosphoglycerate. When these **C₃ plants** close their stomata on hot, dry days to limit water loss, CO_2 concentration in the leaf air spaces falls, slowing the Calvin cycle. As more O_2 than CO_2 accumulates, rubisco adds O_2 in place of CO_2 to RuBP. The product splits and a two-carbon compound leaves the chloroplast and is broken down to release CO_2. This seemingly wasteful process is called **photorespiration.**

■ INTERACTIVE QUESTION   10.10

What possible explanation is there for photorespiration, a process that can result in the loss of as much as 50% of the carbon fixed in the Calvin cycle?

C₄ Plants   In **C₄ plants,** CO_2 is first added to a 3-carbon compound, PEP, with the aid of an enzyme **(PEP carboxylase)** that has a high affinity for CO_2. The resulting four-carbon compound formed in the **meso-phyll cells** of the leaf is transported to **bundle-sheath cells** tightly packed around the veins of the leaf. The compound is broken down to release CO_2, creating concentrations high enough that rubisco will accept CO_2 rather than O_2 and initiate the Calvin cycle.

■ INTERACTIVE QUESTION   10.11

a. Where does the Calvin cycle take place in C_4 plants?

b. How can C_4 plants successfully perform the Calvin cycle in hot, dry conditions when C_3 plants would be undergoing photorespiration?

CAM Plants   Many succulent plants close their stomata during the day to prevent water loss, but open them at night to take up CO_2 and incorporate it into a variety of organic acids. **CAM plants** break these compounds down to release CO_2 during daylight so that the Calvin cycle can proceed. Unlike the C_4 pathway, the **crassulacean acid metabolism (CAM)** pathway does not structurally separate carbon fixation from the Calvin cycle; instead, the two processes are separated in time.

The Importance of Photosynthesis: *A Review*

About 50% of the organic material produced by photosynthesis is used as fuel for cellular respiration in the mitochondria of plant cells; the rest is used as carbon skeletons for synthesis of organic molecules (proteins, lipids, and a great deal of cellulose), stored as starch, or lost through photorespiration. About 160 billion metric tons of carbohydrate per year are produced by photosynthesis.

▶ Word Roots

auto- = self; **-troph** = food (*autotroph:* an organism that obtains organic food molecules without eating other organisms)

chloro- = green; **-phyll** = leaf (*chlorophyll:* photosynthetic pigment in chloroplasts)

electro- = electricity; **magnet-** = magnetic (*electromagnetic spectrum:* the entire spectrum of radiation)

hetero- = other (*heterotroph:* an organism that obtains organic food molecules by eating other organisms or their by-products)

meso- = middle (*mesophyll:* the green tissue in the middle, inside of a leaf)

photo- = light (*photosystem:* cluster of pigment molecules)

▶ Structure Your Knowledge

1. You have already filled in the blanks in several diagrams of photosynthesis. To really understand this process, however, you should create your own representation. In a diagrammatic form, outline the key events of photosynthesis. Trace the flow of electrons through photosystem II and I, the production of ATP and NADPH by the light reactions and their transfer into the Calvin cycle, and the major steps in the production of G3P. Note where these reactions occur in the chloroplast. You can compare your creation to the sketch in the answer section. Then talk a friend through the steps in your diagram. Perhaps your study group can combine several representations into a clear, concise summary of this chapter.

2. Create a concept map to confirm your understanding of the chemiosmotic synthesis of ATP in photophosphorylation.

▶ Test Your Knowledge

MULTIPLE CHOICE: *Choose the one best answer.*

1. Which of the following is mismatched with its location?
   a. light reactions—grana
   b. electron transport chain—thylakoid membrane
   c. Calvin cycle—stroma
   d. ATP synthase—double membrane surrounding chloroplast
   e. splitting of water—thylakoid space

2. Photosynthesis is a redox process in which
   a. CO_2 is reduced and water is oxidized.
   b. $NADP^+$ is reduced and RuBP is oxidized.
   c. CO_2, $NADP^+$, and water are reduced.
   d. O_2 acts as an oxidizing agent and water acts as a reducing agent.
   e. G3P is reduced and the electron transport chain is oxidized.

3. Blue light has more energy than red light. Therefore, blue light
   a. has a longer wavelength than red light.
   b. has a shorter wavelength than red light.
   c. contains more photons than red light.
   d. has a broader electromagnetic spectrum than red light.
   e. is absorbed faster by chlorophyll *a*.

4. A spectrophotometer can be used to measure
   a. the absorption spectrum of a substance.
   b. the action spectrum of a substance.
   c. the amount of energy in a photon.
   d. the wavelength of visible light.
   e. the efficiency of photosynthesis.

5. Accessory pigments within chloroplasts are responsible for
   a. driving the splitting of water molecules.
   b. absorbing photons of different wavelengths of light and passing that energy to P680 or P700.
   c. providing electrons to the reaction-center chlorophyll after photoexcited electrons pass to NADP$^+$.
   d. pumping H$^+$ across the thylakoid membrane to create a proton-motive force.
   e. anchoring chlorophyll *a* within the reaction center.

6. Below is an absorption spectrum for an unknown pigment molecule. What color would this pigment appear to you?

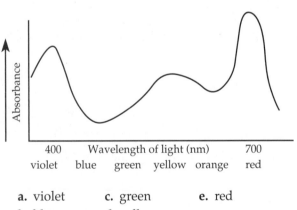

    400   Wavelength of light (nm)   700

violet   blue   green yellow orange   red

   a. violet   c. green   e. red
   b. blue   d. yellow

7. Noncyclic electron flow along with chemiosmosis in the chloroplast results in the production of
   a. ATP only.
   b. ATP and NADPH.
   c. ATP and G3P.
   d. ATP and O$_2$.
   e. ATP, NADPH, and O$_2$.

8. The chlorophyll known as P680 has its electron "holes" filled by electrons from
   a. photosystem I.
   b. photosystem II.
   c. water.
   d. NADPH.
   e. accessory pigments.

9. CAM plants avoid photorespiration by
   a. fixing CO$_2$ into organic acids during the night; these acids then release CO$_2$ during the day.
   b. performing the Calvin cycle at night.
   c. fixing CO$_2$ into four-carbon compounds in the mesophyll, which release CO$_2$ in the bundle-sheath cells.
   d. using PEP carboxylate to fix CO$_2$ to ribulose bisphosphate (RuBP).
   e. keeping their stomata closed during the day.

10. Electrons that flow through the two photosystems have their highest potential energy in
    a. water.
    b. P680.
    c. NADPH.
    d. the electron transport chain.
    e. photoexcited P700.

11. Chloroplasts can make carbohydrate in the dark if provided with
    a. ATP and NADPH and CO$_2$.
    b. an artificially induced proton gradient.
    c. organic acids or four-carbon compounds.
    d. a source of hydrogen.
    e. photons and CO$_2$.

12. In the chemiosmotic synthesis of ATP in chloroplast, H$^+$ diffuses through the ATP synthase
    a. from the stroma into the thylakoid space.
    b. from the thylakoid space into the stroma.
    c. from the intermembrane space into the matrix.
    d. from the cytoplasm into the intermembrane space.
    e. from the matrix into the stroma.

13. In C$_4$ plants, the Calvin cycle
    a. takes place at night.
    b. only occurs when the stomata are closed.
    c. takes place in the mesophyll cells.
    d. takes place in the bundle-sheath cells.
    e. uses PEP carboxylase instead of rubisco because of its greater affinity for CO$_2$.

14. How many "turns" of the Calvin cycle are required to produce one molecule of glucose?
    a. 1   c. 3   e. 12
    b. 2   d. 6

15. In green plants, most of the ATP for synthesis of proteins, cytoplasmic streaming, and other cellular activities comes directly from
    a. photosystem I.
    b. photosystem II.
    c. the Calvin cycle.
    d. oxidative phosphorylation.
    e. photophosphorylation.

16. Six molecules of G3P formed in the Calvin cycle are used to produce
    a. three molecules of glucose.
    b. three molecules of RuBP and one G3P.
    c. one molecule of glucose and four molecules of 3-phosphoglycerate.
    d. one G3P and three four-carbon intermediates.
    e. none of the above, since three molecules of G3P result from three turns of the Calvin cycle.

17. A difference between electron transport in photosynthesis and respiration is that in photosynthesis
    a. NADPH rather than NADH passes electrons to the electron transport chain.
    b. ATP synthase releases ATP into the stroma rather than into the cytosol.
    c. light provides the energy to push electrons to the top of the electron chain, rather than energy from the oxidation of food molecules.
    d. an H^+ concentration gradient rather than a proton-motive force drives the phosphorylation of ATP.
    e. both a and c are correct.

18. NADPH and ATP from the light reactions are both needed
    a. in the carbon fixation stage to provide energy and reducing power to rubisco.
    b. to regenerate three RuBP from five G3P (glyceraldehyde-3-phosphate).
    c. to combine two molecules of G3P to produce glucose.
    d. to reduce 3-phosphoglycerate to G3P.
    e. to reduce the H^+ concentration in the stroma and contribute to the proton-motive force.

19. What portion of an illuminated plant cell would you expect to have the lowest pH?
    a. nucleus
    b. cytosol
    c. chloroplast
    d. stroma of chloroplast
    e. thylakoid space

20. How does cyclic electron flow differ from non-cyclic electron flow?
    a. No NADPH is produced by cyclic electron flow.
    b. No O_2 is produced by cyclic electron flow.
    c. The cytochrome complex in the electron transport chain is not involved in cyclic electron flow.
    d. Both a and b are correct.
    e. a, b, and c are correct.

21. What does rubisco do?
    a. reduces CO_2 to G3P
    b. regenerates RuBP with the aid of ATP
    c. combines electrons and H^+ to reduce $NADP^+$ to NADPH
    d. adds CO_2 to RuBP in the carbon fixation stage
    e. transfers electrons from NADPH to 1,3-bisphosphoglycerate to produce G3P

22. What are the final electron acceptors for the electron transport chains in the light reactions of photosynthesis and in cellular respiration?
    a. O_2 in both
    b. CO_2 in both
    c. H_2O in the light reactions and O_2 in respiration
    d. P700 and $NADP^+$ in the light reactions and NAD^+ or FAD in respiration
    e. $NADP^+$ in the light reactions and O_2 in respiration

Use the following for questions 23 through 28.

Indicate if the following events occur during

    a. respiration
    b. photosynthesis
    c. both respiration and photosynthesis
    d. neither respiration nor photosynthesis

———— 23. Chemiosmotic synthesis of ATP

———— 24. Reduction of oxygen

———— 25. Reduction of CO_2

———— 26. Reduction of NAD^+

———— 27. Oxidation of $NADP^+$

———— 28. Oxidative phosphorylation

Chapter 11
Cell Communication

Framework

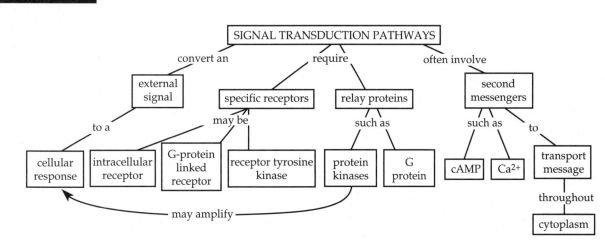

Chapter Review

Cell-to-cell communication is critical to the development and functioning of multicellular organisms and also to communication between unicellular organisms. Some universal mechanisms of cellular interaction provide evidence for the evolutionary relatedness of all life.

11.1 External signals are converted into responses within the cell

Evolution of Cell Signaling The two mating types of yeast secrete chemical factors that bind to receptors on the other cell type, initiating fusion (mating) of the cells. The series of steps involved in the conversion of a cell surface signal to a cellular response is called a **signal transduction pathway.** Similarities among these pathways in bacteria, yeast, plants, and animals suggest an early evolution of cell-signaling mechanisms.

Local and Long-Distance Signaling Chemical signals may be communicated between cells through direct cytoplasmic connections (gap junctions or plasmodesmata) or through contact of surface molecules (cell-cell recognition in animal cells).

In *paracrine signaling* in animals, a signaling cell releases messenger molecules into the extracellular fluid, and these **local regulators** influence nearby cells. In another type of local signaling called *synaptic signaling,* a nerve cell releases neurotransmitter molecules into the narrow synapse separating it from its target cell.

Hormones are chemical signals that travel to more distant cells. In hormonal or endocrine signaling in animals, the circulatory system transports hormones throughout the body to reach target cells with appropriate receptors.

Transmission of signals through the nervous system is also a type of long-distance signaling.

■ INTERACTIVE QUESTION 11.1

a. Do plant cells communicate using hormones?

b. If so, how do those hormones travel between secreting cells and target cells?

The Three Stages of Cell Signaling: A Preview
E. W. Sutherland's work studying epinephrine's effect on the hydrolysis of glycogen in liver cells established that cell signaling involves three stages: **reception** of a chemical signal by binding to a receptor protein either inside the cell or on its surface; **transduction** of the signal, often by a signal transduction pathway—a sequence of changes in relay molecules; and the final **response** of the cell.

11.2 Reception: A signal molecule binds to a receptor protein, causing it to change shape

A signal molecule acts as a **ligand,** which specifically binds to a receptor protein and usually induces a change in the receptor's conformation.

Intracellular Receptors Hydrophobic chemical messengers may cross a cell's plasma membrane and bind to receptors in the cytoplasm or nucleus of target cells. Steroid hormones activate receptors in target cells that function as *transcription factors* to regulate gene expression.

Receptors in the Plasma Membrane There are three major types of membrane receptors that bind with water-soluble signal molecules and transmit information into the cell.

The various receptors that work with the aid of a G protein, called **G-protein-linked receptors,** are structurally similar, with seven α helices spanning the plasma membrane. Binding of the appropriate extracellular signal to a G-protein-linked receptor activates the receptor, which binds to and activates a specific **G protein** located on the cytoplasmic side of the membrane. This activation occurs when a GTP nucleotide replaces the GDP bound to the G protein. The G protein then activates a membrane-bound enzyme,

after which it hydrolyzes its GTP and becomes inactive again. The activated enzyme triggers the next step in the pathway to the cell's response.

G-protein-linked receptor systems are used by many hormones and neurotransmitters and are involved in embryological development and sensory reception. Many bacteria produce toxins that interfere with G-protein systems; up to 60% of all medicines influence G-protein pathways.

■ INTERACTIVE QUESTION 11.2

Explain why G-protein-regulated pathways shut down rapidly in the absence of a signal molecule.

Receptor tyrosine kinases are receptor proteins with enzymatic activity that can trigger several pathways at once. Part of the receptor protein is tyrosine kinase, an enzyme that transfers phosphate groups from ATP to the amino acid tyrosine on a protein. Many receptor tyrosine receptors exist as single transmembrane polypeptides with a signal binding site, a transmembrane α helix, and a cytoplasmic tail with a series of tyrosine amino acids. Ligand binding causes two receptors to form a dimer, which activates the tyrosine kinase portions of the molecules. These then phosphorylate the tyrosines on each other's cytoplasmic tails. Different relay proteins now bind to specific phosphorylated tyrosines and become activated, triggering many different transduction pathways in response to one type of signal.

■ INTERACTIVE QUESTION 11.3

Label the parts in this diagram of an activated receptor tyrosine kinase dimer.

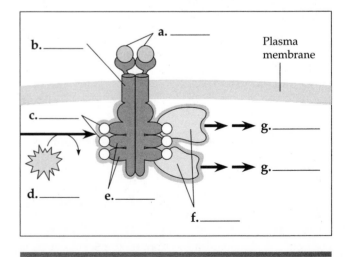

The binding of a chemical signal to a **ligand-gated ion channel** opens or closes the protein pore, thus allowing or blocking the flow of specific ions through the membrane. The resulting change in ion concentration inside the cell triggers a cellular response. Neurotransmitters often bind to ligand-gated ion channels in the transmission of nervous signals.

■ INTERACTIVE QUESTION 11.4

a. What determines whether a cell is a target cell for a particular signal molecule?

b. What determines whether a signal molecule binds to a membrane-surface receptor or an intracellular receptor?

11.3 Transduction: Cascades of molecular interactions relay signals from receptors to target molecules in the cell

Multistep signal pathways allow a small number of extracellular signal molecules to be amplified to produce a large cellular response.

Signal Transduction Pathways The relay molecules in a signal transduction pathway are usually proteins, which interact as they pass the message from the extracellular signal to the protein that produces the cellular response.

Protein Phosphorylation and Dephosphorylation **Protein kinases** are enzymes that transfer phosphate groups from ATP to proteins, often to the amino acid serine or threonine. Relay molecules in signal transduction pathways are often protein kinases, which are sequentially phosphorylated, producing a conformational change that activates each enzyme. Hundreds of different kinds of protein kinases regulate the activity of a cell's proteins.

Protein phosphatases are enzymes that remove phosphate groups from proteins. They effectively shut down signaling pathways when the extracellular signal is no longer present.

■ INTERACTIVE QUESTION 11.5

a. What does a protein kinase do?

b. What does a protein phosphatase do?

c. What is a phosphorylation cascade?

Small Molecules and Ions as Second Messengers
Small, water-soluble molecules or ions often function as **second messengers,** which rapidly relay the signal from the membrane-receptor-bound "first messenger" into a cell's interior.

Binding of an extracellular signal to a G-protein-linked receptor activates a G protein that may activate **adenylyl cyclase,** a membrane protein that converts ATP to cyclic adenosine monophosphate (**cyclic AMP** or **cAMP**). The cAMP often activates *protein kinase A,* which phosphorylates other proteins. A cytoplasmic enzyme, phosphodiesterase, converts cAMP to inactive AMP, thus removing the second messenger.

Some signal molecules may activate an inhibitory G protein that inhibits adenylyl cyclase.

■ INTERACTIVE QUESTION 11.6

Label the components in this diagram of the steps in a signal transduction pathway that uses cAMP as a second messenger.

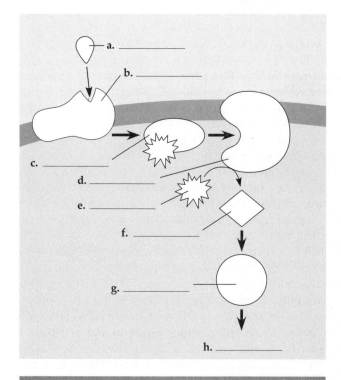

Calcium ions are widely used as a second messenger in many G-protein-linked and receptor tyrosine kinase pathways. Cytosolic concentration of Ca^{2+} is usually kept very low by active transport of calcium out of the cell and into the endoplasmic reticulum. The release of Ca^{2+} from the ER involves another second messenger, **inositol trisphosphate (IP$_3$).** In response to the reception of a signal, the enzyme phospholipase C cleaves a membrane phospholipid into two second messengers, IP$_3$ and **diacylglycerol (DAG).** IP$_3$ binds to and opens ligand-gated calcium channels in the ER. The calcium ions then induce cellular responses.

■ **INTERACTIVE QUESTION   11.7**

Fill in the blanks to review the steps in a signal transduction pathway involving a G-protein-linked receptor and Ca^{2+} as a second messenger.

A a. _____ binds to a G-protein-linked receptor. An activated b. _____ activates the enzyme phospholipase C, which cleaves a c. _____ into DAG and d. _____, which binds to and opens a ligand-gated channel, releasing e. _____ from the f. _____.

11.4   Response: Cell signaling leads to regulation of cytoplasmic activities or transcription

Cytoplasmic and Nuclear Responses   Signal transduction pathways may lead to the activation of cytoplasmic enzymes or other proteins, or may lead to the synthesis of such proteins by affecting gene expression. Growth factors and certain animal and plant hormones may initiate pathways that ultimately activate transcription factors, which regulate the transcription of mRNA from specific genes.

Fine-Tuning of the Response   A signal transduction pathway can amplify a signal in an enzyme cascade, as each successive enzyme in the pathway can process multiple molecules that then activate the next step.

As a result of their particular set of receptor proteins, relay proteins, and effector proteins, different cells can respond to different signals or can exhibit different responses to the same molecular signal. Pathways may branch to produce multiple responses, or two pathways may interact ("cross-talk") to mediate a single response.

Scaffolding proteins are large relay proteins to which other relay proteins attach, increasing the efficiency of signal transduction in a pathway.

Inactivation mechanisms that discontinue a cell's response to a signal are essential in keeping a cell responsive to regulation.

■ **INTERACTIVE QUESTION   11.8**

How do the following mechanisms or molecules maintain a cell's ability to respond to fresh signals?

a. reversible binding of signal molecules

b. GTPase activity of G protein

c. phosphodiesterase

d. protein phosphatases

Word Roots

liga- = bound or tied (*ligand:* a small molecule that specifically binds to a larger one)

trans- = across (*signal transduction pathway:* the process by which a signal on a cell's surface is converted into a specific cellular response inside the cell)

-yl = substance or matter (*adenylyl cyclase:* an enzyme built into the plasma membrane that converts ATP to cAMP)

Structure Your Knowledge

1. Why is cell signaling such an important component of a cell's life?

2. Briefly describe the three stages of cell signaling.

3. Some signal pathways alter a protein's activity; others may result in the production of new proteins. Explain the mechanisms for these two different responses.

4. How does an enzyme cascade produce an amplified response to a signal molecule?

Test Your Knowledge

MULTIPLE CHOICE: *Choose the one best answer.*

1. When epinephrine binds to cardiac (heart) muscle cells, it speeds their contraction. When it binds to muscle cells of the small intestine, it inhibits their contraction. How can the same hormone have different effects on muscle cells?
   a. Cardiac cells have more receptors for epinephrine than do intestinal cells.
   b. Epinephrine circulates to the heart first and thus is in higher concentration around cardiac cells.
   c. The two types of muscle cells have different signal transduction pathways for epinephrine and thus have different cellular responses.
   d. Cardiac muscle is stronger than intestinal muscle and thus has a stronger response to epinephrine.
   e. Epinephrine binds to G-protein-linked receptors in cardiac cells, and these receptors always increase a response to the signal. Epinephrine binds to receptor tyrosine kinases in intestinal cells, and these receptors always inhibit a response to the signal.

2. Which of the following would be used in the type of local signaling called paracrine signaling in animals?
   a. the neurotransmitter acetylcholine
   b. the hormone epinephrine
   c. the neurotransmitter norepinephrine
   d. a local regulator such as a growth factor
   e. Both a and c are correct.

3. A signal molecule that binds to a plasma-membrane protein functions as a
   a. ligand.
   b. second messenger.
   c. protein phosphatase.
   d. protein kinase.
   e. receptor protein.

4. What is a G protein?
   a. a specific type of membrane-receptor protein
   b. a protein on the cytoplasmic side of a membrane that becomes activated by a receptor protein
   c. a membrane-bound enzyme that converts ATP to cAMP
   d. a tyrosine kinase relay protein
   e. a guanine nucleotide that converts between GDP and GTP to activate and inactivate relay proteins

5. How do receptor tyrosine kinases transduce a signal?
   a. They transport the signal molecule into the cell, where it binds to and activates a transcription factor. The transcription factor then alters gene expression.
   b. Signal binding causes a conformational change that activates membrane-bound tyrosine kinase relay proteins that phosphorylate serine and threonine amino acids.
   c. Their activated tyrosine kinases convert ATP to cAMP; cAMP acts as a second messenger to activate other protein kinases.
   d. When activated, they cleave a membrane phospholipid into two second-messenger molecules. One of the molecules opens Ca^{2+} ion channels on the endoplasmic reticulum.
   e. They form a dimer; they phosphorylate each other's tyrosines; specific proteins bind to and are activated by these phosphorylated tyrosines.

6. Which of the following can activate a protein by transferring a phosphate group to it?
   a. cAMP
   b. G protein
   c. phosphodiesterase
   d. protein kinase
   e. protein phosphatase

7. Many signal transduction pathways use second messengers to
   a. transport a signal through the lipid bilayer portion of the plasma membrane.
   b. relay a signal from the outside to the inside of the cell.
   c. relay the message from the inside of the membrane throughout the cytoplasm.
   d. amplify the message by phosphorylating proteins.
   e. dampen the message once the signal molecule has left the receptor.

8. What is a function of the second messenger IP_3?
   a. bind to and activate protein kinase A
   b. activate transcription factors
   c. activate other membrane-bound relay molecules
   d. convert ATP to cAMP
   e. bind to and open ligand-gated calcium channels on the ER

9. Many human diseases, including bacterial infections, and also many medicines used to treat these diseases produce their effects by influencing which of the following?
   a. cAMP concentrations in the cell
   b. Ca^{2+} concentrations in the cell
   c. G-protein pathways
   d. gene expression
   e. receptor tyrosine kinases

10. Signal amplification is most often achieved by
    a. an enzyme cascade involving multiple protein kinases.
    b. the binding of multiple signal molecules.
    c. branching pathways that produce multiple cellular responses.
    d. activating transcription factors that affect gene expression.
    e. the action of adenylyl cyclase in converting ATP to ADP.

11. From studying the effects of epinephrine on liver cells, Sutherland concluded that
    a. there is a one-to-one correlation between the number of epinephrine molecules bound to receptors and the number of glucose molecules released from glycogen.
    b. epinephrine enters liver cells and binds to receptors that function as transcription factors to turn on the gene for glycogen phosphorylase.
    c. there is a "second messenger" that transmits the signal of epinephrine binding on the plasma membrane to the enzymes involved in glycogen breakdown inside the cell.
    d. the signal transduction pathway through which epinephrine signals glycogen breakdown involves receptor tyrosine kinases and the release of calcium ions that activate glycogen phosphorylase.
    e. epinephrine functions as a ligand to open ion channels in the plasma membrane that allow calcium ions to enter and initiate a cellular response.

12. Which of the following is a similarity between G-protein-linked receptors and receptor tyrosine kinases?
    a. signal-binding sites specific for steroid hormones
    b. formation of a dimer following binding of a signal molecule
    c. activation that results from binding of GTP
    d. phosphorylation of specific amino acids in direct response to signal binding
    e. α-helix regions of the receptor that span the plasma membrane

13. Which of the following is *incorrectly* matched with its description?
    a. scaffolding protein—large relay protein that may bind with several other relay proteins to increase the efficiency of a signaling pathway
    b. protein phosphatase—enzyme that transfers a phosphate group from ATP to a protein, causing a conformational change that usually activates that protein
    c. adenylyl cyclase—enzyme attached to plasma membrane that converts ATP to cAMP in response to an extracellular signal
    d. phospholipase C—enzyme that may be activated by a G protein or receptor tyrosine kinase and cleaves a plasma-membrane phospholipid into the second messengers IP_3 and DAG
    e. G protein—relay protein attached to the inside of plasma membrane that, when activated by an activated G-protein-linked receptor, binds GTP and then usually activates another membrane-attached protein

14. Which of the following signal molecules pass through the plasma membrane and bind to intracellular receptors that move into the nucleus and function as transcription factors to regulate gene expression?
    a. epinephrine
    b. growth factors
    c. yeast mating factors α and a
    d. testosterone, a steroid hormone
    e. neurotransmitter released into synapse between nerve cells

Chapter 12

The Cell Cycle

12.1 Cell division results in genetically identical daughter cells

12.2 The mitotic phase alternates with interphase in the cell cycle

12.3 The cell cycle is regulated by a molecular control system

■ Framework

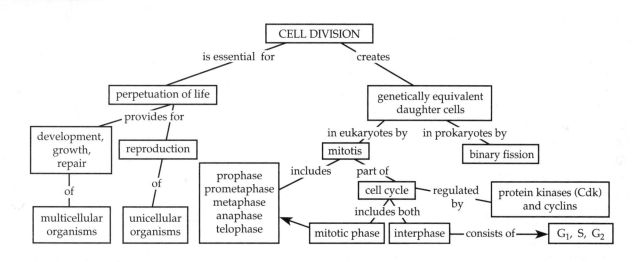

■ Chapter Review

Cell division creates duplicate offspring in unicellular organisms and provides for growth, development, and repair in multicellular organisms. The **cell cycle** extends from the creation of a new cell by the division of its parent cell to its own division into two cells.

12.1 Cell division results in genetically identical daughter cells

The process of re-creating a structure as intricate as a cell necessitates the exact duplication and equal division of the DNA containing the cell's genetic program.

Cellular Organization of the Genetic Material A cell's complete complement of DNA is called its **genome.** Each diploid eukaryotic species has a characteristic number of **chromosomes** in each **somatic cell;** reproductive cells, or **gametes** (egg and sperm), have half that number of chromosomes.

Each chromosome is a very long DNA molecule with associated proteins that help to structure the chromosome and control the activity of the genes. This DNA–protein complex is called **chromatin.**

Distribution of Chromosomes During Cell Division Prior to cell division, a cell copies its DNA and each chromosome densely coils and shortens. Replicated chromosomes consist of two identical **sister chromatids,** attached in their condensed form at a specialized region called a **centromere.** The two sister chromatids separate during **mitosis** (the division of the nucleus), and then the cytoplasm divides during **cytokinesis,** producing two separate, genetically equivalent daughter cells.

A type of cell division called **meiosis** produces daughter cells that have half the number of chromosomes of the parent cell. With the fertilization of egg and sperm, which were formed (in animals) by meiosis, the chromosome number is restored in the somatic cells of the new offspring.

■ INTERACTIVE QUESTION   12.1

a. How many chromosomes do you have in your somatic cells?

b. How many chromosomes in your gametes?

c. How many chromatids in one of your body cells that has duplicated its chromosomes prior to mitosis?

12.2   The mitotic phase alternates with interphase in the cell cycle

Phases of the Cell Cycle   The cell cycle consists of the **mitotic (M) phase,** which includes mitosis and cytokinesis, and **interphase,** during which the cell grows and duplicates its chromosomes. Interphase, usually lasting 90% of the cell cycle, includes the **G₁ phase,** the **S phase,** and the **G₂ phase.** Mitosis is conventionally described in five subphases: **prophase, prometaphase, metaphase, anaphase,** and **telophase.**

■ INTERACTIVE QUESTION   12.2

a. How are the three subphases of interphase alike?

b. What key event happens during the S phase?

The Mitotic Spindle: A Closer Look   The **mitotic spindle** consists of fibers made of microtubules and as-

sociated proteins. The assembly of the spindle begins in the **centrosome,** or microtubule-organizing center. A pair of centrioles is centered in each centrosome of an animal cell, but centrioles are not required for normal spindle operation. The single centrosome replicates during interphase. As spindle microtubules are formed, the two centrosomes move to opposite poles of the cell, and are called spindle poles. Radial arrays of microtubules, called **asters,** extend from the centrosomes in animal cells.

During prophase, the nucleoli disappear and the chromatin fibers coil and fold into visible chromosomes, consisting of sister chromatids joined at the centromere. During prometaphase, some of the spindle microtubules attach to each chromatid's **kinetochore,** a structure of protein associated with DNA located at the centromere region. Nonkinetochore microtubules (or "polar" microtubules) extend out from each centrosome and overlap at the midline. Alternate tugging on the chromosome by opposite kinetochore microtubules moves the chromosome to the midline of the cell. At metaphase, the chromosomes are aligned at the **metaphase plate,** across the midline of the spindle.

The proteins joining sister chromatids are inactivated in anaphase, and the now separate chromosomes move toward the poles. Motor proteins "walk" a chromosome along the kinetochore microtubules as these shorten by depolymerizing at their kinetochore end. The extension of the spindle poles away from each other as an animal cell elongates is probably due to the overlapping nonkinetochore microtubules walking past each other, also using motor proteins.

During telophase, equivalent sets of chromosomes gather at the two poles of the cell. Nuclear envelopes form, nucleoli reappear, and cytokinesis begins.

Cytokinesis: A Closer Look   **Cleavage** is the process that separates the two daughter cells in animals. A **cleavage furrow** on the cell surface forms, as a ring of actin microfilaments interacting with myosin proteins begins to contract on the cytoplasmic side of the membrane. The cleavage furrow deepens until the dividing cell is pinched in two.

In plant cells, a **cell plate** forms from the fusion of membrane vesicles derived from the Golgi apparatus. The membrane of the enlarging cell plate joins with the plasma membrane, separating the two daughter cells. A new cell wall develops between the cells from the contents of the cell plate.

■ **INTERACTIVE QUESTION   12.3**

The following diagrams depict interphase and the five subphases of mitosis in an animal cell. Assuming that this cell has four chromosomes, sketch the chromosomes as they would appear in each subphase. Identify the stages and label the indicated structures.

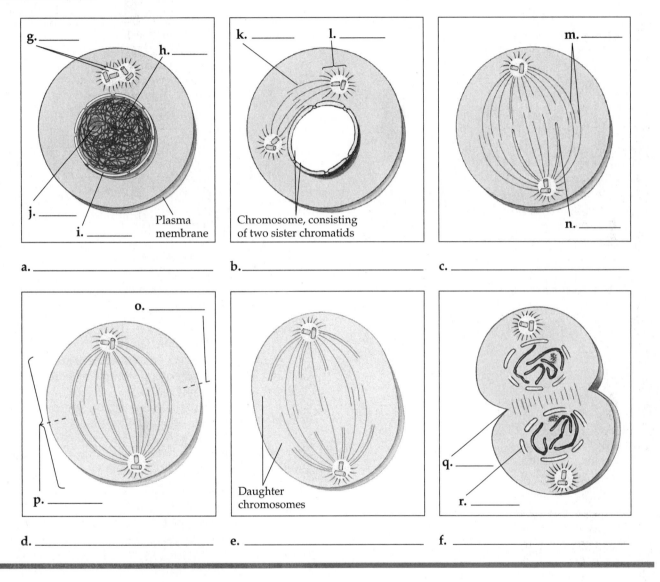

g. _____

h. _____

j. _____

i. _____

Plasma membrane

a. _____

k. _____

l. _____

Chromosome, consisting of two sister chromatids

b. _____

m. _____

n. _____

c. _____

o. _____

p. _____

d. _____

Daughter chromosomes

e. _____

q. _____

r. _____

f. _____

Binary Fission   Prokaryotes reproduce by a process known as **binary fission.** The *bacterial chromosome,* a single circular DNA molecule, begins to replicate at the **origin of replication** and these duplicated origins move apart by an unknown mechanism, ending up apparently anchored at opposite poles of the cell. Replication is completed as the cell doubles in size, and the plasma membrane grows inward to divide the two identical daughter cells.

The Evolution of Mitosis   Evidence for the evolution of mitosis from prokaryotic cell division includes the relatedness of several proteins involved in both types

of division and the possible intermediate stages seen in some unicellular algae, in which chromosomal division takes place within an intact nuclear envelope.

12.3   The cell cycle is regulated by a molecular control system

Normal growth, development, and maintenance depend on proper control of the timing and rate of cell division.

Evidence for Cytoplasmic Signals   Experiments that fuse two cells at different phases of the cell cycle indicate that cytoplasmic chemical signals drive the cell cycle.

The Cell Cycle Control System   A **cell cycle control system,** consisting of a set of molecules that function cyclically, coordinates the events of the cell cycle.

Important internal and external signals are monitored to determine whether the cell cycle will proceed past the three main **checkpoints** in the G_1, G_2, and M phases. If a mammalian cell does not receive a go-ahead signal at the G_1 checkpoint, called the "restriction point," the cell will usually exit the cell cycle to a nondividing state called the **G_0 phase.**

Protein kinases are enzymes that activate or inactivate other proteins by phosphorylating them. The changing concentrations of **cyclins,** regulatory proteins that attach to these kinases, affect the activity of **cyclin-dependent kinases, or Cdks.**

A cyclin–Cdk complex called **MPF,** for maturation or M-phase-promoting factor, triggers passage past the G_2 checkpoint into M phase. In addition to phosphorylating proteins and other kinases that initiate mitotic events, MPF activates a protein breakdown process that destroys its cyclin and thus MPF activity during anaphase. The Cdk portion of the complex remains to associate with new cyclin synthesized during S and G_2 phases of the next cycle.

Other Cdk proteins and cyclins appear to control the movement of a cell past the G_1 checkpoint.

■ **INTERACTIVE QUESTION   12.4**

a. What is MPF?

b. Describe the relative concentrations of MPF and its constituent molecules throughout the cell cycle:

MPF

Cdk

cyclin

An internal signal is required to move past the M-phase checkpoint into anaphase. Kinetochores that are not yet attached to spindle microtubules release a signal molecule that keeps active the proteins holding chromatids together until all chromosomes are attached to the spindle at the metaphase plate.

Growing cells in cell culture has allowed researchers to identify chemical and physical factors that affect cell division. Certain nutrients and regulatory proteins called **growth factors** have been found to be essential for cells to divide in culture. Mammalian fibroblast cells have receptors on their plasma membranes for *platelet-derived growth factor (PDGF)*, which is released from blood platelets at the site of an injury. Binding of PDGF initiates a signal-transduction pathway that stimulates cell division.

Density-dependent inhibition of cell division is related to diminishing supplies of growth factors and essential nutrients. Most animal cells also show **anchorage dependence** and must be attached to a substratum in order to divide.

Loss of Cell Cycle Controls in Cancer Cells   Cancer cells escape from the body's normal control mechanisms. When grown in cell culture, cancer cells do not exhibit density-dependent inhibition and may continue to divide indefinitely instead of stopping after the typical 20 to 50 divisions of normal mammalian cells.

When a normal cell is **transformed** or converted to a cancer cell, the body's immune system usually destroys it. If it proliferates, a mass of abnormal cells develops within a tissue. **Benign tumors** remain at their original site and can be removed by surgery. **Malignant tumors** cause cancer as they invade and disrupt the functions of one or more organs. Malignant tumor cells may have abnormal metabolism and unusual numbers of chromosomes. They lose their attachments to other cells and may **metastasize,** entering the blood and lymph systems and spreading to other sites. Radiation and chemicals are used to treat tumors that metastasize. Much remains to be learned about control of the cell division processes of both normal and cancerous cells.

Word Roots

ana- = up, throughout, again (*anaphase:* the mitotic stage in which the chromatids of each chromosome have separated and the daughter chromosomes are moving to the poles of the cell)

bi- = two (*binary fission:* a type of cell division in which a cell divides in half)

centro- = the center; **-mere** = a part (*centromere:* the narrow "waist" of a condensed chromosome)

chroma- = colored (*chromatin:* DNA and the various associated proteins that form eukaryotic chromosomes)

cyclo- = a circle (*cyclin:* a regulatory protein whose concentration fluctuates cyclically)

cyto- = cell; **-kinet** = move (*cytokinesis:* division of the cytoplasm)

gamet- = a wife or husband (*gamete:* a haploid egg or sperm cell)

gen- = produce (*genome:* a cell's endowment of DNA)

inter- = between (*interphase:* time when a cell metabolizes and performs its various functions)

mal- = bad or evil (*malignant tumor*: a cancerous tumor that is invasive enough to impair functions of one or more organs)

meio- = less (*meiosis*: a variation of cell division that yields daughter cells with half as many chromosomes as the parent cell)

meta- = between (*metaphase*: the mitotic stage in which the chromosomes are aligned in the middle of the cell, at the metaphase plate)

mito- = a thread (*mitosis*: the division of the nucleus)

pro- = before (*prophase*: the first mitotic stage in which the chromatin is condensing)

soma- = body (*centrosome*: a nonmembranous organelle that functions throughout the cell cycle to organize the cell's microtubules)

telos- = an end (*telophase*: the final stage of mitosis in which daughter nuclei are forming and cytokinesis has typically begun)

trans- = across; **-form** = shape (*transformation*: the process that converts a normal cell into a cancer cell)

Structure Your Knowledge

1. Describe the life of one chromosome as it proceeds through an entire cell cycle, starting with G_1 of interphase and ending with telophase of mitosis.

2. Draw a sketch of one half of a mitotic spindle. Identify and list the functions of the components.

3. In this photomicrograph of cells in an onion root tip, identify the cell cycle phases for the indicated cells.

a. _____
b. _____
c. _____
d. _____

Test Your Knowledge

FILL IN THE BLANK: *Identify the appropriate phase of the cell cycle.*

_____ 1. most cells that will no longer divide are in this phase

_____ 2. sister chromatids separate and chromosomes move apart

_____ 3. mitotic spindle begins to form

_____ 4. cell plate forms or cleavage furrow pinches cells apart

_____ 5. chromosomes replicate

_____ 6. chromosomes line up at equatorial plane

_____ 7. nuclear membranes form around separated chromosomes

_____ 8. chromosomes become visible

_____ 9. kinetochore–microtubule interactions move chromosomes to midline

_____ 10. restriction point occurs in this phase

MULTIPLE CHOICE: *Choose the one best answer.*

1. One of the major differences in the cell division of prokaryotic cells compared to eukaryotic cells is that
   a. cytokinesis does not occur in prokaryotic cells.
   b. genes are not replicated on chromosomes in prokaryotic cells.
   c. the duplicated chromosomes are attached to the nuclear membrane in prokaryotic cells and are separated from each other as the membrane grows.
   d. the chromosomes do not separate along a mitotic spindle in prokaryotic cells.
   e. the chromosome number is reduced by half in eukaryotic cells but not prokaryotic cells.

2. A plant cell has 12 chromosomes at the end of mitosis. How many *chromosomes* would it have in the G_2 phase of its next cell cycle?
   a. 6
   b. 9
   c. 12
   d. 24
   e. It depends on whether it is undergoing mitosis or meiosis.

3. How many *chromatids* would this plant cell have in the G_2 phase of its cell cycle?
   a. 6   d. 24
   b. 9   e. 48
   c. 12

4. The longest part of the cell cycle is
   a. prophase.   d. mitosis.
   b. G_1 phase.   e. interphase.
   c. G_2 phase.

5. In animal cells, cytokinesis involves
   a. the separation of sister chromatids.
   b. the contraction of the contractile ring of micro-filaments.
   c. depolymerization of kinetochore microtubules.
   d. a protein kinase that phosphorylates other enzymes.
   e. sliding of nonkinetochore microtubules past each other.

6. Humans have 46 chromosomes. That number of chromosomes will be found in
   a. cells in anaphase.
   b. the egg and sperm cells.
   c. the somatic cells.
   d. all the cells of the body.
   e. only cells in G_1 of interphase.

7. Sister chromatids
   a. have one-half the amount of genetic material as does the original chromosome.
   b. start to move along kinetochore microtubules toward opposite poles during telophase.
   c. each have their own kinetochore.
   d. are formed during prophase.
   e. slide past each other along nonkinetochore microtubules.

8. Which of the following would *not* be exhibited by cancer cells?
   a. changing levels of MPF concentration
   b. passage through the restriction point
   c. density-dependent inhibition
   d. metastasis
   e. mitotic phase of the cell cycle

9. Which of the following is *not* true of a cell plate?
   a. It forms at the site of the metaphase plate.
   b. It results from the fusion of microtubules.
   c. It fuses with the plasma membrane.
   d. A cell wall is laid down between its membranes.
   e. It forms during telophase in plant cells.

10. A cell that passes the restriction point in G_1 will most likely
    a. undergo chromosome duplication.
    b. have just completed cytokinesis.
    c. continue to divide only if it is a cancer cell.

d. show a drop in MPF concentration.
e. move into the G_0 phase.

11. The rhythmic changes in cyclin concentration in a cell cycle are due to
    a. its increased production once the restriction point is passed.
    b. the cascade of increased production once its enzyme is phosphorylated by MPF.
    c. its degradation, which is initiated by active MPF.
    d. the correlation of its production with the production of Cdk.
    e. the binding of the growth factor PDGF.

12. In a plant cell, a centrosome functions in the formation of
    a. the cell plate.
    b. kinetochores.
    c. duplicate chromosomes.
    d. centromeres.
    e. microtubules of the spindle apparatus.

13. A cell in which of the following phases would have the *least* amount of DNA?
    a. G_0 d. metaphase
    b. G_2 e. anaphase
    c. prophase

14. What initiates the separation of sister chromatids in anaphase?
    a. the drop in MPF concentration
    b. a rapid rise in Cdk concentration
    c. movement past the G_2 checkpoint
    d. a signal pathway initiated by the binding of a growth factor
    e. the cessation of delay signals received from unattached kinetochores

15. Cells growing in cell culture that divide and pile up on top of each other are lacking
    a. anchorage dependence.
    b. density independence.
    c. PDGF.
    d. MPF.
    e. nutrients and growth factors.

16. Knowledge of the cell cycle control system will be most beneficial to the area of
    a. human reproduction.
    b. plant genetics.
    c. prokaryotic growth and development.
    d. cancer prevention and treatment.
    e. prevention and treatment of cardiovascular disease.

3 Genetics

Chapter 13
Meiosis and Sexual Life Cycles

Key Concepts

13.1 Offspring acquire genes from parents by inheriting chromosomes

13.2 Fertilization and meiosis alternate in sexual life cycles

13.3 Meiosis reduces the number of chromosome sets from diploid to haploid

13.4 Genetic variation produced in sexual life cycles contributes to evolution

Framework

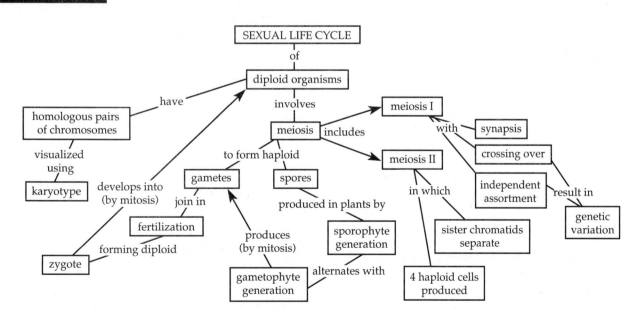

Chapter Review

Genetics is the scientific study of the transmission of traits from parents to offspring (**heredity**) and the **variation** between and within generations.

13.1 Offspring acquire genes from parents by inheriting chromosomes

Inheritance of Genes   The inheritance of traits from parents to offspring involves the transmission of discrete units of information coded in segments of DNA known as **genes.** The collection of inherited genes is called a genome. Specific sequences of the four nucleotides that comprise DNA contain instructions for synthesizing proteins, such as enzymes, that then guide the development of inherited traits.

Precise copies of an organism's genes are packaged into **gametes** (sperm and eggs). Upon fertilization, genes from both parents are passed on to offspring. The DNA of a eukaryotic cell is packaged into a species-specific number of chromosomes, each of which contains hundreds or thousands of genes. A gene's **locus** is its location on a chromosome.

Comparison of Asexual and Sexual Reproduction   In **asexual reproduction,** a single parent passes copies of all its genes on to its offspring. A **clone** is a group of genetically identical offspring of an asexually reproducing individual. In **sexual reproduction,** an individual receives a unique combination of genes inherited from two parents.

13.2 Fertilization and meiosis alternate in sexual life cycles

An organism's **life cycle** is the sequence of stages from conception to production of its own offspring.

Sets of Chromosomes in Human Cells   In **somatic cells,** there are two chromosomes of each type, known as **homologous chromosomes** or homologues. A gene controlling a particular trait is found at the same locus on each chromosome of a homologous pair.

A **karyotype** is an ordered display of an individual's chromosomes. It is usually made using isolated somatic cells, which are stimulated to undergo mitosis, arrested in metaphase, and stained. A computer uses a digital photograph to arrange chromosomes into homologous pairs by size and shape. Karyotyping may be used to identify chromosomal abnormalities associated with inherited disorders.

Sex chromosomes determine the sex of a person: females have two homologous X chromosomes; males have nonhomologous X and Y chromosomes. Chromosomes other than the sex chromosomes are called **autosomes.**

Somatic cells contain a set of chromosomes from each parent. These are **diploid cells,** each with a diploid number of chromosomes, abbreviated $2n$. Gametes, egg and sperm, are **haploid cells** and contain a single set of chromosomes. The haploid number (n) of chromosomes for humans is 23.

■ INTERACTIVE QUESTION   13.1

a. If $2n = 14$, how many chromosomes will be present in somatic cells? _____ How many chromosomes will be found in gametes? _____

b. If $n = 14$, how many chromosomes will be found in diploid somatic cells? _____ How many sets of homologous chromosomes will be found in gametes? _____

c. If $2n = 28$, how many chromatids will be found in a cell in which DNA synthesis has occurred prior to cell division? _____ What is the difference between sister and nonsister chromatids?

Behavior of Chromosome Sets in the Human Life Cycle   **Fertilization,** or fusion of sperm and ovum (egg), produces a **zygote** containing both paternal and maternal sets of chromosomes. The diploid zygote then divides by mitosis to produce the somatic cells of the body, all of which contain the diploid number ($2n$) of chromosomes.

Meiosis is a special type of cell division that halves the chromosome number and provides a haploid set of chromosomes to each gamete. An alternation between diploid and haploid conditions, involving the processes of fertilization and meiosis, is characteristic of the life cycle of all sexually reproducing organisms.

The Variety of Sexual Life Cycles   In most animals, meiosis occurs in the formation of gametes, which are the only haploid cells in the life cycle. In many fungi and some protists, the only diploid stage is the zygote. Meiosis occurs after the gametes fuse, producing haploid cells that divide by mitosis to create a multicellular haploid organism. Gametes are produced by mitosis in these organisms.

Plants and some species of algae have a type of life cycle called **alternation of generations** that includes both diploid and haploid multicellular stages. The multicellular diploid **sporophyte** stage produces haploid **spores** by meiosis. These spores undergo mitosis and develop into a multicellular haploid plant, the **gametophyte,** which produces gametes by mitosis. Gametes fuse to form a diploid zygote that develops into the next sporophyte generation. (See Interactive Question 13.2, p. 97.)

On the test!

■ INTERACTIVE QUESTION 13.2

Complete these three diagrams of sexual life cycles with the names of processes or cells.

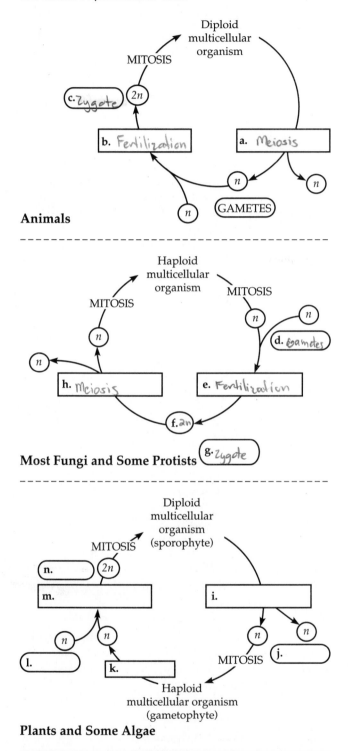

Animals

Most Fungi and Some Protists

Plants and Some Algae

13.3 Meiosis reduces the number of chromosome sets from diploid to haploid

In meiosis, chromosome replication is followed by two consecutive cell divisions: **meiosis I** and **meiosis II**, producing four haploid daughter cells.

The Stages of Meiosis In interphase I, each chromosome replicates, producing two genetically identical sister chromatids that remain attached at their centromeres. During prophase I, homologous chromosomes synapse and crossovers may occur, forming chiasmata.

In metaphase I, tetrads (synapsed chromosomes with four chromatids) line up on the metaphase plate with their kinetochores attached to spindle fibers from opposite poles. The homologous pairs separate in anaphase I, with one homologue moving toward each pole. In telophase I, a haploid set of chromosomes, each composed of two sister chromatids, reaches each pole. Cytokinesis usually occurs during telophase I. There is no replication of genetic material prior to the second division of meiosis.

Meiosis II looks like a regular mitotic division, in which chromosomes line up individually on the metaphase plate, and sister chromatids separate and move apart in anaphase II. At the end of telophase II, there are four haploid daughter cells. (See Interactive Question 13.4, p. 98.)

A Comparison of Mitosis and Meiosis Mitosis produces daughter cells that are genetically identical to the parent cell. Meiosis produces haploid cells that differ genetically from their parent cells and from each other.

The three unique events that produce this result occur during meiosis I: In prophase I, when homologous chromosomes are held together along their lengths by the synaptonemal complex **(synapsis)**, genetic material is rearranged by **crossing over** between nonsister chromatids, visible later in this stage as X-shaped regions called **chiasmata** that hold the **tetrad** together. In metaphase I, chromosomes line up in pairs, not as individuals, on the metaphase plate. During anaphase I, the homologous pairs separate and one homologue goes to each pole.

Meiosis I is called a *reductional division* because it reduces the chromosome sets from two (diploid) to one (haploid). The sister chromatids of each homologue do not separate until meiosis II.

13.4 Genetic variation produced in sexual life cycles contributes to evolution

Origins of Genetic Variation Among Offspring Independent assortment of chromosomes, crossing over, and random fertilization are three mechanisms that generate genetic variation in sexual reproduction.

The first meiotic division results in an assortment of maternal and paternal chromosomes in the two daughter cells. Each homologous pair lines up independently at the metaphase plate—the orientation of the maternal and paternal chromosomes is random. The number of possible combinations of maternal and paternal chromosomes is 2^n, where n is the haploid number.

■ INTERACTIVE QUESTION   13.3

How many assortments of maternal and paternal chromosomes are possible in human gametes?

■ INTERACTIVE QUESTION   13.4

The following diagrams represent some of the stages of meiosis (not in the right order). Label these stages.

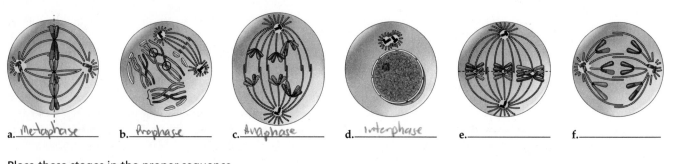

a. Metaphase   b. Prophase   c. Anaphase   d. Interphase   e. _____   f. _____

Place these stages in the proper sequence.

_____   _____   _____   _____   _____   _____   _____

In prophase I, homologous segments of nonsister chromatids exchange or cross over, resulting in new genetic combinations of maternal and paternal genes on the same chromosome. The genetic variability of gametes is greatly increased as the no-longer-equivalent sister chromatids of these **recombinant chromosomes** assort independently during meiosis II.

The random nature of fertilization adds to the genetic variability established in meiosis. Human parents can produce a zygote with any of about 64 trillion (8 million × 8 million) diploid combinations.

Evolutionary Significance of Genetic Variation within Populations   In Darwin's theory of evolution by natural selection, genetic variations present in a population result in adaptation as the individuals best suited to an environment produce the most offspring. The process of sexual reproduction and mutation are the sources of this variation.

Word Roots

-apsis = juncture (*synapsis:* the paring of replicated homologous chromosomes during prophase I of meiosis)

a- = not or without (*asexual:* type of reproduction not involving fertilization)

auto- = self (*autosome:* the chromosomes that do not determine gender)

chiasm- = marked crosswise (*chiasma:* the X-shaped microscopically visible region representing homologous chromosomes that have exchanged genetic material through crossing over during meiosis)

di- = two (*diploid:* cells that contain two homologous sets of chromosomes)

fertil- = fruitful (*fertilization:* process of fusion of a haploid sperm and a haploid egg cell)

haplo- = single (*haploid:* cells that contain only one chromosome of each homologous pair)

homo- = like (*homologous:* like chromosomes that form a pair)

karyo- = nucleus (*karyotype:* a display of the chromosomes of a cell)

meio- = less (*meiosis:* a variation of cell division that yields daughter cells with half as many chromosomes as the parent cell)

soma- = body (*somatic:* body cells with 46 chromosomes in humans)

sporo- = a seed; **-phyte** = a plant (*sporophyte:* the multicellular diploid form in organisms undergoing alternation of generations that results from a union of gametes and that meiotically produces haploid spores that grow into the gametophyte generation)

syn- = together; **gam-** = marriage (*syngamy:* the process of cellular union during fertilization)

tetra- = four (*tetrad:* the four closely associated chromatids of a homologous pair of chromosomes.

Structure Your Knowledge

1. Describe the key events of these stages of meiosis.

a. Interphase
b. Prophase I
c. Metaphase I
d. Anaphase I
e. Metaphase II
f. Anaphase II

2. Create a concept map to help you organize your understanding of the similarities and differences between mitosis and meiosis. Compare your map with those of some classmates to see different ways of organizing this material.

Test Your Knowledge

MULTIPLE CHOICE: *Choose the one best answer.*

1. The restoration of the diploid chromosome number after halving in meiosis is due to
   a. synapsis.
   b. fertilization.
   c. mitosis.
   d. DNA replication.
   e. chiasmata.

2. What is a karyotype?
   a. a genotype of an individual
   b. a unique combination of chromosomes found in a gamete
   c. a blood type determination of an individual
   d. a pictorial display of an individual's chromosomes
   e. a species-specific diploid number of chromosomes

3. What are autosomes?
   a. sex chromosomes
   b. chromosomes that occur singly
   c. chromosomal abnormalities that result in genetic defects
   d. chromosomes found in mitochondria and chloroplasts
   e. none of the above

4. A synaptonemal complex would be found during
   a. prophase I of meiosis.
   b. fertilization or syngamy of gametes.
   c. metaphase II of meiosis.
   d. prophase of mitosis.
   e. anaphase I of meiosis.

5. During the first meiotic division (meiosis I),
   a. homologous chromosomes separate.
   b. the chromosome number becomes haploid.
   c. crossing over between nonsister chromatids occurs.
   d. paternal and maternal chromosomes assort randomly.
   e. all of the above occur.

6. A cell with a diploid number of 6 could produce gametes with how many different combinations of maternal and paternal chromosomes?
   a. 6   c. 12   e. 128
   b. 8   d. 64

7. The DNA content of a diploid cell is measured in the G_1 phase. After meiosis I, the DNA content of one of the two cells produced would be
   a. equal to that of the G_1 cell.
   b. twice that of the G_1 cell.
   c. one-half that of the G_1 cell.
   d. one-fourth that of the G_1 cell.
   e. impossible to estimate due to independent assortment of homologous chromosomes.

8. In most fungi and some protists,
   a. the zygote is the only haploid stage.
   b. gametes are formed by meiosis.
   c. the multicellular organism is haploid.
   d. the gametophyte generation produces gametes by mitosis.
   e. reproduction is exclusively asexual.

9. In the alternation of generations found in plants,
   a. the sporophyte generation produces spores by mitosis.
   b. the gametophyte generation produces gametes by mitosis.
   c. the zygote will develop into a sporophyte generation by meiosis.
   d. spores develop into the haploid sporophyte generation.
   e. the gametophyte generation produces spores by meiosis.

10. Which of the following is least likely to be a source of genetic variation in sexually reproducing organisms?
    a. crossing over
    b. replication of DNA during S phase before meiosis I
    c. independent assortment of chromosomes
    d. random fertilization of gametes
    e. mutation

11. Meiosis II is similar to mitosis because
    a. sister chromatids separate.
    b. homologous chromosomes separate.
    c. DNA replication precedes the division.
    d. they both take the same amount of time.
    e. haploid cells are produced.

12. Pairs of homologous chromosomes
    a. have identical DNA sequences in their genes.
    b. have genes for the same traits at the same loci.
    c. are found in gametes.
    d. separate in meiosis II.
    e. have all of the above characteristics.

13. Asexual reproduction of a diploid organism would
    a. be impossible.
    b. involve meiosis.
    c. produce identical offspring.
    d. show variation among sibling offspring.
    e. involve spores produced by meiosis.

14. In a sexually reproducing species with a diploid number of 8, how many different combinations of paternal and maternal chromosomes would be possible in the *offspring*?
    a. 8   c. 64   e. 512
    b. 16   d. 256

15. The calculation of offspring in Question 14 includes only variation resulting from
    a. crossing over.
    b. random fertilization.
    c. independent assortment of chromosomes.
    d. a, b, and c.
    e. only b and c.

16. How many *chromatids* are present in metaphase II in a cell undergoing meiosis from an organism in which $2n = 24$?
    a. 12   c. 36   e. 96
    b. 24   d. 48

17. Which of the following would *not* be considered a haploid cell?
    a. daughter cell after meiosis II
    b. gamete
    c. daughter cell after mitosis in gametophyte generation of a plant
    d. cell in prophase I
    e. cell in prophase II

18. Which of the following is *not* true of homologous chromosomes?
    a. They behave independently in mitosis.
    b. They synapse during the S phase of meiosis.
    c. They travel together to the metaphase plate in prometaphase of meiosis I.
    d. Each parent contributes one set of homologous chromosomes to an offspring.
    e. Crossing over between nonsister chromatids of homologous chromosomes is indicated by the presence of chiasmata.

19. Which of the following describes why or how recombinant chromosomes add to genetic variability?
    a. They are formed as a result of random fertilization when two sets of chromosomes combine in a zygote.
    b. They are the result of mutations that change alleles.
    c. They randomly orient during metaphase II and the nonequivalent sister chromatids separate in anaphase II.
    d. Genetic material from two parents is combined on the same chromosome.
    e. Both c and d are true.

20. A cell in G_2 before meiosis compared with one of the four cells produced by that meiotic division has
    a. twice as much DNA and twice as many chromosomes.
    b. four times as much DNA and twice as many chromosomes.
    c. four times as much DNA and four times as many chromosomes.
    d. half as much DNA but the same number of chromosomes.
    e. half as much DNA and half as many chromosomes.

Mendel and the Gene Idea

Framework

Through his work with garden peas in the 1860s, Mendel developed the fundamental principles of inheritance and the laws of segregation and independent assortment. This chapter describes the basic monohybrid and dihybrid crosses that Mendel performed to establish that inheritance involves particulate genetic factors (genes) that segregate independently in the formation of gametes and recombine to form offspring. The laws of probability can be applied to predict the outcome of genetic crosses.

The phenotypic expression of genotype may be affected by such factors as incomplete dominance, codominance, multiple alleles, pleiotropy, epistasis, and polygenic inheritance, as well as the environment. Genetic screening and counseling for recessively inherited disorders use new technologies and Mendelian principles to analyze human pedigrees.

Chapter Review

The genetic material of two parents is not blended in offspring but is passed on to future generations as discrete heritable units, or genes. This "particulate" hypothesis of inheritance was developed by Gregor Mendel.

14.1 Mendel used the scientific approach to identify two laws of inheritance

Mendel's Experimental, Quantitative Approach Mendel worked with garden peas, a good choice of study organism because they are available in many varieties, their fertilization is easily controlled, and the characteristics of their offspring can be quantified.

Mendel studied seven **characters,** or heritable features, that occurred in alternative forms called **traits.** He used **true-breeding** varieties of pea plants, which means that self-fertilizing parents always produce offspring with the parental form of the character. To follow the transmission of these well-defined traits, Mendel performed **hybridizations** in which he cross-pollinated contrasting true-breeding varieties, and then allowed the next generation to self-pollinate. The true-breeding parental plants are the **P generation** (parental); the offspring of the first cross are the F_1 **generation** (first filial); and the next generation, from the self-cross of the F_1, is known as the F_2 **generation.**

The Law of Segregation Mendel found that the F_1 offspring did not show a blending of the parental traits. Instead, only one of the parental traits of the character was found in the hybrid offspring. In the F_2 generation, however, the missing parental trait reappeared in the ratio of 3:1—three offspring with the *dominant* trait shown by the F_1 to one offspring with the reappearing *recessive* trait.

Mendel's explanation for this phenomenon contains four parts: (1) alternate forms of genes, now called **alleles,** account for variations in characters; (2) an organism has two alleles for each inherited trait, one received from each parent; (3) when two different alleles occur together, one of them, called the **dominant allele,** determines the organisms appearance while the other, the **recessive allele,** has no observable effect on the organism's appearance; and (4) allele pairs separate (segregate) during the formation of gametes (Mendel's **law of segregation**), so an egg or sperm carries only one allele for each inherited character. This explanation is consistent with the behavior of chromosomes during meiosis.

Mendel's law of segregation explains the 3:1 ratio observed in the F₂ plants. During the segregation of allele pairs in the formation of F₁ gametes, half the gametes receive one allele while the other half receive the alternate allele. Random fertilization of gametes results in one-fourth of the plants having two dominant alleles, half having one dominant and one recessive allele, and one-fourth receiving two recessive alleles, producing a ratio of plants showing the dominant to recessive trait of 3:1. A **Punnett square** can be used to predict the results of simple genetic crosses. Dominant alleles are often symbolized by a capital letter, recessive alleles by a small letter.

■ **INTERACTIVE QUESTION   14.1**

Fill in this diagram of a cross of round- and wrinkled-seeded pea plants. The round allele (*R*) is dominant and the wrinkled allele (*r*) is recessive.

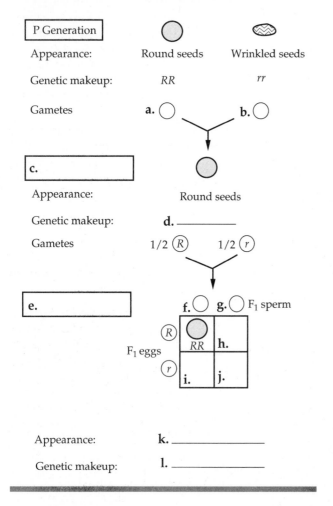

has two different alleles, it is said to be **heterozygous** for that gene. Homozygotes are true-breeding; heterozygotes are not, since they produce gametes with one or the other allele that can combine to produce offspring that are dominant homozygotes, heterozygotes, and recessive homozygotes. **Phenotype** is an organism's expressed traits; **genotype** is its genetic makeup.

In a **testcross,** an organism expressing the dominant phenotype is crossed with a recessive homozygote to determine the genotype of this phenotypically dominant organism.

■ **INTERACTIVE QUESTION   14.2**

A tall pea plant is crossed with a recessive dwarf pea plant. What will the phenotypic and genotypic ratio of offspring be

a. if the tall plant was *TT*?

b. if the tall plant was *Tt*?

The Law of Independent Assortment   In a *monohybrid cross,* the inheritance of a single character is followed through the crossing of **monohybrids,** F₁ offspring that are heterozygous for one character. Mendel used *dihybrid crosses* between F₁ **dihybrids,** which are heterozygous for two characters, to determine whether the two characters were transmitted independently of each other from the parent plants.

If the two pairs of alleles segregate independently, then gametes from an F₁ hybrid generation (*AaBb*) should contain four combinations of alleles in equal quantities (*AB, Ab, aB, ab*). The random fertilization of these four classes of gametes should result in 16 (4 × 4) gamete combinations that produce four phenotypic categories in a ratio of 9:3:3:1 (nine offspring showing both dominant traits, three showing one dominant trait and one recessive, three showing the opposite dominant and recessive traits, and one showing both recessive traits). Mendel obtained these ratios when he categorized the F₂ progeny of dihybrid crosses, providing evidence for his **law of independent assortment.** This principle states that pairs of alleles for each character segregate independently in the formation of gametes. This law applies to genes located on different chromosomes.

An organism that has a pair of identical alleles is said to be **homozygous** for that gene. If the organism

■ INTERACTIVE QUESTION 14.3

A true-breeding tall, purple-flowered pea plant (*TTPP*) is crossed with a true-breeding dwarf, white-flowered plant (*ttpp*).

a. What is the phenotype of the F_1 generation?

b. What is the genotype of the F_1 generation?

c. What four types of gametes are formed by F_1 plants?

_____ _____ _____ _____

d. Fill in the following Punnett square to show the offspring of the F_2 generation. Shade each phenotype a different color so you can see the ratio of offspring.

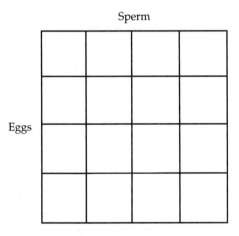

Sperm

Eggs

e. List the phenotypes and ratios found in the F_2 generation.

_____   _____

_____   _____

f. What is the ratio of tall to dwarf plants? _____
Of purple- to white-flowered plants? _____
(Note that the alleles for each individual character segregate as in a monohybrid cross.)

14.2 The laws of probability govern Mendelian inheritance

General rules of probability apply to the laws of segregation and independent assortment. The probability scale goes from 0 to 1; the probabilities of all possible outcomes must add up to 1. The probability of an event occurring is the number of times that event could occur over all the possible events; for example, the probability of drawing an ace from a deck of cards is $\frac{4}{52}$. The outcome of independent events is not affected by previous or simultaneous trials.

The Multiplication and Addition Rules Applied to Monohybrid Crosses The rule of multiplication states that the probability that a certain combination of independent events will occur together is equal to the product of the separate probabilities of the independent events. The probability of a particular genotype being formed by fertilization is equal to the product of the probabilities of forming each type of gamete needed to produce that genotype.

If a genotype can be formed in more than one way, then the rule of addition states that its probability is equal to the sum of the separate probabilities of the different, mutually exclusive ways the event can occur. For example, a heterozygote offspring can occur if the egg contains the dominant allele and the sperm the recessive ($\frac{1}{2} \times \frac{1}{2} = \frac{1}{4}$ probability) or vice versa ($\frac{1}{4}$). Therefore, the heterozygote offspring would be the predicted result from a monohybrid cross half of the time ($\frac{1}{4} + \frac{1}{4} = \frac{1}{2}$).

■ INTERACTIVE QUESTION 14.4

Apply the rule of multiplication to a dihybrid cross. How would you determine the probability of getting an F_2 offspring that is homozygous recessive for both traits?

Solving Complex Genetics Problems with the Rules of Probability Fairly complex genetics problems can be solved by applying the rules of multiplication and addition. The probability of a particular genotype arising from a cross can be determined by considering each gene involved as a separate monohybrid cross and then multiplying the probabilities of all the independent events involved in the final genotype. When more than one outcome is involved, the rule of addition is also used.

The larger the sample size, the more closely the results will conform to statistical predictions.

■ INTERACTIVE QUESTION   14.5

a. In the following cross, what is the probability of obtaining offspring that show all three dominant traits, A_B_C_ (_ indicates that the second allele can be either dominant or recessive without affecting the phenotype determined by the first dominant allele)?

$$AaBbcc \times AabbCC$$

probability of offspring that are A_B_C_ = _____

b. What is the probability that the offspring of this AaBbcc × AabbCC cross will show at least two dominant traits? _____

14.3   Inheritance patterns are often more complex than predicted by simple Mendelian genetics

Extending Mendelian Genetics for a Single Gene   Alleles may show varying degrees of dominance and recessiveness along a *spectrum of dominance.* In the case of an allele showing **complete dominance,** the phenotype of the heterozygote is indistinguishable from that of the dominant homozygote. At the other end of the continuum, alleles that exhibit **codominance** will both affect the phenotype in separate, distinguishable ways, as in the case of the MN blood group.

Intermediate phenotypes are characteristic of alleles showing **incomplete dominance.** The F_1 hybrids have a phenotype intermediate between that of the parents. The F_2 show a 1:2:1 phenotypic and genotypic ratio.

Even when alleles exhibit dominance or incomplete dominance at the phenotypic level, both alleles may be expressed at the molecular level. **Tay-Sachs disease** is a lethal disorder in which brain cells lack a critical enzyme and are unable to metabolize a type of lipid that then accumulates and damages the brain. In a heterozygote, the Tay-Sachs allele is recessive at the organismal level; at the biochemical level, the enzyme activity level is intermediate between both homozygotes; at the molecular level, the alleles are codominant in that each produces its enzyme product, either normal or dysfunctional.

Whether or not an allele is dominant or recessive has no relation to how common it is in a population.

Most genes exist in more than two allelic forms. The gene that determines human blood groups has three alleles. The alleles I^A and I^B are codominant with each other; each codes for an enzyme that attaches a carbohydrate to the surface of red blood cells. The allele i codes for an enzyme that attaches neither the A nor B carbohydrate, and is thus recessive to I^A and I^B. Blood type is critical in transfusions because, if the carbohydrate attached to the donor's blood cells is foreign to the recipient, the recipient's immune system will cause clumping of the donated blood cells.

■ INTERACTIVE QUESTION   14.6

List the possible genotypes for the following blood groups.

a. A _____

b. B _____

c. AB _____

d. O _____

Pleiotropy is the characteristic of a single gene having multiple phenotypic effects in an individual. Certain hereditary diseases with complex sets of symptoms are caused by a single allele.

Extending Mendelian Genetics for Two or More Genes In **epistasis,** a gene at one locus may affect the expression of another gene. F_2 ratios that differ from the typical 9:3:3:1 often indicate epistasis.

■ INTERACTIVE QUESTION   14.7

A dominant allele *M* is necessary for the production of the black pigment melanin; mm individuals are white. A dominant allele *B* results in the deposition of a lot of pigment in an animal's hair, producing a black color. The genotype *bb* results in brown hair. Two black animals heterozygous for both genes are bred. Fill in the following table for the offspring of this MmBb × MmBb cross.

Phenotype	Genotype	Ratio
Black		
	M_bb	

Quantitative characters, such as height or skin color, vary along a continuum in a population. Such phenotypic gradations are usually due to **polygenic inheritance,** in which two or more genes have an additive effect on one character. Each dominant allele contributes one "unit" to the phenotype. A polygenic character may result in a normal distribution (forming a bell-shaped curve) of the character within a population.

■ INTERACTIVE QUESTION 14.8

The height of spike weed is a result of polygenic inheritance involving three genes, each of which can contribute an additional 5 cm to the base height of the plant, which is 10 cm. The tallest plant (*AABBCC*) can reach a height of 40 cm.

a. If a tall plant (*AABBCC*) is crossed with a base-height plant (*aabbcc*), what is the height of the F$_1$ plants?

b. How many phenotypic classes will there be in the F$_2$?

Nature and Nurture: the Environmental Impact on Phenotype   The phenotype of an individual is the result of complex interactions between its genotype and the environment. Genotypes have a phenotypic range called a **norm of reaction** within which the environment influences phenotypic expression. Polygenic characters are often **multifactorial,** meaning that a combination of genetic and environmental factors influences phenotype.

Integrating a Mendelian View of Heredity and Variation   The phenotypic expression of most genes is influenced by other genes and by the environment. The theory of particulate inheritance and Mendel's principles of segregation and independent assortment, however, still form the basis for modern genetics.

14.4   Many human traits follow Mendelian patterns of inheritance

Pedigree Analysis   A family **pedigree** is a family tree with the history of a particular trait shown across the generations. By convention, circles represent females, squares are used for males, and solid symbols indicate individuals that express the trait in question. Parents are joined by a horizontal line, and offspring are listed below parents from left to right in order of birth. The genotype of individuals in the pedigree can often be deduced by following the patterns of inheritance.

■ INTERACTIVE QUESTION 14.9

Consider this pedigree for the trait albinism (lack of skin pigmentation) in three generations of a family. (Solid symbols represent individuals who are albinos.) From your knowledge of Mendelian inheritance, answer the questions that follow.

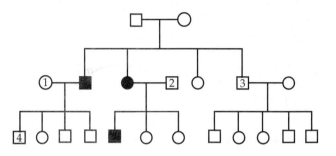

a. Is this trait caused by a dominant or recessive allele? How can you tell?

b. Determine the genotypes of the parents in the first generation. (Let *AA* and *Aa* represent normal pigmentation and *aa* be the albino genotype.) Genotype of father _____ ; of mother _____.

c. Determine the probable genotypes of the mates of the albino offspring in the second generation and the grandson 4 in the third generation. Genotypes: mate 1 _____ mate 2 _____ grandson 4 _____.

d. Can you determine the genotype of son 3 in the second generation? Why or why not?

Recessively Inherited Disorders   Only homozygous recessive individuals express the phenotype for the thousands of genetic disorders that are inherited as simple recessive traits. **Carriers** of the disorder are heterozygotes who are phenotypically normal but may transmit the recessive allele to their offspring.

Cystic fibrosis is the most common lethal genetic disease in the United States; it is found more frequently in people of European descent than in other groups. This recessive allele results in defective chloride channels in certain cell membranes. Accumulating extracellular chloride leads to the buildup of thickened mucus in various organs and a predisposition to bacterial infections.

Sickle-cell disease is the most common inherited disease among African Americans. Due to a single amino acid substitution in the hemoglobin protein, red blood cells deform into a sickle shape when blood-oxygen concentration is low, triggering blood clumping and other pleiotropic effects. Heterozygous individuals are said to have sickle-cell trait but are usually healthy. The resistance to malaria that accompanies the sickle-cell trait may explain why this lethal recessive allele remains in relatively high frequency in areas where malaria is common.

The likelihood of two mating individuals carrying the same rare deleterious allele increases when the individuals have common ancestors. Consanguineous matings, between siblings or close relatives, are indicated on pedigrees by double lines.

■ **INTERACTIVE QUESTION   14.10**

a. What is the probability that a mating between two carriers will produce an offspring with a recessively inherited disorder? _____

b. What is the probability that a phenotypically normal child produced by a mating of two heterozygotes will be a carrier? _____

Dominantly Inherited Disorders   A few human disorders are due to dominant genes. In achondroplasia, dwarfism is due to a single copy of a mutant allele.

Dominant lethal alleles are more rare than recessive lethals because the harmful allele cannot be masked in the heterozygote. A late-acting lethal dominant allele can be passed on if the symptoms do not develop until after reproductive age. Molecular geneticists have developed a method to detect the lethal gene for **Huntington's disease,** a degenerative disease of the nervous system that does not develop until later in life.

Multifactorial Disorders   Many diseases have genetic (usually polygenic) and environmental components. These multifactorial disorders include heart disease, diabetes, cancer, and others.

Genetic Testing and Counseling   The risk of a genetic disorder being transmitted to offspring can sometimes be determined through genetic counseling and testing before a child is conceived or in the early stages of pregnancy. The probability of a child having a genetic defect may be determined by considering the family history of the disease.

■ **INTERACTIVE QUESTION   14.11**

If two prospective parents both have siblings who had a recessive genetic disorder, what is the chance that they would have a child who inherits the disorder?

Determining whether parents are carriers can determine the risk of passing on a genetic disorder. Tests that permit carrier recognition have been developed for a number of heritable disorders.

Amniocentesis is a procedure that extracts a small amount of amniotic fluid from the sac surrounding the fetus. The fluid is analyzed biochemically, and fetal cells present in the fluid are cultured for several weeks and then tested for certain genetic disorders and karyotyped to check for chromosomal defects.

Chorionic villus sampling (CVS) is a technique in which a small amount of fetal tissue is suctioned from the placenta. These rapidly growing cells can be karyotyped immediately, and this procedure can be performed at only 8 to 10 weeks of pregnancy, earlier than an amniocentesis can be performed. A new technique isolates fetal cells from maternal blood, which can then be cultured and tested.

Ultrasound is a simple, noninvasive procedure that can reveal major abnormalities. Fetoscopy, the insertion of a needle-thin viewing scope and light into the uterus, allows the fetus to be checked for anatomical problems.

Some genetic disorders can be detected at birth. Routine screening is now used to test for the recessively inherited disorder phenylketonuria (PKU) in newborns. If detected, dietary adjustments can lead to normal development.

▶ Word Roots

co- = together (*codominance:* phenotype in which both dominant alleles are expressed in the heterozygote)

-centesis = a puncture (*amniocentesis:* a technique for determining genetic abnormalities in a fetus by the presence of certain chemicals or defective fetal cells in the amniotic fluid, obtained by aspiration from a needle inserted into the uterus)

di- = two (*dihybrid cross:* a breeding experiment in which offspring of a cross of parental varieties differing in two traits are mated)

epi- = beside; **-stasis** = standing (*epistasis:* a phenomenon in which one gene alters the expression of another gene that is independently inherited)

geno- = offspring (*genotype:* the genetic makeup of an organism)

hetero- = different (*heterozygous:* having two different alleles for a trait)

homo- = alike (*homozygous:* having two identical alleles for a trait)

mono- = one (*monohybrid cross:* a breeding experiment that crosses offspring of a cross of parental varieties differing in a single character)

pedi- = a child (*pedigree:* a family tree describing the occurrence of heritable characters in parents and offspring across as many generations as possible)

pheno- = appear (*phenotype:* the physical and physiological traits of an organism)

pleio- = more (*pleiotropy:* when a single gene impacts more than one characteristic)

poly- = many; **gene-** = produce (*polygenic:* an additive effect of two or more gene loci on a single phenotypic character)

Structure Your Knowledge

1. Relate Mendel's two laws of inheritance to the behavior of chromosomes in meiosis that you studied in Chapter 13.

2. Mendel worked with characters that exhibited two alternate forms: smooth or wrinkled, green or yellow, tall or short. In his F1 generation, the dominant allele was always expressed, with the recessive trait reappearing in the F2 offspring. But not all allele pairs operate by complete dominance; some show incomplete dominance or codominance. And some that show dominance on the phenotypic level are actually incompletely dominant or codominant when observed at the microscopic or molecular level. Taking into account the mechanisms by which genotype becomes expressed as phenotype, explain this spectrum in dominance.

Genetics Problems

One of the best ways to learn genetics is to work problems. You can't memorize genetic knowledge; you have to practice using it. Work through the problems presented below, methodically setting down the information you are given and what you are to determine. Write down the symbols used for the alleles and genotypes, and the phenotypes resulting from those genotypes. Avoid using Punnett squares; while useful

for learning basic concepts, they are much too laborious and mistake-prone. Instead, break complex crosses into their monohybrid components and rely on the rules of multiplication and addition. Always look at the answers you get to see if they make logical sense. And remember that study groups are great for going over your problems and helping each other "see the light." Answers and explanations are provided at the end of the book.

1. Summer squash are either white or yellow. To get white squash, at least one of the parental plants must be white. The allele for which color is dominant?

2. For the following crosses, determine the probability of obtaining the indicated genotype in an offspring.

Cross	Offspring	Probability
$AAbb \times AaBb$	$AAbb$	**a.**
$AaBB \times AaBb$	$aaBB$	**b.**
$AABbcc \times aabbCC$	$AaBbCc$	**c.**
$AaBbCc \times AaBbcc$	$aabbcc$	**d.**

3. True-breeding tall red-flowered plants are crossed with dwarf white-flowered plants. The resulting F$_1$ generation consists of all tall pink-flowered plants. Assuming that height and flower color are each determined by a single gene locus on different chromosomes, predict the results of an F$_1$ cross of dihybrid plants. Choose appropriate symbols for the alleles of the height and flower color genes. List the phenotypes and predicted ratios for the F$_2$ generation.

4. Blood typing has been used as evidence in paternity cases, when the blood type of the mother and child may indicate that a man alleged to be the father could not possibly have fathered the child. For the following mother and child combinations, indicate which blood groups of potential fathers would be exonerated.

Blood Group of Mother	Blood Group of Child	Man Exonerated if He Belongs to Blood Group(s)
AB	A	**a.**
O	B	**b.**
A	AB	**c.**
O	O	**d.**
B	A	**e.**

5. In rabbits, the homozygous *CC* is normal, *Cc* results in rabbits with deformed legs, and *cc* is lethal. For a gene for coat color, the genotype *BB* produces black, *Bb* brown, and *bb* a white coat. Give the phenotypic ratio of offspring from a cross of a deformed-leg, brown rabbit with a deformed-leg, white rabbit.

6. Polydactyly (extra fingers and toes) is due to a dominant gene. A father is polydactyl, the mother has the normal phenotype, and they have had one normal child. What is the genotype of the father? Of the mother? What is the probability that a second child will have the normal number of digits?

7. In dogs, black (*B*) is dominant to chestnut (*b*), and solid color (*S*) is dominant to spotted (*s*). What are the genotypes of the parents in a mating that produced ⅜ black solid, ⅜ black spotted, ⅛ chestnut solid, and ⅛ chestnut spotted puppies? (*Hint:* First determine what genotypes the offspring must have before you deal with the fractions.)

8. When hairless hamsters are mated with normal-haired hamsters, about one-half the offspring are hairless and one-half are normal. When hairless hamsters are crossed with each other, the ratio of normal-haired to hairless is 1:2. How do you account for the results of the first cross? How would you explain the unusual ratio obtained in the second cross?

9. Two pigs whose tails are exactly 25 cm in length are bred over 10 years and they produce 96 piglets with the following tail lengths: 6 piglets at 15 cm, 25 at 20 cm, 37 at 25 cm, 23 at 30 cm, and 5 at 35 cm.

   a. How many pairs of genes are regulating the tail length character? *Hint:* Count the number of phenotypic classes, or determine the sum of the ratios of the classes. In a monohybrid cross, the F_2 ratios add up to 4 (3:1 or 1:2:1). In a dihybrid cross, the F_2 ratios add up to 16 (9:3:3:1 or some variation if the genes are epistatic or quantitative).

   b. What offspring phenotypes would you expect from a mating between a 15-cm and a 30-cm pig?

10. Fur color in rabbits is determined by a single gene locus for which there are four alleles. Four phenotypes are possible: black, Chinchilla (gray color caused by white hairs with black tips), Himalayan (white with black patches on extremities), and white. The black allele (*C*) is dominant over all other alleles, the Chinchilla allele (C^{ch}) is dominant over Himalayan (C^h), and the white allele (*c*) is recessive to all others.

    a. A black rabbit is crossed with a Himalayan, and the F_1 consists of a ratio of 2 black to 2 Chinchilla. Can you determine the genotypes of the parents?

    b. A second cross was done between a black rabbit and a Chinchilla. The F_1 contained a ratio of 2 black to 1 Chinchilla to 1 Himalayan. Can you determine the genotypes of the parents of this cross?

11. In Labrador retriever dogs, the dominant gene *B* determines black coat color and *bb* produces brown. A separate gene *E*, however, shows dominant epistasis over the *B* and *b* alleles, resulting in a golden coat color. The recessive *e* allows expression of *B* and *b*. A breeder wants to know the genotypes of her three dogs, so she breeds them and makes note of the offspring of several litters. Determine the genotypes of the three dogs.

    a. golden female (Dog 1) × golden male (Dog 2) offspring: 7 golden, 1 black, 1 brown

    b. black female (Dog 3) × golden male (Dog 2) offspring: 8 golden, 5 black, 2 brown

12. The ability to taste phenylthiocarbamide (PTC) is controlled in humans by a single dominant allele (*T*). A woman nontaster married a man taster, and they had three children, two boy tasters and a girl nontaster. All the grandparents were tasters. Create a pedigree for this family for this trait. (Solid symbols should signify nontasters (*tt*).) Where possible, indicate whether tasters are *TT* or *Tt*.

13. Two true-breeding varieties of garden peas are crossed. One parent had red, axial flowers, and the other had white, terminal flowers. All F_1 individuals had red, terminal flowers. If 100 F_2 offspring were counted, how many of them would you expect to have red, axial flowers?

14. You cross true-breeding red-flowered plants with true-breeding white-flowered plants, and the F_1 are all red-flowered plants. The F_2, however, occur in a ratio of 9 red : 6 pale purple : 1 white. How many genes are involved in the inheritance of this color character? Explain why the F_1 are all red and how the 9:6:1 ratio of phenotypes in the F_2 occurred.

Test Your Knowledge

MATCHING: *Match the definition with the correct term.*

_____ **1.** codominance

_____ **2.** homozygote

_____ **3.** heterozygote

_____ **4.** phenotype

_____ **5.** polygenic (quantitative)

_____ **6.** pleiotropy

_____ **7.** epistasis

_____ **8.** testcross

_____ **9.** dihybrid cross

_____ **10.** incomplete dominance

A. true-breeding variety

B. cross between two hybrids

C. cross between hybrids that are heterozygous for two genes

D. cross with recessive homozygote to determine genotype of unknown

E. the physical characteristics of an individual

F. genotype with two different alleles

G. genotype with multiple alleles for same locus

H. one gene influences the expression of another gene

I. both alleles are fully expressed in heterozygote

J. single gene with multiple phenotypic effects

K. heterozygote intermediate between phenotypes of homozygotes

L. two or more genes with additive effect on phenotype

MULTIPLE CHOICE: *Choose the one best answer.*

1. According to Mendel's law of segregation,
   a. there is a 50% probability that a gamete will get a dominant allele.
   b. gene pairs segregate independently of other genes in gamete formation.
   c. allele pairs separate in gamete formation.
   d. the laws of probability determine gamete formation.
   e. there is a 3:1 ratio in the F_2 generation.

2. After obtaining two heads from two tosses of a coin, the probability of tossing the coin and obtaining a head is
   a. ½.   **c.** ⅙.   **e.** ¹⁄₁₆.
   b. ¼.   **d.** ⅛.

3. The probability of tossing three coins simultaneously and obtaining two heads and one tail is
   a. ½.   **c.** ⅛.   **e.** ⅜.
   b. ¾.   **d.** ¹⁄₁₆.

4. A multifactorial disorder
   a. can usually be traced to consanguineous matings.
   b. is caused by recessively inherited lethal genes.
   c. has both genetic and environmental causes.
   d. has a collection of symptoms traceable to an epistatic gene.
   e. is usually associated with quantitative traits.

5. The F_2 generation
   a. has a phenotypic ratio of 3:1.
   b. is the result of the self-fertilization or crossing of F_1 individuals.
   c. can be used to determine the genotype of individuals with the dominant phenotype.
   d. has a phenotypic ratio that equals its genotypic ratio.
   e. has 16 different genotypic possibilities.

6. The base height of the dingdong plant is 10 cm. Four genes contribute to the height of the plant, and each dominant allele contributes 3 cm to height. If you cross a 10-cm plant (quadruply homozygous recessive) with a 34-cm plant, how many phenotypic classes will there be in the F_2?
   a. 4   d. 9
   b. 5   e. 64
   c. 8

7. A 1:1 phenotypic ratio in a testcross indicates that
   a. the alleles are dominant.
   b. one parent must have been homozygous dominant.
   c. the dominant phenotype parent was a heterozygote.
   d. the alleles segregated independently.
   e. the alleles are codominant.

8. Carriers of a genetic disorder
   a. are indicated by solid symbols on a family pedigree.
   b. are involved in consanguineous matings.
   c. will produce children with the disease.
   d. are heterozygotes for the gene that can cause the disorder.
   e. have a homozygous recessive genotype.

9. If both parents are carriers of a lethal recessive gene, the probability that their child will inherit and express the disorder is
   a. ⅛.
   b. ¼.
   c. ½.
   d. ½ × ½ × ¼, or ¹⁄₁₆.
   e. ⅔ × ⅔ × ¼, or ⅑.

10. Which phase of meiosis is most directly related to the law of independent assortment?
    a. prophase I
    b. prophase II
    c. metaphase I
    d. metaphase II
    e. anaphase II

11. You think that two alleles for coat color in mice show incomplete dominance. What is the best and simplest cross to perform in order to support your hypothesis?
    a. a testcross of a homozygous recessive mouse with a mouse of unknown genotype
    b. a cross of F₁ mice to look for a 1:2:1 ratio in the offspring
    c. a reciprocal cross in which the sex of the mice of each coat color is reversed
    d. a cross of two true-breeding mice of different colors to look for an intermediate phenotype in the F₁
    e. a cross of F₁ mice to look for a 9:7 ratio in the offspring

12. Which of the following human diseases is inherited as a simple recessive trait?
    a. Tay-Sachs disease
    b. cancer
    c. diabetes
    d. Alzheimer's disease
    e. cardiovascular disease

13. In a dihybrid cross of heterozygotes, what proportion of the offspring will be phenotypically dominant for both traits?
    a. ¹⁄₁₆
    b. ³⁄₁₆
    c. ¼
    d. ⁹⁄₁₆
    e. ¾

14. A mother with type B blood has two children, one with type A blood and one with type O blood. Her husband has type O blood. Which of the following could you conclude from this information?
    a. The husband could not have fathered either child.
    b. The husband could have fathered both children.
    c. The husband must be the father of the child with type O blood and could be the father of the type A child.
    d. The husband could be the father of the child with type O blood, but not the type A child.
    e. Neither the mother nor the husband could be the biological parent of the type A child.

15. In guinea pigs, the brown coat color allele (*B*) is dominant over red (*b*), and the solid color allele (*S*) is dominant over spotted (*s*). The F₁ offspring of a cross between true-breeding brown, solid-colored guinea pigs and red, spotted pigs are crossed. What proportion of their offspring (F₂) would be expected to be red and solid colored?
    a. ⅑
    b. ¹⁄₁₆
    c. ³⁄₁₆
    d. ⁹⁄₁₆
    e. ¾

16. A dominant allele *P* causes the production of purple pigment; *pp* individuals are white. A dominant allele *C* is also required for color production; *cc* individuals are white. What proportion of offspring will be purple from a *ppCc* × *PpCc* cross?
    a. ⅛
    b. ⅜
    c. ½
    d. ¾
    e. None

Chapter 15
The Chromosomal Basis of Inheritance

Framework

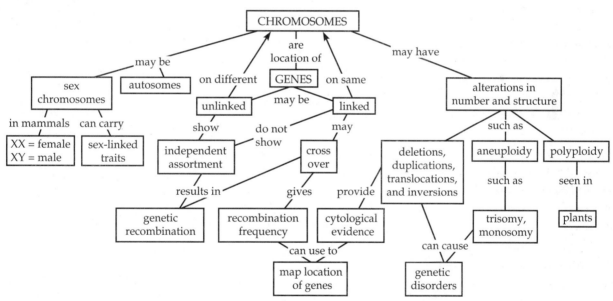

Chapter Review

15.1 Mendelian inheritance has its physical basis in the behavior of chromosomes

Mendel's laws, combined with cytological evidence of the process of meiosis, led to the **chromosome theory of inheritance:** Genes occupy specific positions (loci) on chromosomes, and chromosomes undergo segregation and independent assortment in the process of meiosis in gamete formation.

Morgan's Experimental Evidence: **Scientific Inquiry** T. H. Morgan, working with the fruit fly, *Drosophila melanogaster,* was the first to identify a specific gene with a specific chromosome. Fruit flies are prolific breeders and have only four pairs of chromosomes, which are easily distinguishable with a microscope. The sex chromosomes occur as XX in females and XY in male flies.

The normal phenotype found most commonly in nature for a character is called the **wild type,** whereas alternative traits, assumed to have arisen as mutations, are called *mutant phenotypes.*

Morgan discovered a mutant white-eyed male fly that he mated with a wild-type red-eyed female. The F₁ were all red-eyed. In the F₂, however, all female flies were red-eyed, whereas half of the males were red-eyed and half were white-eyed. Morgan deduced that the gene for eye color was located only on the X chromosome. Males have only one X, so their phenotype is determined by the eye-color allele they inherit from their mother. This association of a specific gene with a chromosome provided evidence for the chromosome theory of inheritance.

■ INTERACTIVE QUESTION   15.1

Complete this summary of Morgan's crosses with the mutant white-eyed fly by filling in the Punnett square and showing the phenotypes and genotypes of the F₂ generation. (*w* stands for the mutant recessive white allele; *w*⁺ for the wild-type red allele.) The Y indicates that male flies, with only one X chromosome, carry only a single allele for this eye color gene.

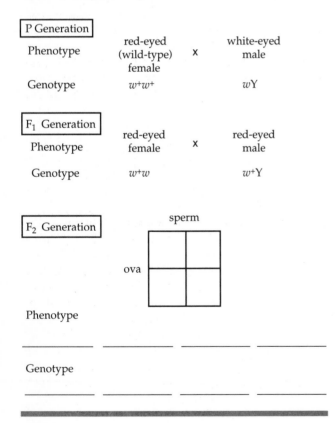

Phenotype

——————   ——————   ——————   ——————

Genotype

——————   ——————   ——————   ——————

15.2   Linked genes tend to be inherited together because they are located near each other on the same chromosome

Genes that are located on the same chromosome and tend to be inherited together are called **linked genes.**

How Linkage Affects Inheritance: **Scientific Inquiry**

Morgan performed a testcross of F₁ dihybrid wild-type flies with flies that were homozygous recessive for black bodies and vestigial wings and found that the offspring were not in the predicted 1:1:1:1 phenotypic classes. Rather, most of the offspring were the same phenotypes as the P generation parents—either wild type (gray, normal wings) or double mutant (black, vestigial). Morgan deduced that these traits were inherited together because their genes were located on the same chromosome.

Genetic recombination results in offspring with combinations of traits that differ from those found in either parent.

Genetic Recombination and Linkage

In a cross between a dihybrid heterozygote and a recessive homozygote, one-half of the offspring will be **parental types** and have phenotypes like one or the other of the P generation parents, and one-half of the offspring, called **recombinant types (recombinants),** will have combinations of the two traits that are unlike the parents. This 50% frequency of recombination is observed when two genes are located on different chromosomes, and it results from the random alignment of homologous chromosomes at metaphase I and the resulting independent assortment of alleles.

■ INTERACTIVE QUESTION   15.2

In a testcross between a heterozygote tall, purple-flowered pea plant and a dwarf, white-flowered plant,

a. what are the phenotypes of offspring that are parental types?

b. what are the phenotypes of offspring that are recombinants?

Linked genes do not assort independently, and one would not expect to see recombination of parental traits in the offspring. Recombination of linked genes does occur, however, due to **crossing over,** the reciprocal trade between nonsister (a maternal and a paternal) chromatids of synapsed homologous chromosomes during prophase of meiosis I. The percentage of recombinant offspring is called the *recombination frequency.*

■ INTERACTIVE QUESTION 15.3

With unlinked genes, an equal number of parental and recombinant offspring are produced. With linked genes, (more/fewer) parentals than recombinants are produced. (Circle and then explain your answer.)

Linkage Mapping Using Recombination Data: **Scientific Inquiry** A **genetic map** is an ordered list of genes on a chromosome. A. H. Sturtevant suggested that recombination frequencies reflect the relative distance between genes; genes located farther apart have a greater probability that a crossover event will occur between them. Sturtevant used recombination data to create a **linkage map,** defining one **map unit** (or centimorgan) as equal to a 1% recombination frequency.

The sequence of genes on a chromosome can be determined by finding the recombination frequency between different pairs of genes. Linkage cannot be determined if genes are so far apart that crossovers between them are almost certain. They would then have the 50% recombination frequency typical of unlinked genes. Such genes are *physically linked* but *genetically unlinked.* Distant genes on the same chromosome may be mapped by adding the recombination frequencies determined between them and intermediate genes.

Sturtevant and his colleagues found that the genes for the various known mutations of *Drosophila* clustered into four groups of linked genes, providing additional evidence that genes are located on chromosomes, in this case, on the four *Drosophila* chromosomes.

The frequency of crossing over may vary along the length of a chromosome, and a linkage map provides the sequence but not the exact location of genes on chromosomes. **Cytogenetic maps** locate gene loci in reference to visible chromosomal features.

■ INTERACTIVE QUESTION 15.4

Recombination frequency is given below for several gene pairs. Create a linkage map for these genes, showing the map unit distance between loci.

j, k	12%	*k, l*	6%
j, m	9%	*l, m*	15%

15.3 Sex-linked genes exhibit unique patterns of inheritance

The Chromosomal Basis of Sex Sex is a phenotypic character usually determined by sex chromosomes. In humans and mammals, females, who are XX, produce ova that each contain an X chromosome. Males, who are XY, produce two kinds of sperm, each with either an X or a Y chromosome. Whether the gonads of an embryo develop into testes or ovaries depends on the presence or absence of the gene *SRY*, found on the Y chromosome, whose protein product regulates many other genes. A few other genes on the Y chromosome are necessary for normal functioning of the testes.

Other sex-determination systems include X-0 (in grasshoppers and some other insects), Z-W (in birds and some fishes and insects), and haplo-diploid (in bees and ants).

Inheritance of Sex-Linked Genes Sex chromosomes may carry genes, called **sex-linked genes,** for traits that are not related to sex. In humans, *sex-linked* typically refers to a gene on the X chromosome. Males inherit their sex-linked alleles from only their mothers; daughters inherit sex-linked alleles from both parents. Recessive sex-linked traits are seen more often in males, since they are *hemizygous* for sex-linked genes.

Duchenne muscular dystrophy is a sex-linked disorder resulting from the lack of a key muscle protein. **Hemophilia** is a sex-linked recessive trait characterized by excessive bleeding due to the absence of one or more blood-clotting proteins.

X Inactivation in Female Mammals Only one of the X chromosomes is fully active in most mammalian female somatic cells. The other X chromosome is contracted into a **Barr body** located inside the nuclear membrane. M. Lyon demonstrated that the selection of which X chromosome is inactivated is a random event occurring independently in embryonic cells. As a result of X-inactivation, both males and females have an equal dosage of genes on the X chromosome. A gene called *XIST* is active on the X chromosome that forms the Barr body. Its RNA product may trigger DNA methylation and X-inactivation.

■ INTERACTIVE QUESTION   15.5

Two normal color-sighted individuals produce the following children and grandchildren. Fill in the probable genotype of the indicated individuals in this pedigree. *Squares are males, circles are females, and solid symbols represent color blindness. Choose an appropriate notation for the genotypes.*

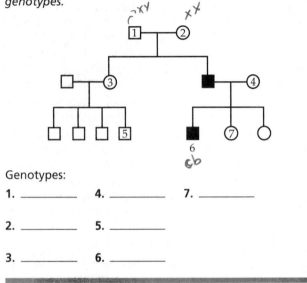

Genotypes:

1. _____   4. _____   7. _____

2. _____   5. _____

3. _____   6. _____

15.4   Alterations of chromosome number or structure cause some genetic disorders

Abnormal Chromosome Number   Nondisjunction occurs when a pair of homologous chromosomes does not separate properly in meiosis I or sister chromatids do not separate in meiosis II. As a result, a gamete receives either two or no copies of that chromosome. A zygote formed with one of these aberrant gametes has a chromosomal alteration known as **aneuploidy**, a nontypical number of a particular chromosome. The zygote will be either **trisomic** for that chromosome (chromosome number is $2n + 1$) or **monosomic** ($2n - 1$). Aneuploid organisms usually have a set of symptoms caused by the abnormal dosage of genes. A mitotic nondisjunction early in embryonic development is likely to be harmful.

Polyploidy is a chromosomal alteration in which an organism has more than two complete chromosomal sets, as in *triploidy* ($3n$) or *tetraploidy* ($4n$). Polyploidy is common in the plant kingdom and has played an important role in the evolution of plants.

■ INTERACTIVE QUESTION   15.6

a. What is the difference between a trisomic and a triploid organism?

b. Which of these is likely to show the most deleterious effects of its chromosomal imbalance?

Alterations of Chromosome Structure   Chromosome breakage can result in chromosome fragments that are lost, called **deletion;** that join to a sister chromatid (during meiosis) or the homologous chromosome, called **duplication;** that rejoin the original chromosome in the reverse orientation, called **inversion;** or that join a nonhomologous chromosome, called **translocation.** A *nonreciprocal* crossover can result in a deletion and duplication in nonsister chromatids, caused by nonequal exchange between chromatids.

A homozygous deletion is usually lethal. Duplications and translocations also are typically harmful. Even though all the genes are present in proper quantities in inversions and translocations, the phenotype may be altered due to the influence of neighboring genes on the expression between the relocated genes.

■ INTERACTIVE QUESTION   15.7

Two nonhomologous chromosomes have gene orders, respectively, of *A-B-C-D-E-F-G-H-I-J* and *M-N-O-P-Q-R-S-T.* What types of chromosome alterations would have occurred if daughter cells were found to have a gene sequence on the first chromosome of *A-B-C-O-P-Q-G-J-I-H*?

Human Disorders Due to Chromosomal Alterations
The frequency of aneuploid zygotes may be fairly high in humans, but development is usually so disrupted that the embryos spontaneously abort. Some genetic disorders, expressed as syndromes of characteristic traits, are the result of aneuploidy. Fetal testing can detect such disorders before birth.

Down syndrome, caused by trisomy of chromosome 21, results in characteristic facial features, short stature, heart defects, and mental retardation. The incidence of Down syndrome increases for older mothers.

XXY males exhibit *Klinefelter syndrome,* a condition in which the individual has abnormally small testes, is sterile, and is usually of normal intelligence.

Males with an extra Y chromosome do not exhibit any well-defined syndrome. Trisomy X results in females who are healthy and distinguishable only by karyotype. Monosomy X individuals (X0) exhibit *Turner syndrome* and are phenotypically female, sterile individuals with short stature and usually normal intelligence.

Structural alterations of chromosomes, such as deletions or translocations, may be associated with specific human disorders. The *cri du chat* syndrome is caused by a deletion in chromosome 5; chronic myelogenous leukemia is a cancer that is associated with a reciprocal chromosomal translocation.

■ INTERACTIVE QUESTION 15.8

What is an explanation for the observation that most sex chromosome aneuploidies have less deleterious effects than do autosomal aneuploidies?

15.5 Some inheritance patterns are exceptions to the standard chromosome theory

Genomic Imprinting A few dozen traits in mammals have been identified that seem to depend on which parent supplied the alleles for the trait. In this **genomic imprinting,** which occurs during gamete formation, certain genes are imprinted or not, depending on whether ova or sperm are being produced. Imprinted alleles are not expressed in the offspring. When this generation makes gametes, old maternal and paternal imprints are removed, and alleles are imprinted according to the sex of the parent. Most of the mammalian genes subject to imprinting identified so far are involved in embryonic development. The addition of methyl groups may inactivate the imprinted gene, assuring that the developing embryo has only one active copy.

Inheritance of Organelle Genes Exceptions to Mendelian inheritance are found in the case of *extranuclear genes* located on small circles of DNA in mitochondria and plant plastids, which are transmitted to offspring in the cytoplasm of the ovum. Some rare human disorders are caused by mitochondrial mutations, and maternally inherited mitochondrial defects may contribute to diabetes, heart disease, and Alzheimer's disease.

Word Roots

aneu- = without (*aneuploidy:* a chromosomal aberration in which certain chromosomes are present in extra copies or are deficient in number)

cyto- = cell (*cytogenetic maps:* charts of chromosomes that locate genes with respect to chromosomal features)

hemo- = blood (*hemophilia:* a human genetic disease caused by a sex-linked recessive allele, characterized by excessive bleeding following injury)

mono- = one (*monosomic:* a chromosomal condition in which a particular cell has only one copy of a chromosome, instead of the normal two; the cell is said to be monosomic for that chromosome)

non- = not; **dis-** = separate (*nondisjunction:* an accident of meiosis or mitosis in which both members of a pair of homologous chromosomes or both sister chromatids fail to move apart properly)

poly- = many (*polyploidy:* a chromosomal alteration in which the organism possesses more than two complete chromosome sets)

re- = again; **com-** = together; **bin-** = two at a time (*recombinant:* an offspring whose phenotype differs from that of the parents)

trans- = across (*translocation:* attachment of a chromosomal fragment to a nonhomologous chromosome)

tri- = three; **soma-** = body (*trisomic:* a chromosomal condition in which a particular cell has an extra copy of one chromosome, instead of the normal two; the cell is said to be trisomic for that chromosome)

Structure Your Knowledge

1. Mendel's law of independent assortment applies to genes that are on different chromosomes. However, two of the genes Mendel studied were actually located on the same chromosome. Explain why genes located more than 50 map units apart behave as though they are not linked. How can one determine whether these genes are linked and what the relative distance is between them?

2. You have found a new mutant phenotype in fruit flies that you suspect is recessive and sex-linked. What is the single, best cross you could make to confirm your predictions?

3. Various human disorders or syndromes are related to chromosomal abnormalities. What explanation can you give for the adverse phenotypic effects associated with these chromosomal alterations?

Genetics Problems

Again, one of the best ways to learn genetics is to do problems.

1. The following pedigree traces the inheritance of a genetic trait.

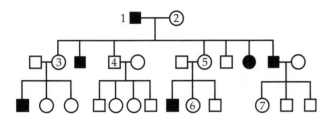

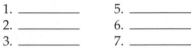

   a. What type of inheritance does this trait show?
   b. Give the predicted genotype for the following individuals:

      1. _____   5. _____
      2. _____   6. _____
      3. _____   7. _____
      4. _____

   c. What is the probability that a child of individual #6 and a phenotypically normal male will have this trait?

2. The following recombination frequencies were found. Determine the order of these genes on the chromosome.

   a, c   10%   b, c   24%
   a, d   30%   b, d   16%

3. In guinea pigs, black (*B*) is dominant to brown (*b*), and solid color (*S*) is dominant to spotted (*s*). A heterozygous black, solid-colored pig is mated with a brown, spotted pig. The total offspring for several litters are black solid = 16, black spotted = 5, brown solid = 5, and brown spotted = 14. Are these genes linked or unlinked? If they are linked, how many map units are they apart?

4. A woman is a carrier for a sex-linked lethal gene that causes an embryo with the gene to spontaneously abort. She has nine children. How many of these children do you expect to be boys?

5. A dominant sex-linked gene *B* produces white bars on black chickens, as seen in the Barred Plymouth Rock breed. A clutch of chicks has equal numbers of black and barred chicks. (Remember that sex is determined by the Z-W system in birds: ZZ are males, ZW are females.)
   a. If only the females are found to be black, what were the genotypes of the parents?
   b. If males and females are evenly represented in the black and barred chicks, what were the genotypes of the parents?

Test Your Knowledge

MULTIPLE CHOICE: *Choose the one best answer.*

1. The chromosomal theory of inheritance states that
   a. genes are located on chromosomes.
   b. chromosomes and their associated genes undergo segregation during meiosis.
   c. chromosomes and their associated genes undergo independent assortment in gamete formation.
   d. Mendel's laws of inheritance relate to the behavior of chromosomes in meiosis.
   e. all of the above are correct.

2. A wild type is
   a. the phenotype found most commonly in nature.
   b. the dominant allele.
   c. designated by a small letter if it is recessive or a capital letter if it is dominant.
   d. a trait found on the X chromosome.
   e. your basic party animal.

3. Sex-linked traits
   a. are carried on an autosome but expressed only in males.
   b. are coded for by genes located on a sex chromosome.
   c. are found in only one or the other sex, depending on the sex-determination system of the species.
   d. are always inherited from the mother in mammals and fruit flies.
   e. depend on whether the gene was inherited from the mother or the father.

4. Linkage and cytogenetic maps for the same chromosome
   a. are both based on mutant phenotypes and recombination data.
   b. may have different orders of genes.
   c. have both the same order of genes and intergenic distances.
   d. have the same order of genes but different intergenic distances.
   e. are created using chromosomal abnormalities.

5. The genetic event that results in Turner syndrome (X0) is probably
   a. nondisjunction.
   b. deletion.
   c. parental imprinting.
   d. monoploidy.
   e. independent assortment.

6. A 1:1:1:1 ratio of offspring from a dihybrid test-cross indicates that
   a. the genes are linked.
   b. the dominant organism was homozygous.
   c. crossing over has occurred.
   d. the genes are 25 map units apart.
   e. the genes are not linked or are more than 50 map units apart.

7. Genes *A* and *B* are linked and 12 map units apart. A heterozygous individual, whose parents were *AAbb* and *aaBB*, would be expected to produce gametes in the following frequencies:
   a. 44% *AB*   6% *Ab*   6% *aB*   44% *ab*
   b. 6% *AB*   44% *Ab*   44% *aB*   6% *ab*
   c. 12% *AB*   38% *Ab*   38% *aB*   12% *ab*
   d. 6% *AB*   6% *Ab*   44% *aB*   44% *ab*
   e. 38% *AB*   12% *Ab*   12% *aB*   38% *ab*

8. A female tortoiseshell cat is heterozygous for the gene that determines black or orange coat color, which is located on the X chromosome. A male tortoiseshell cat
   a. is hemizygous at this locus.
   b. must have had a tortoiseshell mother.
   c. must have resulted from a nondisjunction and has a Barr body in each of his cells.
   d. must have three alleles for coat color, one from his father and two from his mother.
   e. would be hermaphroditic.

9. A color-blind son inherited this trait from his
   a. mother.
   b. father.
   c. mother only if she is color-blind.
   d. father only if he is color-blind.
   e. mother only if she is not color-blind.

10. Genomic imprinting
    a. explains cases in which the phenotypic effect of an allele depends on the gender of the parent from whom that allele is inherited.
    b. may involve the silencing of an allele by methylation so that offspring inherit only one active copy of a gene.
    c. is greatest in females because of the larger maternal contribution of cytoplasm.
    d. is more likely to occur in offspring of older mothers.
    e. involves both a and b.

11. A cross of a wild-type red-eyed female *Drosophila* with a violet-eyed male produces all red-eyed off-spring. If the gene is sex-linked, what should the reciprocal cross (violet-eyed female × red-eyed male) produce? (Assume that the red allele is dominant to the violet allele.)
    a. all violet-eyed flies
    b. 3 red-eyed flies to 1 violet-eyed
    c. a 1:1 ratio of red and violet eyes in both males and females
    d. red-eyed females and violet-eyed males
    e. all red-eyed flies

12. Which of the following chromosomal alterations does not alter genic balance but may alter phenotype because of differences in gene expression?
    a. deletion
    b. inversion
    c. duplication
    d. nondisjunction
    e. genomic imprinting

13. Two true-breeding *Drosophila* are crossed: a normal-winged, red-eyed female and a miniature-winged, vermilion-eyed male. The F_1 all have normal wings and red eyes. F_1 offspring are crossed with miniature-winged, vermilion-eyed flies. The following offspring of that cross were counted:
    233 normal wing, red eye
    247 miniature wing, vermilion eye
    7 normal wing, vermilion eye
    13 miniature wing, red eye
    From these results you could conclude that the alleles for miniature wings and vermilion eyes are
    a. both X-linked and dominant.
    b. located on autosomes and dominant.
    c. recessive, and these genes are located 4 map units apart.
    d. recessive, and these genes are located 20 map units apart.
    e. recessive, and the deviation from the expected 9:3:3:1 ratio is due to epistasis.

14. Tests of social behavior given to Turner's syndrome volunteers (who are X0) found a correlation between scores on "behavioral inhibition" tasks and the source of the lone X chromosome. These test results appear to be an example of
    a. parental imprinting of a gene on the X chromosome.
    b. Y-linked inheritance.
    c. meiotic nondisjunction.
    d. chromosomal reciprocal translocation.
    e. nonreciprocal crossover.

15. Suppose that alleles for a sex-linked character for wing shape in *Drosophila* show incomplete dominance. The X^+ allele codes for pointed wings, the X^r for round wings, and $X^+ X^r$ individuals have oval wings. In a cross between an oval-winged female and a round-winged male, the following offspring were observed: oval-winged females, round-winged females, pointed-winged males, and round-winged males. A rare pointed-winged female was noted. What could account for this unusual offspring?

    a. a crossover between the two X chromosomes

    b. a crossover between the X and Y chromosomes

    c. a nondisjunction in meiosis II between two X^+ chromatids

    d. a nondisjunction in meiosis I between the X^+ and X^r chromosomes

    e. an X^+0 female formed when an X^+ ovum was fertilized by a sperm in which there was no sex chromosome due to a nondisjunction

16. What is the function of the gene *SRY*?

    a. codes for an RNA molecule that coats the X chromosome and initiates X-inactivation

    b. codes for RNA that activates the Y chromosome

    c. codes for a protein that regulates genes that control the development of ovaries

    d. located on the Y chromosome and codes for a protein that regulates genes that control development of testes

    e. located on the Y chromosome and required for production of normal sperm

17. A mutation in a mitochondrial gene has been linked to a rare muscle-wasting disease. This disease is

    a. found more often in males than females.

    b. found more often in females than males.

    c. inherited in a simple Mendelian fashion.

    d. caused by a translocation of a nuclear gene.

    e. inherited from the mother.

18. In which of the following would you expect to find a Barr body?

    a. an ovum

    b. a sperm

    c. a liver cell of a man

    d. a liver cell of a woman

    e. a mitochondrion

19. A cross between a wild-type mouse and a dwarf mouse homozygous for a recessive mutation in the *Igf2* gene produces heterozygotes offspring that are normal if the dwarf parent was the mother, but dwarf if the dwarf parent was the father. Which of the following explains these results?

    a. sex-linked inheritance

    b. monosomy

    c. genomic imprinting by the mother

    d. inheritance of mitochondrial genes

    e. a mutant *XIST* allele

20. Which of the following statements is *not* true about genetic recombination:

    a. Recombination of linked genes occurs by crossing over.

    b. Recombination of unlinked genes occurs by independent assortment of chromosomes.

    c. Genetic recombination results in offspring with combinations of traits that differ from the phenotypes of both parents.

    d. Recombinant offspring outnumber parental-type offspring when two genes are 50 map units apart on a chromosome.

    e. The number of recombinant offspring is proportional to the distance between two gene loci on a chromosome.

Chapter 16

The Molecular Basis of Inheritance

Framework

This chapter outlines the key evidence that was gathered to establish DNA as the molecule of inheritance. Watson and Crick's double helix, with its rungs of specifically paired nitrogenous bases and twisting side ropes of phosphate and sugar groups, provided the three-dimensional model that explained DNA's ability to encode a great variety of information and produce exact copies of itself through semiconservative replication. The replication of DNA is an extremely fast and accurate process involving many enzymes and proteins.

Chapter Review

Deoxyribonucleic acid, or DNA, is the genetic material. Nucleic acids' ability to direct their own replication allows for the precise copying and transmission of DNA to all the cells in the body and from one generation to the next. DNA encodes the blueprints that direct and control the biochemical, anatomical, physiological, and behavioral traits of organisms.

16.1 DNA is the genetic material

The Search for the Genetic Material: **Scientific Inquiry**
Chromosomes were shown to carry hereditary information and to consist of proteins and DNA. Until the 1940s, most scientists believed that proteins were the genetic material because of their specificity and heterogeneity. The role of DNA in heredity was first established through work with microorganisms—bacteria and viruses.

In 1928 F. Griffith was working with two strains of *Streptococcus pneumoniae*. When he mixed the remains of heat-killed pathogenic bacteria with harmless bacteria, some bacteria were changed into disease-causing bacteria. These bacteria had incorporated external genetic material in a process called **transformation,** which results in a change in genotype and phenotype.

O. Avery worked for more than a decade to identify the transforming agent by purifying chemicals from heat-killed pathogenic cells. In 1944 Avery, McCarty, and MacLeod announced that DNA was the molecule that transformed the bacteria.

Viruses consist of little more than DNA, or sometimes RNA, contained in a protein coat. They reproduce by infecting a cell and commandeering that cell's metabolic machinery. **Bacteriophages,** or **phages,** are viruses that infect bacteria. In 1952 A. Hershey and M. Chase showed that DNA was the genetic material of a phage known as T2 that infects the bacterium *Escherichia coli (E. coli).*

E. Chargaff, in 1947, reported that the ratio of nitrogenous bases in the DNA from various organisms was species specific. Chargaff also determined that the number of adenines and thymines was approximately equal, and the number of guanines and cytosines was also equal in the DNA from all the organisms he studied. The A=T and G=C properties of DNA became known as *Chargaff's rules.*

Circumstantial evidence that DNA is the genetic material came from the observation that a eukaryotic cell doubles its DNA content prior to mitosis and that diploid cells have twice as much DNA as haploid gametes of the same organism.

■ INTERACTIVE QUESTION   16.1

Hershey and Chase devised an experiment using radioactive isotopes to determine whether the phage's DNA or protein entered the bacteria and was the genetic material of T2 phage.

a. How did they label phage protein?

b. How did they label phage DNA?

Separate samples of *E. coli* were infected with the differently labeled T2 cells, then blended and centrifuged to isolate the bacterial cells from the lighter viral particles.

c. Where was the radioactivity found in the samples with labeled phage protein?

d. In the samples with labeled phage DNA?

e. What did Hershey and Chase conclude from these results?

Building a Structural Model of DNA: **Scientific Inquiry**
By the early 1950s, the arrangement of covalent bonds in a nucleic acid polymer was established, but the three-dimensional structure of DNA was yet to be determined.

In X-ray crystallography, an X-ray beam passed through a substance produces an X-ray diffraction photo with a pattern of spots that a crystallographer interprets as information about three-dimensional molecular structure. J. Watson saw an X-ray photo produced by R. Franklin that indicated the helical shape of DNA. He deduced that the width of the helix suggested that it consisted of two strands, thus the term **double helix.**

Watson and F. Crick constructed wire models to build a double helix that would conform to the X-ray measurements and the known chemistry of DNA. They arrived at a model that paired the nitrogenous bases on the inside of the helix with the sugar-phosphate chains on the outside. The helix makes one full turn every 3.4 nm; thus ten layers of nucleotide pairs, stacked 0.34 nm apart, are present in each turn of the helix.

To produce the molecule's uniform 2-nm width, a purine base must pair with a pyrimidine. The molecular arrangements of the side groups of the bases permit two hydrogen bonds to form between adenine and thymine and three hydrogen bonds between guanine and cytosine. This complementary pairing explains Chargaff's Rules. Van der Waals attractions between the closely stacked bases help to hold the molecule together.

In 1953 Watson and Crick published a paper in *Nature* reporting the double helix as the molecular model for DNA.

■ **INTERACTIVE QUESTION   16.2**

Review the structure of DNA by labeling the following diagrams.

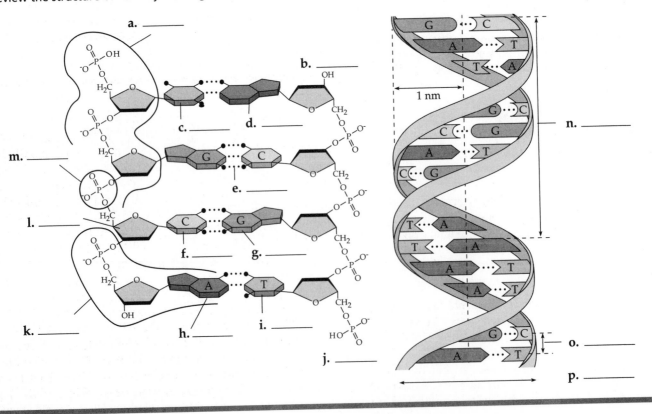

16.2 Many proteins work together in DNA replication and repair

The Basic Principle: Base Pairing to a Template Strand
Watson and Crick noted that the base-pairing rule of DNA sets up a mechanism for its replication. Each side of the double helix is an exact complement to the other. When the two sides of a DNA molecule separate in the replication process, each strand serves as a template for rebuilding a double-stranded molecule.

The **semiconservative model** of DNA replication predicts that the two daughter DNA molecules each have one parental strand and one newly formed strand. In contrast, a conservative model predicts that the parent double helix re-forms and the duplicated molecule is totally new, whereas a dispersive model predicts that all four strands of the two DNA molecules are a mixture of parental and new DNA.

M. Meselson and F. Stahl tested these models by growing *E. coli* in a medium with ^{15}N, a heavy isotope that the bacteria incorporated into their nitrogenous bases. Cells with labeled DNA were transferred to a medium with the lighter isotope, ^{14}N. Samples were removed after one and two generations of bacterial growth, and the DNA was extracted and centrifuged. The resulting locations of the density bands in the centrifuge tubes confirmed the semiconservative model of DNA replication.

■ **INTERACTIVE QUESTION   16.3**

Using different colors for heavy (parental) and light (new) strands of DNA, sketch the results of two replication cycles when *E. coli* were moved from medium containing ^{15}N to ^{14}N medium. Show the resulting density bands in the centrifuge tubes.

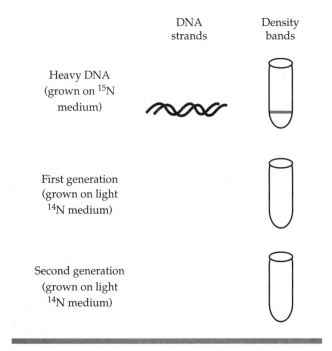

DNA
strands

Density
bands

Heavy DNA
(grown on ^{15}N
medium)

First generation
(grown on light
^{14}N medium)

Second generation
(grown on light
^{14}N medium)

DNA Replication: A Closer Look   Replication begins at special sites, called **origins of replication,** where proteins that initiate replication bind to a specific sequence of nucleotides and separate the two strands to form a replication "bubble." Replication proceeds in both directions in the two Y-shaped **replication forks.**

Enzymes called **DNA polymerases** connect nucleotides to the growing end of the new DNA strand. DNA polymerase III and I are involved in replication in *E. coli;* at least 11 different DNA polymerases have been discovered so far in eukaryotes. A nucleoside triphosphate lines up with its complementary base on the template strand; it loses two phosphate groups and the hydrolysis of this pyrophosphate to two inorganic phosphates ($\text{\textcircled{P}}_i$) provides the energy for polymerization.

The two strands of a DNA molecule are antiparallel; their sugar-phosphate backbones run in opposite directions. The deoxyribose sugar of each nucleotide is connected to its own phosphate group at its 5' carbon and connects to the phosphate group of the adjacent nucleotide by its 3' carbon. Thus a strand of DNA has polarity, with a 5' end where the first nucleotide's phosphate group is exposed, and a 3' end where the nucleotide at the other end has a hydroxyl group attached to its 3' carbon.

■ **INTERACTIVE QUESTION   16.4**

Look back to Interactive Question 16.2 and label the 5' and 3' ends of both strands of the DNA molecule.

DNA polymerases add nucleotides only to the 3' end of growing strands; DNA is replicated in a 5'→3' direction. Because the DNA strands run in an antiparallel direction, the simultaneous synthesis of both strands presents a problem. The **leading strand** is the new 5' → 3' strand being formed along the template by DNA polymerase III (DNA pol III) in the progressing replication fork. The **lagging strand** is created as a series of short segments, called **Okazaki fragments,** that are formed in the 5'→3' direction away from the replication fork. An enzyme called **DNA ligase** joins the sugar-phosphate backbones of the fragments.

DNA polymerases cannot initiate synthesis of a DNA strand; they can only add nucleotides to an existing chain that is attached to a template strand. An enzyme called **primase** joins about 10 RNA nucleotides to form the **primer** needed to start the chain. Each fragment on the lagging strand requires a primer. A continuous strand of DNA is produced after DNA polymerase I (DNA pol I) replaces the RNA primer with DNA nucleotides and ligase joins the fragments.

An enzyme called **helicase** unwinds the helix and separates the parent strand at replication forks, and **single-strand binding proteins** keep the separated strands apart while they serve as templates. **Topoisomerase** helps relieve the strain from the twisting of DNA strands in front of helicase.

The various proteins that function in DNA replication form a large complex. In eukaryotic cells, many such complexes may anchor to the nuclear matrix and the DNA polymerase molecules may pull the parental DNA through them.

■ **INTERACTIVE QUESTION   16.5**

In this diagram showing the replication of DNA, label the following items: leading and lagging strands, Okazaki fragment, DNA pol III, DNA pol I, DNA ligase, helicase, primase, single-strand binding proteins, RNA primer, replication fork, and 5' and 3' ends of parental DNA.

← Overall direction of replication

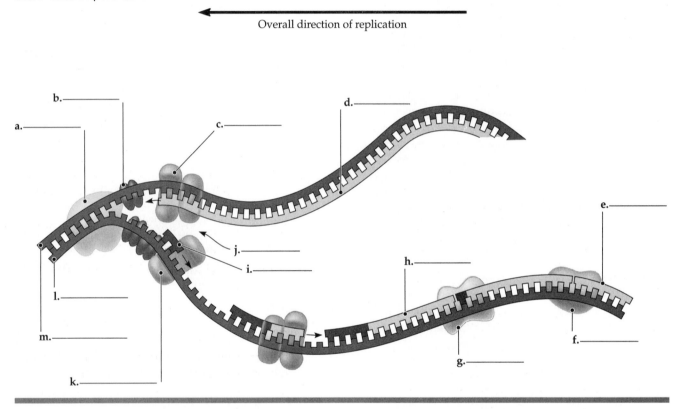

Proofreading and Repairing DNA   Initial pairing errors in nucleotide placement may occur as often as 1 per 100,000 base pairs. The amazing accuracy of DNA replication (one error in ten billion nucleotides) is achieved as DNA polymerases check each newly added nucleotide against its template and replace incorrect nucleotides. Other enzymes also fix incorrectly paired nucleotides, called **mismatch repair.**

Reactive chemicals, radioactive emissions, X-rays, and UV light may alter DNA molecules. Many different types of DNA repair enzymes (130 have been identified in humans so far) may correct these changes. In **nucleotide excision repair,** the damaged strand is cut out by a **nuclease** and the gap is correctly filled through the action of DNA polymerase and ligase. In skin cells, nucleotide excision repair frequently corrects thymine dimers caused by ultraviolet rays of sunlight.

Replicating the Ends of DNA Molecules   Because DNA polymerase cannot attach nucleotides to the 5' end of a daughter DNA strand, repeated replications cause a progressive shortening of linear DNA molecules. Multiple repetitions of a short nucleotide sequence at the ends of chromosomes, called **telomeres,** protect an organism's genes from being shortened during successive DNA replications.

The shortening of telomeres may limit cell division. In germ line cells, however, the enzyme **telomerase,** which contains an RNA template for the telomere sequence, lengthens telomeres. Some somatic cancer cells and "immortal" strains of cultured cells produce telomerase and are thus capable of unregulated cell division.

Word Roots

helic- = a spiral (*helicase:* an enzyme that untwists the double helix of DNA at the replication forks)

liga- = bound or tied (*DNA ligase:* a linking enzyme for DNA replication)

-phage = to eat (*bacteriophages:* viruses that infect bacteria)

semi- = half (*semiconservative model:* type of DNA replication in which the replicated double helix consists of one old strand, derived from the old molecule, and one newly made strand)

telos- = an end (*telomere:* the protective structure at each end of a eukaryotic chromosome)

trans- = across (*transformation:* a phenomenon in which external DNA is assimilated by a cell)

Structure Your Knowledge

1. Summarize the evidence and techniques Watson and Crick used to deduce the double-helix structure of DNA.

2. Review your understanding of DNA replication by describing the key enzymes and proteins (in the order of their functioning) that direct replication.

Test Your Knowledge

MULTIPLE CHOICE: *Choose the one best answer.*

1. One of the reasons most scientists believed proteins were the carriers of genetic information was that
   a. proteins were more heat stable than nucleic acids.
   b. the protein content of duplicating cells always doubled prior to division.
   c. proteins were much more complex and heterogeneous molecules than nucleic acids.
   d. early experimental evidence pointed to proteins as the hereditary material.
   e. proteins were found in DNA.

2. Transformation involves
   a. the uptake of external genetic material, often from one bacterial strain to another.
   b. the creation of a strand of RNA from a DNA molecule.
   c. the infection of bacterial cells by phage.
   d. the type of semiconservative replication shown by DNA.
   e. the replication of DNA along the lagging strand.

3. The DNA of an organism has thymine as 20% of its bases. What percentage of its bases would be guanine?
   a. 20%   c. 40%   e. 80%
   b. 30%   d. 60%

4. In his work with pneumonia-causing bacteria, Griffith found that
   a. DNA was the transforming agent.
   b. the pathogenic and harmless strains mated.
   c. heat-killed harmless cells could cause pneumonia when mixed with heat-killed pathogenic cells.
   d. some heat-stable chemical was transferred to harmless cells to transform them into pathogenic cells.
   e. a T2 phage transformed harmless cells to pathogenic cells.

5. When T2 phages are grown with radioactive sulfur,
   a. their DNA is tagged.
   b. their proteins are tagged.
   c. their DNA is found to be of medium density in a centrifuge tube.
   d. they transfer their radioactivity to *E. coli* chromosomes when they infect the bacteria.
   e. their excision enzymes repair the damage caused by the radiation.

6. Meselson and Stahl
   a. provided evidence for the semiconservative model of DNA replication.
   b. were able to separate phage protein coats from *E. coli* by using a blender.
   c. found that DNA labeled with ^{15}N was of intermediate density.
   d. grew *E. coli* on labeled phosphorus and sulfur.
   e. found that DNA composition was species specific.

7. Watson and Crick concluded that each base could not pair with itself because
   a. there would not be room for the helix to make a full turn every 3.4 nm.
   b. the uniform width of 2 nm would not permit two purines or two pyrimidines to pair together.
   c. the bases could not be stacked 0.34 nm apart.
   d. identical bases could not hydrogen-bond together.
   e. they would be on antiparallel strands.

8. The joining of nucleotides in the polymerization of DNA requires energy from
   a. DNA polymerase.
   b. the hydrolysis of the terminal phosphate group of ATP.
   c. RNA nucleotides.
   d. the hydrolysis of GTP.
   e. the hydrolysis of the pyrophosphates removed from nucleoside triphosphates.

9. Continuous elongation of a new DNA strand along one strand of DNA
   a. requires the action of DNA ligase as well as polymerase.
   b. occurs because DNA ligase can only elongate in the 5′→ 3′ direction.
   c. occurs on the leading strand.
   d. occurs on the lagging strand.
   e. a, b, and c are correct.

10. Which of the following statements about DNA polymerase is incorrect?
    a. It forms the bonds between complementary base pairs.
    b. It is able to proofread and correct for errors in base pairing.
    c. It is unable to initiate synthesis; it requires an RNA primer.
    d. It only works in the 5′→ 3′ direction.
    e. It is found in eukaryotes and prokaryotes.

11. Thymine dimers—covalent links between adjacent thymine bases in DNA—may be induced by UV light. When they occur, they are repaired by
    a. excision enzymes (nucleases).
    b. DNA polymerase.
    c. ligase.
    d. primase.
    e. a, b, and c are all needed.

12. How does DNA synthesis along the lagging strand differ from that on the leading strand?
    a. Nucleotides are added to the 5′ end instead of the 3′ end.
    b. Ligase is the enzyme that polymerizes DNA on the lagging strand.
    c. An RNA primer is needed on the lagging strand but not on the leading strand.
    d. Okazaki fragments, which each grow 5′→ 3′, must be joined along the lagging strand.
    e. Helicase synthesizes Okazaki fragments, which are then joined by ligase.

13. Which of the following enzymes or proteins is paired with an incorrect or inaccurate function?
    a. Helicase—unwind and separate parental double helix
    b. Telomerase—add telomere repetitions to end of chromosomes
    c. Single-strand binding protein—hold strands of unwound DNA apart and straight
    d. Nuclease—cut out (excise) damaged DNA strand
    e. Primase—form DNA primer to start replication

Use the following diagram to answer Questions 14 through 17.

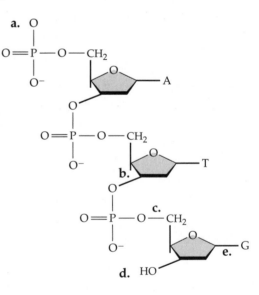

14. Which letter indicates the 5′ end of this single DNA strand?
    a.   b.   c.   d.   e.

15. At which letter would the next nucleotide be added?
    a.   b.   c.   d.   e.

16. Which letter indicates a phosphodiester bond formed by DNA polymerase?

    a.　　　b.　　　c.　　　d.　　　e.

17. The base sequence of the DNA strand made from this template would be (from top to bottom)

    a. A T C.
    b. C G A.
    c. T A C.
    d. U A C.
    e. A T G.

18. T2 phage is grown in *E. coli* with radioactive phosphorus and then allowed to infect other *E. coli*. The culture is blended to separate the viral coats from the bacterial cells and centrifuged. Which of the following best describes the expected results of such an experiment?

    a. Both viral and bacterial DNA are labeled; radioactivity is found in the supernatant.
    b. Both viral and bacterial proteins are labeled; radioactivity is present in both the supernatant and the pellet.
    c. Viral proteins are labeled; radioactivity is found in the supernatant but not in the pellet.
    d. Viral DNA is labeled; radioactivity is found in the pellet.
    e. The virus destroyed the bacteria; no pellet is formed.

19. What are telomeres, and what do they do?

    a. ever-shortening tips of chromosomes that may signal cells to stop dividing at maturity
    b. highly repetitive sequences at tips of chromosomes that protect the lagging strand during replication

    c. repetitive sequences of nucleotides at the centromere region of a chromosome
    d. enzymes that are present in germ-line cells that allow these cells to undergo repeated divisions
    e. Both a and b are correct.

20. You are trying to support your hypothesis that DNA replication is *conservative*; i.e., parental strands separate; complementary strands are made, but these new strands join together to make a new DNA molecule, and the parental strands rejoin. You take *E. coli* that had grown in a medium containing only heavy nitrogen (^{15}N) and transfer a sample to a medium containing light nitrogen (^{14}N). After allowing time for only one DNA replication, you centrifuge a sample and compare the density band(s) formed with control bands for bacteria grown on either normal ^{14}N or ^{15}N medium. Which band location would support your hypothesis of *conservative* DNA replication?

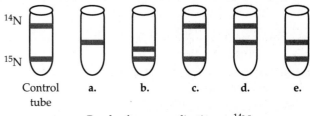

Bands after one replication on ^{14}N

21. Using the experiment explained in question 20, which centrifuge tube would represent the band distribution obtained after one replication showing that DNA replication is *semiconservative*?

Chapter 17
From Gene to Protein

Framework

This chapter deals with the pathway from DNA to RNA to proteins. The instructions of DNA are transcribed to a sequence of codons in mRNA. In eukaryotes, mRNA is processed before it leaves the nucleus. Complexed with ribosomes, mRNA is translated as a sequence of amino acids in a polypeptide as tRNAs match their anticodons to the mRNA codons. Mutations, which alter the base pairs in DNA, usually alter the protein product.

Chapter Review

The information of DNA is contained in specific sequences of nucleotides. These sequences are transcribed and translated to synthesize proteins, which are responsible for the specific traits of an organism and thus form the link between genotype and phenotype.

17.1 Genes specify proteins via transcription and translation

Evidence from the Study of Metabolic Defects   In 1909, A. Garrod first suggested that genes determine phenotype through the action of enzymes, reasoning that inherited diseases were caused by an inability to make certain enzymes.

G. Beadle and E. Tatum, working with mutants of a bread mold, *Neurospora crassa*, demonstrated the relationship between genes and enzymes. They studied several nutritional mutants that could not grow on the minimal medium that sufficed for wild-type mold. By growing nutritional mutants on complete growth medium and then transferring samples to various combinations of minimal medium and one added nutrient, Beadle and Tatum were able to identify the specific metabolic defect for each mutant.

Three classes of *Neurospora* mutants that were unable to synthesize arginine were identified by supplementing different precursors of the pathway. Beadle and Tatum reasoned that the metabolic pathway of each class was blocked at a different enzymatic step. They formulated the *one gene–one enzyme hypothesis*.

Biologists revised Beadle and Tatum's idea to one gene–one protein because not all proteins are enzymes. Because many proteins consist of more than one polypeptide chain, each of which is codified by its own gene, the axiom is now the **one gene–one polypeptide hypothesis**.

Basic Principles of Transcription and Translation   RNA is the link between a gene and the protein for which it codes. RNA differs from DNA in three ways: the sugar component of its nucleotides is ribose, rather than deoxyribose; uracil (U) replaces thymine as one of its nitrogenous bases; and RNA is usually single stranded.

Transcription is the transfer of information from DNA to **messenger RNA (mRNA)** or another type of

RNA, using the "language" of nucleic acids. **Translation** transfers information from mRNA to a polypeptide, changing from the language of nucleotides to that of amino acids. **Ribosomes** are the sites of translation—the synthesis of a polypeptide.

In prokaryotes, which lack a nucleus, transcription and translation can occur simultaneously. In eukaryotes, the mRNA must exit the nucleus before translation can begin. **RNA processing,** the modification of *pre-mRNA* or of the **primary transcript** of any gene (such as one coding for RNA) within the nucleus, occurs only in eukaryotes.

■ **INTERACTIVE QUESTION   17.1**

Fill in the sequence in the synthesis of proteins. Put the name of the process above each arrow.

_____ $\longrightarrow$ _____ $\longrightarrow$ _____

The Genetic Code   A sequence of three nucleotides provides 4^3, or 64, possible unique sequences of nucleotides, more than enough to code for the 20 amino acids. The translation of nucleotides into amino acids uses a **triplet code** to specify each amino acid.

The base triplets along the **template strand** of a gene are transcribed into mRNA **codons.** The same strand of a DNA molecule can be the template strand for one gene and the complementary strand for another. The mRNA is complementary to the DNA template since its bases follow the same base-pairing rules, with the exception that uracil substitutes for thymine in RNA. Codons are read in the 5' → 3' direction. Each codon specifies one of the 20 amino acids.

In the early 1960s, M. Nirenberg added artificial "poly U" mRNA to a test tube containing all the biochemical ingredients necessary for protein synthesis and obtained a polypeptide containing a single amino acid. Molecular biologists had deciphered all 64 codons by the mid-1960s. Three codons function as stop signals, or termination codons. The codon AUG both codes for methionine and functions as an initiation codon, a start signal for translation.

The code is often redundant, meaning that more than one codon may specify a single amino acid. The code is never ambiguous; no codon specifies two different amino acids.

The nucleotide sequence on mRNA is read in the correct **reading frame,** starting at a start codon and reading each triplet sequentially.

■ **INTERACTIVE QUESTION   17.2**

Practice using the dictionary of the genetic code in your textbook. Determine the amino acid sequence for a polypeptide coded for by the following mRNA transcript (written 5' → 3'):

AUGCCUGACUUUAAGUAG

The genetic code of codons and their corresponding amino acids is almost universal. A bacterial cell can translate the genetic messages of human cells. The near universality of a common genetic language provides compelling evidence of the antiquity of the code and the evolutionary connection of all living organisms.

17.2   Transcription is the DNA-directed synthesis of RNA: *a closer look*

Molecular Components of Transcription   The **promoter** is the DNA sequence where **RNA polymerase** attaches and initiates transcription. In prokaryotes, the **terminator** is the sequence that signals the end of transcription. A **transcription unit** is the sequence of DNA that is transcribed into one RNA molecule.

Bacteria have one type of RNA polymerase. Eukaryotes have three types; the one that synthesizes mRNA is called RNA polymerase II.

Synthesis of an RNA Transcript   The promoter is the binding site for RNA polymerase and determines where transcription starts and which DNA strand is used as the template. It includes recognition sequences, such as the **TATA box** common in eukaryotes, upstream from the start point. In eukaryotes, **transcription factors** must first recognize and bind to the promoter before RNA polymerase II can attach, at which point the assembly is called the **transcription initiation complex.**

RNA polymerase untwists the double helix, exposing DNA nucleotides for base pairing with RNA nucleotides, and joins the nucleotides to the 3' end of the growing polymer. The new RNA peels away from the DNA template and the DNA double helix re-forms. Several molecules of RNA polymerase may be transcribing simultaneously from different parts of a single gene, enabling a cell to produce large quantities of mRNA.

In prokaryotes, transcription ends after RNA polymerase transcribes the terminator sequence. In eukaryotes, polymerase continues past a polyadenylation signal sequence (AAUAAA), proteins cut loose the pre-mRNA, and polymerase eventually falls off the DNA after transcribing hundreds more nucleotides.

■ INTERACTIVE QUESTION   17.3

Review the key steps of transcription in eukaryotes:

a.

b.

c.

17.3   Eukaryotic cells modify RNA after transcription

Alteration of mRNA Ends   A modified guanine nucleotide is attached to the 5′ end of a pre-mRNA, and a string of adenine nucleotides, called a **poly-A tail,** is added to the 3′ end. The **5′ cap** and poly-A tail may facilitate transport of mRNA from the nucleus, aid ribosome attachment, and protect the ends of mRNA from hydrolytic enzymes. The cap and tail are attached to the untranslated regions (UTRs) at the 5′ and 3′ ends.

Split Genes and RNA Splicing   Long segments of noncoding base sequences, known as **introns** or intervening sequences, occur within the boundaries of eukaryotic genes. The remaining coding regions are called **exons,** since they are expressed in protein synthesis (except that the 5′ and 3′ UTRs are not translated). A primary transcript is made of the gene—but introns are removed and exons joined before the mRNA leaves the nucleus—in a process called **RNA splicing.**

Signals for RNA splicing are sets of a few nucleotides at either end of each intron. *Small nuclear ribonucleoproteins (snRNPs),* composed of proteins and *small nuclear RNA (snRNA),* are components of a molecular complex called a **spliceosome.** The spliceosome snips an intron out of the RNA transcript and connects the adjoining exons. In addition to splice-site recognition and spliceosome assembly, a function of snRNA may be catalytic in intron removal. RNA molecules that act as enzymes are called **ribozymes.**

In some cases of RNA splicing, intron RNA catalyzes its own removal.

■ INTERACTIVE QUESTION   17.4

How does the mRNA that leaves the nucleus differ from the primary transcript pre-mRNA?

Some introns are involved in regulating gene activity, and splicing is necessary for the export of mRNA from the nucleus. **Alternative RNA splicing** allows some genes to produce different polypeptides. Exons may code for polypeptide **domains,** functional segments of a protein, such as binding and active sites. Introns may facilitate recombination of exons between different alleles or even between different genes. Such exon shuffling can result in novel proteins.

17.4   Translation is the RNA-directed synthesis of a polypeptide: *a closer look*

Molecular Components of Translation   **Transfer RNA (tRNA)** molecules are specific for the amino acid they carry to the ribosomes. They each have a specific base triplet, called an **anticodon,** that base-pairs with a complementary codon on mRNA, thus assuring that amino acids are arranged in the sequence prescribed by the transcription from DNA.

As with other RNAs, transfer RNA is transcribed in the nucleus of a eukaryote and moves into the cytoplasm where it can be used repeatedly. These single-stranded, short RNA molecules are arranged into a clover-leaf shape by hydrogen bonding between complementary base sequences and then folded into a three-dimensional, roughly L-shaped structure. The anticodon is at one end of the L; the 3′ end is the attachment site for its amino acid.

Sixty-one codons for amino acids can be read from mRNA, but there are only about 45 different tRNA molecules. A phenomenon known as **wobble** enables the third nucleotide of some tRNA anticodons to pair with more than one kind of base in the codon. Thus, one tRNA can recognize more than one mRNA codon, all of which code for the same amino acid carried by that tRNA.

■ INTERACTIVE QUESTION   17.5

Using some of the codons and the amino acids you identified in Interactive Question 17.2, fill in the following table.

DNA Triplet $3' \rightarrow 5'$	mRNA Codon $5' \rightarrow 3'$	Anticodon $3' \rightarrow 5'$	Amino Acid
			methionine
		GGA	Stop
TTC	AAG		
ATC	UAG		

Each amino acid has a specific **aminoacyl-tRNA synthetase** that attaches it to its appropriate tRNA molecule to create an aminoacyl tRNA. The hydrolysis of ATP drives this process.

Ribosomes facilitate the specific pairing of tRNA anticodons with mRNA codons during protein synthesis. They consist of a large and a small subunit, each composed of proteins and a form of RNA called **ribosomal RNA (rRNA).** Subunits are constructed in the nucleolus in eukaryotes. Prokaryotic ribosomes are smaller and differ enough in molecular composition that some antibiotics can inhibit them without affecting eukaryotic ribosomes.

A large and small subunit join to form a ribosome when they attach to an mRNA molecule. Ribosomes have a binding site for mRNA, a **P site** (peptidyl-tRNA site) that holds the tRNA carrying the growing polypeptide chain, an **A site** (aminoacyl-tRNA site) that holds the tRNA carrying the next amino acid, and an **E site** (exit site) from which discharged tRNAs leave the ribosome. The attachment of an amino acid to the carboxyl end of the growing polypeptide chain is catalyzed by the ribosome.

According to the recently determined detailed structure of the small and large subunits of bacterial ribosomes, the protein components are mostly on the outside and the rRNA is in the interior at the interface between the subunits and at the A and P sites, supporting the hypothesis that rRNA performs a ribosome's catalytic functions.

Building a Polypeptide   The three stages of protein synthesis—initiation, elongation, and termination—all require the aid of protein "factors." The first two stages also require energy, which is provided by the hydrolysis of GTP (guanosine triphosphate).

The initiation stage begins as the small subunit of the ribosome binds to an mRNA and an initiator tRNA carrying methionine, which attaches to the start codon AUG on the mRNA. With the aid of proteins called *initiation factors* and the expenditure of a GTP, the large subunit of the ribosome attaches to the small one, forming a translation initiation complex. The initiator tRNA fits into the P site.

The addition of amino acids in the elongation stage involves several proteins called *elongation factors* and occurs in a three-step cycle.

In codon recognition, an aminoacyl-tRNA base-pairs with the mRNA codon in the A binding site. This step requires energy from the hydrolysis of 2 GTP.

In the peptide bond formation step, an RNA molecule of the large subunit catalyzes the formation of a peptide bond between the carboxyl end of the polypeptide held in the P site and the amino acid in the A site. The polypeptide is now held by the amino acid in the A site.

In translocation, tRNA carrying the growing polypeptide is translocated to the P site, a process requiring energy from the hydrolysis of a GTP molecule. The empty tRNA from the P site is moved to the E site and released. The next mRNA codon moves into the A site as the mRNA moves through the ribosome.

Termination occurs when a stop codon—UAA, UAG, or UGA—reaches the A site of the ribosome. A *release factor* binds to the stop codon and hydrolyzes the bond between the polypeptide and the tRNA in the P site. The completed polypeptide leaves through the exit tunnel of the large subunit. The two ribosomal subunits and other components dissociate.

An mRNA may be translated simultaneously by several ribosomes in strings called **polyribosomes** (or **polysomes**).

■ INTERACTIVE QUESTION 17.6

In the following diagrams of polypeptide synthesis, name the stages (1–4), identify the components (a–l), and then briefly describe what happens in each stage.

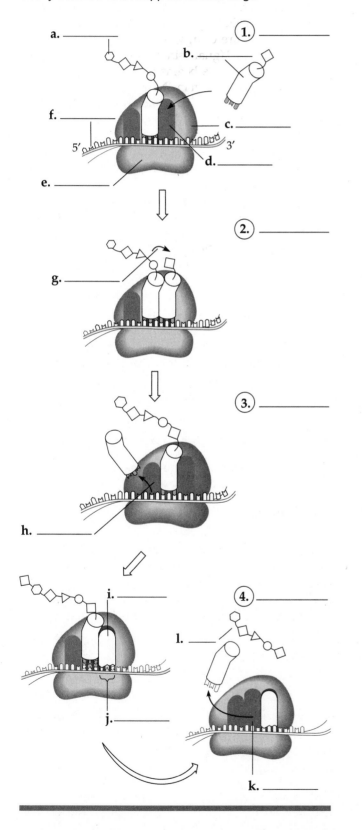

Completing and Targeting the Functional Protein

During and following translation, a polypeptide folds spontaneously into its secondary and tertiary structures. Chaperone proteins often facilitate the correct folding. The protein may need to undergo *post-translational modifications:* Amino acids may be chemically modified; one or more amino acids at the beginning of the chain may be enzymatically removed; segments of the polypeptide may be excised; or several polypeptides may associate into a quaternary structure.

All ribosomes are identical, whether they are free ribosomes that synthesize cytosolic proteins or ER-bound ribosomes that make membrane and secretory proteins. Polypeptide synthesis begins in the cytoplasm. If a protein is destined for the endomembrane system or for secretion, its polypeptide chain will begin with a **signal peptide** that is recognized by a protein-RNA complex called a **signal-recognition particle (SRP),** which attaches the ribosome to a receptor protein that is part of a multiprotein complex on the ER membrane. As the growing polypeptide threads into the ER, the signal peptide is usually removed. Other signal peptides direct some proteins made in the cytosol to specific sites such as mitochondria, chloroplasts, or the interior of the nucleus.

■ INTERACTIVE QUESTION 17.7

What determines if a ribosome becomes bound to the ER?

17.5 RNA plays multiple roles in the cell: *a review*

The versatility of RNA molecules stems from their ability to hydrogen-bond to other RNA or DNA molecules, to form specific three-dimensional shapes by forming intramolecular hydrogen bonds, and to act as a catalyst.

Fill in the functions for the following types of RNA molecules.

a. mRNA

b. tRNA

c. rRNA

d. snRNA

e. SRP RNA

f. snoRNA (small nucleolar RNA)

g. siRNA (small interfering RNA) and miRNA (microRNA)

17.6   Comparing gene expression in prokaryotes and eukaryotes reveals key differences

The basic processes of transcription and translation are the same in prokaryotes and eukaryotes, although the RNA polymerases and ribosomes differ. Transcription factors are required in eukaryotes for RNA polymerase to bind, and transcription is terminated differently. In bacteria, transcription and translation occur almost simultaneously, whereas in eukaryotes, transcription is physically separated from translation by the nuclear envelope, allowing extensive RNA processing to occur before RNA leaves the nucleus. Eukaryotes also can target proteins to particular organelles.

17.7   Point mutations can affect protein structure and function

Mutations, changes in the genetic information of a cell (or virus) may involve large portions of a chromosome or affect just one base pair of nucleotides, as in a **point mutation.** If the mutation is in a cell that gives rise to a gamete, it may be passed on to offspring.

Types of Point Mutations   A **base-pair substitution** replaces one nucleotide and its complementary partner with another pair of nucleotides. Due to the redundancy of the genetic code, some base-pair substitutions in the third nucleotide of a codon do not change the amino acid translation and are called *silent mutations.* A substitution may result in the insertion of a different amino acid without altering the character of the protein if the new amino acid has similar properties or is not located in a region crucial to that protein's function.

A base-pair substitution that results in a different amino acid in a critical portion of a protein, such as the active site of an enzyme, may significantly impair protein function.

A substitution that results in an incorrectly coded amino acid is called a **missense mutation. Nonsense mutations** occur when the point mutation changes an amino acid codon into a stop codon, prematurely halting the translation of the polypeptide chain and usually creating a nonfunctional protein.

Base-pair **insertions** or **deletions** that are not in multiples of three nucleotides alter the reading frame. All nucleotides downstream from the mutation will be improperly grouped into codons, creating extensive missense and usually prematurely ending in nonsense. These **frameshift mutations** almost always produce nonfunctional proteins.

Mutagens   Mutations can occur in a number of ways. *Spontaneous mutations* include base-pair substitutions, insertions, deletions, and longer mutations that occur during DNA replication, repair, or recombination. Physical agents, such as X-rays and UV light, and various chemical agents that cause mutations are called **mutagens.** Chemical mutagens include base analogs that substitute for normal bases and then pair incorrectly in DNA synthesis, chemicals that insert into and distort the double helix, and other agents that chemically change DNA bases. Tests can measure the mutagenic effects of chemicals and thus their potential carcinogenic risk.

Define the following, and explain what type of point mutation could cause each of these mutations.

a. silent mutation

b. missense mutation

c. nonsense mutation

d. frameshift mutation

What is a gene? Revisiting the question

Our definition of a gene has evolved from Mendel's heritable factors, to Morgan's loci along chromosomes, to the one gene–one polypeptide axiom.

Research continually refines our understanding of the structural and functional aspects of genes, which now include introns, promoters, and other regulatory regions. The best working definition of a gene is that it is a region of DNA whose final product is either a polypeptide or an RNA molecule.

poly- = many (*poly-A tail:* the modified end of the 3′ end of an mRNA molecule consisting of the addition of some 50 to 250 adenine nucleotides)

trans- = across; **-script** = write (*transcription:* the synthesis of RNA on a DNA template)

Word Roots

anti- = opposite (*anticodon:* a specialized base triplet on one end of a tRNA molecule that recognizes a particular complementary codon on an mRNA molecule)

exo- = out, outside, without (*exon:* a coding region of a eukaryotic gene that is expressed)

intro- = within (*intron:* a noncoding, intervening sequence within a eukaryotic gene)

muta- = change; **-gen** = producing (*mutagen:* a physical or chemical agent that causes mutations)

Structure Your Knowledge

1. Make sure you understand and can explain the processes of transcription and translation. You may find that filling in the table below (for eukaryotic gene expression) in a study group helps you to review these processes.

2. What is the genetic code? Explain redundancy and the wobble phenomenon. What is the significance of the fact that the genetic code is nearly universal?

3. Prepare a concept map showing the types and consequences of point mutations.

	Transcription	Translation
Template		
Location		
Molecules involved		
Enzymes involved		
Control—start and stop		
Product		
Product processing		
Energy source		

Test Your Knowledge

MULTIPLE CHOICE: *Choose the one best answer.*

1. In Beadle and Tatum's study of *Neurospora*, they were able to identify three classes of mutants that needed arginine added to minimal media in order to grow. The production of arginine includes the following steps: precursor → ornithine → citrulline → arginine. What nutrient(s) had to be supplied for the mutants with a defective enzyme for the precursor → ornithine step to grow?

   a. precursor only
   b. ornithine only
   c. citrulline only
   d. ornithine or citrulline
   e. precursor, ornithine, and citrulline

2. Transcription involves the transfer of information from

   a. DNA to RNA.
   b. RNA to DNA.
   c. mRNA to an amino acid sequence.
   d. DNA to an amino acid sequence.
   e. the nucleus to the cytoplasm.

3. If the 5′ → 3′ nucleotide sequence on the complementary (noncoding) DNA strand is CAT, what is the corresponding codon on mRNA?

   a. UAC
   b. CAU
   c. GUA
   d. GTA
   e. CAT

4. RNA polymerase

   a. is the protein responsible for the production of ribonucleotides.

   b. is the enzyme that creates hydrogen bonds between nucleotides on the DNA template strand and their complementary RNA nucleotides.

   c. is the enzyme that transcribes exons but does not transcribe introns.

   d. is a ribozyme composed of snRNPs.

   e. begins transcription at a promoter sequence and moves along the template strand of DNA, elongating an RNA molecule in a $5' \to 3'$ direction.

5. How is the template strand for a particular gene determined?

   a. It is the DNA strand that runs from the $5' \to 3'$ direction.

   b. It is the DNA strand that runs from the $3' \to 5'$ direction.

   c. It depends on the orientation of RNA polymerase, whose position is determined by particular sequences of nucleotides within the promoter.

   d. It doesn't matter which strand is the template because they are complementary and will produce the same mRNA.

   e. The template strand always contains the TATA box.

6. Which enzyme synthesizes tRNA?

   a. RNA replicase

   b. RNA polymerase

   c. aminoacyl-tRNA synthetase

   d. ribosomal enzymes

   e. ribozymes

7. Which of the following is true of RNA processing?

   a. Exons are excised before the mRNA is translated.

   b. The RNA transcript that leaves the nucleus may be much longer than the original transcript.

   c. Assemblies of protein and snRNPs, called spliceosomes, may catalyze splicing.

   d. Large quantities of rRNA are assembled into ribosomes.

   e. Signal peptides are added to the 5′ end of the transcript.

8. Which of the following is *not* involved in the formation of a eukaryotic transcription initiation complex?

   a. TATA box

   b. transcription factors

   c. snRNA

   d. RNA polymerase II

   e. promoter

9. A prokaryotic gene 600 nucleotides long can code for a polypeptide chain of how many amino acids (at most)?

   a. 100   **c.** 300   **e.** 1,800

   b. 200   **d.** 600

10. All of the following are transcribed from DNA *except*

    a. exons.   **c.** tRNA.   **e.** promoter.

    b. introns.   **d.** rRNA.

11. What might introns have to do with the evolution of new proteins?

    a. The excised introns are transcribed and translated as new proteins by themselves.

    b. Introns are more likely to accumulate mutations than exons, and these mutations then result in the production of novel proteins.

    c. Introns that are self-excising may also function as hydrolytic enzymes for other processes.

    d. Introns provide more area where crossing over may occur (without interfering with the coding sequences) and thus increase the probability of exon shuffling between alleles.

    e. Introns often correspond to domains in proteins that fold independently and have specific functions. Changing domains between nonallelic genes could produce novel proteins.

12. A ribozyme is

    a. an exception to the one gene–one RNA molecule axiom.

    b. an enzyme that adds the 5′ cap and poly-A tail to mRNA.

    c. an example of rearrangement of protein domains caused by RNA splicing.

    d. an RNA molecule that functions as an enzyme.

    e. an enzyme that produces both small and large ribosomal subunits.

13. All of the following would be found in a prokaryotic cell *except*

    a. mRNA.

    b. rRNA.

    c. simultaneous transcription and translation.

    d. snRNA.

    e. RNA polymerase.

14. Which of the following is transcribed and then translated to form a protein product?

    a. gene for tRNA

    b. intron

    c. gene for a transcription factor

    d. 5′ and 3′ UTRs

    e. gene for rRNA

15. Transfer RNA
    a. forms hydrogen bonds between its codon and the anticodon of an mRNA in the A site of a ribosome.
    b. binds to its specific amino acid in the active site of an aminoacyl-tRNA synthetase.
    c. uses GTP as the energy source to bind its amino acid.
    d. is translated from mRNA.
    e. is produced in the nucleolus.

16. Place the following events in the synthesis of a polypeptide in the proper order.
    1. A peptide bond forms.
    2. An aminoacyl tRNA matches its anticodon to the codon in the A site.
    3. A tRNA translocates from the A to the P site, and an unattached tRNA leaves the ribosome from the E site.
    4. The large subunit attaches to the small subunit, with the initiator tRNA in the P site.
    5. A small subunit binds to an mRNA and an initiator tRNA.
    a. 4-5-3-2-1
    b. 4-5-2-1-3
    c. 5-4-3-2-1
    d. 5-4-1-2-3
    e. 5-4-2-1-3

17. Translocation in the process of translation involves
    a. the hydrolysis of a GTP molecule.
    b. the movement of the tRNA in the A site to the P site.
    c. the movement of the mRNA strand one triplet length.
    d. the release of the unattached tRNA from the E site.
    e. all of the above.

18. Which of the following type of molecule catalyzes the formation of a peptide bond?
    a. RNA polymerase
    b. rRNA
    c. mRNA
    d. aminoacyl-tRNA synthetase
    e. protein ribosomal enzyme

19. Which of the following is *not* true of an anticodon?
    a. It consists of three nucleotides.
    b. It lines up in the $5' \rightarrow 3'$ direction along the $5' \rightarrow 3'$ mRNA strand.
    c. It extends from one loop of a tRNA molecule.
    d. It may pair with more than one codon.
    e. Its base uracil base-pairs with adenine.

20. Changes in a polypeptide following translation may involve
    a. the addition of sugars or lipids to certain amino acids.
    b. the action of enzymes to add amino acids at the beginning of the chain.
    c. the removal of poly-A from the end of the chain.
    d. the addition of a 5' cap of a modified guanosine residue.
    e. all of the above.

21. Several proteins may be produced at the same time from a single mRNA by
    a. the action of several ribosomes in a string, called a polyribosome.
    b. several RNA polymerase molecules working sequentially.
    c. signal peptides that associate ribosomes with rough ER.
    d. containing several promoter regions.
    e. the involvement of multiple spliceosomes.

22. A signal peptide
    a. is most likely to be found on cytosolic proteins produced by bacterial cells.
    b. directs an mRNA molecule into the cisternal space of the ER.
    c. is a sign to help bind the small ribosomal unit at the initiation codon.
    d. would be the first 20 or so amino acids of a protein destined for a membrane location or secretion from the cell.
    e. is part of the 5' cap.

23. A base deletion early in the coding sequence of a gene may result in
    a. a nonsense mutation.
    b. a frameshift mutation.
    c. multiple missense mutations.
    d. a nonfunctional protein.
    e. all of the above.

24. Base-pair substitutions may have little effect on the resulting protein for all of the following reasons *except* which one?
    a. The redundancy of the code may result in a silent mutation.
    b. The substitution must involve three nucleotide pairs, otherwise the reading frame is altered.
    c. The missense mutation may not occur in a critical part of the protein.
    d. The new amino acid may have similar properties to the replaced one.
    e. The wobble phenomenon could result in no change in translation.

Chapter 18

The Genetics of Viruses and Bacteria

Framework

A virus is an infectious particle consisting of a genome of single- or double-stranded DNA or RNA enclosed in a protein capsid. Viruses replicate using the metabolic machinery of their bacterial, animal, or plant host. Viral infections may destroy the host cell and cause diseases within the host organism. Viruses may have evolved from plasmids or transposons.

Most bacteria have a circular chromosome and may have additional genes carried on plasmids. In conjunction with the genetic diversity generated by mutation, genetic recombination, and transposable elements, the short generation time of bacteria facilitates their adaptation to changing environments. Individual cells may alter their metabolic response to environmental conditions through feedback inhibition of enzyme activity and the regulation of gene expression.

Chapter Review

The study of viruses and bacteria has provided information about the molecular biology of all organisms and has led to the development of new techniques of manipulating genes that have had an immense impact on basic research and biotechnology. These microbes are called *model systems* because of their use in studies with broad biological applications.

18.1 A virus has a genome but can reproduce only within a host cell

The Discovery of Viruses: **Scientific Inquiry** The search for the cause of tobacco mosaic disease led to the discovery of viruses. The infectious agent could not be removed from infected sap by passing it through a filter designed to remove bacteria. Unlike bacteria, the infectious agent could not be cultivated on nutrient media, but it was able to reproduce within plants. In 1935, W. Stanley crystallized the infectious particle, now known as tobacco mosaic virus (TMV). Since that time, many viruses have been seen with the electron microscope.

Structure of Viruses The ability to crystallize viruses indicated that they were not cells. These infectious particles consist of nucleic acid enclosed in a protein coat, sometimes surrounded by a membranous envelope.

Viral genomes may be single- or double-stranded DNA or single- or double-stranded RNA. Viral genes are contained on a single linear or circular nucleic acid molecule.

The **capsid,** or protein shell, is built from a large number of often identical protein subunits (*capsomeres*) and may be rod-shaped (helical), polyhedral, or more complex in shape. **Viral envelopes,** derived from membranes of the host cell but also including viral proteins and glycoproteins, may cloak the capsids of viruses found in animals. Some viruses also contain a few viral enzymes.

Complex capsids are found among **bacteriophages,** or **phages,** the viruses that infect bacteria. Of the seven phages that infect the bacterium *E. coli,* T2, T4, and T6 have similar capsid structures consisting of a polyhedral head and a protein tail piece with tail fibers for attaching to a bacterium.

General Features of Viral Reproductive Cycles Viruses are obligate intracellular parasites that lack metabolic enzymes and other equipment needed to reproduce. Each virus type has a limited **host range** due to proteins on the outside of the virus that recognize only specific receptor molecules on the host cell surface.

Once the viral genome enters the host cell, the cell's enzymes, nucleotides, amino acids, ribosomes, ATP, and other resources are used to replicate the viral genome and produce capsid proteins. DNA viruses use host DNA polymerases to copy their genome, whereas RNA viruses use virus-encoded polymerases for replicating their RNA genome.

After replication, viral nucleic acid and capsid proteins spontaneously assemble to form new viruses within the host cell, a process called self-assembly. Hundreds or thousands of newly formed virus particles are released, often destroying the host cell in the process.

Reproductive Cycles of Phages   A reproductive cycle of a virus that culminates in lysis of the host cell and release of newly produced phages is known as a **lytic cycle. Virulent phages** reproduce only by a lytic cycle.

The T4 phage uses its tail fibers to stick to a receptor site on the surface of an *E. coli* cell. The sheath of the tail contracts and thrusts its viral DNA into the cell, leaving the empty capsid behind. The *E. coli* cell begins to transcribe and translate phage genes, one of which codes for an enzyme that chops up host cell DNA. Nucleotides from the degraded bacterial DNA are used to create viral DNA. Capsid proteins are made

and assembled into phage tails, tail fibers, and polyhedral heads. The viral components assemble into 100 to 200 phage particles that are released after an enzyme is manufactured that digests the bacterial cell wall.

Mutations that change their receptor sites and **restriction enzymes** that chop up viral DNA once it enters the cell help to defend bacteria against viral infection.

In a **lysogenic cycle,** a virus reproduces its genome without killing its host. **Temperate phages** can reproduce by the lytic and lysogenic cycles.

When the phage lambda (λ) injects its DNA into an *E. coli* cell, it can begin a lytic cycle, or its DNA may be incorporated by genetic recombination into the host cell's chromosome and begin a lysogenic cycle as a **prophage.** Most of the genes of the inserted phage genome are repressed by a protein coded for by a prophage gene. Reproduction of the host cell replicates the phage DNA along with the bacterial DNA. The prophage may exit the bacterial chromosome, usually in response to environmental stimuli, and start the lytic cycle. Complete Interactive Question 18.1 to review the lytic and lysogenic cycles.

Expression of prophage genes may cause a change in the bacterial phenotype. Several disease-causing bacteria would be harmless except for the expression of prophage genes that code for toxins.

■ INTERACTIVE QUESTION   18.1

In this diagram of a lytic and lysogenic cycle, describe steps 1–8 and label structures a–e.

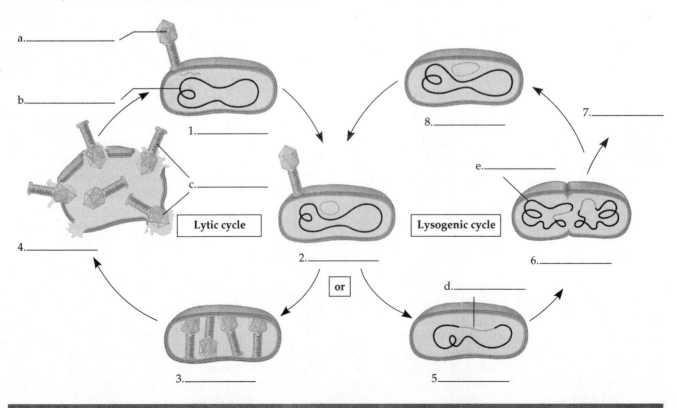

Reproductive Cycles of Animal Viruses Animal viruses are classified based on their type of genome. A viral envelope surrounds the capsid of almost all animal viruses that have RNA genomes. Glycoproteins extending from the viral membrane attach to receptor sites on a host cell plasma membrane, and the two membranes fuse, transporting the capsid into the cell. The viral genome replicates and directs the synthesis of viral proteins. Glycoproteins are produced and embedded in the ER, and then transported to the plasma membrane, where new viruses bud off within an envelope derived from the host's plasma membrane and bearing viral glycoproteins.

The envelopes of herpesviruses come from the host nuclear membrane. The herpesvirus' double-stranded DNA can remain latent as a minichromosome in the host cell nucleus until it initiates herpes infections in times of stress.

RNA viruses infect plants, some bacteria, and animals. The single-stranded RNA of animal class IV viruses can serve directly as mRNA. The RNA genome of class V viruses must first be transcribed into a strand of complementary RNA (using a viral enzyme packaged inside the capsid) that then serves as mRNA and a template for making genome RNA.

In the complicated reproductive cycle of **retroviruses** (class VI), the viral RNA genome is transcribed into double-stranded DNA by a viral enzyme, **reverse transcriptase.** This viral DNA is then integrated into a chromosome, where it is transcribed by the host cell into viral RNA, which acts both as new viral genome and as mRNA for viral proteins. **HIV (human immunodeficiency virus)** is a retrovirus that causes **AIDS (acquired immunodeficiency syndrome).** The integrated viral DNA remains as a **provirus** within the host cell DNA. New viruses, assembled with two copies of the RNA genome and reverse transcriptase within a capsid, bud off covered in host cell plasma membrane studded with viral glycoproteins.

■ **INTERACTIVE QUESTION 18.2**

Summarize the flow of genetic information during replication of a retrovirus. Indicate the enzymes that catalyze this flow.

_____ ⟶ _____ ⟶ _____

Enzymes:

Evolution of Viruses Viruses contain a genetic program, but it can only be expressed within a host cell.

The genomes of viruses are often more similar to those of their host cells than to the genomes of viruses infecting other hosts. Recent sequencing of viral genomes, however, has found genetic similarities among viruses that otherwise seem distantly related. Viruses may have evolved from fragments of cellular nucleic acids that moved from one cell to another and eventually evolved special packaging. Sources of viral genomes may have been plasmids, self-replicating circles of DNA found in bacteria and yeast, and transposons, segments of DNA that can change locations within a cell's genome.

18.2 Viruses, viroids, and prions are formidable pathogens in animals and plants

Viral Diseases in Animals The symptoms of a viral infection may be caused by toxins produced by infected cells, toxic components of the viruses themselves, cells killed or damaged by the virus, or the body's defense mechanisms fighting the infection.

Vaccines are variants or derivatives of pathogens that induce the immune system to react against the actual disease agent. Vaccinations have greatly reduced the incidence of many viral diseases.

Unlike bacteria, viruses use the host's cellular machinery to replicate, and few drugs have been found to treat or cure viral infections. Some antiviral drugs resemble nucleosides and interfere with viral nucleic acid synthesis.

Emerging Viruses The emergence of "new" viral diseases may be linked to the mutation of an existing virus (as in influenza viruses), the spread from one host species to another (as in hantavirus), or the dissemination of an existing virus to a more widespread population (as in HIV).

Viral Diseases in Plants Most plant viruses are RNA viruses. Plant viral diseases may spread through *vertical transmission* from a parent plant or through *horizontal transmission* from an external source. Plant injuries increase susceptibility to viral infections, and insects can act as carriers of viruses.

Viral particles spread easily through the plasmodesmata, the cytoplasmic connections between plant cells. Reducing the spread of disease and breeding resistant varieties are the best preventions of plant viral infections.

Viroids and Prions: The Simplest Infectious Agents **Viroids** are very small infectious molecules of circular RNA that can disrupt the metabolism of a plant cell and severely stunt plant growth.

Prions are protein infectious agents that may be linked to several degenerative brain diseases, such as mad cow disease and Creutzfeldt-Jacob disease in humans. Prions apparently spread disease by converting normal cellular proteins into the defective form of the prion.

■ **INTERACTIVE QUESTION   18.3**

Complete the following concept map that summarizes these sections on viruses.

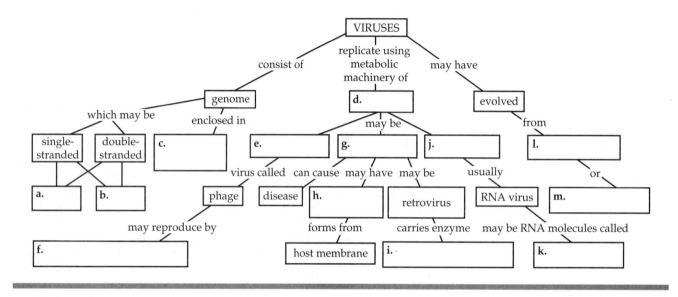

18.3   Rapid reproduction, mutation, and genetic recombination contribute to the genetic diversity of bacteria

The Bacterial Genome and Its Replication   The circular bacterial chromosome contains about 100 times more DNA than a typical virus, but only one-thousandth as much DNA as a typical eukaryotic genome. This double-stranded DNA molecule is tightly packed into a region of the cell called the **nucleoid.** Many bacteria also have plasmids, small rings of DNA with a few genes.

Replication of the bacterial chromosome proceeds bidirectionally from a single origin prior to binary fission.

Mutation and Genetic Recombination as Sources of Genetic Variation   Most bacteria in a colony are genetically identical to the parent cell. Mutations, although statistically rare, produce a great deal of genetic diversity because bacteria reproduce so rapidly. Considering all 4,300 *E. coli* genes, there may be 9 million mutations per day per human host.

Genetic recombination, the combining of genetic material from two sources, also adds to the genetic diversity of bacterial populations. Evidence that genetic recombination occurs in bacteria is provided by growing two mutant strains of *E. coli*, each of which is un-

able to produce a particular amino acid, together in a liquid medium. When later transferred to minimal media, numerous colonies grow that are now able to synthesize both amino acids.

Mechanisms of Gene Transfer and Genetic Recombination in Bacteria   The mechanisms of genetic recombination in bacteria are different from the eukaryotic mechanisms of meiosis and fertilization.

Bacteria can take up segments of naked DNA in a process called **transformation.** The foreign DNA is integrated into the bacterial chromosome by an exchange involving crossing over. Many bacteria have surface proteins specialized for uptake of DNA from closely related species. *E. coli* can be artificially induced to take up foreign DNA, a procedure important to biotechnology.

Phages can transfer genes from one bacterium to another by a process called **transduction.** In *generalized transduction*, a random piece of host DNA is accidentally packaged within a phage capsid and introduced into a new bacterium; in *specialized transduction*, bacterial genes adjacent to a prophage insertion site are excised with the prophage from the bacterial chromosome. By either method, recombination occurs when the newly introduced bacterial genetic material replaces the homologous region of a bacterial chromosome.

■ INTERACTIVE QUESTION 18.4

What type of phage and reproductive cycle would cause specialized transduction?

In **conjugation,** two cells temporarily join by appendages called sex pili and the donor cell transfers DNA to a recipient cell. The ability to form pili and donate DNA usually results from the presence of an **F factor,** a special piece of DNA that is either part of the chromosome or a plasmid.

Plasmids are small, circular, self-replicating DNA molecules. Some plasmids, called **episomes,** can reversibly incorporate into the cell's chromosome. Temperate viruses are also considered to be episomes. Plasmid genes are not required for reproduction and survival under normal conditions, but may confer advantages to bacteria in stressful environments.

Bacterial cells containing the F factor on the **F plasmid** are called F$^+$. The F plasmid replicates in synchrony with the bacterial chromosome, and F$^+$ cells pass the trait to offspring cells. During conjugation, a single strand of the plasmid DNA is transferred through a mating bridge to the recipient cell. In this *rolling-circle replication*, the single parental strands in both cells replicate, and the recipient cell is now also an F$^+$ cell.

Cells in which the F factor is inserted into the bacterial chromosome are called Hfr cells, for "high frequency of recombination." When these cells undergo conjugation, one parental strand of part of the F factor transfers an attached strand of the bacterial chromosome to the recipient cell. Movements disrupt the mating, usually resulting in a partial transfer of genes and F factor. Replication in both donor and recipient cells produces double-stranded DNA. Crossovers between the new DNA and the recipient cell's chromosome produce new genetic combinations in this recombinant F$^-$ bacterium.

R plasmids carry genes that code for antibiotic-destroying enzymes. R plasmids also have genes coding for sex pili and are transferred to nonresistant cells during conjugation, creating the medical problem of antibiotic-resistant pathogens.

Transposition of Genetic Elements *Transposable genetic elements,* or **transposable elements,** are mobile segments of DNA that may move within a chromosome and to and from plasmids. In "cut-and-paste" transposition, the transposable element simply changes location; in "copy-and-paste" transposition, the transposable element first replicates, thus remaining in its original position and also inserting in a new location. Unlike other forms of genetic recombination where alleles exchange between homologous regions, transposable elements can move genes to totally new areas.

Insertion sequences are the simplest transposable element, consisting of only a transposase gene and inverted repeats, which serve as recognition sites for transposase. Transposase is the enzyme responsible for the cutting and ligating of DNA required for transposition. Insertion sequences cause mutations when they insert within a gene or in regulatory regions.

Transposons contain other genes carried between two insertion sequences or inverted repeats. They may confer selective advantage by moving beneficial genes, such as those for antibiotic resistance, about in the bacterial genome.

■ INTERACTIVE QUESTION 18.5

Complete the following concept map that summarizes the genetic characteristics of bacteria.

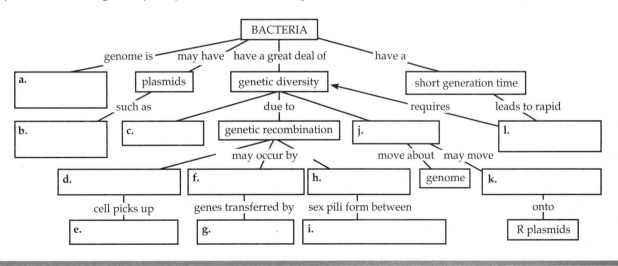

18.4   Individual bacteria respond to environmental change by regulating their gene expression

Feedback inhibition, typical of biosynthetic pathways, allows regulation of enzyme activity in response to short-term environmental fluctuations. Regulation of gene transcription controls enzyme production as metabolic needs change. F. Jacob and J. Monod first described the *operon model* for gene regulation in 1961.

Operons: The Basic Concept   Genes for the different enzymes of a single metabolic pathway may be grouped together into one transcription unit and served by a single promoter. An **operator** is a segment of DNA within the promoter region or between it and the enzyme-coding genes that controls the access of RNA polymerase to the genes. An **operon** is the DNA segment that includes the clustered genes, the promoter, and the operator.

Operons are normally "on." A **repressor** is a protein that binds to a specific operator, blocking attachment of RNA polymerase and thus turning the operon "off." **Regulatory genes** code for repressor proteins. These allosteric proteins, which may assume active or inactive shapes, are usually produced at a slow but continuous rate. The activity of the repressor protein may be determined by the presence or absence of a **corepressor.**

In the *trp* operon, tryptophan is the corepressor that binds to the repressor protein, changing its conformation into its active state, which has a high affinity for the operator and switches off the *trp* operon. Should the tryptophan concentration of the cell fall, repressor proteins are no longer bound with tryptophan and become inactive, the operator is no longer repressed, RNA polymerase attaches to the promoter, and mRNA for the enzymes needed for tryptophan synthesis is produced.

Repressible and Inducible Operons: Two Types of Negative Gene Regulation   The transcription of a *repressible operon,* such as the *trp* operon, is inhibited when a specific small molecule binds to a regulatory protein. The transcription of an *inducible operon* is stimulated when a specific small molecule interacts with a regulatory protein.

The *lac* operon, controlling lactose metabolism in *E. coli,* is an inducible operon that contains three genes. The *lac* repressor, coded for by the regulatory gene, *lacI,* is innately active, binding to the *lac* operator and switching off the operon. Allolactose, an isomer of lactose, acts as an **inducer,** a small molecule that binds to and inactivates the repressor protein, so that the operon can be transcribed.

■ **INTERACTIVE QUESTION   18.6**

In the following diagram of the *lac* operon, an operon for inducible enzymes, identify components a through i.

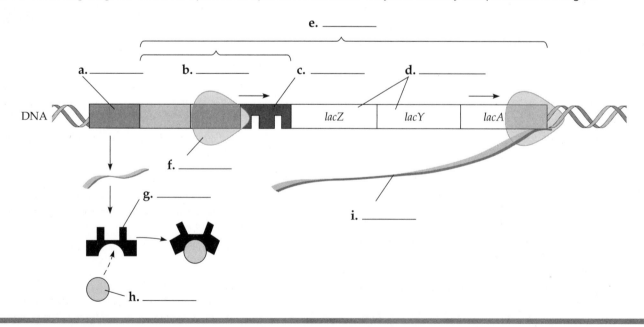

■ INTERACTIVE QUESTION 18.7

a. Repressible enzymes usually function in _____ pathways. The pathway's product serves as a _____ to activate the repressor and turn off enzyme synthesis and prevent overproduction of the product of the pathway. Genes for repressible enzymes are usually switched _____ and the repressor is synthesized in an _____ form.

b. Inducible enzymes usually function in _____ pathways. Nutrient molecules serve as _____ to stimulate production of the enzymes necessary for their breakdown. Genes for inducible enzymes are usually switched _____ and the repressor is synthesized in an _____ form.

Positive Gene Regulation   *E. coli* cells preferentially use glucose as their source of energy and carbon skeletons. Should glucose levels fall, transcription of operons for other catabolic pathways can be increased through the action of the allosteric regulatory protein *catabolite activator protein (CAP)*, which is an **activator.** Cyclic AMP (cAMP) accumulates in the cell when glucose is scarce and binds with CAP, changing it to its active shape. The active CAP attaches above the promoter region and stimulates transcription by facilitating the binding of RNA polymerase.

The regulation of the *lac* operon includes negative control by the repressor protein that is inactivated by the presence of lactose, and positive control by CAP when complexed with cAMP.

▶ Word Roots

capsa- = a box (*capsid:* the protein shell that encloses the viral genome)

conjug- = together (*conjugation:* in bacteria, the transfer of DNA between two cells that are temporarily joined)

lyto- = loosen (*lytic cycle:* a type of viral replication cycle resulting in the release of new phages by death or lysis of the host cell)

-oid = like, form (*nucleoid:* a dense region of DNA in a prokaryotic cell)

-phage = to eat (*bacteriophages:* viruses that infect bacteria)

pro- = before (*provirus:* viral DNA that inserts into a host genome)

retro- = backward (*retrovirus:* an RNA virus that reproduces by transcribing its RNA into DNA and then inserting the DNA into a cellular chromosome)

trans- = across (*transformation:* a phenomenon in which external DNA is assimilated by a cell)

virul- = poisonous (*virulent virus:* a virus that reproduces only by a lytic cycle)

▶ Structure Your Knowledge

1. Create a concept map that describes the lytic and lysogenic cycles of a phage.

2. Create a concept map to develop your understanding of the mechanisms by which bacteria regulate their gene expression in response to varying metabolic needs. Distinguish repressible and inducible operons, which are both examples of negative control, and catabolite activator protein (CAP), which illustrates positive control of gene expression.

▶ Test Your Knowledge

MULTIPLE CHOICE: *Choose the one best answer.*

1. The study of the genetics of viruses and bacteria has done all of the following *except*
   a. provide information on the molecular biology of all organisms.
   b. illuminate the sexual reproductive cycles of viruses.
   c. develop new techniques for manipulating genes.
   d. develop an understanding of the causes of diseases.
   e. show that genetic recombination occurs even in asexual bacteria.

2. Beijerinck concluded that the cause of tobacco mosaic disease was not a filterable toxin because
   a. the infectious agent could not be cultivated on nutrient media.
   b. a plant sprayed with filtered sap would develop the disease.
   c. the infectious agent could be crystallized.
   d. the infectious agent reproduced and could be passed on from a plant infected with filtered sap.
   e. the filtered sap was infectious even though microbes could not be found in it.

3. Viral genomes may be any of the following except
   a. single-stranded DNA.
   b. double-stranded RNA.
   c. misfolded infectious proteins.
   d. a linear single-stranded RNA molecule.
   e. a circular double-stranded DNA molecule.

4. Retroviruses have a gene for reverse transcriptase that
   a. uses viral RNA as a template for making complementary RNA strands.
   b. protects viral DNA from degradation by restriction enzymes.
   c. destroys the host cell DNA.
   d. translates RNA into proteins.
   e. uses viral RNA as a template for DNA synthesis.

5. Virus particles are formed from capsid proteins and nucleic acid molecules
   a. by spontaneous self-assembly.
   b. at the direction of viral enzymes.
   c. by using host cell enzymes.
   d. using energy from ATP stored in the tail piece.
   e. by both b and d.

6. A virus has a base ratio of $(A + G)/(U + C) = 1$. What type of virus is this?
   a. a single-stranded DNA virus
   b. a single-stranded RNA virus
   c. a double-stranded DNA virus
   d. a double-stranded RNA virus
   e. a retrovirus

7. Vertical transmission of a plant viral disease may involve
   a. the movement of viral particles through the plasmodesmata.
   b. the inheritance of an infection from a parent plant.
   c. a bacteriophage transmitting viral particles.
   d. insects carrying viral particles between plants.
   e. the transfer of filtered sap.

8. Bacteria defend against viral infection through the action of
   a. antibiotics that they produce.
   b. restriction nucleases that chop up foreign DNA.
   c. their R plasmids.
   d. reverse transcriptase.
   e. episomes that incorporate viral DNA into the bacterial chromosome.

9. Drugs that are effective in treating viral infections
   a. induce the body to produce antibodies.
   b. inhibit the action of viral ribosomes.
   c. interfere with the synthesis of viral nucleic acid.
   d. change the cell-recognition sites on the host cell.
   e. produce vaccines that stimulate the immune system.

10. Which of the following is true of prions?
    a. They are emerging viruses.
    b. They may spread through vertical or horizontal transmission.
    c. They probably evolved from transposons.
    d. They are infectious proteins that may convert brain proteins into misfolded forms.
    e. They may be transferred between animals by sexual contact.

11. The herpesvirus
    a. acts as a provirus when its DNA becomes incorporated into the host cell's genome.
    b. is a retrovirus that uses restriction enzymes to transcribe DNA from its RNA genome.
    c. has an envelope derived from the host cell's plasma membrane.
    d. is the retrovirus that has been linked to HIV, the virus that causes AIDS.
    e. can be used to vaccinate against hepatitis B.

12. The replication of the genome of an RNA virus uses
    a. DNA polymerase from the host.
    b. RNA replicating enzymes coded for by viral genes.
    c. reverse transcriptase to synthesize RNA.
    d. RNA polymerase from the host.
    e. restriction nucleases from the host.

13. The replication of the genome of a DNA virus uses
    a. DNA polymerase from the host.
    b. RNA replicating enzymes coded for by viral genes.
    c. reverse transcriptase to make an RNA copy from the DNA.
    d. RNA polymerase from the host.
    e. restriction nucleases from the host.

14. Which of the following would never be an episome?
    a. an F plasmid
    b. a prophage
    c. a provirus
    d. a retrovirus
    e. All of the above can be episomes.

15. Tiny molecules of naked RNA that may act as infectious agents are
    a. retroviruses.
    b. transposons.
    c. viroids.
    d. reoviruses.
    e. prions.

Choose from the following types of genetic variation in bacteria to answer questions 16–20.
    a. conjugation
    b. mutation
    c. transduction
    d. transformation
    e. transposon

16. When harmless *Streptococcus pneumoniae* are mixed with heat-killed, broken-open cells of pathogenic bacteria, live pneumonia-causing bacteria are found in the culture.

17. Transfer of genes by viruses is called _____.

18. Transfer of antibiotic-resistant genes to R plasmids may occur this way.

19. The source of most of the genetic variation found in bacterial populations is _____.

20. Two mutant *E. coli* strains, which cannot grow on minimal media, are grown together in complete media (all amino acids supplied). Samples are later transferred to minimal media and numerous colonies are able to grow.

21. Which of the following is *not* descriptive of conjugation between an Hfr and F⁻ bacterium?
    a. The Hfr cell has an F plasmid integrated into its chromosome.
    b. The Hfr cell forms sex pili and transfers one strand of its chromosome into an F⁻ cell.
    c. Random movements often break the conjugation bridge before the entire single strand of its chromosome and F episome is transferred.
    d. Hfr conjugation always transfers antibiotic resistance genes between bacteria.
    e. Crossing over between homologous genes creates a recombinant F⁻ cell.

22. A regulatory gene
    a. has its own promoter.
    b. is transcribed continuously.
    c. is not contained in the operon it controls.

    d. codes for repressor proteins.
    e. is or does all of the above.

23. Inducible enzymes
    a. are usually involved in anabolic pathways.
    b. are produced when a small molecule inactivates the repressor protein.
    c. are produced when an activator molecule enhances the attachment of RNA polymerase with the operator.
    d. are regulated by inherently inactive repressor molecules.
    e. are regulated almost entirely by feedback inhibition.

24. In *E. coli*, tryptophan switches off the *trp* operon by
    a. inactivating the repressor protein.
    b. inactivating the gene for the first enzyme in the pathway by feedback inhibition.
    c. binding to the repressor and increasing the latter's affinity for the operator.
    d. binding to the operator.
    e. binding to the promoter.

25. A mutation that renders nonfunctional the product of a regulatory gene for an inducible operon would result in
    a. continuous transcription of the genes of the operon.
    b. complete blocking of the attachment of RNA polymerase to the promoter.
    c. irreversible binding of the repressor to the operator.
    d. no difference in transcription rate when an activator protein was present.
    e. negative control of transcription.

26. An insertion sequence
    a. carries a gene only for transposase between two inverted repeats.
    b. may transpose genes for antibiotic resistance to R plasmids.
    c. involves the exchange of homologous regions of DNA.
    d. is necessary for the F plasmid to incorporate into the bacterial chromosome to form an Hfr cell.
    e. consists of an inverted repeat sandwiched between two direct repeats.

MATCHING: *Match these components of the lac operon with their functions:*

_____ **1.** β-galactosidase   **A.** is inactivated when attached to allolactose

_____ **2.** active CAP   **B.** codes for repressor protein

_____ **3.** allolactose   **C.** hydrolyzes lactose

_____ **4.** operator   **D.** stimulates gene expression

_____ **5.** promoter   **E.** repressor attaches here

_____ **6.** regulator gene   **F.** RNA polymerase attaches here

_____ **7.** repressor   **G.** acts as inducer that inactivates repressor

_____ **8.** gene in operon   **H.** usually codes for an enzyme

Chapter 19

Eukaryotic Genomes: Organization, Regulation, and Evolution

Framework

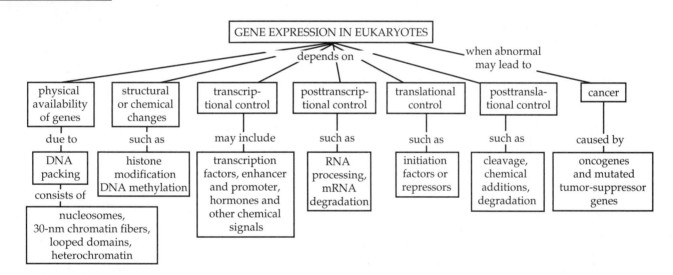

Chapter Review

The control of gene expression is more complex in eukaryotes due to the much greater size of the genome and the cell specialization found in multicellular eukaryotes. **Chromatin** is the DNA-protein complex, which is ordered into multiple levels of structural organization.

19.1 Chromatin structure is based on successive levels of DNA packing

In eukaryotes, each chromosome consists of a single, extremely long DNA double helix precisely complexed with a large amount of protein. In interphase, chromatin is extended in the nucleus, whereas discrete, condensed chromosomes appear during mitosis.

Nucleosomes, or "Beads on a String" **Histones** are small, positively charged proteins that bind tightly to the negatively charged DNA to make up chromatin. Unfolded chromatin appears as a string of beads, each bead a **nucleosome** consisting of the DNA helix wound around a protein core of four pairs of different histone molecules. The amino end (*histone tail*) of each of the eight histones extends outward. Also called the *10-nm fiber* because of its diameter, a nucleosome is the basic unit of DNA packing, remaining intact even during transcription. A fifth histone, H1, attaches to the DNA near the bead for the next level of packing.

Higher Levels of DNA Packing The *30-nm fiber* is a tightly coiled cylinder of nucleosomes organized with the aid of histone H1, the histone tails, and the linker DNA between nucleosomes. The *looped domain* is a loop of the 30-nm chromatin fiber attached to a nonhistone protein scaffold. During prophase, looped domains form a *300-nm fiber*. In a metaphase chromosome, looped domains coil and fold, further compacting the chromatin.

The looped domains of interphase chromosomes appear to be attached to discrete locations of the nuclear lamina inside the nuclear envelope, perhaps helping to organize areas of transcription. Certain regions of chromatin, called **heterochromatin,** are in a highly condensed state during interphase. The more open form of interphase chromatin, called **euchromatin,** is available for transcription.

■ INTERACTIVE QUESTION   19.1

List the multiple levels of packing in a metaphase chromosome in order of increasing complexity.

19.2   Gene expression can be regulated at any stage, but the key step is transcription

Eukaryotic cells in multicellular organisms control the expression of their genes in response to their external and internal environments and also to direct the **cell differentiation** necessary to create specialized cells.

Differential Gene Expression Differences in cells with the same genome are the result of **differential gene expression.** Only a small portion of the DNA codes for proteins; some codes for RNA, and most is noncoding DNA (although some of this is transcribed into RNAs of unknown function). Gene expression can be regulated at various stages, although, as in prokaryotes, it is most commonly regulated at transcription.

Regulation of Chromatin Structure DNA packing and the location of genes relative to nucleosomes and attachment to the protein scaffold may help control which genes are available for transcription.

The attachment of acetyl groups ($-COCH_3$) to amino acids in histone tails, called **histone acetylation,** appears to change the binding of a nucleosome's histones such that transcription proteins may more easily access genes. Some enzymes that acetylate or deacetylate histones are associated with or part of transcription factors. The addition of methyl groups ($-CH_3$) to histone tails leads to condensation of the chromatin. According to the *histone code hypothesis,* the specific patterns of modifications determine chromatin shape and influence transcription.

In a process called DNA methylation, methyl groups are added to DNA bases (usually cytosine) after DNA synthesis. Genes are generally more heavily methylated in cells in which they are not expressed. Some proteins that bind to methylated DNA also interact with histone deacetylation enzymes, reinforcing the transcription repression.

DNA methylation may reinforce gene regulatory decisions of early development as enzymes act during DNA replication to methylate daughter DNA strands and pass on methylation records. Such methylation patterns account for **genomic imprinting** in mammals in which either a maternal or paternal allele of certain genes is permanently regulated.

Epigenic inheritance is the transmission of traits by mechanisms that do not involve changes in DNA sequences, such as chromatin modifications that affect gene expression in future generations of cells.

■ INTERACTIVE QUESTION   19.2

a. Give an example of highly methylated and inactive DNA common in mammalian cells.

b. Would histone tail deacetylation increase or decrease the transcription of a gene located in that nucleosome?

Regulation of Transcription Initiation A typical gene consists of a promoter sequence, where a transcription initiation complex that includes RNA polymerase II attaches, and a sequence of introns interspersed among the coding exons. After transcription, RNA processing of the pre-mRNA removes the introns and adds a 5′ cap and a poly-A tail at the 3′ end. **Control elements** are noncoding sequences to which transcription factors bind that help regulate transcription.

General **transcription factors** bind to RNA polymerase and each other to initiate the transcription of all protein-coding genes. The binding of *specific* transcription factors to control elements, however, greatly increases transcription rates of particular genes.

Proximal control elements are located close to the promoter. **Enhancers** are groups of *distal control elements* located far upstream or downstream of a gene or in an intron. According to a current model, a protein-mediated bend in the DNA brings **activators** bound to enhancers into contact with *mediator proteins*, which interact with proteins at the promoter. These protein-protein interactions assemble the initiation complex. Hundreds of transcription activators have been identified. Most have a DNA-binding domain and one or more activation domains that bind other regulatory proteins.

Some specific transcription factors act as **repressors**, which can inhibit gene expression in several ways, such as blocking the binding of activators or binding to specific control elements in an enhancer. Some activators and repressors influence chromatin structure by recruiting proteins that either acetylate or deacetylate histones. Most eukaryotic gene repression may occur by this *silencing* at the level of chromatin modification.

Only a dozen or so short nucleotide sequences have been found in control elements, but each enhancer contains about ten different control elements. The combination of these elements ensures that specific activator proteins must be present to activate transcription. Thus, the specific activators and repressors made in a cell determine which genes are expressed.

Prokaryotic genes are often organized into operons that are controlled by the same promoter and control elements and transcribed together. In a few cases, coexpressed eukaryotic genes (each with its own promoter) are clustered together and their regulation involves changes in chromatin structure making them available for transcription. Most eukaryotic genes coding for enzymes in the same metabolic pathway are scattered on the chromosomes. The integrated control of these genes may involve a specific collection of control elements associated with each related gene and requiring the same activators.

For example, steroid hormones may activate multiple genes when they bind to specific receptor proteins within a cell that function as transcription activators for several genes. Other signal molecules bind to cell surface receptors and initiate signal transduction pathways that activate particular transcription activators or repressors. The same chemical signal will activate genes with the same control elements and thus coordinate gene expression.

■ **INTERACTIVE QUESTION 19.3**

Label the components of this diagram of how enhancers, mediator proteins, and transcription factors facilitate formation of a transcription initiation complex.

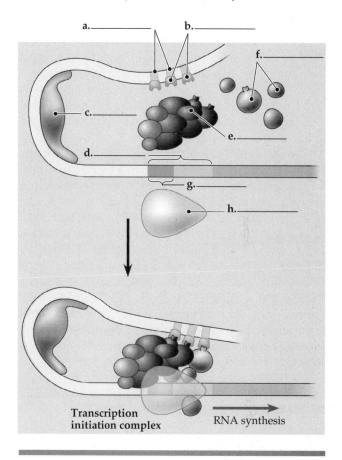

Mechanisms of Post–transcriptional Regulation Gene expression, measured by the amount of functional protein that is made, can be regulated at any posttranscriptional step. At the level of RNA processing, **alternative RNA splicing** may produce different mRNA molecules when regulatory proteins control which segments of the primary transcript are chosen as introns and which as exons.

In contrast to prokaryotic mRNA that is degraded after a few minutes, eukaryotic mRNA can last hours or even weeks. The hydrolysis of mRNA by nucleases is usually preceded by the enzymatic shortening of the poly-A tail and removal of the 5′ cap. The untranslated region (UTR) at the 3′ end of mRNA may contain nucleotide sequences that affect stability.

MicroRNAs (miRNAs) are small, single-stranded RNA molecules that associate with a complex of proteins and then base-pair with complementary sequences on target mRNA. This miRNA-protein complex either blocks translation or degrades the mRNA. Long RNA molecules fold on themselves, and an enzyme called Dicer cuts this double-stranded RNA into short fragments. One strand of the fragment becomes an miRNA. This inhibition of gene expression by RNA molecules was first observed experimentally and called **RNA interference (RNAi).** Injection of double-stranded RNA into a cell produced **small interfering RNAs (siRNAs)** that appear to fuction as miRNAs to inhibit expression of genes with the same sequence.

The translation of eukaryotic mRNA can be delayed by the binding of regulatory proteins to the 5′ untranslated region (5′ UTR) that prevent ribosome binding. A great deal of mRNA is synthesized and stored in egg cells; translation may begin when enzymes add more A residues to poly-A tails or when translation initiation factors are activated following fertilization.

Following translation, polypeptides are often cleaved or chemical groups added to yield an active protein. Some proteins must be transported to target locations. Selective degradation of proteins may serve as a control mechanism in the cell. Molecules of the protein ubiquitin are added to mark proteins for destruction. **Proteasomes** recognize and degrade the marked proteins.

■ **INTERACTIVE QUESTION   19.4**

a. The untranslated regions (UTR) at both the 5′ and 3′ ends of an mRNA may contribute to regulation of gene expression. Describe their different effects.

b. Does the action of microRNAs increase or decrease gene expression? Explain.

19.3   Cancer results from genetic changes that affect cell cycle control

Types of Genes Associated with Cancer   Carcinogens, x-rays, or viruses most often cause the changes in the genes that regulate cell growth and division that lead to cancer. Viruses that can cause cancer in animals are called *tumor viruses.* Cells become transformed into cancer cells when viral nucleic acid becomes integrated into host DNA. Viruses have been linked to certain types of human cancer.

Oncogenes, or cancer-causing genes, were first found in certain retroviruses. Similar genes were later recognized in the genomes of humans and other animals. Cellular **proto-oncogenes,** which code for proteins that stimulate cell growth and division, may become oncogenes by several mechanisms. Mutations may result in more copies of the gene being present than normal (amplification), transposition or chromosomal translocation (both of which may bring the gene under the control of a more active promoter or control element), or a change in a nucleotide sequence in either a promoter or enhancer that increases gene expression or in the gene that creates a more active or resilient protein.

Mutations in **tumor-suppressor genes** can contribute to the onset of cancer when they result in a decrease in the activity of proteins that prevent uncontrolled cell growth. Tumor-suppressor proteins have various functions, such as repair of damaged DNA, control of cell adhesion, and inhibition of the cell cycle.

Interference with Normal Cell–Signaling Pathways
In about 30% of human cancers, the *ras* proto-oncogene is mutated. The *ras* **gene** codes for a G protein that connects a growth-factor receptor on the plasma membrane to a cascade of protein kinases that leads to the production of a cell cycle stimulating protein. A mutation may create a hyperactive version of the Ras protein that relays a signal without binding of a growth factor.

The *p53* **gene** is mutated in about 50% of human cancers. It codes for a tumor-suppressor protein that is involved in the synthesis of several growth-inhibiting proteins. Such proteins may activate the *p21* gene, whose product binds to cyclin-dependent kinases, halting the cell cycle and allowing time for the cell to repair damaged DNA. The p53 transcription factor can also activate genes involved in DNA repair. Should DNA damage be irreparable, p53 activates "suicide genes" that initiate apoptosis.

■ INTERACTIVE QUESTION 19.5

These diagrams represent signaling pathways that stimulate (top diagram) or inhibit (bottom) the cell cycle. Describe the numbered steps and then explain the effect of mutations that make a hyperactive Ras protein (top) or a defective p53 transcription factor (bottom).

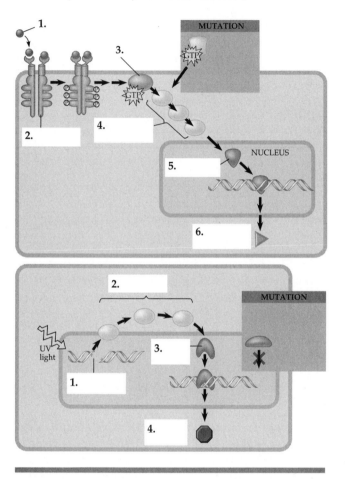

The Multistep Model of Cancer Development More than one somatic mutation appears to be needed to produce a cancerous cell. Mutation of a single proto-oncogene can stimulate cell division, but both alleles for several tumor-suppressor genes must be defective to allow uncontrolled cell growth. In many malignant tumors, the gene for telomerase becomes activated and cells become able to divide indefinitely.

Virus-associated cancers are thought to account for about 15% of human cancer cases. The virus might add an oncogene or its insertion may affect a proto-oncogene or tumor-suppressor gene.

Inherited Predisposition to Cancer A genetic predisposition to certain cancers may involve the inheritance of an oncogene or a recessive mutant allele for a tumor-suppressor gene. Approximately 15% of colorectal cancers involve inherited mutations, often in the tumor-suppressor gene *APC*, which regulates cell migration and adhesion.

The 5–10% of breast cancer cases linked to family history have been linked to mutant alleles for either *BRCA1* or *BRCA2*, both of which appear to be tumor-suppressor genes.

19.4 Eukaryotic genomes can have many noncoding sequences in addition to genes

The Relationship Between Genomic Composition and Organismal Complexity Unlike the prokaryote genome, in which most of the DNA contains uninterrupted codes for proteins (or RNA), the eukaryote genome contains mostly noncoding DNA, including regulatory sequences, introns that interrupt coding sequences, nonfunctional genes, and sequences present in many copies, called **repetitive DNA.** More than half of this repetitive DNA (44% of the human genome) is either made up of or related to transposable elements.

Transposable Elements and Related Sequences **Transposons** move about a genome as a DNA intermediate. **Retrotransposons,** which make up the majority of transposable elements, always move by the "copy and paste" mechanism because they are first transcribed into an RNA intermediate. This RNA transcript is converted back to DNA by reverse transcriptase, coded by the retrotransposon itself. Transposable elements are present as multiple (although not identical) copies of transposons or as related sequences that have lost the ability to move. In primates, perhaps 10% of the genome is made up of *Alu elements,* 300-nucleotide-long sequences, many of which are transcribed into RNA but are of unknown function.

Other Repetitive DNA, Including Simple Sequence DNA About 5% of the human genome is repetitive DNA not related to transposable elements, which consists of large-segment duplications. *Simple-sequence DNA*, on the other hand, makes up 3% and consists of multiple copies of tandemly repeated sequences, often of fewer than 15 nucleotides. Much of this so-called *satellite DNA* is located at centromeres and telomeres.

Genes and Multigene Families   Sequences coding for proteins and structural RNAs make up 1.5% of the human genome. Including introns and regulatory sequences, the amount of gene-related DNA is 25% of the genome. About half of the coding DNA is for genes present in single copies of unique sequences. **Multigene families** are collections of similar or identical genes. With the exception of the genes for histone proteins, identical multigene families code for RNA products. The genes coding for the three largest rRNA molecules are arranged in a single transcription unit repeated in huge tandem arrays, enabling cells to produce the millions of ribosomes needed for protein synthesis.

Examples of multigene families of nonidentical genes are the two families of genes that code for the α and β polypeptide chains of hemoglobin. Different versions of each globin subunit are clustered together on two different chromosomes and are expressed at the appropriate time during development. **Pseudogenes,** nonfunctional DNA sequences similar to real genes, are also found in the globin gene family clusters.

■ **INTERACTIVE QUESTION   19.6**

Match the letter of the description and fill in the percentage (listed below) for each type of DNA sequence found in the human genome.

Types of DNA	Description	%
1. Exons & RNA-coding	_____	_____
2. Introns & regulatory	_____	_____
3. Transposable elements and related repetitive sequences	_____	_____
4. Simple sequence repeats	_____	_____
5. Large-segment duplications	_____	_____
6. Unique noncoding DNA	_____	_____

Descriptions

A. satellite DNA in centromeres and telomeres

B. multiple copies of moveable sequences

C. gene fragments, nonfunctional genes

D. protein and RNA-coding sequences

E. multiple copies of large sequences

F. DNA related to gene expression

Percentages to choose from: 1.5, 3, 5, 15, 24, and 44

19.5   Duplications, rearrangements, and mutations of DNA contribute to genome evolution

Duplication of Chromosome Sets   Extra sets of chromosomes may arise by accidents in meiosis. The resulting extra genes might diverge through mutation, leading to genes with novel functions.

Duplication and Divergence of DNA Segments   Errors such as unequal crossing over during meiosis and slippage of template strands during DNA replication might lead to the duplication of individual genes.

The α-globin and β-globin gene families appear to have evolved from a common ancestral globin, which was duplicated and diverged about 500 million years ago. Multiple duplications and mutations within each family have led to the current family of genes with related functions.

In other cases, mutations of a duplicated gene may lead to a protein product with a new function.

■ **INTERACTIVE QUESTION   19.7**

Lysozyme and α-lactalbumin have similar sequences but different functions. Both genes are found in mammals, but birds have only the gene for lysozyme. What does this observation suggest about the evolution of these genes?

Rearrangements of Parts of Genes: Exon Duplication and Exon Shuffling   Unequal crossing over can lead to a gene with a duplicated exon. Exons often code for domains of a protein. Their duplication could provide a protein with enhanced structure and function. Errors in meiotic recombination could also lead to *exon shuffling* within a gene or between nonallelic genes. For instance, the gene for TPA has three different exons that may have originated in the genes for epidermal growth factor, fibronectin, and plasminogen.

How Transposable Elements Contribute to Genome Evolution   Transposable elements may contribute to the evolution of a genome by promoting recombination, disrupting cellular genes or control elements, or moving genes or exons to new locations. Recombination events can take place between homologous transposable element sequences that are scattered throughout the genome, causing chromosomal mutations that may occasionally be beneficial to the organism.

Transposable elements that insert within a gene may disrupt its functioning; those that insert within regulatory sequences may increase or decrease gene expression. A transposable element can also move a copy of a gene or an exon to a new location. The transposition of an *Alu* element into introns may provide an alternative splice site in the RNA transcript, producing a new portion of a protein that may result in alternative functions. The increased genetic diversity provided by these mechanisms provides raw material for natural selection.

Word Roots

eu- = true (*euchromatin:* the more open, unraveled form of eukaryotic chromatin)

hetero- = different (*heterochromatin:* nontranscribed eukaryotic chromatin that is highly compacted and is visible with a light microscope during interphase)

nucleo- = the nucleus; **-soma** = body (*nucleosome:* the basic beadlike unit of DNA packaging in eukaryotes)

proto- = first, original; **onco-** = tumor (*proto-oncogene:* a normal cellular gene corresponding to an oncogene)

pseudo- = false (*pseudogenes:* DNA segments very similar to real genes but which do not yield functional products)

retro- = backward (*retrotransposons:* transposable elements that move in a genome as an RNA intermediate, a transcript of the retrotransposon DNA)

Structure Your Knowledge

1. Fill in the table below to help you organize the major mechanisms that can regulate the expression of eukaryotic genes.

Level of Control	Examples
Chromatin structure	a.
Transcriptional regulation	b.
Post-transcriptional regulation	c.
Translational regulation	d.
Post-translational regulation	e.

2. **a.** What are proto-oncogenes? How do they become oncogenes?
   b. What is the role of tumor-suppressor genes in the development of cancer?

3. **a.** Describe a retrotransposon.
   b. How do transposable elements contribute to genome evolution?

Test Your Knowledge

MULTIPLE CHOICE: *Choose the one best answer.*

1. The control of gene expression is more complex in eukaryotic cells because
   a. DNA is associated with protein.
   b. gene expression differentiates specialized cells.
   c. the chromosomes are linear and more numerous.
   d. operons are controlled by more than one promoter region.
   e. inhibitory or activating molecules may help regulate transcription.

2. Histones are
   a. small, positively charged proteins that bind tightly to DNA.
   b. small bodies in the nucleus involved in rRNA synthesis.
   c. basic units of DNA packing consisting of DNA wound around a protein core.
   d. DNA bending proteins that facilitate formation of transcription initiation complexes.
   e. proteins responsible for producing repeating sequences at telomeres.

3. Heterochromatin
   a. has a higher degree of packing than does euchromatin.
   b. is visible with a light microscope during interphase.
   c. is not actively involved in transcription.
   d. makes up metaphase chromosomes.
   e. is all of the above.

4. DNA methylation of cytosine residues
   a. can be induced by drugs that reactivate genes.
   b. may be a mechanism of exogenic inheritance when methylation patterns are repeated in daughter cells.
   c. occurs in the promoter region and enhances binding of RNA polymerase.
   d. makes satellite DNA a different density so it can be separated by ultracentrifugation.
   e. may be related to the transformation of proto-oncogenes to oncogenes.

5. Which of the following appears to be attached to the nuclear lamina in a precise and organized fashion?
   a. nucleosomes
   b. heterochromatin
   c. 30-nm fiber
   d. looped domains of interphase chromosomes
   e. enhancer regions of actively transcribed genes

6. Which of the following is *not* true of enhancers?
   a. They may be located thousands of nucleotides upstream from the genes they affect.
   b. When bound with activators, they interact with the promoter region and other transcription factors to increase the activity of a gene.
   c. They may complex with steroid-activated receptor proteins and thus selectively activate specific genes.
   d. They may coordinate the transcription of enzymes involved in the same metabolic pathway.
   e. They are located within the promoter, and when complexed with a steroid or other small molecule, they release an inhibitory protein and thus make DNA more accessible to RNA polymerase.

7. Which of the following is *not* an example of the control of gene expression that occurs after transcription?
   a. mRNA stored in the cytoplasm needing activation of translation initiation factors
   b. the length of time mRNA lasts before it is degraded
   c. rRNA genes amplified in tandem arrays
   d. alternative RNA splicing before mRNA exits from the nucleus
   e. splicing or modification of a polypeptide

8. Pseudogenes are
   a. tandem arrays of rRNA genes that enable actively synthesizing cells to create enough ribosomes.
   b. genes that can become oncogenes when mutated by carcinogens.
   c. genes of multigene families that are expressed at different times during development.
   d. sequences of DNA that are similar to real genes but lack regulatory sequences necessary for gene expression.
   e. both c and d.

9. Which of the following might a proto-oncogene code for?
   a. DNA polymerase
   b. reverse transcriptase
   c. receptor proteins for growth factors
   d. an enhancer
   e. transcription factors that inhibit cell division genes

10. A gene can develop into an oncogene when it
    a. is present in more copies than normal.
    b. undergoes a translocation that removes it from its normal control region.
    c. develops a mutation that creates a more active or resistant protein.
    d. is transposed to a new location where its expression is enhanced.
    e. does any of the above.

11. What is the main reason that prokaryotic genes average 1,000 nucleotide base pairs, whereas human genes average about 27,000 base pairs?
    a. Prokaryotes have smaller, but many more, individual genes.
    b. Prokaryotes are more ancient organisms; longer genes arose later in evolution.
    c. Prokaryotes are much simpler organisms; humans have many types of differentiated cells.
    d. Prokaryotic genes do not have introns; human genes have many.
    e. Human proteins are much larger and more complex than prokaryotic proteins.

12. A tumor-suppressor gene could cause the onset of cancer if
    a. both alleles have mutations that decrease the activity of the gene product.
    b. only one allele has a mutation that alters the gene product.
    c. it is inherited in mutated form from a parent.
    d. a proto-oncogene has also become an oncogene.
    e. both a and d have happened.

13. What is apoptosis?
    a. a cell suicide program that may be initiated by p53 protein in response to DNA damage
    b. metastasis, or the spread of cancer cells to a new location in the body
    c. the transformation of a normal cell to a cancer cell
    d. the mutation of a G protein into a hyperactive form
    e. the transformation of a proto-oncogene to an oncogene by a point mutation

14. Which of the following would you expect to find as part of a receptor protein that binds with a steroid hormone?
    a. a TATA box located within the promoter region
    b. a domain that binds to DNA and protein-binding domains
    c. an activated operator region that allows attachment of RNA polymerase
    d. an enhancer sequence located at some distance upstream or downstream from the promoter
    e. transmembrane domains that facilitate its localization in a plasma membrane

15. A eukaryotic gene typically has all of the following associated with it *except*
    a. a promoter.
    b. an operator.
    c. enhancers.
    d. introns and exons.
    e. control elements.

16. What are proteasomes?
    a. complexes of proteins that excise introns
    b. single-stranded RNA molecules complexed with proteins that block translation of or degrade mRNA
    c. small, positively charged proteins that form the core of nucleosomes
    d. enormous protein complexes that degrade unneeded proteins in the cell
    e. complexes of transcription factors whose protein–protein interactions are required for enhancing gene transcription

17. Tissue plasminogen activator (TPA) is a protein with three types of domains. One of each of these types is found in the protein's epidermal growth factor, fibronectin, and plasminogen. What is a likely explanation for this?
    a. All four genes are members of a multigene family involved in cell signaling.
    b. TPA was the first gene to evolve; the other three genes each lost two domains.
    c. The gene for TPA arose by several instances of exon shuffling from the other three genes.
    d. TPA must have many *Alu* elements that allow for alternative splice sites to incorporate these exons.
    e. Several duplication events led to the evolution of the TPA gene.

18. Which of the following would most likely account for a family history of colorectal cancer?
    a. inheritance of a proto-oncogene
    b. a family diet that is low in fats and high in fiber
    c. a family history of breast cancer
    d. inheritance of the *ras* oncogene that locks the G protein in an active configuration
    e. inheritance of one mutated *APC* allele that regulates cell adhesion and migration

19. Which of the following best describes what pseudogenes and introns have in common?
    a. They are RNA molecules that are not translated into proteins.
    b. They are DNA segments that lack a promoter but do have other control regions.
    c. They are not expressed—they do not produce a functional product.
    d. They code for RNA products, not proteins.
    e. They appear to have arisen from retrotransposons.

20. How is the coordinated transcription of genes involved in the same pathway regulated?
    a. The genes are transcribed in one transcription unit, although each gene has its own promoter.
    b. The genes are located in the same region of the chromosome, and enzymes deacetylate the entire region so that transcription may begin.
    c. The genes all respond to the same general transcription factors, although they may respond to different specific transcription factors.
    d. A steroid hormone selectively binds to the promoters for all the genes.
    e. The genes have the same combination of control elements in the enhancer that bind with the particular activators present in the cell.

Chapter 20

DNA Technology and Genomics

Key Concepts

20.1 DNA cloning permits production of multiple copies of a specific gene or other DNA segment

20.2 Restriction fragment analysis detects DNA differences that affect restriction sites

20.3 Entire genomes can be mapped at the DNA level

20.4 Genome sequences provide clues to important biological questions

20.5 The practical applications of DNA technology affect our lives in many ways

Framework

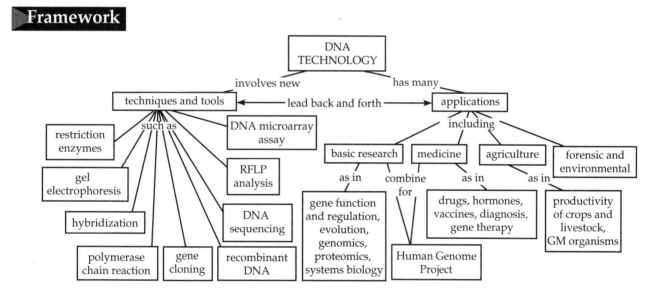

Chapter Review

DNA technology began with techniques for making **recombinant DNA,** which combines nucleotide sequences from different organisms or species into the same DNA molecule. **Genetic engineering,** the direct manipulation of genetic material for practical purposes, has begun an industrial revolution in **biotechnology.** This use of living organisms or their components to manufacture desirable products dates back centuries, but advances in DNA technology have resulted in hundreds of new products. DNA technology has led to major advances in all fields of biology and in our knowledge of the humane genome.

20.1 DNA cloning permits production of multiple copies of a specific gene or other DNA segment

Techniques for **gene cloning** are used to prepare multiple identical copies of pieces of DNA.

DNA Cloning and Its Applications: A Preview One of the approaches used to clone pieces of DNA makes use of the plasmids of bacterial cells. Recombinant DNA may be made by inserting foreign DNA into plasmids. These plasmids are put back into bacterial cells where they will replicate as the *recombinant bacteria* reproduce to form **clones** of identical cells. Such cloned

157

DNA provides multiple copies of the gene and may also be used to produce protein coded for by the foreign DNA.

Using Restriction Enzymes to Make Recombinant DNA   **Restriction enzymes** protect bacteria from the DNA of phages or other bacteria by cutting up foreign DNA in a process called *restriction*. Most restriction enzymes recognize short nucleotide sequences, called **restriction sites,** and cut at specific points within them. The cell protects its own DNA from restriction by methylating nucleotide bases within its own restriction sites.

A restriction site is usually symmetrical, the same sequence of four to eight nucleotides running in opposite directions on the two strands. The most useful restriction enzymes cut phosphodiester bonds in a staggered way, leaving **sticky ends** of short single-stranded sequences on both sides of the resulting **restriction fragment.**

DNA from different sources can be combined in the laboratory when the DNA is cut by the same restriction enzyme and the complementary bases on the sticky ends of the restriction fragments form hydrogen-bonded base pairs. **DNA ligase** is used to seal the strands together.

■ INTERACTIVE QUESTION   20.1

Which of these DNA sequences would most likely function as a restriction site for a restriction enzyme? Why?

··CAGCAG··   ··GTGCTG··   ··GAATTC··

··GTCGTC··   ··CACGAC··   ··CTTAAG··

Cloning a Eukaryotic Gene in a Bacterial Plasmid   **Cloning vectors** are DNA molecules that can move foreign DNA into a cell and replicate there. Recombi-

nant plasmids returned to bacterial cells will replicate the foreign DNA as the bacteria reproduce. The ease with which plasmids can be isolated from and returned to bacterial cells makes bacteria the most common host for gene cloning.

The plasmid method of gene cloning involves treating plasmids containing an antibiotic-resistant gene with a restriction enzyme that cuts the DNA ring at a single restriction site and disrupts a gene whose activity is easily determined, such as *lacZ*, the gene for β-galactosidase. The clipped plasmids are mixed with DNA containing the gene of interest, which has also been treated with the same restriction enzyme, producing many different fragments. The sticky ends form hydrogen bonds with each other, and DNA ligase seals the recombinant molecules.

The plasmids are introduced by transformation into bacterial cells that have a mutation in their *lacZ* gene, and thus are unable to produce β-galactosidase. Bacteria are plated onto a medium containing ampicillin and X-gal, a compound that is cleaved by β-galactosidase and yields a blue product. Colonies that are able to grow on the medium (because they contain a plasmid with the amp^R gene) and are not blue (because the foreign DNA inserted in the middle of the β-galactosidase gene) are carrying a recombinant plasmid.

The colonies are tested to find the ones that contain the gene of interest. One technique detects the gene itself by **nucleic acid hybridization,** using a **nucleic acid probe** that has complementary sequences to segments of the gene. The probe hybridizes with the DNA of the gene after **denaturation** of the cell's DNA by heat or chemicals produces single-stranded DNA. The probe is located by its radioactively labeled molecules or fluorescent tag. Once the desired clone is identified, it can be grown in liquid culture and the gene of interest isolated in large quantities. (Complete Interactive Question 20.2 to review plasmid cloning.)

■ INTERACTIVE QUESTION 20.2

This schematic diagram shows the steps in plasmid cloning of a gene. Identify components a–j. Briefly describe the five steps of the process. How are bacterial clones that have picked up the recombinant plasmid identified?

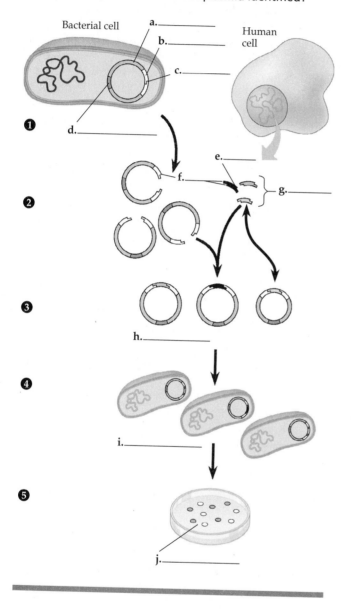

A partial genomic library can be produced using the mRNA molecules isolated from a cell. Using reverse transcriptase from retroviruses, mRNA is used as a template to produce **complementary DNA (cDNA).** Restriction sites are added to the ends and the cDNA is inserted into vector DNA. Such a **cDNA library** contains only the genes that are expressed (transcribed) within the cell.

■ INTERACTIVE QUESTION 20.3

Compare genomic and cDNA libraries with regard to their advantages and disadvantages.

Cloning and Expressing Eukaryotic Genes Clones carrying a particular gene can also be identified based on the presence of its protein product, either by detecting enzyme activity or by antibody binding.

Differences between prokaryotic and eukaryotic mechanisms for gene expression can be overcome by using an **expression vector,** a cloning vector that has an active prokaryotic promoter just upstream from the eukaryotic gene insertion site. Using a cDNA gene removes the problem of long introns, especially since bacteria do not have RNA-splicing machinery.

Yeast cells are eukaryotic hosts used to clone eukaryotic genes; they are easy to grow and have plasmids that serve as vectors. **Yeast artificial chromosomes (YACs)** have also been constructed that carry foreign DNA and contain an origin for DNA replication, a centromere, and two telomeres. These vectors behave normally in mitosis and are able to clone large pieces of DNA.

Plant and animal cells in culture can serve as hosts and may be necessary when a protein must be modified following translation.

More efficient means for introducing DNA into eukaryotic cells include **electroporation,** in which an electric pulse briefly opens holes in the plasma membrane through which DNA can enter, injection into cells using microscopically thin needles, or using the bacterium *Agrobacterium* to introduce DNA into plant cells.

Storing Cloned Genes in DNA Libraries An organism's DNA can be cut into thousands of pieces with restriction enzymes and inserted into plasmids. The collection of the thousands of clones of bacteria containing recombinant plasmids derived from this "shotgun" approach is called a **genomic library.**

Bacteriophages are also used as vectors for creating genomic libraries. DNA fragments are spliced into phage DNA, which is packaged into capsids and used to infect bacteria. The production of new phage clones the foreign DNA.

Amplifying DNA in Vitro: The Polymerase Chain Reaction (PCR) The **polymerase chain reaction (PCR)** can produce billions of copies of a section of DNA *in vitro* in only a few hours. DNA containing the region of interest is incubated with the four nucleotides, a special heat-resistant type of DNA polymerase, and specially synthesized primers that bind upstream from the target sequence. The solution is heated to separate the DNA strands then cooled so the primers can anneal (hydrogen-bond) to complementary sequences. DNA polymerase adds nucleotides to the 3′ ends of the primers. The solution is heated again and the process repeated. The desired DNA segment does not need to be purified from the starting material, and very small samples can be used.

■ **INTERACTIVE QUESTION   20.4**

a. Why is PCR often used prior to cloning a gene in cells?

b. Why even bother cloning genes in cells, since PCR produces so many copies so fast?

20.2   Restriction fragment analysis detects DNA differences that affect restriction sites

Producing cloned segments of DNA allows scientists to study the expression of particular genes and to make comparisons between the genes of different cells, individuals, and species.

Gel Electrophoresis and Southern Blotting   Many of the methods for analyzing and comparing DNA make use of **gel electrophoresis,** a technique that separates nucleic acids and proteins on the basis of their size and electrical charge. Due to the negative charge of their phosphate groups, DNA molecules migrate through the electric field produced in a thin slab of gel toward the positive electrode. Linear molecules of DNA move at a rate inversely proportional to their length, producing band patterns in the gel of fragments of decreasing size.

Cutting a long DNA molecule with a particular restriction enzyme and separating the resulting restriction fragments by gel electrophoresis produces a characteristic pattern of bands, a process called restriction fragment analysis. Pure samples of such bands can be recovered from the gel and retain their biological activity.

Two DNA samples, such as alleles of a gene, will produce different patterns of bands when differences in their DNA sequences add or delete restriction sites. The addition of nucleic acid hybridization with a probe allows two or more unpurified samples of DNA (such as the entire genome) to be compared for the presence and band location of a particular DNA sequence. In a technique known as **Southern blotting,** DNA is treated with a restriction enzyme; the fragments are separated on a gel and then transferred to nitrocellulose paper by blotting; labeled probes of single-stranded DNA are added and

hybridize with complementary DNA sequences; and the restriction fragment bands of interest are identified by autoradiography.

Restriction Fragment Length Differences as Genetic Markers   Samples of noncoding DNA treated with restriction enzymes also produce different band patterns due to differences in nucleotide sequences in restriction sites. Such **restriction fragment length polymorphisms (RFLPs)** can serve as genetic markers on a chromosome and can be used in making linkage maps. A particular RFLP marker often occurs in many variants in a population. RFLPs are detected by Southern blotting, and the entire genome can be used as the DNA starting material.

■ **INTERACTIVE QUESTION   20.5**

A bloody crime has occurred. Police have collected blood samples from the victim, two suspects, and blood found at the scene. Briefly list the steps the lab went through to produce the following autoradiograph.

a.

b.

c.

d.

e.

Which suspect would you charge with the crime?

20.3   Entire genomes can be mapped at the DNA level

The international effort to map the human genome, called the **Human Genome Project,** was largely completed in 2003. It included three stages: genetic (linkage) mapping, physical mapping, and DNA sequencing. The project also included the mapping of important research organisms, such as *E. coli,* yeast, the nematode *C. elegans, Drosophila,* and *Mus musculus* (mouse).

Genetic (Linkage) Mapping: Relative Ordering of Markers   Cytogenetic maps based on banding patterns and fluorescence *in situ* hybridization (FISH) provided the basis for mapping chromosomes. The first stage in mapping a large genome is to develop a **linkage map** of several thousand genetic markers, either genes or other identifiable sequences such as RFLPs or simple sequences repeats. Determining recombination frequencies with these known markers enabled researchers to order genes and locate other markers.

Physical Mapping: Ordering DNA Fragments   In a **physical map,** the actual distances between markers are determined. The DNA of each chromosome is cut into restriction fragments, and their order on the chromosome is determined. First large fragments are cut, cloned, and ordered. Then those fragments are cut and their fragments ordered. Finally, fragments are cut, cloned, and ordered that are small enough to be sequenced.

Cloning vectors include yeast artificial chromosomes, which can carry long fragments, and **bacterial artificial chromosomes (BACs),** artificial bacterial chromosomes that carry shorter inserts.

DNA Sequencing   Pure preparations of DNA fragments of about 800 base pairs are ultimately used to determine the nucleotide sequence of an entire genome. In the *dideoxyribonucleotide chain-termination method* of DNA sequencing, developed by F. Sanger, a sample of a denatured DNA fragment is incubated with a primer, DNA polymerase, the four deoxyribonucleotides, and the four modified nucleotides (dideoxyribonucleotides) that randomly block further synthesis when they are incorporated into a growing DNA strand. The sets of fluorescently tagged strands of varying lengths are separated through a polyacrylamide gel in a capillary tube. A fluorescence detector reads the sequence of their colors, which corresponds to the nucleotide sequence of the fragment.

Faster sequencing machines and better computer software were developed during the Human Genome Project. J. C. Venter, founder of the company Celera Genomics, developed a whole-genome shotgun approach that relies on powerful computer programs to order

the large number of sequenced short sequences cut from a chromosome. In February 2001, Celera (as well as the public consortium) announced that more than 90% of the human genome had been sequenced. Although a few gaps remain, sequencing of the human genome is now basically complete.

■ INTERACTIVE QUESTION   20.6

Compare the strategies of the public consortium of the Human Genome Project and of Celera Genomics for mapping a genome.

20.4   Genome sequences provide clues to important biological questions

Genomics, the study of whole sets of genes and their interactions, is allowing researchers to address questions about gene expression, growth and development, and evolution.

Identifying Protein-Coding Genes in DNA Sequences   Computer data banks of DNA sequences are available to researchers on the Internet. Computer software can scan these sequences for signs of genes, such as start and stop signals; RNA-splicing sites; and *expressed sequence tags,* or *ESTs* (sequences similar to those of known genes).

The number of putative genes revealed by the analysis of the human genome is surprisingly small, perhaps only 25,000. Large amounts of noncoding DNA (in particular, repetitive DNA) and unusually long introns make up the majority of the human genome. Alternative splicing of exons, which creates different polypeptides from the same gene, and greater polypeptide diversity, as well as more complicated control of gene expression, may help explain how this relatively small number of genes can produce the complexity of humans (and vertebrates in general).

The sequences of new gene candidates are compared with sequences of known genes of that species and other species to look for similarities that might indicate the function of the new gene. Many of the putative genes identified so far have never been encountered before.

Determining Gene Function   The function of unknown genes can be studied using *in vitro* **mutagenesis,** in which changes are made to a cloned gene, the gene is returned to the cell, and changes in physiology or developmental patterns that result from the altered gene product are monitored. A new way to stop the expression of selected genes in cells is called **RNA interference (RNAi).** Synthetic double-stranded RNA molecules that match a gene sequence trigger the breakdown or block translation of that gene's mRNA. RNAi was used in one study to silence 86% of the genes of nematode embryos, one at a time.

Studying Expression of Interacting Groups of Genes   In order to study patterns of gene expression, researchers isolate the mRNA made in different cells, create a cDNA library using reverse transcriptase, and then use the cDNA as probes to explore collections of genomic DNA. Using this cDNA in **DNA microarray assays,** scientists can test all the genes expressed in a tissue for hybridization with short, single-stranded DNA fragments from thousands of genes arrayed on a grid (called a DNA chip). The intensity with which the hybridized spots fluoresce indicates the relative amount of mRNA that was in the tissue. Gene expression in different tissues and at different stages of development can be compared.

Comparing Genomes of Different Species   Genomes of about 150 species had been sequenced by the spring of 2004, the majority of these from prokaryotes. Analyses of genome sequences are used to explore evolutionary relationships between both distantly and closely related groups, as well as to answer questions about gene functions.

■ **INTERACTIVE QUESTION   20.7**

a. Give an example of how comparisons with the yeast genome have helped illuminate the human genome.

b. When sequencing the genomes of two closely related species, why is it easier to sequence the smaller genome first?

Future Directions of Genomics   **Proteomics** is the challenging identification and study of entire protein sets coded for by a genome. In humans, the number of proteins greatly exceeds the number of genes. To understand how cells and organisms function requires the study of how proteins interact.

With lists compiled of the parts, researchers are now looking at the functional integration of these parts in biological systems. This systems biology approach is attempting to define gene circuits and protein interaction networks, using computer science and mathematics to process and integrate huge amounts of data.

The human species is comparatively young, and its genetic variation is small. Human DNA sequences are about 99.9% identical. Most variation seems to be one-base-pair variations called **single nucleotide polymorphisms (SNPs),** occurring at about 3 million sites in the human genome. Identifying these sites will provide genetic markers for studying human evolution and differences between human populations, as well as for locating health-related genes.

20.5   The practical applications of DNA technology affect our lives in many ways

Medical Applications   DNA technology is identifying genes responsible for genetic diseases; it is hoped this will lead to new ways to diagnose, treat, and even prevent those disorders. DNA microarray assays allow comparisons of gene expression in healthy and diseased tissues.

PCR and labeled DNA probes are being used to identify pathogens and to diagnose infectious diseases and genetic disorders. Even if a gene has not yet been cloned, a disease allele may be diagnosed when closely linked with an RFLP marker.

Gene therapy may provide the means for correcting genetic disorders in individuals by replacing or supplementing defective genes. New genes would be introduced into somatic cells of the affected tissue. For the correction to be permanent, the cells must be types that actively reproduce within the body, such as bone marrow cells. To date, a few trials have used retroviral vectors to carry a normal allele into bone marrow cells, but results have been mixed.

■ **INTERACTIVE QUESTION   20.8**

a. Why is it easier to perform a test for Huntington's disease now that the gene has been cloned?

b. What are some of the practical and ethical considerations in human gene therapy?

Pharmaceutical Products   Many pharmaceutical proteins are produced using DNA technology. Engineering host cells to secrete a protein as it is made simplifies its purification.

New approaches to disease treatments include genetically engineered proteins that either block or mimic membrane receptor proteins. Recombinant DNA techniques can be used to make vaccines that consist of surface molecules on pathogens, or to modify the genome of a pathogen so as to attenuate it (make it nonpathogenic) so it can be used as a vaccine.

Forensic Evidence   RFLP analysis can be used in criminal cases to compare the **DNA fingerprint,** or specific pattern of RFLP bands, of a victim, suspect, and crime sample. Variations in the number of tandem repeated base sequences (*simple tandem repeats,* or *STRs)* found in various loci are now commonly used in DNA fingerprint analysis. Forensic tests are able to provide a high statistical probability that matching DNA fingerprints come from the same individual.

Environmental Cleanup   Genetically engineered microorganisms that are able to extract heavy metals, such as copper, lead, and nickel, may become important in mining and cleaning up mining waste. Engineering organisms to degrade chlorinated hydrocarbons and other toxic compounds is an area of active research. Environmental disasters such as oil spills and waste dumps are other areas for which detoxifying microbes are being developed.

Agricultural Applications   Vaccines and growth hormones for use in farm animals are being produced with DNA technology.

Transgenic animals containing genes from other organisms are being developed for potential agricultural use. In addition to increasing productivity, another use is as "pharm" animals engineered to produce large quantities of a pharmaceutical product, often by secretion in the animal's milk. Transgenic animals are produced by injecting a foreign gene into egg cells fertilized *in vitro.* The eggs are then transplanted into surrogate mothers.

The ability to regenerate plants from single cells growing in tissue culture has made plant cells easier to genetically manipulate than animal cells. The most commonly used vector is the **Ti plasmid** from the bacterium *Agrobacterium tumefaciens,* which integrates a segment of its DNA into plant chromosomes. Using DNA technology, foreign genes are inserted into Ti plasmids whose disease-causing properties have been eliminated, and the recombinant plasmids are introduced into plant cells growing in culture. When these cells regenerate whole plants, the foreign gene is included in the plant genome.

Many crops have been engineered with bacterial genes for herbicide resistance. Crop plants are being engineered to be resistant to infectious pathogens and insects. Research efforts are also focusing on producing more nutritional transgenic crops, as in the "golden" rice enriched with beta-carotene.

The pharmaceutical industry is working on pharm plants that can produce human proteins for medical use and viral proteins for vaccines.

Safety and Ethical Questions Raised by DNA Technology   Safety regulations have focused on potential hazards of engineered microbes. Most public concern, however, now centers on agricultural **genetically modified (GM) organisms,** which contain artificially acquired genes from the same or different species. One environmental risk of transgenic plants is that their herbicide- or insect-resistant genes might pass to wild plants, creating "superweeds." Some fear that GM foodstuffs may be hazardous to human health. An international Biosafety Protocol requires exporters to label GM organisms in bulk food shipments so that importing countries can decide on potential environmental or health risks. In the United States, several federal agencies evaluate and regulate new products and procedures.

Ethical questions about human genetic information include who should have access to information about a person's genome and how that information should be used. Potential ethical, environmental, and health issues must be considered in the development of these powerful genetic techniques and remarkable products of biotechnology.

Word Roots

liga- = bound, tied (*DNA ligase:* a linking enzyme essential for DNA replication)

electro- = electricity (*electroporation:* a technique to introduce recombinant DNA into cells by applying a brief electrical pulse to a solution containing cells)

muta- = change; **-genesis** = origin, birth (in vitro *mutagenesis:* a technique to discover the function of a gene by introducing specific changes into the sequence of a cloned gene, reinserting the mutated gene into a cell, and studying the phenotype of the mutant)

poly- = many; **morph-** = form (*single nucleotide polymorphisms:* one-base-pair variations in the genome sequence)

Structure Your Knowledge

1. Fill in the table on the next page on the basic tools of gene manipulation used in DNA technology.

Technique or Tool	Brief Description	Some Uses in DNA Technology
Restriction enzymes	**a.**	
Gel electrophoresis	**b.**	
cDNA	**c.**	
Labeled probe	**d.**	
Southern blotting	**e.**	
DNA sequencing	**f.**	
PCR	**g.**	
RFLP analysis	**h.**	
DNA microarray assay	**i.**	

2. Describe several examples of the many possible applications for DNA technology in agriculture and in medicine.

Test Your Knowledge

MULTIPLE CHOICE: *Choose the one best answer.*

1. The role of restriction enzymes in DNA technology is to
   a. provide a vector for the transfer of recombinant DNA.
   b. produce cDNA from mRNA.
   c. produce a cut (usually staggered) at specific restriction sites on DNA.
   d. reseal "sticky ends" after base pairing of complementary bases.
   e. denature DNA into single strands that can hybridize with complementary sequences.

2. The segment of DNA shown at right has restriction sites I and II, which create restriction fragments a, b, and c. Which of the following gels produced by electrophoresis would represent the separation and identity of these fragments?

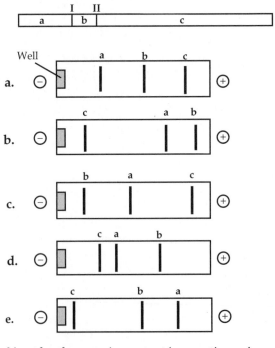

3. Yeast has become important in genetic engineering because it
   a. has RNA splicing machinery.
   b. has plasmids that can be genetically engineered.
   c. allows the study of eukaryotic gene regulation and expression.
   d. grows readily and rapidly in the laboratory.
   e. does all of the above.

4. Which of the following DNA sequences would most likely be a restriction site?

   a. AACCGG
      TTGGCC

   b. GGTTGG
      CCAACC

   c. AAGG
      TTCC

   d. AATTCCGG
      TTAAGGCC

   e. GAATTC
      CTTAAG

5. What is genomics?

   a. the public consortium effort to sequence the human genome

   b. the Celera shotgun approach to sequencing the human genome

   c. the sequencing and systematic study of whole genomes

   d. the use of gene therapy in the treatment of human diseases

   e. the use of nucleotide sequences to determine the function of all proteins encoded by a genome

6. Petroleum-lysing bacteria are being engineered for the treatment of oil spills. What is the most realistic danger of these bacteria to the environment?

   a. mutations leading to the production of a strain pathogenic to humans

   b. extinction of natural microbes due to the competitive advantage of the "petro-bacterium"

   c. destruction of natural oil deposits

   d. poisoning of the food chain

   e. contamination of the water

7. You are attempting to introduce a gene that imparts larval moth resistance to bean plants. Which of the following vectors are you most likely to use?

   a. phage DNA

   b. *E. coli* plasmid

   c. Ti plasmid

   d. yeast plasmid

   e. bacterial artificial chromosome

8. An attenuated virus

   a. is a virus that is nonpathogenic.

   b. is an elongated viral particle.

   c. can transfer recombinant DNA to other viruses.

   d. will not produce an immune response.

   e. is made with cDNA.

9. Which of the following is a difficulty in getting prokaryotic cells to express eukaryotic genes?

   a. The signals that control gene expression are different and prokaryotic promoter regions must be added to the vector.

   b. The genetic code differs because prokaryotes substitute the base uracil for thymine.

   c. Prokaryotic cells cannot transcribe introns because their genes do not have them.

   d. The ribosomes of prokaryotes are not large enough to handle long eukaryotic genes.

   e. The RNA splicing enzymes of bacteria work differently from those of eukaryotes.

10. Complementary DNA does not create as complete a library of genes as the shotgun approach because

    a. it has eliminated introns from the genes.

    b. a cell produces mRNA for only a small portion of its genes.

    c. the shotgun approach produces more restriction fragments.

    d. cDNA is not as easily integrated into plasmids.

    e. restriction enzymes are not used to create cDNA.

11. Which of the following is *not* true of restriction sites?

    a. Modification by methylation of bases within them prevents restriction of bacterial DNA.

    b. They are usually symmetrical sequences of four to eight nucleotides.

    c. They signal the attachment of RNA polymerase.

    d. Each restriction site is cut by a specific restriction enzyme.

    e. Cutting a restriction site in the middle of a functional and identifiable gene is used to screen clones that have taken up foreign DNA.

Use the following choices to answer questions 12–14.

   a. restriction enzyme

   b. reverse transcriptase

   c. ligase

   d. DNA polymerase

   e. RNA polymerase

12. Which is the first enzyme used in the production of cDNA?

13. Which enzyme is used in the polymerase chain reaction?

14. Which is the first enzyme used in the production of DNA fragments for DNA fingerprinting?

15. A plasmid has two antibiotic resistance genes, one for ampicillin and one for tetracycline. It is treated with a restriction enzyme that cuts in the middle of the ampicillin gene. DNA fragments containing a human globin gene were cut with the same enzyme. The plasmids and fragments are mixed, treated with ligase, and used to transform bacterial cells. Clones that have taken up the recombinant DNA are the ones that

    a. are blue and can grow on plates with both antibiotics.

    b. can grow on plates with ampicillin but not with tetracycline.

    c. can grow on plates with tetracycline but not with ampicillin.

    d. cannot grow with any antibiotics.

    e. can grow on plates with tetracycline and are blue.

16. STRs (simple tandem repeats) are a valuable tool for

    a. DNA microarray assays.

    b. infecting plant cells with recombinant DNA.

    c. acting as probes in Southern blots.

    d. DNA fingerprinting.

    e. PCR to produce multiple copies of a DNA segment.

17. You have affixed the chromosomes from a cell onto a microscope slide. Which of the following would *not* make a good radioactively labeled probe to help map a particular gene to one of those chromosomes? (Assume DNA of chromosomes and probes is single stranded.)

    a. cDNA made from the mRNA transcribed from the gene

    b. a portion of the amino acid sequence of that protein

    c. mRNA transcribed from the gene

    d. a piece of the restriction fragment on which the gene is located

    e. a sequence of nucleotide bases determined from the genetic code needed to produce a known sequence of amino acids found in the protein product of the gene

18. If the first three nucleotides in a six-nucleotide restriction site are CTG, what would the next three nucleotides most likely be?

    a. AGG

    b. GTC

    c. CTG

    d. CAG

    e. GAC

19. Computer software is used to identify putative genes in the nucleotide sequences in data banks. Which of the following is the software probably *not* scanning for?

    a. promoters

    b. RNA-splicing sites

    c. STRs (simple tandem repeats)

    d. ESTs (expressed sequence tags)

    e. stop signals for transcription

20. The human genome appears to have only one-third more genes than the simple nematode, *C. elegans*. Which of the following best explains how the more complex humans can have relatively few genes?

    a. The unusually long introns in human genes are involved in regulation of gene expression.

    b. More than one polypeptide can be produced from a gene by alternative splicing.

    c. Human genes code for many more types of domains.

    d. The human genome has a high proportion of noncoding DNA.

    e. The large number of SNPs (single nucleotide polymorphisms) in the human genome provides for a great deal of genetic variability.

21. What is a "pharm" animal?

    a. a transgenic animal that produces large quantities of a pharmaceutical product

    b. an animal used by the pharmaceutical industry to test new medical treatments

    c. a cloned animal that was produced from an adult cell nucleus inserted into an ovum

    d. a genetically engineered animal that produces more meat or milk

    e. a genetically modified organism whose production is permitted in the United States but not in the European Union

22. Which of the following processes or procedures does *not* involve any nucleic acid hybridization?

    a. separation of fragments by gel electrophoresis

    b. Southern blotting

    c. polymerase chain reaction

    d. DNA fingerprinting

    e. DNA microarray assay

23. Which of the following genomes has been completely (or almost completely) sequenced?

    a. nematode *(C. elegans)*

    b. human

    c. *E. coli* and yeast *(Saccharomyces cerevisiae)*

    d. fruit fly *(Drosophila melanogaster)*

    e. all of the above

24. Which of the following techniques can be used to determine the function of a newly identified gene?

    a. comparisons with genes of known functions that have similar sequences and whose products have similar domains

    b. *in vitro* mutagenesis or RNA interference

    c. DNA microarray assay

    d. both a and b

    e. a, b, and c

25. This restriction fragment contains a gene whose recessive allele is lethal. The normal allele has restriction sites for the restriction enzyme *PST*I at sites I and II. The recessive allele lacks restriction site I. An individual who had a sister with the lethal trait is being tested to determine if he is a carrier of that allele. Indicate which of these band patterns would be produced on a gel if he is a carrier (*heterozygous* for the gene)?

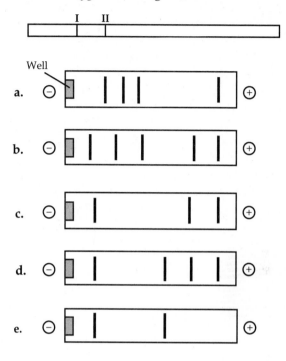

Chapter 21

The Genetic Basis of Development

what causes this?

Framework

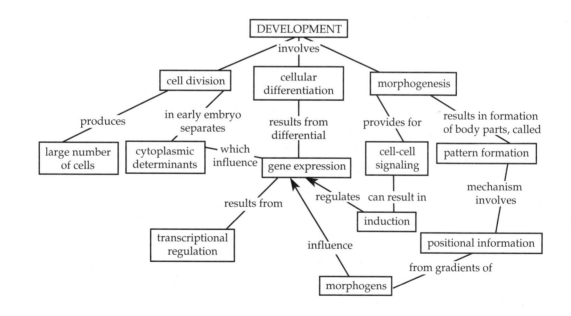

Chapter Review

The application of molecular genetics and DNA technology, along with the study of developmental mutants, has revolutionized the field of development. In order to study broad biological principles, researchers often choose a **model organism** that is representative of a larger group, well suited for answering particular types of research questions, and easy to grow in the lab.

The fruit fly *Drosophila melanogaster* has a long history as a model organism for genetic studies. Its genome is 180×10^6 base pairs long (180 millionbases, Mb) and contains about 13,700 genes. Although its early

169

embryology differs from that of other animals, research on *Drosophila* development has provided key information about animal development.

The transparent nematode *Caenorhabditis elegans* has several advantages as a model organism: its ease of culture and short generation time, its hermaphroditic reproduction that allows for easy detection of recessive mutations, and its small number of cells that has allowed researchers to reconstruct the ancestry of every adult cell. Its genome is 97 Mb long and has about 19,000 genes.

Researchers have already developed a great deal of knowledge of the mouse *Mus musculus* and are now able to manipulate genes to make transgenic mice and mice in which certain genes are "knocked out" by mutation. Their large genome (2,600 Mb with 25,000 genes) and in utero development are disadvantages.

The zebrafish *Danio rerio* reproduces in large numbers and has transparent embryos that develop outside the mother. Its genome is estimated at 1,700 Mb and is still being mapped and sequenced.

Arabidopsis thaliana, a small plant in the mustard family, is easily grown in test tubes, is self-fertilizing, produces large numbers of offspring, and has a relatively small genome—about 118 Mb with an estimated 25,500 genes.

21.1 Embryonic development involves cell division, cell differentiation, and morphogenesis

The three key processes of embryonic development are **cell division,** the production of large numbers of cells; **cell differentiation,** the formation of cells specialized in structure and function; and **morphogenesis,** the physical processes that produce body shape and form.

Morphogenetic events establish body axes early in development. In animals, but not in plants, movements of cells and tissues produce body form. Growth and morphogenesis continue throughout the life of a plant, relying on the perpetually embryonic regions in root and shoot tips known as **apical meristems** to produce new organs and growth.

■ INTERACTIVE QUESTION   21.1

What are some of the important criteria for model organisms chosen for the study of developmental genetics?

21.2 Different cell types result from differential gene expression in cells with the same DNA

Evidence for Genomic Equivalence   Nearly all cells of an organism have the same genes; they have *genomic equivalence.* F. C. Steward demonstrated genomic equivalence in plants by growing new carrot plants from differentiated root cells. Most plant cells remain **totipotent,** retaining the ability to form a complete new organism.

Cloning involves producing genetically identical individuals (**clones**) from a single somatic cell of a multicellular organism.

Early evidence of genomic equivalence in animals was provided by the work of Briggs, King, and Gurdon, who transplanted nuclei from embryonic and tadpole cells into enucleated frog egg cells, a method called *nuclear transplantation.* The ability of the transplanted nucleus to direct normal development was inversely related to its developmental age. The base sequence of an animal cell's DNA usually does not change during differentiation, but its chromatin structure may be modified and thus its nuclear potency restricted.

In 1997, Scottish researchers reported cloning an adult sheep by transplanting a nucleus from a fully differentiated mammary cell into an unfertilized enucleated egg cell, and then implanting the resulting early embryo into a surrogate mother. The mammary cell was induced to dedifferentiate by culturing in a nutrient-poor medium that arrested the cell cycle in the G_0 phase.

The *reproductive cloning* of numerous mammals has shown that cloned animals do not always look and behave identically. Environmental influences and random events play a role in development.

■ INTERACTIVE QUESTION   21.2

Although numerous mammals have now been cloned successfully, most cloned embryos fail to develop normally, and many cloned animals have various defects. What is a likely cause of these developmental failures?

Stem cells are relatively unspecialized cells that continue to reproduce themselves and can, under proper conditions, differentiate into one or more types of cells. *Embryonic stem cells* taken from early embryos can be cultured indefinitely. They are totipotent and can differentiate into cells of any type. Adult stem cells have been isolated from various tissues and grown in culture. Such cells, capable of producing multiple but not all types of cells, are called **pluripotent.** Both types of stem cells can be induced to differentiate into specialized cells. Stem cell research has the potential to provide cells to repair organs that are damaged or diseased. *Therapeutic cloning* of embryonic stem cells, although different from reproductive cloning of humans, still raises ethical debates.

Transcriptional Regulation of Gene Expression During Development
A cell's developmental history leads to its eventual differentiation as a cell with a specific structure and function. **Determination** is a term used to describe the condition when a cell is irreversibly committed to its fate. When a cell becomes differentiated, it expresses genes for *tissue-specific proteins*, and that expression is usually controlled at the level of transcription.

The embryonic precursor cells from which muscle cells arise have the potential to develop into a number of different cell types. Once they become committed to becoming muscle cells, they are called *myoblasts.* Researchers have isolated mRNA from myoblasts, prepared a library of cDNA genes, and inserted them into separate embryonic precursor cells in order to identify "master regulatory genes." One of these is *myoD*, which codes for a transcription factor named MyoD that binds to control elements and initiates transcription of other muscle-specific transcription factors. These secondary transcription factors then activate muscle-protein genes. MyoD also turns on genes that block the cell cycle and stop cell division, and it activates its own transcription, thus maintaining the cell's differentiated state.

■ INTERACTIVE QUESTION 21.3

The MyoD protein has been shown to be able to transform some, but not all, differentiated cells into muscle cells. Why doesn't it work on all kinds of cells?

Cytoplasmic Determinants and Cell-Cell Signals in Cell Differentiation
The cytoplasm of an unfertilized egg cell contains maternal mRNA, proteins, and other substances that are unevenly distributed, and the first few mitotic divisions separate these components and expose cell nuclei to different environments. These maternal components of the egg cell that influence early development by regulating gene expression are called **cytoplasmic determinants.**

The other important source of developmental control comes from signals received from other embryonic cells. Change in the gene expression of target cells resulting from chemical or physical signals from other cells is called **induction.**

21.3 Pattern formation in animals and plants results from similar genetic and cellular mechanisms

Pattern formation is the ordering of cells and tissues into their characteristic structures and locations. In animals, pattern formation takes place in embryos and juveniles; in plants, it occurs continually in the apical meristems.

An animal's three major body axes are laid out early in development. Cells and their progeny develop in response to molecular cues called **positional information** that tell a cell where it is located relative to the body axes and neighboring cells.

Drosophila Development: A Cascade of Gene Activations
A fruit fly's body consists of a series of segments grouped into the head, thorax, and abdomen. The anterior-posterior and dorsal-ventral axes are determined by positional information provided by cytoplasmic determinants localized in the unfertilized egg. After fertilization, positional information establishes the proper number of segments and then triggers the formation of each segment's characteristic structures.

Each egg cell in the mother's ovary is surrounded by nurse cells and follicle cells that supply nutrients and other molecules needed for development. Eggs are laid following fertilization, and the first ten mitotic divisions occur without cytokinesis, yielding a multinucleated embryo. In the blastoderm stage, nuclei migrate to the periphery, and plasma membranes divide the nuclei into separate cells. Segments become visible, organs form, and a wormlike larva hatches out of the egg shell. Following three larval stages and molts, the larva becomes a pupa. Metamorphosis produces an adult fly, with distinct segments bearing characteristic appendages.

In the 1940s, E. B. Lewis studied developmental mutants and was able to map certain mutations that control development in the late embryo to specific genes. In the late 1970s, C. Nüsslein-Volhard and E. Wieschaus undertook a search for the genes that control segment formation. They studied mutations that were **embryonic lethals,** which prevented the development of viable larvae. They exposed flies to a chemical mutagen and then performed many thousands of crosses to detect recessive mutations causing death of embryos or larvae with abnormal segmentation. They identified

1,200 genes essential for development, 120 of which were involved in pattern formation leading to normal segmentation.

Maternal effect genes are genes of the mother that code for proteins or mRNA that are deposited in the unfertilized egg. These genes are also called **egg-polarity genes** because they determine the anterior-posterior and dorsal-ventral axes of the egg and consequently the embryo.

One egg-polarity gene is *bicoid*. The product of the *bicoid* gene is concentrated at one end of the larva and responsible for determining its anterior end. Offspring of a mother defective for this gene have two tail regions and lack the front half of the body. Researchers used the cloned *bicoid* gene as a probe and found *bicoid* mRNA concentrated in the most anterior end of egg cells. Following fertilization, the mRNA is translated into Bicoid protein, which diffuses posteriorly, forming a gradient in the early embryo. Proteins whose gradients determine the posterior end and establish the dorsal-ventral axis have also been identified. Such substances, whose gradients establish an embryo's axes or other features, are called **morphogens.**

The products of the egg-polarity genes are transcription factors. Gradients of these morphogens affect the expression of the embryo's **segmentation genes,** which direct the formation of segments. In a cascade of gene activations, products of the *gap genes* control the localized expression of *pair-rule genes,* which in turn activate specific *segment polarity genes*. The products of many of these segmentation genes are transcription factors; others code for cell-signaling molecules or their receptors, which are necessary for the cell-cell communication needed once plasma membranes have separated the cells of the embryo.

Master regulatory genes called **homeotic genes** next determine each segment's anatomical fate by encoding transcription factors that control the expression of genes that build a segment's characteristic structures.

■ INTERACTIVE QUESTION   21.4

a. What could cause a mutant fly embryo to develop two tail ends but no head?

b. What could cause a mutant embryo to have half the normal number of segments?

c. What could cause a fruit fly to have legs growing out of its head in place of antennae?

C. elegans: *The Role of Cell Signaling* Inductive signaling from one group of cells to nearby cells influences the gene expression that leads to cellular differentiation. Researchers have determined the complete **cell lineage** of *C. elegans*. As early as the four-cell stage, cell signaling is involved in directing the developmental fate of cells. The binding of a cell-surface protein produced by cell 4 to a receptor protein on cell 3 triggers cellular events in cell 3 that lead to different fates for its two daughter cells—the posterior one will form the intestine, the anterior one develops into muscle and gonads.

Later in development, six precursor cells of the second-stage larva become destined to give rise to the vulva. An *anchor cell* of the embryonic gonad secretes a signal protein that reaches the closest precursor cell in high concentration. This inducer stimulates the cell to divide and differentiation to form the inner part of the vulva. Lower concentrations of the signal reach the two adjacent precursor cells. These cells are stimulated to divide and differentiate into the outer vulva. The remaining three cells are too distant to receive the signal; they develop into epidermal cells.

■ INTERACTIVE QUESTION   21.5

Developmental studies of *Drosophila* and *C. elegans* support the following generalizations regarding the role of induction in development:

a. The pathway to organ formation often follows a series of steps that involve _____ _____ .

b. The effect of an inducer on its possible target cells often depends on the _____ of the inducer.

c. Inducers often initiate _____ _____ _____ in the target cell that lead to _____ _____ of important developmental genes.

Lineage analysis of *C. elegans* has established that cell suicide or **apoptosis** occurs numerous times during normal development. Signals activate a cascade of "suicide" proteins, the cell shrinks and forms blebs, and neighboring cells engulf the membrane-bound remains. Genetic screening has identified three key genes involved in cell death: *ced-9* that produces a protein that inhibits the activity of the protein products of *ced-3* and *ced-4.* In the case of apoptosis, protein activity is regulated, not transcription or translation. When a cell receives a "death" signal on its membrane receptor, Ced-9 protein becomes inactivated, and the apoptosis pathway activates enzymes that hydrolyze the cell's proteins and DNA. Ced-3 is the main *caspase* (protease) of apoptosis in the nematode. One of several apoptosis pathways in mammals involves proteins, including cytochrome *c*, that are released from mitochondria.

■ INTERACTIVE QUESTION 21.6

a. What can one conclude from the fact that some mammalian apoptosis proteins are homologous to the Ced-3, Ced-4, and Ced-9 proteins of nematodes?

b. Give some examples of programmed cell death in humans.

Plant Development: Cell Signaling and Transcriptional Regulation The developmental fates of plant cells depend less on cell lineage and more on positional information. Embryonic development occurring in a seed is difficult to study, but research on apical meristems has provided insight into the genetics and control of plant development.

Environmental cues trigger signal transduction pathways through which shoot meristems are transformed to floral meristems. A floral meristem consists of three cell layers, from which the organs of the flower—the carpels, stamens, petals, and sepals-arise. Grafting stems of *fasciated (f)* mutant tomato plants onto wild-type plants produces shoots from the graft site that are **chimeras,** having a mixture of cells from different genetic backgrounds. The number of organs produced in flowers of the chimeras depends on the parental source of the innermost cell layer, which must somehow induce the two outer layers to produce its specified number of organs.

Analogous to homeotic genes in animals are plant **organ identity genes** that determine what type of structure develops from a particular outgrowth from the floral meristem. These genes have been most extensively studied with floral mutants in *Arabidopsis*. The organ identity genes encode transcription factors that control the transcription of many other genes responsible for organ development.

21.4 Comparative studies help explain how the evolution of development leads to morphological diversity

Biologists in the field of evolutionary developmental biology ("evo-devo") compare development processes to understand how they have evolved and how minor changes may lead to diverse forms of life.

Widespread Conservation of Developmental Genes Among Animals A sequence of 180 nucleotides called a **homeobox** has been found in each *Drosophila* homeotic gene. The same or very similar homeobox nucleotide sequences have been identified in homeotic genes of many animals. These are often called *Hox* genes in mammals. Related sequences are found in regulatory genes of yeast, plants, and even prokaryotes. These similarities indicate that the homeobox must have arisen early and been conserved through evolution as part of genes involved in regulation of gene expression and development.

The homeobox sequence is translated into a 60-amino-acid sequence known as a *homeodomain*. This portion of the resulting transcription factor binds to DNA, while other domains interact with other transcription factors to recognize specific enhancers or promoters. Proteins with homeodomains probably coordinate the transcription of groups of developmental genes and thus control pattern formation.

Many other genes involved in development, such as those coding for components of signaling pathways, are highly conserved. The differing patterns of expression of these genes in different body areas may explain the development of animals with different body plans.

■ INTERACTIVE QUESTION 21.7

In *Drosophila*, homeobox sequences have been found not only in the homeotic genes, but also in the egg-polarity gene *bicoid*, in several segmentation genes, and in the master regulatory gene for eye development. Is this just a coincidence? Explain.

Comparison of Animal and Plant Development The processes of development evolved independently in plants and animals due to their ancient divergence. The rigid cell walls of plants restrict movement, and morphogenesis relies more on the orientation of cell divisions and cell enlargement. Some similarities persist from their common ancestral microbe. In both, development involves a cascade of transcriptional regulators, although the master control genes are different in the two groups.

▌Word Roots

apic- = tip (*apical meristem:* embryonic plant tissue in the tips of roots and in the buds of shoots that supplies cells for the plant to grow in length) morph- = form; **-gen** = produce (*morphogen*: a substance that provides positional information in the form of a concentration gradient along an embryonic axis)

toti- = all; **-potent** = powerful (*totipotent*: the ability of a cell to form all parts of the mature organism)

Structure Your Knowledge

1. Describe the cascade of gene activations that leads to the development of a fruit fly.

2. How might the mechanism for transcriptional regulation differ for cytoplasmic determinants and the cell-cell signaling involved in induction?

Test Your Knowledge

MULTIPLE CHOICE: *Choose the one best answer.*

1. Which of the following is descriptive of a cell that is differentiated?
   a. The cell's development fate has been decided, although it may not look any different from another precursor cell.
   b. The cell has definite anterior-posterior and dorsal-ventral axes.
   c. The cell has developed cell surface receptors that allow it to receive signals from other cells.
   d. The cell has been induced to transcribe all its genes.
   e. The cell is producing tissue-specific proteins and has its characteristic structure.

2. Morphogenesis in plants results from
   a. migration of cells from the apical meristems to produce three tissue layers.
   b. differences in the plane of cell division and the direction of cell expansion.
   c. cytoplasmic determinants that were deposited in the egg cell.
   d. the differentiation of cells within the apical meristem.
   e. the production of cell walls and plasmodesmata that communicate between cells.

3. In which of these model organisms has it been possible to create a complete cell lineage?
   a. *Drosophila melanogaster* (fruit fly)
   b. *Danio rerio* (zebrafish)
   c. *Caenorhabditis elegans* (nematode)
   d. *Mus musculus* (mouse)
   e. *Arabidopsis thaliana* (common wall cress)

4. Cytoplasmic determinants are
   a. unevenly distributed cytoplasmic components of an unfertilized egg.
   b. often involved in transcriptional regulation.
   c. often separated in the first few mitotic divisions following fertilization.
   d. maternal contributions that help to direct the initial stages of development.
   e. all of the above.

5. For which of the following can the cloning of Dolly the sheep best be used to provide evidence?
   a. totipotency of most adult animal cells
   b. determination of most adult animal cells
   c. embryonic nature of mammary cells
   d. genomic equivalence of most animal cells
   e. immune tolerance of embryos in mammals

6. The fact that transplanted nuclei from most tadpole cells were unable to direct normal development in an enucleated frog cell gives evidence for
   a. the differentiation of adult cells resulting in the deletion of certain key genes.
   b. the totipotency of adult cells even when they are differentiated.
   c. changes in chromatin during development that may make genes no longer available for transcription.
   d. the dedifferentiation of embryonic cells.
   e. the need for cytoplasmic determinants to direct initial developmental stages.

7. Pattern formation in animals is based on
   a. the first few mitotic divisions.
   b. the induction of cells by the initial fertilized egg.
   c. the locations and activity of apical meristems.
   d. differentiation of cells, which then migrate together to form tissues and organs.
   e. positional information a cell receives from gradients of morphogens.

8. Which of the following developmental processes involves apoptosis?
   a. the development of the vulva in *C. elegans*
   b. the sequential activation of segmentation genes in *Drosophila*
   c. the production of genetically engineered mice with "knockout" genes
   d. the development of separate fingers and toes during mammalian development
   e. the induction of tissue layers in the production of the organs from a floral meristem

9. A fruit fly that has two sets of wings growing from its thorax (instead of a single pair of wings and a pair of small balancing organs) would probably have mutations in its
   a. gap genes.
   b. segment polarity genes.
   c. homeotic genes.
   d. egg-polarity genes.
   e. pair-rule genes.

10. What would be the fate of a *Drosophila* larva that inherits two copies of a mutant *bicoid* gene (one mutant allele from each heterozygous parent)?
    a. It develops two heads, one at each end of the larva.
    b. It develops two tails, one at each end of the larva.
    c. It develops normally but produces mutant larvae that have two tail regions.
    d. It develops into an adult with legs growing out of its head.
    e. It receives no *bicoid* mRNA from the nurse cells of its mother.

11. A highly conserved nucleotide sequence that has been found in master regulatory genes in many diverse organisms is called a
    a. homeobox.
    b. homeodomain.
    c. transcription factor.
    d. homeotic gene.
    e. morphogen.

12. The gene *ced-9* codes for a protein that inactivates the proteins of suicide genes found in the genome of *C. elegans*. For development to proceed normally, the *ced-9* gene
    a. should be activated in all cells, but its protein product must remain inactive.
    b. should be activated in all cells, but its product will be inactivated when cells programmed to die receive the proper signal.
    c. should be inactivated in all cells except when a cell receives a signal to die.
    d. should be inactivated only in those cells that must die for proper development to occur.
    e. should code for a transcription factor that attaches to the enhancers of other suicide genes.

13. Once the developmental fate of a cell is set, the cell is said to be
    a. determined.
    b. differentiated.
    c. totipotent.
    d. genomically equivalent.
    e. induced.

14. Which of the following is *not* true of adult stem cells?
    a. They have been found not only in bone marrow, but also in other tissues, including the adult brain.
    b. Although more difficult to grow than embryonic stem cells, they have been successfully grown in culture and made to differentiate into specialized cells.
    c. They are differentiated cells that can be induced to dedifferentiate and become totipotent.
    d. As pluripotent cells, they are capable of developing into several different types of cells under appropriate conditions.
    e. These relatively unspecialized cells continually reproduce themselves.

15. In this hypothetical embryo, a high concentration of a morphogen called morpho is needed to activate gene *P*; gene *Q* is active at medium concentrations of morpho or above; and gene *R* is expressed as long as there is any quantity of morpho present. A different morphogen called phogen has the following effects: activates gene *S* and inactivates gene *Q* when at medium to high concentrations. If morpho and phogen are diffusing from where they are produced at the opposite ends of the embryo, which genes will be expressed in region 2 of this embryo? (Assume diffusion through the three regions from high at source to medium to low concentration.)

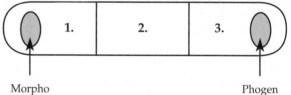

Morpho produced here   Phogen produced here

    a. genes *P*, *Q*, *R*, and *S*
    b. genes *P*, *Q*, and *R*
    c. genes *Q* and *R*
    d. genes *R* and *S*
    e. gene R

16. What do most master regulatory genes do?
    a. produce mRNAs that function as cytoplasmic determinants
    b. code for tissue-specific proteins that differentiate cells
    c. produce proteins that function as inducers to neighboring cells, initiating signaling pathways that activate transcription factors
    d. produce transcription factors that coordinate the expression of other transcription factors and groups of developmental genes
    e. either c or d

Mechanisms of Evolution

Chapter 22

Descent with Modification: A Darwinian View of Life

► Key Concepts

22.1 The Darwinian revolution challenged traditional views of a young Earth inhabited by unchanging species

22.2 In *The Origin of Species*, Darwin proposed that species change through natural selection

22.3 Darwin's theory explains a wide range of observations

► Framework

This chapter describes Darwin's formulation of evolution—all of life has descended from a common ancestor, and the mechanism of natural selection has resulted in the evolution of species adapted to their environments. The scientific and philosophical climate of Darwin's day was quite inhospitable to the implications of evolution, although ideas from taxonomy, geology, and paleontology set the stage for Darwin's work. Evidence for evolution is drawn from homologies (anatomical and molecular), biogeography, and the fossil record, as well as examples of natural selection in action.

► Chapter Review

Charles Darwin presented the first convincing case for evolution in his book *On the Origin of Species by Means of Natural Selection*, published in 1859. Darwin made two major claims: The species present on Earth today descended from ancestral species, and **natural selection** is the mechanism for evolution. Natural selection leads to **evolutionary adaptation,** as individuals with beneficial heritable traits leave more offspring, and the frequency of such traits in a population increases over generations. **Evolution** may be defined as the changes in a population's genetic composition over time.

22.1 The Darwinian revolution challenged traditional views of a young Earth inhabited by unchanging species

Resistance to the Idea of Evolution   Darwin's book challenged both the prevailing scientific views and the world view that had been held for centuries in Western culture.

The Greek philosopher Plato believed in two worlds, an ideal and eternal real world and the illusory world perceived by the senses. Aristotle proposed that all living forms could be arranged on a "scale of nature" of increasing complexity and that each group of organisms was permanent and perfect.

The Old Testament account of creation supported the idea of perfect and fixed species. One of the goals of biology in the 1700s was to classify the species that God had created. Linnaeus developed both a binomial system for naming organisms according to their genus and species, and a hierarchy of classification groupings. **Taxonomy,** the branch of biology that names and classifies organisms, originated in the work of Linnaeus.

Fossils are remnants or impressions of past organisms, usually found in **sedimentary rocks** formed through the compression of layers of sand and mud into superimposed layers called strata. Fossils from strata of different ages reveal some of the organisms that have existed at different periods of Earth's history.

Cuvier may be considered the father of **paleontology,** the study of fossils. Advocating **catastrophism,** he maintained that the differences he observed in the fossils found in different strata were the result of local catastrophic events such as floods or drought and were not indicative of evolution.

Theories of Gradualism   **Gradualism,** the idea that immense change is the cumulative result of slow but continuous processes, was proposed by Hutton in 1795 to explain the geologic state of the Earth. Lyell, a contemporary of Darwin, extended gradualism to a theory of **uniformitarianism,** stating that the rates and effects of geologic processes have remained the same through Earth's history and continue in the present.

Darwin took two ideas from the observations of Hutton and Lyell: The Earth must be very old, and very slow processes can produce substantial change, perhaps even on living species.

Lamarck's Theory of Evolution   Lamarck published a theory of evolution in 1809. He explained the mechanism of evolution with two principles: The use or disuse of body parts leads to their development or deterioration, and acquired characteristics can be inherited. Although present genetic knowledge rejects his mechanism, Lamarck proposed several key evolutionary ideas: that evolution is the best explanation for the fossil record and that adaptation to the environment is the main result of evolution.

■ INTERACTIVE QUESTION   22.1

a. Match the theory or philosophy and its proponent(s) with the following descriptions.

A. catastrophism	a. Aristotle
B. inheritance of acquired characteristics	b. Cuvier
C. gradualism	c. Darwin
D. natural selection	d. Hutton
E. taxonomy	e. Lamarck
F. scale of nature	f. Linnaeus
G. uniformitarianism	g. Lyell

Theory   Proponents

1. _____   _____ Ordering the diversity of God's creations by naming and classifying species

2. _____   _____ History of Earth marked by floods or droughts that resulted in extinctions

3. _____   _____ Early explanation of mechanism of evolution

4. _____   _____ Profound change is the cumulative product of slow but continuous processes

5. _____   _____ Fixed species on a continuum from simple to complex

6. _____   _____ Differential reproductive success leads to adaptation to environment and evolution

7. _____   _____ Geologic processes have constant rates throughout time

b. Now place 1 through 7 in chronological order.

_____ _____ _____ _____ _____ _____ _____

22.2   In *The Origin of Species,* Darwin proposed that species change through natural selection

Darwin's Research   Darwin was 22 years old when he sailed from Great Britain on the HMS *Beagle.* He spent the voyage collecting thousands of specimens of the fauna and flora of South America, observing the various adaptations of organisms living in very diverse habitats, and making special note of the geographic distribution of the distinctly South American species. He was particularly struck by the uniqueness of the fauna of the Galápagos Islands. Darwin also read and was influenced by Lyell's *Principles of Geology.*

Darwin began to link the origin of new species to the process of adaptation to different environments. In 1844 he wrote an essay on the origin of species and natural selection but did not publish it. In 1858 Darwin received Wallace's manuscript describing an identical theory of natural selection. Wallace's paper and extracts of Darwin's unpublished essay were jointly presented to the Linnaean Society, and Darwin published *The Origin of Species* the next year. Within a decade, Darwin's book and its defenders had convinced the majority of biologists that evolution was the best explanation for the diversity of life.

The Origin of Species Darwin's book developed two main points: evolution as the explanation of life's unity and diversity, and natural selection as the mechanism of adaptive evolution.

Darwin's concept of **descent with modification** included the notion that all organisms were related through descent from some unknown ancestor and had developed increasing modifications as they adapted to various habitats. The history of life is analogous to a tree with a common ancestor at the fork of each new branch. The taxonomy developed by Linnaeus provided a hierarchical organization of groups that suggested to Darwin this branching tree of life.

Evolutionary biologist Ernst Mayr described Darwin's theory of natural selection as follows:

Observation 1: Species have the potential for their population size to increase exponentially.
Observation 2: Most population sizes are stable.
Observation 3: Resources are limited.
    Inference 1: There is a struggle for limited resources and only a fraction of offspring survive.
Observation 4: Individuals vary within a population.
Observation 5: Much of this variation is inherited.
    Inference 2: Individuals whose inherited characteristics fit them best to the environment are likely to leave more offspring.
    Inference 3: Unequal reproduction leads to the gradual accumulation of favorable characteristics in a population over generations.

■ **INTERACTIVE QUESTION 22.2**

Summarize in your own words Darwin's theory of natural selection as the mechanism of evolution.

Darwin found support for the struggle for existence and the capacity of organisms to overproduce in the essay on human population growth published by Malthus in 1798.

Artificial selection used in the breeding of domesticated plants and animals provided Darwin with

evidence that selection among the variations present in a population can lead to substantial changes. He reasoned that natural selection, working over thousands of generations, could gradually create the modifications essential for the present diversity of life.

Natural selection results in the evolution of populations, groups of interbreeding individuals of the same species in a common geographic area. Evolution is measured only as change in the relative proportions of variations in a population over time. Natural selection affects only those traits that are heritable—acquired characteristics cannot evolve. And natural selection depends on the specific environmental factors present in a region at a given time. If the environment changes, different adaptations will be favored.

22.3 Darwin's theory explains a wide range of observations

Natural Selection in Action J. Endler and D. Reznick, studying guppy populations in isolated pools in Trinidad over many years, have observed variations in the age and size of sexual maturity that correlate with the type of local predator. They transplanted guppies from pools with pike-cichlids to pools with killifish predators and observed, over 11 years, a change in age and size at maturity in the transplanted population, documenting evolution in a natural setting over a relatively short period of time.

The evolution of drug resistance in HIV illustrates two facets of natural selection: It is an editing, not a creative, mechanism that selects for variations already present in a population. And it is regional and temporal, selecting for traits that fit the local environment at that current time.

■ **INTERACTIVE QUESTION 22.3**

Within a few weeks of treatment with the drug 3TC, a patient's HIV population consists entirely of 3TC-resistant HIV. Explain how this rapid evolution of drug resistance is an example of natural selection.

Homology, Biogeography, and the Fossil Record **Homology** is a similarity resulting from common ancestry. The forelimbs of all mammals are **homologous structures,** containing the same skeletal elements regardless of function or external shape. Comparative anatomy illustrates that evolution is a remodeling process in which ancestral structures become modified for new functions. Some homologous structures that differ greatly in adult form and function are more evident during embryonic development.

Vestigial organs are rudimentary structures, of little or no value to the organism, that are historical remnants of ancestral structures.

Homologies can be seen on a molecular level. DNA, RNA, and an essentially universal genetic code, which have been passed along through all branches of evolution, are important evidence that all forms of life descended from the earliest organisms and are thus related. Homology is evident on different hierarchical levels, reflecting evolutionary history and the degree of relationship among organisms. Closely related species have a larger proportion of DNA and proteins in common than do more distantly related species.

The geographic distribution of species, or **biogeography,** provides evidence for evolution. Islands often have **endemic** species, found nowhere else and usually closely related to species on the nearest island or mainland. Widely separated areas having similar environments are not likely to be populated by closely related species. Rather, each area is more likely to have species that are taxonomically related to those of their region, regardless of environment. Biogeographic distribution patterns are explained by evolution; modern species are found where they are because they evolved from ancestors that inhabited those regions.

The major branches of evolutionary descent established with evidence from homologies and molecular biology are also supported by the sequence of fossil forms found in the fossil record. Paleontologists continue to discover transitional fossils linking modern species to their ancestral forms.

What Is Theoretical about the Darwinian View of Life? A scientific "theory" is a unifying concept with broad explanatory power and predictions that have been and continue to be tested by experiments and observations. Scientists continue to explore whether natural selection is the main mechanism of evolution or whether other factors have contributed to the evolutionary history of life.

■ **INTERACTIVE QUESTION   22.4**

Complete the following concept map that summarizes the main sources of evidence for evolution.

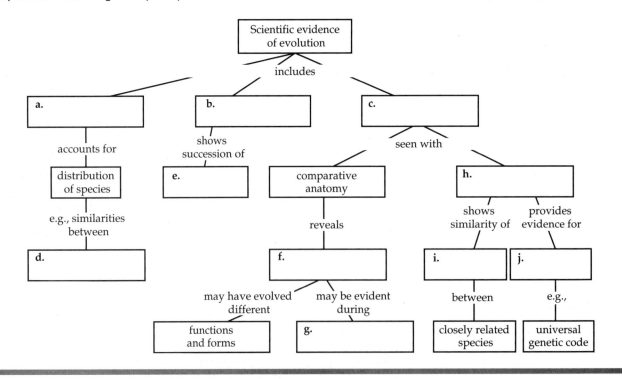

Word Roots

bio- = life; **geo-** = the Earth (*biogeography:* the study of the past and present distribution of species)

end- = within (*endemic:* a type of species that is found only in one region and nowhere else in the world)

homo- = like, resembling (*homology:* similarity in characteristics resulting from a shared ancestry)

paleo- = ancient (*paleontology:* the scientific study of fossils)

taxo- = arrange (*taxonomy:* the branch of biology concerned with naming and classifying the diverse forms of life)

vestigi- = trace (*vestigial organs:* structures of marginal, if any, importance to an organism, historical remnants of structures that had important functions in ancestors)

Structure Your Knowledge

1. Briefly state the main components of Darwin's theory of evolution.

Test Your Knowledge

MULTIPLE CHOICE: *Choose the one best answer.*

1. The classification of organisms into hierarchical groups is called
   a. the scale of nature.
   b. taxonomy.
   c. evolution.
   d. biogeography.
   e. natural selection.

2. The study of fossils is called
   a. homology.
   b. gradualism.
   c. paleontology.
   d. anthropology.
   e. biogeography.

3. To Cuvier, the differences in fossils from different strata were evidence for
   a. changes occurring as a result of cumulative but gradual processes.
   b. divine creation.
   c. evolution by natural selection.
   d. continental drift.
   e. local catastrophic events such as droughts or floods.

4. Darwin proposed that new species evolve from ancestral forms by
   a. the gradual accumulation of adaptations to changing or different environments.
   b. the inheritance of acquired adaptations to the environment.
   c. the struggle for limited resources.
   d. the accumulation of mutations.
   e. the exponential growth of populations.

5. The best description of natural selection is
   a. the survival of the fittest.
   b. the struggle for existence.
   c. the reproductive success of the members of a population best adapted to the environment.
   d. the overproduction of offspring in environments with limited natural resources.
   e. a change in the proportion of inheritable variations within a population.

6. The remnants of pelvic and leg bones in a snake
   a. are vestigial structures.
   b. show that lizards evolved from snakes.
   c. are homologous structures.
   d. provide evidence for inheritance of acquired characteristics.
   e. resulted from artificial selection.

7. The hypothesis that whales evolved from land-dwelling ancestors is supported by
   a. evidence from the biogeographic distribution of whales.
   b. molecular comparisons of whales, fish, and reptiles.
   c. historical accounts of walking whales.
   d. the ability of captive whales to be trained to walk.
   e. fossils of extinct whales found in Egypt and Pakistan that had small hind limbs.

8. Darwin's claim that all of life descended from a common ancestor may best be supported with evidence from
   a. the fossil record.
   b. comparative embryology.
   c. taxonomy.
   d. molecular biology.
   e. comparative anatomy.

9. The smallest unit that can evolve is
   a. a genome.
   b. an individual.
   c. a species.
   d. a population.
   e. a community.

10. Which of the following would *not* be considered part of the process of natural selection?

    a. Many of the variations among individuals in a population are heritable.

    b. More offspring are produced than are able to survive and reproduce.

    c. Individuals with traits best adapted to the environment are likely to leave more offspring.

    d. Many adaptive traits may be acquired during an individual's lifetime, helping that individual to evolve.

    e. Differential reproductive success leads to gradual change in a population.

11. When cytochrome *c* molecules are compared, yeasts and molds are found to differ by approximately 46 amino acids per 100 residues (amino acids making up a protein), whereas insects and vertebrates are found to differ by approximately 29 amino acids per 100 residues. What can one conclude from these data?

    a. Very little can be concluded unless the DNA sequence for the cytochrome *c* genes are compared.

    b. Yeasts evolved from molds, but vertebrates did not evolve from insects.

    c. Insects and vertebrates diverged from a common ancestor more recently than did yeasts and molds.

    d. Yeasts and molds diverged from a common ancestor more recently than did insects and vertebrates.

    e. The evolution of cytochrome *c* occurred more rapidly in yeasts and molds than in insects and vertebrates.

12. All of the following influenced Darwin as he synthesized the theory of evolution by natural selection *except*

    a. the biogeographic distribution of species such as the finches on the Galápagos Islands.

    b. Lyell's book, *Principles of Geology,* on the gradualness of geologic changes.

    c. Linnaeus's hierarchical classification of species, which could be interpreted as evidence of evolutionary relationships.

    d. examples of artificial selection that produce rapid changes in domesticated species.

    e. Mendel's paper in which he described his "laws of inheritance."

13. What might you conclude from the observation that the bones in your arm and hand are similar to the bones that make up a bat's wing?

    a. The bones in the bat's wing may be vestigial structures, no longer useful as "arm" bones.

    b. The bones in a bat's wing may be homologous to your arm and hand bones.

    c. Bats and humans evolved in the same geographic area.

    d. Bats lost their opposable digits during the course of evolution.

    e. Our ancestors could fly.

14. The best description of endemic species are species that are

    a. found only on islands.

    b. found in the same geographic area.

    c. found only on mainlands.

    d. found only in that location and nowhere else on Earth.

    e. disease causing and pesticide resistant.

15. Which of the following is an example of convergent evolution?

    a. the increase in size and age of sexual maturity of guppies transplanted to pools with killifish, predators that prey mainly on small guppies

    b. two very different plants that are found in different habitats, but evolved from a fairly recent common ancestor

    c. similarities between the marsupial sugar glider and the eutherian flying squirrel

    d. the remodeling of a vertebrate forelimb in the evolution of a bird wing

    e. the many different bill sizes and shapes of finches on the Galápagos Islands

16. Chimpanzees and humans share many of the same genes, indicating that most likely

    a. the two groups belong to the same species.

    b. the two groups belong to the same phylum.

    c. the two groups share a relatively recent common ancestor.

    d. humans evolved from chimpanzees.

    e. chimpanzees evolved from humans.

Chapter 23

The Evolution of Populations

Framework

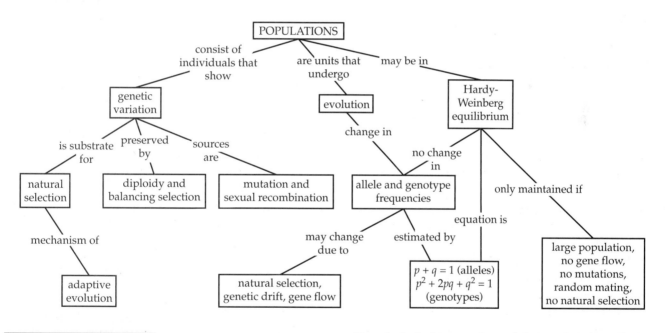

Chapter Review

Although it is individuals that are selected for or against by natural selection, it is populations that actually evolve as some characteristics become more common and others less from one generation to the next.

23.1 Population genetics provides a foundation for studying evolution

Microevolution is defined as changes in the genetic makeup of a population from generation to generation.

Darwin lacked a genetic model to explain how the heritable variations on which natural selection acts are passed on without blending.

The Modern Synthesis When Mendel's work was rediscovered in the early 1900s, many geneticists believed that Darwin's focus on the inheritance of quantitative traits that vary on a continuum could not be explained by the inheritance of discrete Mendelian traits. The discovery that quantitative characters are influenced by multiple genes that each follow Mendelian inheritance helped to reconcile Darwin's and Mendel's ideas. The field of **population genetics,** which studies genetic change in populations over time, gave rise to a comprehensive theory of evolution, known as the **modern synthesis.** Key ideas of the modern synthesis have included the importance of populations as the units of evolution, the essential role of natural selection, and the gradualness of evolution. This paradigm of evolution, however, continues to evolve.

Gene Pools and Allele Frequencies   A **population** is a localized group of individuals that have the ability to interbreed and produce fertile offspring. Populations of a species may be totally isolated or adjoining with their members concentrated in population centers.

The **gene pool** is the term for all the alleles present in a population at any given time. For individuals of a diploid species, the pool includes two alleles for each gene locus. If all individuals are homozygous for the same allele, the allele is said to be *fixed*. More often, two or more alleles are present in the gene pool in some relative proportion or frequency.

■ INTERACTIVE QUESTION   23.1

In a population of 200 mice, 98 are homozygous dominant for brown coat color (*BB*), 84 are heterozygous (*Bb*), and 18 are homozygous recessive (*bb*).

a. The allele frequencies of this population are _____ *B* allele _____ *b* allele.

b. The genotype frequencies of this population are _____ *BB* _____ *Bb* _____ *bb*.

The Hardy-Weinberg Theorem   If only Mendelian segregation and recombination of alleles at fertilization are involved, the gene pool of a population will remain constant from one generation to the next. This stasis is formulated as the **Hardy-Weinberg theorem,** named for its originators. The theorem describes how genetic variation is retained in a population, providing the opportunity for natural selection to act.

The allele frequency within a population determines the proportion of gametes that will contain that allele. The random combination of gametes will yield offspring with genotypes that reflect and reconstitute the allele frequencies of the previous generation. If both gamete recombination and mating are random, the population is in **Hardy-Weinberg equilibrium,** and the frequencies of both alleles and genotypes will remain stable from generation to generation.

With the equation for Hardy-Weinberg equilibrium, the frequencies of genotypes within a population can be calculated from the allele frequencies. In a case with two alleles at a particular gene locus, the letters p and q represent the proportions of the two alleles within the population, and their combined frequencies must equal 1: $p + q = 1$. The frequencies of the genotypes in the offspring reflect the probability of each combination of alleles. According to the rule of multiplication, the probability that two gametes containing the same allele will come together is equal to ($p \times p$) or p^2, or ($q \times q$) or q^2. A p and q allele can combine in two different ways, depending on which parent contributes which allele; therefore, the frequency of a heterozygous offspring is equal to $2pq$. The sum of the frequencies of all possible genotypes in the population adds up to 1: $p^2 + 2pq + q^2 = 1$.

■ INTERACTIVE QUESTION   23.2

Use the allele frequencies you determined in Interactive Question 23.1 to predict the genotype frequencies of the next generation.

Frequencies of

$B (p) =$ _____ $b (q) =$ _____

$BB = p^2 =$ _____ $Bb = 2pq =$ _____ $bb = q^2 =$ _____

Even if a population is self-fertilizing or does not mate randomly, its allele frequencies will remain constant if gametes are produced at random from the gene pool and no other influences are operating.

Hardy-Weinberg equilibrium is maintained only if all of the following five conditions are met: an extremely large population, no gene flow or transfer of alleles between genetically different populations, no mutations, random mating, and no natural selection. Even though natural populations are rarely in true Hardy-Weinberg equilibrium, the rate of evolutionary change may be so slow that allele and genotype frequencies may be estimated using the Hardy-Weinberg equation.

If the frequency of homozygous recessive individuals is known (q^2), then the frequency of q may be estimated as the square root of q^2 (assuming the population is in Hardy-Weinberg equilibrium for that gene). From our example in Interactive Question 23.2, $q = \sqrt{0.09} = 0.3$. If the frequency of individuals with a recessively inherited disease is known (q^2), then the frequency of carriers of that recessive allele in the population can be calculated.

■ INTERACTIVE QUESTION 23.3

Practice using the Hardy-Weinberg equation so that you can easily determine genotype frequencies from allele frequencies and vice versa.

a. The allele frequencies in a population are $A = 0.6$ and $a = 0.4$. Predict the genotype frequencies for the next generation.

   AA _____   *Aa* _____   *aa* _____

b. What would the allele frequencies be for the generation you predicted above in part a.?

   A _____   *a* _____

23.2   Mutation and sexual recombination produce the variation that makes evolution possible

Mutation   New alleles and genes originate by **mutation,** changes in the sequence of nucleotides in DNA. Most mutations occur in somatic cells and cannot be passed on to the next generation. Point mutations that occur in noncoding DNA or do not change the amino acid sequence of a protein are harmless. Those mutations that do alter a protein's function or affect a regulatory region of DNA may have serious effects. Rarely, however, a new mutant allele may increase an individual's fitness, or a mutation already present in the population may be selected for when the environment changes.

Chromosomal mutations are most often deleterious. Occasionally, a translocation of a chromosomal piece may bring alleles together that are beneficial in combination. Duplication of smaller DNA segments, introduced by transposons, may provide an expanded genome with extra loci that could eventually take on new functions by mutation. This gene **duplication** provides an important source of variation. The shuffling of exons may also result in new genes.

Mutation rates in animals and plants average about one in every 100,000 genes per generation. Mutation produces genetic variation very rapidly in HIV because of its short generation span and higher rate of mutation in its RNA genome.

Sexual Recombination   In a sexually reproducing population, the genetic differences that make adaptation possible arise from the reshuffling of alleles into new combinations (recombination) every generation.

■ INTERACTIVE QUESTION 23.4

a. What is a major source of genetic variation for bacteria and viruses?

b. What is the major source of genetic variation for plants and animals?

c. Explain why your answers to a. and b. are different.

23.3   Natural selection, genetic drift, and gene flow can alter a population's genetic composition

Natural Selection   Individuals that are more successful in producing viable, fertile offspring pass their alleles to the next generation in disproportionate numbers. This differential success in reproduction—natural selection—disrupts Hardy-Weinberg equilibrium.

Genetic Drift   Chance deviations from expected results are more likely to occur in a small sample. Chance fluctuations in a small population's allele frequencies from one generation to the next are called **genetic drift** and tend to reduce genetic variation in a population.

The **bottleneck effect** occurs when some disaster or other factor reduces the population size dramatically, and the few surviving individuals are unlikely to represent the genetic makeup of the original population. Genetic drift will remain a factor until the population grows large enough for chance events to be less significant. A bottleneck usually reduces variability because some alleles are lost from the gene pool.

Genetic drift that occurs when only a few individuals colonize a new area is known as the **founder effect.** Allele frequencies in the small sample are unlikely to be representative of the parent population, and genetic drift will affect the gene pool of the new population until it is larger.

Gene Flow   **Gene flow,** the migration of individuals or the transfer of gametes between populations, may change allele frequencies. Differences in allele frequencies between populations tend to be reduced by gene flow.

■ INTERACTIVE QUESTION 23.5

Fill in the following concept map that summarizes three causes of microevolution. Better still, create your own concept map to help you review the ways in which a population's genetic composition may be altered.

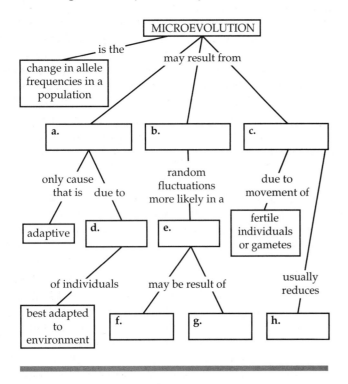

23.4   Natural selection is the primary mechanism of adaptive evolution

Genetic Variation   Individual variation, the slight differences between individuals as a result of their unique genomes, is the raw material for natural selection. *Discrete characters* vary categorically as distinct phenotypes and usually are determined by a single gene locus. *Quantitative characters*, those that are affected by two or more gene loci, provide most of the heritable variation within a population.

A population is **phenotypically polymorphic** when two or more forms of a discrete character—called *morphs*—are evident in a population. Characters that vary along a continuum are influenced by **genetic polymorphisms** for alleles of the genes involved in determining that character.

Measures of genetic variation look at both gene diversity and nucleotide diversity. The **average heterozygosity** of a population is the average percent of loci that are heterozygous. Nucleotide variability measures the average percent of differences in nucleotide sites between individuals of a population.

Geographic variations are regional genetic differences between populations. These variations may be due to differing environmental selection factors and also to genetic drift. A **cline**, or graded variation within a species along a geographic axis, may parallel an environmental gradient.

■ INTERACTIVE QUESTION 23.6

a. Do humans have more or less genetic variation than most species?

b. Two humans differ by about what percentage of their nucleotide bases?

A Closer Look at Natural Selection   **Fitness** is a measure of an individual's relative contribution to the gene pool of the next generation. Population geneticists speak in terms of the **relative fitness** of a genotype as its contribution to the next generation in comparison with the contribution of other genotypes for that locus. The most reproductively successful variants are said to have a relative fitness of 1, whereas the fitness of another genotype is the proportion of offspring it produces compared to the most successful variant.

Natural selection acts on the whole organism. The relative fitness of an allele depends on the entire genetic and environmental context in which it is expressed. Factors that contribute to both survival and fertility determine an individual's evolutionary fitness.

■ INTERACTIVE QUESTION 23.7

a. A gene locus has two alleles, *B* and *b*. The genotype *BB* has a relative fitness of 0.5 and *bb* has a relative fitness of 0.25. What is the relative fitness of the genotype *Bb*?

b. What is the relative fitness of a sterile animal?

The distribution of a trait may be affected by three modes of selection. **Directional selection** occurs most frequently during periods of environmental change, when individuals deviating in one direction from the average for some phenotypic character may be favored. **Disruptive selection** occurs when the environment favors individuals on both extremes of a phenotypic range. **Stabilizing selection** acts against extreme phenotypes and favors more intermediate forms, tending to reduce phenotypic variation.

The Preservation of Genetic Variation The diploidy of most eukaryotes maintains genetic variation by hiding recessive alleles in heterozygotes, enabling them to persist in the population and be selected, should the environment change.

Balancing selection can maintain the frequencies of two of more phenotypes in a population, resulting in **balanced polymorphism.** When individuals heterozygous at a certain gene locus have survival and reproductive advantages, their **heterozygote advantage** tends to maintain two or more alleles at this locus.

In frequency-dependent selection, a morph's reproductive success declines if it becomes too common in the population. Frequency-dependent predation, as predators learn to focus on the more common prey items, may preserve polymorphism in prey populations.

Some of the genetic variation seen in populations may be **neutral variations** that do not confer a selective advantage or disadvantage. In **pseudogenes,** which have become inactivated by mutations, variations are not affected by natural selection and will change randomly by genetic drift. There is no consensus on how much variation is truly neutral.

■ **INTERACTIVE QUESTION 23.8**

a. Why is the highly deleterious sickle-cell allele still present in the gene pool of the U.S. population?

b. Why is this allele at such a relatively high frequency in the gene pool of some African populations?

Sexual Selection **Sexual dimorphism** is the distinction between males and females on the basis of secondary sexual characteristics. **Sexual selection** is the selection for traits that may not be adaptive to the environment but do enhance reproductive success. Such traits may increase an individual's success in competing for **(intrasexual selection)** or attracting **(intersexual selection)** a mate. Also called mate choice, intersexual selection may be based on showy traits that reflect the general health of the male and, thus, the fitness of his alleles.

The Evolutionary Enigma of Sexual Reproduction From a reproductive output standpoint, asexual reproduction is far superior, and sexual reproduction should, in theory, be selected against. One explanation for how sexual reproduction is maintained by natural selection is the selective advantage of genetic variation in disease resistance. For example, maintaining high variation in the cell surface molecules helps a popula-

tion's offspring resist pathogens such as viruses and bacteria that key in on such receptor molecules. The co-evolutionary race between host and rapidly evolving parasites is known as a "Red Queen race."

Why Natural Selection Cannot Fashion Perfect Organisms There are at least four reasons evolution does not produce perfect organisms. First, each species has evolved from a long line of ancestral forms, many of whose structures have been coopted for new situations. Second, adaptations are often compromises between the need to do several different things, such as swim and walk, or be agile and strong. Third, chance events affect a population's evolutionary history. Fourth, natural selection can act only on variations that are available; new alleles do not arise as they are needed.

■ **Word Roots**

inter- = between (*intersexual selection:* individuals of one sex are choosy in selecting their mates from individuals of the other sex, also called mate choice)

intra- = within (*intrasexual selection:* a direct competition among individuals of one sex for mates of the opposite sex)

micro- = small (*microevolution:* a change in the gene pool of a population over a succession of generations)

muta- = change (*mutation:* a change in the DNA of genes that ultimately creates genetic diversity)

poly- = many; **morph-** = form (*polymorphism:* the co-existence of two or more distinct forms of individuals in the same population)

■ **Structure Your Knowledge**

1. a. What is the Hardy-Weinberg theorem?

   b. Define the variables of the equation for Hardy-Weinberg equilibrium. Make sure you can use this equation to determine allele frequencies and predict genotype frequencies.

2. It seems that natural selection would work toward genetic unity; the genotypes that are most fit produce the most offspring, increasing the frequency of adaptive alleles and eliminating less beneficial alleles from the population. Yet there remains a great deal of variability within populations of a species. Describe some of the factors that contribute to this genetic variability.

Test Your Knowledge

MULTIPLE CHOICE: *Choose the one best answer.*

1. Which of the following determines an organism's fitness?
   a. survival
   b. number of matings
   c. adaptation to the environment
   d. successful competition for resources
   e. number of viable offspring

2. According to the Hardy-Weinberg theorem,
   a. the allele frequencies of a population should remain constant from one generation to the next if the population is large and only sexual recombination is involved.
   b. only natural selection, resulting in unequal reproductive success, will cause evolution.
   c. the square root of the frequency of individuals showing the recessive trait will equal the frequency of *p*.
   d. genetic drift, gene flow, and mutations are always maladaptive.
   e. all of the above are correct.

3. If a population has the following genotype frequencies, $AA = 0.42$, $Aa = 0.46$, and $aa = 0.12$, what are the allele frequencies?
   a. $A = 0.42$ $a = 0.12$
   b. $A = 0.6$ $a = 0.4$
   c. $A = 0.65$ $a = 0.35$
   d. $A = 0.76$ $a = 0.24$
   e. $A = 0.88$ $a = 0.12$

4. In a population with two alleles, *B* and *b*, the allele frequency of *b* is 0.4. What would be the frequency of heterozygotes if the population is in Hardy-Weinberg equilibrium?
   a. 0.16
   b. 0.24
   c. 0.48
   d. 0.6
   e. You cannot tell from this information.

5. In a population that is in Hardy-Weinberg equilibrium for two alleles, *C* and *c*, 16% of the population show a recessive trait. Assuming *C* is dominant to *c*, what percent show the dominant trait?
   a. 36%
   b. 48%

   c. 60%
   d. 84%
   e. 96%

6. Genetic drift is likely to be seen in a population
   a. that has a high migration rate.
   b. that has a low mutation rate.
   c. in which natural selection is occurring.
   d. that is very small.
   e. for which environmental conditions are changing.

7. Gene flow often results in
   a. populations that move to better environments.
   b. an increase in randomness in the genetic composition of the next generation.
   c. adaptive microevolution.
   d. a decrease in allele frequencies.
   e. a reduction of the allele frequency differences between populations.

8. The existence of two distinct phenotypic forms in a species is known as
   a. geographic variation.
   b. stabilizing selection.
   c. heterozygote advantage.
   d. polymorphism.
   e. directional selection.

9. The average heterozygosity of *Drosophila* is estimated to be about 14%, which means that
   a. 86% of fruit fly genes are identical.
   b. on average, 14% of a fruit fly's gene loci are heterozygous.
   c. only 14% of nucleotide sites differ between individuals.
   d. nucleotide variability must be very great between individuals.
   e. the fruit fly population never experienced a bottleneck effect.

10. Mutations are rarely the cause of microevolution in eukaryotes because
    a. they are most often harmful and do not get passed on.
    b. they do not directly produce most of the genetic variation present in a diploid population.
    c. they occur very rarely.
    d. they are only passed on when they occur in gametes.
    e. of all of the above.

11. In a study of a population of field mice, you find that 48% of the mice have a coat color that indicates that they are heterozygous for a particular gene. What would be the frequency of the dominant allele in this population?
    a. 0.2.4
    b. 0.48
    c. 0.50
    d. 0.60
    e. You cannot estimate allele frequency from this information.

12. In a random sample of a population of shorthorn cattle, 73 animals were red ($C^R C^R$), 63 were roan, a mixture of red and white ($C^R C^r$), and 13 were white ($C^r C^r$). Estimate the allele frequencies of C^R and C^r, and determine whether the population is in Hardy-Weinberg equilibrium.
    a. $C^R = 0.64$, $C^r = 0.36$; because the population is large and a random sample was chosen, the population is in equilibrium.
    b. $C^R = 0.7$, $C^r = 0.3$; the genotype ratio is not what would be predicted from these frequencies and the population is not in equilibrium.
    c. $C^R = 0.7$, $C^r = 0.3$; the genotype ratio is close to what would be predicted from these frequencies and the population is in equilibrium.
    d. $C^R = 1.04$, $C^r = 0.44$; the allele frequencies add up to greater than 1 and the population is not in equilibrium.
    e. You cannot estimate allele frequency from this information.

13. A scientist observes that the height of a certain species of asters decreases as the altitude on a mountainside increases. She gathers seeds from samples at various altitudes, plants them in a uniform environment, and measures the height of the new plants. All of her experimental asters grow to approximately the same height. From this she concludes that
    a. height is not a quantitative trait.
    b. the cline she observed was due to genetic variations.
    c. the differences in the parent plants' heights were due to directional selection.
    d. the height variation she initially observed was an example of nongenetic environmental influence.
    e. stabilizing selection was responsible for height differences in the parent plants.

14. Sexual selection will
    a. select for traits that enhance an individual's chance of mating.
    b. increase the size of individuals.
    c. result in individuals better adapted to the environment.
    d. produce more offspring.
    e. result in a relative fitness of more than 1.

15. The greatest source of genetic variation in plant and animal populations is from
    a. mutations.
    b. sexual recombination.
    c. selection.
    d. polymorphism.
    e. recessive masking in heterozygotes.

16. A plant population is found in an area that is becoming more arid. The average surface area of leaves has been decreasing over the generations. This trend is an example of
    a. a cline.
    b. directional selection.
    c. disruptive selection.
    d. gene flow.
    e. genetic drift.

17. Mice that are homozygous for a lethal recessive allele die shortly after birth. In a large breeding colony of mice, you find that a surprising 5% of all newborns die from this trait. In checking lab records, you discover that the same proportion of offspring have been dying from this trait in this colony for the past three years. (Mice breed several times a year and have large litters.) How might you explain the persistence of this lethal allele at such a high frequency?
    a. Homozygous recessive mice have a reproductive advantage.
    b. A large mutation rate keeps producing this lethal allele.
    c. There is some sort of heterozygote advantage and perhaps selection against the homozygous dominant trait.
    d. Genetic drift has kept the recessive allele at this high frequency in the population.
    e. Since this is a diploid species, the recessive allele cannot be selected against when it is in the heterozygote.

18. In breeding experiments with *Drosophila*, you count the offspring produced by each of three different genotypes and determine that flies with the genotype *AA* have a relative fitness of 1. What does that mean?

    a. *AA* flies have a lower fitness than do flies that are *Aa* or *aa*.

    b. *AA* flies produce more viable offspring than do *Aa* or *aa* flies.

    c. This fly population must be in Hardy-Weinberg equilibrium.

    d. *AA* flies live longer than do *Aa* or *aa* flies.

    e. The *A* and *a* alleles must both have a frequency of 0.5.

19. All of the following would tend to maintain balanced polymorphism in a population *except*

    a. balancing selection.

    b. directional selection.

    c. diversifying selection.

    d. heterozygote advantage.

    e. frequency-dependent selection.

20. Genetic analysis of a large population of mink inhabiting an island in Michigan revealed an unusual number of loci where one allele was fixed. Which of the following is the most probable explanation for this genetic homogeneity?

    a. The population exhibited nonrandom mating, producing homozygous genotypes.

    b. The gene pool of this population never experienced mutation or gene flow.

    c. A very small number of mink may have colonized this island, and this founder effect and subsequent genetic drift could have fixed many alleles.

    d. Natural selection has selected for and fixed the best adapted alleles at these loci.

    e. The colonizing population may have had much more genetic diversity, but genetic drift in the last year or two may have fixed these alleles by chance.

21. Directional selection would be most likely to occur when

    a. a population's environment has undergone a change.

    b. a population's environment has two very different habitats.

    c. frequency-dependent selection is acting on a population.

    d. a population's environment is very harsh.

    e. a population is small and its environment is stable.

22. If an allele is recessive and lethal in homozygotes,

    a. the allele is present in the population at a frequency of 0.001.

    b. the allele will be removed from the population by natural selection in approximately 1,000 years.

    c. the relative fitness of the homozygous recessive genotype is 0.

    d. the allele will most likely remain in the population at a low frequency because it cannot be selected against when in a heterozygote.

    e. Both c and d are correct.

23. Sexual reproduction may be maintained by natural selection because

    a. it produces the greatest number of offspring.

    b. intrasexual selection produces the strongest males.

    c. intersexual selection allows females to choose their mates.

    d. maintaining a high variability in a population for traits such as cell surface markers protects against pathogens such as viruses and bacteria.

    e. it maintains high genetic variability in a population's gene pool so that the population can adapt should environmental conditions change.

24. Humans have an estimated 1,000 olfactory receptor genes. This is most likely an example of

    a. gene flow.

    b. frequency-dependent selection.

    c. gene duplication.

    d. neutral variation.

    e. a quantitative character.

25. Which of these types of selection is mismatched with its example?

    a. disruptive—a population of black-bellied seedcrackers consists of birds with either small bills (more effective at eating soft seeds) or large bills (able to crack hard seeds)

    b. Intrasexual—elephant seal males are more than four times larger than females; males fight over areas of beach where females congregate during breeding season

    c. Intersexual—female zebra finches chose males with the brightest bills; bill color correlates with high levels of carotenoids (antioxidants)

    d. frequency-dependent—as fish of one coloration become more numerous, predator fish form a "search image" and preferentially feed on them

    e. stabilizing—the frequency of A, B, AB, and O blood groups remains constant in a population

Chapter 24
The Origin of Species

Framework

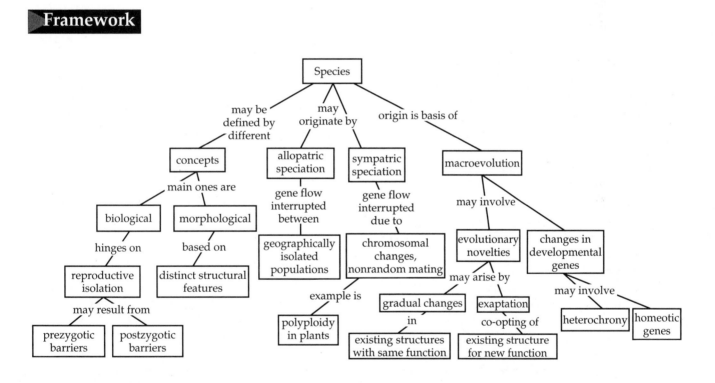

Chapter Review

Microevolution explains evolutionary changes within the gene pool of a population. **Macroevolution** considers the origin of new taxonomic groups (from species to higher taxons). The key process of macroevolution is **speciation,** the origin of a new species. Anagenesis, or phyletic evolution, involves the transformation of one species into a new species. In cladogenesis, or branching evolution, new species arise when a gene pool is split into separate pools and the parent species continues to exist. Cladogenesis is the process that increases biological diversity.

24.1 The biological species concept emphasizes reproductive isolation

Species are most often characterized by their physical form or morphology, although differences in physiology, biochemistry, behavior, and genetics also support the existence of distinct species.

The Biological Species Concept According to the **biological species concept,** developed by Ernst Mayr in 1942, a species is a population or group of populations of individuals that have the potential to interbreed in nature and produce viable, fertile offspring,

191

but which do not successfully interbreed with other species. The **reproductive isolation** that preserves the genetic integrity of a biological species results from barriers that prevent individuals of different species from producing viable, fertile hybrids.

Prezygotic barriers function before the formation of a zygote by preventing mating between species or successful fertilization should gametes meet. Prezygotic barriers include habitat isolation, in which two species that live in the same area occupy different habitats; temporal isolation, in which two species breed at different times; behavioral isolation, in which courtship rituals and behavioral signals are species specific; mechanical isolation, in which anatomical incompatibilities prevent mating with members of other species; and gametic isolation, in which the gametes of different species fail to fuse, due to mechanisms such as the inability of sperm to penetrate the membrane around the egg.

Should a hybrid zygote form, **postzygotic barriers** prevent it from developing into a viable, fertile adult. Postzygotic barriers include reduced hybrid viability, in which a hybrid zygote fails to survive embryonic development; reduced hybrid fertility, in which a viable hybrid individual is sterile, often due to the inability to produce normal gametes in meiosis; or hybrid breakdown, in which the hybrids are viable and fertile, but their offspring are feeble or sterile.

■ INTERACTIVE QUESTION 24.1

Name the type of reproductive barrier and whether it is pre- or postzygotic for the following examples.

Type of Barrier	Pre- or Post-	Example
a.	b.	Two species of frogs are mated in the lab and produce viable, but sterile, offspring.
c.	d.	Two species of sea urchins release gametes at the same time, but no cross fertilization occurs.
e.	f.	Two orchid species with different length nectar tubes are pollinated by different moths.
g.	h.	Two species of mayflies emerge during different weeks in spring.
i.	j.	Two species of salamanders mate and produce offspring, but the hybrid's offspring are sterile.
k.	l.	Two similar species of birds have different mating rituals.
m.	n.	Embryos of two species of mice bred in the lab usually abort.
o.	p.	Peepers breed in woodland ponds; leopard frogs breed in swamps.

The biological species concept does not work for species that are asexual, such as bacteria. Extinct species also cannot be grouped based on the criterion of interbreeding. Even for many living species, it is difficult to obtain data on whether interbreeding occurs. Alternative species concepts may be useful in various contexts.

Other Definitions of Species   Most species have been identified on the basis of physical characteristics, an approach called the **morphological species concept.** The **paleontological species concept** applies to species that are known only from the fossil record. The **ecological species concept** defines species on the basis of their ecological niche, the role they play and resources they use in the specific environments in which they are found. The emphasis of the **phylogenetic species concept** is on evolutionary lineages through which each species has a unique genetic history, producing distinct physical characteristics or molecular sequences. Sibling species, which are morphologically indistinguishable, may be identified in this approach, and the existence of reproductive isolation can confirm their distinction.

■ INTERACTIVE QUESTION 24.2

Fill in the following table to review the five approaches that biologists have proposed for conceptualizing a species.

Concept	Emphasis
biological	a.
b.	anatomical differences, most commonly used
c.	unique roles in specific environments
d.	morphologically discrete fossil species
phylogenetic	e.

24.2 Speciation can take place with or without geographic separation

Allopatric ("Other Country") Speciation **Allopatric speciation** occurs when geographic isolation interrupts gene flow between two subpopulations. The extent of the geographic barrier necessary to maintain genetic separation depends on the ability of the organisms to disperse. Either geologic change or colonization of a new area may geographically isolate populations.

Geographic separation alone is not a reproductive barrier in the biological sense. Intrinsic reproductive barriers may arise coincidentally as allopatric populations go down separate evolutionary paths due to genetic drift and selection. Biologists may assess allopatric speciation by bringing together members of separated populations in a laboratory setting or observing situations in which individuals of the two populations interact in the wild.

Sympatric ("Same Country") Speciation In **sympatric speciation,** reproductive barriers prevent gene flow between overlapping populations.

Mistakes during cell division may lead to **polyploidy** in plants. An **autopolyploid** has more than two sets of chromosomes that have all come from the same species. Failures in cell division can produce tetraploids (4*n*), which can fertilize themselves or other tetraploids but cannot successfully breed with diploids from the parent population, resulting in reproductive isolation in just one generation.

Polyploid species arise more commonly through **allopolyploidy.** Interspecific hybrids may propagate asexually, but are usually sterile due to difficulties in the meiotic production of gametes. Future mitotic or meiotic mistakes can result in the production of a fertile polyploid, which cannot interbreed with either parent species.

Speciation of polyploids has been frequent and important in plant evolution. Many of our agricultural plants are polyploids, and plant geneticists now hybridize plants by inducing meiotic and mitotic errors to create new species.

■ INTERACTIVE QUESTION 24.3

a. A new plant species B forms by autopolyploidy from species A, which had a chromosome number of 2*n* = 10. How many chromosomes would species B have?

b. If species A were to hybridize by allopolyploidy with species C (2*n* = 14) and produce a new, fertile species, D, how many chromosomes would species D have?

Sympatric speciation in animals may involve isolation within the geographic range of the parent population based on different resource usage. Nonrandom mating in a polymorphic population could also lead to sympatric speciation. In a laboratory study of two sympatric species of cichlids, mate choice based on coloration was shown to be the reproductive barrier that normally separates the two species.

■ INTERACTIVE QUESTION 24.4

a. Differentiate between allopatric and sympatric speciation.

b. How might reproductive barriers arise in each type of speciation?

Adaptive Radiation **Adaptive radiation** is the evolution of numerous, variously adapted species from a common ancestor introduced into an environment with many new ecological niches. Colonization of newly formed islands or the opening of numerous niches following mass extinctions may provide the opportunity for multiple speciation events.

■ INTERACTIVE QUESTION 24.5

What factors have contributed to the adaptive radiation of the thousands of endemic species of the Hawaiian Archipelago?

Studying the Genetics of Speciation Researchers have been able to identify genes that play a key role in speciation in some organisms. Two species of *Mimulus,* one pollinated by bees, the other by hummingbirds, have been mated in the greenhouse. The progeny of the hybrids, which have flowers that vary in color and shape, have been tested for pollinator choice. Two gene loci, which influence flower color and nectar production, have been identified as playing a key role in speciation by pollinator preference.

The Tempo of Speciation In the fossil record, new forms often appear rather suddenly, persist unchanged for a long time, and then disappear. According to the model of evolution known as **punctuated equilibrium,** long periods of stasis are punctuated by episodes of relatively rapid speciation and change. In geologic time, the thousands of years during which a species evolves is small compared with the millions of years a successful species may exist, and this short period of divergence may not be captured in the fossil record. Proponents of the gradualism model suggest that the apparently long periods of stasis may be only in external anatomy, while changes in internal anatomy, physiology, and behavior have gone unrecorded.

■ **INTERACTIVE QUESTION   24.6**

Compare the gradual and punctuated equilibrium models of evolution.

24.3   Macroevolutionary changes can accumulate through many speciation events

Evolutionary Novelties   Often very complex organs, such as the eyes of vertebrates and molluscs, have evolved gradually from simpler structures that served and continue to serve similar needs in ancestral species.

Evolutionary novelties may also evolve by the gradual modification of existing structures for new functions. *Exaptation* is the term for structures that evolved and functioned in one setting and were then co-opted for a new function.

■ **INTERACTIVE QUESTION   24.7**

Give examples of reptilian structures that were exaptations for flight in birds.

Evolution of the Genes That Control Development
The combination of the fields of evolutionary and developmental biology, called "evo-devo," explores how slight changes in developmental genes can result in major morphological differences between species.

Heterochrony is an evolutionary change in the rate or timing of development. **Allometric growth,** the dif-

fering rates of growth of various parts of the body, leads to the final shape of the organism. A minor genetic alteration that affects allometric growth can produce a very differently proportioned adult form.

Paedomorphosis is the retention in the adult of juvenile traits of ancestral organisms and can occur when genetic changes speed up the development of reproductive organs relative to the development of body form.

■ **INTERACTIVE QUESTION   24.8**

a. Fetal skulls of humans and chimpanzees have similar shapes. The quite distinctive differences in adult skull shape results from different patterns of _____.

b. A salamander species that retains its gills (a larval trait) when it is full grown and sexually mature is an example of _____.

c. The shorter feet of tree-dwelling salamanders may have resulted from an evolutionary change in a regulatory gene that switches off growth of the foot sooner than in ground-dwelling salamanders. These three cases (a., b., and c.) are all examples of

_____.

Changes in genes that control the spatial arrangement of body parts have also been important in macroevolution. **Homeotic genes** determine where basic body features develop. Mutations in homeotic genes called *Hox* genes, whose products provide positional information in embryos, can drastically alter body form. Duplications of the *Hox* complex of invertebrates may have been central to the evolution of vertebrates.

Evolution Is Not Goal-Oriented   *Equus,* the modern horse, descended from its much smaller, browsing, multitoed ancestor, *Hyracotherium,* through a series of speciation episodes that produced many different species and diverging trends, as documented in the fossil record.

According to S. Stanley's model of **species selection,** an evolutionary trend is analogous to a trend in a population produced by natural selection. The successful species that last the longest before extinction and generate the most new species will determine the direction of the trend. Evolutionary trends are ultimately dictated by environmental conditions; if conditions change, an evolutionary trend may end or change.

Word Roots

allo- = other; **-metron** = measure (*allometric growth:* the variation in the relative rates of growth of various parts of the body, which helps shape the organism)

ana- = up; **-genesis** = origin, birth (*anagenesis:* a pattern of evolutionary change involving the transformation of an entire population, sometimes to a state different enough from the ancestral population to justify renaming it as a separate species)

auto- = self; **poly-** = many (*autopolyploid:* a type of polyploid species resulting from one species doubling its chromosome number to become tetraploid)

clado- = branch (*cladogenesis:* a pattern of evolutionary change that produces biological diversity by budding one or more new species from a parent species that continues to exist)

hetero- = different (*heterochrony:* evolutionary changes in the timing or rate of development)

macro- = large (*macroevolution:* evolutionary change beginning with speciation, encompassing the origin of novel designs, evolutionary trends, adaptive radiation, and mass extinction)

paedo- = child (*paedomorphosis:* the retention in the adult organism of the juvenile features of its evolutionary ancestors)

post- = after (*postzygotic barrier:* any of several species-isolating mechanisms that prevent hybrids produced by two different species from developing into viable, fertile adults)

sym- = together; **-patri** = father (*sympatric speciation:* a mode of speciation occurring as a result of a radical change in the genome that produces a reproductively isolated subpopulation in the midst of its parent population)

Structure Your Knowledge

1. How are speciation and microevolution different?
2. Describe two major mechanisms through which evolutionary novelties may originate.

Test Your Knowledge

MULTIPLE CHOICE: *Choose the one best answer.*

1. The type of evolution that results in the greatest increase in biological diversity is
   a. anagenesis.
   b. cladogenesis.
   c. phyletic evolution.
   d. microevolution.
   e. Both a and c are correct.

2. Which of the following is *not* a type of intrinsic reproductive isolation?
   a. mechanical isolation
   b. behavioral isolation
   c. geographic isolation
   d. gametic isolation
   e. temporal isolation

3. Two species of frogs occasionally mate, but the offspring do not complete development. What type of barrier isolates these gene pools?
   a. gametic isolation
   b. prezygotic barrier
   c. hybrid breakdown
   d. reduced hybrid viability
   e. reduced hybrid fertility

4. For which of the following is the biological species concept least appropriate?
   a. plants
   b. animals
   c. bacteria
   d. fossils
   e. both c and d

5. A horse ($2n = 64$) and a donkey ($2n = 62$) can mate and produce a mule. How many chromosomes would there be in a mule's cells?
   a. 31   c. 63   e. 126
   b. 62   d. 64

6. What prevents horses and donkeys from hybridizing to form a new species?
   a. reduced hybrid fertility
   b. reduced hybrid viability
   c. mechanical isolation
   d. gametic isolation
   e. behavioral isolation

7. Which of the following species concepts identifies species based on their similarities resulting from their unique niche or role in the environment?
   a. biological
   b. ecological
   c. paleontological
   d. morphological
   e. phylogenetic

8. Allopatric speciation is more likely to occur when an isolated population
   a. is large and thus has more genetic variation.
   b. is reintroduced to its original homeland.
   c. is small and exposed to different selection pressures in its new habitat.
   d. inhabits an island close to its parent species' mainland.
   e. All of the above contribute to allopatric speciation.

9. A tetraploid plant species (with four identical sets of chromosomes) is probably the result of
   a. allopolyploidy.
   b. autopolyploidy.
   c. hybridization and nondisjunction.
   d. allopatric speciation.
   e. a and c.

10. All of the following would help to identify sister species *except*
    a. genetic analysis.
    b. test for reproductive incompatibility.
    c. phylogenetic analysis.
    d. morphological comparisons.
    e. All of the above are essential to distinguishing sister species.

11. There are 28 morphologically diverse species of a group of sunflowers called silverswords found on the Hawaiian Archipelago. These species are an example of
    a. a geographical cline.
    b. adaptive radiation.
    c. allopolyploidy.
    d. the bottleneck effect.
    e. sympatric speciation.

12. Which of the following is descriptive of the punctuated equilibrium model?
    a. Long periods of stasis are punctuated by episodes of relatively rapid speciation and change.
    b. Microevolution is the driving force of speciation.
    c. Most rapid speciation events involve polyploidy in plants.
    d. Evolution occurs gradually as the environment gradually changes.
    e. In the framework of geologic time periods, speciation events occur very slowly and the equilibrium of species is punctuated by frequent extinctions.

13. A new plant species C formed from hybridization of species A ($2n = 18$) with species B ($2n = 12$) would probably produce gametes with a chromosome number of
    a. 12.   c. 18.   e. 60.
    b. 15.   d. 30.

14. Which of the following would *not* contribute to allopatric speciation?
    a. geographic separation
    b. genetic drift
    c. gene flow
    d. different selection pressures
    e. founder effect

15. What is meant by the concept of species selection?
    a. Reproductive isolating mechanisms (both prezygotic and postzygotic) are responsible for maintaining the integrity of individual species.
    b. Characteristics that increase the probability of selection as a mate or success in competition for mates will be selected for and increase in frequency in a population.
    c. The species that last the longest and speciate the most often determine the direction of evolutionary trends.
    d. A new species accumulates most of its unique features as it originates and then maintains long periods of stasis in body form.
    e. The colonization of new and diverse habitats can lead to the adaptive radiation of numerous species.

16. Allometric growth
    a. results in paedomorphosis.
    b. results in an evolutionary trend of increasing body size.
    c. results from differences in the locations of expression of *Hox* genes and thus the placement of body parts.
    d. is usually associated with polyploidy in plants.
    e. is the relative differences in growth rates of different body parts.

17. The evolution of the swim bladder from lungs of an ancestral fish is an example of
    a. heterochrony.
    b. paedomorphosis.
    c. exaptation.
    d. changes in homeotic gene expression.
    e. punctuated equilibrium.

18. Which of the following is thought to be a critical event in the evolution of vertebrates from an invertebrate ancestor?
    a. exaptation of body segments to vertebrae
    b. duplication of the *Hox* complex (homeotic genes)
    c. allometric growth of tail segments
    d. allopatric speciation
    e. sympatric speciation

19. A botanist identifies a new species of plant that has 32 chromosomes. It grows in the same habitat with three similar species: species A ($2n = 14$), species B ($2n = 16$), and species C ($2n = 18$). Suggest a possible speciation mechanism for the new species.
    a. allopatric divergence by development of a reproductive isolating mechanism
    b. change in a key developmental gene that causes the plants to flower at different times
    c. autopolyploidy, perhaps due to a nondisjunction in the formation of gametes of species B
    d. allopolyploidy, a hybrid formed from species A and C
    e. Either answer c or d could account for the formation of this new plant species.

20. Which concept of species would be most useful to a field biologist identifying new species in a tropical rain forest?
    a. biological
    b. ecological
    c. paleontological
    d. morphological
    e. phylogenetic

Chapter 25
Phylogeny and Systematics

Key Concepts

25.1 Phylogenies are based on common ancestries inferred from fossil, morphological, and molecular evidence

25.2 Phylogenetic systematics connects classification with evolutionary history

25.3 Phylogenetic systematics informs the construction of phylogenetic trees based on shared characters

25.4 Much of an organism's evolutionary history is documented in its genome

25.5 Molecular clocks help track evolutionary time

Framework

A goal of systematics is to reconstruct the phylogenetic history of species. Cladistic analysis is used to identify monophyletic taxa based on shared derived characters. The best phylogenetic hypotheses are parsimonious trees based on molecular homologies as well as on morphological and fossil evidence.

Chapter Review

Phylogeny is the evolutionary history of a species. The analytical study of the diversity of life and its phylogenetic history is called **systematics. Molecular systematics** uses DNA and RNA comparisons to infer relationships between genes and whole genomes.

25.1 Phylogenies are based on common ancestries inferred from fossil, morphological, and molecular evidence

The Fossil Record Fossils are preserved impressions or remnants of past organisms. The strata of sedimen-

tary rocks form from the compression of layers of sand and silt deposits. The succession of organisms found in strata comprise the **fossil record.**

Paleontologists analyze a variety of fossils: preserved bones, teeth, or shells; mineralized fossils; sedimentary fossils that retain organic material that can be analyzed; molds or casts of organisms; trace fossils, such as footprints or burrows that give clues to an animal's behavior; and occasionally, an entire body of an organism preserved in ice, peat, or amber. The formation of a fossil is an unlikely occurrence. A large number of species that lived probably left no fossils; geologic processes destroy many fossils; and only a fraction of existing fossils have been found.

■ **INTERACTIVE QUESTION   25.1**

What types of organisms are most likely to appear in the fossil record?

Morphological and Molecular Homologies Homologies are similarities due to shared ancestry. Morphological and molecular homologies among living organisms can be used to infer phylogenetic history. The more similar morphologies or DNA sequences are, the more likely that organisms are closely related.

Sorting homology from analogy is essential in constructing phylogenies. **Analogy** is similarity due to convergent evolution, in which unrelated species develop similar features because they have similar ecological roles and natural selection has led to similar adaptations. Analogous structures are sometimes called **homoplasies.** In general, when two similar structures are very complex, they are more likely to be homologous and have a shared origin.

■ INTERACTIVE QUESTION 25.2

What two complications may make it difficult to determine phylogenetic relationships based on morphological similarities between species? Give examples.

In molecular comparisons, the more nucleotide sequences shared between comparable genes, the more likely the genes are to be homologous. Molecular comparisons are complicated by insertion or deletion mutations that change the lengths of homologous regions of DNA. Computer programs are used to identify similar sequences and then insert gaps in order to align homologous DNA segments properly for nucleotide comparisons. The matching bases between organisms that are not closely related may be molecular homoplasies. Mathematical tools have been developed to identify "distant" homologies between extremely divergent sequences.

■ INTERACTIVE QUESTION 25.3

Base deletions have changed the lengths and alignments of these two homologous regions of DNA. Determine the best possible fit between these two DNA sequences. How many deletions and base changes have occurred in these DNA segments?

A C G T G C A C G

A G T G A G G

25.2   Phylogenetic systematics connects classification with evolutionary history

Systematics dates back to Linnaeus, who named and classified organisms. **Taxonomy** divides organisms into categories based on a set of characteristics used to determine similarities and differences. Although not based on genealogy, parts of his system are useful in phylogenetic systematics.

Binomial Nomenclature   Each species is assigned a two-part latinized name—a **binomial**—consisting of the **genus** and the **specific epithet.**

Hierarchical Classification   Linnaeus also organized species into broader categories. Related genera are grouped into **families,** which are then grouped into **orders, classes, phyla,** and **kingdoms,** and more recently, into **domains.**

A taxonomic unit at any level is called a **taxon.** Both words of the binomial are italicized, and the names for all taxa at the genus level and higher are capitalized. Taxa across different lineages are not comparable because grouping into higher classification levels is ultimately arbitrary.

Linking Classification and Phylogeny   A **phylogenetic tree** represents hypotheses about evolutionary relationships. Each dichotomous branch point represents the divergence of two species from a common ancestor. Phylogenetic systematics originated with Darwin, who recognized the evolutionary implications of the classification hierarchy.

25.3   Phylogenetic systematics informs the construction of phylogenetic trees based on shared characters

Cladistics   A **cladogram** is a phylogenetic tree that shows the pattern of homologous shared characteristics among taxa. A **clade** consists of an ancestral species and all of its descendant species. Such a clade is **monophyletic.** A **polyphyletic** clade includes several clades but not their common ancestor, and a **paraphyletic** clade excludes some species that share a common ancestor with other species in the clade. **Cladistics** is the study of the relationships among clades.

After separating similar characters that are homologous from those that are analogous, systematists identify **shared primitive characters,** which are common to more inclusive taxa and evolved at an earlier branch point, and **shared derived characters,** which are evolutionary novelties that are unique to a particular clade.

To determine the branching sequence of a group of related species, the group is compared to an **outgroup,** a species or group of species that is closely related to the group being studied. Any homologies that are common to both the outgroup and the **ingroup,** the taxa to be grouped, are shared primitive characters that were present in an ancestor common to both groups. A comparison of the numbers of characters that are present in each taxon of the ingroup indicates the sequence in which shared derived characters evolved and determines the branch points used to produce a cladogram.

■ INTERACTIVE QUESTION 25.4

Place the taxa (outgroup, A, B, C, and D) on the clado-gram based on the presence or absence of the characters 1–4 as shown in this table. Indicate before each branch point the number for the shared derived character that evolved in the ancestor of the clade.

	Taxa Outgroup O	A	B	C	D
1	0	1	1	1	1
2	0	0	1	0	1
3	0	1	1	0	1
4	0	0	1	0	0

Characters

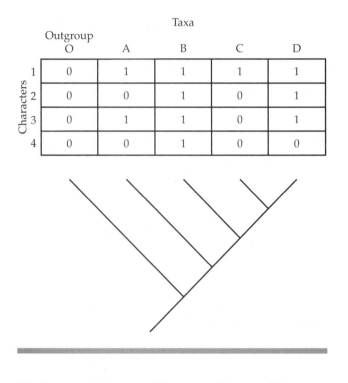

Phylogenetic Trees and Timing   The branching pattern of a phylogenetic tree is relative rather than absolute; it indicates only the order in which members of each clade last shared a common ancestor.

In **phylograms,** which are based on homologous nucleotide sequences, the length of a branch correlates with the number of evolutionary changes (in a nucleotide sequence) that have occurred in the lineage. **Ultrametric trees** use the fossil record to locate branch points in the context of geologic time and also show all the branches from a common ancestor to the present as the same length.

Maximum Parsimony and Maximum Likelihood   Systematists use morphological characters or molecular comparisons to choose among many possible phylogenetic trees using the principle of **maximum parsimony**—the smallest number of evolutionary changes is the simplest explanation and thus the best hypothesis to consider first. According to the principle of **maximum likelihood,** a tree can be found that rep-

resents the most likely sequence of events, given a certain set of assumptions. Computer programs search for the most parsimonious and most likely trees, using either "distance" methods that minimize the total of the percent differences among the sequences or "character-state" methods that minimize the total number of base changes or the most likely pattern of base changes among the sequences.

Phylogenetic Trees as Hypotheses   The most parsimonious tree based on the fewest number of evolutionary changes represents the best hypothesis of the relationships among a set of species. A hypothesis of phylogenetic relationships becomes more reliable with the more data that can be compared (a large database of DNA sequence comparisons and several derived characters used to define each clade).

■ INTERACTIVE QUESTION 25.5

According to the principle of parsimony, the evolution of the four-chambered heart should place birds and mammals in the same clade. Why does the most accepted evolutionary tree show them as separate branches from the reptilian line?

25.4 Much of an organism's evolutionary history is documented in its genome

Molecular systematics uses nucleic acid or other molecular comparisons to uncover phylogenetic relationships, making it possible both to compare genetic divergence among individuals within a species and to reconstruct phylogeny among very distantly related species. The ability to span this length of time depends on genes that evolve both very slowly and very rapidly, such as rRNA sequences and mtDNA.

Gene Duplications and Gene Families   Gene duplications are important evolutionary mutations, increasing the number of genes in the genome and the opportunity for change. **Orthologous genes** are homologous genes that diverge following speciation. **Paralogous genes** result from gene duplication and can diverge while in the same gene pool, providing new opportunities for evolutionary change.

Genome Evolution   Widespread orthologous genes indicate that all organisms share many biochemical and developmental pathways. There is a relatively small difference in the numbers of genes in organisms of very different complexity.

■ **INTERACTIVE QUESTION   25.6**

a. Give an example of genes that would be compared to uncover phylogenetic relationships among the earliest branches on the tree of life.

b. Give an example of genes that evolve very rapidly and are used to discriminate among closely related species.

c. Give an example of paralogous genes.

25.5   Molecular clocks help track evolutionary time

Molecular Clocks   Some regions of DNA appear to evolve at constant rates, and comparisons of the number of nucleotide substitutions in orthologous genes can serve as **molecular clocks** to estimate the time since species branched from their common ancestor. The number of differences in paralogous genes is proportional to the time since the genes were duplicated.

Some genes appear to have a reliable average rate of evolution. Graphs that plot nucleotide or amino acid differences against times for known evolutionary branch points can be used to estimate phylogenetic branchings that are not evident from the fossil record.

According to the **neutral theory,** much of the evolutionary change in genes does not affect fitness. Harmful mutations are removed quickly from the gene pool, but neutral mutational changes should become fixed at a constant rate.

■ **INTERACTIVE QUESTION   25.7**

Using the neutral theory of molecular evolution, explain why different genes might have a different molecular clock rate.

Natural selection, which favors some DNA changes over others, should disrupt the smooth running of the molecular clock. Evolutionists disagree about the extent of neutral mutations, and thus about whether molecular clocks are reliable for timing evolution. Many are skeptical when molecular clocks are used to date evolutionary divergences that occurred billions of years ago.

By comparing sequences of HIV viruses from samples taken at various times during the epidemic, including one from 1959, researchers have observed a remarkably consistent rate of evolution. They estimated that the HIV-1 M strain first infected humans in the 1930s.

The Universal Tree of Life   The rRNA genes have evolved so slowly that they have been used as the basis for constructing a universal tree of life. The tree consists of the domains Bacteria, Archaea, and Eukarya. Genome comparisons from the three domains indicate that **horizontal gene transfer,** perhaps through fusions of different organisms, occurred during the early history of life.

▶ **Word Roots**

analog- = proportion (*analogy:* similarity due to convergence)

bi- = two; **nom-** = name (*binomial:* a two-part latinized name of a species)

clado- = branch (*cladogram:* a dichotomous phylogenetic tree that branches repeatedly)

homo- = like, resembling (*homology:* similarity in characteristics resulting from a shared ancestry)

mono- = one (*monophyletic:* pertaining to a taxon derived from a single ancestral species that gave rise to no species in any other taxa)

parsi- = few (*principle of parsimony:* the premise that a theory about nature should be the simplest explanation that is consistent with the facts)

phylo- = tribe; **-geny** = origin (*phylogeny:* the evolutionary history of a taxon)

Structure Your Knowledge

1. Complete this concept map to help you review some key ideas about the field of systematics.

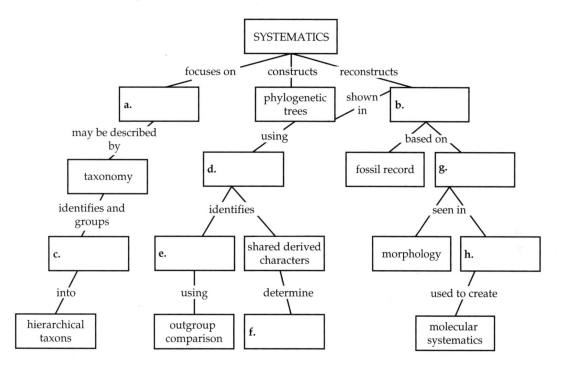

Test Your Knowledge

MULTIPLE CHOICE: *Choose the one best answer.*

1. The richest source of fossils is found
   a. in coal and peat moss.
   b. along gorges.
   c. within sedimentary rock strata.
   d. encased in volcanic rocks.
   e. in amber.

2. Which of the following is *least* likely to leave a fossil?
   a. a soft-bodied land organism such as a slug
   b. a marine organism with a shell such as a mussel
   c. a vascular plant embedded in layers of mud
   d. a freshwater snake
   e. a human

3. Which of the following is the best example of structures that are homoplasies?
   a. forelimbs of bat and mole
   b. silversword plants of Hawaii
   c. four-chambered hearts of birds and bats
   d. skulls of apes and humans
   e. hindlegs of Australian and North American moles

4. Which of these taxonomic units are comparable across all eukaryotes?
   a. genera
   b. species
   c. phyla
   d. kingdoms
   e. All of these units are determined by set criteria, and all are comparable across different lineages of eukaryotes.

5. Which of the following is a shared derived character for marsupial and eutherian mammals?
   a. parental care
   b. internal fertilization
   c. amniote egg
   d. production of milk for young
   e. complete embryonic development inside a uterus

6. Related families are grouped into the next-highest taxon called a
   a. class.
   b. order.
   c. phylum.
   d. genus.
   e. kingdom.

7. Which of the following would provide the best data for determining the phylogeny of three very similar extant species?
   a. the fossil record
   b. a comparison of embryological development
   c. a quantitative analysis of their morphological similarities and differences
   d. comparisons of DNA sequences
   e. comparison of their taxonomic assignments

8. Convergent evolution may result
   a. when older structures are co-opted for new functions.
   b. when homologous structures are adapted for different functions.
   c. from adaptive radiation.
   d. when species are widely separated geographically.
   e. when species have similar ecological roles.

9. Which of the following is the best description of our current hypothesis of the universal tree of life?
   a. The tree of life consists of three great domains: Bacteria, Archaea, and Eukarya.
   b. The base of the tree of life is still uncertain because the molecular clock is not accurate for evolutionary events that occurred that long ago.
   c. The Domain Archaea is known to be the first branch; domains Bacteria and Eukarya are more closely related to each other.
   d. There was substantial horizontal gene transfer and perhaps even fusion of different organisms during the early history of life.
   e. Both a and d represent our current hypothesis.

10. How could a molecular clock be useful for studying orthologous genes?
    a. It could help determine when the original gene duplication occurred.
    b. It could help date the divergence of different lineages that all contain that homologous gene.
    c. Because orthologous genes do not have neutral mutations, the molecular clock could not be used to study them.
    d. It could date when the gene first arose.
    e. Because the evolution of orthologous genes does not occur at a consistent rate, a molecular clock would not be useful.

11. Shared derived characters are
    a. those that characterize all the species on a branch of a dichotomous phylogenetic tree.
    b. determined through a computer comparison of orthologous genes in a group of species.
    c. homologous structures that develop during adaptive radiation.
    d. characters found in the outgroup but not in the species to be classified.
    e. used to identify species but not higher taxa.

12. A comparative study of which of the following would provide the best data on the early evolution of fungi and plants?
    a. mitochondrial DNA
    b. DNA from a chloroplast
    c. the amino acid sequence of chlorophyll
    d. DNA for ribosomal RNA
    e. the morphology of present-day specimens

13. A comparative study of which of the following would provide the best data on the ancestry of people from Germany, Italy, and Spain?
    a. mitochondrial DNA
    b. orthologous genes
    c. ribosomal RNA
    d. paralogous genes
    e. the fossil record

14. A taxon such as the class Reptilia, which does not include its relatives, the birds, is
    a. really an order.
    b. a clade.
    c. monophyletic.
    d. polyphyletic.
    e. paraphyletic.

15.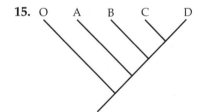

According to this cladogram, C and D are most closely related because they
    a. do not share a common ancestor with O, A, or B.
    b. are monophyletic.
    c. evolved from a common ancestor a long time ago.
    d. have the most shared derived characters in common.
    e. have the youngest species.

16. A biologist is studying the evolution of four similar species of birds. Which approach would allow her to choose the best phylogenetic tree from all possible phylogenies?

    a. Draw the simplest tree and choose that one.
    b. From a comparison of nucleotide sequences, determine the number of evolutionary events required for each tree and choose the most parsimonious tree.
    c. Compare the entire genomes of each species; the two most similar genomes are the two species that are most closely related.
    d. Determine which species can interbreed; those that can evolved from a common ancestor most recently.
    e. Choose the tree that has the most evolutionary changes required as the most probable explanation for why these similar birds have evolved into four distinct species.

TRUE OR FALSE: *Indicate T or F, and then correct the false statements.*

_____ 1. A monophyletic taxon includes an ancestral species and all its descendants.

_____ 2. The more the sequences of amino acids in homologous proteins vary, the more recently the two species have diverged.

_____ 3. Phylogenetic trees determined on the basis of similar structures may be inaccurate when adaptive radiations have created large differences or when convergent evolution has created misleading analogies.

_____ 4. Phylogenetic trees designed with the concept of parsimony should show the most likely evolutionary relationships.

_____ 5. The strongest support for a phylogenetic hypothesis comes from the concurrence of molecular, morphological, and fossil evidence.

_____ 6. Phlyograms are phylogenetic trees that indicate chronological branch points as found in the fossil record.

The Evolutionary History of Biological Diversity

Chapter 26

The Tree of Life: An Introduction to Biological Diversity

▶ Key Concepts

26.1 Conditions on early Earth made the origin of life possible

26.2 The fossil record chronicles life on Earth

26.3 As prokaryotes evolved, they exploited and changed young Earth

26.4 Eukaryotic cells arose from symbioses and genetic exchanges between prokaryotes

26.5 Multicellularity evolved several times in eukaryotes

26.6 New information has revised our understanding of the tree of life

▶ Framework

The evolutionary history of life is chronicled in the fossil record. Division between eras in the geologic record are marked by major biological transitions, often linked with continental drift and mass extinctions. This chapter presents hypotheses on how life began on Earth between 4.0 and 3.5 billion years ago, how prokaryotes evolved and changed Earth, how eukaryotic cells arose from endosymbioses and genetic exchange between prokaryotes, and how multicellularity may have evolved.

From these proposed beginnings, an incredible biological diversity has evolved. These diverse organisms are grouped into multiple kingdoms within a three-domain system—Archaea, Bacteria, and Eukarya.

▶ Chapter Review

26.1 Conditions on early Earth made the origin of life possible

Synthesis of Organic Compounds on Early Earth Earth formed about 4.6 billion years ago and was bombarded by huge rocks until about 3.9 billion years ago. As Earth cooled, water vapor condensed into oceans, and the atmosphere probably contained nitrogen and its oxides, carbon dioxide, methane, ammonia, and hydrogen sulfide.

In the 1920s, A. Oparin and J. Haldane independently hypothesized that conditions on the primitive Earth, in particular the reducing atmosphere, lightning, and intense ultraviolet radiation, favored the synthesis of organic compounds from inorganic precursors available in the atmosphere and seas.

In 1953, S. Miller and H. Urey tested the Oparin-Haldane hypothesis with an apparatus that simulated the hypothetical conditions of early Earth. After a week of applying sparks (lightning) to a warmed flask of water (the primeval sea) in an atmosphere of

H_2O, H_2, CH_4, and NH_3, Miller and Urey found a variety of amino acids and other organic molecules in the flask. Numerous laboratory replications using various combinations of atmospheric gases have been able to produce organic compounds. It is not clear, however, that the atmosphere was reducing enough or that it played a significant role in producing organic molecules.

Another hypothesis focuses on volcanoes and deep-sea vents as the possible locations of these early reactions and sources of chemical precursors.

It is also possible that organic compounds could reach Earth on meteorites. Carbonaceous chondrites are rocks from meteorites that contain carbon compounds, including amino acids similar to those produced in the Miller-Urey experiment. Scientists continue to look for signs of liquid water, oxygen in the atmosphere, and the possibility of life on other planets.

Abiotic Synthesis of Polymers Macromolecules are also needed for life. Researchers have created polypeptides by dripping dilute solutions of organic monomers onto hot sand or rock in the laboratory. Such polymers may have acted as catalysts on early Earth.

Protobionts Life requires accurate replication and metabolism, and the two are interdependent. **Protobionts** are aggregates of abiotically produced organic molecules surrounded by a membrane. Laboratory experiments indicate that protobionts could have formed spontaneously. When a mixture of organic ingredients includes lipids, droplets, called liposomes, become surrounded by a lipid bilayer. This bilayer acts as a selectively permeable membrane. If enzymes are added, liposomes can carry out simple metabolic reactions. An energy-storing membrane potential may develop across this membrane.

The "RNA World" and the Dawn of Natural Selection Once protobionts containing self-replicating molecules that carried genetic information had formed, natural selection could begin to shape the properties of protobionts.

RNA probably functioned as the first hereditary material. The discovery of RNA catalysts, called **ribozymes,** indicates that RNA molecules may have been capable of ribozyme-catalyzed replication. Ribozymes have been produced that can make complementary copies of parts of their sequence. Natural selection has been observed with RNA in the laboratory, in which sequences that are most stable and autocatalytic prevail in the population.

RNA-directed protein synthesis may have begun with the weak binding of specific amino acids to bases along RNA molecules and their linkage to form a short polypeptide. This polypeptide may then have behaved as an enzyme to help the RNA molecule replicate.

Protobionts with self-replicating, catalytic RNA could grow, split, pass some of their RNA to daughters, and be acted on by natural selection. Differential reproductive success of these offspring would have accumulated metabolic and hereditary improvements. DNA, a more stable genetic molecule, eventually replaced RNA as the carrier of genetic information.

■ **INTERACTIVE QUESTION   26.1**

Why do we say that for life to have begun required both accurate replication and metabolism?

26.2   The fossil record chronicles life on Earth

How Rocks and Fossils Are Dated The order in which fossils appear in the strata of sedimentary rocks indicates their relative age. Index fossils, such as shells of widespread animals, are used to correlate strata from different locations. Gaps may appear in the sequence, indicating that an area was above sea level or underwent erosion during some period of geologic time.

Radiometric dating is used to determine the ages of rocks and fossils. Each radioactive isotope has a fixed rate of decay—its **half-life,** which is the number of years it takes for 50% of an original sample to decay. During an organism's lifetime, it accumulates radioactive isotopes in proportions equal to the relative abundance of the isotopes in the environment. After the organism dies, the isotopes decay at their fixed rate. Carbon-14, with a half-life of 5,730 years, is used for determining the age of fossils up to 75,000 years old; potassium-40, which decays to the gas argon-40, can be used to date volcanic rock and thus infer the age of fossils associated with those rocks.

Patterns of **magnetic reversals,** when the orientation of Earth's magnetic field has reversed, can also be used to date rocks.

■ **INTERACTIVE QUESTION   26.2**

A fossil has one-eighth of the atmospheric ratio of C-14 to C-12. Estimate the age of this fossil.

The Geologic Record Geologists have developed a geologic record of Earth's history. The Archaean and the Proterozoic eons lasted approximately 4 billion years and are collectively referred to as the Precambrian. The Phanerozoic eon covers the last half billion years and is divided into three eras: the Paleozoic, Mesozoic, and Cenozoic. These eras are delineated by major transitions, such as mass extinctions and extensive radiations. The eras are subdivided into periods and epochs, which are often associated with lesser extinctions shown in the fossil record.

Mass Extinctions A species may become extinct due to a change in its physical or biological environment. Mass extinctions have occurred during periods of major environmental change.

The Permian mass extinction, occurring at the boundary between the Paleozoic and Mesozoic eras about 250 mya, claimed about 96% of marine animal species and many terrestrial ones. These extinctions may be related to massive volcanic eruptions in Siberia, which probably warmed the global climate and slowed the mixing of oceans, reducing the oxygen supply to marine organisms.

The Cretaceous mass extinction, which marks the boundary between the Mesozoic and Cenozoic eras about 65 mya, claimed more than one-half of the marine species, many families of terrestrial plants and animals, and most of the dinosaurs. Possibly an asteroid or comet collided with Earth and caused the Cretaceous extinction. The thin layer of clay rich in iridium (an element common in meteorites) that separates Mesozoic from Cenozoic sediments may have been the fallout from a huge cloud of dust created by the collision. This cloud would have blocked sunlight and severely affected weather. The huge Chicxulub crater on the Yucatán coast of Mexico may have been the site of the impact.

■ INTERACTIVE QUESTION 26.3

Why do extensive adaptive radiations often follow mass extinctions?

26.3 As prokaryotes evolved, they exploited and changed young Earth

The oldest known fossils are 3.5-billion-year-old **stromatolites,** which are rocklike layers of bacteria and sediment. They resemble the banded mats of sediment that form around certain modern-day bacterial colonies.

It is possible that the earliest life forms may have emerged as early as 3.9 billion years ago.

The First Prokaryotes Early protobionts must have used molecules available in the primitive soup for their growth and replication. They were eventually replaced with autotrophs that could produce their needed compounds from molecules in the environment. Some of these may have diversified and become able to use light energy. Heterotrophs then emerged that could live on products of the autotrophs or on the autotrophs themselves. These autotrophs and heterotrophs—the prokaryotes that diverged into the bacteria and the archaea—were the sole inhabitants of Earth from 3.5 to about 2 billion years ago.

Electron Transport Systems The chemiosmotic synthesis of ATP is common to all three domains. Transmembrane proton pumps probably used ATP to expel H^+ to regulate pH. Electron transport chains that coupled the oxidation of organic acids to transport of H^+ out of the cell would have saved the cell ATP. When electron transport systems extruded extra H^+, the diffusion of H^+ back into the cell could reverse the proton pumps and generate ATP.

Photosynthesis and the Oxygen Revolution Prokaryotes capable of non-oxygen-producing photosynthesis probably appeared early in prokaryotic history. Cyanobacteria may have evolved as early as 3.5 billion years ago and are the only living photosynthetic prokaryotes that generate O_2. Banded iron formations in marine sediments and the rusting of iron in terrestrial rocks that began about 2.7 billion years ago provide evidence of oxygen accumulation from the photosynthesis of cyanobacteria. The relatively rapid accumulation of atmospheric oxygen around 2.2 billion years ago caused the extinction of many prokaryotic groups and relegated others to anaerobic habitats.

■ INTERACTIVE QUESTION 26.4

What adaptation evolved in some prokaryotes that enabled them to thrive in this new oxygen-rich atmosphere?

26.4 Eukaryotic cells arose from symbioses and genetic exchanges between prokaryotes

The First Eukaryotes Chemical traces of eukaryotes may date back to 2.7 billion years ago. Fossils that are generally accepted as eukaryotic date from 2.1 billion years ago.

Endosymbiotic Origin of Mitochondria and Plastids

The cytoskeleton of eukaryotic cells allows them to change shape and engulf other cells. The first eukaryotes may have been predators of other cells. The cytoskeleton is also necessary for the movement of chromosomes in mitosis and meiosis.

The first eukaryotic cells may have engulfed an aerobic heterotrophic prokaryote and packaged it inside a vacuole. The prokaryote became an *endosymbiont*, living within the host cell. Eventually the relationship became mutualistic, and finally the endosymbiont evolved into a mitochondrion. Plastids, including chloroplasts, may have evolved through a similar endosymbiotic process. In this **serial endosymbiosis,** mitochondria probably arose first because they or their genetic remnants are present in all eukaryotes.

A great deal of evidence supports the endosymbiotic origin of plastids and mitochondria. Comparisons of small-subunit ribosomal RNA (SSU-rRNA) indicate that the alpha proteobacteria are the closest relatives of mitochondria, and that cyanobacteria are the closest relatives of plastids. Transposable elements probably transferred some of the genes present in mitochondria and plastids to the nucleus of the host cell.

■ **INTERACTIVE QUESTION   26.5**

List some of the evidence supporting an endosymbiotic origin of mitochondria and plastids?

Eukaryotic Cells as Genetic Chimeras

Some researchers propose that the nucleus evolved from an archaeal endosymbiont. Genes with close relatives in both bacteria and archaea have been found in nuclei, perhaps as a result of **genetic annealing,** involving horizontal gene transfers between many bacterial and archaeal lineages. According to the "you are what you eat" hypothesis, evolving eukaryotes occasionally incorporated some of the genes of their various bacterial and archaeal prey into their nucleus.

The Golgi apparatus and endoplasmic reticulum perhaps originated from infoldings of the plasma membrane. Homologs of cytoskeletal proteins have been found in bacteria. Despite the lack of the 9 + 2 microtubule apparatus in any prokaryote, some researchers speculate that flagella and cilia may have evolved from symbiotic bacteria.

26.5 Multicellularity evolved several times in eukaryotes

The Earliest Multicellular Eukaryotes

Although molecular clocks estimate that the common multicellular ancestor arose 1.5 billion years ago, the oldest known fossils are small algae from 1.2 billion years ago. Larger animals appear in the fossil record in the late Proterozoic. A diversity of algae and animals, including embryolike structures, has been found in a Chinese site dating from 570 million years ago.

According to the **snowball Earth hypothesis,** a severe ice age that occurred from 750 to 570 million years ago accounted for the limited diversity and distribution of multicellular eukaryotes, with the first major diversification occurring after the Earth thawed.

The Colonial Connection

The first multicellular organisms were **colonies** of autonomously replicating cells in which some cells became specialized. Some prokaryotes show cellular specialization, such as the nitrogen-fixing cells called heterocysts found in the filamentous cyanobacterium *Nostoc*.

A multicellular organism generally develops from a single cell, which divides to form an organism with many specialized cells. Multicellularity evolved several times among early eukaryotes.

The "Cambrian Explosion"

Fossils of Porifera (sponges) and Cnidaria date from the late Proterozoic. The appearance of most of the major phyla of animals in the first 20 million years of the Cambrian period, beginning 542 mya, is referred to as the "Cambrian explosion." But molecular evidence indicates that many animal phyla began to diverge between 1 billion and 700 million years ago.

Colonization of Land by Plants, Fungi, and Animals

Cyanobacteria coated damp terrestrial surfaces more than a billion years ago, but plants, fungi, and animals colonized land only about 500 million years ago, during the early Paleozoic era. The move onto land was associated with adaptations that helped prevent dehydration and permitted reproduction on land. Plants and fungi appear to have colonized the land together in symbiotic associations. Arthropods and tetrapod vertebrates are the most widespread and diverse land animals.

Continental Drift

The continents rest on great plates of crust that float on the molten mantle. Earthquakes, islands, and mountains are formed in regions where these shifting plates abut. Large-scale continental drift brought all the landmasses together into a supercontinent named **Pangaea** about 250 million years ago (mya), near the end of the Paleozoic era. This tremendous change had a great environmental impact as shorelines were eliminated, shallow coastal areas were drained, and the majority of the land became part of the continental interior. Many species became extinct, and new opportunities for remaining species became available. About 180 mya, during the Mesozoic era, Pangaea broke up and the continents drifted apart, creating a huge geographic isolation event.

■ INTERACTIVE QUESTION 26.6

Marsupials evolved in what is now North America, yet their greatest diversity is found in Australia. How can you account for this biogeographic distribution?

■ INTERACTIVE QUESTION 26.7

Explain the snowball Earth hypothesis.

26.6 New information has revised our understanding of the tree of life

Previous Taxonomic Systems Intuitively and historically, we have divided the diversity of life into two kingdoms—plants and animals. R. Whittaker proposed a five-kingdom system in 1969. The prokaryotes were set apart from the eukaryotes and placed in the kingdom Monera. The kingdoms Plantae, Fungi, and Animalia consist of multicellular eukaryotes, defined partly by characteristics of nutrition. The kingdom Protista included unicellular eukaryotes or simple multicellular organisms that are believed to have descended from them.

Reconstructing the Tree of Life: A Work in Progress
As systematists increase their understanding of phylogenetic relationships, alternative classification systems are proposed. The current **three-domain system** places the three major evolutionary lineages into a superkingdom taxon. The domains Bacteria and Archaea represent the early evolutionary divergence within the prokaryotes, and the domain Eukarya includes all the kingdoms of eukaryotes. Using molecular systematics and cladistics, systematists propose splitting the Protista into five or more kingdoms. New data will continue to be used in the challenge of constructing a classification system that reflects the evolutionary history of life.

▶ Word Roots

proto- = first (*protobionts:* aggregates of abiotically produced molecules)
stromato- = something spread out; **-lite** = a stone (*stromatolite:* rocks made of banded domes of sediment in which are found the most ancient forms of life)

▶ Structure Your Knowledge

1. What do scientists think the primitive Earth was like? How could life possibly arise in such an inhospitable environment?

2. Develop a simple concept map of Whittaker's five-kingdom system, indicating the key characteristics that are used to differentiate the kingdoms.

3. Now draw a map that shows the current three-domain system. Why do systematists keep changing the classification system?

▶ Test Your Knowledge

MULTIPLE CHOICE: *Choose the one best answer.*

1. Index fossils are fossils of
   a. unique organisms that are used to determine the relative rates of evolution in different areas.
   b. widespread organisms that allow geologists to correlate strata of rocks from different locations.
   c. transitional forms that link major evolutionary groups.
   d. extinct organisms that indicate the separation of different eras.
   e. fossils that have been dated using magnetic reversal data.

2. The half-life of carbon-14 is 5,730 years. A fossil that is 22,920 years old would have what amount of the normal proportion of C-14 to C-12?
   a. ½   c. ⅙   e. ¹⁄₁₆
   b. ¼   d. ⅛

3. The Permian mass extinction between the Paleozoic and Mesozoic eras
   a. affected a large percentage of marine and terrestrial species.
   b. coincided with the breaking apart of Pangaea.
   c. made way for the adaptive radiation of mammals, birds, and pollinating insects.
   d. appears to have been caused by a large asteroid striking Earth.
   e. did all of the above.

4. The primitive atmosphere of Earth may have favored the synthesis of organic molecules because it
   a. was highly oxidative.
   b. was reducing and had energy sources in the form of lightning and UV radiation.
   c. had a great deal of methane and organic fuels.
   d. had plenty of water vapor, carbon, oxygen, and nitrogen, providing the C, H, O, and N needed for organic molecules.
   e. consisted almost entirely of hydrogen gas.

5. Life on Earth is thought to have begun
   a. 540 million years ago, at the beginning of the Paleozoic era.
   b. 700 million years ago, during the Precambrian.
   c. 2.7 billion years ago, when oxygen began to accumulate in the atmosphere.
   d. between 3.5 and 4.0 billion years ago.
   e. 4.5 billion years ago, when the Earth formed.

6. Stromatolites are
   a. early prokaryotic fossils found in sediments around hydrothermal vents.
   b. a protobiont that forms when lipids assemble into a bilayer surrounding organic molecules.
   c. members of the kingdom Protista that are both motile and photosynthetic.
   d. fossils appearing to be eukaryotes that are about twice the size of bacteria.
   e. layered communities of bacteria, fossils of which represent the oldest known prokaryotes.

7. Liposomes are
   a. spontaneously forming droplets surrounded by a lipid membrane.
   b. prokaryotes that lived as endosymbionts in larger prokaryotic cells.
   c. polypeptides formed in the laboratory by dripping organic monomers onto hot rocks.
   d. abiotically produced lipids.
   e. RNA self-replicating molecules.

8. Which of the following is a proposed hypothesis for the origin of genetic information?
   a. Early DNA molecules coded for RNA, which then catalyzed the production of proteins.
   b. Early polypeptides became associated with RNA bases and catalyzed their linkage into RNA molecules.
   c. Short RNA strands were capable of self-replication and evolved by natural selection.
   d. As protobionts grew and split, they distributed copies of their molecules to their offspring.
   e. Early RNA molecules coded for DNA, which then dictated the order of amino acids in a polypeptide.

9. Which of the following provides the poorest evidence that life may have formed spontaneously?
   a. the discovery of ribozymes, showing that prebiotic RNA molecules may have been autocatalytic.
   b. the laboratory synthesis of liposomes.
   c. the fossil record.
   d. the abiotic synthesis of polymers from monomers dripped onto hot rocks.
   e. the production of organic compounds in the Miller-Urey experimental apparatus.

10. Banded iron formations in marine sediments provide evidence of
    a. the first prokaryotes around 3.5 billion years ago.
    b. oxidized iron layers in terrestrial rocks.
    c. the accumulation of oxygen in the seas from the photosynthesis of cyanobacteria.
    d. the evolution of photosynthetic archaea near deep-sea vents.
    e. the crashing of meteorites onto Earth, possibly transporting abiotically produced organic molecules from space.

11. Simple animals such as jellies and worms first appear in the fossil record around 600 million years ago. This time period may correspond to
    a. the thawing of snowball Earth.
    b. the first appearance of multicellular eukaryotes in the fossil record.
    c. the Cambrian explosion of animal forms.
    d. the colonization of land by plants and fungi, providing the food source for animals to follow.
    e. the rapid increase in atmospheric oxygen that allowed for the active lifestyle of animals.

12. The first generally accepted fossil evidence of eukaryotic cells
    a. dates from 1.2 billion years ago.
    b. appears in sediments around deep-sea vents.
    c. is found in stromatolites.
    d. dates from 2.7 billion years ago when oxygen accumulated in the atmosphere.
    e. appears after prokaryotes had been evolving on Earth for 1.5 billion years.

13. According to the theory of serial endosymbiosis,
    a. multicellularity evolved when colonies of cells became specialized and began to reproduce from eggs.
    b. plants were able to colonize land with fungi as symbionts in their roots.
    c. the mitochondria and plastids of eukaryotic cells were originally prokaryotic endosymbionts that became permanent parts of the host cell.
    d. the infoldings and specializations of the plasma membrane led to the evolution of the endomembrane system.
    e. the nuclear membrane evolved first, then plastids, and then mitochondria.

14. Evidence that the evolution of eukaryotic cells may have involved genetic annealing from multiple lineages comes from
    a. the universal presence of ribosomes in all cells.
    b. evidence of the transfer of genes from endosymbiotic prokaryotes to the host cell's nucleus.
    c. the presence of genes related to both bacteria and to archaea in eukaryotic nuclei.
    d. evidence of "chemical signatures" of eukaryotes that date from 2.7 billion years ago.
    e. Both b and c provide such evidence.

15. Mitochondria and plastids contain DNA and ribosomes and make some, but not all, of their proteins. Some of their proteins are coded for by nuclear DNA and produced in the cytoplasm. What may explain this division of labor?
    a. Over the course of evolution, some of the original endosymbiont's genes were transferred to the host cell's nucleus.
    b. The host cell's genome always included genes for making mitochondrial and plastid proteins.
    c. These organelles do not have sufficient resources to make all their proteins and rely on the help of the cell.
    d. Some mitochondria and plastid genes were contributed by early bacterial prokaryotes that shared genes with other primitive cells.
    e. The smaller prokaryotic ribosomes in these organelles cannot produce the eukaryotic proteins required for their functions.

16. Why is the diversity of life now organized into three domains?
    a. Molecular evidence indicates that the protists really include three separate lineages.
    b. The domains Bacteria and Archaea reflect the early evolutionary divergence of these two lineages, and they differ from eukaryotes.
    c. The eukaryotes are more alike than are the prokaryotes and thus belong in one big group.
    d. The division into plants, animals, and bacteria is more intuitive and accessible to most people.
    e. The origin of life involved three distinct stages—protobiont, prokaryote, and eukaryote—and each domain represents one of those stages.

Key Concepts

27.1 Structural, functional, and genetic adaptations contribute to prokaryotic success

27.2 A great diversity of nutritional and metabolic adaptations have evolved in prokaryotes

27.3 Molecular systematics is illuminating prokaryotic phylogeny

27.4 Prokaryotes play crucial roles in the biosphere

27.5 Prokaryotes have both harmful and beneficial impacts on humans

Framework

This chapter presents the morphology, phylogeny, and ecology of prokaryotes. These generally single-celled organisms greatly outnumber and outweigh all eukaryotes combined and flourish in all habitats, including ones that are too harsh for any other forms of life. The collective impact of these microscopic organisms is huge.

Chapter Review

27.1 Structural, functional, and genetic adaptations contribute to prokaryotic success

Prokaryotic cells are usually 1–5 μm in diameter. The three most common cell shapes are spheres (cocci), rods (bacilli), and spirals.

Cell-Surface Structures Prokaryotic cell walls maintain cell shape and protect the cell from bursting in a hypotonic surrounding. Most prokaryotes plasmolyze in a hypertonic environment. The cell walls of bacteria (but not archaea) contain **peptidoglycan,** a matrix composed of modified-sugar polymers cross-linked by short polypeptides.

The **Gram stain** is an important tool for identifying bacteria as **gram-positive** (bacteria with walls containing a thicker layer of peptidoglycan) or **gram-negative** (bacteria with more complex walls including an outer lipopolysaccharide membrane).

Many prokaryotes secrete a sticky **capsule** outside the cell wall that serves as protection from attack by a host's immune system and as glue for adhering to a substrate or other prokaryotes. Some bacteria may also attach by means of surface hairlike appendages called **fimbriae** and **pili.** Fimbriae are shorter and more numerous, and pili may be specialized to hold bacteria together during conjugation.

Motility Many bacteria are equipped with flagella, either scattered over the cell surface or concentrated at one or both ends of the cell. These flagella differ from eukaryotic flagella in their lack of a plasma membrane cover, structure, and function. ATP-driven pumps and the diffusion of H^+ back into the cell power the basal apparatus, embedded in the cell wall, which turns a hook attached to the protein filament of the flagellum.

Many motile bacteria exhibit **taxis,** an oriented movement in response to chemical, light, magnetic, or other stimuli.

Internal and Genomic Organization Extensive internal compartmentalization is not found in prokaryotic cells, although some may have membranes, usually infoldings of the plasma membrane that function in respiration or photosynthesis.

The circular DNA chromosome is concentrated in a **nucleoid region.** It contains one one-thousandth as much DNA as a eukaryotic genome and has relatively little protein associated with it. Smaller rings of DNA, called **plasmids,** may carry genes for antibiotic resistance, metabolism of unusual nutrients, or other functions. They replicate independently and may be transferred between bacteria during conjugation.

The ribosomes of prokaryotes are smaller than eukaryotic ribosomes and differ in their protein and RNA content. Some antibiotics work by binding to prokaryotic ribosomes and blocking protein synthesis.

Reproduction and Adaptation   With generation times from 20 minutes to 3 hours, prokaryotes have a huge reproductive potential. The growth of colonies usually stops due to the exhaustion of nutrients or the toxic accumulation of wastes. Some microorganisms release antibiotics, which inhibit the growth of competitors.

Some bacteria produce **endospores,** which are tough-walled dormant cells formed in response to harsh conditions. Endospores are so durable that microbiologists must use autoclaves to sterilize laboratory equipment and media.

Due to short generation times, favorable mutations are rapidly propagated, facilitating adaptive evolution to environmental changes. Scientists have studied such evolution in colonies of *E. coli* since 1988, documenting mutations and changes in gene expression through more than 20,000 generations. Horizontal gene transfer contributes to this rapid evolution.

Property	Description
Cell shape	**a.**
Cell size	**b.**
Cell surface	**c.**
Motility	**d.**
Internal membranes	**e.**
Genome	**f.**
Reproduction and growth	**g.**

■ **INTERACTIVE QUESTION   27.1**

Fill in the table in the next column with a brief description of the characteristics of prokaryotic cells.

27.2   A great diversity of nutritional and metabolic adaptations have evolved in prokaryotes

Nutrition refers to how an organism obtains energy (photo- or chemotroph) and the carbon it uses for synthesizing organic compounds (auto- or heterotroph).

■ **INTERACTIVE QUESTION   27.2**

Complete the following concept map that summarizes the four categories of nutrition.

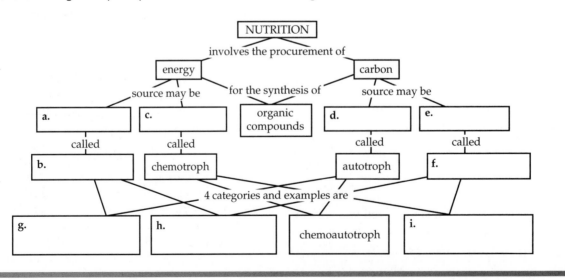

Thus, there are four major nutritional categories: (1) **photoautotrophs,** which use light energy and CO_2 to synthesize organic compounds; (2) **chemoautotrophs,** which obtain energy by oxidizing inorganic substances (such as H_2S, NH_3, or Fe^{2+}) and need only CO_2 as a carbon source; (3) **photoheterotrophs,** which use light energy but must obtain carbon in organic form; and (4) **chemoheterotrophs,** which use organic molecules as both an energy and a carbon source.

Metabolic Relationships to Oxygen **Obligate aerobes** need oxygen for cellular respiration; **facultative anaerobes** can use oxygen but also can grow in anaerobic conditions using fermentation; **obligate anaerobes** are poisoned by oxygen. Some obligate anaerobes use **anaerobic respiration** to break down nutrients, with inorganic molecules other than oxygen serving as the final electron acceptor.

Nitrogen Metabolism Nitrogen is an essential component of proteins and nucleic acids. Prokaryotes are able to metabolize various nitrogenous compounds. Some cyanobacteria and a few other prokaryotes obtain nitrogen through **nitrogen fixation,** converting atmospheric N_2 to ammonium.

Metabolic Cooperation The cyanobacterium *Anabaena* forms filaments in which most cells carry out photosynthesis, while a few cells called *heterocysts* perform nitrogen fixation. Members of the colony share nutrients through intercellular connections. **Biofilms** are surface-coating colonies characterized by intercellular signaling, proteins that adhere cells to each other and to the substrate, and channels in the colony for movement of nutrients and wastes.

27.3 Molecular systematics is illuminating prokaryotic phylogeny

While shape, staining characteristics, nutritional mode, and motility have been helpful characters for clinically identifying bacteria, they have not provided a phylogenetic classification.

Molecular Systematics Molecular comparisons of base sequences of SSU-rRNA indicate that domains Bacteria and Archaea diverged early in the history of life. A tentative phylogeny of taxa has been developed using molecular systematics. Continuing analysis of prokaryotic genomes shows that the genetic diversity of prokaryotes is huge and that horizontal gene transfer has been and continues to be significant in the evolution of prokaryotes.

Bacteria Bacteria show great diversity in their modes of nutrition and metabolism. Five major groups are described in the text.

Proteobacteria is a nutritionally diverse group of aerobic and anaerobic gram-negative bacteria. Its five subgroups are **Alpha Proteobacteria,** many of which are mutual symbionts or parasites of eukaryotes such as *Rhizobium* and *Agrobacterium;* **Beta Proteobacteria,** which include the important soil bacteria *Nitrosomonas;* **Gamma Proteobacteria,** which include photosynthetic sulfur bacteria and some serious pathogens and enterics (intestinal inhabitants such as *E. coli*); **Delta Proteobacteria,** including the colony-forming myxobacteria and the predacious bdellovibrio; and **Epsilon Proteobacteria,** the mostly pathogenic group that includes the stomach ulcer–causing *Helicobacter pylori.*

Chlamydias are obligate intracellular animal parasites. One of these gram-negative species is the most common cause of blindness and nongonococcal urethritis, a sexually transmitted disease.

Spirochetes are helical heterotrophs that move in a corkscrew fashion. Spirochetes cause syphilis and Lyme disease.

The **gram-positive bacteria** are a diverse group. The subgroup actinomycetes, colonial bacteria once mistaken for fungi, includes the species that cause tuberculosis and leprosy. Most actinomycetes are soil bacteria, some of which are cultured to produce antibiotics. There are diverse solitary species of gram-positive bacteria, including those responsible for anthrax and botulism, and the species of *Staphylococcus* and *Streptococcus.* Mycoplasmas, the smallest of all cells, are the only bacteria that lack cell walls.

The **cyanobacteria** have plantlike, oxygenic photosynthesis and are important producers in freshwater and marine ecosystems. Some filamentous cyanobacteria have cells specialized for nitrogen fixation.

■ INTERACTIVE QUESTION 27.3

a. Identify the large structure inside this *Bacillus anthracis* cell. What is its significance?

Bacillus anthracis

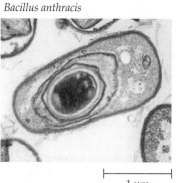

1 μm

b. Identify the major bacterial group to which this filamentous, photosynthetic species belongs. What is the function of the spherical cell indicated by the arrow?

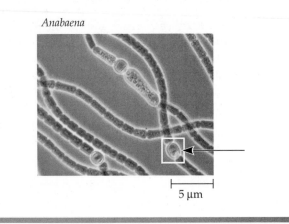

Anabaena

5 μm

Archaea   Archaea have some characteristics in common with bacteria, some with eukaryotes, and some unique characteristics. Many archaea are **extremophiles,** species that live in extreme habitats.

Extreme thermophiles may be found in hot sulfur springs and near deep-sea hydrothermal vents. The **extreme halophiles** (salt-lovers) live in extremely saline waters. **Methanogens** have a unique energy metabolism in which CO_2 is used to oxidize H_2, producing the waste product methane (CH_4). These strict anaerobes often live in swamps and marshes, are important decomposers in sewage treatment, and are gut inhabitants that contribute to the nutrition of cattle and other herbivores.

The methanogens and halophiles are placed into a clade called Euryarchaeota. Most thermophiles fit into the Crenarchaeota. Not all archaea are extremeophiles; numerous archaea have been found in more moderate habitats through genetic prospecting.

Prokaryote phylogeny continues to change as new groups are discovered. Examples include the oldest lineage, Korarchaeota, and the tiny Nanoarchaeota.

■ **INTERACTIVE QUESTION   27.4**

What is genetic prospecting? Why does it contribute to prokaryotic phylogeny?

27.4   Prokaryotes play crucial roles in the biosphere

Chemical Recycling   Prokaryotes are indispensable in recycling chemical elements. As **decomposers,** they return carbon, nitrogen, and other elements to the environment for assimilation into new living forms.

Symbiotic Relationships   **Symbiosis** is an ecological relationship involving direct contact between organisms of different species. If one organism is much larger than the other, it is called the **host** and the smaller is known as the **symbiont.** In **mutualism,** both symbiotic organisms benefit. Many of the 500–1,000 species of bacteria living in the human intestines are mutualists, digesting food and synthesizing important nutrients. In **commensalism,** one organism benefits while the other is neither harmed nor helped. In **parasitism,** a **parasite** benefits at the expense of the host.

■ **INTERACTIVE QUESTION   27.5**

Explain how cyanobacteria play an important role in chemical recycling.

27.5   Prokaryotes have both harmful and beneficial impacts of humans

Pathogenic Prokaryotes   About one-half of all human diseases are caused by pathogenic prokaryotes. Pathogens most commonly cause disease by producing toxins. **Exotoxins,** are proteins secreted by prokaryotes that cause such diseases as botulism and cholera. **Endotoxins,** which are lipopolysaccharides released from the outer membrane of gram-negative bacteria that have died, cause such diseases as typhoid fever and *Salmonella* food poisoning.

Improved sanitation and the development of antibiotics have decreased the incidence of bacterial disease in developed countries. The evolution of antibiotic-resistant strains of pathogenic bacteria, however, poses a serious health threat. Horizontal gene transfer also spreads genes connected with virulence, as in the emergence of the dangerous *E. coli* strain, O157:H7. Pathogenic prokaryotes such as *Bacillus anthracis*, *C. botulinum,* and *Yersinia pestis*, which causes plague, are potential bioterrorism weapons.

Prokaryotes in Research and Technology   **Bioremediation** is the use of organisms to remove environmental pollutants. Prokaryotes are used to treat sewage and clean up oil spills. They are used in the mining industry to recover metals from ores. Prokaryotes have been engineered to make vitamins, antibiotics, and other products.

Word Roots

-gen = produce (*methanogen:* microorganisms that obtain energy by using carbon dioxide to oxidize hydrogen, producing methane as a waste product)

-oid = like, form (*nucleoid:* a dense region of DNA in a prokaryotic cell)

an- = without, not; **aero-** = the air (*anaerobic:* lacking oxygen; referring to an organism, environment, or cellular process that lacks oxygen and may be poisoned by it)

anti- = against; **-biot** = life (*antibiotic:* a chemical that kills bacteria or inhibits their growth)

bi- = two (*binary fission:* the type of cell division by which prokaryotes reproduce; each dividing daughter cell receives a copy of the single parental chromosome)

chemo- = chemical; **hetero-** = different (*chemoheterotroph:* an organism that must consume organic molecules for both energy and carbon)

endo- = inner, within (*endotoxin:* a component of the outer membranes of certain gram-negative bacteria responsible for generalized symptoms of fever and ache)

exo- = outside (*exotoxin:* a toxic protein secreted by a bacterial cell that produces specific symptoms even in the absence of the bacterium)

halo- = salt; **-philos** = loving (*halophile:* microorganisms that live in unusually highly saline environments such as the Great Salt Lake or the Dead Sea)

mutu- = reciprocal (*mutualism:* a symbiotic relationship in which both the host and the symbiont benefit)

photo- = light; **auto-** = self; **-troph** = food, nourish (*photoautotroph:* an organism that harnesses light energy to drive the synthesis of organic compounds from carbon dioxide)

sym- = with, together; **-bios** = life (*symbiosis:* an ecological relationship between organisms of two different species that live together in direct contact)

thermo- = temperature (*thermophiles:* microorganisms that thrive in hot environments, often 60–80°C)

Structure Your Knowledge

1. What is the rationale for separating the archaea, the bacteria, and all the eukaryotes into three domains?

2. Describe four positive ways in which prokaryotes have an impact on our lives and on the world around us.

Test Your Knowledge

MULTIPLE CHOICE: *Choose the one best answer.*

1. Which of the following is *not* true of plasmids?
   a. They replicate independently of the main chromosome.
   b. They may carry genes for antibiotic resistance.
   c. They are essential for the existence of bacterial cells.
   d. They may be transferred between bacteria during conjugation.
   e. They may carry genes for special metabolic pathways.

2. Gram-positive bacteria
   a. have peptidoglycan in their cell walls, whereas gram-negative bacteria do not.
   b. have an outer membrane around their cell walls.
   c. have lipopolysaccharides in their cell walls and thus may be more pathogenic than gram-negative bacteria.
   d. are members of domain Archaea.
   e. have simpler, thick peptidoglycan cell walls.

3. Many prokaryotes secrete a sticky capsule outside the cell wall that
   a. allows them to glide along a slime thread.
   b. serves as protection from host defenses and glue for adherence.
   c. reacts with the Gram stain.
   d. is used for attaching cells during conjugation.
   e. is composed of peptidoglycan, polymers of modified sugars cross-linked by short polypeptides.

4. Chemoautotrophs
   a. are photosynthetic.
   b. use organic molecules for an energy and carbon source.
   c. oxidize inorganic substances for energy and use CO_2 as a carbon source.
   d. use light to generate ATP but need organic molecules for a carbon source.
   e. use light energy to extract electrons from H_2S.

5. A major source of genetic variation in prokaryotes is
   a. horizontal gene transfer.
   b. mutation.
   c. conjugation.
   d. plasmid exchange.
   e. all of the above.

6. A relationship in which one organism benefits and the host is neither harmed nor helped is called
   a. symbiosis.
   b. metabolic cooperation.
   c. mutualism.
   d. commensalism.
   e. parasitism.

7. Facultative anaerobes
   a. can survive with or without oxygen.
   b. are poisoned by oxygen.
   c. are anaerobic chemoautotrophs.
   d. are able to fix atmospheric nitrogen to make NH_4^+.
   e. include the methanogens that reduce CO_2 to CH_4.

8. Some prokaryotes have specialized membranes that
   a. contain ribosomes and function in protein synthesis.
   b. arose from endosymbiosis of smaller prokaryotes, creating mitochondria and chloroplasts.
   c. form from infoldings of the plasma membrane and may function in cellular respiration or photosynthesis.
   d. are produced by the endoplasmic reticulum, but differ in composition from eukaryotic membranes.
   e. enclose the nucleoid region and separate plasmids from the single prokaryotic chromosome.

9. Organizing prokaryotes into major clades within the domains Archaea and Bacteria has been based on
   a. fossil records of prokaryotes that have been recently discovered.
   b. molecular comparisons.
   c. the Gram stain and colony characteristics when grown on solid media.
   d. shape and motility.
   e. nutritional modes and ecology.

10. Some cyanobacteria are capable of nitrogen fixation. This process
    a. oxidizes nitrogen-containing compounds to produce ATP.
    b. allows these bacteria to live in anaerobic environments.
    c. removes soil nitrogen and returns N_2 gas to the atmosphere.
    d. converts N_2 to ammonium, making nitrogen available to plants for incorporation into proteins and nucleic acids.
    e. is an essential part of the nitrogen cycle that rescues nitrogen trapped in mineral deposits and makes it available to organisms for their nitrogen metabolism.

11. Archaea
    a. are believed to be more closely related to eukaryotes than to bacteria.
    b. have cell walls that lack peptidoglycan.
    c. are often found in harsh habitats, reminiscent of the environment of early Earth.
    d. include the methanogens, extreme halophiles, and extreme thermophiles, as well as groups found in more moderate habitats.
    e. are all of the above.

12. Which of the following groups is matched with an incorrect description?
    a. Proteobacteria—diverse gram-negative bacteria including pathogens such as *Salmonella* and *Helicobacter pylori*, as well as beneficial species such as *Rhizobium*
    b. Chlamydias—intracellular parasites, including a species that causes blindness and nongonococcal urethritis
    c. Spirochetes—helical heterotrophs, including pathogens causing syphilis and Lyme disease
    d. Gram-positive bacteria—diverse group that includes actinomycetes, mycoplasmas, and pathogens that cause anthrax, botulism, and tuberculosis
    e. Cyanobacteria—photosynthetic filamentous colonies; some solitary species use the pigment bacteriorhodopsin and inhabit saline waters

FILL IN THE BLANKS

_____ 1. the name for spherical prokaryotes

_____ 2. region in which the prokaryotic chromosome is found

_____ 3. common laboratory technique that identifies two groups of bacteria

_____ 4. surface appendages of prokaryotes used for adherence to substrate

_____ 5. an oriented movement in response to light or chemical stimuli

_____ 6. resistant cell that can survive harsh conditions

_____ 7. surface-coating cooperative colonies of prokaryotes

_____ 8. proteins that are secreted by pathogens and are potent poisons

_____ 9. use of organisms to remove environmental pollutants

_____ 10. type of respiration that uses molecules other than O_2 as final electron acceptor

Chapter 28
Protists

Framework

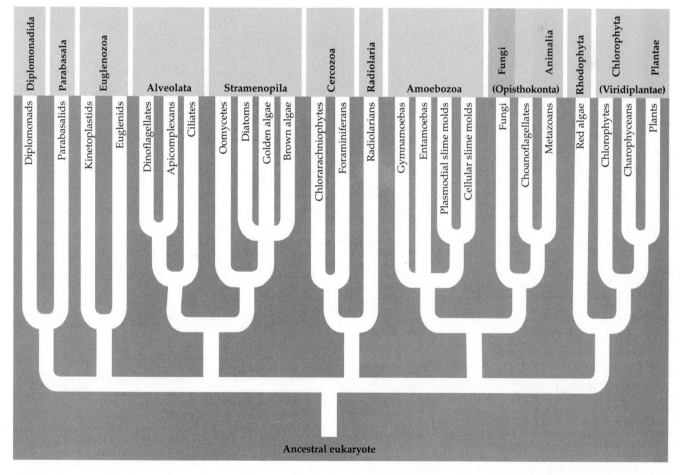

Ancestral eukaryote

This chapter surveys the incredible diversity of protists—eukaryotes that are not fungi, plants, or animals. A tentative phylogeny of eukaryotes is on page 221.

▶ Chapter Review

The traditional kingdom Protista has been abandoned; many lineages of this paraphyletic group may be recognized as their own kingdoms. The informal term *protist* is still used for eukaryotes that are not plants, animals, or fungi.

28.1   Protists are an extremely diverse assortment of eukaryotes

Protists include eukaryotic primarily unicellular organisms, with a few multicellular forms. Nutritionally, protists can be photoautotrophs, heterotrophs, or **mixotrophs,** which are both photosynthetic and heterotrophic. These modes of nutrition are spread throughout the various protist lineages. Based on their nutrition and not phylogeny, protists are often grouped for convenience into three categories: photosynthetic, plant-like protists (algae); ingestive, animal-like protists (protozoans); and absorptive, funguslike protists.

Protists abound almost anywhere there is water—they occupy freshwater, marine, and moist terrestrial habitats or live symbiotically within the bodies of hosts. They form an important part of plankton, the drifting community of mostly microscopic organisms found in bodies of water. Most aquatic food chains are based on phytoplankton, planktonic algae and cyanobacteria.

Reproduction and life cycles are highly diverse.

Endosymbiosis in Eukaryote Evolution   Mitochondria evolved from endosymbiotic alpha proteobacteria within the earliest eukaryotes. All eukaryotes studied thus far have mitochondria, or signs of having had them in the past. Plastids evolved from photosynthetic cyanobacteria that became endosymbionts within eukaryotic cells. This lineage of cells eventually gave rise to red and green algae. In several instances of **secondary endosymbiosis,** a red or green alga was engulfed by a heterotrophic eukaryote, leading to new protist lineages.

■ INTERACTIVE QUESTION   28.1

The plastids of chlorarachniophytes are surrounded by four membranes and enclose a vestigial nucleus called a *nucleomorph.* What does this evidence suggest?

28.2   Diplomonads and parabasalids have modified mitochondria

Diplomonads and parabasalids are found in anaerobic environments. They do not have plastids and their reduced mitochondria lack DNA, electron transport chains, and enzymes of the citric acid cycle.

Diplomonads   **Diplomonads,** such as the intestinal parasite *Giardia intestinalis,* have two nuclei and multiple flagella.

Parabasalids   The **parabasalids** include protists called trichomonads, of which *Trichomonas vaginalis* is the best known. This common vaginal inhabitant moves with flagella and an undulating part of its plasma membrane.

28.3   Euglenozoans have flagella with a unique internal structure

The diverse clade Euglenozoa includes predatory heterotrophs, autotrophs, and pathogenic parasites, all of which have a spiral or crystalline rod inside their flagella. Most have disk-shaped mitochondrial cristae.

Kinetoplastids   A single large mitochondrion containing a mass of DNA called a kinetoplast is characteristic of the group named **kinetoplastids.** These protists include free-living heterotrophs as well as a number of parasites of plants, animals, and other protists.

Euglenids   The **euglenids** are characterized by one or two flagella that emerge from an anterior pocket and by the storage polymer paramylon. Many species of the photosynthetic *Euglena* switch to heterotrophy in the absence of sunlight.

■ INTERACTIVE QUESTION   28.2

Sleeping sickness, caused by the kinetoplastid *Trypanosoma* and transmitted by the African tsetse fly, is fatal if untreated. Why is it so difficult for the immune system to attack this parasite?

28.4   Alveolates have sacs beneath the plasma membrane

Membrane-bounded alveoli under the plasma membrane are characteristic of the groups in the clade Alveolata. These sacs may help to stabilize the cell surface or to osmoregulate.

Dinoflagellates Most species of **dinoflagellates** are unicellular, each with a characteristic shape. The beating of two flagella in perpendicular grooves on the cell surface between the internal cellulose plates of the cell produces a characteristic spinning movement.

Dinoflagellates make up a large proportion of marine and freshwater phytoplankton. Blooms of some populations are responsible for the often-harmful red tides. Photosynthesis by symbiotic dinoflagellates, living in small coral polyps, provides the main food source for coral reef communities.

Apicomplexans All **apicomplexans** are animal parasites, and most have complex life cycles that often include sexual and asexual stages and several host species. The parasites spread by tiny infectious cells called **sporozoites.** They have an apical complex of organelles specialized for invading host cells and a non-photosynthetic plastid called an apicoplast.

The control of malaria, caused by *Plasmodium,* is complicated by multiple factors: development of insecticide resistance in *Anopheles* mosquitoes, which spread the disease; drug resistance in *Plasmodium;* the sequestering of the parasite within human liver and blood cells; and the ability of the parasite to change its surface proteins. The genome of *Plasmodium* has been sequenced and the expression of most genes has been tracked. These developments may help development of vaccines.

Ciliates These protists are characterized by the cilia they use to move and feed. Their numerous cilia, coordinated by a submembrane system of microtubules, may be widespread or clumped.

Ciliates have two types of nuclei: Large macronuclei control everyday functions of the cell, have multiple copies of the genome, and split during binary fission (asexual reproduction); small micronuclei are exchanged in a process called **conjugation,** which provides for genetic recombination. Foreign genetic material is also excised from the genome during reproduction following conjugation.

■ **INTERACTIVE QUESTION   28.3**

Label the indicated structures in this diagram of a *Paramecium.* To which group and protist clade does this organism belong?

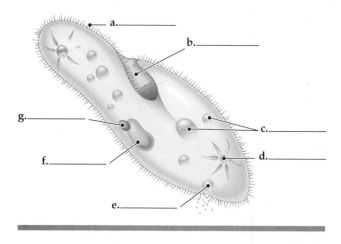

28.5 Stramenopiles have "hairy" and smooth flagella

The clade Stramenopila includes several heterotrophic groups and various groups of algae, all characterized by hairlike projections on a flagellum. In some cases, only reproductive cells are flagellated, having one smooth and one "hairy" flagellum.

Oomycetes (Water Molds and Their Relatives) **Oomycetes** include water molds, white rusts, and downy mildews. Although many resemble fungi in appearance, oomycetes have cell walls made of cellulose, as well as a predominant diploid stage and flagellated cells in their life cycles. Molecular systematics confirms that oomycetes and fungi are not closely related.

The large egg cell of a water mold is fertilized by a smaller sperm nucleus. Zygotes, following dormancy, germinate to form hyphae, which are tipped by zoosporangia that asexually produce biflagellated zoospores. Water molds are important decomposers in aquatic ecosystems. White rusts and downy mildews can be destructive parasites of land plants. Genetic studies of *Phytophthora infestans,* the oomycete that causes potato late blight, indicate that it has acquired genes in recent decades that make it more aggressive and resistant to pesticides.

Diatoms Diatoms (baccilariophytes) are a major component of marine and freshwater phytoplankton with unique, protective, boxlike silica walls. These unicellular protists usually reproduce asexually, although eggs and sperm may be produced. Food reserves are in the form of laminarin, a glucose polymer. Massive quantities of fossilized diatom walls make up diatomaceous earth. Diatoms may make a contribution to the field of nanotechnology as nanoengineers study the self-assembly of their intricate shells.

Golden Algae The color of **golden algae** (chrysophytes) results from yellow and brown carotenoids. Most golden algae are unicellular, biflagellated, and found among marine and freshwater plankton. Some species are mixotrophic. Many species form long-lasting resistant cysts when population densities become too high.

Brown Algae The mostly marine **brown algae** (phaeophytes) include some of the largest and most complex algae. Their color is due to accessory pigments in their plastids, which are homologous to the plastids of golden algae and diatoms.

Large marine brown algae (and some red and green algae) are called seaweeds. Some seaweeds have tissues and organs that are analogous to those found in plants. Seaweeds may have a plantlike **thallus,** or body, consisting of a rootlike **holdfast** and a stemlike **stipe** that supports leaflike **blades.** Some brown algae have floats, which keep the photosynthesizing blades near the surface. The giant, fast-growing brown algal kelps live in deeper waters. Humans use some seaweeds for food and as thickeners in processed foods.

Some algae have an **alternation of generations.** The multicellular, haploid gametophyte produces gametes. After syngamy, the diploid zygote grows into the multicellular sporophyte, which then produces spores by meiosis. The gametophyte and sporophyte may be similar in appearance **(isomorphic)** or distinct **(heteromorphic).**

■ **INTERACTIVE QUESTION   28.4**

What adaptations help brown algal seaweeds inhabit intertidal zones?

28.6   Cercozoans and radiolarians have threadlike pseudopodia

Amoebas, originally defined on the basis of their cellular extensions called **pseudopodia,** are spread across many eukaryotic taxa. Cercozoa, a newly recognized clade, contains species of amoebas that have threadlike pseudopodia. Chlorarachniophytes and foraminiferans are cercozoans; radiolarians, also with threadlike pseudopodia, are closely related.

Foraminiferans (Forams) **Foraminiferans,** or **forams,** are known for their porous, multichambered shells, called **tests,** made of organic material and calcium carbonate. Pseudopodia extending through the pores function in swimming, test formation, and feeding. Many forams obtain nourishment from symbiotic algae.

Radiolarians Slender pseudopodia called axopodia help these organisms phagocytize microscopic food organisms. **Radiolarians** are primarily marine and have delicate silica shells.

■ **INTERACTIVE QUESTION   28.5**

How are foram fossils used?

28.7   Amoebozoans have lobe-shaped pseudopodia

Gymnamoebas Most amoebas of the diverse group gymnamoebas are found free-living in freshwater, marine, or soil habitats, where they consume bacteria and other protists.

Entamoebas Amoebas of the genus *Entamoeba* are parasites of vertebrates and some invertebrates. *E. histolytica* causes amebic dysentery in humans.

Slime Molds The resemblance of the slime molds (or mycetozoans) to fungi is the result of convergent evolution to similar lifestyles. Molecular comparisons place them in the clade Amoebozoa.

Plasmodial slime molds engulf food particles by phagocytosis as they grow through leaf litter or rotting logs. This multinucleate mass called a **plasmodium** is the feeding stage. Under harsh conditions, sporangia on erect stalks produce resistant spores by meiosis. Spores germinate, the amoeboid or flagellated haploid cells fuse, and the diploid nucleus repeatedly divides to form a new plasmodium.

During the feeding stage of the life cycle, **cellular slime molds** consist of haploid solitary amoeboid cells. As food is depleted, the individual cells congregate into a mass. Asexual fruiting bodies produce resistant spores. In sexual reproduction, two amoebas fuse to form a zygote that develops into a giant cell enclosed in a protective wall. Following meiosis and mitosis, haploid amoebas are released. The cellular slime mold *Dictyostelium discoideum* is an experimental model for studying the evolution of multicellularity. Research indicates that a recognition system allows noncheater cells to aggregate differentially from "cheater" cells (mutant cells that always become spore cells).

■ INTERACTIVE QUESTION 28.6

Compare the following aspects of the life cycles of plasmodial slime molds and cellular slime molds.

Group	Haploid or Diploid	Common Body Form	Gametes or Sexual Cells
Plasmodial slime molds	a.	b.	c.
Cellular slime molds	d.	e.	f.

28.8 Red algae and green algae are the closest relatives of land plants

Descendants of the heterotrophic protist that acquired a cyanobacterium endosymbiont evolved into red algae and green algae more than a billion years ago. Land plants arose from the green algal lineage at least 475 million years ago.

Red Algae The accessory pigment phycoerythrin produces the color of **red algae**, and allows them to absorb the wavelengths of light that penetrate into deep water. Most red algae species are marine and multicellular, often with filamentous, delicately branched thalli. Some coralline algae have cell walls hardened by calcium carbonate and are members of coral reef communities. Alternation of generations is common, although life cycles are diverse and they have no flagellated stages in their life cycle.

Green Algae The chloroplasts of **green algae** resemble those of plants, and molecular systematics indicates that green algae and plants share a common ancestor. Some systematists favor including green algae in an extended kingdom, Viridiplantae. The two main green algal groups are chlorophytes and charophyceans. The charophyceans are very closely related to land plants.

Most chlorophytes live in fresh water, although many are marine. Unicellular forms may be planktonic, inhabitants of damp soil, symbionts in other eukaryotes, or mutualistic partners with fungi—forming associations known as lichens.

Size and complexity have increased in the green algae in three ways: the formation of colonies; repeated nuclear divisions to produce multinucleated filaments; and cell division and differentiation to produce true multicellular forms. Some large marine forms are considered seaweeds.

The life cycles of most green algae include sexual (with biflagellated gametes) and asexual stages. Some multicellular green algae have an alternation of generations, as in the seaweed *Ulva*.

Word Roots

-phyte = plant (*gametophyte*: the multicellular haploid form in organisms undergoing alternation of generations)

con- = with, together (*conjugation*: in ciliates, the transfer of micronuclei between two cells that are temporarily joined)

hetero- = different; **-morph** = form (*heteromorphic*: a condition in the life cycle of all modern plants in which the sporophyte and gametophyte generations differ in morphology)

iso- = same (*isomorphic*: alternating generations in which the sporophytes and gametophytes look alike, although they differ in chromosome number)

pseudo- = false; **-podium** = foot (*pseudopodium*: a cellular extension of amoeboid cells used in moving and feeding)

thallos- = sprout (*thallus*: a seaweed body that is plantlike but lacks true roots, stems, and leaves)

Structure Your Knowledge

1. Evidence indicates that all plastids evolved from a cyanobacterium that was engulfed by an ancestral heterotrophic eukaryote. Fill in the blanks in the figure on page 226, which diagrams the diversification of this ancestral eukaryote into red algae and green algae as well as several instances of secondary endosymbiosis that led to various protist groups.

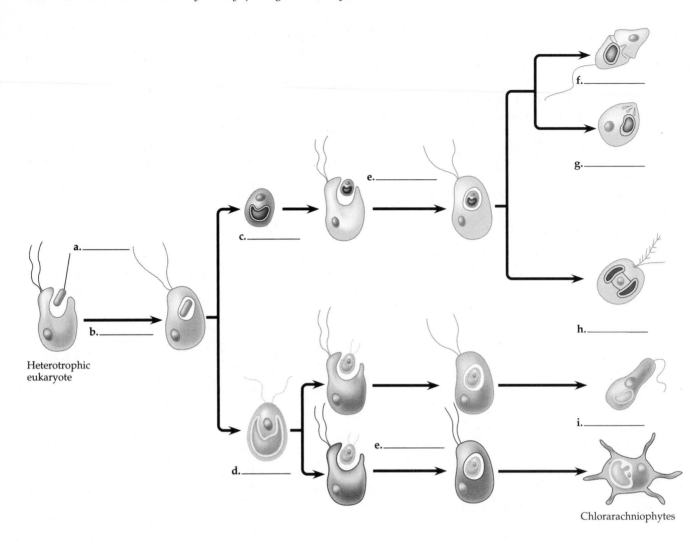

Heterotrophic
eukaryote

a. _____

b. _____

c. _____

d. _____

e. _____

e. _____

f. _____

g. _____

h. _____

i. _____

Chlorarachniophytes

▶ Test Your Knowledge

MATCHING: *Match the protist clades with their characteristics and examples.*

_____ **1.** modified mitochondria, two nuclei; *Giardia*

_____ **2.** plant-type chloroplasts; *Chlamydomonas, Ulva*

_____ **3.** lobe-shaped pseudopodia; free-living amoeba, parasites, slime molds

_____ **4.** phycoerythrin, no flagellated stage; red algae

_____ **5.** hairy and smooth flagella; water molds, diatoms, golden and brown algae

_____ **6.** amoebas with threadlike pseudopodia; forams, radiolarians

_____ **7.** subsurface sacs; dinoflagellates, apicomplexans, ciliates

_____ **8.** flagella with spiral or crystalline rod; *Trypanosoma, Euglena*

_____ **9.** undulating membrane, modified mitochondria; *Trichomonas*

A. Alveolata

B. Amoebozoa

C. Cercozoa & Radiolaria

D. Chlorophyta

E. Diplomonadida

F. Euglenozoa

G. Parabasala

H. Rhodophyta

I. Stramenopila

MULTIPLE CHOICE: *Choose the one best answer.*

1. Mixotrophs are
   a. plasmodial slime molds.
   b. cellular slime molds.
   c. gametophyte and sporophyte generations that differ in morphology.
   d. organisms that can be both heterotrophic and autotrophic.
   e. organisms that can be both parasitic and free-living.

2. Phytoplankton
   a. form the basis of most marine food chains.
   b. include the multicellular green, red, and brown algae.
   c. are mutualistic symbionts that provide food for coral reef communities.
   d. are unicellular heterotrophs that float near the ocean surface.
   e. Both a and d are correct.

3. According to the theory of secondary endosymbiosis,
   a. multicellularity evolved when primitive cells incorporated prokaryotic cells that then took on specialized functions.
   b. the symbiotic associations found in lichens resulted from the incorporation of algal protists into the ancestors of fungi.
   c. cells that had obtained their plastids through endosymbiosis were engulfed and themselves became plastids in heterotrophic eukaryotic cells.
   d. the infoldings and specializations of the plasma membrane led to the evolution of the endomembrane system.
   e. the nuclear membrane evolved first, then mitochondria, and then plastids.

4. The diplomonads and parabasalids are unique among eukaryotes in that
   a. they lack nuclear membranes, even though diplomonads have two nuclei.
   b. they lack ribosomes.
   c. they lack plastids, and all other protist lineages have at least some members that are autotrophic.
   d. they have modified mitochondria that lack electron transport chains and citric acid-cycle enzymes.
   e. they were an early branch from the bacteria rather than the archaea.

5. Genetic variation is generated in the ciliate *Paramecium* when
   a. a micronucleus replicates its genome many times and becomes a macronucleus.
   b. zoospores, which are gametes, fuse.
   c. micronuclei are exchanged in conjugation.
   d. mutations occur in the many copies of the genome contained in the macronucleus and are passed on to offspring.
   e. plasmids are exchanged in conjugation.

6. The Chlorophyta, or green algae, are believed to share a common ancestor with plants because
   a. they are the only multicellular algal protists.
   b. they do not have flagellated gametes.
   c. they are the only protists whose plastids evolved from cyanobacteria.
   d. their chloroplasts are similar in ultrastructure and pigment composition to those of plants.
   e. they are the only algae that exhibit alternation of generations.

7. Which of the following is *not* true of seaweeds?
   a. They are found in the red, green, and brown algal groups.
   b. They have true roots that anchor them tightly to withstand the turbulence of waves.
   c. They are a source of food and commercial products.
   d. The gel-forming polysaccharides in their cell walls protect them from the abrasive action of waves.
   e. They are multicellular photoautotrophs.

8. Examples of mutualistic symbiotic relationships include
   a. dinoflagellates producing red tides.
   b. dinoflagellates living within cnidarians in coral reefs.
   c. sporozoans in their multiple hosts.
   d. lichens.
   e. both b and d.

9. The protist clade Alveolata includes
   a. the diatoms, golden algae, and brown algae.
   b. the dinoflagellates, apicomplexans, and ciliates.
   c. the slime molds and water molds that are the ancestors of fungi.
   d. the diplomonads and parabasalids.
   e. the amoebas and other protists that move using pseudopods.

10. Which of the following is *not* an autotroph?
    a. dinoflagellate
    b. diatom
    c. kinetoplastid
    d. brown algae
    e. chlorarachniophyte

Chapter 29

Plant Diversity I: How Plants Colonized Land

Key Concepts

29.1 Land plants evolved from green algae

29.2 Land plants possess a set of derived terrestrial adaptations

29.3 The life cycles of mosses and other bryophytes are dominated by the gametophyte stage

29.4 Ferns and other seedless vascular plants formed the first forests

Framework

The evolution of plants involved adaptations to terrestrial habitats. The kingdom Plantae includes multicellular, photoautotrophic eukaryotes that develop from embryos nourished by the parent plant. Plants exhibit an alternation of generations in which the diploid sporophyte is the more conspicuous stage in all groups except the bryophytes. This chapter outlines the major periods of plant evolution in which bryophytes and seedless vascular plants appeared and radiated.

Chapter Review

For more than the first 3 billion years of Earth's history, life was confined to aquatic environments. Plants began to move onto land about 500 million years ago, and have now diversified into 290,000 species.

29.1 Land plants evolved from green algae

Several lines of evidence indicate that a few lineages of green algae, called charophyceans, are the closet relatives of land plants.

Morphological and Biochemical Evidence Land plants have chloroplasts containing chlorophylls *a* and *b* and cell walls of cellulose. These characteristics, however, are shared with several algal groups and cannot be used to distinguish plants from algae. Two features that do link plants to the charophyceans are **rosette cellulose-synthesizing complexes,** which produce the cellulose microfibrils of their cell walls, and enzymes within peroxisomes that help minimize photorespiration losses. Other derived homologies in charophyceans and land plants include similarities in sperm cells (in those land plants that have flagellated sperm) and the formation of a **phragmoplast** in mitosis. This alignment of cytoskeletal elements and vesicles in the synthesis of a cell plate is found only in plants and certain charophyceans, such as *Chara* and *Coleochaete.*

Genetic Evidence Results from the analysis of nuclear and chloroplast genes done during the international initiative called "Deep Green" confirm that charophyceans species such as *Chara* and *Coleochaete* are the closest living relatives of land plants.

Adaptations Enabling the Move to Land The tough polymer **sporopollenin** protects charophycean zygotes during fluctuations in water levels. With the accumulation of traits such as sporopollenin, ancient charophyceans living along the edges of ponds and lakes may have given rise to the first plants to colonize land.

■ **INTERACTIVE QUESTION 29.1**

List some of the lines of evidence that indicate that charophyceans and land plants shared a common ancestor.

29.2   Land plants possess a set of derived terrestrial adaptations

Defining the Plant Kingdom   Systematists are debating three different ways in which to establish the boundaries of the plant kingdom: the traditional kingdom Plantae containing only embryophytes (plants with embryos); the kingdom Streptophyta, which includes the charophyceans; or the kingdom Viridiplantae, which includes the chlorophytes as well as the charophyceans with the embryophytes. This textbook uses the traditional embryophyte definition of kingdom Plantae.

Derived Traits of Plants   The following derived traits distinguish land plants as a clade: apical meristems, alternations of generations, walled spores produced in sporangia, multicellular gametangia, and multicellular dependent embryos.

Regions of cell division at the tips of roots and shoots, called **apical meristems,** produce linear growth and cells that differentiate into a protective epidermis and internal plant tissues. The elongation of roots and shoots provides increasing access to environmental resources.

Alternation of generations is the life cycle found in all land plants. Whereas this type of reproductive cycle did evolve in various groups of algae, it does not occur in the charophyceans and appears to be a derived character of land plants. In this life cycle, the multicellular haploid **gametophyte** alternates with the multicellular diploid **sporophyte.** The sporophyte produces **spores,** which are reproductive cells that directly develop into new organisms. Review the alternation of generations in Interactive Question 29.2.

■ INTERACTIVE QUESTION   29.2

The gametophyte produces a. _____ by b. _____. Following fertilization, the c. _____ divides by d. _____ to develop into the e. _____. The sporophyte produces f. _____ by g. _____. The spores germinate and develop into the h. _____.

The tough polymer sporopollenin protects spores as they disperse on land. Spores develop and are protected within multicellular **sporangia** on the sporophyte plant.

Diploid cells called **sporocytes,** or spore mother cells, undergo meiosis and produce haploid spores in sporangia.

Gametes are produced within multicellular **gametangia.** Sperm are produced in gametangia called **antheridia.** A single egg cell is produced and fertilized within an **archegonium,** where the zygote develops into an embryo. In the reduced gametophytes of seed plants, the antheridia and archegonia have been lost in most lineages.

Plant embryos, which are retained in the female plant, have **placental transfer cells** that facilitate the transfer of nutrients from parental tissues. This derived trait is the basis for referring to land plants as **embryophytes.**

Many land plants have a waxy **cuticle** coating their leaves and stems, which protects from water loss and microbial attack. Many plants produce secondary compounds through side branches of their main metabolic pathways. These compounds, which include alkaloids, terpenes, tannins, and phenolics, may discourage herbivory or microbial attack, protect from UV radiation, or serve as signals.

The Origin and Diversification of Plants   Fossil spores have been found that date back 475 million years. In 2003 a "molecular clock" study estimated that the common ancestor of living plants lived 490 to 425 million years ago.

Most plants have **vascular tissue,** a transport system composed of cells joined into tubes, and are called **vascular plants.** Liverworts, hornworts, and mosses do not have an extensive transport system and are referred to as nonvascular plants and informally called **bryophytes.** Whether or not these groups form a clade is still debated.

Vascular plants, about 93% of plant species, do form a clade, and are divided into three smaller clades. The **lycophytes** (club mosses and their relatives) and the **pterophytes** (ferns and their relatives) are two clades informally called **seedless vascular plants.** A **seed** is an embryo enclosed with a supply of nutrients in a protective coat. The third clade, the seed plants, contains two groups: The **gymnosperms** are called "naked seed" plants because their seeds are not enclosed; **angiosperms** are the flowering plants, in which seeds develop inside chambers called ovaries in a flower. Taxonomists recognize ten phyla of extant plants.

■ INTERACTIVE QUESTION 29.3

Fill in this overview diagram of the major plant groupings. (Note that the plants in b. do not constitute a clade—they are not monophyletic. The plants in a. may or may not form a clade.)

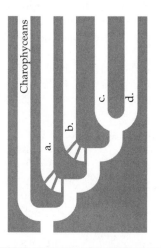

29.3 The life cycles of mosses and other bryophytes are dominated by the gametophyte stage

The three separate phyla of nonvascular plants are **liverworts** (phylum Hepatophyta), **hornworts** (phylum Anthocerophyta), and **mosses** (phylum Bryophyta). The evolutionary relationships of these groups continue to be debated.

Bryophyte Gametophytes   In the bryophytes, the gametophyte is the prevalent generation; sporophytes are present only part of the time. Liverwort and hornwort gametophytes grow more horizontally than do moss gametophyte. When moss spores germinate in a moist habitat, they grow into a mass of green filaments called a **protonema.** Meristem-containing buds grow into gamete-producing upright "stems" called **gametophores.** Bryophyte gametophytes are generally only a few cells thick and low growing. They are anchored by **rhizoids.**

Some mosses, such as *Polytrichum*, have leaf ridges coated with cuticle and conducting tissues in their "stems." Plant biologists have yet to determine whether these tissues are homologous or analogous with vascular tissue.

Gametangia are enclosed in jackets of protective tissue. Sperm are produced in antheridia, from which they swim to eggs retained in archegonia. Zygotes develop into embryos, which rely on nutrients transported by placental transfer cells to develop into sporophytes.

Bryophyte Sporophytes   Although they are usually photosynthetic when young, bryophyte sporophytes remain attached to and dependent on their parental gametophytes. A sporophyte has a **foot** that obtains nutrients from the gametophyte, a stalk **(seta)** that elongates for spore dispersal, and a sporangium or **capsule** in which millions of spores are produced by meiosis. The protective **calyptra** covers the immature capsule. The specialized "toothed" **peristome** gradually releases spores to be dispersed by wind currents.

Liverworts have tiny, simple sporophytes. The sporophytes of hornworts and mosses have **stomata,** pores through which gases are exchanged and water evaporates. There are three possible evolutionary origins of stomata: in the ancestor of all three groups and later lost in the liverwort lineage; in the ancestor of the branch leading to hornworts, mosses, and vascular plants; or in the ancestor of mosses and vascular plants, with an independent origin in hornworts.

■ INTERACTIVE QUESTION 29.4

Review the life cycle of a typical moss plant by filling in the following blanks.

The dominant generation is the a. _____. Female gametophytes produce eggs in b. _____. Male gametophytes produce sperm in c. _____. Sperm d. _____ through the damp environment to fertilize the egg. The zygote remains in the archegonium and grows into the e. _____ still attached to the female gametophyte. Spores are formed by the process of f. _____ in the g. _____. When shed, spores develop into the h. _____.

Ecological and Economic Importance of Mosses
Bryophytes are widely distributed and common in moist habitats. Some are adapted to very cold or dry habitats due to their ability to dehydrate and rehydrate without dying. Phenolic compounds absorb damaging radiation in harsh environments. *Sphagnum*, or peat moss, is an abundant wetland moss that forms undecayed organic deposits known as **peat.** The decay-resistant compounds in peat bogs can preserve organisms for thousands of years. Peat serves as huge stores of organic carbon and helps to stabilize atmospheric carbon dioxide concentrations. Peat is harvested for use as fuel, a soil conditioner, and plant packing material.

29.4   Ferns and other seedless vascular plants formed the first forests

Bryophytes were the dominant plants in the first 100 million years of plant evolution. Seedless vascular plants began to spread during the Carboniferous period, but remained limited to damp habitats because of their swimming sperm and fragile gametophytes.

Origins and Traits of Vascular Plants   Fossils of ancient relatives of modern vascular plants date back 420 million years. These tiny plants had branched, independent sporophytes and terminal sporangia. The branching made possible more complex bodies and multiple sporangia. They lacked other derived traits of vascular plants.

In modern vascular plants, the diploid sporophyte is the larger and more complex plant. Gametophytes are tiny structures growing on or under the soil surface. Trace the life cycle of a fern in Interactive Question 29.5.

■ **INTERACTIVE QUESTION   29.5**

In the following diagram of the life cycle of a fern, label the processes (in the boxes) and structures (on the lines). Indicate which portion of the life cycle is haploid and which is diploid.

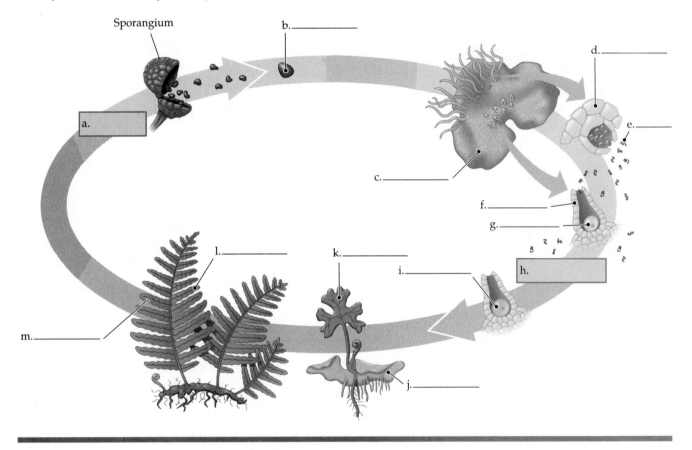

Water and minerals are conducted up from roots in **xylem.** The xylem of all vascular plants includes tube-shaped cells called **tracheids** with cell walls strengthened by **lignin.** Vascular plants are also called tracheophytes. **Phloem** transports sugars and other organic nutrients through living cells arranged into tubes.

Unlike the rhizoids of bryophytes, **roots** both anchor plants and absorb water and nutrients. Roots may have evolved from subterranean stems, either once in a common ancestor or independently in different lineages of vascular plants.

Leaves increase a plants photosynthetic surface area. The single-veined leaves of lycophytes, known as **microphylls,** are usually small and spine-shaped. The branching vascular systems of **megaphylls,** typical of most other vascular plants, support larger leaves with greater photosynthetic capacity. Megaphylls appear in the fossil record at the end of the Devonian period, 40 million years after the appearance of microphylls.

Microphylls may have originated as small outgrowths of stems, supported by single vascular tissue strands, whereas megaphylls may have evolved by the joining of closely lying branches.

Sporophylls are modified leaves bearing sporangia. In ferns, clusters of sporangia called **sori** are produced on the undersides of sporophylls. Cones, or **strobili,** are formed from groups of sporophylls in gymnosperms and many lycophytes. Most seedless vascular plants are **homosporous,** producing only one kind of spore, which develops into bisexual gametophytes. **Heterosporous** plants include all seed and some seedless vascular plants and produce two kinds of spores: **megaspores** that develop into female gametophytes and **microspores** that develop into male gametophytes.

Classification of Seedless Vascular Plants The two clades of living seedless vascular plant are classified into two phyla: phylum Lycophyta and phylum Pterophyta. Some systematists still retain three separate phyla in the pterophyte clade: phylum Pterophyta (ferns), phylum Sphenophyta (horsetails), and phylum Psilophyta (whisk ferns).

Lycophytes were a major part of the landscape during the Carboniferous period. One evolutionary line, the giant lycophytes, became extinct when the climate became drier. The other line of small lycophytes is represented today by the club mosses, spike mosses, and quillworts. Many tropical species grow on trees as epiphytes—plants that anchor to other organisms but are not parasites. Sporophytes grow upright with many small leaves; horizontal stems grow along the surface and produce roots.

Ferns were also found in the great forests of the Carboniferous period and are the most numerous seedless vascular plants in the modern flora. The large fern leaves (megaphylls), or fronds, often grow from horizontal stems. Most fern species are homosporous; gametophytes are small.

Horsetails grew as tall plants during the Carboniferous period. A few species of the genus *Equisetum* are the only representatives found today. These small, upright plants grow in damp locations and many have cones at the tips of some of the green, jointed stems.

Because of their dichotomous branching and lack of true leaves and roots, plant biologists initially considered *Psilotum,* the whisk fern, to be "living fossils." Molecular analysis and sperm structure comparisons now indicate that whisk ferns are closely related to ferns, and the true roots and leaves of the ancestor of whisk ferns were lost during evolution.

The Significance of Seedless Vascular Plants The seedless vascular plants of the Carboniferous forests decreased CO_2 levels in the atmosphere, perhaps contributing to the global cooling at the end of the Carboniferous. Dead plant material did not completely decay in the stagnant swamp waters, and great accumulations of peat developed. When the sea later covered the swamps and marine sediments piled on top, heat and pressure converted the peat to coal.

Word Roots

-angio = vessel (*gametangia:* the reproductive organ of bryophytes, consisting of the male antheridium and female archegonium; a multichambered jacket of sterile cells in which gametes are formed)

-phore = bearer (*gametophore:* the mature gamete-producing structure of a gametophyte body of a moss)

bryo- = moss; **-phyte** = plant (*bryophytes:* the mosses, liverworts, and hornworts; a group of nonvascular plants that inhabit the land but lack many of the terrestrial adaptations of vascular plants)

gymno- = naked; **-sperm** = seed (*gymnosperm:* a vascular plant that bears naked seeds not enclosed in any specialized chambers)

hetero- = different; **-sporo** = a seed (*heterosporous:* referring to plants in which the sporophyte produces two kinds of spores that develop into unisexual gametophytes, either female or male)

homo- = like (*homosporous:* referring to plants in which a single type of spore develops into a bisexual gametophyte having both male and female sex organs)

mega- = large (*megaspores:* a spore from a heterosporous plant that develops into a female gametophyte bearing archegonia)

micro- = small; **-phyll** = leaf (*microphylls:* the small leaves of lycophytes that have only a single, unbranched vein)

peri- = around; **-stoma** = mouth (*peristome:* the upper part of the moss capsule often specialized for gradual spore discharge)

phragmo- = a partition; **-plast** = formed, molded (*phragmoplast:* an alignment of cytoskeletal elements and Golgi-derived vesicles across the midline of a dividing plant cell)

proto- = first; **-nema** = thread (*protonema:* a mass of green, branched, one-cell-thick filaments produced by germinating moss spores)

pter- = fern (*pterophytes:* seedless plants with true roots with lignified vascular tissue; the group includes ferns, whisk ferns, and horsetails)

rhizo- = root; **-oid** = like, form (*rhizoids:* long tubular single cells or filaments of cells that anchor bryophytes to the ground)

Structure Your Knowledge

1. The evolution of land plants involved adaptations to the terrestrial habitat. List some of these adaptations that evolved in the bryophytes and the seedless vascular plants.

2. The diagram below presents a widely held view of plant phylogeny. Fill in blanks a–e with the informal names for the major groups and blanks f–j with the common names for the modern plant groups (give the 3 groups included in clade j).

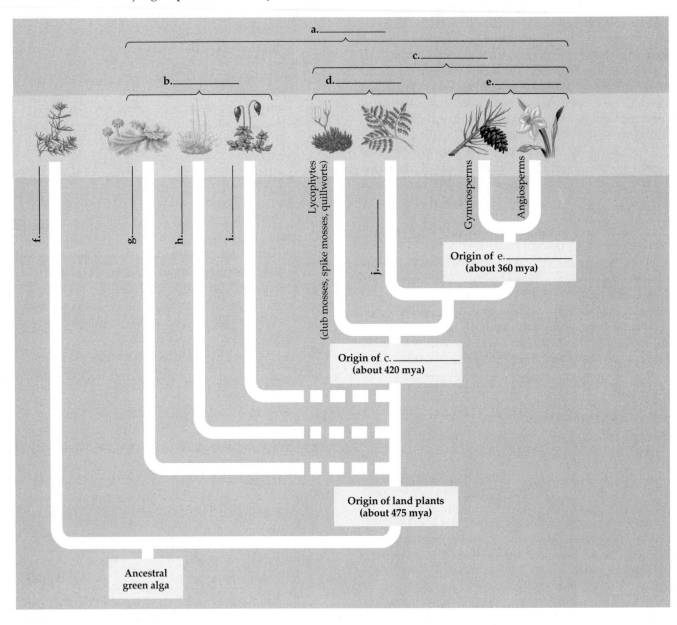

Test Your Knowledge

MULTIPLE CHOICE: *Choose the one best answer.*

1. Adaptations for terrestrial life seen in all plants are
   a. chlorophylls *a* and *b*.
   b. cell walls of cellulose and lignin.
   c. sporopollenin, protection and nourishment of embryo by gametophyte.
   d. vascular tissue and stomata.
   e. alternation of generations.

2. Plants are thought to be most closely related to charophyceans based on
   a. peroxisome enzymes that minimize photorespiration.
   b. rosette cellulose-synthesizing complexes in their plasma membranes.
   c. phragmoplast formation of cell plates in mitosis.
   d. similarities in sperm cells (in those land plants with flagellated sperm).
   e. all of the above.

3. Megaphylls
   a. are large leaves.
   b. are gametophyte plants that develop from megaspores.
   c. are leaves specialized for reproduction.
   d. are leaves with branching vascular systems.
   e. were a dominant group of the great coal forests.

4. Bryophytes differ from all other land plant groups because
   a. their gametophyte generation is dominant.
   b. they are lacking gametangia.
   c. they have flagellated sperm.
   d. they are not embryophytes.
   e. of all of the above.

5. Which of the following plant groups is *incorrectly* paired with its gametophyte generation?
   a. charophycean—no alternation of generations
   b. lycophyte—either green or nonphotosynthetic underground plants
   c. moss—green matlike plant
   d. fern—frond growing from horizontal stem
   e. liverwort—flattened, low-growing green plant

6. Which of the following is most likely the closest relative of the vascular plants?
   a. charophyceans
   b. mosses
   c. liverworts
   d. pterophytes
   e. lycophytes

7. The evolution of sporopollenin was important to the movement of plants onto land because it
   a. provided a mechanism for the dispersal of spores and pollen.
   b. provided the structural support necessary to withstand gravity.
   c. enclosed developing gametes and embryos in maternal tissue and prevented desiccation.
   d. provided a tough coating for spores so they could disperse on land.
   e. initiated the alternation of generations that is characteristic of all plants.

8. If a plant's life cycle includes both a male and female gametophyte, the sporophyte plant must be
   a. heterosporous.
   b. homosporous.
   c. homologous.
   d. analogous.
   e. megasporous.

9. Xylem and phloem are found in
   a. all plants.
   b. bryophytes, ferns, conifers, and angiosperms.
   c. only the gametophytes of vascular plants.
   d. the vascular plants, which include ferns, conifers, and angiosperms.
   e. only the vascular plants with seeds.

10. Which of the following functions may secondary compounds serve?
    a. protect plant from ultraviolet radiation
    b. support plant cell walls (lignin)
    c. discourage herbivory or inhibit microbes
    d. function as signaling molecules
    e. all of the above

11. Which of the following is most likely the closet relative of the seed plants?
    a. bryophytes
    b. moss
    c. lycophytes
    d. pterophytes
    e. gymnosperms

12. Which of the following has been proposed by plant biologists as the clade that establishes the boundary of the plant kingdom?
    a. kingdom Plantae, which includes only the embryophyte clade
    b. kingdom Streptophyta, which also includes the charophyceans
    c. kingdom Viridiplantae, which also includes all green algae (chlorophytes)
    d. kingdom Vasculata, which includes only the vascular plants
    e. a, b, and c have all been proposed as the clade that defines the plant kingdom

13. Large stores of organic carbon that help to stabilize atmospheric carbon dioxide concentrations are found in
    a. huge tropical swamps dominated by mosses and ferns.
    b. peat.
    c. coal deposits left by Cambrian forests.
    d. the abundant epiphytic lycophytes found in tropical regions.
    e. large tracts of small seedless vascular plants found in boreal regions.

14. Which of the following is synonymous with tracheophytes?
    a. bryophytes
    b. pterophytes
    c. lycophytes
    d. vascular plants
    e. seed plants

15. If you could take a time machine back to the Carboniferous period, which of the following scenarios would you most likely encounter?
    a. creeping mats of low-growing bryophytes
    b. fields of tall grasses swaying in the wind
    c. swampy forests dominated by large lycophytes, horsetails, and ferns
    d. huge forests of naked-seed trees filling the air with pollen
    e. the dominance of flowering plants

Chapter 30

Plant Diversity II: The Evolution of Seed Plants

Framework

The reduction and protection of the gametophyte within the parent sporophyte, the use of seeds as a means of dispersal, and the development of pollen for the transfer of sperm are reproductive adaptations of seed plants that enhanced their success on land.

Gymnosperms bear their seeds "naked" on modified sporophylls. Angiosperms are the most diverse and widespread of plants, owing much of their success to their efficient reproductive apparatus—the flower— and their dispersal mechanism—the fruit. Agriculture is based almost entirely on angiosperms.

Chapter Review

30.1 The reduced gametophytes of seed plants are protected in ovules and pollen grains

Seed plants originated about 360 million years ago. The key adaptations that enabled seed plants to become the dominant members of terrestrial ecosystems include seeds, reduced gametophytes, heterospory, ovules, and pollen.

Advantages of Reduced Gametophytes The extremely reduced gametophytes of seed plants develop within the sporangium, thus protected and nourished by the sporophyte plant.

Heterospory: The Rule Among Seed Plants All seed plants are heterosporous, with megaspores that give rise to female gametophytes, and microspores that give rise to male gametophytes. A megaspongium produces a single functional megaspore; a microsporangium produces many microspores.

Ovules and Production of Eggs The megasporangium is surrounded by protective **integuments** derived from sporophyte tissue. An unfertilized **ovule** consists of the integuments, megasporangium, and megaspore. The female gametophyte develops from the megaspore and produces one or more eggs.

Pollen and Production of Sperm Microspores develop into male gametophytes contained in sporopollenin-protected **pollen grains.** Pollen grains may be dispersed over long distances by wind or animals to an ovule, where they release sperm to the female gametophyte. This transfer of pollen to the part of a plant containing the ovules is called **pollination.**

The Evolutionary Advantage of Seeds An ovule develops into a **seed,** consisting of a sporophyte embryo surrounded by a food supply and enclosed in a protective coat derived from the integuments of the ovule. Seeds enable seed plants to withstand harsh environmental conditions and to disperse offspring. Unicellular spores perform those functions for bryophytes and seedless vascular plants.

■ INTERACTIVE QUESTION 30.1

Label the parts in this generalized diagram of an unfertilized ovule and seed of a gymnosperm. Indicate whether structures are diploid or haploid tissues.

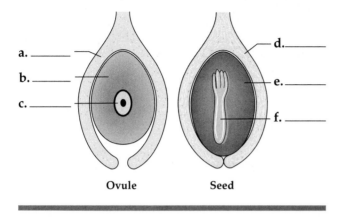

Ovule   Seed

■ INTERACTIVE QUESTION 30.2

Why is the evolution of pollen an important terrestrial adaptation?

30.2   Gymnosperms bear "naked" seeds, typically on cones

Gymnosperms are named for their naked seeds that develop on the surface of modified leaves, which usually form cones (strobili).

There are four gymnosperm phyla. Cycadophyta includes cycads with large cones and fern-like or palm-like leaves. They thrived during the Mesozoic era. Ginkgophyta has only one extant species, the deciduous, ornamental ginkgo tree. Gnetophyta includes three genera: the bizarre *Welwitschia*, the tropical *Gnetum*, and the desert shrub *Ephedra*. The largest gymnosperm phylum is Coniferophyta. The name **conifer** refers to the reproductive structure, the cone. Most conifers are evergreens with needle-shaped leaves. Coniferous trees are some of the largest and oldest living organisms.

Gymnosperm Evolution   Transitional species of heterosporous seedless vascular plants are sometimes called **progymnosperms.** Around 360 million years ago, the first seed plants appear in the fossil record. These early gymnosperm lineages became extinct, and their relationships to the two clades of surviving seed plants, the gymnosperms and angiosperms, is not clear. During the Carboniferous, early gymnosperms lived in ecosystems dominated by seedless vascular plants. Drier conditions during the Permian gave gymnosperms a selective advantage. Gymnosperms supported the giant dinosaurs of the Mesozoic. With the environmental changes at the end of the Mesozoic, most of the dinosaurs became extinct, but many gymnosperms persisted.

A Closer Look at the Life Cycle of a Pine   The pine tree is a heterosporous sporophyte. Pollen cones consist of many scalelike structures that bear sporangia. Meiosis gives rise to microspores that develop into pollen grains enclosing the male gametophytes. Scales of the ovulate cone hold ovules, each of which contains a megasporangium. A megasporocyte undergoes meiosis, and one of the resulting megaspores undergoes repeated divisions to produce a female gametophyte in which a few archegonia develop.

Following pollination, a pollen grain enters through the micropyle, an opening in the integuments. A pollen tube grows and digests its way through the megasporangium. Fertilization may occur more than a year after pollination. The zygote develops into a sporophyte embryo, which is nourished by the remaining female gametophyte tissue and is enclosed in a seed coat derived from integuments of the parent sporophyte.

a. Describe a pollen cone and the formation of a male gametophyte.

b. Describe an ovulate cone and the formation of a female gametophyte.

30.3 The reproductive adaptations of angiosperms include flowers and fruits

Angiosperms, the vascular seed plants that produce flowers and fruits, are the most diverse and widespread of modern plants.

Characteristics of Angiosperms The phylum Anthophyta contains more than 90% of plants species, more than 250,000 species.

The **flower,** the reproductive structure of an angiosperm, has four whorls of modified leaves, which attach at the **receptacle,** or base of the flower. The outer **sepals** are usually green, whereas **petals** are brightly colored in most flowers that are pollinated by insects and birds. **Stamens,** the microsporophylls, produce microspores, which develop into pollen grains containing male gametophytes. A stamen has a stalk, called a **filament,** and a terminal **anther,** in which pollen is produced. **Carpels,** the megasporophylls, produce megaspores, which develop into female gametophytes. The carpel has a sticky **stigma,** which receives pollen, and a **style,** which leads to the **ovary.** The ovary contains ovules, which develop into seeds. A carpel or group of fused carpels may be called a **pistil.**

A **fruit** is a mature ovary that functions in the protection and dispersal of seeds. Hormonal changes following pollination cause the ovary to enlarge, its wall becoming the **pericarp,** or thickened wall of the fruit. Fruits may be fleshy or dry, with several layers of pericarp. The dry fruits of grasses are the major foods for humans. Fruits may be modified in various ways to disperse seeds, enlisting the aid of wind or animals.

a. Name the four whorls of modified leaves that make up a flower.

b. What does a seed consist of?

c. What is a fruit?

In the angiosperm life cycle, the two highly reduced gametophytes are produced within the flower. Pollen grains, which develop from microspores within the anthers, contain the male gametophytes, consisting of two haploid cells. Ovules contain the female gametophyte, called an **embryo sac,** which consists of a few cells.

Most flowers have some mechanism to ensure **cross-pollination.** A pollen grain germinates on the stigma and extends a pollen tube down the style, through the **micropyle,** and into the ovule, where it releases two sperm cells into the embryo sac. In a process called **double fertilization,** one sperm unites with the egg to form the zygote and the other sperm fuses with the two polar nuclei in the large cell in the center of the female gametophyte.

The zygote divides and forms the sporophyte embryo, consisting of a rudimentary root and one or two seed leaves, called **cotyledons.** The **endosperm,** which develops from the triploid central cell, serves as a food reserve for the embryo.

■ INTERACTIVE QUESTION   30.5

In this diagram of the life cycle of an angiosperm, label the indicated structures and processes. Which structures represent the male and female gametophyte generations?

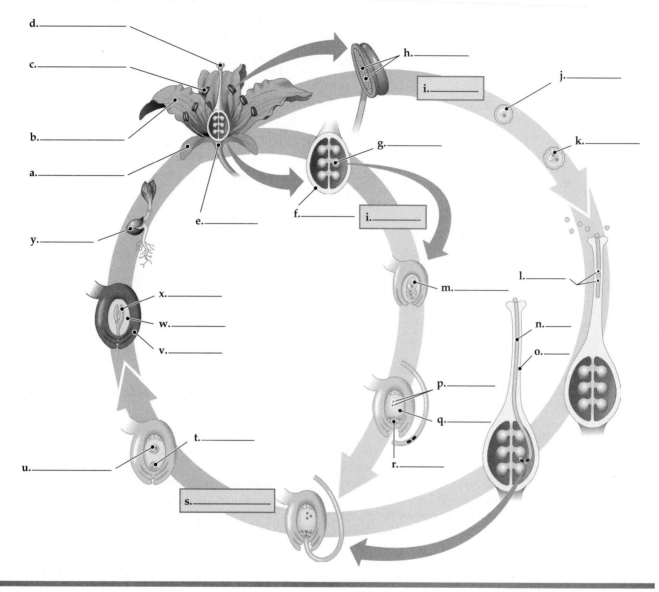

Angiosperm Evolution Angiosperms originated at least 140 million years ago. By the end of the Mesozoic era, 65 million years ago, angiosperms were becoming the dominant plants. The origin of angiosperms and flowers is still a mystery. A phylogenetic comparison of recently discovered fossils of 125-million-year-old angiosperms, called *Archaefructus,* have shown this group to be most closely related to all living angiosperms. These fossils suggest that proto-angiosperms may have been herbaceous, aquatic plants, although some paleobotanists dispute this interpretation.

An "evo-devo" approach is being used to hypothesize how pollen-producing and ovule-producing structures, which are separate in gymnosperms, became combined into a single flower. M. Frohlich's "mostly male" hypothesis is based on the relationship between flower-development genes and pollen-producing gymnosperm genes.

Angiosperm Diversity The angiosperms were traditionally divided into two main groups, the **monocots** and **dicots.** Recent DNA comparisons revealed that, although the monocots are a monophyletic group, the dicots are not. The clade **eudicots** includes most previously classified dicots, but several other small lineages have been identified. The most primitive flowering plants are represented by three lineages, informally called **basal angiosperms:** water lilies, star anise, and the oldest branch represented today by one species, *Amborella trichopoda. Amborella* lacks the xylem vessels found in more derived angiosperms. A lineage known as the **magnoliids** is more closely related to monocots and eudicots than to the basal angiosperms.

■ INTERACTIVE QUESTION 30.6

List some of the differences between monocots and eudicots for the following characteristics.

Characteristic	Monocot	Eudicot
Number of cotyledons		
Leaf venation		
Vascular bundles in stems		
Root system		
Flower parts		
Openings in pollen grain		

Evolutionary Links Between Angiosperms and Animals Animals influenced the evolution of plants and vice versa. Animals on the forest floor may have provided selective pressure for plants to keep spores and vulnerable gametophytes off the ground. As flowers and fruits evolved, some animals became beneficial as pollinators and seed dispersers. The mutual evolutionary influence of angiosperms and their pollinators is seen in the diversity of flowers, whose color, fragrance, and shape are often matched to the sense of sight and smell or to the particular morphology of a group of pollinators. Angiosperms have contributed to the variety of insect species, and, as grasslands expanded during the past 65 million years, to the diversity of grazing mammals.

30.4 Human welfare depends greatly on seed plants

Products from Seed Plants Our fruit and vegetable crops are angiosperms. The dramatic evolution of do-

mesticated plants is a result of artificial selection in plant breeding. Seed plants also supply wood and medicines.

Threats to Plant Diversity The growing human population with its demand for space, food, and natural resources is leading to the extinction of hundreds of species each year. Tropical rain forests, where plant diversity is greatest, are rapidly being destroyed. Humans are losing their potential supply of new food crops and medicines. Preserving plant diversity is an urgent problem.

▶ Word Roots

co- = with, together (*coevolution:* the mutual influence on the evolution of two different species interacting with each other and reciprocally influencing each other's adaptations)

endo- = inner (*endosperm:* a nutrient-rich tissue formed by the union of a sperm cell with two polar nuclei during double fertilization, which provides nourishment to the developing embryo in angiosperm seeds)

peri- = around; **-carp** = fruit (*pericarp:* the thickened wall of a fruit)

pro- = before; **gymno-** = naked; **-sperm** = seed (*progymnosperm:* an extinct group of plants that is probably ancestral to gymnosperms and angiosperms)

▶ Structure Your Knowledge

1. **a.** List the four phyla that are considered gymnosperms.
   b. The angiosperms are grouped into a single phylum, Anthophyta, which traditionally contained the classes, monocots and dicots. List five of the clades that have recently been distinguished in this phylum.

2. List the characteristics of seed plants that were evolutionary adaptations to their terrestrial habitat.

3. What adaptations helped angiosperms to become the most successful and widespread land plants?

Test Your Knowledge

MULTIPLE CHOICE: *Choose the one best answer.*

1. In which of the following groups do sperm no longer have to swim to reach the female gametophyte?
   a. bryophytes
   b. ferns
   c. gymnosperms
   d. angiosperms
   e. Both c and d are correct.

2. What provides food for a developing sporophyte embryo in a gymnosperm seed?
   a. endosperm
   b. female gametophyte tissue
   c. female sporophyte tissue
   d. male gametophyte tissue
   e. pine cone

3. Which of the following is the correct path that a pollen tube takes to reach the female gametophyte in an angiosperm?
   a. stigma, style, ovary, ovule, embryo sac
   b. anther, stigma, filament, ovule, ovum
   c. stigma, filament, carpel, ovary, ovule
   d. carpel, pistil, ovary, ovule, embryo sac
   e. stigma, style, pistil, ovule, ovary

4. Which correctly states the number and describes the generations that are represented in a pine seed?
   a. one: the new sporophyte generation
   b. two: seed coat and food supply from female gametophyte and sporophyte embryo
   c. two: seed coat from integuments of parent sporophyte and new sporophyte embryo
   d. three: seed coat from parent sporophyte, food supply from gametophyte, and sporophyte embryo
   e. three: seed coat from female gametophyte, food supply from parent sporophyte, and sporophyte embryo

5. Gymnosperms rose to dominance during which of the following periods?
   a. Carboniferous, when they formed important components of the great "coal forests"
   b. Devonian, when they successfully competed with the short-statured bryophytes
   c. Permian, when they replaced the seedless vascular plants as the climate became drier with the formation of the supercontinent Pangaea
   d. Cretaceous, when cooler climates favored their growth over the ferns and other seedless vascular plants
   e. Cretaceous, when they replaced angiosperms as the largest and most widespread group of plants

6. If the angiosperm gametophyte generation has been reduced to so few cells, why hasn't it been eliminated from the life cycle?
   a. The gametophyte generation produces the resistant spores that allow flowering plants to withstand harsh conditions and to disperse on land.
   b. In the ancestors of angiosperms, the gametophyte was the dominant generation, and thus natural selection has retained it.
   c. The gametophyte generation produces both the protective seed coat and nourishment for the developing embryo.
   d. The gametophyte may allow for elimination of harmful mutations that are expressed in these haploid cells, and it helps to provide nourishment during early development of the embryo.
   e. Natural selection is not effective at eliminating such microscopic structures.

7. Which of the following is a key difference between seedless vascular plants and plants with seeds?
   a. The gametophyte generation is dominant in the seedless plants, whereas the sporophyte is dominant in the seed plants.
   b. The spore is the agent of dispersal in the first, whereas the seed functions in dispersal in the second.
   c. Seedless plants are heterosporous; seed plants are homosporous.
   d. The embryo is unprotected in the seedless plants but retained within the female reproductive structure in the seed plants.
   e. Sporopollenin is not found in the seedless plants.

8. An example of coevolution is
   a. wind pollination in conifers.
   b. a flower with a nectar tube that is the length of its pollinator's proboscis.
   c. the synchronization of nutrient development and fertilization resulting from double fertilization.
   d. the evolution of seeds in both gymnosperms and angiosperms.
   e. the development of alternation of generations independently in land plants and some algal groups.

9. Where would you find a microsporangium in the life cycle of a pine?
   a. within the embryo sac in an ovule
   b. in the pollen sacs in an anther
   c. at the base of a sporophyll in a pollen cone
   d. on a scalelike sporophyll found in an ovulate cone
   e. forming a seed coat surrounding a pine seed

10. Which of these plants is believed to be the only survivor of the oldest branch of the angiosperm lineage?
    a. the ornamental ginkgo tree
    b. the water lilies
    c. the eudicot poppies
    d. the star anise
    e. the small shrub *Amborella*

TRUE OR FALSE: *Indicate T or F, and then correct the false statements.*

_____ 1. All photoautotrophic, multicellular eukaryotes are plants.

_____ 2. Heterosporous plants produce male and female spores.

_____ 3. The gametophyte generation is most reduced in the gymnosperms.

_____ 4. The *Ginkgo,* cycads, and conifers are naked-seed plants.

_____ 5. A sporangium produces spores, no matter what group it is found in.

_____ 6. A fruit consists of an embryo, nutritive material, and a protective coat.

_____ 7. According to the "Mostly Male" hypothesis, bisexual flowers evolved from the male cone of a gymnosperm ancestor.

_____ 8. A stamen consists of a filament and anther in which microspores are produced, which give rise to pollen grains.

_____ 9. The female gametophyte in angiosperms consists of haploid cells in which a few archegonia develop.

_____ 10. The male gametophyte in angiosperms is contained within a pollen grain.

Chapter 31

Fungi

Framework

This chapter describes the morphology, life cycles, evolutionary history, diversity, and economic and ecological importance of the kingdom Fungi. Fungi play an essential ecological role, both as decomposers and by their mycorrhizal association with plant roots. A flagellated protistan may have been the common ancestor to fungi and animals.

Chapter Review

31.1 Fungi are heterotrophs that feed by absorption

Nutrition and Fungal Lifestyles Fungi are heterotrophs that obtain their nutrients by absorption; they secrete digestive enzymes, called **exoenzymes,** into the surrounding food and absorb the resulting small organic molecules. Fungi may be saprobes that decompose nonliving organic material; parasites that absorb nutrients from living cells; or mutualists that feed on, but also benefit, their hosts.

Body Structure Most fungi are composed of **hyphae** that form a network called a **mycelium.** This body form provides an extensive surface area for absorption. The tubular cell walls are composed of **chitin.** Although fungi are nonmotile, they rapidly enter into new food territory by the fast growth of their hyphae.

The hyphae of most fungi are divided into cells by cross-walls called **septa,** which usually have pores through which nutrients and cell organelles can pass. Some hyphae are aseptate, and these **coenocytic** fungi consist of a continuous cytoplasmic mass with many nuclei.

Mutualistic and parasitic fungi may penetrate plant host cells with specialized hyphae called **haustoria.** Mutualistic relationships with plant roots are known as **mycorrhizae. Ectomycorrhizal** fungi form hyphal sheaths around a root. **Endomycorrhizal** fungi extend their hyphae through root cell walls and into membrane-lined tubes within root cells.

■ INTERACTIVE QUESTION 31.1

What do mycorrhizal fungi do for plants? What do plants do for the fungi?

31.2 Fungi produce spores through sexual or asexual life cycles

Fungi release huge quantities of spores, produced either sexually or asexually, that aid in dispersal.

Sexual Reproduction Nuclei of hyphae and spores of most species are haploid. Sexual reproduction involves the release of sexual signaling **pheromones,** which may cause a receiving hypha of a suitable mating type to grow toward and fuse with the signaling hypha. Such cytoplasmic fusion is called **plasmogamy,** forming a mycelium with genetically different nuclei, or a **het-erokaryon.** The haploid nuclei may stay in separate regions of the mycelia, or they may mingle and exchange genes in a process similar to crossing over. The two nuclei may pair up in cells and divide in tandem, forming **dikaryotic** cells. After a period of time, the nuclei fuse (**karyogamy**), forming the only diploid stage in most fungi. Meiosis produces haploid spores.

Asexual Reproduction   Many fungi reproduce asexually by spores; some species reproduce only asexually.

Molds are rapidly growing mycelia and reproduce asexually by spores. Many molds can reproduce sexually if they encounter other mating types. **Yeasts** are unicellular fungi that grow in liquid or moist habitats. Reproduction is asexual by cell division or budding. Some yeasts can also grow as mycelia, and some also reproduce sexually.

Molds and yeasts for which no sexual stage is known are called **deuteromycetes,** or **imperfect fungi.** When mycologists discover a sexual stage in one of these fungi, they classify it accordingly.

■ INTERACTIVE QUESTION   31.2

Briefly define the following terms that relate to the structure and reproduction of fungi:

a. mycelium

b. septa

c. coenocytic

d. heterokaryon

e. plasmogamy

f. dikaryon

g. karyogamy

31.3   Fungi descended from an aquatic, single-celled flagellated protist

Fungi and Animalia are now recognized as sister kingdoms, more closely related to each other than to plants or other eukaryotes.

The Origin of Fungi   Although most fungi lack flagella, phylogenetic systematics indicates that they evolved from a flagellated ancestor. The fungi, animals, and their protistan relatives are called **opisthokonts.** Evidence also suggests that the ancestor of fungi was unicellular. Fungi and animals may have evolved multicellularity independently from different unicellular ancestors.

The molecular clock estimates that animals and fungi diverged 1.5 billion years ago. The oldest undisputed fossils of fungi date back 460 million years.

The Move to Land   The first vascular plant fossils have evidence of mycorrhizae. The adaptive radiation of fungi may have begun with this move to land.

31.4   Fungi have radiated into a diverse set of lineages

Chytrids   The **chytrids,** phylum Chytridiomycota, are unique among fungi because of their flagellated **zoospores.** Molecular evidence indicates they diverged earliest in fungal evolution. Chytrids are primarily aquatic saprobes or parasites. Some form colonies; others are unicellular.

Zygomycetes   **Zygomycetes** of phylum Zygomycota have diverse life histories and include molds, parasites, and commensal symbionts.

Rhizopus stolonifer, the black bread mold, is a common zygomycete that spreads horizontal coenocytic hyphae across its food substrate and erects hyphae with bulbous black sporangia containing hundreds of asexual haploid spores. Sexual reproduction between mycelia of opposite mating types produces a heterokaryotic **zygosporangium** with a tough protective coat that remains dormant until conditions are favorable. Karyogamy occurs between paired nuclei, followed by meiosis to produce haploid spores.

Microsporidia are unicellular animal and protist parasites that have tiny organelles derived from mitochondria. Molecular data suggest that microsporidia should be classified as zygomycetes.

Glomeromycetes   Anaylsis of fungal genomes has identified the **glomeromycetes** as a separate clade (phylum Glomeromycota). All glomeromycetes form **arbuscular mycorrhizae.** Hyphae that invade plant roots branch into tiny treelike arbuscles. About 90% of all plants form mycorrhizae with glomeromycetes.

Ascomycetes   The **ascomycetes** (phylum Ascomycota), or **sac fungi,** are found in a wide variety of habitats. They produce sexual spores in saclike **asci,** usually contained in fruiting bodies called **ascocarps.**

Ascomycetes range in complexity from yeasts to elaborate cup fungi and morels. Some are serious plant pathogens; others are important saprobes. Many species are in symbiotic associations with algae or cyanobacteria called lichens or form mycorrhizae with plants.

Asexual reproduction involves spores called **conidia,** which are produced in chains or clusters at the ends of hyphae called conidiophores. Plasmogamy between different mating types produces a heterokaryotic ascogonium. Karyogamy occurs in terminal cells of dikaryotic hyphae, which develop into asci. Meiosis yields haploid spores called ascospores.

The genome of the *Neurospora crassa* has 10,000 genes with few stretches of noncoding DNA.

Basidiomycetes   The **basidiomycetes** (phylum Basidiomycota), which include the mushrooms, shelf fungi, and puffballs, produce a club-shaped **basidium** during a short diploid stage in the life cycle. These **club fungi** include important saprobes, mycorrhizae-forming mutualists, and the plant parasites rusts and smuts.

In response to environmental stimuli, an elaborate fruiting body called a **basidiocarp** is formed from the long-lived dikaryotic mycelium. Karyogamy and meiosis occur in numerous basidia in the basidiocarp, producing huge numbers of sexual spores.

■ INTERACTIVE QUESTION   31.3

Indicate whether the following diagrams (1, 2, 3) are from a zygomycete, ascomycete, or basidiomycete life cycle. Identify the labeled structures.

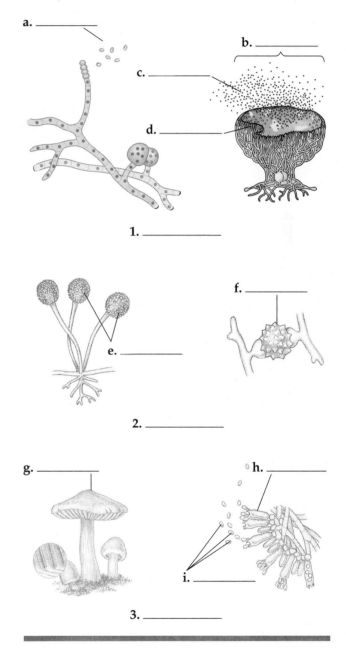

a. _____

b. _____

c. _____

d. _____

1. _____

e. _____

f. _____

2. _____

g. _____

h. _____

i. _____

3. _____

31.5 Fungi have a powerful impact on ecosystems and human welfare

Decomposers Fungi and bacteria are decomposers of organic matter, making possible the essential recycling of chemical elements for plant growth.

Symbionts Mycorrhizae are very common and important in natural ecosystems and agriculture. Fungal inhabitants of the guts of grazing mammals help digest plant material. Farmer insects raise fungal crops by gathering and feeding leaves to the fungus.

Lichens are symbiotic associations of millions of photosynthetic organisms in a lattice of fungal hyphae. The fungus is usually an ascomycete, and the partner is usually unicellular or filamentous green algae or cyanobacteria. The alga provides the fungus with food and, in the case of cyanobacteria, nitrogen. The fungus creates most of the mass of the lichen and provides protection for the alga and absorbs water and minerals.

Lichens reproduce asexually, either as fragments or as tiny clusters called soredia. In addition, it is common for the fungal component to reproduce sexually and for the algal component independently to reproduce asexually.

Molecular studies trace all lichens to three original associations between a fungus and photosynthetic symbionts. Numerous free-living fungi appear to have descended from lichen-forming ancestors.

Lichens are important colonizers of bare rock and soil and can withstand desiccation and great cold. Lichens cannot, however, tolerate air pollution.

Pathogens Fungal parasites of plants are common, killing elm and chestnut trees and spoiling grain crops. Fungi spoil a large proportion of the world's fruit harvest. An ascomycete forms ergots on rye that can cause serious symptoms when accidentally milled into flour. Lysergic acid, the raw material of LSD, is one of the toxins in the ergots.

Fungal infections in animals are called mycoses. Parasitic fungi cause ringworm, athlete's foot, vaginal yeast infections, and lung infections. Systemic mycoses are very serious. Many mycoses are opportunistic.

Practical Uses of Fungi Commercially cultivated mushrooms are eaten, and fungi are used to ripen some cheeses. *Saccharomyces cerevisiae* is the yeast used in baking, brewing, and molecular research and biotechnology. Some fungi are the source of medical compounds, including antibiotics.

Word Roots

-osis = a condition of (*mycosis:* the general term for a fungal infection)

coeno- = common; **-cyto** = cell (*coenocytic:* referring to a multinucleated condition resulting from the repeated division of nuclei without cytoplasmic division)

di- = two; **-karyo** = nucleus (*dikaryotic:* a mycelium with two haploid nuclei per cell)

exo- = out, outside (*exoenzymes:* powerful hydrolytic enzymes secreted by a fungus outside its body to digest food)

hetero- = different (*heterokaryon:* a mycelium formed by the fusion of two hyphae that have genetically different nuclei)

myco- = fungus; **rhizo-** = root (*mycorrhizae:* mutualistic associations of plant roots and fungi)

plasmo- plasm; **-gamy** = marriage (*plasmogamy:* the fusion of the cytoplasm of cells from two individuals; occurs as one stage of syngamy)

Structure Your Knowledge

1. Fill in the table on the next page that summarizes the characteristics of the five fungal phyla.

Phylum	Examples	Key Features and Reproduction
Chytridiomycota	a.	b.
Zygomycota	c.	d.
Glomeromycota	e.	f.
Ascomycota	g.	h.
Basidiomycota	i.	j.

2. The kingdom Fungi contains members with saprobic, parasitic, and mutualistic modes of nutrition. How do these types of nutrition relate to the ecological and economic importance of this group?

3. What is the basis for saying that fungi and animals evolved from a common protistan ancestor?

Test Your Knowledge

FILL IN THE BLANKS

_____ 1. division between cells in fungal hyphae

_____ 2. hyphae with nuclei from two parents

_____ 3. component of cell walls in most fungi

_____ 4. produced by fungi for absorptive nutrition

_____ 5. club-shaped reproductive structure found in mushrooms

_____ 6. sacs that contain sexual spores in sac fungi

_____ 7. mutualistic associations between plant roots and fungi

_____ 8. resistant structures produced by zygomycetes

_____ 9. hyphae with many nuclei

_____ 10. fungal group that diverged earliest

_____ 11. distinctive feature of glomeromycetes

_____ 12. clade that includes fungi, animals, and related flagellated protists

MULTIPLE CHOICE: *Choose the one best answer.*

1. The major difference between fungi and plants is that fungi
   a. have an absorptive form of nutrition.
   b. do not have a cell wall.
   c. are not eukaryotic.
   d. are multinucleate but not multicellular.
   e. reproduce by spores.

2. Chytrids had previously been classified with protists because they
   a. do not have chitin in their cell walls.
   b. do not have absorptive nutrition.
   c. have flagellated zoospores, whereas most fungi lack flagella.
   d. have metabolic pathways like those of protists.
   e. are aquatic, and fungi are terrestrial.

3. A fungus that is both a parasite and a saprobe is one that
   a. digests the nonliving portions of its host's body.
   b. lives off the sap within its host's body.
   c. first lives as a parasite but then consumes the host after it dies.
   d. lives as a mutualistic symbiont on its host.
   e. causes athlete's foot and vaginal infections.

4. The fact that karyogamy does not immediately follow plasmogamy
   a. is necessary to create coenocytic hyphae.
   b. allows for more genetic variation.
   c. allows fungi to reproduce asexually most often.
   d. results in heterokaryotic cells that may benefit from the variation present in two genomes.
   e. is characteristic of yeasts.

5. Imperfect fungi
   a. represent the most ancient lineage of fungi.
   b. include the fungal components of lichens.
   c. have abnormal forms of sexual reproduction.
   d. are fungi that are predatory.
   e. are fungi whose sexual stage is lacking or unknown.

6. In the Ascomycota,
   a. sexual reproduction occurs by conjugation.
   b. spores often line up in a sac in the order in which they were formed by meiosis.
   c. asexual spores form in sporangia on erect hyphae.
   d. most hyphae are dikaryotic.
   e. reproduction is always sexual.

7. Lichens are symbiotic associations that
   a. usually involve an ascomycete and a green alga or cyanobacterium.
   b. can reproduce sexually by forming soredia.
   c. require moist environments to grow.
   d. fix nitrogen for absorption by plant roots.
   e. are unusually resistant to air pollution.

8. The name given to three of the fungal phyla is based on
   a. the structure in which karyogamy occurs during sexual reproduction.
   b. the location of plasmogamy during sexual reproduction.
   c. the location of the heterokaryotic stage in the life cycle.
   d. the structure that produces asexual spores.
   e. their ancestral origin.

9. Fungi and animals appear to have evolved from a common ancestor
   a. because neither of them are photosynthetic.
   b. based on similarities in cell structure.
   c. based on molecular analyses.
   d. about the time that fungi and plants moved onto land.
   e. based on homologous ultrastructure of their flagella.

10. How can fungi be classified as opisthokonts, which means "posterior flagella," when most fungi lack flagella?
    a. Most fungal lineages appear to have lost their flagella during their evolution.
    b. The most primitive lineage of fungi, the chytrids, have flagellated zoospores.
    c. Fungi are in the same lineage as all flagellated protists.
    d. Fungi are closely related to animals, and since animals, which do have flagella, are opisthokonts, fungi must be, too.
    e. Both a and b are the best reasons given here for their classification.

Chapter 32

An Introduction to Animal Diversity

Key Concepts

32.1 Animals are multicellular, heterotrophic eukaryotes with tissues that develop from embryonic layers

32.2 The history of animals may span more than a billion years

32.3 Animals can be characterized by "body plans"

32.4 Leading hypotheses agree on major features of the animal phylogenetic tree

Framework

Animals are multicellular eukaryotic heterotrophs that ingest their food. Fossil evidence for the origin and rapid divergence of animals is limited. A colonial flagellatedprotist is the probable ancestor, and most animal phyla diversified during the 20 million years of the Cambrian explosion. The traditional phylogenetic tree of animals is based on body plan grades and includes such characteristics as symmetry, tissue layers, body cavities, and development. Molecular systematics splits the protostomes into clades Ecdysozoa and Lophotrochozoa.

Chapter Review

32.1 Animals are multicellular, heterotrophic eukaryotes with tissues that develop from embryonic layers

Nutritional Mode Most animals ingest their food, eating either other living organisms or nonliving organic matter.

Cell Structure and Specialization Animal cells lack walls and may have desmosomes, gap junctions, and tight junctions connecting adjacent cells. Structural proteins, such as collagen, hold animal bodies together. Muscle and nervous tissues are unique to animals.

Reproduction and Development The diploid stage is usually dominant, and reproduction is primarily sexual, with a flagellated sperm fertilizing a larger, nonmotile egg. The zygote undergoes a series of mitotic divisions, called **cleavage,** usually passing through a **blastula** stage during embryonic development. The process of **gastrulation** produces layers of embryonic tissues, resulting in the **gastrula** stage. The life cycle of many animals includes a **larva**—a free-living, sexually immature form. **Metamorphosis** transforms a larva into an adult.

While other eukaryotes have regulatory genes, many of which contain DNA sequences called homeoboxes, animals share a unique homeobox-containing family of genes, called *Hox* genes, that are involved in regulating gene expression in embryonic development.

32.2 The history of animals may span more than a billion years

The diversity of the animal kingdom includes the estimated 99% of species that are extinct. Some molecular clock calculations indicate that the ancestors of fungi and animals diverged 1.5 billion years ago. The common ancestor of all living animals may have lived 0.8–1.2 billion years ago and probably resembled choanoflagellates—colonial, flagellated protists.

■ **INTERACTIVE QUESTION 32.1**

List a hypothetical sequence of events that could have led to the origin of animals from a colonial protist.

Neoproterozoic Era (1 Billion–542 Million Years Ago)
The earliest generally accepted fossils of animals, known as the **Ediacaran fauna,** are only about 575 million years old. These appear to be related to cnidarians and soft-bodied molluscs, although fossilized burrows indicate that worms may have evolved in this period. Fossilized animal embryos date from 570 million years ago.

Paleozoic Era (542–251 Million Years Ago) Fossils from the **Cambrian explosion** (542 to 525 mya) represent about half of all extant phyla and include the first animals with hard skeletons. There are three main hypotheses to explain the Cambrian explosion: The emergence of new predator-prey relationships may have changed community dynamics and triggered various evolutionary adaptations. The accumulation of atmospheric oxygen may have supported the active metabolisms of mobile and larger animals. The evolution of the *Hox* complex of regulatory genes, and then changes in their spatial and temporal expression during embryonic development, may have produced the body plan differences that appear during the Cambrian explosion.

During the Silurian and Devonian periods, animal diversity continued to increase. Fishes became the top predators of the seas. Arthropods appeared on land by 460 million years ago. Vertebrates began to adapt to land around 360 million years ago. Of the several lineages that diversified, the amphibians and amniotes survive today.

Mesozoic Era (251–65.5 Million Years Ago) The first coral reefs formed; reptiles returned to water; and flight appeared in pterosaurs and birds. Large predatory and herbivorous dinosaurs emerged, as did the first mammals.

Cenozoic Era (65.6 Million Years Ago to the Present)
Mass extinctions marked the beginning of this period and included the nonflying dinosaurs and marine reptiles. Mammals began to exploit the vacated niches.

■ **INTERACTIVE QUESTION   32.2**

Review the three hypotheses for the rapid radiation of animal phyla during the Cambrian period.

32.3   Animals can be characterized by "body plans"

A group of animals that share a similar organizational complexity is called a **grade.** The sets of morphological and developmental traits that define a grade are often referred to as a **body plan.**

Symmetry Sponges lack symmetry. The parts of an animal with **radial symmetry** radiate from the center. Animals with **bilateral symmetry** have distinct **anterior** (head) and **posterior** (tail) ends, and left and right sides. Bilateral animals also have **dorsal** (top) and **ventral** (bottom) sides. Bilateral symmetry is associated with **cephalization,** the concentration of sensory organs and a central nervous system in the head end, which is an adaptation for unidirectional movement.

Tissues True tissues are groups of specialized cells separated from other tissues by membranous layers. Sponges lack true tissues. During gastrulation, an embryo develops concentric layers of cells called **germ layers: Ectoderm** develops into the outer body covering and, in some phyla, into the central nervous system; **endoderm** lines the developing digestive tube, or **archenteron,** and gives rise to the lining of the digestive tract and associated organs. Cnidarians and comb jellies are **diploblastic,** forming only these two germ layers. The bilateria animals are **triploblastic,** producing a middle layer, the **mesoderm,** from which arise muscles and most other organs.

Body Cavities A fluid-filled **body cavity** separating the digestive tract from the outer body wall is called a **coelom.** A "true" coelom is completely lined by mesodermally derived tissue, with mesenteries extending from the dorsal and ventral sides that suspend internal organs. Animals with a true coelom are called **coelomates. Pseudocoelomates** are animals whose body cavity, called a pseudocoelom, is formed from the blastocoel rather than from mesoderm. Triploblastic animals that have solid bodies are called **acoelomates.**

A fluid-filled body cavity cushions internal organs, allows organs to grow and move independently of the outer body wall, and also functions as a hydrostatic skeleton in soft-bodied animals. During animal evolution, coeloms and pseudocoeloms have evolved and been lost many times and cannot be used to distinguish clades.

Protostome and Deuterostome Development Three features distinguish **protostome development** and **deuterostome development.**

Protostome development, typical of molluscs, annelids, and arthropods, is characterized by **spiral cleavage,** in which the planes of cell division are diagonal and newly formed cells fit in the grooves between cells of adjacent tiers. **Determinate cleavage** of some of these animals sets the developmental fate of each embryonic cell very early. Deuterostome development, typical of echinoderms and chordates, involves **radial**

cleavage, in which parallel and perpendicular cleavage planes result in aligned tiers of cells. **Indeterminate cleavage,** present in most animals with deuterostome development, means that cells from early cleavage divisions retain the capacity to develop into complete embryos.

In protostome development, the coelom forms from splits within solid masses of mesoderm, called **schizocoelous** development. In deuterostome development, the mesoderm begins as buds from the archenteron, called **enterocoelous** development.

The **blastopore** is the opening during gastrulation leading to the developing archenteron. In protostome development ("first mouth"), the blastopore develops into the mouth, and a second opening forms at the end of the archenteron to produce an anus. In deuterostome development, the blastopore becomes the anus, and the second opening develops into the mouth.

■ **INTERACTIVE QUESTION 32.3**

Fill in this table to review some of the differences in the early embryological development of protostomes and deuterostomes.

	Protostomes	Deuterostomes
Cleavage	a.	b.
Coelom formation	c.	d.
Blastopore fate	e.	f.

32.4 Leading hypotheses agree on major features of the animal phylogenetic tree

The relationships of the 35 recognized animal phyla continue to be debated. The traditional animal phylogenetic tree was based on anatomical features. Modern phylogenetic systematics is based on identifying a hierarchy of clades nested within larger clades. Molecular systematics, new studies of lesser-known phyla, and analyses of fossils are contributing to the identification of such clades, defined by shared derived characters unique to the taxa in the clade and their common ancestor. Two current phylogenetic hypotheses are based on systematic analyses of morphological characters or on results from recent molecular studies.

Points of Agreement There are five major points of agreement between the two hypotheses: (1) All animals share a common ancestor. (2) Sponges are basal animals, exhibiting a **parazoan** grade of organization. Phylum Porifera may be paraphyletic. (3) Eumetazoa is a clade of animals with true tissues. The **eumetazoans** include all animals except sponges. Basal members of the clade include the diploblastic, radially symmetrical phylum Cnidaria (which includes jellies) and Ctenophora (comb jellies). These two are often placed in the informal grade called Radiata. (4) Most animal phyla belong to the clade Bilateria. The **bilaterians** are placed in a clade defined by the shared derived character of bilateral symmetry. (5) Vertebrates and some other phyla belong to the clade Deuterostomia.

Disagreement over the Bilaterians The morphology-based tree divides the bilaterians into the deuterostomes and protostomes, assuming that these developmental differences reflect phylogeny. This tree also groups arthropods with annelids, both of which have segmented bodies. Recent molecular studies distinguish two protostome sister taxa: **lophotrochozoans,** which include the annelids and molluscs, and **ecdysozoans,** which include the arthropods and nematodes. The name *Ecdysozoa* refers to the shedding of an exoskeleton, called ecdysis, a trait shared by some edcysozoan phyla. The name *Lophotrochozoa* is based on the feeding apparatus called a **lophophore,** found in the ectoprocts, and the larval stage called the **trochophore larva** shared by other members of the lophotrochozoans.

Future Directions in Animal Systematics Large-scale comparisons of multiple genes across many animal phyla currently being conducted may help systematists to test hypotheses about animal phylogeny.

▶ Word Roots

a- = without; **-koilos** = a hollow (*acoelomate:* the condition of lacking a coelom)

arch- = ancient, beginning (*archenteron:* the endoderm-lined cavity, formed during the gastrulation process, that develops into the digestive tract of an animal)

bi- = two (*Bilateria:* the branch of eumetazoans possessing bilateral symmetry)

blast- = bud, sprout; **-pore** = a passage (*blastopore:* the opening of the archenteron in the gastrula that develops into the mouth in protostomes and the anus in deuterostomes)

cephal- = head (*cephalization:* an evolutionary trend toward the concentration of sensory equipment on the anterior end of the body)

deutero- = second (*deuterostome:* one of two lines of coelomates characterized by radial, indeterminate cleavage, enterocoelous formation of the coelom, and development of the anus from the blastopore)

di- = two (*diploblastic:* having two germ layers)

ecdys- = an escape (*Ecdysozoa:* one of two proposed clades within the protostomes; it includes the arthropods)

ecto- = outside; **-derm** = skin (*ectoderm:* the outermost of the three primary germ layers in animal embryos)

endo- = within (*endoderm:* the innermost of the three primary germ layers in animal embryos)

entero- = the intestine, gut (*enterocoelous:* the type of development found in deuterostomes; the coelomic cavities form when mesoderm buds from the wall of the archenteron and hollows out)

gastro- = stomach, belly (*gastrulation:* the formation of a gastrula from a blastula)

in- = into; **-gest** = carried (*ingestion:* a heterotrophic mode of nutrition in which other organisms or detritus are eaten whole or in pieces)

lopho- = a crest, tuft; **-trocho** = a wheel; (*Lophotrochozoa:* one of two proposed clades within the protostomes that includes annelids and molluscs)

meso- = middle (*mesoderm:* the middle primary germ layer of an early embryo)

meta- = boundary, turning point; **-morph** = form (*metamorphosis:* the resurgence of development in an animal larva that transforms it into a sexually mature adult)

para- = beside; **-zoan** = animal (*parazoan:* grade of body form lacking symmetry and tissues; describes the sponges)

proto- = first; **-stoma** = mouth (*protostomes:* a member of one of two distinct evolutionary lines of coelomates characterized by spiral, determinate cleavage, schizocoelous formation of the coelom, and development of the mouth from the blastopore)

pseudo- = false (*pseudocoelom:* a body cavity that is not completely lined by mesoderm)

radia- = a spoke, ray (*Radiata:* the radially symmetrical animal phyla, including cnidarians)

schizo- = split (*schizocoelous:* the type of development found in protostomes; initially, solid masses of mesoderm split to form coelomic cavities)

tri- = three (*triploblastic:* having three germ layers)

▶ Structure Your Knowledge

1. These two very abbreviated trees represent two hypotheses of animal phylogeny. In the boxes, fill in the clades or body plan grades for these two trees. In the table on the next page, list some of the shared derived characters that define each group.

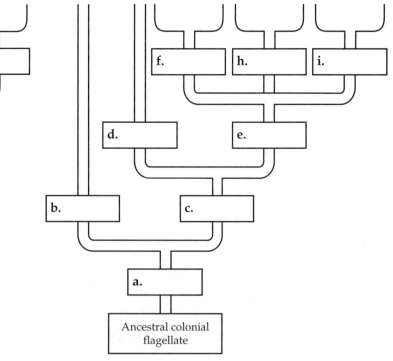

Clade or Body Plan Grade	Shared Derived Characters
a.	
b.	
c.	
d.	
e.	
f.	
g.	
h.	
i.	

Test Your Knowledge

MULTIPLE CHOICE: *Choose the one best answer.*

1. Sponges differ from the rest of the animals because
   a. they are completely sessile.
   b. they have radial symmetry and are suspension feeders.
   c. their simple body structure has no true tissues, and they have no symmetry.
   d. they are not multicellular.
   e. they have no flagellated cells.

2. An insect larva
   a. is a miniature version of the adult.
   b. is transformed into an adult by molting.
   c. ensures more genetic variation in the insect life cycle.
   d. is a sexually immature organism specialized for eating and growth.
   e. is all of the above.

3. Cephalization
   a. is the development of bilateral symmetry.
   b. is the formation of a coelom by budding from the archenteron.
   c. is a diagnostic characteristic of deuterostomes.
   d. is common in radially symmetrical animals.
   e. is associated with motile animals that concentrate sensory organs in a head region.

4. A true coelom
   a. has mesenteries extending from the dorsal and ventral sides that support internal organs.
   b. allows organs to grow and move independently of the outer body wall.
   c. is a fluid-filled cavity completely lined by mesoderm.
   d. may be used as a hydrostatic skeleton by soft-bodied coelomates.
   e. is all of the above.

5. Which of the following is descriptive of protostome development?
   a. radial and determinate cleavage, blastopore becomes mouth
   b. spiral and indeterminate cleavage, coelom forms as split in solid mass of mesoderm
   c. spiral and determinate cleavage, blastopore becomes mouth, schizocoelous development
   d. spiral and indeterminate cleavage, blastopore becomes mouth, enterocoelous development
   e. radial and determinate cleavage, enterocoelous development, blastopore becomes anus

6. Which of the following is *not* descriptive of a pseudocoelomate?
   a. a body cavity incompletely lined by mesoderm
   b. bilateral symmetry
   c. triploblastic
   d. true tissues
   e. schizocoelous formation of body cavity

7. Which of the following characteristics is found only in animals?
   a. homeobox-containing genes
   b. flagellated sperm
   c. heterotrophic nutrition
   d. *Hox* genes
   e. All of the above are exclusive animal traits.

8. Some of the oldest known animal fossils are
   a. colonies of flagellated protists.
   b. worms that are 1 billion years old.
   c. sponges, because they were the first animals.
   d. soft-bodied cnidarians of the Ediacaran period.
   e. bizarre-looking animals from the Cambrian explosion.

9. The grade-based and molecular-based phylogenetic trees agree in which of the following ways?
   a. the acoelomate condition as a branch point before the divergence of the two protostome clades
   b. the deepest branch points of the parazoa–eumetazoa and radiata–bilateria, and the deuterostomes as a monophyletic clade
   c. parazoa as the probable ancestor of the animals
   d. the grouping of the segmented annelids and arthropods
   e. the origin of the *Hox* complex as the probable cause of the Cambrian explosion

10. The rapid diversification of animal phyla during the Cambrian explosion is most likely linked with
    a. the movement of animals onto land.
    b. the origin of the triploblastic body plan that allowed for the development of organs.
    c. the origin of the first homeobox-containing genes.
    d. the evolution of bilateral symmetry and cephalization that permitted more efficient movement and processing of sensory data.
    e. ecological changes such as the development of predator-prey relationships and geologic causes such as the accumulation of atmospheric oxygen.

Chapter 33

Invertebrates

Framework

This chapter surveys the amazing diversity of invertebrate animals. Characteristics and representatives of the major animal phyla are presented. The groups covered are the sponges, cnidarians, flatworms, rotifers, three lophophorate phyla, nemerteans, molluscs, annelids, nematodes, arthropods, echinoderms, and chordates.

Chapter Summary

More than 95% of known animal species, and all except one of the 35 described phyla of animals, are **invertebrates,** animals that lack backbones.

33.1 Sponges are sessile and have a porous body and choanocytes

Sponges (phylum Porifera) are sessile animals found in fresh and marine waters. Water is drawn through pores in the body wall of this saclike animal into a central cavity, the **spongocoel,** and flows out through the **osculum.** Sponges are **suspension feeders,** collecting food particles by the action of collared, flagellated **choanocytes** lining the spongocoel or internal water chambers. Choanocytes resemble the cells of choanoflagellates. Molecular evidence indicates that animals originated from a choanoflagellate-like ancestor.

Sponges lack true tissues. In the **mesohyl,** or gelatinous matrix between the two body-wall layers, are **amoebocytes.** These cells take up food from the water and from choanocytes, digest it, and carry nutrients to other cells. Amoebocytes also form skeletal fibers, which may be sharp spicules or flexible fibers.

Most sponges are **hermaphrodites,** producing both eggs and sperm, although usually sequentially functioning as one sex or the other. Sperm, carried out through the osculum, fertilize eggs retained in the mesohyl of neighboring sponges. Flagellated larvae disperse to a suitable substratum and develop into sessile adults. Sponges produce defensive compounds that may have pharmaceutical uses.

■ INTERACTIVE QUESTION 33.1

Give the locations and functions of the following:

a. choanocytes

b. amoebocytes

33.2 Cnidarians have radial symmetry, a gastrovascular cavity, and cnidocytes

One of the oldest groups of the clade Eumetazoa, animals with true tissues, is the phylum Cnidaria. The cnidarians include hydras, jellies, and corals. Their simple anatomy consists of a sac with a central **gastrovascular cavity** and a single opening serving as both mouth and anus. This body plan has two forms: **polyp,** which is a sessile, cylindrical form with mouth and tentacles extending upward, and **medusa,** which is a flattened, mouth-down polyp that moves by passive drifting and weak body contractions. Both these body forms occur in the life histories of some cnidarians.

Cnidarians use their ring of tentacles, armed with **cnidocytes,** to capture prey. Cnidocytes contain cnidae, which are capsules that can evert and discharge long threads; stinging capsules are called **nematocysts.** A gelatinous mesoglea is sandwiched between the epidermis and gastrodermis. Cells of these layers have bundles of microfilaments arranged into contractile fibers acting as simple muscles. A nerve net is associated with simple sensory receptors and coordinates the contraction of cells against the hydrostatic skeleton of the gastrovascular cavity, producing movement.

Hydrozoans Most hydrozoans alternate between an asexually reproducing polyp and a sexually reproducing medusa form. The common, freshwater hydras exist only in polyp form.

Scyphozoans The medusa stage is more prevalent in the scyphozoans. The sessile polyp stage often does not occur in the jellies of the open ocean.

Cubozoans Cubozoans have a box-shaped medusa stage and complex eyes in the fringe of their medusae. Many species, such as the sea wasp, have highly toxic cnidocytes.

Anthozoans Sea anemones and corals occur only as polyps. Corals secrete calcified external skeletons, and the accumulation of such skeletons produces coral.

■ **INTERACTIVE QUESTION 33.2**

Name these two cnidarian body plans and identify the indicated structures.

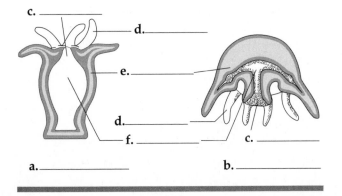

c. _____

d. _____

e. _____

d. _____

f. _____

c. _____

a. _____

b. _____

33.3 Most animals have bilateral symmetry

Clade Bilateria consists of bilaterally symmetrical animals with triploblastic development. Most are coelomates.

Flatworms Free-living flatworms live in marine, freshwater, and damp terrestrial habitats, and many are parasitic. They are triploblastic acoelomates. Most flatworms have a branching gastrovascular cavity, which functions in both digestion and distribution of food. Gas exchange and diffusion of nitrogenous wastes occur across the body wall. Ciliated flame bulbs associated with branched ducts function in osmoregulation. Phylum Platyhelminthes is divided into four classes.

Turbellarians include freshwater **planarians** and other, mostly marine, free-living flatworms. Planarians move using cilia to glide on secreted mucus or may use body undulations to swim. Eyespots on the head detect light, and lateral head flaps detect chemicals. Their nervous system consists of pairs of anterior ganglia and ventral nerve cords. Planarians reproduce asexually by regeneration or sexually by copulation between hermaphroditic worms.

Adapted to live as parasites in or on other animals, flukes (trematodes) have a tough outer covering, suckers, and extensive reproductive organs. The life cycles of trematodes are complex, usually including asexual and sexual stages and intermediate hosts in which larvae develop. Most monogeneans are external parasites of fishes and have a simple life cycle with ciliated larvae dispersing to new hosts.

As parasites, mostly of vertebrates, tapeworms (class Cestoidea) consist of a scolex, with suckers and hooks for attaching to the host's intestinal lining, and a ribbon of proglottids packed with reproductive organs. Predigested food is absorbed from the host. The life cycles of tapeworms may include intermediate hosts.

■ **INTERACTIVE QUESTION 33.3**

a. Describe the digestive system of a planarian.

b. Why do tapeworms, which are also flatworms, lack a digestive system?

Rotifers Rotifers are smaller than many protists but have an **alimentary canal,** with separate mouth and anus, and other organ systems. A pseudocoelom, a body cavity not completely lined by mesoderm, functions as a hydrostatic skeleton and a circulatory system. A crown of cilia draws microscopic food into the mouth. The pharynx bears jaws that grind the ingested microorganisms. Some species reproduce by **parthenogenesis,** in which female offspring develop from unfertilized eggs. In other species, two types of eggs develop parthenogenically: one type forming females and the other developing into simplified males that produce sperm. The resulting resistant zygotes survive harsh conditions in a dormant state.

Lophophorates: Ectoprocts, Phoronids, and Brachiopods The three phyla of lophophorates all have a lophophore, a horseshoe-shaped or circular fold bearing ciliated tentacles, which surrounds the mouth and functions in feeding. This complex structure suggests that all three phyla are related. They have a true coelom.

    Ectoprocts, commonly called bryozoans, are tiny, mostly marine animals living in colonies that are often encased in a hard exoskeleton, with pores through which their lophophores extend. Some species are important reef builders. **Phoronids** are marine, tube-dwelling worms that often live buried in sand, extending and withdrawing their lophophore from the tube opening. **Brachiopods,** or lamp shells, attach to the sea floor by a stalk and open their hinged shell to allow water to flow through the lophophore.

Nemerteans Most ribbon or proboscis worms are marine. Although their body is structurally acoelomate, a fluid-filled sac allows these worms to hydraulically operate an extensible proboscis to capture prey. Their excretory, sensory, and nervous systems are similar to those of flatworms, but they differ in having an alimentary canal and a **closed circulatory system.**

33.4 Molluscs have a muscular foot, a visceral mass, and a mantle

Molluscs are soft-bodied, mostly marine animals, most of which are protected by a shell. The molluscan body plan has three main parts: a muscular **foot** used for movement, a **visceral mass** containing the internal organs, and a **mantle** that covers the visceral mass and may secrete a shell. In many molluscs, a **mantle cavity,** a water-filled chamber formed by the extension of the mantle, encloses the gills, anus, and excretory pores. A rasping **radula** is used for feeding by many molluscs.

    Most molluscs have separate sexes. Many marine molluscs have a life cycle that includes a ciliated larva called the **trochophore,** also found in marine annelids. Four of the eight molluscan classes are discussed in the text.

Chitons Chitons are oval marine animals with shells that are divided into eight dorsal plates. Chitons cling to and creep slowly over rocks, where they feed on algae.

Gastropods Most members of this largest molluscan class are marine, although there are many freshwater species, and some snails and slugs are terrestrial. A distinctive feature of this class is **torsion,** the embryonic rotation of the visceral mass that results in the anus and mantle cavity being above the head. Most gastropods have single coiled shells. Many gastropods have distinct heads with eyes at the tips of tentacles. Moving by the rippling of the foot, most gastropods graze on plant material with their radula. Land snails lack gills; the lining of the mantle cavity functions as a lung.

Bivalves Clams, oysters, mussels, and scallops have the two halves of their shell hinged at the mid-dorsal line. Most bivalves are suspension feeders; water flows into and out of the mantle cavity through siphons, and food particles are trapped in the mucus that coats the gills and then are swept to the mouth by cilia.

Cephalopods Squids and octopuses are rapid-moving carnivores. The mouth has beaklike jaws to bite prey and is surrounded by tentacles. The shell is reduced and internal in squids, absent in octopuses, and external only in the chambered nautilus. In squid, the foot has been modified to form parts of the tentacles and head and the muscular siphon used to jet-propel the squid when water from the mantle cavity is expelled.

    Cephalopods are the only molluscs with a closed circulatory system. They have a well-developed nervous system, sense organs, and a complex brain—important features for active predators. The ancestors of octopuses and squids were probably shelled, predaceous molluscs. Shelled **ammonites** were the dominant invertebrate predators until their extinction at the end of the Cretaceous period.

■ **INTERACTIVE QUESTION 33.4**

a. Describe the three parts of the molluscan body plan.

   1.

   2.

   3.

b. Compare the feeding behaviors and activity levels of snails, clams, and squid.

   1. Snails:

   2. Clams:

   3. Squid:

33.5   Annelids are segmented worms

Annelids are segmented worms found in marine, freshwater, and damp soil habitats.

Oligochaetes   This class includes the earthworms and several aquatic species. Earthworm castings improve soil texture. The earthworm has a closed circulatory system. Respiration occurs across the moist, highly vascularized skin. Septa partition the coelom into segments, in each of which is found a pair of excretory metanephridia, which filter metabolic wastes from the blood and coelomic fluid. The nervous system consists of a pair of cerebral ganglia, a subpharyngeal ganglion, and fused segmental ganglia along the ventral nerve cord. Earthworms are hermaphrodites; sperm are exchanged between worms during mating. A mucous cocoon, secreted by the clitellum, slides off the worm after picking up its eggs and stored sperm.

Noncompressible coelomic fluid within the body segments serves as a hydrostatic skeleton; circular and longitudinal muscles contract alternately to move. Pairs of chaetae provide traction for burrowing.

■ INTERACTIVE QUESTION   33.5

Identify the structures shown in this body segment of an earthworm.

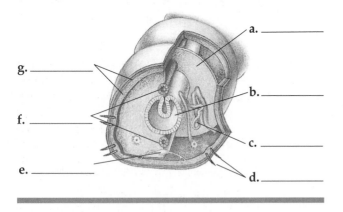

a. _____
g. _____
b. _____
f. _____
c. _____
e. _____
d. _____

Polychaetes   These mostly marine worms have parapodia on each segment that may function in gas exchange and locomotion. Polychaetes may be planktonic, bottom burrowers, or tube dwellers.

Leeches   Most leeches inhabit fresh water. Many feed on small invertebrates, whereas others are parasites that temporarily attach to animals, slit or digest a hole through the skin, and suck the blood of their host.

33.6   Nematodes are nonsegmented pseudocoelomates covered by a tough cuticle

Roundworms are among the most widespread of all animals, found inhabiting water, soil, and the bodies of plants and animals. These cylindrical worms have a tough covering called a **cuticle,** which is periodically shed as they grow. They have a complete digestive tract. Their thrashing movement is produced by contraction of longitudinal muscles. Fluid in the pseudocoelom circulates nutrients. Reproduction is usually sexual, fertilization is internal, and most zygotes form resistant cells.

Numerous nematode species are ecologically important decomposers. Other nematodes are serious agricultural pests and animal parasites. Parasitic nematodes can enlist the cells of their host to supply nutrients or to expand to house them.

■ INTERACTIVE QUESTION   33.6

Compare the locomotion of an earthworm and a nematode.

33.7   Arthropods are segmented coelomates that have an exoskeleton and jointed appendages

In terms of species diversity, distribution, and vast numbers, arthropods are the most successful group of animals.

General Characteristics of Arthropods   Characteristics of **arthropods** include segmentation, which allows for regional specialization; a hard exoskeleton; and jointed appendages. **Trilobites** were early arthropods with uniform appendages. Appendages have become modified for walking, feeding, sensing, mating, and defense. A cuticle of chitin and protein completely covers the body as an **exoskeleton,** providing protection and points of attachment for the muscles that move the appendages. To grow, an arthropod must **molt,** shedding its exoskeleton and secreting a larger one. The exoskeleton probably first evolved in the seas as protection and anchorage for muscles, but as arthropods diversified on land, its functions came to include protection from desiccation and support.

Arthropods have extensive cephalization and well-developed sensory organs. A heart pumps hemolymph through an **open circulatory system** consisting of short arteries and a network of sinuses known as the hemocoel. The embryonic coelom becomes reduced during development, and the hemocoel becomes the main body cavity. Gas exchange in most aquatic species occurs through gills, whereas terrestrial arthropods have tracheal systems of branching internal ducts.

Molecular systematics suggests that arthropods diverged into four evolutionary lineages: **cheliceriforms, myriapods, hexapods,** and **crustaceans.**

Cheliceriforms Cheliceriforms (horseshoe crabs, scorpions, spiders, mites, and ticks) have clawlike **chelicerae** that serve as pincers or fangs used for feeding. They have an anterior cephalothorax and an abdomen. The earliest cheliceriforms were large, predatory **eurypterids.** The horseshoe crab and sea spiders are the few marine cheliceriforms that survive today.

Most modern cheliceriforms are arachnids. Most ticks are blood-sucking parasites on reptiles or mammals. Many mites are also parasites. In arachnids, the cephalothorax has six pairs of appendages: four pairs of walking legs, the chelicerae, and sensing or feeding appendages called pedipalps. In most spiders, **book lungs,** consisting of stacked internal plates, function in gas exchange. Many spiders spin characteristic webs of silk from special abdominal glands.

■ **INTERACTIVE QUESTION   33.7**

a. What are chelicerae?

b. How do spiders trap, kill, and eat their prey?

Myriapods Millipedes and centipedes are in the subphylum Myriapoda. The millipedes of class Diplopoda are wormlike, segmented vegetarians with two pairs of walking legs per segment. Millipedes may have been one of the first land animals. The centipedes of class Chilopoda are terrestrial carnivores with appendages modified as jaw-like **mandibles** and poison claws. Each segment of the trunk has one pair of legs.

Insects Insects and their relatives in the subphylum Hexapoda have more known species than all other forms of life combined. In a typical insect, the head has fused segments and has one pair of antennae, a pair of compound eyes, and several pairs of mouthparts modified for various types of ingestion. The thorax and abdomen have obvious segments. The nervous system consists of a cerebral ganglion and paired ventral nerve cords with segmental ganglia. Malpighian tubules, outpocketings of the digestive tract, function in excretion. Tracheal tubes form the respiratory system.

The oldest insect fossils are from the Devonian period (about 416 mya), but a major diversification occurred in the Carboniferous and Permian periods with the evolution of flight and modification of mouthparts for specialized feeding on plants. The major diversification of insects appears to have preceded and probably influenced the radiation of flowering plants.

Flight is a major key to the success of insects. Many insects have one or two pairs of wings that are extensions of the cuticle of the dorsal thorax. Dragonflies were among the first flying insects.

In **incomplete metamorphosis,** the young are smaller versions of the adult and pass through several molts before developing wings and becoming sexually mature. In **complete metamorphosis,** the larvae look entirely different from the adult. The larvae eat and grow; adults primarily reproduce and disperse. Metamorphosis occurs in a pupal stage. Reproduction is usually sexual; fertilization is usually internal.

Insects are classified into about 26 orders. Insects affect humans as pollinators of crops, vectors of disease, and competitors for food.

Crustaceans The mostly aquatic crustaceans have two pairs of antennae; three or more pairs of mouthpart appendages, including mandibles; walking legs on the thorax; and appendages on the abdomen. Larger crustaceans have gills. Nitrogenous wastes pass by diffusion through thin areas of the cuticle, and a pair of glands regulates the salt balance of the hemolymph. Sexes usually are separate. One or more swimming larval stages occur in most aquatic crustaceans.

Isopods are mostly small aquatic crustaceans but include terrestrial pill bugs. Lobsters, crayfish, crabs, and shrimp are large crustaceans called **decapods.** Their cuticle is hardened by calcium carbonate, and a carapace covers the dorsal side of their cephalothorax. The small, very numerous **copepods** are important members of marine and freshwater plankton communities. Barnacles are sessile crustaceans that strain food from the water with their appendages.

■ INTERACTIVE QUESTION   33.8

Compare and contrast the following features for insects and crustaceans.

Feature	Insects	Crustaceans
Habitat	a.	b.
Locomotion	c.	d.
Respiration	e.	f.
Excretion	g.	h.
Antennae Number	i.	j.
Appendages	k.	l.

33.8   Echinoderms and chordates are deuterostomes

The echinoderms and chordates are grouped together based on their common embryological traits of radial cleavage, coelom formation from the archenteron, and origin of the mouth opposite the blastopore. Molecular systematics also supports Deuterostomia as a clade.

Echinoderms   Most **echinoderms** are sessile or slow-moving marine animals. They have a thin skin covering an endoskeleton of calcareous plates. A **water vascular system** with a network of hydraulic canals controls extensions called **tube feet** that function in locomotion, feeding, and gas exchange. Sexual reproduction usually involves separate sexes and external fertilization. Bilateral larvae metamorphose into radial adults.

Sea stars (class Asteroidea) have multiple arms radiating from a central disk and use tube feet lining the undersurfaces of these arms to creep slowly and to grasp and open prey. They evert their stomach through their mouth and slip it into a slightly opened bivalve shell. Sea stars can regenerate lost arms.

Brittle stars (class Ophiuroidea) have distinct central disks and move by lashing their long, flexible arms. They may be predators, scavengers, or suspension feeders.

Sea urchins and sand dollars, members of class Echinoidea, have no arms but are able to move slowly using their five rows of tube feet. Long spines also aid a sea urchin's movement. A sea urchin's mouth is ringed by complex jaw-like structures used to eat seaweeds and other foods.

Sea lilies live attached to the substrate by stalks; feather stars crawl about. Their long, flexible arms extend upward from around the mouth and are used in suspension feeding. Members of class Crinoidea have changed little in 500 million years of evolution.

Sea cucumbers (Class Holothuroidea) are elongated animals that bear little resemblance to other echinoderms other than having five rows of tube feet.

Discovered in 1986, sea daisies (class Concentricycloidea) have an armless, disk-shaped body with a five-fold symmetry.

■ INTERACTIVE QUESTION   33.9

List the key characteristics that distinguish the phylum Echinodermata.

a.

b.

c.

Chordates   Echinoderms and chordates have existed as separate phyla for at least 500 million years. Phylum Chordata contains two subphyla of invertebrates in addition to the hagfishes and vertebrates.

Word Roots

arthro- = jointed; **-pod** = foot (*Arthropoda:* segmented coelomates with exoskeletons and jointed appendages)

arachn- = spider (*Arachnida:* the arthropod group that includes scorpions, spiders, ticks, and mites)

brachio- = the arm (*brachiopod:* also called lamp shells, these animals superficially resemble clams and other bivalve molluscs, but the two halves of the brachiopod shell are dorsal and ventral to the animal rather than lateral, as in clams)

bryo- = moss; **-zoa** = animal (*bryozoan:* colonial animals (phylum Ectoprocta) that superficially resemble mosses)

cheli- = a claw (*chelicerae:* clawlike feeding appendages characteristic of the cheliceriform group)

choano- = a funnel; **-cyte** = cell (*choanocyte:* flagellated collar cells of a sponge)

cnido- = a nettle (*cnidocytes:* unique cells that function in defense and prey capture in cnidarians)

-coel = hollow (*spongocoel:* the central cavity of a sponge)

cope- = an oar (*copepods:* a group of small crustaceans that are important members of marine and freshwater plankton communities)

cuti- = the skin (*cuticle:* the exoskeleton of an arthropod)

deca- = ten (*decapod:* a large group of crustaceans that includes lobsters, crayfish, crabs, and shrimp)

diplo- = double (*Diplopoda:* the millipede class)

echino- = spiny; **-derm** = skin (*echinoderm:* sessile or slow-moving animals with a thin skin that covers an exoskeleton; the group includes sea stars, sea urchins, brittle stars, crinoids, sea cucumbers, and sea daisies)

eury- = broad, wide; **-pter** = a wing, a feather, a fin (*eurypterid:* mainly marine and freshwater, extinct cheliceriforms; these predators, also called water scorpions, ranged up to 3 meters long)

exo- = outside (*exoskeleton:* a hard encasement on the surface of an animal)

gastro- = stomach; **-vascula** = a little vessel (*gastrovascular cavity:* the central digestive compartment, usually with a single opening that functions as both mouth and anus)

hermaphrod- = with both male and female organs (*hermaphrodite:* an individual that functions as both male and female in sexual reproduction by producing both sperm and eggs)

in- = without (*invertebrates:* animals without a backbone)

iso- = equal (*isopods:* one of the largest groups of crustaceans, primarily marine, but including pill bugs common under logs and moist vegetation next to the ground)

lopho- = a crest, tuft; **-phora** = to carry (*lophophore:* a horseshoe-shaped or circular fold of the body wall bearing ciliated tentacles that surround the mouth)

meso- = the middle; **-hyl** = matter (*mesohyl:* a gelatinous region between the two layers of cells of a sponge.)

meta- = change; **-morph** = shape (*metamorphosis:* the resurgence of development in an animal larva that transforms it into a sexually mature adult)

nemato- = a thread; **-cyst** = a bag (*nematocysts:* the stinging capsules in cnidocytes, unique cells that function in defense and capture of prey)

nephri- = the kidney (*metanephridium:* in annelids, a type of excretory tubule with internal openings called nephrostomes that collect body fluids)

oscul- = a little mouth (*osculum:* a large opening in a sponge that connects the spongocoel to the environment)

partheno- = without fertilization; **-genesis** = producing (*parthenogenesis:* a type of reproduction in which females produce offspring from unfertilized eggs)

plan- = flat or wandering (*planarians:* flatworms that prey on smaller animals or feed on dead animals)

tri- = three; **-lobi** = a lobe (*trilobite:* an extinct group of arthropods with pronounced segmentation)

trocho- = a wheel (*trochophore:* a ciliated larva common to the life cycle of many molluscs; it is also characteristic of marine annelids and some other groups)

Structure Your Knowledge

1. Recall from Chapter 32 that there are several hypotheses of animal phylogeny, and the relationships among the groups may become more evident with further molecular comparisons. To help you review the incredible diversity of animal phyla, let's use the simplified tree on page 264 that shows the groupings agreed upon by both the body plan grade and molecular hypotheses. The small sketches at the top represent the phyla surveyed in this chapter. (The three lophophorate phyla, Ectoprocta, Brachiopoda, and Phoronida, are represented by a figure of a phoronid.) Identify and give common name examples for each of these phyla. The following list of phyla should help:

Annelida	Nematoda
Arthropoda	Nemertea
Chordata	Platyhelminthes
Cnidaria	Porifera
Echinodermata	Rotifera
Mollusca	

Lophophorates: Ectoprocta, Brachiopoda, Phoronida

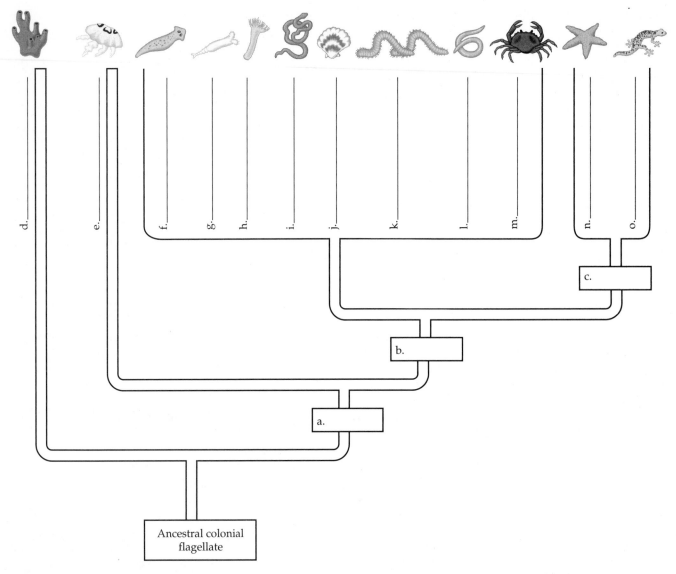

Ancestral colonial
flagellate

Test Your Knowledge

MATCHING: *Match the following organisms with their phylum and class. Answers may be used more than once or not at all.*

Organism	Phylum	Class
1. jelly (jellyfish)	_____	_____
2. crayfish	_____	_____
3. snail	_____	_____
4. leech	_____	_____
5. tapeworm	_____	_____
6. cricket	_____	_____
7. scallop	_____	_____
8. tick	_____	_____
9. sea urchin	_____	_____
10. hydra	_____	_____
11. planaria	_____	_____
12. chambered nautilus	_____	_____

Phyla	Classes
A. Annelida	**a.** Arachnida
B. Arthropoda	**b.** Bivalvia
C. Cnidaria	**c.** Cephalopoda
D. Echinodermata	**d.** Cestoidea
E. Mollusca	**e.** Crustacea
F. Nematoda	**f.** Echinoidea
G. Nemertea	**g.** Gastropoda
H. Platyhelminthes	**h.** Hirudinea
I. Porifera	**i.** Hydrozoa
J. Rotifera	**j.** Insects (Hexopoda)
	k. Oligochaeta
	l. Scyphozoa
	m. Turbellaria

MULTIPLE CHOICE: *Choose the one best answer.*

1. Invertebrates include
   a. all animals except for the phylum Vertebrata.
   b. all animals without backbones.
   c. only animals that use hydrostatic skeletons.
   d. members of the parazoa, radiata, and proto-stomes, but not of the deuterostomes.
   e. all animals without an endoskeleton.

2. Which of the following is the best description of the phylum Porifera?
   a. no real symmetry, diploblastic, cnidocytes for capturing prey
   b. radial symmetry, triploblastic, nematocysts
   c. no real symmetry, without true tissues, choanocytes for trapping food particles
   d. bilateral symmetry, pseudocoel, flame bulbs for excretion
   e. bilateral symmetry, osculum and spongocoel for filtering water

3. Which of the following does *not* have a gastrovascular cavity for digestion?
   a. flatworm
   b. hydra
   c. polychaete worm
   d. sea anemone
   e. fluke

4. The oldest and least-derived group in clade Eumetazoa is
   a. Porifera.
   b. Lophophorates.
   c. Platyhelminthes.
   d. Parazoa.
   e. Cnidaria.

5. Hermaphrodites
   a. contain male and female sex organs but usually cross-fertilize.
   b. include sponges, earthworms, and most insects.
   c. are characteristically parthenogenic rotifers.
   d. are both a and b.
   e. are a, b, and c.

6. Which of the following is *not* true of cnidarians?
   a. An alternation of medusa and polyp stage is common in class Hydrozoa.
   b. They use a ring of tentacles armed with stinging cells to capture prey.
   c. They include hydras, jellies, sponges, and sea anemones.
   d. They have a nerve net that coordinates contraction of microfilaments for movement.
   e. They have a gastrovascular cavity.

7. Which of the following combinations of phylum and characteristics is *incorrect*?
   a. Nemertea—proboscis worm, complete digestive tract
   b. Rotifera—parthenogenesis, crown of cilia, microscopic animals
   c. Nematoda—gastrovascular cavity, tough cuticle, ubiquitous
   d. Annelida—segmentation, closed circulatory system, hydrostatic skeleton
   e. Echinodermata—radial anatomy, endoskeleton, water vascular system

8. Which of the following is an excretory or osmoregulatory structure that is *incorrectly* matched with its class?
   a. metanephridia—Oligochaeta
   b. Malpighian tubules—Echinoidea
   c. flame cells—Turbellaria
   d. thin region of cuticle—Crustacea
   e. diffusion across cell membranes—Hydrozoa

9. Torsion
   a. is embryonic rotation of the visceral mass that results in a U-shaped digestive tract in gastropods.
   b. is characteristic of molluscs.
   c. is responsible for the spiral growth of bivalve shells.
   d. describes the thrashing movement of nematodes.
   e. is responsible for the metamorphosis of insects.

10. Bivalves differ from other molluscs in that they
    a. are predaceous.
    b. have no heads and are suspension feeders.
    c. have shells.
    d. have open circulatory systems.
    e. use a radula to feed as they burrow through sand.

11. The exoskeleton of arthropods
    a. functions in protection and anchorage for muscles.
    b. is composed of chitin and cellulose.
    c. is absent in millipedes and centipedes.
    d. expands at the joints when the arthropod grows.
    e. functions in respiration and movement.

12. Which of the following does *not* function in suspension feeding?
    a. lophophore of brachiopods
    b. radula of snails
    c. choanocytes of sponges
    d. mucus-coated gills of clams
    e. crown of cilia of rotifers

13. What do nematodes and arthropods have in common?
    a. They are both segmented.
    b. They are both pseudocoelomates.
    c. They include important members of plankton communities.
    d. They both have exoskeletons and undergo ecdysis (molting).
    e. Both a and d are correct.

14. Which of the following structures is *not* associated with prey capture?
    a. chaetae of earthworm
    b. mandibles of centipedes
    c. cnidocytes of hydra
    d. proboscis of ribbon or proboscis worm
    e. tube feet of sea star

15. Many animals are parasitic. Which of the following is an *incorrect* description of one of these parasites?
    a. Ticks are blood-sucking parasites belonging to class Arachnida.
    b. Some roundworms (Nematoda) are internal parasites of humans.
    c. Lice are wingless ectoparasites in class Insecta.
    d. Flukes are flatworms and may have complex life cycles.
    e. Tapeworms are annelids that reproduce by shedding proglottids.

Chapter 34

Vertebrates

Key Concepts

34.1 Chordates have a notochord and a dorsal, hollow nerve cord

34.2 Craniates are chordates that have a head

34.3 Vertebrates are craniates that have a backbone

34.4 Gnathostomes are vertebrates that have jaws

34.5 Tetrapods are gnathostomes that have limbs and feet

34.6 Amniotes are tetrapods that have a terrestrially adapted egg

34.7 Mammals are amniotes that have hair and produce milk

34.8 Humans are bipedal hominoids with a large brain

Framework

This chapter focuses on the phylogeny of chordates, introducing the derived characters that define the major clades. Fill in the cladogram of chordates on p. 269 as you work through this chapter. Fossil evidence and hypotheses about human ancestry are also described.

Chapter Review

34.1 Chordates have a notochord and a dorsal, hollow nerve cord

The deuterostome clade of the animal kingdom includes the echinoderms and **chordates.** The chordates include two invertebrate groups—the urochordates and the cephalochordates, the hagfishes, and the **vertebrates,** which have a backbone, or vertebral column.

Derived Characters of Chordates Chordates have four distinguishing anatomical features, which may appear only during the embryonic stage: (1) a flexible rod called a **notochord** that provides skeletal support along the length of the animal; (2) a dorsal, hollow nerve cord that develops into the brain and spinal cord; (3) **pharyngeal clefts,** or grooves that may develop into **pharyngeal slits** that open from the pharynx to the outside and function in suspension feeding or are modified for gas exchange and other functions; and (4) a post-anal tail containing skeletal elements and muscles.

Tunicates Tunicates (subphylum Urochordata) belong to the deepest-branching lineage of chordates. The larval traits of notochord, nerve cord, and tail are lost in the adult animal. In this saclike, sessile animal, commonly called a sea squirt, water flows in the incurrent siphon, through pharyngeal slits where food is filtered by a mucous net, and out the excurrent siphon.

Lancelets Lancelets (subphylum Cephalochordata) are tiny marine animals that retain all four chordate characteristics into the adult stage. These suspension feeders burrow backward into the sand and filter food particles through a mucous net secreted across the pharyngeal slits. The lancelet swims by coordinated contractions of serial muscles that flex the notochord from side to side. Blocks of mesoderm, called **somites,** develop into these muscle segments.

■ INTERACTIVE QUESTION   34.1

What is the name for the animal shown in this diagram, and to which subphylum does it belong? Identify the labeled structures and indicate the four chordate characters.

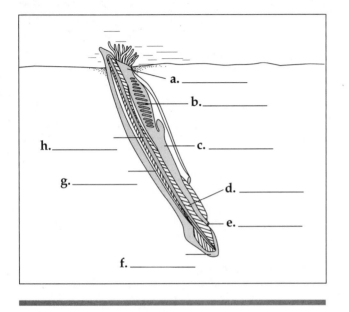

a. _____

b. _____

h. _____   c. _____

g. _____   d. _____

e. _____

f. _____

Early Chordate Evolution   An early hypothesis was that lancelets evolved by paedogenesis from a tunicate-like larva. Current evidence indicates that both the larval and degenerate adult stage of tunicates are derived traits. Studies have shown that several *Hox* genes are expressed in the vertebrate brain and in the anterior end of the lancelet dorsal nerve cord in the same anterior-to-posterior pattern.

34.2   Craniates are chordates that have a head

Derived Characteristics of Craniates   The origin of a head with eyes and other sensory organs and a skull enclosing a brain was a major evolutionary transition seen in **craniates.**

Derived characters that distinguish craniates from other chordates include two clusters of *Hox* genes instead of only one, and a duplicated family of genes for signaling molecules and transcription factors. A group of embryonic cells called the **neural crest** is found along the margin of the embryonic folds that form the nerve cord. These cells migrate in the embryo and contribute to various craniate structures, including bones and cartilage of the cranium (skull). The pharyngeal clefts of craniates evolved into gill slits with muscles and nerves that facilitate sucking in food and water for gas exchange. Craniates have a higher metabolism, more extensive muscles, a heart with at least two chambers, red blood cells, hemoglobin, and kidneys.

The Origin of Craniates   Newly discovered 530 million-year-old fossils (during the Cambrian explosion) appear to be transitional stages: an animal called *Haikouella*, which resembled lancelets, but with craniate characters such as eyes, mineralized toothlike structures, and a small but well-formed brain. Another fossil from this period, *Haikouichthys*, had a skull and is considered to be a true craniate.

Hagfishes   Hagfishes, class Myxini, are the most primitive surviving craniate lineage. Their cartilaginous skeleton includes a skull and a notochord; they lack vertebrae and jaws. The marine hagfishes are mostly bottom-dwelling scavengers that produce slime to repulse predators and competing scavengers.

■ INTERACTIVE QUESTION   34.2

List the derived characters of craniates.

34.3   Vertebrates are craniates that have a backbone

Derived Characters of Vertebrates   The gene duplication involving the *Dlx* family of transcription factor genes was associated with nervous system innovations and the more extensive skull and backbone of vertebrates. In most vertebrates, the vertebrae enclose the spinal cord. The dorsal, ventral, and anal fins supported by fin rays of aquatic vertebrates provided thrust and steering control for swimming.

Lampreys   The jawless lampreys of class Cephalaspidomorphi represent the oldest lineage of vertebrates. Lampreys are suspension feeders as larvae and blood-sucking parasites as adults. The notochord is the main axial skeleton and is surrounded by a cartilaginous pipe with pairs of projections that partially enclose the nerve cord.

Fossils of Early Vertebrates   **Conodants** were soft-bodied vertebrates with prominent eyes and barbed, mineralized hooks in their mouth, and dental elements in the pharynx for processing food. Conodonts appeared in the late Cambrian and were abundant for more than 300 million years.

During the Ordovician, Silurian, and Devonian periods, vertebrates with paired fins and an inner ear with semicircular canals emerged. These animals lacked jaws but had a muscular pharynx and were armored with mineralized bone. This paraphyletic group of diverse armored jawless vertebrates, previously called "ostracoderms," disappeared by the end of the Devonian period.

Origins of Bone and Teeth   One hypothesis for mineralization in vertebrates is that it is associated with the switch from suspension-feeding to scavenging and predation. The earliest known mineralized structures are conodont dental elements. The armor of jawless vertebrates was derived from dental mineralization.

■ INTERACTIVE QUESTION   34.3

List the derived characters of vertebrates.

■ INTERACTIVE QUESTION   34.4

This diagram shows the major clades of chordates with some of the derived characters that define them. As you work through this chapter, fill in the eight large clades across the top and representative animals of each group.

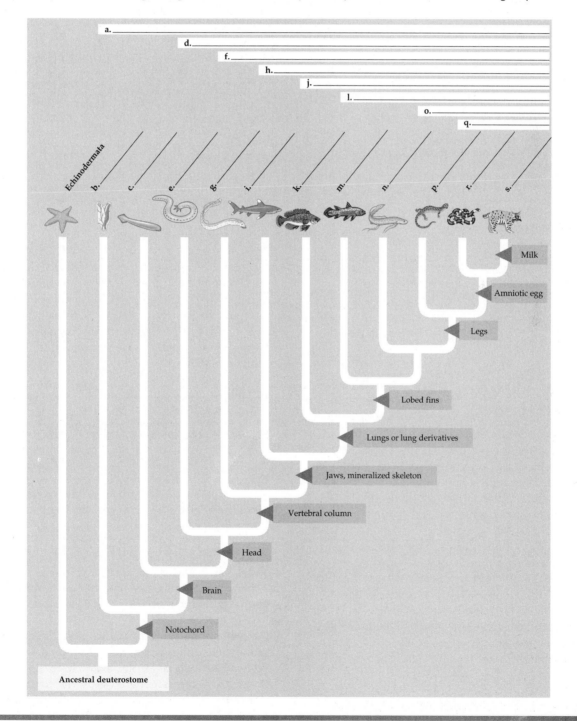

34.4   Gnathostomes are vertebrates that have jaws

Derived Characters of Gnathostomes   The hinged jaws of **gnathostomes** may have evolved from the skeletal rods of the anterior pharyngeal slits, with the remaining gill slits functioning in gas exchange.

Other derived characters of gnathostomes include an addition duplication of *Hox* genes, duplication of other gene clusters, an enlarged forebrain associated with enhanced senses of smell and vision, a pressure-sensing **lateral line system,** and a mineralized endoskeleton.

Fossil Gnathostomes   Beginning in the mid-Ordovician period, about 470 mya, gnathostomes became increasingly diverse. Their paired fins, tail, and hinged jaws facilitated a predatory lifestyle. The earliest gnathostomes were the armored **placoderms. Acanthodians** radiated during the Devonian period and were related to the ancestors of osteichthyans (ray-finned fishes and lobe-fins).

Chondrichthyans (Sharks, Rays, and Their Relatives)   The sharks, rays, and their relatives of class Chondrichthyes have a skeleton made predominantly of cartilage. The cartilaginous skeleton of **chondrichthyans** appears to be a derived condition.

The largest subclass consists of sharks and rays. Sharks swim using powerful muscles in their caudal fin. The paired pectoral and pelvic fins provide lift. Swimming maintains buoyancy and moves water past their gills. Although most sharks are carnivores, the largest sharks and rays are suspension feeders on plankton.

Sharks have keen senses: sharp vision, nostrils, and skin regions in the head that can detect electric fields generated by muscle contractions of nearby animals. The shark's body transmits sound waves to the inner ear.

Following internal fertilization, **oviparous** species of sharks lay eggs that hatch outside the mother; **ovoviviparous** species retain the fertilized eggs until they hatch; and, in the few **viviparous** species, developing young are nourished by a yolk sac placenta until born.

Rays, which propel themselves with their enlarged pectoral fins, are primarily bottom dwellers that feed on mollusks and crustaceans.

■ INTERACTIVE QUESTION   34.5

a. List the derived characters of gnathostomes.

b. What is the function of the **spiral valve** in a shark's intestine?

c. What is a **cloaca?**

Ray-Finned Fishes and Lobe-Fins   Nearly all vertebrates belong to a clade of gnathostomes called Osteichthyes, which historically included only the "bony fishes" but now includes the tetrapods. Almost all living **osteichthyans** have an ossified endoskeleton with a hard calcium phosphate matrix.

Aquatic osteichthyans (informally called fishes) move water through the mouth and out between the gills by the contraction of muscles in the gill chambers and movement of the protective bony flap called the **operculum.** Most aquatic osteichthyans also have a **swim bladder,** an air sac that controls buoyancy. It appears that lungs arose first as a supplement to gas exchange in the gills, and later evolved into swim bladders in some lineages. The flattened, bony scales that cover the skin of some aquatic osteichthyans are coated with mucus to reduce drag. They have a lateral line system similar to that of sharks.

Most aquatic osteichthyans have external fertilization and are oviparous.

Ray-finned fishes (class Actinopterygii) probably originated in fresh water during the Devonian period. They include most of the familiar modern fish, many of which spread to the seas during their evolution and some of which returned to fresh water.

The **lobe-fins** (Sarcopterygii), known mainly from the fossil record, have paired muscular fins supported by a series of rod-shaped bones that may have enabled these lobe-fins to "walk." Only three lobe-fin lineages survive: Two populations of coelacanths (class Actinistia) found in deep marine waters, three genera of lungfishes (class Dipnoi) found in stagnant ponds and swamps in the Southern Hemisphere, and the tetrapods found on land.

■ INTERACTIVE QUESTION   34.6

Go back to the cladogram in Interactive Question 34.4 and make sure you have filled in the information up to letter *n.*

34.5   Tetrapods are gnathostomes that have limbs and feet

About 360 million years ago, the fins of some lobe-fins evolved into the limbs and feet of tetrapods.

Derived Characters of Tetrapods   **Tetrapods** have limbs that support their weight on land and feet with digits. The pelvic girdle is fused to the backbone, transferring forces generated by the hind legs to the rest of the body. Pharyngeal clefts formed during development give rise to parts of the ears, glands, and other structures.

The Origin of Tetrapods During the Devonian, a wide range of lobe-fins had evolved, probably using their stout fins to crawl through shallow waters and supplementing their gas exchange with their lungs. In one lineage, the fins became more limb-like while the rest of the body retained adaptations for the water. A diversity of tetrapods emerged during the Devonian and Carboniferous periods, most of which probably remained tied to the water. Many newly discovered fossils from 400 to 350 million years ago document the origin of tetrapods.

Amphibians **Amphibians** (class Amphibia) include salamanders (order Urodela), frogs (order Anura), and caecilians (order Apoda).

Salamanders, which may be aquatic or terrestrial, move with a lateral bending of the body. Frogs are more specialized for moving on land, using their powerful hind legs for hopping. Protective adaptations of frogs include camouflage coloring and distasteful or poisonous skin mucus that may be advertised by bright coloration. Tropical caecilians are burrowing, legless, nearly blind, wormlike animals.

Many frogs undergo a metamorphosis from a tadpole—the aquatic, herbivorous larval form with gills, a lateral line system, and a finned tail—to the carnivorous, terrestrial adult form with legs and lungs. Some amphibians are exclusively aquatic or terrestrial. Amphibians are most abundant in damp habitats, since some of their gas exchange occurs across their moist skin or mouth.

Fertilization is generally external. Oviparous species lay their eggs in aquatic or moist environments. Some species show varying amounts of parental care. During breeding season, some frogs communicate with vocalizations. Some species migrate to specific breeding sites, using various forms of communication and navigation.

■ INTERACTIVE QUESTION 34.7

a. What are the three lineages of lobe-fins?

b. Describe *Acanthostega*, a Devonian relative of tetrapods.

34.6 Amniotes are tetrapods that have a terrestrially adapted egg

Amniotes include reptiles (and birds) and mammals.

Derived Characters of Amniotes The amniotic egg contains **extraembryonic membranes** that are involved in gas exchange, waste storage, and transfer of nutrients to the embryo. The amnion encases the embryo in fluid. Most reptiles and some mammals have shelled amniotic eggs. The shells of bird eggs are calcareous; those of nonbird reptiles are leathery and flexible. Most mammals retain the embryo within the mother. Other terrestrial adaptations of the amniotes include a waterproof skin and increased use of the rib cage to ventilate the lungs.

■ INTERACTIVE QUESTION 34.8

Identify the four extraembryonic membranes in this sketch of an amniotic egg. What are the functions of these membranes?

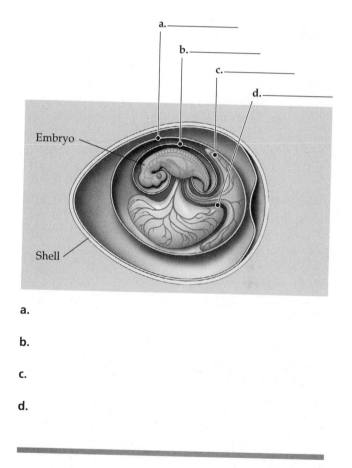

a.

b.

c.

d.

Early Amniotes The most recent common ancestor of living amphibians and amniotes likely lived in the late Devonian period. Early amniotes lived in drier environments than earlier tetrapods. They included herbivores and predators.

Reptiles The **reptile** clade includes a number of extinct groups (such as dinosaurs), and tuatara, lizards, snakes, turtles, crocodilians, and birds. We must infer the derived characters that distinguished early reptiles from other tetrapods. The skin of reptiles is covered with keratinized, waterproof scales. Lungs are the vehicle for gas exchange. Most species lay shelled amniotic eggs, and fertilization is internal. Some species are viviparous, in which case the extraembryonic membranes form a placenta.

Many reptiles are **ectothermic.** They absorb external heat rather than generating their own, although they may regulate their body temperature through behavioral adaptations. Birds are **endothermic,** warming their bodies with metabolic heat.

The oldest reptilian fossils date from the end of the Carboniferous period, 300 million years ago. The first major branch to emerge was the **parareptiles,** large herbivores with dermal plates on their skin. The relationship of turtles to parareptiles remains controversial.

The **diapsids** diversified next. Their most obvious derived character is a pair of holes on each side of the skull. The two main lineages of the diapsids are the **lepidosaurs,** which include tuatara, lizards, and snakes (and a number of now extinct huge marine reptiles) and the **archosaurs,** which produced the crocodilians and the extinct pterosaurs and dinosaurs. The **pterosaurs** had wings formed from a vascularized, muscled membrane stretched between an elongated finger and the hind leg. The highly diverse **dinosaurs** included the largest animals ever to live on land. Two main branches of dinosaurs were the herbivorous ornithischians and the saurischians, which gave rise to long-necked giants and the **theropods,** which included the ancestors of birds.

Scientists continue to debate whether dinosaurs were ectothermic or endothermic. The ancestor of birds was certainly endothermic. Evidence indicates that many dinosaurs were active, agile, and, in some species, social. At the end of the Cretaceous, all dinosaurs (except birds) became extinct.

The lizardlike tuatara represent one lineage of lepidosaurs, surviving only on islands off New Zealand. The other living lineage of lepidosaurs are the squamates. Lizards are the most numerous and diverse reptiles (aside from birds). Snakes are limbless, although some species have vestigial pelvic and limb bones. Snakes are carnivorous, with adaptations for locating (heat-detecting and olfactory organs), killing (sharp teeth, sometimes that inject toxins), and swallowing (loosely articulated jaws and elastic skin) their prey.

The earliest turtle fossils, from 220 mya, already had fully developed shells, so the origin of the turtle shell has not been established. This protective, usually hard shell has an upper and lower shield fused to the vertebrae, clavicles, and ribs.

Crocodiles and alligators, collectively called crocodilians, belong to an archosaur lineage. Living in warm regions, they are adapted to aquatic habitats.

■ **INTERACTIVE QUESTION   34.9**
The two main lineages of diapsids are archosaurs and lepidosaurs. List several representatives, both living and extinct, of these two lineages.

Birds   Birds are archosaurs, whose reptilian features have been modified as they adapted to flight. Many of the derived characters of birds are adaptations that reduce weight: only one ovary, honeycombed bones, and lack of teeth.

Feathers made of β-keratin are extremely light and strong, and shape the wing into an airfoil. Large pectoral muscles attached to a keel on the sternum flap the aerodynamic wings. The evolution of strong flying ability enhanced hunting, foraging, escape from predators, and migration to favorable habitats.

Birds are endothermic; feathers and a fat layer in some species help retain metabolic heat. The four-chambered heart and efficient respiratory system facilitate a high metabolic rate.

Birds have excellent vision. Their relatively large brains permit detailed visual processing and motor coordination. Birds display complex behaviors, particularly in courtship rituals and parenting. Eggs are brooded during development by the female and/or male bird.

Cladistic analysis indicates that birds are theropods. Feathered, but flightless theropod fossils have recently been found in Chinese sediments. Several species had feathers with vanes, others had filamentous feathers. Functions of early feathers may have been insulation and courtship displays.

Flight may have evolved as ground-running dinosaurs used feathers to gain extra lift or as tree-climbing dinosaurs used feathers to glide.

Feathered theropods had evolved into birds by 150 million years ago. *Archaeopteryx,* the oldest bird known, had feathered wings with clawed digits, teeth, and a long tail. Fossils from the Cretaceous period show a gradual loss of dinosaur features and acquisition of bird innovations.

The clade Neornithes, which includes the 28 orders of living birds, appeared about 65 million years ago. Most birds can fly. Flightless birds, collectively called **ratites,** such as the ostrich, kiwi, and emu, lack a keeled breastbone and enlarged pectoral muscles. The general body form of many flying birds is similar. The beak has assumed a great variety of shapes associated with different diets. Foot structure also shows variation.

■ **INTERACTIVE QUESTION 34.10**

List several adaptations for flight found in birds.

34.7 Mammals are amniotes that have hair and produce milk

Derived Characters of Mammals **Mammals** are the other amniote lineage. Their derived characters include mammary glands, which produce milk to nourish the young, and hair, which (along with a layer of fat under the skin) helps to insulate these endothermic animals. Active metabolism is provided for by an efficient respiratory system that uses a diaphragm to help ventilate the lungs and a circulatory system with a four-chambered heart.

Mammals generally have large brains and many are capable of learning. Extended parental care of the young provides time to learn survival skills.

Mammalian teeth come in a diverse assortment of shapes and sizes that are specialized for eating a variety of foods.

Early Evolution of Mammals Mammals are part of the amniote group known as **synapsids,** distinguished by the temporal fenestra, a hole behind each eye socket. During the evolution of mammals from non-mammalian synapsids, the jaw was remodeled, with a stronger, single-bone lower jaw and the incorporation of two jaw bones into the middle ear. Synapsids were the dominant tetrapods during a time in the Permian period, but many became extinct at the end of that period. Mammal-like synapsids emerged 200 million years ago at the end of the Triassic.

The first true mammals arose and diversified during the Jurassic period. The still small Mesozoic mammals, which were probably nocturnal and insectivorous, co-existed with dinosaurs. Mammals underwent an extensive radiation in the wake of the Cretaceous extinctions. There are three major lineages of mammals today.

Monotremes The platypus and echidnas (spiny anteaters) are the only extant **monotremes,** egg-laying mammals. Although egg-laying is a primitive character for amniotes, monotremes have hair and produce milk for their young. Monotremes may have diverged from other mammals about 200 mya.

Marsupials **Marsupials,** including opossums, kangaroos, and koalas, share derived characters with eutherians including a higher metabolic rate, nipples that provide milk, and giving birth to live young. The embryo develops in the uterus. A **placenta,** formed from the uterine lining and extraembryonic membranes of the embryo, nourishes the developing embryo. A marsupial is born very early in development and completes its embryonic development in a maternal pouch called a marsupium.

Australian marsupials have radiated and filled the niches occupied by eutherian mammals in other parts of the world. After the breakup of Pangaea, South America and Australia became island continents where marsupials diversified in isolation from eutherians. Eutherians reached South America in several migrations, whereas Australia has remained isolated since about 65 mya.

Eutherians (Placental Mammals) **Eutherians** complete development attached to a placenta within the maternal uterus. Eutherians and marsupials may have diverged about 180 million years ago.

Molecular systematics is helping to establish the phylogeny of eutherian orders, although there is still no consensus. There appear to be at least four main clades. One clade evolved in Africa and includes the elephants, manatees, hyraxes, and aardvarks. A second clade radiated in South America and includes the sloths, anteaters, and armadillos. The third and largest clade includes the lagomorphs (rabbits), rodents, and primates. The fourth diverse clade includes many groups: the bats, shrews and moles, carnivores, the hoofed herbivores, and the cetaceans (dolphins and whales).

Primates include the lemurs, tarsiers, monkeys, and apes (including humans). Derived characters include grasping hands and feet, and flat nails on their digits. Compared to other mammals, primates have a large brain, short jaws, forward-looking eyes with overlapping visual fields that enhance depth perception, and exhibit complex social behavior and extensive parental care. Many of these traits were shaped by natural selection in the tree-dwelling early primates. All primates, except humans, have a wide separation between the big toe and the other toes, allowing them to grasp branches with their feet. The anthropoids have an **opposable thumb,** which functions in a grasping "power grip" in monkeys and apes other than humans, but is adapted for precise manipulation in humans.

There are three main groups of living primates: the lemurs, lorises, and pottos; the tarsiers; and the anthropoids. The lemurs, lorises, and pottos probably resemble early arboreal primates. The **anthropoids** include monkeys and apes. The oldest known anthropoid fossils are 45 million years old and point to tarsiers as the closest relative. Monkeys appeared in the New World during the Oligocene, after first evolving in the Old World. New World monkeys are strictly arboreal. Most monkeys in both groups are diurnal and live in social bands.

The **hominoids,** primates informally called apes, include gibbons, orangutans, gorillas, chimpanzees and bonobos, and humans. Hominoids diverged from Old World monkeys about 20–25 million years ago. Nonhuman hominoids live in the Old World tropics. Living hominoids, with the exception of gibbons, are larger than monkeys. All living hominoids have relatively long arms, short legs, and no tails. Hominoids have a larger brain and their behavior is more flexible than that of other primates. Gorillas and chimpanzees are highly social.

■ **INTERACTIVE QUESTION   34.11**

a. List the derived characteristics of mammals.

b. Go back to the cladogram in Interactive Question 34.4 and make sure you have filled in all the information.

34.8   Humans are bipedal hominoids with a large brain

Derived Characters of Humans   The species *Homo sapiens* is about 160,000 years old. Characters that distinguish humans from other hominoids are upright stance and bipedal locomotion, larger brain with the capacity for language and symbolic thought, and the manufacture and use of complex tools. Humans also have reduced jaws and a shorter digestive tract.

The Earliest Hominids   **Paleoanthropology** is the study of human origins. Approximately 20 extinct species of **hominids,** species that are more closely related to humans than to chimpanzees, have been identified. The oldest of these lived about 7–6 million years ago. These early hominids shared some derived characters of humans, such as reduced canine teeth, relatively flat faces, and upright stance, as indicated by the location of the foramen magnum underneath the skull. Different human features have evolved at different rates, known as **mosaic evolution.** Thus, for example, bipedalism evolved before an enlarged brain developed.

These early hominids and chimpanzees represent divergent branches from a common, anthropoid ancestor. Human evolution has not occurred within an unbranched hominid line; there have been times when several different human species coexisted.

Australopiths   Hominid diversity increased between 4 and 2 million years ago, and many of the hominids from this period are called australopiths, although their phylogeny is unresolved. The name came from the 1924 discovery of *Australopithecus africanus.* It appears that *A. africanus* lived between 3 and 2.4 million years ago and was an upright hominid, with humanlike hands and teeth but a small brain.

In 1974, a 3.24-million-year-old *Australopithecus* skeleton was discovered in the Afar region of Ethiopia. "Lucy" and similar fossils have been designated *Australopithecus afarensis,* a species that appears to have existed for about 1 million years. Although bipedal, *A. afarensis* had a brain size similar to that of a chimpanzee, an apelike skull with a long lower jaw, and long arms.

Another lineage from that time was the "robust" australopiths, such as *Paranthropus boisei,* which had sturdy skulls with powerful jaws and teeth.

Bipedalism   By about 10 million years ago, the Himalayan range had formed, the climate became drier, and the forests of Africa and Asia contracted, presenting an increased savannah habitat that may have influenced anthropoid evolution. But all recently discovered fossils of early hominids show signs of bipedalism, and these hominids lived in mixed forests and open woodlands. About 1.9 million years ago hominids that lived in more arid environments began to walk long distances on two legs.

Tool Use   The first evidence of tool use is 2.5-million-year-old cuts on animal bones, suggesting that hominids used stone tools. *Australopithecus garhi,* whose fossils were found nearby, had a relatively small brain.

***Early* Homo**   Larger-brained fossils dating from 2.4 to 1.6 million years old are the first to be placed in the genus *Homo.* Sometimes found with sharp stone tools, *Homo habilis,* or "handy man," had some derived hominid characters. Fossils from 1.9 to 1.5 years ago are recognized by a number of paleoanthropologists as the species *Homo ergaster,* which had a larger brain, long legs and hip joints adapted for long-distance walking, relatively short and straight fingers, and more sophisticated stone tools.

The sexual dimorphism of size difference between the sexes was also less in early *Homo* than that in *Australopithecus afarensis,* a trend that continued in humans, with males now only about 1.2 times the weight of females. In living hominoids, reduced sexual dimorphism is associated with more pair-bonding. Thus, *H. ergaster* may have engaged in pair-bonding, perhaps associated with long-term biparental care of babies.

Some paleoanthropologists still consider *H. ergaster* to be early members of *Homo erectus,* the first hominid to migrate out of Africa. The oldest fossils of hominids outside of Africa date back 1.8 million years. Most paleoanthropologists conclude that *H. erectus* became extinct around 200,000 years ago.

Neanderthals   The Neanderthals appear to be descendants of *Homo heidelbergensis,* which originated about 600,000 years ago in Africa. Neanderthals lived in Europe from about 200,000 to 30,000 years ago. They had large brains and made tools from stone and wood. Comparisons of DNA from four Neanderthal fossils with DNA of living humans show that Neanderthals form a clade, while Europeans are more closely related to Africans and Asians.

Homo sapiens   Evidence indicates that the ancestors of humans originated in Africa. The oldest known fossils of *H. sapiens* date from 195,000 years ago. These early humans lacked heavy brow ridges and were more slender than other hominids. DNA comparisons show that Europeans and Asians share a relatively recent common ancestor, with many African lineages branching off earlier. Comparisons of mitochondrial DNA and Y chromosomes of various populations support a common African ancestor of all *Homo sapiens.*

The oldest fossils outside Africa date back about 50,000 years. Y chromosome studies suggest that one wave of humans first spread into Asia, then to Europe and Australia. The date of arrival in the New World may be 15,000 years ago. The rapid expansion of our species may be tied to the evolution of human cognition. Symbolic thought may have emerged along with human language, enabling the construction of new tools and the ability to teach others to build them.

The gene *FOXP2* has been identified as essential for human language. Comparisons with homologous genes in other mammals indicate that the gene underwent intense natural selection within the past 200,000 years.

■ **INTERACTIVE QUESTION   34.12**

List some of the derived characters of humans.

Word Roots

arch- = ancient (*archosaurs:* the reptilian group that includes crocodiles, alligators, dinosaurs, and birds)
cephalo- = head (*cephalochordates:* a chordate without a backbone, represented by lancelets)

aktin- = a ray; **-pterygi** = a fin (*Actinopterygii:* the class of ray-finned fishes)
crani- = the skull (*craniata:* the chordate clade that possess a cranium)
crocodil- = a crocodile (*Crocodilia:* the reptile group that includes crocodiles and alligators)
di- = two (*diapsids:* a group of amniotes distinguished by a pair of holes on each side of the skull)
dino- = terrible; **-saur** = lizard (*dinosaurs:* an extremely diverse group of ancient reptiles varying in body shape, size, and habitat)
endo- = inner; **-therm** = heat (*endotherm:* an animal that uses metabolic energy to maintain a constant body temperature, such as a bird or mammal)
eu- = good (*eutherians:* placental mammals; those whose young complete their embryonic development within the uterus, joined to the mother by the placenta)
extra- = outside, more (*extraembryonic membranes:* four membranes that support the developing embryo in reptiles and mammals)
gnantho- = the jaw; **-stoma** = the mouth (*gnathostomes:* the vertebrate clade that possesses jaws)
homin- = man (*hominid:* a term that refers to mammals that are more closely related to humans than to any other living species)
lepido- = a scale (*lepidosaurs:* the reptilian group that includes lizards, snakes, and tuatara)
marsupi- = a bag, pouch (*marsupial:* a mammal, such as a koala, kangaroo, or opossum, whose young complete their embryonic development inside a maternal pouch called the marsupium)
mono- = one (*monotremes:* an egg-laying mammal, represented by the platypus and echidna)
neuro- = nerve (*neural crest:* a band of cells along the border where the neural tube pinches off from the ectoderm)
noto- = the back; **-chord** = a string (*notochord:* a longitudinal, flexible rod formed from dorsal mesoderm and located between the gut and the nerve cord in all chordate embryos)
opercul- = a covering, lid (*operculum:* a protective flap that covers the gills of fishes)
osteo- = bone; **-ichthy** = fish (*Osteichthyans:* the vertebrate clade that includes the ray-finned fishes and lobe-fins)
ostraco- = a shell; **-derm** = skin (*ostracoderm:* an extinct paraphyletic group of armored fish-like vertebrates)
ovi- = an egg; **-parous** = bearing (*oviparous:* referring to a type of development in which young hatch from eggs laid outside the mother's body)
paedo- = a child; **-genic** = producing (*paedogenesis:* the precocious development of sexual maturity in a larva)

paleo- = ancient; **anthrop-** = man; **-ology** = the science of (*paleoanthropology:* the study of human origins and evolution)

placo- = a plate (*placoderm:* a member of an extinct group of gnathostomes that had jaws and were enclosed in a tough, outer armor)

ptero- = a wing (*pterosaurs:* winged reptiles that lived during the time of dinosaurs)

ratit- = flat-bottomed (*ratites:* the group of flightless birds)

soma- = body (*somites:* blocks of mesoderm that give rise to muscle segments in chordates)

syn- = together (*synapsids:* an amniote group distinguished by a single hole behind each eye socket)

tetra- = four; **-podi** = foot (*tetrapod:* a terrestrial lobe-fin, possessing two pairs of limbs, such as amphibians, reptiles, and mammals)

tunic- = a covering (*tunicates:* members of the subphylum Urochordata)

uro- = tail (*urochordate:* a chordate without a backbone, commonly called a tunicate)

uro- = (note) tail; **-del** = visible (*Urodela:* the order of salamanders that includes amphibians with tails)

vivi- = alive (*ovoviviparous:* referring to a type of development in which young hatch from eggs that are retained in the mother's uterus)

Structure Your Knowledge

1. Review the cladogram on page 269. Make note of the derived characters that define each clade.

2. Describe several examples from vertebrate evolution that illustrate the common evolutionary theme that new adaptations usually evolve from preexisting structures.

3. How do various vertebrate groups meet the challenges of a terrestrial habitat?

Test Your Knowledge

FILL IN THE BLANKS

_____ 1. chordate subphylum of sessile, suspension-feeding marine animals

_____ 2. chordate subphylum that includes lancelets

_____ 3. blocks of mesoderm along notochord that develop into muscles

_____ 4. clade of jawed vertebrates

_____ 5. flap over the gills of bony fishes

_____ 6. adaptation that allows reptiles to reproduce on land

_____ 7. egg-laying mammals

_____ 8. structure that helps mammals ventilate their lungs

_____ 9. group of primates that includes monkeys and apes

_____ 10. genus in which the fossil Lucy is placed

MULTIPLE CHOICE: *Choose the one best answer.*

1. Pharyngeal slits appear to have functioned first as
   a. suspension-feeding devices.
   b. gill slits for respiration.
   c. components of the jaw.
   d. portions of the inner ear.
   e. mouth openings.

2. Which of the following is *not* a derived character of craniates?
   a. cranium or skull
   b. neural crest in embryonic development
   c. a mineralized endoskeleton
   d. heart with at least two chambers
   e. cephalization with sensory organs

3. Which of these represents the oldest lineage of vertebrates?
   a. caecilians
   b. sharks and rays
   c. lancelet
   d. hagfishes
   e. lampreys

4. Which of the following is in the lobe-fin clade?
   a. lampreys
   b. sharks and rays
   c. ray-finned fishes
   d. hagfishes
   e. tetrapods

5. Which of the following is not in the same lineage as the others?
   a. lizards
   b. birds
   c. dinosaurs
   d. crocodilians
   e. pterosaurs

6. Which of the following is *incorrectly* paired with its gas exchange mechanism?
   a. amphibians—skin and lungs
   b. lungfishes—gills and lungs
   c. reptiles—lungs
   d. bony fishes—swim bladder
   e. mammals—lungs with diaphragm to ventilate

7. Nonbird reptiles have lower caloric needs than do mammals of comparable size because they
   a. are ectotherms.
   b. have waterproof scales.
   c. have a longer digestive tract and obtain more nutrients from their food.
   d. move by bending their vertebral column back and forth.
   e. have a more efficient respiratory system.

8. Which of the following best describes the earliest mammals?
   a. large, herbivorous
   b. large, carnivorous
   c. small, insectivorous
   d. small, herbivorous
   e. small, carnivorous

9. Oviparity is a reproductive strategy that
   a. allows mammals to bear well-developed young.
   b. is used by both reptiles and some sharks.
   c. is necessary for vertebrates to reproduce on land.
   d. is a necessity for all flying vertebrates.
   e. protects the embryo inside the mother and uses the food resources of the egg.

10. In Australia, marsupials fill the niches that eutherians (placental mammals) fill in other parts of the world because
    a. they are better adapted and have outcompeted eutherians.
    b. their offspring complete their development attached to a nipple in a marsupium.
    c. they originated in Australia.
    d. they evolved from monotremes that migrated to Australia about 65 million years ago.
    e. after Pangaea broke up, they diversified in isolation from eutherians.

11. Which of the following is closely associated with language development?
    a. four clusters of *Hox* genes
    b. gene duplication involving *Dlx* genes
    c. reduced sexual dimorphism and postnatal care
    d. *FOXP2* gene
    e. shorter hinged jaws and a hole in the skull behind each eye socket through which jaw muscles pass

12. In accordance with the model of mosaic evolution,
    a. biogeography explains the adaptive radiation of many groups.
    b. erect posture preceded the enlargement of the brain in human evolution.
    c. modern humans evolved in parallel in different parts of the world, laying the groundwork for geographic differences more than a million years ago.
    d. modern humans first evolved in Africa and later dispersed to other regions.
    e. the rapid expansion of *H. sapiens* may be tied to the evolution of human cognition, including symbolic thought, language, and complex tool construction.

Plant Form and Function

Chapter 35
Plant Structure, Growth, and Development

▶ Framework

A rather large new vocabulary is needed to name the specialized cells and structures in a study of plant structure and growth. Focus your attention on how the roots, stems, and leaves of a plant are specialized to function in absorption, support, transport, protection, and photosynthesis. The plant body is composed of dermal, vascular, and ground tissue systems. Apical meristems at the tips of roots and shoots create primary growth. The lateral meristems, vascular cambium and cork cambium, create secondary growth that adds girth to stems and roots. New techniques and model systems such as *Arabidopsis* are allowing researchers to explore the molecular bases for plant growth, morphogenesis, and cellular differentiation.

▶ Chapter Review

Plant development and structure are controlled by both genetic and environmental factors. Due to their developmental **plasticity,** no two plants are alike. Evolutionary adaptation to terrestrial environments has shaped distinctive plant **morphologies.**

35.1 The plant body has a hierarchy of organs, tissues, and cells

A plant **organ** performs a specific function and is composed of several types of **tissues,** groups of cells with a particular function.

The Three Basic Plant Organs: Roots, Stems, and Leaves As an evolutionary adaptation to the dispersed resources in a terrestrial environment, vascular plants have an underground **root system** for obtaining water and minerals from the soil and an aerial **shoot system** of stems and leaves for absorbing light and carbon dioxide for photosynthesis. The shoot system includes vegetative shoots and reproductive shoots, which, in angiosperms, are flowers.

The functions of **roots** include anchorage, absorption, and storage of food. A **taproot system** is found commonly in eudicots and gymnosperms. The one main deep root gives rise to **lateral roots** and often stores nutrients. Monocots and seedless vascular plants generally have a more shallow **fibrous root system** in which many small roots grow from the stem. Such roots are called **adventitious,** as they grow in an unusual location.

Most absorption of water and minerals occurs through tiny **root hairs,** extensions of epidermal cells that are clustered near root tips. Symbiotic associations between roots and fungi and bacteria may contribute to water and mineral absorption. Modified roots may serve various functions, including support, storage, and oxygen absorption.

A **stem** consists of alternating **nodes,** the points at which leaves are attached, and segments between nodes, called **internodes.** An **axillary bud** is found in the angle (axil) between a leaf and the stem. A **terminal bud,** consisting of developing leaves and compacted nodes and internodes, is found at the tip or apex of a shoot. The terminal bud may exhibit **apical dominance,** inhibiting the growth of axillary buds and producing a taller plant. Axillary buds may develop into lateral shoots, or branches, with their own terminal buds, leaves, and axillary buds.

Modifications of shoots include stolons, rhizomes, tubers, and bulbs. These structures may function in asexual reproduction and food storage.

Leaves, the main photosynthetic organs of most plants, usually consist of a flattened **blade** and a **petiole,** or stalk. Many monocots lack petioles. Monocot leaves usually have parallel major **veins** (vascular tissue of leaves), whereas eudicot leaves have networks of branched veins. Leaves may be simple or compound. Modified leaves may function in support, storage, and reproduction.

■ **INTERACTIVE QUESTION   35.1**

Label the parts in this diagram of a flowering plant.

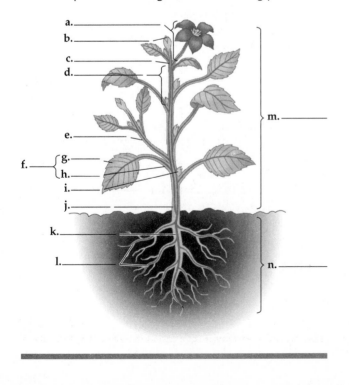

a. _____
b. _____
c. _____
d. _____
m. _____
e. _____
f. _____
g. _____
h. _____
i. _____
j. _____
k. _____
l. _____
n. _____

The Three Tissue Systems: Dermal, Vascular, and Ground **Tissue systems** are functional units of tissues that are continuous throughout the plant but have specific characteristics in each plant organ (root, stem, and leaf).

The **dermal tissue system** forms a protective outer layer. The **epidermis** is a single layer of cells that covers nonwoody plants. Older stems and roots of woody plants are covered by a protective **periderm.** Root hairs are extensions of epidermal cells near root tips. The epidermis of leaves and most stems is covered with a **cuticle,** a waxy coating that prevents excess water loss.

The **vascular tissue system** consists of **xylem** and **phloem** and functions in long-distance transport. In angiosperms, the **stele,** or vascular tissue of a root or stem, is arranged as a solid central **vascular cylinder** in roots, but as **vascular bundles** in stems and leaves.

The **ground tissue system** contains cells that function in photosynthesis, support, and storage. Ground tissue internal to vascular tissue is called **pith;** outside the vascular tissue it is called **cortex.**

Common Types of Plant Cells   Cellular differentiation of plant cells may produce modifications of the cell wall and/or of the **protoplast,** the cell contents excluding the cell wall, that specialize cells for different functions.

Parenchyma cells carry on most of a plant's metabolic functions, such as photosynthesis and food storage. They usually lack secondary walls and have large central vacuoles. Most retain the ability to divide and differentiate into other types of plant cells.

Collenchyma cells lack secondary walls but have thickened primary walls. Strands or cylinders of these cells function in flexible support for young parts of the plant and elongate along with the plant.

Sclerenchyma cells have thick secondary walls strengthened with lignin. These specialized supporting cells often lose their protoplasts at maturity. **Fibers** are long, tapered cells that usually occur in threads. **Sclereids** are shorter and irregular in shape, with very thick, lignified cell walls.

The water-conducting cells of xylem die at functional maturity, leaving behind their secondary walls, which are interrupted only by scattered pits. **Tracheids,** found in all vascular plants, are long, thin, tapered cells with lignin-strengthened walls. Water passes through pits from cell to cell. **Vessel elements,** found only in angiosperms, are wider, shorter, and thinner walled, with perforations in their end walls. They align to form long tubes known as **vessels.**

In the phloem of angiosperms, sugars flow through chains of cells called **sieve-tube members,** which remain alive at functional maturity but lack nuclei, ribosomes, and vacuoles. Fluid flows through pores in the **sieve plates** in the end walls between cells. The nucleus and ribosomes of an adjacent **companion cell,** which is connected to a sieve-tube member by numerous plasmodesmata, may serve both cells.

■ **INTERACTIVE QUESTION 35.2**

a. Which types of plant cells are dead at functional maturity?

b. Which types of plant cells lack nuclei at functional maturity?

35.2 Meristems generate cells for new organs

Most plants exhibit **indeterminate growth,** continuing to grow as long as they live. Plant organs such as leaves and flowers, as well as animals, have **determinate growth** and stop growing after reaching a certain size. Plants may be **annuals,** which complete their life cycle in a year or less; **biennials,** which have a life cycle spanning two years; or **perennials,** which live many years.

Plants have tissues called **meristems,** which remain embryonic and perpetually divide to form new cells. **Apical meristems,** located at the tips of roots and in the buds of shoots, produce **primary growth,** resulting in the elongation of roots and shoots. **Herbaceous** (nonwoody) plants have primary growth. **Secondary growth** is an increase in diameter as new cells are produced by **lateral meristems. Vascular cambium** produces secondary xylem and phloem; **cork cambium** produces the periderm. Cells that remain to divide in a meristem are called **initials,** whereas cells that are displaced from the meristem and become specialized in developing tissues are called **derivatives.**

■ **INTERACTIVE QUESTION 35.3**

What types of growth occur in woody plants?

35.3 Primary growth lengthens roots and shoots

The **primary plant body,** produced by primary growth, is the entire herbaceous plant (usually) and the youngest parts of a woody plant.

Primary Growth of Roots The apical meristem of the root tip is protected by a **root cap,** which secretes a polysaccharide slime to lubricate the growth route. The **zone of cell division** includes the apical meristem and its derivatives. In the **zone of elongation,** cells lengthen to many times their original size, which pushes the root tip through the soil. Cells specialize in structure and function as the zone of elongation grades into the **zone of maturation.**

The primary tissues of a root are the epidermis, ground tissue, and vascular tissue. In most roots, xylem cells radiate from the center in spokes, with phloem in between. The vascular tissue of a monocot may have pith, a central core of undifferentiated parenchyma cells, inside rings of xylem and phloem.

The ground tissue consists of parenchyma cells in the cortex. The innermost layer of cortex is the one-cell-thick **endodermis,** which regulates the passage of materials into the vascular cylinder.

Lateral roots may develop from the **pericycle,** the outer layer of cells in the vascular cylinder. A lateral root pushes through the cortex and retains its connection with the vascular cylinder.

■ **INTERACTIVE QUESTION 35.4**

Label the tissues in the cross sections of a eudicot root and its vascular cylinder. Identify the functions of these tissues.

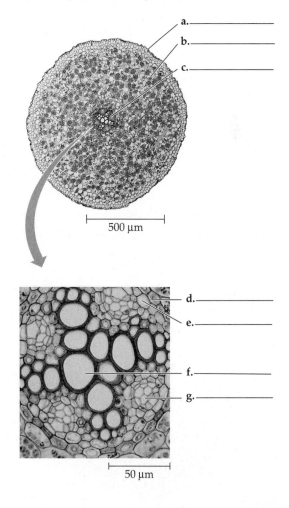

a._____
b._____
c._____

500 μm

d._____
e._____
f._____
g._____

50 μm

Primary Growth of Shoots   The dome-shaped mass at the tip of the terminal bud is the shoot apical meristem. **Leaf primordia** form on the sides of the apical meristem, and axillary buds develop from clumps of meristematic cells left at the bases of the leaf primordia.

Elongation of the shoot occurs by cell division and cell elongation within young internodes. In grasses and some other plants, internodes have intercalary meristems that enable shoots to continue elongation.

The epidermis of the dermal tissue system covers stems. In gymnosperms and most eudicots, vascular bundles may be arranged in a ring, with the ground tissues pith inside and cortex outside the ring. Xylem is located internal to the phloem in the vascular bundles. In most monocot stems, the vascular bundles are scattered throughout the ground tissue. A layer of collenchyma just beneath the epidermis and fiber cells of sclerenchyma within vascular bundles may strengthen the stem.

The leaf is covered by the wax-coated, tightly interlocking cells of the epidermis. **Stomata,** tiny pores flanked by **guard cells,** permit both gas exchange and transpiration, the evaporation of water from the leaf.

Mesophyll consists of parenchyma ground tissue cells containing chloroplasts. In many eudicot leaves, columnar **palisade mesophyll** is located above **spongy mesophyll,** which has loosely packed cells surrounding many air spaces.

A **leaf trace,** which is a branch of a vascular bundle in the stem, continues into the petiole and divides repeatedly within the blade of the leaf, providing support and vascular tissue to the photosynthetic mesophyll. Each vein is enclosed in a ring of protective cells called a **bundle sheath.**

■ **INTERACTIVE QUESTION   35.5**

Name the indicated structures in this diagram of a leaf.

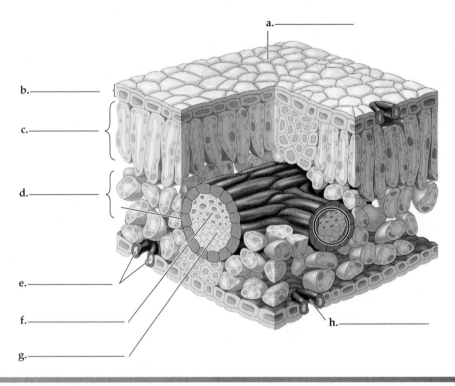

a.—————————

b.—————————

c.—————————

d.—————————

e.—————————

f.—————————

g.—————————

h.—————————

35.4   Secondary growth adds girth to stems and roots in woody plants

The vascular cambium and cork cambium produce the **secondary plant body.** Secondary growth occurs in all gymnosperms and many eudicots but is rare in monocots.

■ **INTERACTIVE QUESTION   35.6**

Explain how primary and secondary growth can occur simultaneously in a woody stem.

The Vascular Cambium and Secondary Vascular Tissue
Vascular cambium forms a continuous cylinder from a band of parenchyma cells that become meristematic, located between the primary xylem and phloem of each vascular bundle, and in the ground tissue between bundles. **Fusiform initials** produce secondary xylem to the inside and secondary phloem to the outside. **Ray initials** produce radial lines of parenchyma cells called xylem rays and phloem rays, which function in storage and lateral transport of water and nutrients.

Wood is the accumulation of secondary xylem cells with thick, lignified walls. Annual growth rings result from the seasonal cycle of cambium dormancy, early wood production (usually in the spring), and late wood production (in the summer) in temperate regions. In older trees, a central column of **heartwood** consists of older xylem with resin-filled cell cavities; the **sapwood** consists of secondary xylem that still functions in transport.

Older secondary phloem is sloughed off as bark; only that closest to the vascular cambium functions in sugar transport.

Cork Cambia and the Production of Periderm The epidermis splits off during secondary growth and is replaced by new protective tissues produced by the cork cambium, a meristematic cylinder that first forms in the outer cortex in stems and the outer pericycle in roots. The cork cambium produces a thin layered phelloderm to the inside and cork cells to the exterior, which develop suberin-impregnated walls. The protective coat formed by a cork cambium and its tissues is a layer of periderm. As secondary growth continually splits the outer layers of periderm, new cork cambia develop, eventually forming from parenchyma cells in the secondary phloem. **Lenticels** are spaces between cork cells through which gas exchange occurs. **Bark** refers to phloem and periderm.

■ **INTERACTIVE QUESTION 35.7**

Starting from the outside, place the letters of the tissues in the order in which they are located in a woody tree trunk.

A. primary phloem   E. pith

B. secondary phloem   F. cork cambium

C. primary xylem   G. vascular cambium

D. secondary xylem   H. cork cells

___ ___ ___ ___ ___ ___ ___ ___

35.5 Growth, morphogenesis, and differentiation produce the plant body

Growth, or the increase in mass, results from cell division and expansion. The organization of cells to create body form is called **morphogenesis.** The third developmental process, differentiation, creates cells with specific structural and functional features.

Molecular Biology: Revolutionizing the Study of Plants Modern plant biology is using new methods and new research organisms, such as *Arabidopsis thaliana,* to study the genetic control of plant development. The entire genome of *Arabidopsis* has been sequenced, and researchers are working to determine the functions of all of its 26,000 (of 15,000 different types) genes by 2010. A major goal of **systems biology** is to track which genes are activated and where during a plants' development.

Growth: Cell Division and Cell Expansion The plane and symmetry of plant cell division is important in determining form. In **asymmetrical cell division,** one daughter cell receives more cytoplasm than the other, often leading to a key developmental event.

The plane of cell division is determined by the **preprophase band,** a ring of cytoskeletal microtubules that forms during late interphase. The microtubules disperse but leave behind an ordered array of actin microfilaments that indicate the plane of cell division.

About 90% of plant cell growth is due to the uptake of water into the central vacuole. This economical means of cell elongation produces the rapid growth of shoots and roots. When enzymes break the crosslinks in the cell wall, the restraint on the turgid cell is reduced and water enters the cell by osmosis. Cells expand in a direction perpendicular to the orientation of the cellulose microfibrils in the inner layers of the cell wall. This orientation of microfibrils was determined by the orientation of microtubules in the outer cytoplasm.

In the squat-bodied *fass* mutants of *Arabidopsis,* preprophase bands are not formed in mitosis and the random distribution of cortical microtubules leads to the random arrangement of cellulose microfibrils. The defects in microtubule organization result in cells that divide in a random arrangement and expand equally in all directions.

■ **INTERACTIVE QUESTION 35.8**

Review the role of microtubules in the orientation of plant cell division and expansion.

Morphogenesis and Pattern Formation   **Pattern formation** is the characteristic development of structures in specific locations. Pattern formation appears to depend on **positional information,** signals that indicate a cell's location within a developing organ and thus direct its growth and differentiation. An embryonic cell may detect its location by gradients of molecules, probably proteins or mRNAs, that diffuse from specific locations in a developing structure.

Plants typically have an axial **polarity** with a root end and a shoot end. The asymmetric first division of the zygote establishes the shoot–root polarity.

Plants have homeotic genes that regulate key developmental events. In many plants, the *KNOTTED-1* homeotic gene influences leaf morphology. Its overexpression in tomato plants results in "super-compound" leaves.

Gene Expression and Control of Cellular Differentiation   Differential gene expression within genetically identical cells leads to their differentiation into the diverse cell types in a plant. In the root epidermis of *Arabidopsis*, the homeotic gene *GLABRA-2* is expressed in epidermal cells that are in contact with only one underlying cortical cell, and these cells do not develop root hairs. The gene is not expressed in those cells in contact with two cortical cells, and they differentiate into root hair cells. A mutant gene results in every root epidermal cell developing a root hair.

Location and a Cell's Developmental Fate   Using clonal analysis, researchers are able to identify cells in the apical meristem and follow their lineages to pinpoint when the developmental fate of a cell becomes determined. Apparently the cells of the meristem are not committed to the formation of specific organs and tissues. A cell's final position in a developing organ determines what type of cell it will become.

Shifts in Development: Phase Changes   In a process known as a **phase change,** the apical meristem can switch from one developmental phase to another, such as from producing juvenile vegetative nodes and internodes to laying down mature nodes. A node's developmental phase is fixed when it is produced, and thus both juvenile and mature leaves can coexist along the shoots of a plant.

Genetic Control of Flowering   In the transition from a vegetative shoot to a reproductive shoot (flower), **meristem identity genes** are expressed that code for transcription factors that activate the genes necessary for floral meristem development.

The four floral organs develop from leaf primordial in four concentric whorls: sepals, petals, stamens, and carpels. Some of the **organ identity genes** that determine this characteristic floral pattern have been identi-

fied, and their expression seems to be influenced by positional information. Mutations of organ-identity genes result in the placement of one type of floral organ where another type would normally develop. The **ABC model** of flower formation proposes three classes of organ identity genes, each of which affects two adjacent whorls. This model explains the phenotypes of mutants lacking *A, B,* or *C* gene expression, with the additional observation that activity of the *A* or *C* gene inhibits the expression of the other. If either *A* or *C* is nonfunctional, the other gene is expressed in its place.

■ INTERACTIVE QUESTION   35.9

This diagram of the ABC model shows the active genes in each whorl and the resulting anatomy of a wild-type flower.

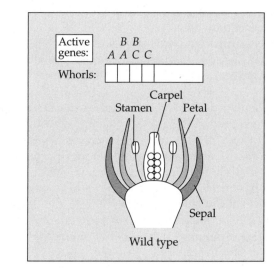

a. Fill in the table below to show which organs are produced in the whorls in a normal flower. In a mutant that lacks a functional gene *A*, what gene expression pattern and resulting flower organ arrangement would be produced? (Remember that a lack of *A* activity removes the inhibition of gene *C*.)

Whorl	Genes Active	Organs in Normal	Genes Active in Mutant *A*	Organs in Mutant *A*
1	*A*			
2	*AB*			
3	*BC*			
4	*C*			

b. If you had a double-mutant plant that had no gene activity for *B* or *C*, what would the resulting flower look like?

Word Roots

apic- = the tip; **meristo-** = divided (*apical meristems:* embryonic plant tissue on the tips of roots and in the buds of shoots that supplies cells for the plant to grow)

a- = not, without; **-symmetr** = symmetrical (*asymmetric cell division:* cell division in which one daughter cell receives more cytoplasm than the other during mitosis)

bienn- = every 2 years (*biennial:* a plant that requires two years to complete its life cycle)

coll- = glue; **-enchyma** = an infusion (*collenchyma cell:* a flexible plant cell type that occurs in strands or cylinders that support young parts of the plant without restraining growth)

endo- = inner; **derm-** = skin (*endodermis:* the innermost layer of the cortex in plants roots)

epi- = over (*epidermis:* the dermal tissue system in plants; the outer covering of animals)

fusi- = a spindle (*fusiform initials:* the cambium cells within the vascular bundles; the name refers to the tapered ends of these elongated cells)

inter- = between (*internode:* the segment of a plant stem between the points where leaves are attached)

meso- = middle; **-phyll** = a leaf (*mesophyll:* the ground tissue of a leaf, sandwiched between the upper and lower epidermis and specialized for photosynthesis)

morpho- = form; **-genesis** = origin (*morphogenesis:* the development of body shape and organization during ontogeny)

perenni- = through the year (*perennial:* a plant that lives for many years)

peri- = around; **-cycle** = a circle (*pericycle:* a layer of cells just inside the endodermis of a root that may become meristematic and begin dividing again)

phloe- = the bark of a tree (*phloem:* the portion of the vascular system in plants consisting of living cells arranged into elongated tubes that transport sugar and other organic nutrients throughout the plant)

pro- = before (*procambium:* a primary meristem of roots and shoots that forms the vascular tissue)

proto- = first; **-plast** = formed, molded (*protoplast:* the contents of a plant cell exclusive of the cell wall)

sclero- = hard (*sclereid:* a short, irregular sclerenchyma cell in nutshells and seed coats and scattered through the parenchyma of some plants)

trachei- = the windpipe (*tracheids:* a water-conducting and supportive element of xylem composed of long, thin cells with tapered ends and walls hardened with lignin)

trans- = across (*transpiration:* the evaporative loss of water from a plant)

vascula- = a little vessel (*vascular tissue:* plant tissue consisting of cells joined into tubes that transport water and nutrients throughout the plant body)

xyl- = wood (*xylem:* the tube-shaped, nonliving portion of the vascular system in plants that carries water and minerals from the roots to the rest of the plant)

Structure Your Knowledge

1. How does the indeterminate growth pattern of plants lead to developmental plasticity? What is the adaptive advantage of such plasticity?

2. In this cross section of a young stem, label the indicated structures. Is this a stem of a monocot or a eudicot? How can you tell?

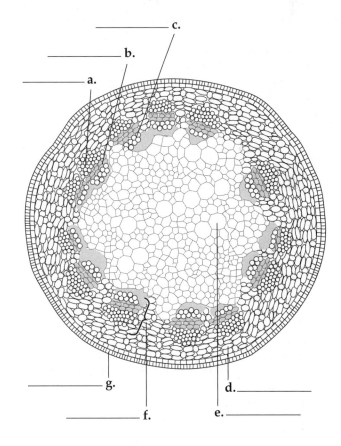

Test Your Knowledge

MATCHING: *Match the plant tissue with its description.*

_____ **1.** sclerenchyma

_____ **2.** collenchyma

_____ **3.** tracheids

_____ **4.** fibers

_____ **5.** fusiform initials

_____ **6.** pericycle

_____ **7.** mesophyll

_____ **8.** periderm

_____ **9.** endodermis

_____ **10.** pith

A. layer from which lateral roots originate

B. tapered xylem cells with lignin in cell walls

C. parenchyma cells with chloroplasts in leaves

D. protective coat made of cork and cork cambium

E. bundles of long sclerenchyma cells

F. supporting cells with thickened primary walls

G. parenchyma cells inside vascular ring in eudicot stem

H. supporting cells with thick secondary walls

I. cambium cells that produce secondary xylem and phloem

J. cell layer in root regulating movement into vascular cylinder

MULTIPLE CHOICE: *Choose the one best answer.*

1. Which of the following is *incorrect?* Monocots typically have
   a. a taproot rather than a fibrous root system.
   b. leaves with parallel veins rather than branching venation.
   c. no secondary growth.
   d. scattered vascular bundles in the stem rather than bundles in a ring.
   e. pith in the center of the vascular cylinder in the root.

2. Which of the following is *incorrectly* paired with its function?
   a. ray initials—form radial xylem and phloem rays
   b. lenticels—gas exchange in woody stem
   c. root hairs—absorption of water and dissolved minerals
   d. root cap—protects root as it pushes through soil
   e. procambium—meristematic tissue that forms protective layer of cork

3. Axillary buds
   a. may exhibit apical dominance over the terminal bud.
   b. form at nodes in the angle where leaves join the stem.
   c. grow out from the pericycle layer.
   d. are formed from intercalary meristems.
   e. only develop into vegetative shoots.

4. A leaf trace is
   a. a petiole.
   b. the outline of the vascular bundles in a leaf.
   c. vascular bundle that extends into a leaf.
   d. a tiny bulge on the flank of the apical dome that grows into a leaf.
   e. a system of plant identification based on leaf morphology.

5. Which of the following structures of a plant shows determinate growth?
   a. roots
   b. vegetative shoots
   c. adventitious roots
   d. leaves
   e. No parts of a plant show determinate growth.

6. Morphogenesis in plants results from
   a. differences in the plane of cell division and the direction of cell expansion.
   b. migration of cells from the apical meristems to the dermal, vascular, and ground tissue systems.
   c. the differentiation of cells within the apical meristem.
   d. cells moving from the zone of elongation to the zone of maturation.
   e. the expression of meristem identity genes.

7. Which of the following is *incorrectly* paired with the length of its life cycle?
   a. pine tree—perennial
   b. rose bush—perennial
   c. carrot—biennial
   d. marigold—annual
   e. corn—biennial

8. Secondary xylem and phloem are produced in a root by the
   a. pericycle.
   b. endodermis.
   c. vascular cambium.
   d. apical meristem.
   e. cork cambium.

9. Bark consists of
   a. secondary phloem.
   b. periderm.
   c. cork cells.
   d. cork cambium.
   e. all of the above.

10. Sieve-tube members
    a. are responsible for lateral transport through a woody stem.
    b. control the activities of phloem cells that have no nuclei or ribosomes.
    c. have spiral thickenings that allow the cell to elongate along with a young shoot.
    d. are transport cells with sieve plates in the end walls between cells.
    e. are tapered water transport cells with pits.

11. Which of the following cells are dead at functional maturity?
    a. tracheids
    b. cork cells
    c. vessel elements
    d. sclerenchyma cells
    e. all of the above

12. Clonal analysis of cells of the shoot apex indicates that
    a. organ-identity gene mutations substitute one body part for another.
    b. a cell's developmental fate is more influenced by position effects than by its meristematic lineage.
    c. cellular differentiation results from regulation of gene expression resulting in production of different proteins in different cells.
    d. all cells were derived from the same parent cell.
    e. the lateral meristem produces the dermal, vascular, and ground tissue systems.

13. In what direction does a plant cell enlarge?
    a. toward the basal end as a result of positional information
    b. parallel to the orientation of the preprophase band of microtubules
    c. perpendicular to the orientation of cellulose microfibrils in the cell wall
    d. in the direction from which water flows into the cell
    e. perpendicular to the internodes

14. What do organ identity genes code for?
    a. the three tissue systems
    b. pollen and egg cells
    c. transcription factors that control development of floral organs
    d. signals that change a vegetative shoot into a floral meristem
    e. the root and shoot meristems

15. What is a usual sign of the change of the apical meristem from the juvenile to the mature phase?
    a. the production of a flower
    b. the initiation of secondary growth
    c. the production of longer internodes
    d. the activation of axillary buds
    e. a change in the morphology of leaves produced at nodes

16. Which of the following is essential to establishing the axial polarity of a plant?
    a. expression of different homeotic genes in the shoot and root meristems
    b. orientation of cortical microtubules perpendicular to the ground
    c. the expression of the *GLABRA-2* gene in the shoot but not the root
    d. the asymmetric first division of the zygote
    e. the proper orientation of microtubules that then orient cellulose-microfibrils

Chapter 36
Transport in Vascular Plants

Framework

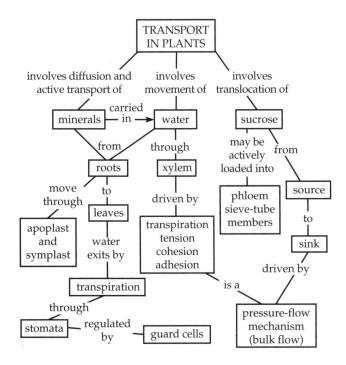

Chapter Review

Adaptations to life on land involved the specialization of roots to absorb water and minerals, shoots to absorb CO_2 and light, and vascular tissue for the transport of materials between roots and shoots.

36.1 Physical forces drive the transport of materials in plants over a range of distances

Transport in plants involves the transport of water and solutes by individual cells, the short-distance movement of substances from cell to cell within tissues, and the long-distance transport of sap in xylem and phloem.

Selective Permeability of Membranes: A Review Solutes may move across the selectively permeable plasma membrane by **passive transport** when they diffuse down their concentration gradients. **Active transport** requires the cell to expend energy to move a solute against its electrochemical gradient. **Transport proteins** speed passive transport and may be either proteins that selectively bind and transport a solute, or selective channels through a membrane. Some channels are gated: They open and close in response to certain stimuli.

The Central Role of Proton Pumps A plant **proton pump** uses ATP to pump hydrogen ions (H^+) out of the cell, generating an energy-storing proton gradient and a **membrane potential** resulting from the separation of charges. The membrane potential helps drive positively charged ions (such as K^+) down their electrochemical gradient and into the negatively charged cell. In **cotransport,** a solute can be pumped against its concentration gradient when a transport protein links its passage to the "downhill" movement of H^+. **Chemiosmosis** links energy-releasing and energy-consuming processes using a transmembrane proton gradient like that created by the proton pump.

Effects of Differences in Water Potential   The passive transport of water across a membrane is called **osmosis**. **Water potential,** designated by Ψ *(psi),* is a useful measurement for predicting the direction that water will move when a plant cell is surrounded by a particular solution. It takes into account both solute concentration and the physical pressure exerted by the plant cell wall. Free water, which is not bound to solutes or surfaces, will flow from a region of higher water potential to one of lower water potential. Water potential is measured in **megapascals (MPa)**; 1 MPa is equal to about 10 atmospheres of pressure. The water potential of pure water in an open container under standard conditions is assigned the value of 0 ($\Psi = 0$ MPa).

The combined effect of solute concentration and pressure is shown by the *water potential equation:* $\Psi = \Psi_S + \Psi_P$. **Solute potential (Ψ_S),** also called **osmotic potential,** is proportional to the number of solute molecules. Solutes bind water, reducing the number of free water molecules and the capacity of the water to do work. Solutes lower the water potential, and a solution's Ψ_S is always negative.

Pressure potential (Ψ_P) measures the pressure on a solution and can be a positive or negative value. Negative pressure is tension. The positive pressure of a cell pushing against the cell wall is called **turgor pressure.**

A **flaccid** plant cell ($\Psi_P = 0$) bathed in a solution more concentrated than the cell will lose water by osmosis because the solution has a lower (more negative) Ψ. The cell's protoplast will shrink and **plasmolyze.** When bathed in pure water, the cell has the lower Ψ. Water will enter the cell until enough turgor pressure builds up so that Ψ_P and Ψ_S are equal and opposite in magnitude, and $\Psi = 0$ both inside and outside the cell. Net movement of water will then stop.

Plant cells are usually **turgid**; they have a greater solute concentration than their extracellular environment and turgor pressure keeps them firm. The loss of turgor can be seen when a plant's leaves and stems **wilt.**

Although small water molecules can move relatively freely across membranes, the rate at which they move is too rapid to be attributed entirely to diffusion through the lipid bilayer. Water-specific transport proteins called **aquaporins** increase the rate of water diffusion.

■ **INTERACTIVE QUESTION   36.1**

a. A flaccid plant cell has a water potential of −0.6 MPa. Fill in the water potential equation for this cell.

$\Psi_P =$

$+\Psi_S =$

$\Psi =$

b. The cell is then placed in a beaker of distilled water ($\Psi = 0$). Fill in the equation for the cell after it reaches equilibrium in pure water. Explain what happens to the water potential of this cell.

$\Psi_P =$

$+\Psi_S =$

$\Psi =$

c. Explain what would happen to the same cell if it is placed in a solution that has a water potential of −0.8 MPa. Fill in the equation for the cell after it is moved to this more concentrated solution and reaches equilibrium.

$\Psi_P =$

$+\Psi_S =$

$\Psi =$

Three Major Compartments of Vacuolated Plant Cells   The three compartments of most mature plant cells are the cell wall, cytosol, and the vacuole. The **vacuolar membrane,** or **tonoplast,** regulates solute movement between the cytosol and the cell sap of the vacuole. Its proton pumps move H^+ from the cytosol into the vacuole, creating a pH gradient that is used to move ions.

The cytosolic compartments of plant cells are connected by plasmodesmata, forming a cytosolic continuum called the **symplast.** The continuum of cell walls and intercellular spaces is called the **apoplast.**

Functions of the Symplast and Apoplast in Transport   Short-distance or lateral transport of water and solutes within plant tissues can occur by three routes: transmembrane, by crossing plasma membranes and cell walls; symplastic, moving through plasmodesmata; and apoplastic, the extracellular pathway along cell walls and extracellular spaces. Solutes and water may change routes during transit.

■ **INTERACTIVE QUESTION   36.2**

Label the diagram of the cell layers and routes of transport of water and minerals from the soil through the root. (Letters a and b refer to routes of transport; letters c–i refer to cell layers or structures.)

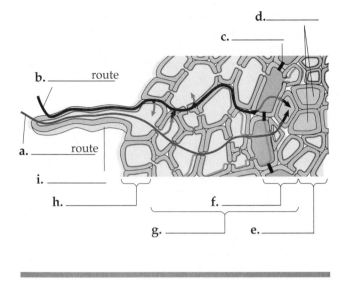

Bulk Flow in Long-Distance Transport Long-distance transport throughout plants occurs by **bulk flow,** the movement of fluid driven by pressure. Water and minerals move through tracheids and vessels of xylem as the result of tension created by transpiration. Sap is forced through phloem sieve tubes by positive pressure.

Sieve-tube members lack most cellular organelles, and vessel elements and tracheids are dead (and empty) at maturity, both facilitating more efficient bulk flow.

36.2 Roots absorb water and minerals from the soil

The Roles of Root Hairs, Mycorrhizae, and Cortical Cells Much of the absorption of water and mineral salts occurs along young root tips where root hairs are located. The soil solution soaks into the hydrophilic walls of epidermal cells and moves along the apoplast into the root cortex, exposing a large surface area of plasma membrane for the uptake of water and minerals. Active transport of minerals allows cells to accumulate essential minerals. **Mycorrhizae,** symbiotic associations of plant roots and fungal hyphae, greatly increase the surface area for absorption of water and selected minerals.

The Endodermis: A Selective Sentry The **endodermis,** the innermost layer of cortex cells surrounding the vascular cylinder, selectively screens all minerals entering the vascular tissue. A ring of suberin around each endodermal cell, called the **Casparian strip,** prevents water from the apoplast from entering the vascular cylinder without passing through the selectively permeable plasma membrane of an endodermal cell. Water and minerals that had already entered the symplast through a cortex or epidermal cell pass through plasmodesmata of endodermal cells into the vascular cylinder.

By a combination of diffusion and active transport, minerals move from endodermal cells to the apoplast and enter the nonliving xylem tracheids and vessel elements along with water.

36.3 Water and minerals ascend from roots to shoot through the xylem

The branching of xylem veins provides water to the cells of each leaf. Through **transpiration,** the loss of water vapor from leaves, plants lose a tremendous amount of water that must be replaced by water transported up from the roots.

Factors Affecting the Ascent of Xylem Sap As minerals are actively pumped into the vascular cylinder and prevented from leaking out by the endodermis, water potential in the vascular cylinder is lowered. Water flows in from the cortex, resulting in **root pressure,** which pushes xylem sap upward. Root pressure may cause **guttation,** the exudation of water droplets from leaves when more water is forced up the xylem than is transpired by the plant.

By transpiration, water vapor from saturated air spaces within a leaf exits to drier air outside the leaf by way of stomata. The thin layer of water that coats the mesophyll cells lining the air spaces begins to evaporate. The adhesion of the remaining water to the hydrophilic walls and the cohesion between water molecules causes the water film to form a concave shape. This meniscus increases the tension of the water layer. This negative pressure pulls water from the xylem and through the apoplast and symplast of the mesophyll to the cells and surface film lining the air spaces. Water moves along a gradient toward the most negative water potential, from xylem to neighboring cells to air spaces to the drier air outside the leaf.

The transpirational pull on xylem sap is transmitted from the leaves to the root tips by the cohesiveness of water that results from hydrogen bonding between molecules. Adhesion of water molecules to the hydrophilic walls of the narrow xylem elements and tracheids also contributes to overcoming the downward pull of gravity.

The upward transpirational pull on the cohesive sap creates tension within the xylem, lowering the water potential so that water flows passively from the soil, across the cortex, and into the vascular cylinder.

A break in the chain of water molecules by the formation of a water vapor pocket in a xylem vessel, called cavitation, breaks the transpirational pull, and the vessel cannot function in transport.

Xylem Sap Ascent by Bulk Flow: A Review The transpiration-cohesion-tension mechanism results in the bulk flow of water and solutes from roots to leaves. Water potential differences caused by solute concentration and pressure contribute to the movement of water from cell to cell, but solar-powered tension caused by transpiration is responsible for long-distance transport of water and minerals.

■ INTERACTIVE QUESTION   36.3

Explain the contribution of each of the following to the long-distance transport of water.

a. Transpiration:

b. Cohesion:

c. Adhesion:

d. Tension:

36.4   Stomata help regulate the rate of transpiration

A plant's tremendous requirement for water is partly a consequence of making food by photosynthesis. To obtain sufficient CO_2 for photosynthesis, leaves must exchange gases through the stomata and provide a large internal surface area for CO_2 uptake, but also from which water may evaporate.

Effects of Transpiration on Wilting and Leaf Temperature   When transpiration exceeds the water available, leaves wilt as cells lose turgor pressure. Transpiration also produces evaporative cooling, maintaining a cooler temperature in leaf cells for critical enzymes.

Stomata: Major Pathways for Water Loss   A plant loses most of its water through stomata. Guard cells regulate the size of stomatal openings and control the rate of transpiration. Stomatal densities in many plant species are both genetically and environmentally influenced; densities have been shown to relate to CO_2 levels, providing a measure of such levels in past climates.

When the kidney-shaped guard cells of eudicots become turgid and swell, their radially oriented microfibrils cause them to increase in length and buckle outward, increasing the size of the gap between them. When the guard cells become flaccid, they sag and close the space.

Guard cells can actively accumulate potassium ions (K^+), which lowers water potential and leads to the osmotic inflow of water and an increase in turgor pressure. The exodus of K^+ (with water following) leads to a loss of turgor. The regulation of aquaporins may also vary the membrane's permeability to water. The movement of K^+ across the guard cell membrane is probably coupled with the generation of membrane potentials by proton pumps that transport H^+ out of the cell.

The opening of stomata at dawn is related to at least three factors. First, light stimulates guard cells to accumulate K^+, perhaps triggered by the illumination of blue-light receptors that activate the proton pumps. Second, stomata are stimulated to open when CO_2 within air spaces of the leaf is depleted as photosynthesis begins in the mesophyll. The third factor is a daily rhythm of opening and closing that is endogenous to guard cells. Cycles that have intervals of approximately 24 hours are called **circadian rhythms.**

Environmental stress can cause stomata to close during the day. Guard cells lose turgor when water is in short supply. Also, a hormone called abscisic acid, produced in the roots in response to a lack of water, signals guard cells to close stomata.

■ INTERACTIVE QUESTION   36.4

What is meant by the photosynthesis-transpiration compromise? How might a sunny, windy, dry day influence this compromise in a plant?

Xerophyte Adaptations That Reduce Transpiration   Many **xerophytes,** plants adapted to arid climates, have leaves that are small and thick, limiting water loss by reducing their surface-to-volume ratio. Leaf cuticles may be thick, and the stomata may be sheltered in depressions. Some desert plants lose their leaves in the driest months.

Succulent plants of the family Crassulaceae and some other plant families assimilate CO_2 into organic acids during the night by a pathway known as CAM (crassulacean acid metabolism) and then release it for photosynthesis during the day. Thus, the stomata are open at night and can be closed during the day, reducing water loss.

36.5   Organic nutrients are translocated through the phloem

Movement from Sugar Sources to Sugar Sinks   **Translocation,** the transport of photosynthetic products throughout the plant, occurs in the sieve-tube members of phloem. Phloem sap may have a sucrose concentration as high as 30% and may also contain minerals, amino acids, and hormones.

Phloem sap flows from a **sugar source,** where it is produced by photosynthesis or the breakdown of starch, to a **sugar sink,** an organ that consumes or stores sugar. The direction of transport in any one sieve tube depends on the location of the source and sink connected by that tube, and the direction may change with the season or needs of the plant.

■ INTERACTIVE QUESTION 36.5

Explain how a root or tuber can serve as both a sugar source and a sugar sink.

In some species, sugar in the leaf moves through the symplast of the mesophyll cells to sieve-tube members. In other species, sugar first moves through the symplast and then into the apoplast in the vicinity of sieve-tube members and companion cells, which actively accumulate sugar. In some plants, companion cells are specialized as **transfer cells** with ingrowths of their wall that increase surface area for movement of solutes from apoplast to symplast.

Phloem loading requires active transport in plants that accumulate high sugar concentration in the sieve tubes. Proton pumps and the cotransport of sucrose through membrane proteins along with the returning protons is the mechanism used for active transport. Sugar is moved by various mechanisms out of sieve tubes at the sink end. The concentration gradient favors this movement because sugar is either being used or converted into starch within sink cells.

■ INTERACTIVE QUESTION 36.6

Label the components of this diagram of the chemiosmotic mechanism used to actively transport sucrose into companion cells or sieve-tube members.

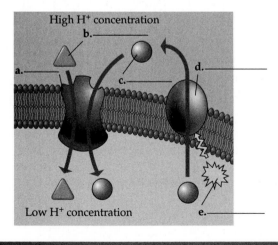

Pressure Flow: The Mechanism of Translocation in Angiosperms The rapid movement of phloem sap from source to sink is due to a pressure flow mechanism. High solute concentration at the source lowers the water potential, and the resulting movement of water into the sieve tube produces positive pressure. At the sink end, the osmotic loss of water following the exodus of sucrose into the surrounding tissue results in a lower pressure. The difference in these pressures causes sap to move by bulk flow from source to sink. Innovative tests of this hypothesis support it as the explanation for the flow of sap in the phloem of angiosperms.

▐ Word Roots

apo- = off, away; **-plast** = formed, molded (*apoplast:* in plants, the nonliving continuum formed by the extracellular pathway provided by the continuous matrix of cell walls)

aqua- = water; **-pori** = a pore, small opening (*aquaporin:* a transport protein in the plasma membranes of a plant or animal cell that specifically facilitates the diffusion of water across the membrane)

chemo- = chemical (*chemiosmosis:* the production of ATP using the energy of hydrogen-ion gradients across membranes to phosphorylate ADP)

circa- = a circle (*circadian rhythm:* a physiological cycle of about 24 hours, present in all eukaryotic organisms, that persists even in the absence of external cues)

co- = together; **trans-** = across; **-port** = a gate, door (*cotransport:* the coupling of the "downhill" diffusion of one substance to the "uphill" transport of another against its own concentration gradient)

endo- = within, inner; **-derm** = skin (*endodermis:* the innermost layer of the cortex in plant roots)

gutt- = a drop (*guttation:* the exudation of water droplets caused by root pressure in certain plants)

mega- = large, great (*megapascal:* a unit of pressure equivalent to 10 atmospheres of pressure)

myco- = a fungus; **-rhizo** = a root (*mycorrhizae:* mutualistic associations of plant roots and fungi)

osmo- = pushing (*osmosis:* the diffusion of water across a selectively permeable membrane)

sym- = with, together (*symplast:* in plants, the continuum of cytoplasm connected by plasmodesmata between cells)

turg- = swollen (*turgor pressure:* the force directed against a cell wall after the influx of water and the swelling of a walled cell due to osmosis)

xero- = dry; **-phyto** = a plant (*xerophytes:* plants adapted to arid climates)

Structure Your Knowledge

1. Describe the ways in which solutes may move across the plasma membrane in plants.
2. Both xylem sap and phloem sap move by bulk flow in an angiosperm. Compare and contrast the mechanisms for their movement.

Test Your Knowledge

MULTIPLE CHOICE: *Choose the one best answer.*

1. Which of the following is *not* a component of the symplast?
   a. sieve-tube members
   b. xylem tracheids
   c. endodermal cells
   d. cortex cells
   e. companion cells

2. Proton pumps in the plasma membranes of plant cells may
   a. generate a membrane potential that helps drive cations into the cell through their specific carriers.
   b. be coupled to the movement of K^+ into guard cells.
   c. drive the accumulation of sucrose in sieve-tube members.
   d. contribute to the movement of anions through a cotransport mechanism.
   e. be involved in all of the above.

3. The Casparian strip prevents water and minerals from entering the vascular cylinder through the
   a. plasmodesmata.
   b. endodermal cells.
   c. symplast.
   d. apoplast.
   e. xylem vessels.

4. The water potential of a plant cell
   a. is equal to 0 MPa when the cell is in pure water and is turgid.
   b. is equal to that of air.
   c. is equal to -0.23 MPa.
   d. becomes greater when K^+ ions are actively moved into the cell.
   e. becomes 0 MPa due to loss of turgor pressure in a concentrated sugar solution.

5. Guttation results from
   a. the pressure flow of sap through phloem.
   b. a water vapor break in the column of xylem sap.
   c. root pressure causing water to flow up through xylem faster than it can be lost by transpiration.
   d. a higher water potential of the leaves than of the roots.
   e. specialized structures in transport cells that accumulate sucrose.

6. Which of these is *not* a major factor in the movement of xylem sap up a tall tree?
   a. transpiration
   b. plasmodesmata
   c. adhesion
   d. cohesion
   e. tension

7. Adhesion is a result of
   a. hydrogen bonding between water molecules.
   b. the pull on the water column as water evaporates from the surfaces of mesophyll cells.
   c. tension within the xylem caused by a negative pressure.
   d. attraction of water molecules to hydrophilic walls of narrow xylem tubes.
   e. the high surface tension of water.

8. What is a function of the tonoplast?
   a. regulate movement of solutes between cells joined by plasmodesmata
   b. help move water and minerals past the Casparian strip into the vascular cylinder
   c. help maintain low cytosolic H^+ concentration by pumping H^+ into the vacuole
   d. increase the surface area for pumping H^+ out of transfer cells
   e. control the expansion of turgid guard cells so that the space between them opens up

9. Which of these does *not* stimulate the opening of stomata?
   a. daylight in a CAM plant
   b. depletion of CO_2 in the air spaces of the leaf
   c. stimulation of proton pumps that results in the movement of K^+ into the guard cells
   d. circadian rhythm of guard cell opening
   e. an increase in the turgor of guard cells

10. Your favorite spider plant is wilting. What is the most likely cause and remedy for its declining condition?
    a. Water potential is too low; apply sugar water.
    b. The stomata won't open; no remedy available.
    c. Plasmolysis of its cells; water the plant.
    d. Cavitation; perform a xylem vessel bypass.
    e. Circadian rhythm has stomata closed; place it in bright light.

11. A turgid plant cell placed in a solution in an open beaker becomes flaccid.
    a. The water potential of the cell was initially higher than that of the solution.
    b. The water potential of the cell was initially equal to that of the solution.
    c. The pressure (Ψ_P) of the cell was initially lower than that of the solution.
    d. The Ψ_S of the cell was initially more negative than that of the solution.
    e. Turgor pressure disappeared because the cell no longer needed support.

12. What facilitates the movement of K^+ into epidermal cells of the root?
    a. cotransport through a membrane protein
    b. bulk flow of water into the root
    c. passage through selective channels, aided by the membrane potential created by proton pumps
    d. active transport through a potassium pump
    e. simple diffusion across the cell membrane down its concentration gradient

13. What are aquaporins?
    a. cytoplasmic connections between cortical cells
    b. pores through the ends of sieve-tube members through which phloem sap flows
    c. openings in the lower epidermis of leaves through which water vapor escapes
    d. openings into root hairs through which water enters
    e. water-specific channels in membranes that may speed the rate of osmosis

14. The formation of a meniscus along the cell walls surrounding the air space of a leaf contributes to water transport by
    a. creating a more positive water potential than in the surrounding mesophyll cells.
    b. creating tension, thus lowering pressure and the water potential of the leaf.
    c. raising the water potential of the surrounding saturated air.
    d. increasing the adhesion of water molecules to the cell walls.
    e. increasing the rate of transpiration from the leaf.

15. Considering an animal cell (first group) and a plant cell (second group) placed in test solutions, which of the following choices gives the *correct* direction for water flow by osmosis?
    a. hypertonic → hypotonic; higher Ψ → lower Ψ
    b. hypertonic → hypotonic; lower Ψ → higher Ψ
    c. hypotonic → hypertonic; higher Ψ → lower Ψ
    d. hypotonic → hypertonic; lower Ψ → higher Ψ
    e. One cannot tell unless told the Ψ_p for the plant cell.

16. All of the following increase the surface area available for absorption of water and minerals by a root *except*
    a. mycorrhizae.
    b. numerous branch roots.
    c. root hairs.
    d. cytoplasmic extensions of the endodermis.
    e. the large surface area of cortical cells.

17. By what method do most mineral anions (negatively charged ions) enter root cells?
    a. apoplastic route
    b. symplastic route
    c. diffusion
    d. cotransport using a proton gradient
    e. bulk flow

18. What mechanism explains the movement of sucrose from source to sink?
    a. evaporation of water and active transport of sucrose from the sink
    b. osmotic movement of water into the sucrose-loaded sieve-tube members creating a higher pressure in the source than in the sink
    c. tension created by the differences in pressure in the source and sink
    d. active transport of sucrose through the sieve-tube cells driven by proton pumps
    e. the hydrolysis of starch to sucrose in the mesophyll cells that raises their water potential and drives the bulk flow of sap to the sink

Chapter 37

Plant Nutrition

Framework

The nutritional requirements of plants include essential macronutrients and micronutrients. Carbon dioxide enters the plant through the leaves, but water and minerals must be absorbed through the roots. Soil fertility is influenced by its texture and composition. Nitrogen assimilation by plants is made possible by the decomposition of humus by microbes and nitrogen fixation by bacteria. Mycorrhizae are important associations between fungi and plant roots that increase mineral and water absorption.

Chapter Review

Plants, as photoautotrophs, make their own organic compounds. Their roots, shoots, and leaves are structurally adapted to obtain water, minerals, and carbon dioxide from the soil and air.

37.1 Plants require certain chemical elements to complete their life cycle

Early scientists speculated on whether soil, water, or air provides the substance for plant growth. **Mineral nutrients,** essential inorganic ions absorbed from the soil, make only a small contribution to the mass of a plant. Water makes up more than three-fourths of the weight of a plant; supplies most of the hydrogen incorporated into organic compounds; and is used as a solvent, for growth by cell elongation, or for support through turgor pressure. By weight, CO_2 from the air is the source of most of the organic material of a plant. Carbon, oxygen, and hydrogen, as the main components of organic compounds such as cellulose, are the most abundant elements making up the dry weight of a plant. Nitrogen, sulfur, and phosphorus—ingredients of some organic compounds—are relatively abundant.

Macronutrients and micronutrients

Essential elements are those required for a plant to complete its life cycle from a seed to an adult that produces more seeds. **Hydroponic culture** has been used to determine which of the mineral elements found in plants are essential nutrients. Seventeen elements have been identified as essential in all plants.

Nine **macronutrients** are required by plants in relatively large amounts and include the six major elements of organic compounds as well as calcium, potassium, and magnesium.

Eight **micronutrients** have been identified as needed by plants in very small amounts, functioning mainly as cofactors of enzymatic reactions.

■ **INTERACTIVE QUESTION 37.1**

a. What is a function of the macronutrient magnesium in plants?

b. What is a function of the macronutrient phosphorus?

c. What is a function of the micronutrient iron?

Symptoms of Mineral Deficiency A mobile nutrient will move to young, growing tissues, so that a deficiency will show up first in older parts of the plant. Symptoms of a mineral deficiency may be distinctive enough for the cause to be diagnosed by a plant physiologist or farmer. Soil and plant analysis can confirm a specific deficiency. Nitrogen, potassium, and phosphorus deficiencies are most common.

■ **INTERACTIVE QUESTION   37.2**

Where would you expect a deficiency of a relatively immobile element to be seen first?

37.2   Soil quality is a major determinant of plant distribution and growth

Texture and Composition of Soils   The formation of soil begins with the weathering of rock and accelerates with the secretion of acids by lichens, fungi, bacteria, and plant roots. **Topsoil** is a mixture of broken-down rock, living organisms, and **humus** (decomposing organic matter). Several other distinct soil layers, or **horizons,** are found under the topsoil layer.

The texture of topsoil depends on particle size, which varies from coarse sand to fine clay. **Loams,** made up of a mixture of sand, silt, and clay, are often the most fertile soils, having enough fine particles to provide a large surface area for retaining water and minerals but enough coarse particles to provide air spaces with oxygen for respiring roots.

The activities of the numerous soil inhabitants, such as bacteria, fungi, algae, other protists, insects, worms, nematodes, and plant roots, affect the physical and chemical properties of soil.

Humus builds a crumbly soil that retains water, provides good aeration of roots, and supplies mineral nutrients.

Water containing dissolved minerals binds to hydrophilic soil particles and is held there in small spaces, available for uptake by plant roots. Positively charged minerals, such as K^+, Ca^{2+}, and Mg^{2+}, adhere to the negatively charged surfaces of finely divided clay particles. Negatively charged minerals, such as nitrate (NO_3^-), phosphate ($H_2PO_4^-$), and sulfate (SO_4^{2-}), tend to leach away more quickly. The release of H^+ by roots facilitates **cation exchange,** in which hydrogen ions displace positively charged mineral ions from the clay particles, making the ions available for absorption.

■ **INTERACTIVE QUESTION   37.3**

Describe the characteristics of a fertile soil.

Soil Conservation and Sustainable Agriculture   Without good soil conservation, agriculture can quickly destroy the fertility of a soil that has built up over centuries. Agriculture diverts essential elements from the chemical cycles when crops are harvested, and many crops use more water than the natural vegetation.

Historically, farmers used manure to fertilize their crops. Today in developed nations, commercially produced fertilizers, usually containing nitrogen, phosphorus, and potassium, are used. Manure, fishmeal, and compost are called organic fertilizers because they contain organic material that is in the process of decomposing. These fertilizers decompose into inorganic nutrients, which are taken up by the plant in the same form supplied by commercial fertilizers. Commercial fertilizers may be rapidly leached from the soil, polluting streams and lakes.

The acidity of the soil affects cation exchange and can alter the chemical form of minerals and thus their ability to be absorbed by the plant. Managing the pH of soil is an important aspect of maintaining fertility.

Irrigation can make farming possible in arid regions, but it places a huge drain on water resources and raises soil salinity. New methods of irrigation and new varieties of plants that can tolerate less water may reduce some of these problems.

In the United States, topsoil from thousands of acres of farmland is lost to water and wind erosion each year. Agricultural use of cover crops, windbreaks, and terracing can minimize erosion.

■ **INTERACTIVE QUESTION   37.4**

a. How could a genetically engineered "smart plant" help reduce fertilizer use?

b. What is **sustainable agriculture?**

The use of plants to extract heavy metals and other pollutants from contaminated soils is an emerging technology known as **phytoremediation.**

37.3 Nitrogen is often the mineral that has the greatest effect on plant growth

Soil Bacteria and Nitrogen Availability To be absorbed by plants, nitrogen must be converted to nitrate (NO_3^-) or ammonium (NH_4^+) by the action of microbes, such as ammonifying bacteria, that decompose humus. Some nitrate is lost to the atmosphere by the action of denitrifying bacteria. **Nitrogen-fixing bacteria** convert atmospheric nitrogen into ammonia through the process of **nitrogen fixation.**

Bacteria capable of nitrogen fixation contain **nitrogenase,** an enzyme complex that reduces N_2 by adding H^+ and electrons to form ammonia (NH_3). This process is energetically expensive, and nitrogen-fixing bacteria rely on organic material in soils or symbiotic relationships with plant roots to supply fuel for their cellular respiration. In the soil solution, ammonia forms ammonium, which plants can absorb. Nitrifying bacteria oxidize ammonium, producing nitrate, the form of nitrogen most readily absorbed by roots. Most plants incorporate nitrogen into amino acids or other organic compounds in the roots and then transport it through the xylem to shoots (see Interactive Question 37.5).

Improving the Protein Yield of Crops Protein deficiency is the most common form of human malnutrition. Agricultural research attempts to improve the quality and quantity of proteins in crops. Varieties of corn, wheat, and rice have been developed that are enriched in protein, but they require the addition of large quantities of expensive nitrogen fertilizer.

37.4 Plant nutritional adaptations often involve relationships with other organisms

The Role of Bacteria in Symbiotic Nitrogen Fixation Plants of the legume family have root swellings, called **nodules,** composed of plant cells with nitrogen-fixing *Rhizobium* bacteria in a form called **bacteroids** contained in vesicles. Nitrogen fixation requires an anaerobic environment. Lignified external layers of nodule cells may limit gas exchange. In some nodules, the presence of oxygen-binding leghemoglobin keeps the concentration of free O_2 low and regulates the oxygen supply for the bacterial respiration required to provide ATP for nitrogen fixation.

Mutual signaling between roots and bacteria lead to root hair elongation and formation of an infection thread within a root hair. Bacteria form bacteroids as they bud into dividing cortical cells from the infection thread. Dividing cells of the cortex and pericycle fuse to form the nodule, which develops vascular tissue into the vascular cylinder. The plant provides the bacteria with carbohydrates and other organic molecules. Nodule cells use most of the fixed nitrogen to make amino acids, which are then transported throughout the plant.

The plant initiates the chemical dialogue with *Rhizobium* by secreting flavonoids that are detected and absorbed only by its particular *Rhizobium* species. The plant signal activates a gene-regulating protein that turns on the bacterial genes called *nod* (for "nodulation") genes. The enzymes produced from these genes catalyze the production of Nod factors that are secreted by the bacterial cells. Nod factors signal the

---

■ INTERACTIVE QUESTION 37.5

Fill in the types of bacteria (a–d) that participate in the nitrogen nutrition of plants. Indicate the form (e) in which nitrogen is transported in xylem to the shoot system.

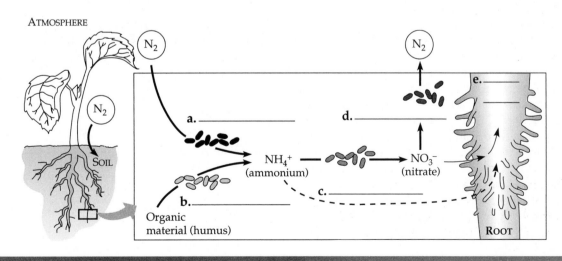

activation of a plant's early nodulin genes, which initiate formation of an infection thread.

Symbiotic nitrogen fixation benefits agriculture. Legume seeds are coated with their specific *Rhizobium* before planting. Rice farmers culture a water fern that has symbiotic cyanobacteria that fix nitrogen, improving the fertility of rice paddies.

■ INTERACTIVE QUESTION   37.6

a. What is **crop rotation?**

b. What is "green manure"?

Mycorrhizae and Plant Nutrition   Most plants have symbiotic associations of roots and fungi called **mycorrhizae.** The fungus receives food from the plant, while providing a large surface area for the absorption of water and minerals, which it provides to the plant. The fungus secretes growth hormones that stimulate root growth and branching and may help protect the plant from some soil pathogens.

Especially common in woody plants, **ectomycorrhizae** form a dense sheath or mantle of mycelium over the root surface. Hyphae extending from the mantle provide a huge surface area for the absorption of water and minerals, especially phosphate. Hyphae also grow into extracellular spaces in the root cortex and facilitate exchange between plant and fungus.

Endomycorrhizae do not form a sheath but extend fine fungal hyphae into the soil. Hyphae also penetrate into root cell walls and form tubes that invaginate the root cell's membrane. Some of these tubes form dense branched structures called arbuscles that facilitate nutrient exchange. Eighty-five percent of plant species have endomycorrhizae.

Most plants form mycorrhizae when they grow in their natural habitat. When seeds are planted in foreign soil where their particular species of fungus may not be found, or when soil fungi are poisoned, plants show signs of malnutrition. Foresters now inoculate seeds with spores of their mycorrhizal fungi to improve the growth of seedlings.

■ INTERACTIVE QUESTION   37.7

Why is it thought that mycorrhizae were important to the colonization of land by the first plants?

Epiphytes, Parasitic Plants, and Carnivorous Plants
Epiphytes are plants that grow on the surface of another plant but do not take nourishment from it. Parasitic plants, such as the mistletoe or dodder, produce haustoria that may invade a host plant and siphon xylem or phloem sap from its vascular tissue.

Living in acid bogs or other nutrient-poor soils, carnivorous plants obtain nitrogen and minerals by killing and digesting insects that are caught in traps formed from modified leaves.

▶ Word Roots

ecto- = outside; **-myco-** = a fungus; **-rhizo** = a root (*ectomycorrhizae:* a type of mycorrhizae in which the mycelium forms a dense sheath, or mantle, over the surface of the root; hyphae extend from the mantle into the soil, greatly increasing the surface area for water and mineral absorption)

endo- = inside (*endomycorrhizae:* a type of mycorrhizae that unlike ectomycorrhizae, do not have a dense mantle ensheathing the root; instead, microscopic fungal hyphae extend from the root into the soil)

macro- = large (*macronutrient:* elements required by plants and animals in relatively large amounts)

micro- = small (*micronutrient:* elements required by plants and animals in very small amounts)

-phyto = a plant (*phytoremediation:* an emerging, nondestructive technology that seeks to cheaply reclaim contaminated areas by taking advantage of the remarkable ability of some plant species to extract heavy metals and other pollutants from the soil and to concentrate them in easily harvested portions of the plant)

▶ Structure Your Knowledge

1. Develop a concept map that organizes your understanding of the basic nutritional requirements of plants.
2. What are the differences between root nodules and mycorrhizae? How are each beneficial to plants?
3. What is a similarity between root nodules and mycorrhizae?

Test Your Knowledge

MULTIPLE CHOICE: *Choose the one best answer.*

1. The inorganic compound that contributes most of the mass to a plant's organic matter is
   a. H_2O.
   b. CO_2.
   c. NO_3^-.
   d. O_2.
   e. $C_6H_{12}O_6$.

2. The effects of mineral deficiencies involving fairly mobile nutrients will first be observed in
   a. older portions of the plant.
   b. new leaves and shoots.
   c. the root system.
   d. the color of the leaves.
   e. the flowers.

3. Micronutrients may be
   a. cofactors in enzymes.
   b. required in very minute quantities.
   c. components of cytochromes.
   d. identified by hydroponic culture.
   e. all of the above.

4. The most fertile type of soil is usually
   a. sand because its large particles allow room for air spaces.
   b. loam, which has a mixture of fine and coarse particles.
   c. clay, because the fine particles provide much surface area to which minerals and water adhere.
   d. humus, which is decomposing organic material.
   e. wet and alkaline.

5. Chlorosis is
   a. a symptom of a mineral deficiency indicated by yellowing leaves due to decreased chlorophyll production.
   b. the uptake of the micronutrient chlorine by a plant, which is facilitated by symbiotic bacteria.
   c. the production of chlorophyll within the thylakoid membranes of a plant.
   d. a contamination of glassware in hydroponic culture.
   e. a mold of roots caused by wet soil conditions.

6. Negatively charged minerals
   a. are released from clay particles by cation exchange.
   b. are reduced by cation exchange before they can be absorbed.
   c. are converted into amino acids before they are transported through the plant.
   d. are bound when roots release acids into the soil.
   e. are leached away by the action of rainwater more easily than positively charged minerals.

7. Nitrogenase
   a. is an enzyme complex that reduces atmospheric nitrogen to ammonia.
   b. is found in *Rhizobium* and other nitrogen-fixing bacteria.
   c. catalyzes the energy-expensive fixation of nitrogen.
   d. provides a model for chemical engineers to design catalysts to make nitrogen fertilizers.
   e. is or does all of the above.

8. Epiphytes
   a. have haustoria for anchoring to their host plants and obtaining xylem or phloem sap.
   b. are symbiotic relationships between leaves and fungi.
   c. live in poor soil and digest insects to obtain nitrogen.
   d. grow on other plants but do not obtain nutrients from their hosts.
   e. are able to fix their own nitrogen.

9. The nitrogen content of some agricultural soils may be improved by
   a. the synthesis of leghemoglobin by ammonifying bacteria.
   b. mycorrhizae on legumes.
   c. water ferns with symbiotic cyanobacteria, or other plants with nitrogen-fixing bacteria.
   d. cation exchange.
   e. the action of denitrifying bacteria.

10. An advantage of organic fertilizers over chemical fertilizers is that they
    a. are more natural.
    b. release their nutrients over a longer period of time and are less likely to be lost to runoff.
    c. provide nutrients in the forms most readily absorbed by plants.
    d. are easier to mass produce and transport.
    e. are all of the above.

11. The early nodulin genes
    a. code for cytokinin plant hormones.
    b. produce Nod factors that are signals released from bacteria to roots.
    c. are turned on by Nod factor signals from *Rhizobium* and initiate nodule development in roots.
    d. code for receptors for signals released by *Rhizobium*.
    e. are fungal genes that control the development of arbuscles within root cells.

12. Which of the following takes the form of a mycelial sheath over plant roots with hyphae extending out that increase the surface area for absorption?
    a. root nodules
    b. haustoria
    c. endomycorrhizae
    d. ectomycorrhizae
    e. infection threads

13. Which of the following is an example of phytoremediation?
    a. dusting legume seeds with spores of *Rhizobium* to increase nodule formation
    b. inoculating seeds with fungal spores to ensure mycorrhizal formation
    c. using plants to remove toxic heavy metals from contaminated soils
    d. genetic engineering of plants to increase protein content
    e. seeding the ocean with iron to create algal blooms that reduce atmospheric CO_2 levels

14. What created the Dust Bowl in the Great Plains in the 1930s?
    a. several years of drought
    b. building dams that removed the normal supply of water to the region
    c. removal of prairie grasses for wheat and cattle farming
    d. lack of crop rotation so that soil fertility was destroyed
    e. Both a and c were important factors.

15. Place these steps in the sequence of root nodule formation in proper order:
    1. production of infection thread through which bacteria enter root
    2. secretion of flavonoids by root
    3. continuation of nodule growth and connection to vascular stele of root
    4. activation of gene regulating protein, which turns on bacterial *nod* genes
    5. activation of plant early nodulin genes by bacterial Nod factors
    6. bacteria wrapped in vesicles in cortical cells, becoming bacteroids

    a. 3, 5, 6, 1, 4, 2
    b. 2, 4, 5, 1, 6, 3
    c. 2, 1, 4, 5, 6, 3
    d. 5, 4, 2, 1, 3, 6
    e. 1, 2, 4, 5, 6, 3

Chapter 38

Angiosperm Reproduction and Biotechnology

Key Concepts

38.1 Pollination enables gametes to come together within a flower

38.2 After fertilization, ovules develop into seeds and ovaries into fruits

38.3 Many flowering plants clone themselves by asexual reproduction

38.4 Plant biotechnology is transforming agriculture

Framework

This chapter describes the sexual and asexual reproduction of flowering plants. The flower produces spores that grow into the haploid gametophyte stages of the life cycle: Microspores in the anther develop into pollen grains, and a megaspore in the ovule produces an embryo sac. Pollination and the double fertilization of egg and polar nuclei are followed by the development of a seed with a quiescent embryo and endosperm, protected in a seed coat and housed within a fruit. Seed dormancy is broken following proper environmental cues and the imbibition of water.

Vegetative propagation allows successful plants to clone themselves. Agriculture makes extensive use of this type of plant reproduction by using cuttings, grafts, and test-tube cloning.

Plant biotechnologists are creating genetically modified (GM or transgenic) plants that have such traits as insect and disease resistance, herbicide tolerance, and improved nutritional value. Opposition to the development of GM organisms focuses on human health concerns and the unknown dangers of introducing transgenic plants into the environment.

Chapter Review

38.1 Pollination enables gametes to come together within a flower

Plants exhibit an alternation of generations between haploid (n) and diploid ($2n$) generations. The diploid plant, the sporophyte, produces haploid spores by meiosis. Spores develop into multicellular haploid male and female gametophytes, which produce gametes by mitosis. Pollination brings pollen to the stigma of a flower, and pollen germination brings sperm to the female gametophyte. Fertilization yields diploid zygotes that grow into new sporophyte plants. In angiosperms, the male and female gametophytes have become reduced to only a few cells that develop within the anthers and ovules of the flower.

Flower Structure   Flowers are determinant, reproductive shoots that usually contain four whorls of modified leaves called floral organs: **sepals, petals, stamens,** and **carpels,** which attach to the stem at the **receptacle.** Sepals enclose and protect the unopened floral bud. Petals are generally more brightly colored and may attract pollinators. Stamens consist of a filament and an **anther,** which contains pollen sacs. A carpel consists of a sticky **stigma** at the top of a slender **style,** which leads to an **ovary.** The ovary encloses one or more **ovules.** A flower may have a single carpel or multiple fused carpels; either may be referred to as a **pistil.**

A **complete flower** has sepals, petals, stamens, and carpels. **Incomplete flowers** lack one or more of these floral organs. Floral variations include fused or separated floral organs; bilateral or radial symmetry; an ovary that is superior, semi-inferior, or inferior; individual flowers or clusters of flowers called **inflorescences;** as well as diverse shapes, colors, and odors adapted to attract different pollinators.

All complete flowers and some incomplete flowers have both stamens and carpels. Most incomplete flowers are either staminate or carpellate. In **monoecious** plant species, both staminate and carpellate flowers are on the same plant; in **dioecious** species, these flowers are on separate plants.

■ **INTERACTIVE QUESTION   38.1**

Identify the flower parts in the following diagram.

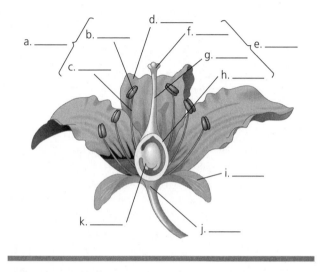

a. _____
b. _____
c. _____
d. _____
e. _____
f. _____
g. _____
h. _____
i. _____
j. _____
k. _____

Gametophyte Development and Pollination   Anthers and ovules contain the sporangia in which spores and gametophytes develop. A pollen grain is the sperm-producing male gametophyte surrounded by a spore wall, and embryo sacs are the egg-producing female gametophytes.

Pollination is the transfer of pollen from an anther to a stigma. The pollen grain grows a tube down the style, releasing its sperm within the embryo sac.

Following fertilization, the zygote develops into an embryo as the ovule containing the embryo develops into a seed. The entire ovary forms a fruit, which aids in seed dispersal.

Within the microsporangium (pollen sac) diploid cells called microsporocytes undergo meiosis to form four haploid **microspores.** A microspore divides once by mitosis to produce a generative cell and a tube cell. The wall surrounding the two cells thickens into the sculptured coat of the pollen grain. A pollen grain develops into a mature male gametophyte when the generative cell, which has moved into the tube cell, divides to form two sperm cells, usually after the tube cell begins to form the pollen tube. After growing through

the style, the pollen tube releases the sperm cells near an embryo sac.

The megasporocyte in the single megasporangium of each ovule undergoes meiosis to form four haploid **megaspores,** only one of which usually survives. This megaspore grows and divides by mitosis three times, forming the female gametophyte, called the embryo sac, which typically consists of eight nuclei contained in seven cells. At one end of the embryo sac, an egg cell is lodged between two cells called synergids; three antipodal cells are at the other end; and two nuclei, called polar nuclei, are in a large central cell. The ovule consists of the embryo sac and its surrounding protective sporophyte layers called integuments.

Pollination is accomplished by wind, water, or animals. In Interactive Question 38.1, indicate where pollen is produced and where pollination and fertilization occur.

■ **INTERACTIVE QUESTION   38.2**

a. Describe the male gametophyte.

b. Describe the female gametophyte.

Mechanisms That Prevent Self-Fertilization   Some flowers self-fertilize; this "selfing" ensures that a seed will develop within the fruit but does not increase the genetic diversity of offspring. Self-fertilization may be prevented by temporal or structural mechanisms. In flowers that are **self-incompatible,** a biochemical block prevents the development of pollen that does land on a stigma of the same plant.

The ability of flowers to reject their own pollen, or that of closely related individuals, depends on genes for self-incompatibility called S-genes. A plant population may have dozens of alleles of the S-gene, and if the pollen's allele matches an allele of the stigma, a pollen tube does not develop. In gametophytic self-incompatibility, RNA-hydrolyzing enzymes produced by the style destroy the RNA of developing pollen tubes that have a matching allele. In sporophytic self-incompatibility, self-recognition activates a signal-transduction pathway in stigma cells that blocks pollen germination.

Further research on the molecular basis of self-incompatibility may allow plant breeders to manipulate crop species to assure hybridization.

38.2 After fertilization, ovules develop into seeds and ovaries into fruits

Double Fertilization   A pollen grain that lands on a receptive stigma absorbs moisture and germinates. The pollen tube grows through the style, and the generative cell divides to form two sperm. The pollen tube probes through the micropyle, an opening through the integuments of the ovule, and releases its two sperm within the embryo sac. By **double fertilization,** one sperm fertilizes the egg to form the zygote, and the other combines with the polar nuclei to form a triploid nucleus, which will develop into a food-storing tissue called the **endosperm.** Gamete fusion is immediately followed by an increase in cytoplasmic Ca^{2+} levels in the egg and the establishment of a block to polyspermy.

■ **INTERACTIVE QUESTION   38.3**

What function does double fertilization serve?

From Ovule to Seed   The triploid nucleus divides to form the endosperm, a multicellular mass rich in nutrients (proteins, oils, and starch) that are provided to the developing embryo and may be stored for later use by the seedling. In many eudicots, the food reserves of the endosperm are transferred to the cotyledons before the seed matures.

In the zygote, the transverse first mitotic division creates a basal cell and a terminal cell. The basal cell divides to produce a thread of cells, called the suspensor, that anchors the embryo and transfers nutrients to it. The terminal cell divides to form a spherical proembryo, on which the cotyledons (two in eudicots, one in monocots) begin to form as bumps. The embryo elongates and apical meristems develop at the apexes of the embryonic shoot and root.

As it matures, the seed dehydrates and the embryo becomes dormant. The embryo and its food supply, the endosperm and/or enlarged cotyledons, are enclosed in a **seed coat** formed from the ovule integuments.

In a eudicot seed, such as a bean, the embryo is an elongated embryonic axis attached to fleshy cotyledons. (Some eudicots have thin cotyledons.) The axis below the cotyledonary attachment is called the **hypocotyl;** it terminates in the **radicle,** or embryonic root. The upper axis is the **epicotyl;** it terminates as a shoot tip with a pair of leaves.

The monocot seed found in members of the grass family has a single thin cotyledon, called a **scutellum,** which absorbs nutrients from the endosperm during germination. A sheath called a **coleorhiza** covers the root, and a **coleoptile** encloses the young shoot.

■ **INTERACTIVE QUESTION   38.4**

Label the parts in these diagrams of a bean and a corn seed.

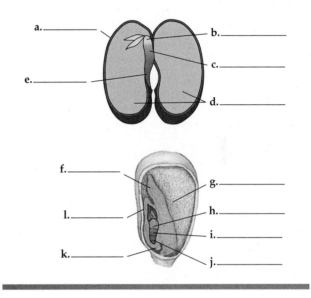

From Ovary to Fruit   The ovary of the flower develops into a **fruit,** which both protects and helps to disperse the seeds. Hormonal changes following fertilization cause the ovary to enlarge, its wall becoming the pericarp, or thickened wall of the fruit. Fruit usually does not set if a flower has not been pollinated.

A **simple fruit** is derived from a single carpel or several fused carpels; an **aggregate fruit** results from a flower with more than one seperate carpel; a **multiple fruit** develops from an inflorescence, whose many ovaries fuse together to become one fruit. Other floral parts (such as the receptacle) may contribute to what we commonly call a fruit. Such fruits, such as apples, are called accessory fruits.

Fruits usually ripen as the seeds are completing their development.

■ **INTERACTIVE QUESTION   38.5**

What changes usually occur when a fleshy fruit ripens?

Seed Germination   Growth and development are suspended when a seed matures and enters **dormancy.**

Dormancy increases the chances that the seed will germinate when and where the seedling has a good chance of surviving. The specific cues for breaking dormancy vary with the environment and may include heavy rain, intense heat from fires, cold, light, or chemical breakdown of the seed coat. The viability of a dormant seed may vary from a few days to decades or longer.

Imbibition, the absorption of water by the dry seed, causes the seed to expand, rupture its coat, and begin a series of metabolic changes. Enzymes digest stored compounds, and nutrients are sent to growing regions.

The radicle emerges from the seed first, followed by the shoot tip. In many eudicots, a hook that forms in the hypocotyl is pushed up through the ground, pulling the delicate shoot and cotyledons behind it. Light stimulates the straightening of the hook, and the first foliage leaves begin photosynthesis.

■ **INTERACTIVE QUESTION   38.6**

How does the shoot tip break through the soil in maize and other grasses?

38.3   Many flowering plants clone themselves by asexual reproduction

Many plant species produce genetically identical copies of themselves through **asexual reproduction.** Advantages of asexual or **vegetative reproduction** include the production of clones of plants that are well suited to a certain environment and of progeny that are usually not as frail as seedlings. Sexual reproduction generates variation in a population, an advantage when the environment changes. Seeds, which are almost always produced sexually, provide a means of dispersal to new locations and dormancy during harsh conditions.

Mechanisms of Asexual Reproduction   Asexual reproduction is an extension of the indeterminate growth of plants in which meristematic tissues can grow indefinitely and parenchyma cells can divide and differentiate into specialized cells. A common type of vegetative reproduction is **fragmentation,** the formation of whole plants from parts of a parent plant. In some species, the root system gives rise to many adventitious shoots that develop into a clone with separate shoot systems. Some plants, such as dandelions, can produce seeds asexually, a process called **apomixis.**

■ **INTERACTIVE QUESTION   38.7**

What is an advantage of apomixis?

Vegetative Propagation and Agriculture   New plants may develop from stem cuttings when a **callus,** or mass of dividing cells, forms at the cut end of the shoot and adventitious roots develop from the callus. Adventitious roots can also form from a node in the shoot fragment.

Twigs or buds of one plant can be grafted onto a plant of a different variety or closely related species. The plant that provides the root system is called the **stock,** and the twig is called the **scion.** Grafting can combine the best qualities of different plants.

In test-tube cloning, whole plants can develop from pieces of tissue, called explants, or even from single parenchyma cells. A single plant can be cloned into thousands of plants by subdividing the undifferentiated calluses as they grow in tissue culture. Stimulated by proper hormone balances, calluses sprout shoots and roots and develop into plantlets, which can be transferred to soil to develop.

Foreign DNA may be inserted into individual plant cells, which then grow into **transgenic** or genetically modified (GM) plants by test-tube culture.

A technique called **protoplast fusion** is being coupled with tissue culture to create new plant varieties. Protoplasts are cells whose cell walls have been enzymatically removed. Protoplasts from different species can be fused and cultured to form hybrid plantlets.

38.4   Plant biotechnology is transforming agriculture

Plant biotechnology refers both to the age-old use of plants to make products for human use and to the use of GM organisms in agriculture.

Artificial Selection   Almost all our crop species were first domesticated by Neolithic (late Stone Age) humans about 10,000 years ago. Natural hybridization between different species of plants is common, and humans have exploited such genetic variations using selective breeding and artificial selection to develop and improve crops.

Reducing World Hunger and Malnutrition   The serious malnutrition of millions of people may be due to inequities in distribution of food resources or may signal that the world is already overpopulated.

The planting of transgenic crops has increased dramatically over the past several years. Cotton, maize, and potatoes have been engineered to contain genes from *Bacillus thuringiensis* that code for *Bt* toxin, reducing the need for spraying chemical insecticides. Other transgenic crops have been developed that are resistant to a number of herbicides, allowing farmers to "weed" crops without heavy tillage. Some transgenic plants are more resistant to disease or have improved nutritional quality.

The Debate over Plant Biotechnology GM organisms (GMOs) may present an unknown risk to human health or the environment. One concern is the transfer of allergens to a food source. Some GM crops may be safer; *Bt* maize contains less of a cancer-causing mycotoxin.

Many ecologists are concerned about the effect of GM crops on nontarget organisms. All GM crops need to be field tested for nontarget effects.

A serious concern is the escape of herbicide- or disease-resistance genes through crop-to-weed hybridization that may create "superweeds." Various techniques are being developed to help reduce the ability of transgenic crops to hybridize, such as developing male sterility so that transgenic pollen is not produced; introducing genes into chloroplast DNA, which is not present in pollen; and developing "terminator technology" through which seeds are rendered inviable when they mature.

Word Roots

a- = without; **-pomo** = fruit (*apomixis:* the asexual production of seeds)

anth- = a flower (*anther:* the terminal pollen sac of a stamen, inside which pollen grains with male gametes form in the flower of an angiosperm)

bi- = two (*bisexual flower:* a flower equipped with both stamens and carpels)

carp- = a fruit (*carpel:* The female reproductive organ of a flower, consisting of the stigma, style, and ovary)

coleo- = a sheath; **-rhiza** = a root (*coleorhiza:* the covering of the young root of the embryo of a grass seed)

di- = two (*dioecious:* referring to a plant species that has staminate and carpellate flowers on separate plants)

dorm- = sleep (*dormancy:* a condition typified by extremely low metabolic rate and a suspension of growth and development)

endo- = within (*endosperm:* a nutrient-rich tissue formed by the union of a sperm cell with two polar nuclei during double fertilization, which provides nourishment to the developing embryo in angiosperm seeds)

epi- = on, over (*epicotyl:* the embryonic axis above the point at which the cotyledons are attached)

gamet- = a wife or husband (*gametophyte:* the multicellular haploid form in organisms undergoing alternation of generations, which mitotically produces haploid gametes that unite and grow into the sporophyte generation)

hypo- = under (*hypocotyl:* the embryonic axis below the point at which the cotyledons are attached)

mega- = large (*megaspore:* a large, haploid spore that can continue to grow to eventually produce a female gametophyte)

micro- = small (*microspore:* a small, haploid spore that can give rise to a haploid male gametophyte)

mono- = one; **-ecious** = house (*monoecious:* referring to a plant species that has both staminate and carpellate flowers on the same individual)

peri- = around; **-carp** = a fruit (*pericarp:* the thickened wall of fruit)

proto- = first; **-plast** = formed, molded (*protoplast:* the contents of a plant cell exclusive of the cell wall)

scutell- = a little shield (*scutellum:* a specialized type of cotyledon found in the grass family)

sporo- = a seed; **-phyto** = a plant (*sporophyte:* the multicellular diploid form in organisms undergoing alternation of generations that results from a union of gametes and that meiotically produces haploid spores that grow into the gametophyte generation)

stam- = standing upright (*stamen:* the pollen-producing male reproductive organ of a flower, consisting of an anther and filament)

uni- = one (*unisexual flower:* a flower missing either stamens or carpels)

Structure Your Knowledge

1. Draw yourself a diagram of the major events in the life cycle of an angiosperm.

2. List the advantages and disadvantages of sexual and asexual reproduction in plants.

3. List some of the potential benefits and dangers of plant biotechnology.

Test Your Knowledge

FILL IN THE BLANKS

_____ 1. structure from which fruit typically develops

_____ 2. generation that produces spores by meiosis

_____ 3. species with staminate and carpellate flowers on the same plant

_____ 4. female gametophyte of angiosperms

_____ 5. embryonic root

_____ 6. embryonic axis above attachment of cotyledon

_____ 7. protects grass shoot as it breaks through the soil

_____ 8. twig or stem portion of a graft

_____ 9. plant cell from which cell wall is removed

_____ 10. mass of dividing cells at cut end of a shoot

MULTIPLE CHOICE: *Choose the one best answer.*

1. A flower on a dioecious plant would be
   a. complete.
   b. biennial.
   c. incomplete.
   d. staminate or carpellate.
   e. both c and d.

2. Which of the following structures is haploid?
   a. embryo sac
   b. anther
   c. endosperm
   d. microsporocyte
   e. both a and b

3. The terminal cell of an early plant embryo
   a. develops into the shoot apex of the embryo.
   b. forms the suspensor that anchors the embryo and transfers nutrients.
   c. develops into the endosperm when fertilized by a sperm nucleus.
   d. divides to form the proembryo.
   e. develops into the cotyledons.

4. In angiosperms, sperm are formed by
   a. meiosis in the anther.
   b. meiosis in the pollen grain.
   c. mitosis in the anther.
   d. mitosis in the pollen tube.
   e. double fertilization in the embryo sac.

5. The endosperm
   a. may have its nutrients absorbed by the cotyledons in the seeds of eudicots.
   b. is usually a triploid tissue.
   c. is digested by enzymes in monocot seeds following hydration.
   d. develops in concert with the embryo as a result of double fertilization.
   e. is or does all of the above.

6. A seed consists of
   a. an embryo, a seed coat, and a nutrient supply.
   b. an embryo sac.
   c. a gametophyte and a nutrient supply.
   d. an enlarged ovary.
   e. an immature ovule.

7. Which structure protects a bean shoot as it breaks through the soil?
   a. hypocotyl hook
   b. radicle
   c. coleoptile
   d. coleorhiza
   e. seed coat

8. Which of the following is a form of asexual or vegetative reproduction?
   a. apomixis
   b. grafting
   c. test-tube cloning
   d. fragmentation
   e. all of the above

9. Protoplast fusion
   a. is used to study the fertilization of plant egg and sperm.
   b. is the method used to produce test-tube plantlets.
   c. can be used to form new plant species.
   d. occurs within a callus.
   e. is done with a gene gun.

10. In the plant embryo, the suspensor
    a. is produced by vertical mitotic divisions within the proembryo.
    b. connects the early root and shoot apexes.
    c. develops into the endosperm.
    d. is analogous to the umbilical cord in mammals.
    e. is the point of attachment of the cotyledons.

11. Flower organs have evolved from modified
    a. leaves.
    b. buds.
    c. sporangia.
    d. sporophytes.
    e. apical meristems.

12. What does self-incompatibility provide for a plant?
    a. means of transferring pollen to another plant
    b. a means of coordinating the fertilization of an egg with the development of stored nutrients
    c. a means of destroying foreign pollen before it fertilizes the egg cell
    d. a biochemical block to self-fertilization so that cross-fertilization is assured
    e. a means of producing seeds without the need for fertilization

13. Into what does a microspore develop in an angiosperm?
    a. the male gametophyte
    b. a pollen grain
    c. the male sporophyte
    d. the embryo sac
    e. Both a and b are correct.

14. Many plants form clones that develop after shoots emerge from the same root system. What is an advantage of forming such clones?
    a. provide a strong start for new plants
    b. disperse offspring to new habitats
    c. provide a period of dormancy until specific environmental cues signal regrowth
    d. provide a strong root system for a genetically altered shoot system
    e. allow for the production of seeds without fertilization

15. Why did it take nearly 20 years for plant breeders to convert the *opaque-2* mutant maize (with higher levels of two essential amino acids) into a variety that had a more durable endosperm?
    a. Such genetic recombination between species was restricted by government regulations.
    b. Its development was delayed because of concern that the new variety, intended for swine feed, would get mixed with maize intended for human consumption.
    c. Traditional plant breeding using hybridization and artificial selection is a time-intensive process.
    d. Plant breeders were trying to combine two varieties that were not closely enough related.
    e. Few people saw the benefit of improving the protein content of maize, and funding was severely lacking for such research.

16. Which of the following is a technique being developed to reduce the threat of introduced genes for herbicide or insect resistance escaping to closely related weed species?
    a. planting a nontransgenic plant border around crop fields to reduce crop-to-weed gene transfer
    b. breeding male sterility into transgenic plants so that they have no pollen to be transferred to nearby weeds
    c. engineering the gene of interest into chloroplast DNA, which is inherited from the maternal plant and is not transferred by pollen
    d. engineering crops, such as soybeans, that have no weedy relatives nearby, or introducing genes for beneficial crop traits that would actually reduce the fitness of hybrid weeds
    e. All of the above would reduce the risk of crop-to-weed transgene escape.

Chapter 39
Plant Responses to Internal and External Signals

Framework

Environmental stimuli and internal signals are linked by signal-transduction pathways to cellular responses such as changes in gene expression and activation of enzymes. Plant hormones—auxin, cytokinins, gibberellins, brassinosteroids, abscisic acid, and ethylene—control growth, development, flowering, and senescence, as plants respond and adapt to their environments. Plant movements in response to environmental stimuli include phototropism, gravitropism, and thigmotropism. The biological clock of plants controls circadian rhythms, such as stomatal opening and sleep movements. Phytochromes function as photoreceptors and are involved in the photoperiodic control of flowering. Plants have various physiological responses to environmental stresses and pathogens.

Chapter Review

Plants are able to sense and adaptively respond to their environments, generally by altering their patterns of growth and development.

39.1 Signal transduction pathways link signal reception to response

Plants have cellular receptors that initiate signal transduction pathways, which couple reception of a stimulus to a response. The growth pattern of a sprouting potato shoot growing in darkness, called **etiolation,** facilitates the shoot breaking ground. When the shoot reaches sunlight, stem elongation slows, leaves expand, the root system elongates, and chlorophyll production begins—all part of a process known as **de-etiolation** or greening.

Reception Signals are detected by receptors, proteins that change shape in response to a specific stimulus. A *phytochrome* is the photoreceptor involved in de-etiolation, and, unlike receptors built into the plasma membrane, it is located in the cytosol. Researchers have studied the role of phytochrome in de-etiolation using the tomato mutant *aurea,* which has lower than normal levels of phytochrome.

Transduction **Second messengers** are small molecules that amplify the signal from the receptor and transfer it to proteins that produce the specific response. Each light-activated phytochrome may lead to the production of hundreds of second messenger molecules, each of which may activate hundreds of specific enzymes. The second messenger cyclic GMP affects membrane ion channels or activates protein kinases. Phytochrome signal transduction also opens calcium channels. Increases in cytoplasmic Ca^{2+} also activate certain protein kinases.

Response The response to a signal-transduction pathway usually involves the activation of specific enzymes, either by stimulating gene expression for those enzymes or by activating existing enzymes.

Several transcription factors are activated during phytochrome-induced de-etiolation, some by cGMP and others by Ca^{2+}. Changes in gene expression may involve the activation of positive transcriptional factors or negative transcriptional factors.

Post-translational modification of existing proteins usually involves phosphorylations, catalyzed by protein kinases. Cascades of protein kinase activations may lead to the phosphorylation of transcription factors, and thus, ultimately, to changes in gene expression. Protein phosphatases are enzymes that dephosphorylate specific proteins, allowing signal pathways to be turned off when a signal is no longer present.

■ INTERACTIVE QUESTION   39.1

In the process known as de-etiolation,

a. what is the signal and the receptor?

b. Briefly describe some of the steps in the transduction of this signal.

c. What is the plant's response?

39.2   Plant hormones help coordinate growth, development, and responses to stimuli

Hormones, chemical signals that coordinate the parts of an organism, are transported through the plant body, where minute concentrations are able to trigger responses in target cells and tissues.

The Discovery of Plant Hormones   A **tropism** is a growth response of plant organs toward or away from stimuli. The growth of a shoot toward light is called positive **phototropism.** A coleoptile, enclosing the shoot of a grass seedling, bends toward the light when illuminated from one side because of the elongation of cells on the darker side.

Darwin and his son observed that a grass seedling would not bend toward light if its tip were removed or covered by an opaque cap. They postulated that a signal must be transmitted from the tip to the elongating region of the coleoptile. P. Boysen-Jensen demonstrated that the signal was a mobile substance, capable of being transmitted through a block of gelatin separating the tip from the rest of the coleoptile.

In 1926 F. Went placed coleoptile tips on blocks of agar to extract the chemical messenger. From his studies he concluded that the chemical produced in the tip, which he called auxin, promoted growth and that it was in higher concentration on the side away from the light.

Researchers have not found a light-induced asymmetrical distribution of auxin in eudicots, but certain substances that may act as growth inhibitors have been shown to be more concentrated on the lighted sides of such stems.

A Survey of Plant Hormones   Several major classes of plant hormones have been identified. These small molecules, which may move from cell to cell across cell walls, usually affect cell division, elongation, and differentiation. Depending on the site of action, the developmental stage of the plant, and relative hormone concentrations, the effect of a hormone will vary. Very low concentrations of hormones, acting through signal-transduction pathways, may affect the expression of genes, the activity of enzymes, or the properties of membranes.

Auxins include any substance that stimulates elongation of coleoptiles. The natural auxin extracted from plants is indoleacetic acid (IAA).

Auxin is transported through parenchyma tissue from the shoot tip down the shoot. This polar transport involves auxin transporters located only at the basal ends of cells.

Auxin is synthesized in the apical meristems of a shoot. According to the acid growth hypothesis, auxin initiates cell growth in the region of elongation by binding to a plasma membrane receptor and stimulating proton pumps. The proton pumps lower pH in the cell wall, activating enzymes called **expansins** that break cross-links between cellulose microfibrils. The proton pumps also increase the membrane potential, enhancing ion uptake and the resulting osmotic uptake of water. For continued growth after this relatively fast elongation, the cell must produce more cytoplasm and wall material, processes that rely on changes in gene expression that are also stimulated by auxin.

Auxin is involved in root branching and is used commercially to enhance formation of adventitious roots at the cut base of stems. Synthetic auxins, such as 2,4-D, are used as herbicides, killing eudicot (broadleaf) weeds with a hormonal overdose. Auxin stimulates cell division in the vascular cambium and differentiation of secondary xylem. Auxin produced by developing seeds promotes fruit growth; synthetic auxins can induce seedless fruit development.

In tissue culture, coconut milk and degraded DNA were found to induce plant cell growth; later, cytokinins were identified as the active ingredients. **Cytokinins** are modified forms of adenine, named because they stimulate cytokinesis. Zeatin is the most common naturally occurring cytokinin.

Cytokinins are produced in actively growing roots, embryos, and fruits. Acting with auxin, they stimulate cell division and affect differentiation.

According to the direct inhibition hypothesis, the control of apical dominance involves the interaction between auxin, transported down from the terminal bud, which restrains axillary bud development, and cytokinins, transported up from the roots, which stimulate bud growth. Several lines of evidence support this hypothesis. Biochemical analyses, however, have shown that removal of the apical bud leads to an increase in auxin levels in the axillary buds, exactly opposite the prediction of the direct inhibition hypothesis.

Cytokinins can retard aging of some plant organs, because they stimulate RNA and protein synthesis, mobilize nutrients, and inhibit protein breakdown.

In the 1930s Japanese scientists determined that the fungus *Gibberella* secreted a chemical that caused the hyperelongation of rice stems or "foolish seedling disease." More than 100 different naturally occurring gibberellins have now been identified.

Gibberellins, which are produced by roots and young leaves, stimulate growth in both leaves and stem, affecting cell division and elongation in stems. Gibberellins may promote cell elongation by stimulating cell wall-loosening enzymes, thus facilitating the penetration of expansins into the cell wall.

Gibberellin applied to dwarf plants may cause them to grow to normal height. Bolting, the growth of an elongated floral stalk, is caused by a surge of gibberellins. In many plants, both auxin and gibberellins contribute to fruit set. Gibberellins are sprayed in the production of Thompson seedless grapes. The release of gibberellins from the embryo signals seeds of many plants to break dormancy.

Similar to cholesterol and animal sex hormones, **brassinosteroids** have effects very similar to those of auxin: They promote cell elongation and division, retard leaf abscission, and promote xylem differentiation. Identification of a brassinosteroid-deficient mutant of *Arabidopsis* helped to establish these compounds as nonauxin plant hormones.

The hormone **abscisic acid (ABA)** generally slows growth. The high concentration of ABA in maturing seeds inhibits germination and stimulates production of proteins that protect the seeds during dehydration. For dormancy to be broken in some seeds, ABA must be removed or inactivated, or the ratio of gibberellins to ABA must increase.

ABA also reduces drought stress. In a wilting plant, ABA causes stomata in the leaves to close. ABA may be produced in the roots in response to water shortage and transported to the leaves.

Plants produce the gas **ethylene** in response to stress and during fruit ripening and programmed cell death. Ethylene production may be induced by a high concentration of auxin.

The mechanical stress of a seedling pushing against an obstacle as it grows upward through the soil induces the production of ethylene. Ethylene then initiates a growth pattern called the **triple response,** consisting of a slowing of stem elongation, a thickening of the stem, and initiation of horizontal growth. When the growing tip no longer detects a solid object above it, ethylene production decreases and normal upward growth resumes. Researchers have identified *Arabidopsis* mutants that are ethylene insensitive *(ein)*, ethylene overproducing *(eto)*, and that undergo the triple response in the absence of ethylene. In these latter constitutive triple response *(ctr)* mutants, the ethylene signal-transduction pathway is permanently turned on. Their mutant gene codes for a protein kinase, suggesting that the normal kinase product is a negative regulator of ethylene signal transduction. Binding of ethylene to the ethylene receptor may normally lead to the inactivation of the negative kinase, which allows the synthesis of the proteins involved in the triple response.

Apoptosis, or programmed cell death, requires the synthesis of new enzymes that break down many cellular components, which the plant may salvage. Ethylene is almost always associated with this programmed death of cells or organs, or of the entire plant.

Deciduous leaf loss protects against winter desiccation. Before leaves abscise in the autumn, many of their compounds are stored in the stem awaiting recycling to new leaves. A change in the balance of auxin and ethylene initiates changes in the abscission layer located near the base of the petiole, including the production of enzymes that hydrolyze polysaccharides in cell walls. A layer of cork forms a protective covering on the twig side of the abscission layer.

Ethylene initiates the breakdown of cell walls and conversion of starches to sugars associated with fruit ripening. In a rare example of positive feedback, ethylene triggers ripening, and ripening triggers even more ethylene production. Many commercial fruits are ripened in huge containers perfused with ethylene gas.

■ INTERACTIVE QUESTION   39.2

Fill in the name of the hormone that is responsible for each of the following functions:

a. _____ promotes fruit ripening; initiates triple response; involved in apoptosis

b. _____ stimulate cell division, growth, and germination; anti-aging

c. _____ inhibits growth; maintains dormancy; closes stomata during water stress

d. _____ stimulates stem elongation, root branching, fruit development; apical dominance

e. _____ promote cell elongation and division, xylem differentiation; retard leaf abscission

f. _____ promote stem elongation, seed germination; contributes to fruit set

Systems Biology and Hormone Interactions   Systems biology studies properties that emerge from interactions of many system elements. Biologists can identify all the genes in a plant, and microarray and proteomic techniques can indicate which genes are activated or inactivated in response to environmental stimuli or during development. A systems-based approach may allow biologists to determine how these gene products interact in producing plant responses and to predict the results of genetic manipulation.

39.3   Responses to light are critical for plant success

The effect of light on plant growth and development is called **photomorphogenesis.** An **action spectrum** graphs a physiological response across different wavelengths of light. An absorption spectrum shows the wavelengths of light a pigment absorbs. Close correlation between an action spectrum for a plant response and the absorption spectrum of a pigment may indicate that the pigment is the photoreceptor involved in the response. Red and blue light have the most influence on a plant's photomorphogenesis. **Blue-light photoreceptors** and **phytochromes** that absorb mostly red light are the two major classes of light receptors.

Blue-Light Photoreceptors   Molecular biologists have determined that plants use at least three different types of pigments to detect blue light: *cryptochrome* (for inhibition of hypocotyl elongation when a seedling breaks ground), *phototropin* (for phototropism), and *zeaxanthin* (for stomatal opening).

Phytochromes as Photoreceptors   Studies of lettuce seed germination in the 1930s determined that red light (660 nm wavelength) increased germination the most and far-red light (730 nm) inhibited germination. The effects of red and far-red light are reversible, with the seed's response determined by the last flash of light it receives.

Five different phytochromes have been identified in *Arabidopsis.* A phytochrome is a photoreceptor that consists of a protein component bonded to a nonprotein light-absorbing chromophore. The chromophore alternates between two isomers, one of which absorbs red light and the other, far-red light. These two variations of phytochrome are photoreversible; the P_r to P_{fr} interconversion acts as a switch, controlling various events in the life of the plant.

Phytochrome tells the plant that light is present by the conversion of P_r, which is the form the plant synthesizes, to P_{fr} in the presence of sunlight. P_{fr} triggers many plant responses to light, such as breaking seed dormancy.

■ INTERACTIVE QUESTION   39.3

Explain how phytochrome can also indicate the quality of light available to a plant. What effect might this have on a shaded tree?

Biological Clocks and Circadian Rhythms   A **circadian rhythm** is a physiological cycle with about a 24-hour frequency. These rhythms persist, even when the organism is sheltered from environmental cues. Research indicates that the circadian clock is endogenous, although it is set (entrained) to a 24-hour period by daily environmental signals.

■ INTERACTIVE QUESTION   39.4

What are free-running periods?

The molecular mechanism of the biological clock, which may be common to all eukaryotes, may be the cyclical changes in the concentration of a protein that is a transcription factor that inhibits the expression of its own gene.

Researchers have identified clock mutants of *Arabidopsis* by splicing the gene for luciferase to the promoter for genes for photosynthesis-related proteins that follow a circadian rhythm in their production. When the biological clock turned on the promoter for these genes, the plants glowed, and plants that glowed for a nonnormal amount of time were isolated as clock mutants. Some of these mutants had defects in proteins that normally bind photoreceptors, pointing to a light-dependent mechanism that sets the biological clock.

The Effect of Light on the Biological Clock Both blue-light photoreceptors and phytochrome can entrain the biological clock in plants. In darkness, the phytochrome ratio shifts toward P_r, in part because P_{fr} converts slowly to P_r in some plants, and also because P_{fr} is degraded and new pigment is synthesized as P_r. When the sun rises, P_r is rapidly converted to P_{fr}, resetting the biological clock each day at dawn.

Photoperiodism and Responses to Seasons Seasonal events in the life cycle of plants usually are cued by photoperiod, the relative length of night and day. A physiological response to photoperiod is called **photoperiodism.**

Researchers discovered that a variety of tobacco plant flowered only when the day length was 14 hours or shorter. They termed it a **short-day plant. Long-day plants** flower when days are longer than a certain number of hours. The flowering of **day-neutral plants** is unaffected by photoperiod.

Researchers have found that it is night length, not day length, that controls flowering and other photoperiod responses. If the dark period is interrupted by even a few minutes of light, a short-day (long-night) plant such as the cocklebur will not flower. Photoperiodic responses thus depend on a critical night length.

Red light was found to be the most effective in interrupting a plant's perception of night length. A brief exposure to red light breaks a dark period of sufficient length and prevents short-day plants from flowering, whereas a flash of red light during a dark period longer than the critical length will induce flowering in a long-day plant. Due to the photoreversibility of phytochrome, a subsequent flash of far-red light negates the effect of the red light.

Some plants bloom after a single exposure to the required photoperiod. Others respond to photoperiod only after exposure to another environmental stimulus. The need for pretreatment with cold before flowering is called **vernalization.**

Leaves detect the photoperiod. The signal for flowering, called **florigen,** travels from leaves to buds. The signal is believed to be a hormone or change in the concentration of two or more hormones, but it has yet to be identified.

In order for flowering to occur, the bud meristem must transition from a vegetative to a flowering state, a change that requires the activation of meristem-identity genes, followed by organ-identity genes. Researchers are looking for the signal-transduction pathways that link photoperiod and hormonal cues to such changes in gene expression.

■ INTERACTIVE QUESTION 39.5

Indicate whether a short-day plant (a–e) and a long-day plant (f–j) would flower or not flower under the indicated light conditions.

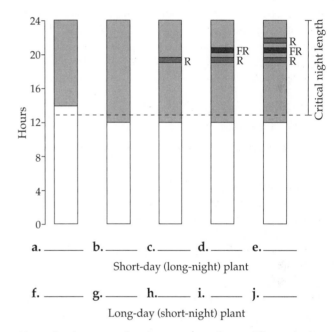

a. _____ b. _____ c. _____ d. _____ e. _____

Short-day (long-night) plant

f. _____ g. _____ h. _____ i. _____ j. _____

Long-day (short-night) plant

How do these results demonstrate the red/far-red photoreversibility of phytochrome?

39.4 Plants respond to a wide variety of stimuli other than light

Gravity Roots exhibit positive **gravitropism,** whereas shoots show negative gravitropism. According to one hypothesis, the settling of **statoliths,** plastids containing dense starch grains, in cells of the root cap triggers movement of calcium, which causes the lateral transport of auxin. Both of these accumulate on the lower side of the growing root, where the high concentration of auxin inhibits cell elongation, causing the root to curve downward. The settling of the protoplast and large organelles may distort the cytoskeleton and also signal gravitational direction.

Mechanical Stimuli The stunting of growth in height and increase in girth of plants that are exposed to wind is an example of **thigmomorphogenesis,** changes in plant form in response to mechanical stimulation. Mechanical stimuli initiate a signal-transduction pathway that results in specific gene activations that alter growth patterns and cell wall properties.

Most climbing plants have tendrils that coil rapidly around supports, exhibiting **thigmotropism** or directional growth in response to touch.

The sensitive plant *Mimosa* folds its leaves after being touched due to the rapid loss of turgor by cells in specialized motor organs called pulvini, located at the joints of the leaf. The message travels through the plant from the point of stimulation, perhaps as the result of electrical impulses, called **action potentials.** These electrical messages may be used in plants as a form of internal communication.

■ INTERACTIVE QUESTION 39.6

a. Name the types of tropisms that a stem exhibits.

b. Name and describe the growth mechanism that produces the coiling of a tendril.

Environmental Stress An **abiotic** environmental stress may be severe enough to threaten a plant's growth, reproduction, and survival. Plants also have defensive responses to **biotic** stresses such as pathogens and herbivores.

Mechanisms that reduce transpiration help a plant respond to water deficit. Guard cells lose turgor and stomata close. Abscisic acid, produced by leaves in response to water deficit, acts on guard cell membranes to keep stomata closed. Growth of young leaves is inhibited by the lack of cell-expanding water, and wilted leaves may roll up, further reducing transpiration. All of these responses, however, reduce photosynthesis. During a drought, root growth in shallow, dry soil decreases while deeper roots in moist soil continue to grow.

Plants adapted to wet habitats may have aerial roots that provide oxygen to their submerged roots. When roots of other plants are in waterlogged soils, oxygen deprivation may stimulate ethylene production, causing some root cortical cells to undergo apoptosis, opening up air tubes within the roots.

Excess salts in the soil may lower the water potential of the soil solution below that of roots, causing roots to lose water. The plasma membranes of root cells can reduce the uptake of sodium and some other ions that are toxic to plants in high concentrations. Plants may respond to moderate soil salinity by producing organic solutes that lower the water potential of root cells. Special adaptations for dealing with high soil salinity have evolved in halophytes.

Transpiration creates evaporative cooling for a plant, but this effect may be lost on hot, dry days when stomata close to reduce water loss. Above critical temperatures, plant cells produce **heat-shock proteins** that may provide temporary support to reduce protein denaturation.

Plants respond to cold stress by increasing the proportion of unsaturated fatty acids in membrane lipids in order to maintain the fluidity of cell membranes. Subfreezing temperatures cause ice to form in cell walls, lowering the extracellular water potential and causing cells to dehydrate. Plants adapted to cold winters have adaptations to deal with freezing stress, such as changing the solute composition of the cytosol.

■ INTERACTIVE QUESTION 39.7

a. A plant is exposed to a spell of very hot and dry weather. How might it survive this stress?

b. How might a plant adapt to an unusually cold and very wet fall?

39.5 Plants defend themselves against herbivores and pathogens

Defenses Against Herbivores Physical defenses, such as thorns, and chemical defenses, such as distasteful or toxic compounds, help plants cope with herbivory. Some plants produce *canavanine*, which resembles arginine and may be incorporated into an insect's proteins, altering protein conformation and causing death.

Some plants recruit predators of their herbivores. A combination of a compound in a caterpillar's saliva and the damage caused by its eating stimulates leaves to release volatile compounds that attract parasitoid wasps. The wasps lay their eggs inside the caterpillar, and the wasp larvae eat their host. These volatile compounds may also signal neighboring plants to activate defense genes. These gene activations are similar in pattern to those produced by exposure to the important plant defense molecule **jasmonic acid.**

Defenses Against Pathogens The physical barrier of the plant's epidermis and/or periderm is the first line of defense against pathogenic viruses, bacteria, and fungi. Should pathogens enter the plant due to injuries or through openings such as stomata, the plant mounts a chemical defense.

Virulent pathogens are those to which the plant has no specific defense. Plants are generally resistant to most pathogens. These **avirulent** pathogens do not extensively harm or kill the plant.

Specific resistance is based on **gene-for-gene recognition** between the protein products of a plant's disease resistance (*R*) genes and a pathogen molecule coded for by an *Avr* gene. According to the receptor-ligand model, binding of an AVR protein to an R protein receptor triggers a signal-transduction pathway that produces a plant defense response. Plants have many different *R* genes, corresponding to the many different potential pathogens.

In response to chemical signals released from plant cells damaged by a pathogen, the plant mounts a localized attack. **Elicitors,** often cellulose fragments called **oligosaccharins,** stimulate the production of antimicrobial **phytoalexins.** Infection also activates genes for **PR proteins** (some of which are antimicrobial, while others serve as alarm signals to neighboring cells) and stimulates strengthening of the cell walls to slow the spread of the pathogen. A **hypersensitive response (HR)** is triggered when the pathogen is avirulent and there is an *R-Avr* match. This more vigorous defense includes increased production of phytoalexins and PR proteins and a more effective sealing of the area.

A hypersensitive response also triggers release of alarm hormones, which stimulate phytoalexin and PR protein production throughout the plant, creating a **systemic acquired resistance (SAR).** This generalized defense response helps protect uninfected tissue. One of the hormones involved in activating SAR is probably **salicylic acid,** which humans have modified as the active ingredient in aspirin.

Word Roots

aux- = grow, enlarge (*auxins:* a class of plant hormones, including indoleacetic acid, having a variety of effects, such as phototropic response through the stimulation of cell elongation, stimulation of secondary growth, and the development of leaf traces and fruit)

circ- = a circle (*circadian rhythm:* a physiological cycle of about 24 hours, present in all eukaryotic organisms, that persists even in the absence of external cues)

crypto- = hidden; **-chromo** = color (*cryptochrome:* the name given to the unidentified blue-light photoreceptor)

cyto- = cell; **-kine** = moving (*cytokinins:* a class of related plant hormones that retard aging and act in concert with auxins to stimulate cell division, influence the pathway of differentiation, and control apical dominance)

gibb- = humped (*gibberellins:* a class of related plant hormones that stimulate growth in the stem and leaves, trigger the germination of seeds and breaking of bud dormancy, and stimulate fruit development with auxin)

hyper- = excessive (*hypersensitive response:* a vigorous, localized defense response to a pathogen that is avirulent based on an *R-Avr* match)

photo- = light; **-trop** = turn, change (*phototropism:* growth of a plant shoot toward or away from light)

phyto- = a plant; **-alexi** to ward off (*phytoalexin:* an antibiotic, produced by plants, that destroys microorganisms or inhibits their growth)

stato- = standing, placed; **-lith** = a stone (*statolith:* specialized plastids that help a plant tell up from down)

thigmo- = a touch; **morpho-** = form; **-genesis** = origin (*thigmomorphogenesis:* a response in plants to chronic mechanical stimulation, resulting from increased ethylene production; an example is thickening stems in response to strong winds)

zea- = a grain; **-xantho** = yellow (*zeaxanthin:* a blue-light photoreceptor involved in stomatal opening)

Structure Your Knowledge

1. List some agricultural uses of plant hormones.

2. Develop a concept map to illustrate your understanding of photoperiodism and the control of flowering. Do not forget to include the role of the biological clock.

3. Briefly describe the steps in the hypersensitive response and systemic acquired resistance that result from a plant's encounter with an avirulent pathogen.

▶ Test Your Knowledge

TRUE OR FALSE: *Indicate T or F and then correct the false statements.*

_____ 1. Abscisic acid is necessary for a seed to break dormancy.

_____ 2. Roots exhibit negative phototropism and positive gravitropism.

_____ 3. Gibberellins, synthesized in the root, counteract apical dominance.

_____ 4. Thigmomorphogenesis is a growth response of a seedling that encounters an obstacle while pushing upward through the soil.

_____ 5. The application of gibberellins or auxin may induce the development of seedless fruit.

_____ 6. Action potentials are electrical impulses that may be used as internal communication in plants.

_____ 7. A physiological response to day or night length is called a circadian rhythm.

_____ 8. Vernalization is the need for pretreatment with cold before seed germination.

_____ 9. A virulent pathogen is one to which a plant has a specific resistance based on gene-for-gene recognition.

_____ 10. A cryptochrome is the light-absorbing portion of a phytochrome that reverts between two isomeric forms.

MULTIPLE CHOICE: *Choose the one best answer.*

1. The body form within a species of plants may vary more than that within a species of animals because
   a. growth in animals is indeterminate.
   b. plants respond adaptively to their environments by altering their patterns of growth and development.
   c. plant growth and development are governed by many hormones.
   d. plants can respond to environmental stress.
   e. all of the above are true.

2. Polar transport of auxin involves
   a. the accumulation of higher concentrations on the side of a shoot away from light.
   b. the movement of auxin from roots to shoots.
   c. movement of auxin ions through carrier proteins located at the basal end of cells.
   d. the unidirectional active transport of auxin into and out of parenchyma cells.
   e. the settling of statoliths.

3. According to the acid growth hypothesis,
   a. auxin stimulates membrane proton pumps.
   b. a lowered pH outside the cell activates expansins that break cross-links between cellulose microfibrils.
   c. the membrane potential created by the proton pumps enhances ion uptake, increasing the osmotic movement of water into cells.
   d. cells elongate when they take up water by osmosis.
   e. all of the above are involved in cell elongation.

4. Which of the following situations would most likely stimulate the development of axillary buds?
   a. a large quantity of auxin traveling down from the shoot and a small amount of cytokinin produced by the roots
   b. a small amount of auxin traveling down from the shoot and a large amount of cytokinin traveling up from the roots
   c. an equal ratio of gibberellins to auxin
   d. the absence of cytokinins caused by the removal of the terminal bud
   e. an increase in the concentration of brassinosteroids produced by leaves

5. The growth inhibitor in seeds is usually
   a. abscisic acid.
   b. ethylene.
   c. gibberellin.
   d. a small amount of ABA combined with a larger concentration of gibberellins.
   e. a high cytokinin-to-auxin ratio.

6. A circadian rhythm
   a. is controlled by an external oscillator.
   b. is a physiological cycle of approximately a 24-hour frequency.
   c. involves an internal biological clock that is not set by daily environmental signals.
   d. provides the signal for seasonal flowering.
   e. involves all of the above.

7. Which of the following is *not* a component of the signal-transduction pathway involved in the deetiolation process when a shoot breaks ground?
   a. reception of light by a phytochrome located in the cytosol
   b. inhibition of the triple response and initiation of phototropism as the shoot bends toward light
   c. production of second messengers such as cGMP and Ca^{2+}
   d. cascades of protein kinase activations that phosphorylate transcription factors
   e. activation of genes that code for photosynthesis-related enzymes

8. Injecting phytochrome into cells of the tomato mutant *aurea*
   a. causes the plant to undergo the triple response above ground when exposed to ethylene.
   b. causes the plant to undergo the triple response above ground without exposure to ethylene.
   c. helps researchers identify clock mutants.
   d. initiates flowering even when nights are longer than a critical period.
   e. causes the plant to develop a normal de-etiolation response to light.

9. A flash of far-red light during a critical-length dark period
   a. will induce flowering in a long-day plant.
   b. will induce flowering in a short-day plant.
   c. will not influence flowering.
   d. will increase the P_{fr} level suddenly.
   e. will be negated by a second flash of far-red light.

10. The conversion of P_r to P_{fr}
    a. occurs slowly at night.
    b. may be the way in which plants sense daybreak and serves to entrain the biological clock.
    c. is the molecular mechanism responsible for the biological clock.
    d. occurs when P_r absorbs far-red light.
    e. occurs in flower buds and induces flowering.

11. A plant may withstand salt stress by
    a. releasing abscisic acid that closes stomata to salt accumulation.
    b. the production of organic solutes that lower the water potential of root cells.
    c. wilting, which reduces water and salt uptake by reducing transpiration.
    d. producing canavanine that reduces the toxic effect of sodium ions.
    e. releasing ethylene that leads to apoptosis of damaged cells.

12. Which of the following hormones would be sprayed on barley seeds to speed germination in the production of malt for making beer?
    a. abscisic acid
    b. auxin
    c. cytokinin
    d. ethylene
    e. gibberellin

13. Many plants will flower in response to a specific
    a. flowering hormone that is produced in the apical bud.
    b. elicitor, which is a cellulose fragment.
    c. minimal temperature that appears to signal a seasonal change.
    d. photoperiod, which seems to be measured by the length of darkness to which the leaves of the plant are exposed.
    e. combination of a high level of auxin and a low level of cytokinin.

14. Which of the following is *not* a plant defense against herbivory?
    a. production of distasteful compounds.
    b. production of toxic compounds such as canavanine.
    c. physical defenses such as thorns.
    d. initiation of a hypersensitive response with production of phytoalexins and PR proteins.
    e. release of volatile compounds that recruit parasitoid wasps.

15. Which of the following is *not* true of the hypersensitive response?
    a. It relies on an *R-Avr* recognition between a specific plant cell receptor protein and a pathogen molecule.
    b. It increases the production of PR proteins that may have antimicrobial or signaling functions.
    c. It enhances the production of ethylene that serves as a signal molecule transported throughout the plant to activate SAR (systemic acquired resistance).
    d. It contains an infection by stimulating cross-linking of cell wall molecules and production of lignin.
    e. The plant cells involved in the defense destroy themselves, leaving lesions that indicate the site of the contained infection.

UNIT 7

Animal Form and Function

Chapter 40

Basic Principles of Animal Form and Function

Framework

Compact animal bodies have highly folded exchange surfaces and a circulatory system that distributes materials throughout the body. The body structures of an animal are built from four basic tissues: epithelial, connective, muscle, and nervous. Organs function together in organ systems. Structure correlates with function in these hierarchical levels of organization.

The functions of an animal are powered by chemical energy derived from food. Metabolic rate, the amount of energy used in a unit of time, is higher for endothermic animals and inversely related to body size.

The internal environment is carefully regulated by mechanisms of homeostasis. Ectotherms may use behavioral mechanisms to control body temperature. Thermoregulatory mechanisms of endotherms include insulation, circulatory adaptations, evaporative heat loss, behavioral responses, and regulation of metabolic heat production.

Chapter Review

The comparative study of animals illustrates several key biological themes: evolution, regulation, bioenergetics, and the correlation of structure and function. **Anatomy** is the study of an organism's structure; **physiology** is the study of function.

40.1 Physical laws and the environment constrain animal size and shape

Physical Laws and Animal Form Physical laws constrain the evolution of an animal's body plan or design. The laws of hydrodynamics drove the natural selection of a fusiform body shape for both fast-swimming fishes and mammals, an example of convergent evolution.

Exchange with the Environment   Every cell must be in an aqueous medium to maintain the integrity of the plasma membrane and allow for exchange across it. Single-celled organisms or animals with two-layered, saclike bodies or thin, flat bodies can maintain sufficient cellular contact with the aqueous environment.

Animals with complex bodies, however, must provide extensively branched or folded internal membranes for exchanging materials with the environment. The circulatory system connects these exchange surfaces with the aqueous environment bathing the body's cells.

■ INTERACTIVE QUESTION 40.1

List some advantages of a compact, complex body form in which most of the body cells are not in contact with the external environment.

40.2 Animal form and function are correlated at all levels of organization

Hierarchical levels of organization characterize life. Multicellular organisms have specialized cells grouped into tissues, which may be combined into organs. Various organs may function together in organ systems.

Tissue Structure and Function   **Tissues** are collections of cells with a common structure and function. Tissues are classified into four categories.

Epithelial tissue lines the outer and inner surfaces of the body in protective sheets of tightly packed cells. Cells at the base of an epithelium are attached to a **basement membrane,** a dense layer of extracellular matrix. A **simple epithelium** has one layer, whereas a **stratified epithelium** has multiple layers of cells. The shape of cells at the free surface may be **squamous** (flat), **cuboidal** (boxlike), or **columnar** (pillarlike).

Glandular epithelia are specialized for absorption or secretion. **Mucous membranes** lining the digestive and respiratory tracts secrete mucus. Cilia on the epithelium lining the air passages sweep particles trapped in mucus away from the lungs.

■ INTERACTIVE QUESTION 40.2

Name the two types of epithelium illustrated below. One of these epithelial types lines the esophagus, and the other lines the digestive tract. Explain why each would be found in its location.

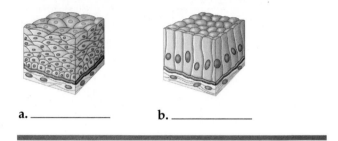

a. _____   b. _____

Connective tissue connects and supports other tissues and is characterized by relatively few cells suspended in an extracellular matrix of fibers, which may be embedded in a liquid, jellylike, or solid substance.

Connective tissue fibers are of three types. **Collagenous fibers** are made of collagen and have nonelastic strength. **Elastic fibers,** made of the protein elastin, can stretch and provide resilience. Branched and thin **reticular fibers** are composed of collagen and form a tightly woven connection with adjacent tissues.

Loose connective tissue, made of loosely woven fibers, attaches epithelia to underlying tissues and holds organs in place. The most common types of cells enmeshed in loose connective tissue are **fibroblasts,** which secrete the protein of the extracellular fibers, and **macrophages,** amoeboid cells that engulf bacteria and cellular debris.

Adipose tissue is a special form of loose connective tissue that pads and insulates the body and stores fat. Adipose cells each contain a large fat droplet.

Fibrous connective tissue, with its dense arrangement of parallel collagenous fibers, is found in **tendons,** which attach muscles to bones, and in **ligaments,** which join bones together at joints.

Cartilage is composed of collagenous fibers embedded in a rubbery substance called chondroitin sulfate, both secreted by **chondrocytes.** Cartilage is a strong but somewhat flexible support material, making up the skeleton of vertebrate embryos.

Bone is a mineralized connective tissue formed by **osteoblasts** that deposit a matrix of collagen and calcium, magnesium, and phosphate ions, which hardens into hydroxyapatite. **Osteons** (Haversian systems) consist of concentric layers of matrix around a central canal containing blood vessels and nerves.

Blood is a connective tissue that has a liquid extracellular matrix called plasma. Erythrocytes (red blood cells) carry oxygen; leukocytes (white blood cells) function in defense; and cell fragments called platelets are involved in the clotting of blood.

Nervous tissue senses stimuli and transmits electrical signals. The **neuron**, or nerve cell, consists of a cell body and two or more processes that conduct impulses toward (dendrites) and away from (axons) the rest of the neuron.

■ INTERACTIVE QUESTION 40.3

Identify the types of connective tissue and their components in the following three micrographs.

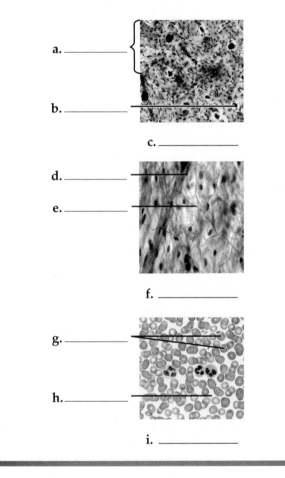

a. _____

b. _____

c. _____

d. _____

e. _____

f. _____

g. _____

h. _____

i. _____

■ INTERACTIVE QUESTION 40.4

Identify these types of vertebrate muscle. What are the dark bands at the end of the line in figure **a,** and what is their function?

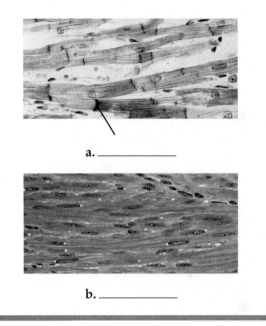

a. _____

b. _____

Muscle tissue consists of long, contractile cells called muscle fibers that are packed with myofibrils of actin and myosin. **Skeletal muscle**—also called **striated muscle**—is responsible for voluntary body movements. **Cardiac muscle**, forming the wall of the heart, is also striated, but its cells are branched and joined by intercalated discs. **Smooth muscle** is composed of spindle-shaped cells lacking striations. It is found in the walls of the digestive tract, arteries, and other internal organs, and is under involuntary control.

Organs and Organ Systems In all animals but sponges and some cnidarians, tissues are organized into **organs**. Organs often consist of a layered arrangement of tissues. Many vertebrate organs are suspended by **mesenteries** in moist or fluid-filled body cavities. Mammals have a **thoracic cavity** separated by a muscular diaphragm from an **abdominal cavity.**

Groups of organs are integrated into **organ systems,** which perform the major functions required for life. The organ systems are coordinated to create the functional integration essential to an organism.

■ INTERACTIVE QUESTION 40.5

There are 11 organ systems in mammals. How many of them can you name?

40.3   Animals use the chemical energy in food to sustain form and function

Bioenergetics   An animal's **bioenergetics,** its use of energy resources, determines its behavior, growth, reproduction, and food needs. The fuel molecules obtained from the digestion of food are absorbed into body cells and used to generate ATP for cellular work and as carbon skeletons for biosynthesis.

The total energy an animal uses in a unit of time is its **metabolic rate.** Energy is measured in calories (cal) or kilocalories (kcal). Metabolic rate can be measured by placing an animal in a calorimeter and measuring heat loss. The rate of oxygen consumption, also a measure of metabolic rate, can be determined with a respirometer.

Birds and mammals are **endothermic,** warming their bodies with metabolic heat. This high-energy bioenergetic strategy allows for high activity levels over a range of environmental temperatures. **Ectothermic** animals, such as most fishes, amphibians, reptiles other than birds, and invertebrates, require less energy and gain heat from external sources.

Influences on Metabolic Rate   The energy required to maintain each gram of body weight is inversely related to body size. Smaller animals have higher metabolic rates, breathing rates, volume of blood, and heart rates, and require more food per unit of body mass. With a greater surface-to-volume ratio, small endotherms may have a higher energy cost to maintain a stable body temperature. But the factors that contribute to this inverse relationship among ectotherms as well as endotherms are not fully understood.

The minimal metabolic rate for a nongrowing endotherm at rest, fasting, and nonstressed is called the **basal metabolic rate (BMR).** A human adult male's rate is about 1,600–1,800 kcal/day; a female's is about 1,300–1,500 kcal/day. The **standard metabolic rate (SMR)** is the metabolic rate of a resting, fasting, nonstressed ectotherm determined at a specific temperature.

Maximal metabolic rates occur during intense activity and are inversely related to the duration of the activity. Metabolic rates are influenced by age, sex, size, activity level, time of day, and other variables. Most terrestrial animals have an average daily rate of energy consumption that is 2–4 times BMR or SMR.

■ INTERACTIVE QUESTION   40.6

Why can an endotherm sustain intense activity for a much longer time period than an ectotherm can?

Energy Budgets   Different species vary in the allocation of their food energy for BMR or SMR, activity, growth, reproduction, and temperature regulation. For most animals, the majority of food is used to produce ATP, with little going to growth and reproduction. Endotherms require much more energy per kilogram than do ectotherms; and a small animal requires more energy per kilogram than does a large animal of the same taxonomic class.

40.4   Many animals regulate their internal environment within relatively narrow limits

The internal environment of vertebrates is the **interstitial fluid** surrounding the cells, through which oxygen, nutrients, and wastes are exchanged with blood in capillaries. The "steady state" or dynamic internal balance of this environment is called **homeostasis.**

Regulating and Conforming   For any given environmental variable, an animal may be a **regulator,** using homeostatic mechanisms to control internal fluctuations, or a **conformer,** allowing internal conditions to vary with certain environmental changes.

Mechanisms of Homeostasis   The mechanisms by which animals maintain homeostasis involve a *receptor* that detects a change in the internal environment and a *control center* that processes information and directs an *effector* to respond. Most homeostatic control mechanisms operate by **negative feedback.** When some variable moves above or below a set point, a control mechanism is turned on or off to counteract further change in that direction.

Positive feedback in a physiological function is a mechanism in which a change in a variable serves to amplify rather than reverse the activity. The stimulation of uterine contractions during childbirth is an example.

Homeostatic mechanisms allow for *regulated change* in the body's internal environment when necessary.

40.5   Thermoregulation contributes to homeostasis and involves anatomy, physiology, and behavior

Changes in temperature affect most biochemical and physiological processes. Animals maintain internal temperatures within an optimal range through the process of **thermoregulation.**

Ectotherms and Endotherms   The use of metabolic heat in regulating body temperature differentiates **endotherms** (birds, mammals, some fishes, a few other reptiles, and some insects) from **ectotherms** (most invertebrates, fishes, amphibians, lizards, snakes, and turtles). The environmental temperature usually determines the body temperature of an ectotherm because its low metabolic rate does not generate much heat. The high metabolic rate of an endotherm contributes enough heat to maintain a body temperature warmer than the environment.

Poikilotherm refers to animals with varying internal temperatures; homeotherm refers to animals with stable internal temperatures. These terms are no longer commonly used.

Endotherms can sustain vigorous activity because their high body temperature produces a high aerobic metabolism. Endothermy also facilitates functioning within the more severe temperature fluctuations found on land. The advantages of endothermy, however, are quite energetically expensive.

■ INTERACTIVE QUESTION 40.7

Explain the following statement: Some ectotherms have body temperatures that are more constant, and some even have higher body temperatures, than those of endotherms.

Modes of Heat Exchange The flow of heat within an organism and between an organism and the external environment is affected by the physical processes of **conduction,** the direct transfer of thermal motion between objects in contact; **convection,** the transfer of heat by the flow of air or water past a surface; **radiation,** the emission of electromagnetic waves by all objects warmer than absolute zero; and **evaporation,** the loss of heat due to the conversion of the surface molecules of a liquid to a gas.

Balancing Heat Loss and Gain Five categories of adaptations enable endotherms and those ectotherms that thermoregulate to balance their heat budgets.

Insulation reduces the rate of heat exchange with the environment. Hair, feathers, and fat layers have evolved in mammals and birds. Fur and feathers insulate mainly by trapping still air. A very thick layer of blubber insulates marine mammals. The mammalian **integumentary system** consists of skin, hair, and nails. The skin serves both thermoregulatory and protective functions. The epidermis consists mostly of dead epithelial cells. The dermis contains hair follicles, oil and sweat glands, muscles, nerves, and blood vessels. The hypodermis has adipose tissue and blood vessels.

Circulatory adaptations contribute to thermoregulation. **Vasodilation** of superficial blood vessels in-

creases blood flow in the skin and the transfer of body heat to the environment. **Vasoconstriction** reduces blood flow and heat transfer. In a **countercurrent heat exchanger,** common in marine mammals and birds, the close association of blood vessels servicing the extremities allows heat in arterial blood leaving the body core to be transferred to returning venous blood. Some reptiles can conserve heat by vasoconstriction. Many endothermic insects maintain high temperatures for the flight muscles in the thorax with a countercurrent heat exchanger.

■ INTERACTIVE QUESTION 40.8

What adaptation allows large, powerful swimming fish such as bluefin tuna, swordfish, and the great white shark to be endothermic?

Mammals and birds may also use the cooling mechanism of evaporative heat loss. Evaporation across the skin and respiratory surface can prevent body temperatures from rising when environmental temperatures are above body temperature. Panting, sweating, and bathing may enhance evaporative cooling. Some amphibians vary their mucus secretion to affect evaporative cooling.

Behavioral responses contribute to thermoregulation for both endotherms and ectotherms. Amphibians easily lose heat by evaporation from their moist skin. Behavioral adaptations usually enable them to maintain internal temperatures within their particular optimal range. Reptiles other than birds often thermoregulate by orienting the body to the sun or finding suitable microclimates. Many terrestrial invertebrates use similar behavioral mechanisms. The social organization of honeybees enables them to retain heat in cold temperatures by huddling together and to cool their hive during hot weather by bringing in water and fanning their wings.

Mammals and birds are able to regulate the rate of metabolic heat production through activity and shivering. Some mammals generate heat through **nonshivering thermogenesis (NST),** a rise in metabolic rate and the production of heat instead of ATP. **Brown fat** is specialized for heat production by NST.

A few large reptiles use shivering to generate heat and become endothermic while incubating eggs. Endothermic flying insects may warm up before taking off by contracting their powerful flight muscles.

Feedback Mechanisms in Thermoregulation The complex system of temperature regulation in mammals depends on feedback mechanisms. Both warm and cold temperature-sensing nerve cells are located in the skin, hypothalamus, and some other body regions. A group of neurons in the hypothalamus functions as the thermostat.

■ INTERACTIVE QUESTION   40.9

a. What cooling mechanisms are activated when the hypothalamus senses a body temperature above the set point?

b. What mechanisms are activated when body temperature falls below the set point?

Adjustment to Changing Temperatures Many animals are capable of **acclimatization,** a physiological adjustment to a different temperature range.

Ectotherms and endotherms acclimatize in different ways. Birds and mammals adjust the thickness of their insulating coats or vary their capacity for metabolic heat production to meet seasonal changes. Ectotherms must adjust their cellular physiology to acclimatize to changes in environmental and thus body temperature. They may produce more enzymes to compensate for reduced enzyme activity or enzymes with different temperature optima, or change the proportions of saturated and unsaturated lipids to help maintain membrane fluidity. Ectotherms in subzero environments may produce cryoprotectants to prevent ice formation in their cells.

Cells can make rapid adjustments to temperature increases and other stresses by producing **stress-induced proteins,** including **heat-shock proteins.** These molecules protect proteins that might otherwise be denatured and may prevent cell death when an organism faces severe environmental changes.

Torpor and Energy Conservation **Torpor** is a physiological state characterized by decreases in metabolism and activity. In **hibernation,** the body temperature is maintained at a lower level, greatly conserving energy and allowing the animal to withstand long periods of cold temperature and decreased food supply. A Belding's ground squirrel reduces its energy needs from

150 to an average of 5–8 kcal per day during its eight-month hibernation.

In **estivation,** animals survive long stretches of elevated temperature and diminished water supply by entering a period of inactivity and lowered metabolism. Hibernation and estivation may be triggered by seasonal changes in the length of daylight.

Many small mammals and birds with very high metabolic rates enter a **daily torpor** controlled by the biological clock during periods when they are not feeding.

Word Roots

chondro- = cartilage; **-cyte** = cell (*chondrocytes:* cartilage cells)

con- = with; **-vect** = carried (*convection:* the mass movement of warmed air or liquid to or from the surface of a body or object)

counter- = opposite (*countercurrent heat exchanger:* a special arrangement of blood vessels that helps trap heat in the body core and is important in reducing heat loss in many endotherms)

-dilat = expanded (*vasodilation:* an increase in the diameter of superficial blood vessels triggered by nerve signals that relax the muscles of the vessel walls)

ecto- = outside; **-therm** = heat (*ectotherm:* an animal, such as a reptile, fish, or amphibian, that must use environmental energy and behavioral adaptations to regulate its body temperature)

endo- = inner (*endotherm:* an animal, such as a bird or mammal, that uses metabolic energy to maintain a constant body temperature)

fibro- = a fiber (*fibroblast:* a type of cell in loose connective tissue that secretes the protein ingredients of the extracellular fibers)

homeo- = same; **-stasis** = standing, posture (*homeostasis:* the steady-state physiological condition of the body)

inter- = between (*interstitial fluid:* the internal environment of vertebrates, consisting of the fluid filling the space between cells)

macro- = large (*macrophage:* an amoeboid cell that moves through tissue fibers, engulfing bacteria and dead cells by phagocytosis)

osteo- = bone; **-blast** = a bud, sprout (*osteoblasts:* bone-forming cells that deposit a matrix of collagen)

Structure Your Knowledge

1. Fill in the table below on the structure and function of the four types of animal tissues.

Tissue	Structural Characteristics	General Functions	Specific Examples

2. Organs are composed of layers of several different tissues. Which of the four animal tissues do you think would be included in all organs? In what types of organs would you predict the remaining tissues would be included? Give an example of an organ composed of several layers of tissues.

3. Explain this statement: Endotherms can tolerate wider external fluctuations in temperature; ectotherms may be able to tolerate wider internal temperature fluctuations.

Test Your Knowledge

MULTIPLE CHOICE: *Choose the one best answer.*

1. Which of the following is *not* an organ system?
   a. skeletal
   b. connective
   c. digestive
   d. excretory
   e. immune and lymphatic

2. A stratified squamous epithelium would be composed of
   a. several layers of cells, the outer one being flat, attached to a basement membrane.
   b. a layer of ciliated, mucus-secreting, flattened cells.
   c. a hierarchical arrangement of boxlike cells.
   d. an irregularly arranged layer of pillarlike cells.
   e. several layers of flat cells beneath columnar cells.

3. Which of the following is *not* true of connective tissue?
   a. It consists of few cells surrounded by fibers in a matrix.
   b. It includes such diverse tissues as bone, cartilage, tendons, adipose, and loose connective tissue.
   c. It connects and supports other tissues.
   d. It forms the internal and external lining of many organs.
   e. It can have a matrix that is a liquid, gel, or solid.

4. Which of the following are *incorrectly* paired?
   a. blood—erythrocytes, leukocytes, and platelets in plasma
   b. bone—osteocytes embedded in hydroxyapatite in osteon
   c. loose connective tissue—collagenous, elastic, reticular fibers
   d. adipose tissue—loose connective tissue with fat-storing cells
   e. fibrous connective tissue—chondrocytes embedded in chondroitin sulfate

5. Which of the following is the best description of smooth muscle?
   a. striated, branching cells; involuntary control
   b. spindle-shaped cells; involuntary control
   c. spindle-shaped cells connected by intercalated disks; voluntary control
   d. striated cells containing overlapping filaments; involuntary control
   e. spindle-shaped striated cells; voluntary control

6. The diaphragm
    a. is a mesentery.
    b. increases the surface area of the lungs.
    c. is part of the mammalian reproductive system.
    d. separates the thoracic and abdominal cavities in mammals.
    e. separates the lungs from the heart cavity.

7. The interstitial fluid of vertebrates
    a. is the internal environment within cells.
    b. bathes cells and provides for the exchange of nutrients and wastes.
    c. makes up the plasma of blood.
    d. surrounds unicellular and flat, thin animals.
    e. is less abundant in ectotherms than in endotherms.

8. Negative feedback circuits are
    a. mechanisms that most commonly maintain homeostasis.
    b. activated only when a physiological variable rises above a set point.
    c. analogous to a radiator that heats a room.
    d. involved in maintaining contractions during childbirth.
    e. found in endotherms but not in ectotherms.

9. The basal metabolic rate
    a. is constant for each species.
    b. may vary depending on the sex or size of an organism.
    c. is highest when an animal is actively exercising.
    d. is lower than the standard metabolic rate for ectotherms.
    e. may be measured from the quantity of food an animal eats after exercising.

10. Which of the following would most likely have the highest metabolic rate?
    a. whale
    b. dog
    c. snake
    d. tuna
    e. bat

11. Dolphins, penguins, sharks, and seals all have a fusiform body shape. What is the best explanation for this similarity?
    a. This shape is an example of divergent evolution.
    b. They are all vertebrates and evolved from a common ancestor.
    c. The physical laws of hydrodynamics drove the natural selection for this shape.
    d. All aquatic animals have this shape, an example of convergent evolution.
    e. This shape is determined by the thermodynamics of maintaining body temperature in cold water.

12. Of the following animals, whose percentage of the energy budget available for growth would be the largest?
    a. deer mouse from temperate forest
    b. penguin from Antarctica
    c. human from temperate climate
    d. python from tropical India
    e. They would all be the same, because most energy is used to maintain their basic metabolic rate.

13. Which of the following is used as a mechanism to dissipate heat?
    a. hibernation
    b. countercurrent heat exchange between vessels that service the extremities
    c. raising fur or feathers
    d. vasodilation of surface vessels
    e. vasoconstriction of surface vessels

14. The rate of metabolic heat production can be increased by
    a. nonshivering thermogenesis.
    b. storage of brown fat.
    c. vasoconstriction.
    d. thick layers of blubber and countercurrent heat exchangers.
    e. all of the above.

15. The function of stress-induced proteins is to
    a. provide proteins that function at a lower temperature.
    b. change the composition of plasma membranes to maintain fluidity.
    c. prevent ice formation and protect cells from freezing.
    d. regulate daily torpor in small mammals with high metabolic rates.
    e. protect cellular proteins during rapid temperature increases.

16. Which of these animals would be likely to experience a daily torpor?
    a. bear
    b. hummingbird
    c. lizard
    d. Belding's ground squirrel
    e. crow

17. Which of the following would be *least* true of an animal that is a regulator?
    a. It can live in a variable climate because of its homeostatic mechanisms.
    b. It may have a larger geographic range than a conformer.
    c. Much of its energy budget can be allocated to reproduction.
    d. It may acclimatize to winter by increasing the thickness of its insulating coat.
    e. It has behavioral as well as physiological mechanisms for responding to changing conditions.

18. For most terrestrial animals, the average daily rate of energy consumption is 2–4 times BMR or SMR. Which of the following animals has an unusually low average daily metabolic rate?
    a. human in developed country
    b. wild turkey
    c. domesticated turkey
    d. forest-dwelling salamander
    e. desert snake

19. Which of the following is *not* descriptive of endotherms?
    a. ability to sustain vigorous activity
    b. efficient circulatory and gas exchange systems
    c. high metabolic rate and retained metabolic heat
    d. energetically efficient and relatively inexpensive
    e. ability to deal with terrestrial temperature fluctuations

20. Which of the following would be likely to produce cryoprotectants?
    a. polar bear or artic wolf
    b. hibernating mammal
    c. estivating amphibian
    d. ectotherm in subzero environment
    e. chickadee in very cold winter

Chapter 41

Animal Nutrition

41.1 Homeostatic mechanisms manage an animal's energy budget

41.2 An animal's diet must supply carbon skeletons and essential nutrients

41.3 The main stages of food processing are ingestion, digestion, absorption, and elimination

41.4 Each organ of the mammalian digestive system has specialized food-processing functions

41.5 Evolutionary adaptations of vertebrate digestive systems are often associated with diet

Framework

Animals eat other organisms to obtain fuel for respiration, organic raw materials for biosynthesis, and essential nutrients. Digestion is the enzymatic hydrolysis of macromolecules into monomers that can be absorbed across cell membranes. A nutritionally adequate diet provides sufficient calories, essential amino acids, vitamins, and minerals.

Gastrovascular cavities are digestive sacs with only one opening, in which some extracellular digestion takes place. Alimentary canals are one-way tracts with specialized regions for mechanical breakdown of food, storage, digestion, absorption of nutrients, and elimination of wastes.

This chapter details the structures, functions, enzymes, and hormones of the human digestive tract.

Chapter Review

Herbivores eat autotrophs; **carnivores** eat animals; and **omnivores** consume both autotrophs and animals. Most animals are opportunistic and eat foods from different categories when they are available.

Many aquatic animals are **suspension feeders,** sifting small food particles from the water. **Substrate feeders** live in or on their food, eating their way through it. **Fluid feeders** suck fluids from a living plant or animal host. Most animals are **bulk feeders,** eating relatively large pieces of food.

A nutritionally adequate diet provides chemical energy for cellular work, raw materials for biosynthesis, and essential nutrients in prefabricated form.

41.1 Homeostatic mechanisms manage an animal's energy budget

The monomers of carbohydrates, fats, and proteins can be used as fuel to produce ATP by cellular respiration, although the first two are used preferentially.

Glucose Regulation as an Example of Homeostasis When an animal consumes more calories than are needed to meet its energy requirements, the excess can be used for biosynthesis, or stored in the liver and muscles as glycogen or in adipose tissue as fat when the glycogen stores are full. The level of glucose in the blood is carefully regulated by the pancreatic hormones insulin, which lowers blood glucose, and glucagon, which raises blood glucose.

Caloric Imbalance An **undernourished** person or other animal has a diet insufficient in calories. With severe deficiency, the body breaks down its own proteins for energy, eventually causing irreversible damage.

Overnourishment, or obesity, is becoming increasingly common in affluent nations. Excess fat calories in the diet are readily converted into fat stores.

Obesity is a major global health problem; it is associated with a number of health problems. In the United States, 30% of people are obese (*very* overweight) and 35% are overweight.

Appetite-regulating hormones act on a region of the brain that regulates the "satiety center," which generates impulses that make one feel hungry or satiated. These hormones include ghrelin, which is secreted by the stomach and triggers hunger, and three hormones that suppress appetite: insulin from the pancreas, PHY from the small intestine, and leptin produced by adipose tissue. The human body seems to have complex feedback mechanisms that stabilize weight. Inheritance plays a role in obesity, and defects in weight-controlling mechanisms may someday be treated with new drugs.

Sometimes obesity is adaptive, as in the obese chicks of some petrel species.

■ **INTERACTIVE QUESTION   41.1**

How might the human craving for fatty foods, which is helping to fuel the obesity crisis, have evolved through natural selection?

41.2 An animal's diet must supply carbon skeletons and essential nutrients

Animals can fabricate most of the organic molecules they need from the carbon skeletons and organic nitrogen acquired from food.

Molecules that an animal requires but cannot make are called **essential nutrients.** These requirements vary from species to species. When the diet is lacking one or more essential nutrients, the animal is said to be **malnourished.**

Essential Amino Acids   Eight of the 20 amino acids required to make proteins are **essential amino acids** in the adult human diet. Protein deficiency develops from a diet that lacks one or more essential amino acids and produces retarded physical and perhaps mental development in children.

Meat, eggs, and cheese contain complete proteins with all essential amino acids in proportions that meet human requirements. Most plant proteins are incomplete, and diets built on a single staple, such as only corn, beans, or rice, can result in protein deficiency.

■ **INTERACTIVE QUESTION   41.2**

How can vegetarians avoid protein deficiencies?

Essential Fatty Acids   Animals are able to make most of the fatty acids they need. Linoleic acid is required in the human diet. Deficiencies of **essential fatty acids** are rare.

Vitamins   **Vitamins** are essential organic molecules required in small amounts in the diet. Thirteen vitamins essential to humans have been identified. Water-soluble vitamins include the B-complex, most of which function as coenzymes, and vitamin C, required for the production of connective tissue. The fat-soluble vitamins are A, incorporated into visual pigments; D, aiding in calcium absorption and bone formation; E, seeming to protect phospholipids in membranes from oxidation; and K, required for blood clotting.

Minerals   **Minerals** are inorganic nutrients, usually needed in very small amounts. Vertebrates require relatively large quantities of calcium and phosphorus for bone construction. Calcium is also needed for normal nerve and muscle function, and phosphorus is needed for ATP and nucleic acids. Iron is a component of the cytochromes and hemoglobin. Other minerals function as cofactors of enzymes. Iodine is needed by vertebrates to make the metabolism-regulating thyroid hormones. Sodium, potassium, and chlorine are important in nerve function and osmotic balance.

■ **INTERACTIVE QUESTION   41.3**

What are RDAs?

41.3 The main stages of food processing are ingestion, digestion, absorption, and elimination

Ingestion is the act of eating. **Digestion** splits macromolecules into monomers by **enzymatic hydrolysis,** the addition of a water molecule when the bond between monomers is broken. Mechanical fragmentation often precedes chemical digestion.

In the last two stages of food processing, monomers and small molecules are **absorbed** into the cells of the animal, and the undigested remainder of the food is **eliminated.**

■ INTERACTIVE QUESTION 41.4

Why must macromolecules be digested into monomers?

Digestive Compartments **Intracellular digestion** of food molecules in food vacuoles is typical of sponges. Food vacuoles fuse with enzyme-containing lysosomes, and digestion occurs safely within a membrane-enclosed compartment.

Most animals break down their food, at least initially, by **extracellular digestion** within a separate compartment of the body that connects to the external environment.

A single-opening **gastrovascular cavity** functions in both digestion and transport of nutrients throughout the body. In the small cnidarian hydra, digestive enzymes secreted into the gastrovascular cavity initiate food breakdown. Gastrodermal cells take in food particles by phagocytosis, and hydrolysis of macromolecules occurs within food vacuoles. Undigested materials are expelled through the mouth/anus.

Most animals have **complete digestive tracts** or **alimentary canals,** with two openings, a mouth and an anus, and specialized regions allowing for the sequential digestion and absorption of nutrients. Food, ingested through the mouth and pharynx, passes through an esophagus that leads to a crop, stomach, or gizzard—organs specialized for storing or grinding food. In the intestine, digestive enzymes hydrolyze macromolecules, and nutrients are absorbed across the tube lining. Undigested material exits through the anus.

■ INTERACTIVE QUESTION 41.5

Label and list the functions for the five organs of an earthworm's digestive tract. What is the function of the typhlosole?

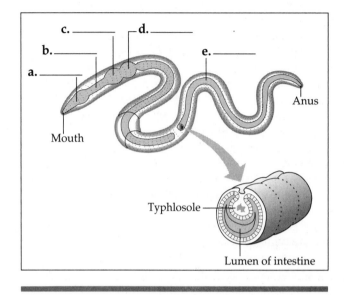

41.4 Each organ of the mammalian digestive system has specialized food-processing functions

The mammalian digestive system consists of the alimentary canal and accessory glands. Rhythmic waves of muscular contraction called **peristalsis** push food through the tract. Ringlike valves called **sphincters** regulate the passage of material between some segments. Accessory glands, the **salivary glands, pancreas,** and **liver** with its **gallbladder,** secrete digestive juices into the alimentary canal through ducts.

The Oral Cavity, Pharynx, and Esophagus Physical and chemical digestion begins in the mouth, where teeth grind food to expose a greater surface area to enzyme action. The presence of food in the **oral cavity** triggers the release of saliva.

Saliva has several components: mucin, a glycoprotein that protects the mouth lining from abrasion and lubricates food for swallowing; buffers to neutralize acidity; antibacterial agents; and **salivary amylase,** which begins the hydrolysis of starch and glycogen into smaller polysaccharides and maltose. The tongue tastes and manipulates food, and pushes the food ball, or **bolus,** into the pharynx for swallowing.

The **pharynx** is the intersection leading to both the esophagus and the trachea. During swallowing, the top of the windpipe moves up so that its opening is blocked by the cartilaginous **epiglottis.**

Food moves down through the **esophagus** to the stomach, squeezed along by a wave of peristalsis.

The Stomach   The expandable **stomach** stores food so we do not have to eat constantly.

The epithelium lining deep pits in the stomach secretes **gastric juice,** a digestive fluid containing hydrochloric acid that breaks down food tissues, kills bacteria, and denatures proteins, and **pepsin,** an enzyme that hydrolyzes specific peptide bonds in proteins. Pepsin is synthesized and secreted by chief cells in an inactive form called **pepsinogen.** It is activated by hydrochloric acid (secreted by parietal cells) and by pepsin itself—an example of positive feedback.

The mucous coating secreted by the epithelium protects the stomach lining from digestion. Gastric ulcers, mainly caused by the bacterium *Helicobacter pylori,* may worsen when the lining is eroded faster than it can be regenerated.

Smooth muscles mix the contents of the stomach. **Acid chyme** is the nutrient broth produced by the action of the stomach and its secretions on ingested food. The stomach is usually closed off by two sphincters: One at the cardiac orifice prevents backflow into the esophagus, and the **pyloric sphincter** regulates passage of acid chyme into the intestine.

■ INTERACTIVE QUESTION   41.6

List the two types of macromolecules that have been partially digested by the time acid chyme moves into the intestine. Where did this digestion take place, and what enzymes were involved?

a.

b.

The Small Intestine   Most enzymatic hydrolysis of macromolecules and nutrient absorption into the blood takes place in the **small intestine.**

Digestive juices (from the pancreas, liver, gallbladder, and gland cells of the intestinal wall) are mixed with the chyme in the **duodenum,** the first section of the small intestine. The pancreas produces digestive enzymes and a bicarbonate-rich alkaline solution that offsets the acidity of the chyme. The liver produces **bile,** which is stored in the gallbladder until needed. Bile aids in the digestion of fats and contains pigments that are by-products of the

breakdown of red blood cells in the liver. The brush border of the duodenum secretes some enzymes; others are bound to the surfaces of these epithelial cells.

The digestion of starch and glycogen into disaccharides is continued by pancreatic amylases. Disaccharidases, enzymes of the brush border, digest them into monosaccharides.

Protein digestion is completed in the small intestine by pancreatic trypsin and chymotrypsin, enzymes specific for peptide bonds adjacent to certain amino acids; carboxypeptidase and aminopeptidase, which split amino acids off from opposite ends of a polypeptide; and dipeptidases of the brush border, which split small peptides. The pancreatic protein-digesting enzymes are secreted in inactive form and activated by enteropeptidase.

Pancreatic nucleases are a group of enzymes that hydrolyze DNA and RNA into their nucleotide monomers. Other enzymes dismantle nucleotides.

The digestion of fats is aided by bile salts, which coat or emulsify tiny fat droplets so they do not coalesce, leaving a greater surface area for lipase to hydrolyze the fat molecules into glycerol, fatty acids, and glycerides.

Hormones regulate the release of digestive secretions. **Gastrin,** released by the stomach into the circulatory system, stimulates secretion of gastric juice. **Secretin** is released from the duodenum and stimulates the pancreas to release bicarbonate. **Cholecystokinin (CCK),** produced in response to amino acids or fatty acids, stimulates the release of bile from the gallbladder and the release of pancreatic enzymes. **Enterogastrone** is secreted by the duodenum in response to an acid chyme rich in fats and inhibits peristalsis in the stomach, thus slowing digestion.

Most digestion is completed in the duodenum. The *jejunum* and *ileum* are regions of the small intestine that function in nutrient and water absorption.

■ INTERACTIVE QUESTION   41.7

List the enzymes that hydrolyze the following macromolecules in the small intestine.

a. polysaccharides

b. polypeptides

c. DNA and RNA

d. fats

Circular folds of the small intestine lining are covered with fingerlike projections called **villi,** on which the epithelial cells have microscopic extensions called **microvilli,** creating a huge surface area for absorption.

The core of each villus has a net of capillaries and a lymph vessel called a **lacteal.** Nutrients are absorbed across the epithelium of the villus and then across the single-celled epithelium of the capillaries or lacteal. Transport may be passive by diffusion or active by pumping against a gradient.

Glycerol and fatty acids are absorbed by epithelial cells where they recombine to form fats and are mixed with cholesterol and coated with proteins to make tiny globules called **chylomicrons.** These packages move out of the epithelial cells by exocytosis and into a lacteal, and they are then transported by the lymphatic system to veins near the heart.

Absorbed amino acids and sugars enter capillaries. The nutrient-laden blood from the small intestine is carried directly to the liver by the **hepatic portal vein.** The liver interconverts molecules and regulates the nutrient content of the blood.

The Large Intestine   The small intestine leads into the **large intestine,** or **colon,** at a T-shaped junction with a sphincter. A pouch, called the **cecum,** with a fingerlike extension in humans, called the **appendix,** attaches at this juncture. The colon finishes the reabsorption of the large quantity of water secreted with digestive enzymes into the digestive tract.

Escherichia coli and other mostly harmless bacteria live on unabsorbed organic material in the large intestine. Some of these bacteria produce vitamins such as several B vitamins and vitamin K, which are absorbed by the host. The **feces** contain cellulose, other undigested materials, and a large proportion of intestinal bacteria. Feces are stored in the **rectum.**

41.5 Evolutionary adaptations of vertebrate digestive systems are often associated with diet

Some Dental Adaptations   Dentition, the type and arrangement of teeth, correlates with diet.

Stomach and Intestinal Adaptations   Herbivores have longer alimentary canals because plant material is more difficult to digest than meat.

Symbiotic Adaptations   Many herbivorous mammals have fermentation chambers filled with symbiotic bacteria and protists. These microorganisms, often housed in the cecum, digest cellulose to simple sugars and produce a variety of essential nutrients.

Ruminants have an elaborate system involving several stomach chambers, regurgitation and rechewing of the cud, and digestion of their symbiotic bacteria to maximize the nutrient yield of their grass or hay diet.

■ INTERACTIVE QUESTION   41.8

Compare and contrast the dentition and alimentary canals of carnivores and herbivores.

▶ Word Roots

chylo- = juice; **micro-** = small (*chylomicron:* small globules composed of fats that are mixed with cholesterol and coated with special proteins)

chymo- = juice; **-trypsi** = wearing out (*chymotrypsin:* an enzyme found in the duodenum; it is specific for peptide bonds adjacent to certain amino acids)

di- = two (*dipeptidase:* an enzyme found attached to the intestinal lining; it splits small peptides)

entero- = the intestines (*enterogastrones:* a category of hormones secreted by the wall of the duodenum)

epi- = over; **-glotti** = the tongue (*epiglottis:* a cartilaginous flap that blocks the top of the windpipe, the glottis, during swallowing)

extra- = outside (*extracellular digestion:* the breakdown of food outside cells)

gastro- = stomach; **-vascula** = a little vessel (*gastrovascular cavities:* pouch that serves as the site of extracellular digestion and a passageway to disperse materials throughout most of an animal's body)

herb- = grass; **-vora** = eat (*herbivore:* a heterotrophic animal that eats plants)

hydro- = water; **-lysis** = to loosen (*hydrolysis:* a chemical process that lyses or splits molecules by the addition of water)

intra- = inside (*intracellular digestion:* the joining of food vacuoles and lysosomes to allow chemical digestion to occur within the cytoplasm of a cell)

micro- = small; **-villi** = shaggy hair (*microvilli:* many fine, fingerlike projections of the epithelial cells in the lumen of the small intestine that increase its surface area)

omni- = all (*omnivore:* a heterotrophic animal that consumes both meat and plant material)

peri- = around; **-stalsis** = a constriction (*peristalsis:* rhythmic waves of contraction of smooth muscle that push food along the digestive tract)

Structure Your Knowledge

1. Food provides fuel, organic raw materials, and essential nutrients. Complete the following concept map that summarizes the nutritional needs of animals.

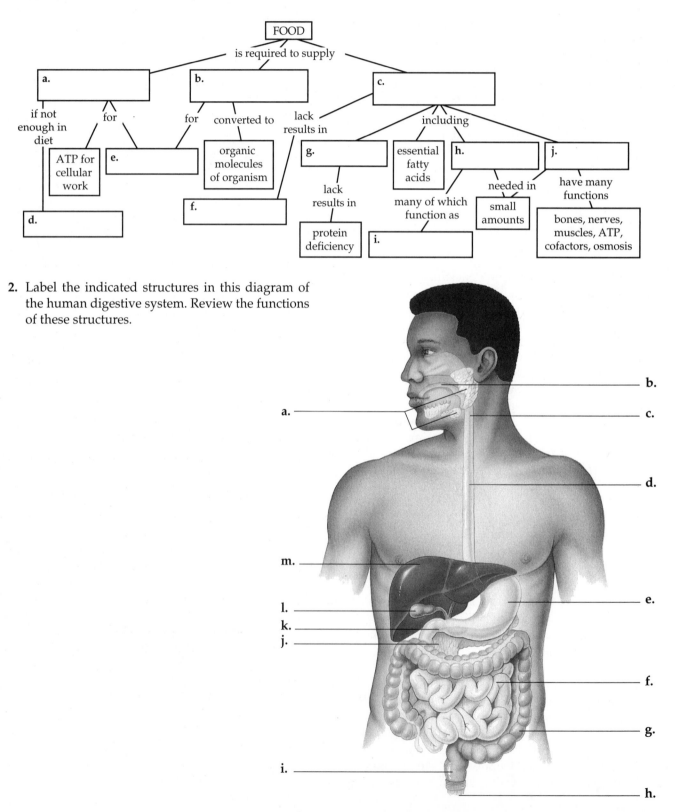

2. Label the indicated structures in this diagram of the human digestive system. Review the functions of these structures.

Test Your Knowledge

MATCHING: *Match the description with the correct enzyme or hormone.*

_____ 1. enzyme that hydrolyzes peptide bonds, works in the stomach

_____ 2. hormone that stimulates secretion of gastric juice

_____ 3. hormone that stimulates release of bile and pancreatic enzymes

_____ 4. enzyme that begins digestion of starch in the mouth

_____ 5. enzymes specific for hydrolyzing various disaccharides

_____ 6. hormone that inhibits peristalsis in the stomach

_____ 7. enzyme that hydrolyzes fats

_____ 8. intestinal enzyme that activates protein-digesting enzymes

_____ 9. hormone that stimulates secretion of bicarbonate ions from the pancreas

_____ 10. intestinal enzyme specific for peptide bonds adjacent to specific amino acids

A. aminopeptidase
B. amylase
C. bile salts
D. cholecystokinin
E. chymotrypsin
F. disaccharidases
G. enterogastrone
H. enteropeptidase
I. gastrin
J. lipase
K. maltase
L. nuclease
M. pepsin
N. secretin

MULTIPLE CHOICE: *Choose the one best answer.*

1. Substrate feeders such as earthworms
   a. eat mostly mineral substrates.
   b. filter small organisms from water.
   c. eat autotrophs.
   d. move and eat through their food.
   e. are bulk feeders.

2. The energy content of fats
   a. is released by bile salts.
   b. may be lost unless an herbivore eats some of its feces.
   c. is approximately two times that of carbohydrates or proteins.
   d. can reverse the effects of malnutrition.
   e. Both c and d are correct.

3. Nucleosidases are
   a. hormones that stimulate release of pancreatic enzymes.
   b. enzymes attached to the intestinal epithelium that hydrolyze nucleosides.
   c. hydrolytic enzymes manufactured in inactive forms to protect the cells that produce them.
   d. protein-digesting enzymes that are activated by hydrochloric acid.
   e. enzymes that hydrolyze DNA.

4. Which of the following statements is *false?*
   a. The average human has enough stored fat to supply calories for several weeks.
   b. An increase in leptin levels leads to an increase in appetite and weight gain.
   c. Conversion of glucose and glycogen takes place in the liver.
   d. After glycogen stores are filled, excessive calories are stored as fat, regardless of their original food source.
   e. Carbohydrates and fats are preferentially used as fuel before proteins are used.

5. The purpose of antidiarrhea medicine would most likely be to
   a. speed up peristalsis in the small intestine.
   b. speed up peristalsis in the large intestine.
   c. kill *E. coli* in the intestine.
   d. increase water reabsorption in the large intestine.
   e. increase salt secretion into the feces.

6. Incomplete proteins are
   a. lacking in essential vitamins.
   b. a cause of undernourishment.
   c. found in meat, eggs, and cheese.
   d. lacking in one or more essential amino acids.
   e. a result of overcooking vegetables.

7. Which of the following is a *true* statement about vitamins?
   a. They may be produced by intestinal microorganisms.
   b. They are the same from one species to the next.
   c. They are stored in large quantities in the liver.
   d. They are inorganic nutrients, needed in small amounts, that usually function as cofactors.
   e. They are all water soluble and must be replaced every day.

8. In animals, a distinct advantage of extracellular digestion over intracellular digestion is that
   a. polymers are hydrolyzed to monomers by digestive enzymes.
   b. there is a greater surface area for absorption of digested nutrients.
   c. larger pieces of food can be ingested and then digested.
   d. all four types of macromolecules can be digested instead of just glucose.
   e. the products of extracellular digestion can be absorbed into all body cells, without the need for a transport system.

9. Ruminants
   a. have teeth adapted for an omnivorous diet.
   b. use microorganisms to digest cellulose.
   c. eat their feces to obtain nutrients digested from cellulose by microorganisms.
   d. house symbiotic bacteria and protists in a cecum.
   e. get all of their nutrition from digested plant material.

10. How do molting penguins obtain enough protein to produce new feathers?
    a. They become obese right before molting in order to store protein.
    b. They combine different protein sources to make up for the incomplete proteins in their food.
    c. They become malnourished during the period of new feather production.
    d. They feed on fish, which are an excellent protein source.
    e. They stockpile extra muscle proteins prior to molting, which then supply amino acids for protein synthesis.

11. Which of the following is the most widespread dietary health problem in the United States?
    a. undernourishment
    b. obesity
    c. anorexia
    d. malnourishment
    e. pesticides in food

12. After a meal of greasy french fries, which enzymes would you expect to be most active?
    a. salivary and pancreatic amylase, disaccharidases, lipase
    b. lipase, lactase, maltase
    c. pepsin, trypsin, chymotrypsin, dipeptidases
    d. gastric juice, bile, bicarbonate
    e. sucrase, lipase, bile

13. Chylomicrons are
    a. lipoproteins transported by the circulatory system.
    b. small branches of the lymphatic system.
    c. protein-coated fat globules excreted out of epithelial cells into a lacteal.
    d. small peptides acted upon by chymotrypsin.
    e. fats emulsified by bile salts.

14. Which of the following is *not* a common component of feces?
    a. intestinal bacteria
    b. cellulose
    c. saturated fats
    d. bile pigments
    e. undigested materials

15. The hepatic portal vein
    a. supplies the capillaries of the intestines.
    b. carries absorbed nutrients to the liver for processing.
    c. carries blood from the liver to the heart.
    d. drains the lacteals of the villi.
    e. supplies oxygenated blood to the liver.

16. An organism that exclusively uses extracellular digestion would
    a. be a sponge.
    b. have a gastrovascular cavity.
    c. not need a circulatory system.
    d. have to be very thin and elongated.
    e. have a specialized body compartment with digestive enzymes.

17. One would expect to find a gastrovascular cavity in a(n)
    a. hydra.
    b. fish.
    c. insect.
    d. earthworm.
    e. bird.

18. What moves a bolus of food down the esophagus to the stomach?
    a. the closing of the epiglottis over the glottis
    b. a lower pressure in the abdominal cavity compared to the oral cavity
    c. the contraction of the diaphragm and the opening of the pyloric sphincter
    d. peristalsis caused by waves of smooth muscle contraction
    e. the action of cilia coated by mucus

19. What are villi, and what do they do?
    a. folds in the stomach that allow the stomach to expand
    b. extensions of the lymphatic system that pick up digested fats for transport to the circulatory system and then to the liver
    c. fingerlike projections of the small intestine lining that increase the surface area for absorption
    d. microscopic extensions of epithelial cells lining the small intestine that provide more surface area for digestion
    e. projections of the intestinal capillaries that join to form the hepatic portal vein

20. Which of the following is mismatched with its function?
    a. most B vitamins—coenzymes
    b. vitamin E—antioxidant
    c. vitamin K—blood clotting
    d. iron—component of thyroid hormones
    e. phosphorus—bone formation, nucleotide synthesis

21. Why does salivary amylase not hydrolyze starch in the duodenum?
    a. Starch is completely hydrolyzed into maltose in the oral cavity.
    b. The acid pH of the stomach denatures salivary amylase, and pepsin begins hydrolyzing it.
    c. Salivary amylase is produced by salivary glands and never leaves the oral cavity.
    d. Pancreatic amylase is a more effective enzyme in the pH of the duodenum.
    e. Salivary amylase can hydrolyze glycogen but not starch.

22. Why do many vegetarians combine different protein sources in the same meal or eat some animal products such as eggs or milk products?
    a. to make sure they obtain sufficient calories
    b. to provide sufficient vitamins
    c. to make sure they ingest all essential fatty acids
    d. to make their diet more interesting
    e. to provide all essential amino acids at the same time

Chapter 42
Circulation and Gas Exchange

Framework

This chapter surveys the basic approaches to circulation and gas exchange found in the animal kingdom, with special attention to the human systems and the problems of cardiovascular disease. The following concept map organizes some of the chapter's key ideas.

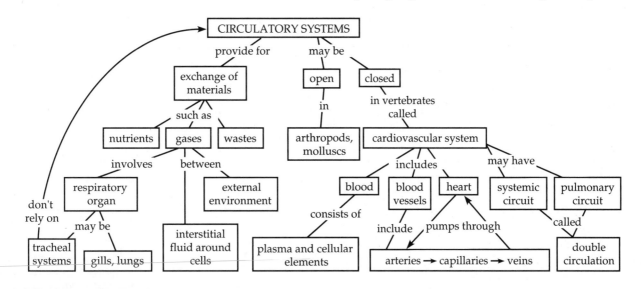

Chapter Review

Cells must reside in an aqueous environment, which provides oxygen and nutrients and permits disposal of carbon dioxide and metabolic wastes. Most animals have organ systems that exchange materials with the environment, and many have an internal transport system to service the body's cells.

42.1   Circulatory systems reflect phylogeny

The circulatory system provides an internal transport system to move substances across long distances in animals. Blood exchanges materials within organs specialized for gas exchange, nutrient absorption, and waste removal and then carries oxygen and nutrients throughout the body for delivery to the interstitial fluid that bathes the body cells.

Invertebrate Circulation   The central gastrovascular cavity inside the two-cell-thick body wall of cnidarians serves for both digestion and transport of materials. The internal fluid exchanges directly with the aqueous environment through the single opening. Flatworms also have gastrovascular cavities that branch throughout the thin and flat body.

The circulatory systems of more complex animals consist of **blood** pumped by a **heart** through **blood vessels. Blood pressure** is the hydrostatic force that moves blood.

In the **open circulatory system** found in insects, other arthropods, and in most molluscs, **hemolymph** (body fluid) in **sinuses,** or spaces between organs, bathes the internal tissues, providing for chemical exchange.

Annelids, some molluscs, and vertebrates have **closed circulatory systems,** in which the blood remains in vessels and exchanges materials with the interstitial fluid bathing the cells.

■ INTERACTIVE QUESTION   42.1

a. How is the hemolymph of an open circulatory system moved throughout the body?

b. What are some advantages of open circulatory systems?

Survey of Vertebrate Circulation   In the **cardiovascular system** of vertebrates, the heart has one or two **atria,** which receive blood, and one or more **ventricles,** which pump blood out of the heart.

Arteries, carrying blood away from the heart, branch into tiny **arterioles** within organs, which then divide into microscopic **capillaries,** which infiltrate tissues in networks called **capillary beds.** Exchange of substances between blood and interstitial fluid occurs across capillary walls. Capillaries converge to form **venules,** which meet to form the **veins** that return blood to the heart.

The ventricle of the two-chambered heart of a fish pumps blood first to the capillary beds of the gills (the **gill circulation**), from which the oxygen-rich blood flows through a vessel to the capillary beds in the other organs (the **systemic circulation**). Veins return the oxygen-poor blood to the atrium. Passage through two capillary beds slows the flow of blood, but body movements help to maintain circulation.

In the three-chambered heart of amphibians, the single ventricle pumps blood through a forked artery into the **pulmocutaneous circuit,** which leads to lungs and skin capillaries and then back to the left atrium, and the **systemic circuit,** which carries blood to the rest of the body and back to the right atrium. This **double circulation** repumps blood after it returns from the capillary beds of the lungs or skin, ensuring a strong flow of oxygen-rich blood to the brain, muscles, and body organs. A ridge in the ventricle helps to direct oxygen-rich blood from the left atrium into the systemic circuit and oxygen-poor blood into the pulmocutaneous circuit.

The three-chambered heart of lizards, snakes, and turtles has a partially divided ventricle that helps to separate blood flow through a **pulmonary circuit** to the gas exchange tissues in the lungs and through a systematic circuit to the body tissues.

Delivery of oxygen for cellular respiration is most efficient in birds and mammals, which, as endotherms, have high oxygen and energy demands. The left side of the large and powerful four-chambered heart handles only oxygen-rich blood, whereas the right side receives and pumps oxygen-poor blood.

42.2   Double circulation in mammals depends on the anatomy and pumping cycle of the heart

Mammalian Circulation: **The Pathway**   Review the circulation of blood through a mammalian cardiovascular system by completing Interactive Question 42.2.

The Mammalian Heart: **A Closer Look**   The human heart, located just beneath the sternum, is composed mostly of cardiac muscle. The atria have relatively thin muscular walls, whereas the ventricles have thicker walls. The **cardiac cycle** consists of **systole,** during which the cardiac muscle contracts and the chambers pump blood, and **diastole,** when the heart chambers are relaxed and filling with blood.

The **cardiac output,** or volume of blood pumped per minute into the systemic circuit, depends on **heart rate** and **stroke volume,** or quantity of blood pumped by each contraction of the left ventricle.

Atrioventricular (AV) valves between each atrium and ventricle are snapped shut when blood is forced against them as the ventricles contract. The **semilunar valves** at the exit of the aorta and pulmonary artery are forced open by ventricular contraction and close when the ventricles relax and blood starts to flow back toward the ventricles.

Heart rate can be measured by taking the **pulse,** which is caused by the rhythmic stretching of arteries as the left ventricle contracts and pumps blood through them.

The heartbeat sounds are caused by the recoil of blood against the closed heart valves. A **heart murmur** is the detectable hissing sound of blood leaking back through a defective valve.

■ **INTERACTIVE QUESTION 42.2**

The diagram below will help you review the flow of blood through a mammalian circulatory system. Label the indicated parts, color the vessels that carry oxygen-rich blood red, and then trace the flow of blood by numbering the circles from 1–11. Start with number 1 in the right ventricle.

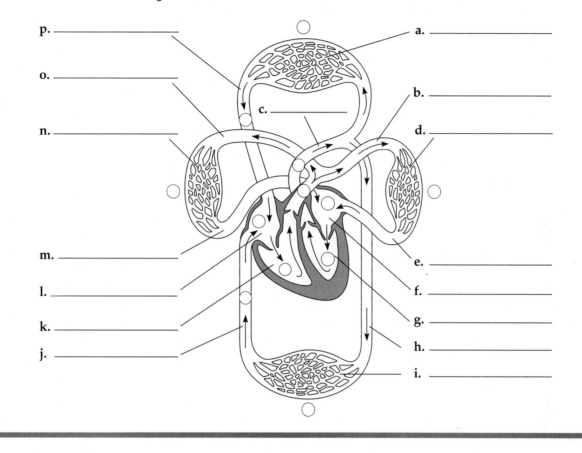

p. _____   a. _____

o. _____   b. _____

c. _____   d. _____

n. _____

m. _____   e. _____

l. _____   f. _____

k. _____   g. _____

j. _____   h. _____

i. _____

Maintaining the Heart's Rhythmic Beat Certain cells of cardiac muscle are self-excitable; they have an intrinsic ability to contract. The rhythm of contractions is coordinated by the **sinoatrial (SA) node,** or **pacemaker,** a region of specialized muscle tissue located in the wall of the right atrium. This arrangement of the pacemaker within the heart of vertebrates is called a **myogenic heart.** The **neurogenic heart** of most arthropods is controlled by motor nerves originating outside the heart.

The SA node initiates an electrical impulse that spreads via the intercalated disks of the cardiac muscle cells, and the two atria contract. The **atrioventricular (AV) node,** located between the right atrium and ventricle, relays the impulse after a 0.1-second delay through fibers to the apex of the heart and then through all parts of the ventricular walls, causing contraction. The electrical currents produced during the heart cycle can be detected by electrodes placed on the skin and recorded in an **electrocardiogram (ECG or EKG).**

The SA node is controlled by two sets of nerves with antagonistic signals and is influenced by hormones, temperature, and exercise.

■ **INTERACTIVE QUESTION 42.3**

a. The valve between an atrium and ventricle is

_____.

b. The valve between a ventricle and the aorta or pulmonary artery is _____.

c. The node that directly controls contraction of the atria is _____.

d. The node that directly controls contraction of the ventricles is _____.

42.3   Physical principles govern blood circulation

Blood Vessel Structure and Function   The wall of an artery or vein consists of three layers: an outer connective tissue layer with elastic fibers that allow the vessel to stretch and recoil; a middle layer of smooth muscle and more elastic fibers; and an inner lining of **endothelium,** a single layer of flattened cells. The outer two layers are thicker in arteries, which must be stronger and more elastic. One-way valves in the larger veins assure that blood flows back toward the heart. Capillaries have only the endothelial layer.

Blood Flow Velocity   Following the *law of continuity,* the flow of blood decelerates with the increasing cross-sectional area of the branching vessel system. The enormous number of capillaries creates a large total diameter, and blood flow is very slow, improving opportunity for exchange with interstitial fluid. Blood flow speeds up within venules and veins, due to the decrease in total cross-sectional area.

Blood Pressure   Blood pressure, the hydrostatic force exerted against the wall of a blood vessel, is much greater in arteries than in veins and is greatest during systole **(systolic pressure).** This hydrostatic pressure drives blood from the heart to the capillary beds. **Peripheral resistance,** caused by the narrow openings of the arterioles impeding the exit of blood from arteries, causes the swelling of the arteries during systole. The recoiling of the stretched elastic arteries during diastole creates **diastolic pressure** and maintains a continuous blood flow into arterioles and capillaries.

Arterial blood pressure may be measured with a sphygmomanometer. The first number is the pressure during systole (when blood first spurts through the artery that was closed off by the cuff), and the second is the pressure during diastole.

Together, cardiac output and peripheral resistance determine blood pressure. Contraction of smooth muscles in arteriole walls increases resistance and thus increases blood pressure, whereas dilation of arterioles as smooth muscles relax increases blood flow into the arterioles and thus lowers blood pressure. Nervous and hormonal signals control these muscles.

Blood pressure is negligible by the time blood exits the capillary beds. Contraction of the smooth muscle walls of venules and veins and the contraction of skeletal muscles between which veins are embedded force blood to flow back to the heart. Pressure changes during breathing also draw blood into the large veins in the thoracic cavity.

■ INTERACTIVE QUESTION   42.4

Complete the following concept map to help you organize your understanding of blood pressure.

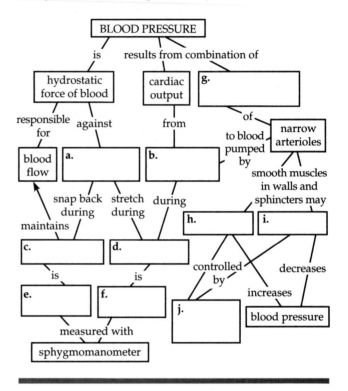

Capillary Function   Only about 5% to 10% of the body's capillaries have blood flowing through them at any one time. Capillaries branch off thoroughfare channels that are direct connections between arterioles and venules. The distribution of blood to capillary beds varies with need and is regulated by nerve signals and hormones that control contraction of smooth muscle in arteriole walls and precapillary sphincters at the entrance to capillary beds.

The exchange of substances between blood and interstitial fluid may involve transport by endocytosis and exocytosis by the endothelial cells, passive diffusion of small molecules, and bulk flow through the clefts between cells due to hydrostatic pressure. Water, sugars, salts, oxygen, and urea move through capillary clefts, whereas blood cells and proteins are too large to fit through. Blood pressure at the upstream end of a capillary forces fluid out, whereas osmotic pressure at the downstream end tends to draw about 85% of that fluid back into the capillary.

Fluid Return by the Lymphatic System   The fluid that does not return to the capillary and any proteins that may have leaked out are returned to the blood through the **lymphatic system.** Fluid diffuses into lymph capillaries intermingled in the blood capillary net. The fluid, called **lymph,** moves through lymph vessels with one-way valves, mainly as a result of the movement of skeletal muscles, and is returned to the circulatory system near the junction of the venae cavae with the right atrium. In **lymph nodes,** the lymph is filtered, and white blood cells attack viruses and bacteria.

42.4 Blood is a connective tissue with cells suspended in plasma

Blood Composition and Function Cells and cell fragments make up about 45% of the volume of blood; the rest is a liquid matrix called **plasma.**

The plasma consists of a large variety of solutes dissolved in water. The collective and individual concentrations of the dissolved ions of inorganic salts are important to osmotic balance, pH levels, and the functioning of muscles and nerves. The kidney is responsible for the homeostatic regulation of these electrolytes.

Plasma proteins function in the osmotic balance of blood and as buffers, antibodies, escorts for lipids, and clotting factors called fibrinogens. Blood plasma from which clotting factors have been removed is called serum. Nutrients, metabolic wastes, gases, and hormones are transported in the plasma.

Red blood cells (erythrocytes) transport oxygen. They lack mitochondria and generate ATP by anaerobic metabolism. The red blood cells of mammals lack nuclei. Their small size and biconcave shape create a large surface area of plasma membrane across which oxygen can diffuse. Erythrocytes are packed with **hemoglobin,** an iron-containing protein that binds oxygen. Hemoglobin also binds nitric oxide (NO). The release of NO relaxes the walls of capillaries, probably helping to deliver O_2 from the expanded capillaries.

There are five major types of **leukocytes,** or **white blood cells,** all of which fight infections. Some white cells are phagocytes that engulf bacteria and cellular debris; lymphocytes give rise to cells that produce the immune response. The majority of leukocytes are located in interstitial fluid and lymph nodes.

Platelets, pinched-off fragments of large cells in the bone marrow, are involved in the blood-clotting mechanism.

Erythrocytes circulate for about 3 to 4 months before they are phagocytosed by cells in the liver and spleen and their components recycled. Red and white blood cells and platelets are produced from pluripotent **stem cells** in red bone marrow. The production of red blood cells is controlled by a negative feedback mechanism involving the hormone **erythropoietin (EPO),** which is produced by the kidney in response to low oxygen supply in tissues. Isolated and purified pluripotent stem cells grown in tissue culture may soon be used to treat human diseases such as leukemia.

The clotting process usually begins when platelets, clumped together along a damaged endothelium, release clotting factors. By a series of steps, the plasma protein **fibrinogen** is converted to its active form, **fibrin.** The threads of fibrin form a patch. An inherited defect in any step of the complex clotting process causes **hemophilia.** Anticlotting factors normally prevent clotting of blood in the absence of injury. A **thrombus** is a clot that occurs within a blood vessel and blocks the flow of blood.

■ **INTERACTIVE QUESTION 42.5**

Complete this concept map of the components of vertebrate blood and their functions.

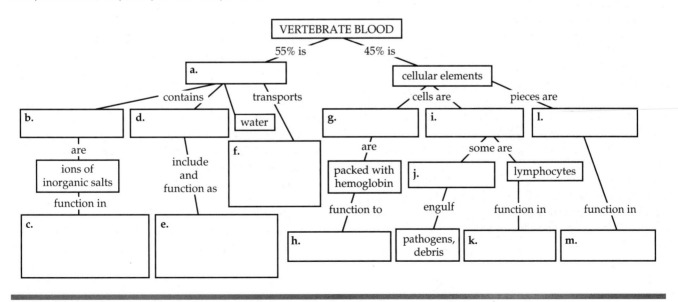

Cardiovascular Disease   Diseases of the heart and blood vessels are called **cardiovascular diseases.** Both lifestyle and genetics contribute to these diseases.

Cholesterol is carried in the blood as **low-density lipoproteins (LDLs)** and this type is associated with cholesterol deposits in plaques, growths that develop on the inner walls of arteries. **High-density lipoproteins (HDLs)** are cholesterol particles that appear to reduce cholesterol deposition in plaques. In the chronic cardiovascular disease known as **atherosclerosis,** plaques narrow the arteries, whose rough lining may trigger the clotting process.

Hypertension, or high blood pressure, is thought to damage the endothelium and initiate plaque formation, promoting atherosclerosis and increasing the risk of heart attack and stroke. This condition can be easily diagnosed and controlled by drugs, diet, and exercise.

Blockage of coronary arteries leads to the death of cardiac muscle in a **heart attack;** blockage of arteries in the head leads to a **stroke,** the death of nervous tissue in the brain. A heart attack or stroke may result from blockage by a thrombus or an embolus, which is a clot formed elsewhere that moves through the circulatory system. The accumulation of LDLs in the artery lining may trigger an inflammatory response that can cause plaques to rupture and a thrombus to form. Occasional chest pains, known as angina pectoris, may warn that a coronary artery is partially blocked.

■ **INTERACTIVE QUESTION   42.6**

What lifestyle choices have been correlated with increased risk of cardiovascular disease?

42.5   Gas exchange occurs across specialized respiratory surfaces

Gas exchange, or respiration, is the uptake of O_2 and the discharge of CO_2 and usually involves both the respiratory and the circulatory system. The **respiratory medium** is air for terrestrial animals and water for aquatic ones. The **respiratory surface,** where gas exchange with the respiratory medium occurs, must be moist, thin, and large enough to supply the whole body.

In sponges, cnidarians, and flatworms, gas exchange takes place across the entire surface of the body. In animals with denser bodies, a localized region of the body surface is usually specialized as a respiratory surface with a thin, moist epithelium separating a rich blood supply and the respiratory medium. Most animals have an extensively branched or folded respiratory organ.

Gills in Aquatic Animals   **Gills** are outfoldings of the body surface, ranging from simple bumps on echinoderms to the more complex gills of molluscs, crustaceans, and fishes. Due to the low oxygen concentration in a water environment, gills usually require **ventilation,** or movement of the respiratory medium across the respiratory surface—often an energy-intensive process due to the density of water.

In the gills of a fish, blood flows through capillaries in a direction opposite the flow of water. This arrangement sets up a **countercurrent exchange,** in which the diffusion gradient favors the movement of oxygen into the blood along the length of the capillary.

Tracheal Systems in Insects   The advantages of air as a respiratory medium include its higher concentration of oxygen, faster diffusion rate of O_2 and CO_2, and lower density. To prevent the disadvantageous water loss from large, moist surfaces, the respiratory surfaces of most terrestrial organisms are invaginated.

The **tracheal systems** of insects are tiny air tubes that branch throughout the body to come into contact with nearly every cell. Thus, the open circulatory system does not function in O_2 and CO_2 transport. Rhythmic body movements of large insects and the contraction and relaxation of flight muscles serve to ventilate tracheal systems.

Lungs   **Lungs** are invaginated respiratory surfaces restricted to one location from which oxygen is transported to the rest of the body by the circulatory system. The lungs of amphibians are small, and gas exchange is supplemented by diffusion across the skin. Birds, most other reptiles, and mammals rely totally on lungs for gas exchange.

The lungs of mammals, located in the thoracic cavity, have a honeycombed moist epithelium that provides a large surface area for gas exchange.

Air, entering through the nostrils, is filtered, warmed, humidified, and smelled in the nasal cavity. Air passes through the pharynx and enters the respiratory tract through the glottis. The **larynx** moves up and tips the epiglottis over the glottis when food is swallowed. The larynx functions as a voicebox in most mammals; exhaled air vibrates a pair of **vocal cords.** The **trachea,** or windpipe, branches into two **bronchi,** which then branch repeatedly into **bronchioles** within each lung. Ciliated, mucus-coated epithelium lines the major branches of the respiratory tree. Multilobed air sacs encased in a web of capillaries are at the tips of the tiniest bronchioles. Gas exchange takes place across the thin, moist epithelium of these **alveoli.**

42.6   Breathing ventilates the lungs

Vertebrate lungs are ventilated by **breathing,** the alternate inhalation and exhalation of air.

How an Amphibian Breathes   A frog uses **positive pressure breathing.** It lowers the floor of the oral cavity, expanding it and drawing air into the mouth; raising the floor of the oral cavity then pushes air into the lungs.

How a Mammal Breathes   Mammals ventilate their lungs by **negative pressure breathing.** The lungs are enclosed in a double-walled sac. Surface tension holds the two layers together, and they adhere to the lungs and the chest-cavity wall. Contraction of the rib muscles and the **diaphragm** expands the chest cavity and increases the volume of the lungs. Air pressure is reduced within this increased volume, and air is drawn through the nostrils and mouth down to the lungs. Relaxation of the rib muscles and diaphragm compresses the lungs, increasing the pressure and forcing air out.

Tidal volume is the volume of air inhaled and exhaled by an animal during normal breathing. The maximum volume that can be inhaled and exhaled by forced breathing is called **vital capacity.** The **residual volume** is the air that remains in the alveoli and lungs after forceful exhaling.

How a Bird Breathes   Birds have air sacs penetrating their abdomen, neck, and wings that act as bellows to maintain air flow through the lungs. The air sacs and lungs are ventilated through a circuit that includes a one-way passage through tiny, gas-exchange channels, called **parabronchi,** in the lungs.

Control of Breathing in Humans   Breathing is under automatic regulation by the **breathing control centers** in the medulla oblongata and the pons. Nerve impulses from the medulla instruct the rib muscles and diaphragm to contract. A drop in the pH of the blood and cerebrospinal fluid when CO_2 concentration increases is sensed by the medulla's control center, which increases the depth and rate of breathing. Although breathing centers respond primarily to CO_2 levels, they are alerted by oxygen sensors in the aorta and carotid arteries that react to severe deficiencies of O_2.

42.7   Respiratory pigments bind and transport gases

The Role of Partial Pressure Gradients   The concentration of a gas in air or dissolved in water is measured as **partial pressure.** At sea level, the partial pressure of O_2, which makes up 21% of the atmosphere, is 160 mm Hg (0.21×760 mm—atmospheric pressure at sea level). The partial pressure of CO_2 is 0.23 mm Hg. A gas will diffuse from a region of higher partial pressure to lower partial pressure. Blood entering the capillaries of the lungs has a lower P_{O_2} and a higher P_{CO_2} than does the air in the alveoli, so oxygen diffuses into the capillaries and carbon dioxide diffuses out. In the tissue capillaries, pressure differences favor the diffusion of O_2 out of the blood into the interstitial fluid and CO_2 into the blood.

Respiratory Pigments   In most animals, O_2 is carried by **respiratory pigments** in the blood. Copper is the oxygen-binding component in the blue respiratory protein **hemocyanin,** common in arthropods and many molluscs.

Hemoglobin, contained in red blood cells, is the respiratory pigment of almost all vertebrates. Hemoglobin is composed of four subunits, each of which has a cofactor called a heme group with iron at its center. The binding of O_2 to the iron atom of one subunit induces a change in shape of the other subunits, and their affinity for oxygen increases. Likewise, the unloading of the first O_2 lowers the other subunits' affinity for oxygen.

The **dissociation curve** for hemoglobin shows the relative amounts of O_2 bound to hemoglobin under

---

■ **INTERACTIVE QUESTION   42.7**

For the following animals, indicate the respiratory medium, respiratory surface, and means of ventilation.

Animal	Respiratory Medium	Respiratory Surface	Ventilation
Fish	a.	b.	c.
Grasshopper	d.	e.	f.
Frog	g.	h.	i.
Human	j.	k.	l.

varying O_2 partial pressures. In the steep part of this S-shaped curve, a slight change in partial pressure will cause hemoglobin to load or unload a substantial amount of O_2. Because of the **Bohr shift** in response to the lower pH of active tissues, hemoglobin releases more O_2 in such tissues.

About 70% of CO_2 is transported in blood as bicarbonate ions. Carbon dioxide enters red blood cells, where it first combines with H_2O to form carbonic acid (catalyzed by carbonic anhydrase) and then dissociates into H^+ and HCO_3^-. Bicarbonate moves into the plasma for transport. The H^+ binds to hemoglobin and other proteins, which reduces changes in pH of the blood during the transport of CO_2. In the lungs, the diffusion of CO_2 out of the blood shifts the equilibrium in favor of the conversion of HCO_3^- back to CO_2, which is unloaded from the blood.

■ INTERACTIVE QUESTION   42.8

The following graph shows dissociation curves for hemoglobin at two different pH values. Explain the significance of these two curves.

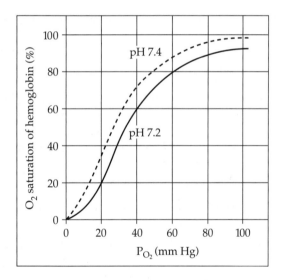

Elite Animal Athletes   Pronghorns are very fast long-distance runners. Adaptations for speed and endurance include a large exchange surface in the lungs, high cardiac output, large muscle mass, and large numbers of mitochondria, all of which contribute to a very high rate of O_2 consumption.

Special physiological adaptations have enabled some air-breathing animals to make sustained underwater dives. Weddell seals store twice the amount of O_2/kg body weight as do humans, mostly by having a larger volume of blood, a huge spleen that stores blood, and a higher concentration of **myoglobin,** an oxygen-storing muscle protein. Oxygen-conserving adaptations during a dive include decreasing the heart rate, rerouting blood to the brain and essential organs, and restricting blood supply to the muscles, which can use fermentation to produce ATP during long dives.

▶ Word Roots

alveol- = a cavity (*alveoli:* one of the dead-end, multi-lobed air sacs that constitute the gas exchange surface of the lungs)

atrio- = a vestibule; **-ventriculo** = ventricle (*atrioventricular node:* a region of specialized muscle tissue between the right atrium and right ventricle; it generates electrical impulses that primarily cause the ventricles to contract)

cardi- = heart; **-vascula** = a little vessel (*cardiovascular system:* the closed circulatory system characteristic of vertebrates)

counter- = opposite (*countercurrent exchange:* opposite flow of adjacent fluids that maximizes transfer rates)

endo- = inner (*endothelium:* the innermost, simple squamous layer of cells lining the blood vessels; the only constituent structure of capillaries)

erythro- = red; **-poiet** = produce (*erythropoietin:* a hormone produced in the kidney when tissues of the body do not receive enough oxygen. This hormone stimulates the production of erythrocytes)

fibrino- = a fiber; **-gen** = produce (*fibrinogen:* the inactive form of the plasma protein that is converted to the active form fibrin, which aggregates into threads that form the framework of a blood clot)

hemo- = blood; **-philia** = loving (*hemophilia:* a human genetic disease caused by a sex-linked recessive allele, characterized by excessive bleeding following injury)

leuko- = white; **-cyte** = cell (*leukocyte:* a white blood cell)

myo- = muscle (*myoglobin:* an oxygen-storing, pigmented protein in muscle cells)

para- = beside, near (*parabronchi:* the sites of gas exchange in bird lungs; they allow air to flow past the respiratory surface in just one direction)

pluri- = more, several; **-potent** = powerful (*pluripotent stem cell:* a cell within bone marrow that is a progenitor for any kind of blood cell)

pulmo- = a lung; **-cutane** = skin (*pulmocutaneous:* the route of circulation that directs blood to the skin and lungs)

semi- = half; **-luna** = moon (*semilunar valve:* a valve located at the two exits of the heart, where the aorta leaves the left ventricle and the pulmonary artery leaves the right ventricle)

thrombo- = a clot (*thrombus:* a clump of platelets and fibrin that block the blood flow through a vessel)

Structure Your Knowledge

1. Identify the labeled structures in this diagram of a human heart. Draw arrows to trace the flow of blood.

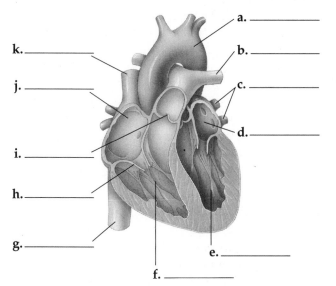

2. Trace the path of a molecule of oxygen being inhaled through the nasal cavity to delivery to the kidney.

Test Your Knowledge

MULTIPLE CHOICE: *Choose the one best answer.*

1. A gastrovascular cavity
   a. is found in cnidarians and annelids.
   b. functions to pump fluid in short vessels from which it spreads throughout the body.
   c. functions in both digestion and distribution of nutrients.
   d. has a single opening for ingestion and elimination, but a separate opening for gas exchange.
   e. involves all of the above.

2. Which of the following is *not* a similarity between open and closed circulatory systems?
   a. Some sort of pumping device helps to move blood through the body.
   b. Some of the circulation of blood is a result of movements of the body.
   c. The blood and interstitial fluid are distinguishable from each other.
   d. All tissues come into close contact with the circulating body fluid so that the exchange of nutrients and wastes can take place.
   e. All of these apply to both open and closed circulatory systems.

3. In a system with double circulation,
   a. blood is usually pumped at two separate locations as it circulates through the body.
   b. there is a countercurrent exchange in the gills.
   c. hemolymph circulates both through a pumping blood vessel and through body sinuses.
   d. blood is pumped to the gas-exchange organ and returning blood is then pumped through systemic vessels.
   e. there are always two ventricles and often two atria.

4. During diastole,
   a. the atria fill with blood.
   b. blood flows passively into the ventricles.
   c. the elastic recoil of the arteries maintains hydrostatic pressure on the blood.
   d. semilunar valves are closed, but atrioventricular valves are open.
   e. all of the above are occurring.

5. An atrioventricular valve prevents the backflow or leakage of blood from
   a. the right ventricle into the right atrium.
   b. the left atrium into the left ventricle.
   c. the aorta into the left ventricle.
   d. the pulmonary vein into the right atrium.
   e. the right atrium into the vena cava.

6. Breathing rate will increase when _____ CO_2 in your blood causes a _____ in pH.
   a. increased/rise
   b. increased/drop
   c. decreased/rise
   d. decreased/drop
   e. It is the level of O_2 in the blood that normally signals changes in breathing rate.

7. A heartbeat is initiated in humans by the
   a. SA node.
   b. AV node.
   c. right atrium.
   d. anterior and posterior venae cavae.
   e. left ventricle.

8. Blood flows more slowly in the arterioles than in the arteries because the arterioles
   a. have thoroughfare channels to venules that are often closed off.
   b. collectively have a larger cross-sectional area than do the arteries.
   c. must provide opportunity for exchange with the interstitial fluid.
   d. have sphincters that restrict flow to capillary beds.
   e. are narrower than arteries.

9. If all the body's capillaries were open at one time,
   a. blood pressure would fall dramatically.
   b. peripheral resistance would increase.
   c. blood would move too rapidly through the capillary beds.
   d. the amount of blood returning to the heart would increase.
   e. the increased gas exchange would allow for strenuous exercise.

10. Which of the following is *not* a factor in the exchange of substances in capillary beds?
    a. endocytosis and exocytosis
    b. passive diffusion
    c. hydrostatic pressure
    d. bulk flow through clefts between endothelial cells
    e. active transport by white blood cells

11. Edema involves the swelling of body tissues. It can result from
    a. swollen lymph glands.
    b. a loss of interstitial fluid.
    c. too much lymph being returned to the circulatory system.
    d. too low a concentration of blood proteins.
    e. swollen feet.

12. The functions of the kidney include
    a. production of the hormone erythropoietin to increase production of red blood cells.
    b. maintenance of electrolyte balance.
    c. destruction and recycling of red blood cells.
    d. both a and b.
    e. a, b, and c.

13. Fibrinogen is
    a. a blood protein that escorts lipids through the circulatory system.
    b. a cell fragment involved in the blood-clotting mechanism.
    c. a blood protein that is converted to fibrin to form a blood clot.
    d. a leukocyte involved in trapping viruses and bacteria.
    e. a lymph protein that regulates osmotic balance in the tissues.

14. A thrombus
    a. may form at a region of plaque in an artery.
    b. is a traveling embolism that may cause a heart attack or stroke.
    c. may cause hypertension.
    d. narrows the diameter of arteries due to the deposition of cholesterol.
    e. is often associated with high levels of HDLs in the blood.

15. In countercurrent exchange,
    a. the flow of fluids in opposite directions maintains a favorable diffusion gradient along the length of an exchange surface.
    b. oxygen is exchanged for carbon dioxide.
    c. double circulation keeps oxygenated and deoxygenated blood separate.
    d. oxygen moves from a region of high partial pressure to one of low partial pressure, but carbon dioxide moves in the opposite direction.
    e. the capillaries of the lung pick up more oxygen than do tissue capillaries.

16. The tracheae of insects
    a. are stiffened with chitinous rings and lead into the lungs.
    b. are filled by positive pressure breathing.
    c. ramify along capillaries of the circulatory system for gas exchange.
    d. are highly branched, coming into contact with almost every cell for gas exchange.
    e. provide an extensive, fluid-filled system of tubes for gas exchange.

17. Which of the following is *not* involved in speeding up breathing?

   a. nervous and chemical signals

   b. stretch receptors in the lungs

   c. impulses from the breathing centers in the medulla

   d. severe deficiencies of oxygen

   e. a drop in the pH of cerebrospinal fluid

18. The binding of an O_2 atom to the first iron atom of a hemoglobin molecule

   a. occurs only at a very high partial pressure of oxygen.

   b. produces a conformational change that lowers the other subunits' affinity for oxygen.

   c. is an example of cooperativity as O_2 readily binds to the other subunits.

   d. occurs more readily at a lower pH value.

   e. is the signal for hemoglobin to release H^+ so that bicarbonate converts to CO_2.

19. Fluid moves back into the capillaries at the venous end of a capillary bed as a result of

   a. active transport.

   b. a blood pressure that is higher than the osmotic pressure.

   c. an osmotic pressure that is higher than the blood pressure.

   d. a partial pressure that is lower than that at the arteriole end.

   e. the squeezing of surrounding muscles on the interstitial fluid.

20. As a general rule, blood leaving the right ventricle of a mammal's heart will pass through how many capillary beds before it returns to the right ventricle?

   a. one

   b. two

   c. three

   d. one or two, depending on the circuit it takes

   e. at least two, but almost always three

21. Gas exchange between maternal and fetal blood takes place in the placenta. Fetal and adult hemoglobin differ slightly in composition, and their dissociation curves are different so that oxygen can be transferred from maternal blood to fetal blood. If curve *c* in the graph that follows represents the dissociation curve of adult hemoglobin, which curve would represent fetal hemoglobin? (*Hint:* Must fetal hemoglobin have a higher or lower affinity for oxygen than maternal hemoglobin has?)

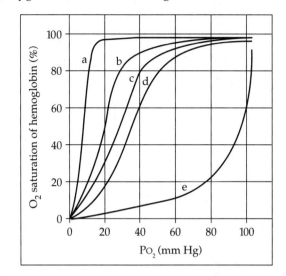

Use the following to answer questions 22–25.

   1. vena cava   4. right atrium

   2. left ventricle   5. aorta

   3. pulmonary vein   6. pulmonary capillaries

22. Which of the following sequences represents the flow of blood through the human body? (Obviously, many structures are not included.)

   a. 1-5-3-6-2-4

   b. 1-2-6-3-4-5

   c. 1-4-3-6-2-5

   d. 1-2-3-6-4-5

   e. 1-4-6-3-2-5

23. In which vessel would blood pressure be the highest?

   a. 1   **c.** 4   **e.** 6

   b. 3   **d.** 5

24. In which location would velocity of blood flow be the slowest?

   a. 1   **c.** 4   **e.** 6

   b. 3   **d.** 5

25. Of these structures, which would have the thickest muscle layer?

   a. 1   **c.** 3   **e.** 5

   b. 2   **d.** 4

26. The nurse tells you that your blood pressure is 112/70. What does the 70 refer to?
    a. your heart rate
    b. the velocity of blood during diastole
    c. the systolic pressure from ventricular contraction
    d. the diastolic pressure from the recoil of the arteries
    e. the venous pressure caused by the compression of the blood pressure cuff

27. What do the alveoli of mammalian lungs, the gill filaments of fish, and the tracheoles of insects all have in common?
    a. use of a circulatory system to transport absorbed O_2
    b. respiratory surfaces that are invaginated
    c. countercurrent exchange
    d. a large, thin surface area for exchange
    e. all of the above

28. Which of the following would *not* be a possible function of a plasma protein in a mammal?
    a. pH buffer
    b. osmotic balance
    c. clotting agent
    d. oxygen transport
    e. antibody

29. What is the function of the cilia in the trachea and bronchi?
    a. movement of air into and out of the lungs
    b. increase the surface area for gas exchange
    c. vibrate when air rushes past them to produce sounds
    d. filter the air that rushes through them
    e. sweep mucus with its trapped particles up and out of the respiratory tract

30. Both birds and mammals have four-chambered hearts. Which of the following is the best explanation for this similarity?
    a. They shared a common ancestor that had a four-chambered heart.
    b. They are the only vertebrates with double circulation, and this double pumping requires four chambers.
    c. They are both endotherms, and the evolution of efficient circulatory systems was necessary to support the high metabolic rate of endotherms.
    d. This is an example of convergent evolution, because animals that obtain their oxygen from air require both a pulmonary and a systemic circuit.
    e. The more inefficient single atrium of amphibians and many reptiles could not supply the higher oxygen needs of these endotherms.

31. What causes a heart murmur?
    a. a weak systole
    b. the closing of the atrioventricular valves
    c. a leaking heart valve
    d. low blood pressure
    e. an interruption of the impulse from the SA to the AV node

32. All of the following correctly describe how an arthropod heart or circulatory system differs from that of a vertebrate except one. Identify the incorrect comparison.
    a. two-chambered heart rather than three- or four-chambered heart
    b. neurogenic heart rather than myogenic
    c. heart pumps hemolymph rather than blood
    d. open circulatory system rather than closed
    e. hemocyanin as respiratory pigment rather than hemoglobin

33. More O_2 is unloaded from hemoglobin in an actively metabolizing tissue than in a resting tissue. Why?
    a. The pH of an active tissue is higher as more CO_2 combines with water to produce H^+ and HCO_3^-.
    b. The conformation of hemoglobin changes at the lower pH of an actively metabolizing tissue and O_2 dissociates more easily.
    c. Hemoglobin can bind four O_2 due to a change in conformation caused by the lower pH of the active tissue.
    d. More oxygen is bound to myoglobin at lower levels of oxygen concentration.
    e. More CO_2 is drawn out of tissue cells as it is converted to bicarbonate in red blood cells.

34. Which of the following does *not* contribute to the respiratory adaptations of deep-diving animals?
    a. an unusually large volume of blood
    b. large quantity of myoglobin in the muscles
    c. reduction in blood flow to noncritical body areas
    d. extra-large lungs
    e. large spleen that stores blood and releases it during a dive

Chapter 43

The Immune System

Framework

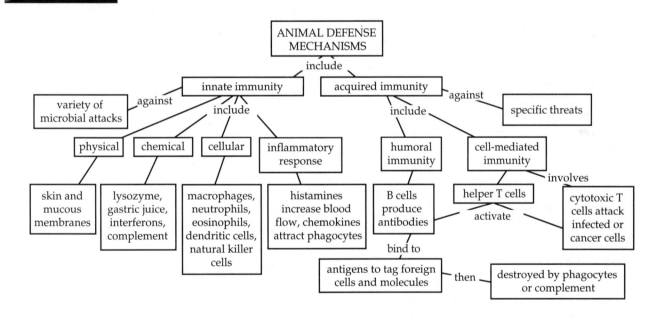

Chapter Review

The defense system protects an animal from bacteria, viruses, other pathogens, and early-stage cancer cells. **Innate immunity** involves nonspecific defense mechanisms, including both the external physical barriers of the skin and mucous membranes and the internal defenses of chemicals and phagocytic cells. **Acquired immunity,** also called *adaptive immunity,* is a line of defense in which **lymphocytes** react specifically to threat. In the humoral response, **antibodies** produced by cells derived from B lymphocytes mark microbes for destruction. In the cell-mediated response, cytotoxic lymphocytes destroy infected body cells, cancer cells, and foreign tissue.

■ **INTERACTIVE QUESTION   43.1**

How do the words *innate* and *acquired* relate to *when* these types of immunity develop in an animal's body?

43.1   Innate immunity provides broad defenses against infection

External Defenses   Skin and the mucous membranes lining the digestive, respiratory, and genitourinary tracts are physical barriers to microbes. Skin secretions maintain a low pH, which discourages colonization by microbes. The acidity of gastric juice kills most bacteria that reach the stomach. The ciliated, *mucus*-coated epithelial lining of the respiratory tract traps and removes microbes. **Lysozyme,** an enzyme that attacks bacterial cell walls, is present in tears, saliva, and mucus.

Internal Cellular and Chemical Defenses   Internal innate defense relies on **phagocytosis** of microbes by white blood cells that are *phagocytes.* The release of antimicrobial proteins by these cells helps initiate inflammation. Natural killer cells are nonphagocytic cells that also participate in innate defenses.

Surface receptors on phagocytes bind to microbial surface molecules. Microbes are engulfed and the resulting vacuoles fuse with lysosomes, which contain toxic forms of oxygen (such as nitric oxide) and enzymes. Some microbes have outer capsules that hide their surface polysaccharides or resist destruction once engulfed.

Short-lived **neutrophils** are the most numerous phagocytic white blood cells (leukocytes). **Monocytes** migrate into tissues and develop into **macrophages,** large, long-lived phagocytic cells. Macrophages may migrate through the body or become permanently attached in various organs, including lymph nodes, spleen, and other tissues of the lymphatic system. These macrophages attack microbes filtered from the blood in the spleen and from interstitial fluid that flows as lymph through lymph nodes.

Eosinophils are leukocytes that attack multicellular parasitic invaders with destructive enzymes. The primary role of phagocytic **dendritic cells** is to stimulate acquired immunity.

The **complement system** is a group of about 30 serum proteins that, when activated, may lyse microbes, trigger inflammation, or assist in acquired defenses. Virus-infected cells produce two types of **interferons** that stimulate neighboring cells to produce substances that inhibit viral reproduction in those cells. Another type of interferon activates macrophages. Interferons produced by recombinant DNA technology are being tested for their effectiveness in treating viral infections and cancer. *Defensins* are antimicrobial proteins secreted by macrophages that affect broad groups of pathogens.

Chemicals released in response to physical injury or pathogen entry can trigger an **inflammatory response,** characterized by redness, swelling, and heat. Damaged **mast cells** in connective tissue release **histamine,** which triggers dilation and leakiness of blood vessels; activated macrophages and other cells release prostaglandins and other chemicals that promote blood flow to the damaged area. Blood-clotting elements delivered to the area begin vessel repair and help seal off infections. Vasodilation as well as signaling proteins called **chemokines** result in the congregation of phagocytic cells and the production of antimicrobial compounds.

A systemic response to an infection may include an increase in the number of circulating white blood cells and a fever. Fever may be triggered by toxins produced by pathogens or by chemicals released by macrophages. Fevers stimulate phagocytosis and may speed tissue repair. *Septic shock* is a dangerous condition resulting from an overwhelming systemic inflammatory response.

Natural killer (NK) cells recognize general features of virus-infected or cancer cells, attaching to them and triggering **apoptosis,** or programmed cell death.

■ INTERACTIVE QUESTION   43.2

Complete the following table that summarizes the functions of the cells and chemicals of the innate defense mechanisms.

Cells or Compounds	Functions
Neutrophils	a.
Monocytes	b.
Macrophages	c.
Eosinophils	d.
Natural killer (NK) cells	e.
Mast cells	f.
Histamine	g.
Interferons	h.
Complement system	i.
Lysozyme	j.
Prostaglandins	k.
Chemokines	l.

Invertebrate Immune Mechanisms Insect defenses include their exoskeleton, various *hemocytes* that can phagocytose bacteria, secrete antimicrobial peptides, and wall off parasites and wounded tissue with aggregates produced by the cascade activation of phenoloxidase. Some invertebrate defenses have characteristics of acquired immunity, such as the ability to distinguish self from nonself and to respond more quickly to a second encounter with a particular microbe or foreign tissue.

43.2 In acquired immunity, lymphocytes provide specific defenses against infection

Lymphocytes, the key cells of acquired immunity, are activated by contact with microbes and by **cytokines,** proteins secreted by macrophages and dendritic cells after they engulf microbes. **Antigens** are foreign molecules recognized by lymphocytes. Most antigens are large proteins or polysaccharides, often protruding from the surfaces of microbes or transplanted cells. The small region of an antigen to which a lymphocyte or secreted antibody binds is called an **epitope,** or antigenic determinant.

Antigen Recognition by Lymphocytes **B lymphocytes (B cells)** and **T lymphocytes (T cells)** circulate in blood and lymph and reside in the spleen and lymph nodes. B and T cells both have membrane-bound **antigen receptors** that allow them to recognize specific epitopes. Each B or T lymphocyte carries about 100,000 identical receptors.

Each Y-shaped **B cell receptor** consists of four polypeptide chains: two identical **light chains** and two identical **heavy chains,** linked together by disulfide bridges. Both heavy and light chains have *variable (V) regions* at the ends of the two arms of the Y, which form two identical antigen-binding sites. The *constant (C) regions* of the molecule vary little from cell to cell. Secreted antibodies, or **immunoglobulins,** lack anchoring transmembrane regions but are otherwise structurally similar to B cell receptors, which are often called *membrane antibodies.*

A **T cell receptor,** also anchored by a transmembrane region, consists of one α chain and one β chain, linked by a disulfide bridge. The variable regions at the tip of the molecule form a single antigen-binding site. T cells recognize small fragments of antigens complexed with MHC molecules. A family of genes called the **major histocompatibility complex (MHC)** codes for these cell-surface proteins. In a process called **antigen presentation,** newly formed MHC molecules bind with antigen fragments within the cell and then display them on the cell's surface, where a T cell receptor can recognize the antigen and MHC molecule complex.

Class I MHC molecules are found on almost all nucleated cells. They bind foreign antigens that an in-fected or cancerous cell has produced. **Cytotoxic T cells** recognize class I MHC molecules displaying peptide antigens.

Class II MHC molecules are made by dendritic cells, macrophages, and B cells. These cells, called **antigen-presenting cells,** engulf and fragment microbes and display their antigens in class II MHC molecules. **Helper T cells** recognize class II MHC molecule-peptide antigen complexes.

The hundreds of different alleles for each class I and class II MHC gene result in a unique biochemical fingerprint for each individual (except identical twins).

■ INTERACTIVE QUESTION 43.3

Describe the differences between the antigens that B cell receptors and antibodies recognize and the antigens that T cell receptors on cytotoxic T cells and helper T cells recognize.

Lymphocyte Development Lymphocytes, like all blood cells, develop from pluripotent stem cells in the bone marrow. Lymphocytes either migrate to the **thymus** and differentiate into T cells or continue to develop in the bone marrow as B cells.

Each person may have as many as a million different B cells and 10 million different T cells. The genes coding for this diversity have numerous coding *gene segments* that are randomly and permanently rearranged. The genes for the α and β chains of the T cell receptor, and encoding the light and heavy chains of the B cell receptor all undergo rearrangement.

Immunoglobulin (Ig) genes code for both the B cell receptor and secreted antibodies. In the Ig light-chain gene, 40 variable *(V)* gene segments and five joining *(J)* gene segments are separated by a long stretch of DNA. The *J* segments are followed by an intron and a *C* exon that codes for the constant region. Early in B cell development, one *V* gene segment is randomly linked to one *J* segment by a set of enzymes called recombinase, producing one of 200 possible light chains. After transcription, the pre-mRNA is processed, and the mRNA is translated into a light chain with a variable and a constant region. These light chains combine with heavy chains that were produced the same way.

In the bone marrow and thymus, respectively, the antigen receptors of maturing B and T cells are tested for self-reactivity. Lymphocytes with receptors specific for some of the body's own molecules are either inactivated or destroyed by apoptosis. This critical *self-tolerance* means that normally there are no mature lymphocytes that react against self components.

When an antigen interacts with receptors on B cells or T cells that are specific for epitopes of that antigen, those particular lymphocytes are activated to divide repeatedly and differentiate into two clones—a large number of short-lived **effector cells,** which combat the antigen, and a clone of long-lived **memory cells,** all of which carry receptors for that antigen. By this **clonal selection,** a small number of cells is selected by their interaction with a specific antigen to produce thousands of cells keyed to that particular antigen.

The body mounts a **primary immune response** upon first exposure to an antigen. About 10 to 17 days are required for selected lymphocytes to proliferate and differentiate to yield the maximum response produced by effector T cells and the antibody-producing effector B cells, called **plasma cells.** Should the body reencounter the same antigen, the **secondary immune response** is more rapid, effective, and prolonged. The long-lived T and B memory cells are responsible for this *immunological memory.* This secondary immune response provides long-term protection against a previously encountered pathogen.

43.3   Humoral and cell-mediated immunity defend against different types of threats

The **humoral immune response** involves B cell activation and production of antibodies that circulate in the blood and lymph and defend against pathogens and toxins in extracellular fluid. The **cell-mediated immune response** involves cytotoxic T cells that destroy infected body cells, cancer cells, and transplanted tissues. Helper T cells activate both B cells and cytotoxic T cells.

Helper T Cells: A Response to Nearly All Antigens
Helper T cells recognize specific class II MHC molecule-antigen complexes on antigen-presenting cells. A T cell surface protein called **CD4** enhances the binding between an antigen-presenting cell and helper T cell, which results in the proliferation and differentiation of a clone of activated helper T cells and memory cells. Activated helper T cells secrete cytokines, which stimulate both the cell-mediated and humoral responses.

Dendritic cells are located in the epidermis and other tissues. They migrate from sites of infection to lymphoid tissues and present antigen in their class II MHC molecules to *naive* helper T cells, triggering a primary immune response. Macrophages are involved in a secondary immune response, presenting antigen to memory helper T cells. B cells interact with helper T cells in the humoral response.

■ **INTERACTIVE QUESTION   43.4**

Answer the following questions concerning the three major stages in the development of lymphocytes.

a. How is the great diversity of B and T cells produced?

b. What prevents B and T cells from reacting against the body's own molecules?

c. Describe clonal selection.

■ **INTERACTIVE QUESTION   43.5**

Label the components in this diagram that shows a helper T cell being activated by interaction with a dendritic cell and the central role of the helper T cell in activating both humoral and cell-mediated immunity.

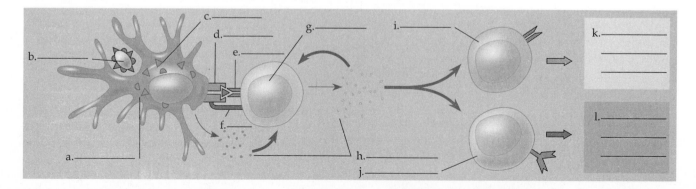

Cytotoxic T Cells: A Response to Infected Cells and Cancer Cells   Cytotoxic T cells recognize nonself proteins synthesized in infected cells and displayed with class I MHC molecules. **CD8** surface proteins on cytotoxic T cells bind to the side of class I MHC molecules and enhance the interaction between the cells, while the cytotoxic T cells, also stimulated by cytokines from nearby helper T cells, differentiate into active killers. The activated cell secretes proteins that kill the target cell. Pathogens released from the destroyed cell are marked by circulating antibodies for destruction.

Class I MHC molecules on tumor cells present fragments of tumor antigen to cytotoxic T cells, which recognize them as foreign and destroy them.

■ **INTERACTIVE QUESTION   43.6**

a. What surface molecule of a helper T cell facilitates the interaction with a class II MHC molecule of an antigen-presenting cell and the helper T cell?

b. What surface molecule on a cytotoxic T cell assists in the interaction with a class I MHC molecule displayed on infected cells?

c. What does an activated helper T cell release?

d. What does a cytotoxic T cell attached to an infected body cell release?

B Cells: A Response to Extracellular Pathogens
B cells are selectively activated by antigens (usually proteins or polysaccharides) on the surface of bacteria or transplanted cells. The activation is aided by cytokines released from helper T cells also activated by that antigen. Upon first binding antigen, the B cell takes in a few foreign molecules by receptor-mediated endocytosis and presents antigen fragments in its class II MHC molecules to helper T cells. Most protein antigens are *T-dependent antigens* that require the aid of helper T cells to stimulate antibody production. The activated B cell then proliferates into a clone of plasma cells and a clone of memory B cells.

T-independent antigens trigger antibody production by B cells without the aid of helper T cells. Found in bacterial capsules and flagella, these antigenic molecules usually have long chains of repeating subunits that apparently bind to enough antigen receptors on a B cell to activate the cell.

A variety of B cells will be stimulated by the different epitopes of an antigen, each giving rise to thousands of plasma cells. And each plasma cell can secrete about 2,000 antibody molecules per second during its four- to five-day life span.

The antigen-binding sites on the arms of an antibody allow it to identify a specific antigen. The heavy-chain constant regions in the tail determine the antibody's distribution in the body and its function. There are five major types of constant regions, creating five classes of antibodies: IgM, IgG, IgA, IgD, and IgE. IgG is the most abundant antibody in blood and confers passive immunity on a fetus; IgA is present in tears, saliva, mucus, and breast milk.

Researchers can use the specificity of antigen-antibody binding for biological research, clinical testing, and medical applications. Some antibody tools are *polyclonal* because they were formed by several different B cell clones, each specific for a different epitope. A technique for making **monoclonal antibodies** can supply quantities of identical antibodies.

Antibodies label antigens for disposal by one of several mechanisms. In *viral neutralization*, antibodies may block the ability of a virus to infect a host cell by binding to its surface. In *opsonization*, antibodies coat microbes and enhance phagocytosis by macrophages. Because each antibody molecule has at least two antigen-binding sites, the formation of antigen-antibody complexes may lead to the *agglutination* of bacteria or viruses, or the *precipitation* of soluble antigen molecules. The resulting clumps are then engulfed by phagocytes.

Antigen-antibody complexes on microbes or transplanted body cells may activate the complement system by binding with complement proteins and triggering the generation of a **membrane attack complex (MAC),** which produces a pore in the membrane and causes the cell to lyse. Complement proteins can be activated as part of the innate or acquired defenses. In addition to lysing microbes, activated complement proteins promote inflammation and phagocytosis.

■ **INTERACTIVE QUESTION   43.7**

List four ways in which antibodies mediate the disposal of antigens. Which of these enhance phagocytosis by macrophages?

a.

b.

c.

d.

Active and Passive Immunization   **Active immunity** can be acquired when the body produces antibodies and develops immunological memory from either exposure to an infectious agent or from **immunization,** also called **vaccination.** A vaccine may be an inactivated toxin, a killed or weakened microbe, a portion of a microbe, or even genes for microbial proteins. In **passive immunity,** temporary immunity is provided by antibodies supplied through the placenta to a fetus, through milk to a nursing infant, or by an antibody injection.

43.4   The immune system's ability to distinguish self from nonself limits tissue transplantation

Blood Groups and Transfusion   The immune response to the chemical markers that determine *ABO blood groups* must be considered in blood transfusions. Antibodies to other blood group antigens (other than the individual's own antigens) arise in response to normal bacterial flora and circulate in the blood plasma, where they will induce a devastating transfusion reaction to transfused blood cells with matching antigens. Blood group antigens are polysaccharides that induce primary immune responses in which no memory cells are produced. Thus, anti-blood group antibodies are in the IgM class and do not cross the placenta.

An Rh-negative mother may develop antibodies against the **Rh factor,** a protein red blood cell antigen, if fetal blood from an Rh-positive child leaks across the placenta. Should she carry a second Rh-positive fetus, her immunological memory may result in the production of IgG antibodies that cross the placenta and destroy fetal red blood cells. Treatment of the mother with anti-Rh antibodies late in pregnancy and just after delivery destroys any Rh antigen on fetal blood cells that may have leaked into her circulation and prevents the mother's immunological response to and memory of the antigen.

■ INTERACTIVE QUESTION   43.8

Fill in the following table to review your understanding of the antigens and antibodies of the ABO blood groups. Remember to compare the antigens on the donor cells with the antibodies in the recipient's plasma. (Assume these are packed cell transfusions.)

Blood Type	Antigens on RBCs	Antibodies in Plasma	Can Receive Blood from	Can Donate Blood to

Tissue and Organ Transplants   Transplanted tissues and organs are rejected because the foreign MHC molecules are antigenic and trigger immune responses. The use of closely related donors, as well as drugs that suppress immune responses, helps to minimize rejection.

In bone marrow transplants, used to treat leukemia and blood cell diseases, the graft itself may be the source of immune rejection. The recipient's bone marrow cells are destroyed by irradiation, eliminating the recipient's immune system. The lymphocytes in the bone marrow transplant, however, may produce a **graft versus host reaction** if the MHC molecules of donor and recipient are not well matched.

43.5   Exaggerated, self-directed, or diminished immune responses can cause disease

Allergies   Allergies are hypersensitivities to certain environmental antigens, or *allergens.* IgE antibodies created in response to an allergen may bind to mast cells in connective tissue. When allergens then bind to these cell surface antibodies, the mast cells, in a process called *degranulation,* release histamines, which create an inflammatory response that may include sneezing, a runny nose, and difficulty in breathing due to smooth muscle contractions. Antihistamines are drugs that combat these symptoms by blocking receptors for histamine. **Anaphylactic shock** is a severe allergic response in which the abrupt dilation of peripheral blood vessels caused by widespread mast cell degranulation leads to a life-threatening drop in blood pressure.

Autoimmune Diseases   Sometimes the immune system turns against self, leading to autoimmune diseases, such as *lupus, rheumatoid arthritis, insulin-dependent diabetes mellitus,* and *multiple sclerosis.* These diseases may be caused by a failure in the regulation of self-reactive lymphocytes.

Immunodeficiency Diseases   A genetic or developmental defect in the immune system is called an *inborn* or *primary immunodeficiency;* a defect that arises later in response to chemical or biological agents is called an *acquired* or *secondary immunodeficiency.* An inborn immunodeficiency may occur in any of the components of the immune system. In the rare congenital disease known as *severe combined immunodeficiency (SCID),* both humoral and cell-mediated immune systems are nonfunctional. Gene therapy has been used to treat individuals with a type of SCID caused by a deficiency of the enzyme adenosine deaminase. Acquired immunodeficiency may be caused by drugs used against autoimmune diseases or to suppress transplant rejection. Certain cancers, such as Hodgkin's disease, and **acquired immunodeficiency syndrome,** or **AIDS,** suppress the immune system.

There is growing evidence that general emotional health and immunity are related. Hormones secreted during stress affect the number of leukocytes; nerve fibers penetrate deep into the thymus, and receptors for chemical signals from nerve cells have been found on lymphocytes.

Individuals with AIDS are highly susceptible to opportunistic infections and cancers that take advantage of a suppressed immune system. The infectious agent responsible for AIDS is **HIV (human immunodeficiency virus).** HIV infects cells with surface CD4 molecules, including helper T cells, macrophages, and brain cells. CD4 and a protein *co-receptor,* which normally functions in chemokine reception, are required for viral entry. HIV RNA is reverse transcribed into DNA, which is integrated into the host cell genome, from where it directs production of new viral particles. HIV infection kills helper T cells through the effects of virus reproduction or by undergoing apoptosis, triggered by the virus.

While not able to cure HIV, new drug combinations are slowing the progression to AIDS. The frequent mutational changes during replication generate drug-resistant strains of HIV. And frequent mutational changes in surface antigens have made development of an effective vaccine difficult. HIV is transmitted by transfer of body fluids containing infected cells, such as blood or semen. Education may be the best approach to slowing the continuing spread of HIV.

■ INTERACTIVE QUESTION   43.9

a. Why is AIDS such a deadly disease?

b. Why has it proved so difficult to prevent and cure this disease?

Word Roots

agglutinat- = glued together (*agglutination:* an antibody-mediated immune response in which bacteria or viruses are clumped together)

an- = without; **-aphy** = suck (*anaphylactic shock:* an acute, life-threatening, allergic response)

anti- = against; **-gen** = produce (*antigen:* a foreign macromolecule that does not belong to the host organism and that elicits an immune response)

chemo- = chemistry; **-kine** = movement (*chemokine:* a group of about 50 different proteins secreted by blood vessel endothelial cells and monocytes; these molecules bind to receptors on many types of leukocytes and induce numerous changes central to inflammation)

cyto- = cell (*cytokines:* in the vertebrate immune system, protein factors secreted by macrophages and helper T cells as regulators of neighboring cells)

epi- = over; **-topo** = place (*epitope:* a localized region on the surface of an antigen that is chemically recognized by antibodies)

immuno- = safe, free; **-glob** = globe, sphere (*immunoglobulin:* one of the class of proteins comprising the antibodies)

macro- = large; **-phage** = eat (*macrophage:* an amoe-boid cell that moves through tissue fibers, engulf-ing bacteria and dead cells by phagocytosis)

mono- = one (*monocyte:* an agranular leukocyte that is able to migrate into tissues and transform into a macrophage)

neutro- = neutral; **-phil** = loving (*neutrophil:* the most abundant type of leukocyte; neutrophils tend to self-destruct as they destroy foreign invaders, lim-iting their lifespan to but a few days)

perfora- = bore through (*perforin:* a protein that forms pores in a target cell's membrane)

Structure Your Knowledge

This chapter contains a wealth of information that is proba-bly fairly new to you. If you take a little time and pull out the key players of the immune system and organize them first into very basic concept clusters and then develop more inter-related concept maps, you will find that this information is both understandable and fascinating.

1. Create a concept map outlining acquired immunity, showing the cells involved in humoral and cell-mediated responses and their functions.

2. Describe the structure of an antibody molecule and relate this structure to its function.

3. While we presented innate and acquired immunity separately, these defense mechanisms interact in several ways. Describe a few of the chemical and cellular players they share.

Test Your Knowledge

MULTIPLE CHOICE: *Choose the one best answer.*

1. Which of the following is *incorrectly* paired with its effect?
   a. gastric juice—kills bacteria in the stomach
   b. fever—stimulates phagocytosis
   c. histamine—causes blood vessels to dilate
   d. vaccination—creates passive immunity
   e. lysozyme—attacks cell walls of bacteria

2. Which of the following would release interferon?
   a. a macrophage that has become an antigen-presenting cell
   b. an injured epithelial cell of a blood vessel
   c. a cell infected by a virus
   d. a mast cell that has bound an antigen
   e. a helper T cell bound to an antigen-presenting cell

3. Antibodies are
   a. proteins or polysaccharides usually found on the cell surface of invading bacteria or viruses.
   b. proteins embedded in T cell membranes.
   c. proteins circulating in the blood that tag foreign cells for complement destruction.
   d. proteins that consist of two light and two heavy polypeptide chains.
   e. c and d are both correct.

4. A secondary immune response is more rapid and greater in effect than a primary immune response because
   a. histamines and prostaglandins cause rapid va-sodilation.
   b. the second response is an active immunity, whereas the primary one was a passive immunity.
   c. helper T cells are available to activate other blood cells.
   d. chemokines cause the rapid accumulation of phagocytic cells.
   e. memory cells respond to the pathogen and rap-idly clone more effector cells.

5. Lymphocytes capable of reacting against "self" molecules
   a. are usually not a problem until a woman's second pregnancy.
   b. are usually inactivated or destroyed as they mature.
   c. are usually kept separate from the immune system.
   d. contribute to immunodeficiency diseases.
   e. are characterized by class I MHC molecules.

6. Major histocompatibility complex molecules
   a. are involved in the ability to distinguish self from nonself.
   b. are a collection of cell surface proteins.
   c. may trigger T cell responses after transplant operations.
   d. present antigen fragments on infected cells.
   e. All of the above are correct.

7. In opsonization,
   a. antibodies coat microorganisms and help phago-cytes bind to and engulf the foreign cell.
   b. a set of complement proteins lyses a hole in a foreign cell's membrane.
   c. antibodies precipitate soluble antigens.
   d. a flood of histamines is released that may result in anaphylactic shock.
   e. *V* gene segments and *J* gene segments are joined by recombinase.

8. Severe combined immunodeficiency
   a. is an inborn autoimmune disease.
   b. is a form of cancer in which the membrane surface of the cell has changed.
   c. is a disease in which both T and B cells are absent or inactive.
   d. is an immune disorder in which the number of helper T cells is greatly reduced.
   e. results from a few types of cancers, such as Hodgkin's disease.

9. A transfusion of type B blood given to a person who has type A blood would result in
   a. the recipient's anti-B antibodies reacting with the donated red blood cells.
   b. the recipient's B antigens reacting with the donated anti-B antibodies.
   c. the recipient forming both anti-A and anti-B antibodies.
   d. no reaction, because B is a universal donor type of blood.
   e. the introduced blood cells being destroyed by nonspecific defense mechanisms.

10. Which of the following are *incorrectly* paired?
    a. variable region—determines antibody specificity for an epitope
    b. immunoglobulins—glycoproteins that form epitopes
    c. constant region—determines class and function of antibody
    d. IgG—most abundant circulating antibodies, confer passive immunity to fetus
    e. IgE—antibody molecules attached to mast cells

11. Which of the following destroys a target cell by phagocytosis?
    a. neutrophil   d. complement proteins
    b. cytotoxic T cell   e. plasma cell
    c. natural killer cell

12. A T-independent antigen
    a. does not need the aid of helper T cells to bind to antibody molecules, whereas T-dependent antigens do.
    b. needs to bind to helper T cells to activate antibody production.
    c. is often a large molecule that simultaneously binds with several antigenic receptors on a B cell at one time.
    d. will result in the production of both plasma cells and memory cells.
    e. is responsible for the secondary immune response.

13. What do IgE antibodies, T cell receptors, and MHC molecules have in common?
    a. They are found exclusively in cells of the immune system.
    b. They are all part of the complement system.
    c. They are antigen-presenting molecules.
    d. They are or can be membrane-bound proteins.
    e. They are involved in the cell-mediated portion of the immune system.

14. Which of the following is an effective defense against bacteria but does *not* work against viral particles?
    a. secretion of interferon by an infected cell
    b. neutralization by antibodies
    c. the enzyme lysozyme
    d. a secondary immune response
    e. humoral immunity

15. How are antibodies and complement related?
    a. They are both coded for by genes that have hundreds of alleles.
    b. They are both used in nonspecific defenses.
    c. They are both produced by plasma cells.
    d. Antibodies bound to antigens on a pathogen's membrane may activate complement proteins to form a membrane attack complex.
    e. Complement proteins tag foreign cells for destruction; antibodies destroy cells by agglutination.

16. Which of the following describes the main difference between an inflammatory response and an immune response?
    a. The inflammatory response responds only to free pathogens in a localized area; the immune response responds only to pathogens that have entered body cells.
    b. The inflammatory response involves only leukocytes, whereas the immune response involves only lymphocytes.
    c. The inflammatory response relies on phagocytes to destroy pathogens, whereas the immune response does not involve phagocytes.
    d. The inflammatory response is nonspecific, whereas the immune response reacts to specific microbes on the basis of their different antigens.
    e. Complement proteins participate in the immune response but not in the inflammatory response.

17. Clonal selection is responsible for the
    a. proliferation of clones of effector and memory cells specific for an encountered antigen.
    b. recognition of class I MHC molecules by cytotoxic T cells.
    c. rearrangement of antibody genes for the light and heavy chains.
    d. formation of cell cultures in the commercial production of monoclonal antibodies.
    e. transformation of a clone of helper T cells into cytotoxic T cells keyed to a specific antigen.

18. What role does a macrophage play in the immune response?
    a. activates complement proteins to form a membrane attack complex
    b. binds to the CD8 receptors on cytotoxic T cells to activate their production of perforin
    c. releases cytokines to activate B cells to produce clones of plasma cells
    d. activates both humoral and cell-mediated immunity by releasing interferons after it has ingested a virus
    e. presents antigens of an engulfed pathogen in its class II MHC molecules to helper T cells, and releases cytokines

19. All of the following are involved with innate immunity *except*
    a. the inflammatory response.
    b. the complement system.
    c. antimicrobial proteins such as lysozyme.
    d. chemokines that attract phagocytes.
    e. plasma cells.

20. Which of the following is *not* a description of invertebrate defense systems?
    a. They seem able to distinguish self from nonself.
    b. They may have phagocytic cells that engulf foreign matter.
    c. They produce specific antibodies against bacterial antigens, but not against viral antigens.
    d. Immunological memory against tissue grafts has been documented in earthworms.
    e. Certain hemocytes in insects contain phenoloxidase, which can create large aggregates that surround parasites and wounded tissue.

21. Helper T cells play which of the following roles in the immune response?
    a. bind to class I MHC molecules and activate complement proteins to attack and lyse cancer cells.
    b. bind to the antigens presented in the CD8 receptors on cytotoxic T cells and release perforin.
    c. produce interferons and histamines that help initiate a specialized inflammatory response.
    d. present antigens of an engulfed pathogen in its class II MHC molecules to B cells, which are then stimulated to develop into a clone of plasma cells.
    e. activate both the humoral and cell-mediated immunities by releasing cytokines after recognizing class II MHC molecule-antigen complexes on an antigen-presenting cell.

22. Which of the following statements about humoral immunity is correct?
    a. It is a form of passive immunity produced by vaccination.
    b. It defends against free pathogens with effector mechanisms such as neutralization, agglutination, precipitation, opsonization, or complement activation.
    c. It protects the body against pathogens that have invaded body cells as well as against abnormal body cells.
    d. It is mounted by lymphocytes that have matured in the thymus.
    e. It depends on the recognition of class I MHC molecules that are bound to a specific antigen to activate its effector mechanism.

23. What accounts for the huge diversity of antigens to which B cells can respond?
    a. The antibody genes have millions of alleles.
    b. The rearrangement of the antibody genes during development results in millions of possible combinations of randomly constructed light and heavy polypeptide chains.
    c. The antigen-binding sites at the arms of the molecule can assume a huge diversity of shapes in response to the specific antigen encountered.
    d. B cells have thousands of copies of antibodies bound to their plasma membranes.
    e. B cells can be antigen-presenting cells when they take in antigens by endocytosis and display fragments in their class II MHC molecules.

24. What is the function of CD4?

    a. a surface molecule on a cytotoxic T cell that enhances its binding to a class I MHC molecule displaying a foreign antigen

    b. a membrane protein on an antigen-presenting cell that helps a helper T cell recognize the MHC molecule-antigen complex

    c. a receptor that normally functions for cytokines, but which HIV uses as its receptor

    d. a surface molecule on a helper T cell that enhances its binding to a class II MHC molecule displaying a foreign antigen

    e. a portion of the class I MHC molecule found on all nucleated cells that identify cells as "self"

25. From which of the following would an AIDS patient be *least* likely to suffer?

    a. Kaposi's sarcoma or other cancers

    b. tuberculosis

    c. rheumatoid arthritis

    d. pneumonia

    e. yeast infections of mucous membranes

26. In which circumstance do B cells display antigens to T cells?

    a. They take in a few antigen molecules by endocytosis and display them in class II MHC molecules to helper T cells.

    b. They phagocytose bacteria and display bacterial peptide antigens in class II MHC molecules to helper T cells.

    c. After being infected by a virus, they display viral peptides they have synthesized in class I MHC molecules to cytotoxic T cells.

    d. They bind free antigens and display them to helper T cells attached to their B cell receptors.

    e. Both a and c are possible ways that B cells can display antigens.

27. All of the following are considered diseases or malfunctions of the immune system *except*

    a. MHC-induced transplant rejection.

    b. SCID (severe combined immunodeficiency).

    c. lupus, multiple sclerosis, and insulin-dependent diabetes.

    d. AIDS.

    e. allergic anaphylactic shock.

28. Which of the following may induce a graft-versus-host reaction?

    a. organ transplant

    b. blood transfusion

    c. bone marrow transplant

    d. skin transplant

    e. gene therapy

29. How does the immune system recognize malignant tumor cells?

    a. They do not display class I MHC molecules.

    b. They display fragments of tumor antigen in their MHC molecules.

    c. They display cancerous viral fragments in their class II MHC molecules.

    d. They have abnormal amounts of polysaccharides in their extracellular matrix that trigger a T-independent immune response.

    e. They undergo opsonization by complement proteins so that they are recognized by phagocytes.

30. Place the following steps in the helper T cell activation of cell-mediated and humoral immunity in the correct order:

    1. Helper T cell secretes cytokines.

    2. Macrophage engulfs pathogen and presents antigen in class II MHC.

    3. Plasma cells secrete antibodies and cytotoxic T cells attack cells with class I MHC molecule-antigen complex.

    4. T cell receptor recognizes class II MHC molecule-antigen complex.

    5. Macrophage secretes cytokines.

    6. Activated B cells form plasma and memory cells, activated T cells form cytotoxic T cells and memory cells.

    a. 1, 3, 5, 6, 2, 4

    b. 5, 1, 2, 6, 4, 3

    c. 2, 4, 5, 1, 6, 3

    d. 5, 2, 4, 1, 3, 6

    e. 2, 1, 4, 5, 6, 3

Chapter 44
Osmoregulation and Excretion

Framework

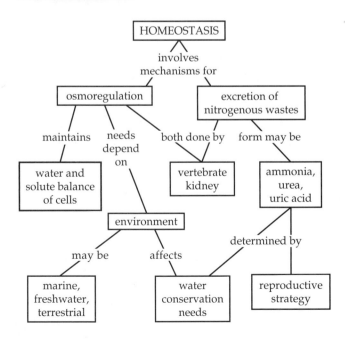

Chapter Review

Animals are able to survive large fluctuations in their external environment by regulating solute and water balance (**osmoregulation**). Animals must also dispose of nitrogen-containing wastes (**excretion**).

44.1 Osmoregulation balances the uptake and loss of water and solutes

Osmosis   Whatever an animal's habitat or nitrogenous waste product, its water gain must balance water loss.

Osmosis is the diffusion of water across a selectively permeable membrane that separates two solutions differing in **osmolarity** (moles of solute per liter). Osmolarity is expressed in units of milliosmoles per liter (mosm/L). *Isoosmotic* solutions are equal in osmolarity, and there is no net osmosis between them. There is a net flow of water from a *hypoosmotic* (more dilute) to a *hyperosmotic* (more concentrated) solution.

Osmotic Challenges   **Osmoconformers** are isoosmotic with their surroundings and do not regulate their osmolarity. The body fluids of **osmoregulators** are not isoosmotic with their external environment. Osmoregulators must get rid of excess water if they live in a hypoosmotic medium or take in water to offset osmotic loss if they inhabit a hyperosmotic environment. Osmoregulation is energetically costly because animals must actively transport solutes in order to maintain osmotic gradients needed to gain or lose water.

Most animals are **stenohaline,** able to tolerate only small changes in external osmolarity. Animals that are **euryhaline** can survive large differences in the osmotic environment.

Most marine invertebrates are osmoconformers, whereas marine vertebrates are osmoregulators. Marine bony fishes are hypoosmotic to seawater. They gain salt by diffusion, from food, and from the large quantities of seawater they drink to replace the water they lose by osmosis through their skin and gills. Excess salt is pumped out through the gills and other ions are excreted in the scanty urine.

Sharks have an internal salt concentration lower than that of seawater. Salt tends to diffuse into the body and is excreted by the rectal glands and kidneys. A shark's osmolarity is slightly higher than that of seawater because it maintains high concentrations of urea in its tissues, as well as trimethylamine oxide (TMAO), which protects proteins from the damaging effects of urea. Water that enters a shark's body by osmosis or in food is disposed of in urine produced in the kidneys.

Freshwater animals constantly take in water by osmosis and lose salts by diffusion. Freshwater fishes excrete large quantities of dilute urine. Salt supplies are replaced from their food or by active uptake of ions across the gills.

Some animals are capable of **anhydrobiosis,** surviving dehydration in a dormant state. Anhydrobiotic roundworms produce large quantities of trehalose, a disaccharide that replaces water around membranes and proteins during dehydration. Many insects survive freezing in the same manner.

Adaptations to prevent dehydration in terrestrial animals include water-impervious coverings and behavioral adaptations such as nocturnal lifestyles. Water is lost in urine and feces and through evaporation, and gained by drinking and eating moist food and by metabolic production during cellular respiration.

■ INTERACTIVE QUESTION 44.1

Indicate whether the following animals are isoosmotic, hyperosmotic, or hypoosmotic to their environment. Then briefly list their mechanisms of osmoregulation.

Animal	Osmotic Relation to Environment	Osmoregulatory Mechanisms
Marine invertebrate	a.	b.
Shark	c.	d.
Marine bony fish	e.	f.
Freshwater fish	g.	h.
Terrestrial animal	i.	j.

Transport Epithelia   The transport of solutes across a **transport epithelium** is essential for both osmotic regulation and disposing of metabolic wastes. The cells of this epithelium are linked by impermeable tight junctions and regulate the passage of specific solutes between the internal fluid and the environment. Transport epithelium is usually arranged in tubular networks that provide large surface areas for exchange.

44.2 An animal's nitrogenous wastes reflect its phylogeny and habitat

Ammonia, a small and toxic molecule, is produced when proteins and nucleic acids are metabolized. Some animals excrete ammonia; others expend energy to convert it to less toxic wastes such as urea or uric acid.

Forms of Nitrogenous Waste   Aquatic animals can excrete nitrogenous wastes as ammonia because it is very soluble and easily permeates membranes. Many invertebrates lose ammonia across their whole body surface. In fishes, most of the ammonia is passed across the epithelia of the gills. In freshwater fishes, the gills exchange NH_4^+ for Na^+, helping to maintain Na^+ concentrations in body fluids.

Mammals, most adult amphibians, sharks, and some marine bony fishes and turtles produce **urea,** a much less toxic compound than ammonia. Urea can be tolerated in a more concentrated form and excreted with less loss of water. Ammonia and carbon dioxide are combined in the liver to produce urea, which is then carried by the circulatory system to the kidneys. The production of urea requires energy.

Insects, land snails, and many reptiles, including birds, produce **uric acid,** a compound of low solubility in water that can be excreted as a semisolid with very little water loss. Its synthesis, however, is energetically expensive.

The Influence of Evolution and Environment on Nitrogenous Wastes   The mode of reproduction of animal groups seems to have determined whether they excrete uric acid or urea as their nitrogenous waste product.

The form of nitrogenous wastes also relates to habitat. Terrestrial turtles excrete mainly uric acid, whereas aquatic turtles excrete both urea and ammonia. Some animals actually shift their nitrogenous waste product depending on environmental conditions.

■ INTERACTIVE QUESTION 44.2

a. Which would produce more nitrogenous waste for the same size animal?

—an endotherm or an ectotherm?

—an herbivore or a carnivore?

Explain your answers.

b. Vertebrates that produce shelled eggs excrete _____. Mammals produce _____. What adaptive advantage do these types of nitrogenous wastes provide?

44.3 Diverse excretory systems are variations on a tubular theme

Excretory Processes During **filtration,** water and small solutes are forced out of the blood or body fluids into the excretory system. Two mechanisms transform this **filtrate** to urine: Valuable solutes are returned from the filtrate in **selective reabsorption.** In selective **secretion,** excess salts, toxins, and other solutes are added to the filtrate. The osmotic movement of water into or out of the filtrate follows the pumping of solutes.

Survey of Excretory Systems The **protonephridia** of flatworms are branched systems of closed tubules. Water and solutes from the interstitial fluid pass into the flame bulbs at the ends of the tubules, propelled by cilia. Urine empties into the external environment by way of nephridiopores. The flame-bulb system of freshwater flatworms is primarily osmoregulatory; metabolic wastes diffuse through the body surface or are excreted through the gastrovascular cavity. In parasitic flatworms, however, protonephridia function mainly in excretion of nitrogenous wastes.

Metanephridia are found in most annelids. They occur in pairs in each segment of an earthworm. An open ciliated funnel called a nephrostome collects coelomic fluid, which then moves through a folded tubule encased in capillaries. The transport epithelium of the tubule reabsorbs most solutes, which then reenter the blood. Hypoosmotic urine, carrying nitrogenous wastes and excess water absorbed by osmosis, exits by way of nephridiopores.

Malpighian tubules are excretory organs in insects and other terrestrial arthropods. Transport epithelia lining these blind sacs, which open into the digestive tract, secrete solutes from the hemolymph into the tubule. The fluid passes through the hindgut and into the rectum, where most of the solutes are pumped back into the hemolymph, and water follows by osmosis. In this water-conserving system, nitrogenous wastes are eliminated as dry matter along with the feces.

The numerous excretory tubules of most vertebrates are enclosed in a network of capillaries and arranged into compact kidneys. The kidneys, their associated blood vessels, and the structures that carry urine out of the body comprise the vertebrate excretory system.

■ INTERACTIVE QUESTION 44.3

Indicate whether these tubular systems function in osmoregulation, excretion, or both. If they function in osmoregulation, do they help conserve water or remove excess water that had entered by osmosis?

a. Protonephridia of freshwater planaria

b. Metanephridia of earthworms

c. Malpighian tubules of insects

d. Kidneys of terrestrial mammals

44.4 Nephrons and associated blood vessels are the functional units of the mammalian kidney

Blood enters each of the pair of bean-shaped kidneys through a **renal artery** and leaves by way of a **renal vein.** Urine exits through a **ureter** and is temporarily stored in the **urinary bladder.** Urine exits the body through the **urethra.**

Structure and Function of the Nephron and Associated Structures   The outer **renal cortex** and inner **renal medulla** of the kidney are packed with excretory tubules and blood vessels. The **nephron** consists of a long tubule with a cuplike **Bowman's capsule** at the blind end that encloses a ball of capillaries called the **glomerulus.**

Filtration of the blood occurs as blood pressure forces water, urea, salts, and small molecules through the porous capillary walls and specialized capsule cells called podocytes into Bowman's capsule, forming the filtrate. The filtrate passes through the **proximal tubule,** the **loop of Henle** with a descending limb and an ascending limb, and the **distal tubule. Collecting ducts** receive processed filtrate from many nephrons and pass the urine into the **renal pelvis.**

In the human kidney, 80% of the nephrons are **cortical nephrons** located entirely in the renal cortex. **Juxtamedullary nephrons,** found only in mammals and birds, have long loops of Henle that extend into the renal medulla and allow mammals to form hyperosmotic urine. The transport epithelium lining the nephrons and collecting ducts processes about 180 L of filtrate each day to produce about 1.5 L of urine.

An **afferent arteriole** supplies each nephron, subdividing to form the capillaries of the glomerulus, which then converge to form an **efferent arteriole.** This arteriole divides into a second capillary network—the **peritubular capillaries**—that surrounds the proximal and distal tubules. The **vasa recta** includes descending and ascending capillaries that parallel the loop of Henle. Exchange between the tubules and capillaries is via the interstitial fluid.

■ INTERACTIVE QUESTION   44.4

a. What is found in the filtrate that moves from Bowman's capsule into the tubule?

b. What remains in the capillaries?

From Blood Filtrate to Urine: A Closer Look   (1) Reabsorption and secretion both take place in the proximal tubule. Its transport epithelium selectively secretes H^+, ammonia, and drugs and poisons (processed by the liver and delivered via peritubular capillaries) into the filtrate. Glucose, amino acids, potassium, and the buffer bicarbonate are actively or passively reabsorbed. Salt diffuses from the filtrate into the cells of the proximal tubule, which then actively transport Na^+ across the membrane into the interstitial fluid; Cl^- is transported passively out of the cells to balance the positive charge; and water follows by osmosis. Salt and water diffuse from the interstitial fluid into the peritubular capillaries.

(2) The descending limb of the loop of Henle is freely permeable to water but not to salt or other small solutes. The interstitial fluid is increasingly hyperosmotic toward the inner medulla, so water exits by osmosis, leaving behind a filtrate with a high solute concentration.

(3) The ascending limb of the loop of Henle is not permeable to water but is permeable to salt, which diffuses out of the thin lower segment of the loop, increasing the osmolarity of the medulla. NaCl is actively transported out of the thick upper portion of the ascending limb. As salt leaves, the filtrate becomes less concentrated.

(4) The distal tubule regulates K^+ secretion and NaCl reabsorption, and it helps to regulate pH by secreting H^+ and reabsorbing bicarbonate.

(5) As the filtrate moves in the collecting duct through the osmotic gradient of the medulla, more and more water exits by osmosis. Urea diffuses out of the lower portion of the duct and helps to maintain the osmotic gradient. Salt excretion is controlled by the transport epithelium actively reabsorbing NaCl.

Review these stages in Interactive Question 44.5.

44.5   The mammalian kidney's ability to conserve water is a key adaptation to terrestrial environments

Solute Gradients and Water Conservation   The osmolarity gradient of NaCl and urea in the renal medulla is produced by the loops of Henle and collecting ducts and enables the human kidney to produce urine up to four times as concentrated as blood and normal interstitial fluid.

Filtrate leaving Bowman's capsule has an osmolarity of about 300 mosm/L, the same as blood. Both water and salt are reabsorbed in the proximal tubule; thus, the volume decreases but the osmolarity remains the same. In the trip down the descending limb of the loop of Henle, water exits by osmosis, and the filtrate becomes more concentrated. Salt, which is now in high concentration in the filtrate, diffuses out as the filtrate moves up the salt-permeable but water-impermeable ascending limb—helping to create the osmolarity gradient.

The flow of filtrate in the loop of Henle, along with the energy-consuming active transport of NaCl from the thick segment of the ascending limb, maintain the steep osmotic gradient between the cortex and medulla. This **countercurrent multiplier system** expends energy to create the gradient of the high salt concentration.

■ INTERACTIVE QUESTION 44.5

In this diagram of a nephron and collecting tubule, label the parts on the indicated lines. Label the numbered arrows to indicate the movement of salt, water, nutrients, K^+, bicarbonate (HCO_3^-), H^+, and urea into or out of the tubule.

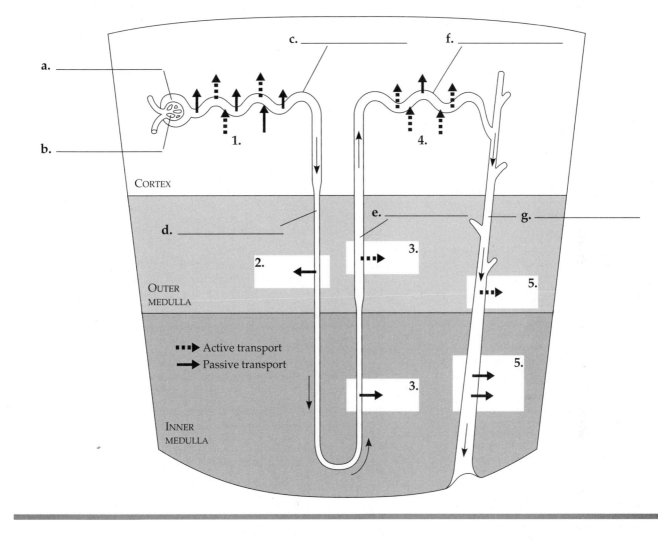

The vasa recta also has a countercurrent flow through the loop of Henle. As the blood in these vessels moves down into the inner medulla, water leaves by osmosis and salt enters; as the blood moves back up toward the cortex, water moves back in and salt diffuses out. Thus the capillaries can carry nutrients and other supplies to the medulla without disrupting the osmolarity gradient.

The filtrate makes one final pass through the medulla, this time in the collecting duct, which is permeable to water and urea but not to salt. Water flows out by osmosis, and as the filtrate becomes more concentrated, urea leaks into the interstitial fluid, adding to the high osmolarity of the inner medulla. The resulting concentrated urine may achieve an osmolarity as high as 1,200 mosm/L, isoosmotic to the interstitial fluid of the inner medulla.

Regulation of Kidney Function   Osmolarity of urine in humans can vary from 70 to 1,200 mosm/L, depending on the body's water and salt balance and urea production.

Antidiuretic hormone (ADH) is produced in the hypothalamus and stored in the pituitary gland. When blood osmolarity rises above a set point of 300 mosm/L, osmoreceptor cells in the hypothalamus trigger the release of ADH. This hormone increases water permeability of the distal tubules and collecting ducts, increasing water reabsorption from the urine. After consuming water in food or drink, negative feedback decreases the release of ADH. When blood osmolarity is low, little ADH is released, resulting in the production of large volumes of dilute urine, called diuresis. Alcohol inhibits ADH release and can cause dehydration.

The **juxtaglomerular apparatus (JGA),** located near the afferent arteriole leading to the glomerulus, responds to a drop in blood pressure or volume by releasing renin, an enzyme that converts the plasma protein angiotensinogen to **angiotensin II.** Angiotensin II functions as a hormone and constricts arterioles, stimulates the proximal tubules to reabsorb more NaCl and water, and stimulates the adrenal glands to release **aldosterone.** This hormone stimulates Na^+ and water reabsorption in the distal tubules. The **renin-angiotensin-aldosterone system (RAAS)** is a homeostatic feedback circuit that maintains adequate blood pressure and volume.

Atrial natriuretic factor (ANF) counters the RAAS. ANF, released by the atria of the heart in response to increased blood volume and pressure, inhibits the release of renin from the JGA, inhibits NaCl reabsorption by the collecting ducts, and reduces the release of aldosterone from the adrenals.

■ INTERACTIVE QUESTION   44.6

This concept map may look overwhelming, but if you work through it, it should help organize your understanding of the nervous and hormonal control of water and salt reabsorption in the kidneys.

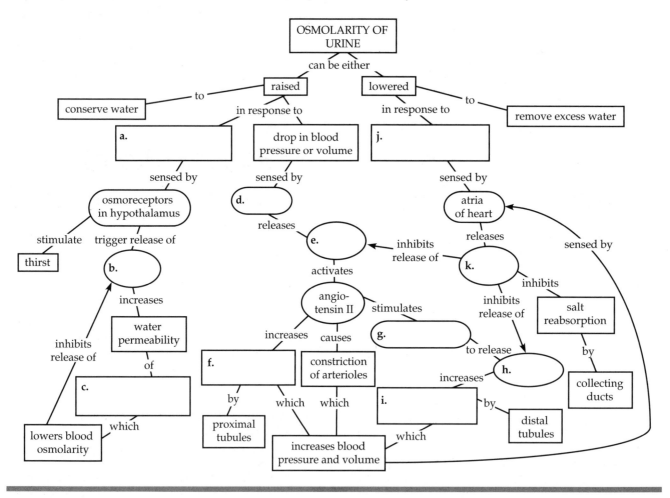

44.6   Diverse adaptations of the vertebrate kidney have evolved in different environments

Modifications of nephron structure and function correspond to osmoregulatory requirements in various habitats. Mammals in dry habitats, needing to conserve water and excrete hyperosmotic urine, have exceptionally long loops of Henle. Mammals that spend much of their time in fresh water have nephrons with very short loops. Birds also have juxtamedullary nephrons, but their shorter loops of Henle prevent them from producing urine as hyperosmotic as that of mammals. The production of uric acid is their main means of water conservation. The kidneys of other reptiles have only cortical nephrons, and their urine is isoosmotic to body fluids; water is conserved, however, through reabsorption in the cloaca and the production of uric acid as their nitrogenous waste.

Freshwater fishes are hyperosmotic to their environment and must excrete large quantities of very dilute urine. Salts are conserved by the efficient reabsorption of ions from the filtrate. When in fresh water, frogs and other amphibians actively absorb certain salts across their skin and excrete a dilute urine. When on land, water is reabsorbed from the urinary bladder. The kidneys of marine bony fishes excrete very little urine and function mainly to get rid of divalent ions taken in by drinking seawater.

■ INTERACTIVE QUESTION 44.7

a. What type of nephron allows the production of hyperosmotic urine?

b. Which animal groups have this type of nephron?

▶ Word Roots

an- = without; **hydro-** = water; **-bios** = life (*anhydrobiosis:* the ability to survive in a dormant state when an organism's habitat dries up)

anti- = against; **-diure** = urinate (*antidiuretic hormone:* a hormone that helps regulate water balance)

eury- = broad, wide; **-halin** = salt (*euryhaline:* organisms that can tolerate substantial changes in external osmolarity)

glomer- = a ball (*glomerulus:* a ball of capillaries surrounded by Bowman's capsule in the nephron and serving as the site of filtration in the vertebrate kidney)

homeo- = like, same; **-stasis** = standing (*homeostasis:* the steady-state physiological condition of the body)

juxta- = near to (*juxtaglomerular apparatus:* a specialized tissue located near the afferent arteriole that supplies blood to the glomerulus)

meta- = with; **-nephri** = kidney (*metanephridium:* in annelid worms, a type of excretory tubule with internal openings called nephrostomes that collect body fluids and external openings called nephridiopores)

osmo- = pushing; **-regula** = regular (*osmoregulation:* adaptations to control the water balance in organisms living in hyperosmotic, hypoosmotic, or terrestrial environments)

peri- = around (*peritubular capillaries:* the network of tiny blood vessels that surrounds the proximal and distal tubules in the kidney)

podo- = foot; **-cyte** = cell (*podocytes:* specialized cells of Bowman's capsule that are permeable to water and small solutes but not to blood cells or large molecules such as plasma proteins)

proto- = first (*protonephridium:* an excretory system, such as the flame-cell system of flatworms, consisting of a network of closed tubules having external openings called nephridiopores and lacking internal openings)

reni- = a kidney; **-angio** = a vessel; **-tens** = stretched (*renin-angiotensin-aldosterone system:* a part of a complex feedback circuit that normally partners with antidiuretic hormone in osmoregulation)

steno- = narrow (*stenohaline:* organisms that cannot tolerate substantial changes in external osmolarity)

vasa- = a vessel; **-recta** = straight (*vasa recta:* the capillary system that serves the loop of Henle)

▶ Structure Your Knowledge

1. List the substances that are filtered, secreted, and reabsorbed in the production of urine in mammals.
   a. Substances filtered:

   b. Substances secreted:

   c. Substances reabsorbed:

2. Referring back to the diagram in Interactive Question 44.5, describe the two-solute model that allows the production of concentrated urine.

▶ Test Your Knowledge

MULTIPLE CHOICE: *Choose the one best answer.*

1. A unicellular protist may use a contractile vacuole to expel excess water. Contractile vacuoles most likely would be found in protists that are
   a. in a freshwater environment.
   b. in a marine environment.
   c. internal parasites.
   d. hypoosmotic to their environment.
   e. isoosmotic to their environment.

2. Transport epithelia are responsible for
   a. pumping water across a membrane.
   b. forming an impermeable boundary at an interface with the environment.
   c. the movement of solutes for osmoregulation or excretion.
   d. transporting urine in the ureter and urethra.
   e. the passive transport of H^+ and HCO_3^- for the regulation of pH.

3. Which of the following is *incorrectly* paired with its excretory system?
   a. insect—Malpighian tubules
   b. flatworm—flame-bulb system
   c. earthworm—protonephridia
   d. amphibian—kidneys
   e. fish—kidneys

4. A freshwater fish would be expected to
   a. pump salt out through salt glands in the gills.
   b. produce copious quantities of dilute urine.
   c. diffuse urea across the epithelium of the gills.
   d. have scales that reduce water loss to the environment.
   e. do all of the above.

5. Which of the following is *not* part of the filtrate entering Bowman's capsule?
   a. water, salt, and electrolytes
   b. glucose
   c. urea
   d. plasma proteins
   e. amino acids

6. Which is the correct pathway for the passage of urine in vertebrates?
   a. collecting tubule → ureter → bladder → urethra
   b. renal vein → renal ureter → bladder → urethra
   c. nephron → urethra → bladder → ureter
   d. cortex → medulla → bladder → ureter
   e. renal pelvis → medulla → bladder → urethra

7. Aldosterone
   a. is a hormone that stimulates thirst.
   b. is secreted by the adrenal glands in response to a high osmolarity of the blood.
   c. stimulates the active reabsorption of Na^+ in the distal tubules.
   d. causes diuresis.
   e. is converted from a blood protein by the action of renin.

8. Which of the following statements is *incorrect?*
   a. Long loops of Henle are associated with steep osmotic gradients and the production of hyperosmotic urine.
   b. Ammonia is a toxic nitrogenous waste molecule that passively diffuses out of the bodies of aquatic invertebrates.
   c. Uric acid is the form of nitrogenous waste that requires the least amount of water to excrete.
   d. In the mammalian kidney, urea diffuses out of the collecting duct and contributes to the osmotic gradient within the medulla.
   e. Uric acid is produced by a mammalian fetus and removed through the placenta to the mother's excretory system.

9. The process of reabsorption in the formation of urine insures that
   a. excess H^+ is removed from the blood.
   b. drugs and other poisons are removed from the blood.
   c. urine is always hyperosmotic to interstitial fluid.
   d. glucose, salts, and water are returned to the blood.
   e. pH is maintained with a balance of hydrogen ions and bicarbonate.

10. The peritubular capillaries
    a. form the ball of capillaries inside Bowman's capsule from which filtrate is forced by blood pressure into the renal tubule.
    b. intertwine with the proximal and distal tubules and exchange solutes with the interstitial fluid.
    c. form a countercurrent flow of blood through the medulla to supply nutrients without interfering with the osmolarity gradient.
    d. surround the collecting ducts and reabsorb water, helping to create a hyperosmotic urine.
    e. rejoin to form the efferent arteriole.

11. ADH and the RAAS both increase water reabsorption, but they respond to different osmoregulatory problems. Which two of the following statements are true?
    1. ADH will be released in response to high alcohol consumption.
    2. ADH is released when osmoregulatory cells in the hypothalamus sense an increase in blood osmolarity.
    3. The RAAS will increase the osmolarity of urine due to the cooperative action of renin, aldosterone, and ANP.
    4. The RAAS is a response to a rise in blood pressure or volume.
    5. The RAAS is most likely to respond following an accident or severe case of diarrhea.
    a. 1 and 4
    b. 1 and 5
    c. 2 and 3
    d. 2 and 4
    e. 2 and 5

12. Which of the following sections of the mammalian nephron is *incorrectly* paired with its function?
    a. Bowman's capsule and glomerulus—filtration of blood
    b. proximal tubule—secretion of ammonia and H^+ into filtrate and transport of glucose and amino acids out of tubule
    c. descending limb of loop of Henle—diffusion of urea out of filtrate
    d. ascending limb of loop of Henle—diffusion and pumping of NaCl out of filtrate
    e. distal tubule—regulation of pH and K^+

13. One of the advantages of the production of uric acid by birds is that uric acid
    a. has low toxicity and can be safely stored in the egg.
    b. is very soluble in water and takes very little energy to produce.
    c. contributes to the production of the egg shell and can thus serve two purposes.
    d. takes less energy to produce than urea and is nontoxic.
    e. requires a moderate amount of water to excrete and is isoosmotic to body fluids.

14. What is the mechanism for the filtration of blood within the nephron?
    a. the active transport of Na^+ and glucose, followed by osmosis
    b. both active and passive secretion of ions, toxins, and NH_3 into the tubule
    c. high hydrostatic pressure of blood forcing water and small molecules out of the capillary
    d. the high osmolarity of the medulla that was created by active and passive transport of salt from the tubule and passive diffusion of urea from the collecting duct
    e. a lower osmotic pressure in Bowman's capsule compared to that in the glomerulus

15. What stimulus causes the atria of the heart to release ANF (atrial natriuretic factor)?
    a. a rise in blood pressure
    b. a drop in blood pressure
    c. a drop in blood pH
    d. a rise in blood osmolarity
    e. a drop in blood osmolarity

16. What stimulus causes the juxtaglomerular apparatus to release renin?
    a. a drop in blood pH
    b. a rise in blood pH
    c. a drop in blood pressure
    d. a rise in blood osmolarity
    e. the release of ANF by the atria of the heart

17. Which of the following would be most likely to produce the sugar trehalose?
    a. desert mammal
    b. euryhaline fish
    c. stenohaline protist
    d. anhydrobiotic roundworm
    e. vampire bat

18. To produce urine that is hyperosmotic to blood and interstitial fluids requires
    a. juxtamedullary nephrons.
    b. an increasing osmotic concentration down through the medulla.
    c. the inhibition of ADH release.
    d. all three of the above.
    e. only a and b above.

Chapter 45

Hormones and the Endocrine System

Framework

This chapter introduces the intricate system of chemical control and communication within animals. Endocrine cells or neurosecretory cells produce hormones that regulate the activity of other cells and organs. Steroid hormones bind with receptors within their target cells and influence gene expression. The signal-transduction pathways of most peptide/protein and amine hormones involve binding with cell surface receptors and triggering metabolic reactions in target cells.

The hypothalamus and pituitary gland play coordinating roles by integrating the nervous and endocrine systems and producing many tropic hormones that control the synthesis and secretion of hormones in other endocrine glands and organs.

Chapter Review

An animal **hormone** is a chemical signal usually transported through the bloodstream that elicits a specific response from target cells.

45.1 The endocrine system and the nervous system act individually and together in regulating an animal's physiology

Coordination and communication among the specialized parts of complex animals are achieved by the nervous system, which conveys high-speed messages along neurons, and the **endocrine system,** which produces hormones that travel more slowly and regulate many biological processes. **Endocrine glands** are ductless secretory organs that release hormones, which diffuse into the circulation.

Overlap Between Endocrine and Nervous Regulation The endocrine system and nervous system function together in regulating many physiological processes. **Neurosecretory cells** are specialized nerve cells that secrete hormones (sometimes called *neurohormones*). Several chemicals are used by both systems, as hormones in one and neurotransmitters in the other.

Control Pathways and Feedback Loops In a biological control system, a receptor/sensor detects a change (the stimulus) and notifies a control center, which sends out an efferent signal to direct an effector's response. An endocrine cell acts both as sensor and control center and sends out either a hormone or neurohormone signal. Three simple types of hormonal pathways are endocrine, neurohormone, and neuroendocrine pathways.

Both endocrine and nervous pathways may be controlled by **negative feedback.**

Explain the difference between negative and positive feedback in the control of hormonal pathways.

45.2   Hormones and other chemical signals bind to target cell receptors, initiating pathways that culminate in specific cell responses

Hormones are signals that travel throughout the body; local regulators act on neighboring cells; and pheromones communicate between different individuals.

The *reception* of a signal is its binding to a specific protein receptor on or in the target cell. *Signal transduction* leads to a *response* in the target cell. Three classes of hormones include peptide/protein and amine hormones, which are water soluble, and steroid hormones.

Cell-Surface Receptors for Water-Soluble Hormones
Most water-soluble hormones bind to plasma membrane receptors, initiating **signal-transduction pathways,** which convert extracellular signals to specific intracellular responses. Responses may vary depending on the type of receptor, signal-transduction pathway, and/or proteins for carrying out the response.

Intracellular Receptors for Lipid-Soluble Hormones
The protein receptors for steroid and thyroid hormones and the hormonal form of vitamin D are located within target cells. These small nonpolar hormones cross the membrane and bind to an intracellular receptor; the hormone-receptor complex is usually a transcription factor that binds to regulatory sites on DNA, either inducing or suppressing gene expression. Most receptors are located in the nucleus; some remain in the cytoplasm until bound to their hormone. A given signal can have different effects on different target cells and can produce different effects in different species.

Paracrine Signaling by Local Regulators   Local regulators are chemical signals that affect nearby target cells, called paracrine signaling. The neurotransmitter released from a neuron to its target cell is a type of local regulator. **Cytokines** are peptide/protein regulators involved in immune responses. **Growth factors** are peptides and proteins that are required in the extracellular environment for many types of cells to divide and develop.

Nitric oxide (NO) is a gas that can serve various roles as a smooth muscle relaxant that increases blood flow, a neurotransmitter, and a defense chemical.

Prostaglandins (PGs) are modified fatty acids that are released from most cells and have a wide range of effects on nearby target cells. In mammals, prostaglandins help to induce labor. Aspirin and ibuprofen inhibit the synthesis of prostaglandins and thus reduce their fever-and inflammation-inducing and pain-intensifying actions. The balance of antagonistic signals, such as the prostaglandins PGE and PGF that regulate blood flow to the lungs, is an important regulatory mechanism.

Briefly review the characteristics of the following:

a. hormone

b. local regulator

c. endocrine cell

d. neurosecretory cell

45.3   The hypothalamus and pituitary integrate many functions of the vertebrate endocrine system

Many organs whose main functions are nonendocrine, such as the heart, digestive tract, kidney, and liver, also secrete important hormones.

Relationship Between the Hypothalamus and Pituitary Gland   The **hypothalamus,** situated in the lower brain, plays a key role in integrating the endocrine and nervous systems. The hypothalamus receives nerve signals from throughout the body, and its neurosecretory cells release hormones that are stored in or regulate the pituitary gland at the base of the hypothalamus.

The **pituitary gland** has two discrete parts that develop separately and have different functions. The **posterior pituitary,** or **neurohypophysis,** is an extension of the hypothalamus that grows downward during embryonic development. It stores and secretes two hormones that are produced by and delivered from neurosecretory cells of the hypothalamus.

The **anterior pituitary,** or **adenohypophysis,** develops from the roof of the embryonic mouth. It synthesizes at least six hormones, several of which are **tropic hormones** that target other endocrine glands and are particularly important to chemical coordination. Some hypothalamic neurosecretory cells produce releasing hormones and inhibiting hormones that regulate the anterior pituitary. These tropic hormones are released into capillaries at the base of the hypothalamus and travel via a short portal vessel to capillary beds in the anterior pituitary.

Posterior Pituitary Hormones **Oxytocin** induces uterine contractions during birth and milk ejection during nursing. **Antidiuretic hormone (ADH)** functions in osmoregulation, increasing water retention by the kidneys and thus decreasing urine volume. Oxytocin signaling exhibits positive feedback, ADH is controlled by negative feedback.

■ **INTERACTIVE QUESTION   45.3**

What hormones are produced by the hypothalamus? To where and how are they transported?

Anterior Pituitary Hormones   Three of the tropic hormones produced by the anterior pituitary are similar glycoproteins. **Thyroid-stimulating hormone (TSH)** regulates production of thyroid hormones. **Follicle-stimulating hormone (FSH)** and **luteinizing hormone (LH),** also called **gonadotropins,** stimulate gonad activity. **Adrenocorticotropic hormone (ACTH)** is cleaved from a large precursor protein and stimulates the adrenal cortex to produce and secrete its steroid hormones. All four tropic hormones are part of complex neuroendocrine pathways.

The hormone **prolactin (PRL)** is a protein with diverse effects in different vertebrate species, ranging from milk production and secretion in mammals, to delay of metamorphosis in amphibians, to osmoregulation in fish.

Melanocyte-stimulating hormone (MSH) regulates the activity of pigment-containing cells in the skin of some vertebrates and appears to inhibit hunger in mammals. β-endorphin is classified as an **endorphin,** chemical signals that inhibit pain perception. Endorphins are also produced by certain neurons in the brain. MSH, β-endorphin, and ACTH all come from fragments of the same precursor protein.

Growth hormone (GH) affects a variety of target tissues. Its main tropic action is signalling the release of **insulin–like growth factors (IGFs)** that are produced by the liver and stimulate bone and cartilage growth. Gigantism and acromegaly are human growth disorders caused by excessive GH. Pituitary dwarfism can now be treated with genetically engineered GH.

■ **INTERACTIVE QUESTION   45.4**

Fill in the following table to review the hormones stored and released by the posterior pituitary (a and b) and secreted by the anterior pituitary (c–i).

Hormone	Main Actions
Oxytocin	a.
ADH	b.
TSH	c.
FSH and LH	d.
ACTH	e.
Prolactin	f.
MSH	g.
β-endorphin	h.
GH	i.

45.4 Nonpituitary hormones help regulate metabolism, homeostasis, development, and behavior

Thyroid Hormones   The **thyroid gland** produces two hormones, **triiodothyronine (T_3)** and **thyroxine (T_4).** A negative feedback loop controls secretion of thyroid hormones: Secretion of TSH-releasing hormone, or TRH, by the hypothalamus stimulates the anterior pituitary to secrete thyroid-stimulating hormone. TSH stimulates the thyroid gland, triggering the synthesis and release of thyroid hormones. High levels of T_3 and T_4 inhibit the secretion of TRH and TSH.

Thyroid hormones are critical to vertebrate development and maturation; an inherited deficiency in humans results in retarded skeletal and mental development, a condition known as cretinism.

The thyroid gland contributes to homeostasis in mammals, helping to maintain normal blood pressure, heart rate, muscle tone, digestion, and reproductive functions. T_3 and T_4 increase cellular metabolism. Excess thyroid hormone results in hyperthyroidism, with symptoms of weight loss, irritability, and high body temperature and blood pressure. Hypothyroidism can cause cretinism in infants or weight gain and lethargy in adults.

■ **INTERACTIVE QUESTION   45.5**

Explain how a lack of iodine in the diet may result in goiter, an enlarged thyroid gland.

Parathyroid Hormone and Calcitonin: Control of Blood Calcium   A balance between antagonistic hormones maintains blood calcium homeostasis. The four **parathyroid glands** secrete **parathyroid hormone (PTH),** which stimulates Ca^{2+} reabsorption in the kidney, and its release from bone to raise blood calcium levels. PTH also activates **vitamin D,** which then acts as a hormone to increase Ca^{2+} uptake from food in the intestines. If blood Ca^{2+} rises above the set point, the thyroid gland secretes **calcitonin,** a hormone that lowers calcium levels in the blood.

Insulin and Glucagon: Control of Blood Glucose   Scattered within the exocrine tissue of the **pancreas** are clusters of endocrine cells known as the **islets of Langerhans.** Within each islet are *alpha cells* that secrete the hormone **glucagon** and *beta cells* that secrete the hormone **insulin.** These antagonistic hormones regulate glucose concentration in the blood, and negative feedback controls their simple endocrine pathways.

Insulin lowers blood glucose levels by promoting the movement of glucose from the blood into body cells, by slowing the breakdown of glycogen in the liver, and by inhibiting the conversion of amino acids and glycerol (from fats) to sugar. Glucagon raises glucose concentrations by stimulating the liver to increase glycogen hydrolysis, convert amino acids and glycerol to glucose, and release glucose to the blood.

In **diabetes mellitus** the absence of insulin in the bloodstream or the loss of response to insulin in target tissues reduces glucose uptake by cells. Glucose accumulates in the blood and is excreted in the urine, with an accompanying loss of water. The body must use fats for fuel, and acids from fat breakdown may lower blood pH.

Type I diabetes mellitus, also known as insulin-dependent diabetes, is an autoimmune disorder in which pancreatic cells are destroyed. This type of diabetes is treated by regular injections of genetically engineered human insulin. More than 90% of diabetics have *type II diabetes mellitus,* or noninsulin-dependent diabetes, characterized either by insulin deficiency or reduced responsiveness of target cells. Exercise and dietary control are often sufficient to manage this disease.

Adrenal Hormones: Response to Stress   In mammals the **adrenal glands** consist of two different glands: the outer *adrenal cortex* and the central *adrenal medulla.* The adrenal medulla produces **epinephrine** (adrenaline) and **norepinephrine** (noradrenaline)—both of which are **catecholamines,** a class of compounds synthesized from the amino acid tyrosine.

Epinephrine and norepinephrine, released in response to positive or negative stress, increase the availability of energy sources by stimulating glycogen hydrolysis in skeletal muscle and the liver, glucose release from the liver, and fatty acid release from fat cells.

■ **INTERACTIVE QUESTION   45.6**

Complete this concept map on the regulation of blood glucose levels by hormones of the pancreas.

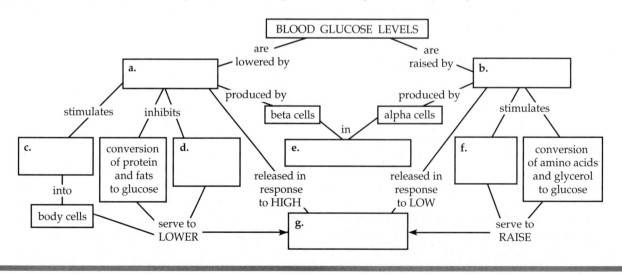

These hormones increase metabolic rate and the rate and volume of the heartbeat, dilate bronchioles in the lungs, and influence the contraction or relaxation of smooth muscles to increase blood supply to the heart, brain, and skeletal muscles, while reducing the supply to other organs. Norepinephrine's primary role is in sustaining blood pressure. The sympathetic division of the autonomic nervous system stimulates the release of epinephrine from the adrenal medulla.

The adrenal cortex responds to endocrine signals released in response to stress. A releasing hormone from the hypothalamus causes the anterior pituitary to release ACTH. This tropic hormone stimulates the adrenal cortex to synthesize and secrete **corticosteroids,** a group of steroid hormones.

The two main human corticosteroids are the **glucocorticoids,** such as cortisol, and the **mineralocorticoids,** such as aldosterone. Glucocorticoids promote synthesis of glucose from noncarbohydrates (such as muscle proteins) and thus increase energy supplies during stress. Cortisol has been used to treat serious inflammatory conditions such as arthritis, but its immunosuppressive effects can be dangerous.

Mineralocorticoids affect salt and water balance. Aldosterone stimulates kidney cells to reabsorb sodium ions and water from the filtrate. ACTH secreted by the anterior pituitary in response to severe stress increases aldosterone secretion. Both glucocorticoids and mineralocorticoids appear to help maintain homeostasis during extended periods of stress.

A third group of corticosteroids produced by the adrenal cortex are sex hormones, in particular androgens, which appear to influence female sex drive.

■ INTERACTIVE QUESTION   45.7

Describe how the adrenal gland is signaled by the hypothalamus and responds to short-term and long-term stress.

a. Short-term stress:

b. Long-term stress:

Gonadal Sex Hormones   The testes of males and ovaries of females produce steroids that affect growth, development, and reproductive cycles and behaviors. The three major categories of gonadal steroids—androgens, estrogens, and progestins—are found in different proportions in males and females.

The testes primarily synthesize **androgens,** such as **testosterone,** which determine the gender of the developing embryo and stimulate development of the male reproductive system and secondary sex characteristics. **Estrogens** regulate the development and maintenance of the female reproductive system and secondary sex characteristics. In mammals **progestins** help prepare and maintain the uterus for the growth of an embryo.

In a complex neuroendocrine pathway, a hypothalamic releasing hormone, GnRH, controls secretion of FSH and LH, gonadotropins from the anterior pituitary gland that control the synthesis of estrogens and androgens.

Melatonin and Biorhythms   The **pineal gland,** located near the center of the mammalian brain, secretes **melatonin,** which regulates functions related to light and changes in day length. The secretion of melatonin at night may function with a biological clock for daily or seasonal activities such as reproduction.

45.5   Invertebrate regulatory systems also involve endocrine and nervous system interactions

In invertebrates, homeostasis, reproduction, development, and behavior are usually controlled by an integration of endocrine and nervous systems.

Arthropods have extensive endocrine systems. Insects have three hormones that interact to control molting and development of adult characteristics. **Brain hormone,** produced by neurosecretory cells in the brain, stimulates the prothoracic glands to secrete ecdysone. **Ecdysone** functions to trigger molts and to promote development of adult characteristics. **Juvenile hormone** counters the action of ecdysone and promotes retention of larval characteristics during molting. When the JH level is high, ecdysone-induced molting produces larger larval stages; only after the JH level has decreased does molting result in a pupa.

▶ Word Roots

adeno- = gland; **-hypo** = below (*adenohypophysis:* also called the anterior pituitary, a gland positioned at the base of the hypothalamus)

andro- = male; **-gen** = produce (*androgens:* the principal male steroid hormones, such as testosterone, which stimulate the development and maintenance of the male reproductive system and secondary sex characteristics)

anti- = against; **-diure** = urinate (*antidiuretic hormone:* a hormone that helps regulate water balance)

cata- = down; **-chol** = anger (*catecholamines:* a class of compounds, including epinephrine and norepinephrine, synthesized from the amino acid tyrosine)

-cortico = the shell; **-tropic** = to turn or change (*adrenocorticotropic hormone:* a peptide hormone released from the anterior pituitary, it stimulates the production and secretion of steroid hormones by the adrenal cortex)

ecdys- = an escape (*ecdysone:* a steroid hormone that triggers molting in arthropods)

endo- = inside (*endorphin:* a hormone produced in the brain and anterior pituitary that inhibits pain perception)

epi- = above, over (*epinephrine:* a hormone produced as a response to stress; also called adrenaline)

gluco- = sweet (*glucagon:* a peptide hormone secreted by pancreatic endocrine cells that raises blood glucose levels; an antagonistic hormone to insulin)

lut- = yellow (*luteinizing hormone:* a gonadotropin secreted by the anterior pituitary)

melan- = black (*melatonin:* a modified amino acid hormone secreted by the pineal gland)

neuro- = nerve (*neurohypophysis:* also called the posterior pituitary, it is an extension of the brain)

oxy- = sharp, acid (*oxytocin:* a hormone that induces contractions of the uterine muscles and causes the mammary glands to eject milk during nursing)

para- = beside, near (*parathyroid glands:* four endocrine glands, embedded in the surface of the thyroid gland, that secrete parathyroid hormone and raise blood calcium levels)

pro- = before; **-lact** = milk (*prolactin:* a hormone produced by the anterior pituitary gland, it stimulates milk synthesis in mammals)

tri- = three; **-iodo** = violet (*triiodothyrodine:* one of two very similar hormones produced by the thyroid gland and derived from the amino acid tyrosine)

Structure Your Knowledge

1. Briefly describe the two general mechanisms by which chemical signals trigger responses in target cells, depending on the location of the receptor.

2. Most homeostatic functions are maintained by a balance between hormones with antagonistic effects. Describe how the thyroid and parathyroid glands regulate calcium levels in the blood.

Test Your Knowledge

MATCHING: *Match the hormone and gland or organ that produces it to the descriptions. Choices may be used more than once, and not all choices are used.*

Hormones	Gland or Organ
A. ACTH	a. adrenal cortex
B. androgens	b. adrenal medulla
C. ADH	c. hypothalamus
D. calcitonin	d. pancreas
E. epinephrine	e. parathyroid
F. glucagon	f. pineal
G. glucocorticoids	g. pituitary
H. insulin	h. testis
I. melatonin	i. thymus
J. oxytocin	j. thyroid
K. PTH	
L. thyroxine	

Hormone	Gland	Hormone Action
_____	_____	1. involved in biological clock and seasonal activities
_____	_____	2. break down muscle protein for conversion to glucose
_____	_____	3. increase blood sugar, glycogen breakdown in liver
_____	_____	4. stimulate development of male reproductive system
_____	_____	5. stimulate adrenal cortex to synthesize corticosteroids
_____	_____	6. increase available energy, heart rate, metabolism
_____	_____	7. regulate metabolism, growth, and development
_____	_____	8. lower blood calcium levels
_____	_____	9. increase reabsorption of water by kidney
_____	_____	10. stimulate contraction of uterus, milk secretion

MULTIPLE CHOICE: *Choose the one best answer.*

1. Which of the following is *not* an accurate statement about hormones?
   a. Not all hormones are secreted by endocrine glands.
   b. Most hormones move through the circulatory system to their destination.
   c. Target cells have specific protein receptors for hormones.
   d. Hormones are essential to homeostasis.
   e. Steroid hormones often function as neurotransmitters.

2. The best description of the difference between pheromones and hormones is that
   a. pheromones are small, volatile molecules, whereas hormones are steroids.
   b. pheromones are involved in reproduction, whereas hormones are not.
   c. pheromones are a form of neural communication; hormones are a form of chemical communication.
   d. pheromones are signals that function between organisms, whereas hormones communicate among the parts within an organism.
   e. pheromones are local regulators, whereas hormones travel greater distances.

3. Which one of the following hormones is *incorrectly* paired with its origin?
   a. releasing hormones—hypothalamus
   b. growth hormone—anterior pituitary
   c. progestins—ovary
   d. TSH—thyroid
   e. mineralocorticoids—adrenal cortex

4. Which of the following is an example of a positive feedback mechanism?
   a. the liver's production of insulin-like growth factors in response to growth hormone, which promotes skeletal growth
   b. the ability of the neurotransmitter acetylcholine to cause skeletal muscle to contract, heart muscle to relax, and cells of the adrenal medulla to secrete epinephrine
   c. prostaglandins released from placental cells promoting muscle contraction during childbirth, with muscle contractions stimulating more prostaglandin release
   d. the interplay of prostaglandin E and prostaglandin F on blood vessels servicing the lungs
   e. elevated levels of stress resulting in neural stimulation of the adrenal medulla and hormonal stimulation of the adrenal cortex

5. Ecdysone
   a. is a steroid hormone produced in insects that promotes retention of larval characteristics.
   b. is responsible for color changes in amphibians.
   c. is a hormone secreted from specialized neurons that stimulates egg laying in the mollusk *Aplysia*.
   d. is secreted by prothoracic glands in insects and triggers molts and development of adult characteristics.
   e. is involved in metamorphosis in amphibians.

6. Which of the following local regulators amplifies the sensation of pain?
   a. prostaglandins
   b. PGE and PGF
   c. growth factors
   d. cytokines
   e. nitric oxide

7. Antidiuretic hormone (ADH)
   a. is produced by cells in the kidney and liver in response to low blood volume or pressure.
   b. inhibits the reabsorption of Na^+ from the urine.
   c. is released from the posterior pituitary and increases water reabsorption in the kidneys.
   d. is a steroid hormone produced by the adrenal cortex.
   e. is part of the system that regulates salt balance.

8. The anterior pituitary
   a. stores oxytocin and ADH produced by the hypothalamus.
   b. receives releasing and inhibiting hormones from the hypothalamus through portal vessels connecting capillary beds.
   c. produces several releasing and inhibiting hormones.
   d. is responsible for nervous and hormonal stimulation of the adrenal glands.
   e. produces only tropic hormones.

9. Acromegaly, the abnormal growth of bones of the hands, feet, and head, is caused by
   a. an autoimmune disorder.
   b. an excess of thyroxine.
   c. an excess of glucocorticoids.
   d. an abnormally high androgen-to-estrogen ratio.
   e. an excess of growth hormone.

10. Which of the following hormones is *not* involved with increasing the blood glucose concentration?
    a. glucagon
    b. epinephrine
    c. glucocorticoids
    d. ACTH (adrenocorticotropic hormone)
    e. insulin

11. MSH (melanocyte-stimulating hormone) causes frog skin cells to darken when added to the interstitial fluid but has no effect on color when injected into the cells. Which of the following does this evidence support?
    a. MSH is a tropic hormone.
    b. MSH is a neurohormone.
    c. MSH is a steroid hormone.
    d. MSH binds to cell surface receptors.
    e. Both c and d are reasonable conclusions.

12. Which of the following is *not* true of norepinephrine?
    a. It is secreted by the adrenal medulla.
    b. It maintains blood pressure.
    c. Its release is stimulated by ACTH.
    d. It serves as a neurotransmitter.
    e. It is part of the flight-or-fight response to stress.

Choose from the following hormones to answer questions 13–15.
    a. ACTH
    b. parathyroid hormone
    c. epinephrine
    d. estrogen
    e. insulin

13. Which of the above is a tropic hormone?

14. Which hormone is a steroid hormone?

15. Which hormone is released in response to a nervous impulse?

16. A lack of iodine in the diet can lead to formation of a goiter because
    a. excess thyroxine production causes the thyroid gland to enlarge.
    b. with limited iodine available, triiodothyronine (T_3) rather than thyroxine (T_4) is produced.
    c. iodine is a key ingredient of the growth hormones that control growth of the thyroid.
    d. little thyroxine is produced, TRH and TSH production is not inhibited, and thyroid stimulation continues.
    e. the hypothalamus is stimulated to increase its production of TRH.

17. A tropic hormone is a hormone
    a. whose target tissue is another endocrine gland.
    b. that is produced by the hypothalamus but stored and released from the posterior pituitary.
    c. that acts by negative feedback to regulate its own level in the body.
    d. from the hypothalamus that regulates the synthesis and secretion of hormones of the posterior pituitary.
    e. that is released in response to nervous stimulation.

18. What is the best description of the mechanism of action of steroid hormones?
    a. transported by neurosecretory cells directly to target tissues
    b. form a hormone-receptor complex inside the cell that regulates gene expression
    c. amplified response using second messengers
    d. bind to membrane-bound receptors and initiate a signal-transduction pathway
    e. secreted into the interstitial fluid and act as a local regulator

Chapter 46
Animal Reproduction

Framework

This chapter covers the patterns and mechanisms of animal reproduction. In sexual reproduction, fertilization may occur externally or internally. Development of the zygote may take place internally within the female; externally in a moist environment; or in a protective, resistant egg. Placental mammals provide nourishment as well as shelter for the embryo, and they nurse their young.

The human reproductive system is described, including its organs, glands, hormones, gamete formation, and sexual response. The chapter also covers pregnancy and birth, contraception, and recently developed reproductive technologies.

Chapter Review

46.1 Both asexual and sexual reproduction occur in the animal kingdom

In **asexual reproduction,** a single individual produces offspring, usually using only mitotic cell division. In **sexual reproduction,** two meiotically formed haploid **gametes** fuse to form a diploid **zygote,** which develops into an offspring. Gametes are usually a relatively large, nonmotile **egg** or **ovum** and a small, motile **sperm.** Sexual reproduction combines genes from two parents and produces offspring with varying phenotypes.

Mechanisms of Asexual Reproduction Many invertebrates can reproduce asexually by **fission,** in which a parent is separated into two or more individuals of equal size; by **budding,** in which a new individual grows out from the parent's body; or by **fragmentation,** in which the body is broken into several pieces, each of which develops into a complete animal. **Regeneration** is necessary for a fragment to develop into a new organism.

Reproductive Cycles and Patterns Periodic reproductive cycles may be linked to favorable conditions or energy supplies. A combination of environmental and hormonal cues controls the timing of these cycles.

The freshwater crustacean *Daphnia*, like aphids and rotifers, produces eggs that develop by **parthenogenesis,** as well as eggs that are fertilized. In bees, wasps, and ants, males are produced parthenogenetically, whereas sterile worker females and reproductive females are produced from fertilized eggs. In a few parthenogenetic fishes, amphibians, and lizards, doubling of chromosomes after meiosis creates diploid "zygotes."

In **hermaphroditism,** found in some sessile, burrowing, and parasitic animals, each individual has functioning male and female reproductive systems. Mating results in fertilization of both individuals. In some fishes and oysters, individuals reverse their sex during their lifetime, a pattern known as **sequential hermaphroditism.** Sex reversal may be related to age or to the relative advantage conferred by size. Thus some sequential hermaphrodites are females first; others are males first.

■ INTERACTIVE QUESTION   46.1

a. What adaptive advantages would asexual reproduction provide?

b. What adaptive advantage would sexual reproduction provide?

46.2   Fertilization depends on mechanisms that help sperm meet eggs of the same species

Fertilization is the union of sperm and egg. In **external fertilization,** eggs and sperm are shed, and fertilization occurs in the environment. **Internal fertilization** involves the placement of sperm in or near the female reproductive tract so that the egg and sperm unite internally.

External fertilization occurs almost exclusively in moist habitats, where the gametes and developing zygote are not in danger of desiccation.

Behavioral cooperation, as well as copulatory organs and sperm receptacles, are required for internal fertilization.

Pheromones are small, volatile chemical signals that may function as mate attractants.

■ INTERACTIVE QUESTION   46.2

List three mechanisms that may help to ensure that gamete release is synchronized when fertilization is external.

a.

b.

c.

Ensuring the Survival of Offspring   Developing embryos may receive some type of protection. The shells of the amniote eggs of birds, reptiles, and monotremes protect the embryo from water loss or injury. Embryos of eutherian (placental) mammals develop in the uterus, nourished from the mother's blood through the placenta. Parental care of young is widespread among vertebrates and even among many invertebrates.

Gamete Production and Delivery   The simplest reproductive systems do not even have **gonads** to produce gametes. In polychaete annelids, eggs or sperm develop from cells lining the coelom. Parasitic flatworms are hermaphroditic and have complex reproductive systems.

Most insects have separate sexes and complex reproductive systems. Sperm develop in the testes, are stored in the seminal vesicles, and are ejaculated into the female. Eggs are produced in the ovaries and fertilized in the vagina. Females may have a **spermatheca,** or sperm-storing sac.

With the exception of most mammals, many vertebrates have a common opening, a **cloaca,** for the digestive, excretory, and reproductive systems. Many non-mammalian vertebrates evert the cloaca to ejaculate.

Most mammals have a separate opening for the digestive tract. The urethra is used by both the excretory and reproductive systems in males; most female mammals have a separate vagina and urethra.

46.3   Reproductive organs produce and transport gametes: focus on humans

Female Reproductive Anatomy   The external reproductive structures include the clitoris and two sets of labia. The female **ovaries** contain many **follicles,** which are sacs of cells that nourish and protect the egg cell contained within each of them. During each menstrual cycle, a maturing follicle produces estrogens, the primary female sex hormones. Following **ovulation,** the follicle forms a solid mass called the **corpus luteum,** which secretes progesterone and estrogens.

The egg cell is expelled into the abdominal cavity and swept by cilia into the **oviduct,** or fallopian tube, through which it is transported to the **uterus.** The **endometrium,** or lining of the uterus, is highly vascularized. The neck of the uterus, the **cervix,** opens into the **vagina,** the thin-walled birth canal and repository for sperm during copulation. It opens at the **vulva.**

The vaginal opening is initially covered by a membrane called the **hymen.** The separate openings of the vagina and urethra are in a region called the **vestibule,** enclosed by two pairs of skin folds, the inner **labia minora** and the outer **labia majora.** The **clitoris,** composed of erectile tissue, is at the top of the vestibule. **Bartholin's glands** secrete mucus into the vestibule during sexual arousal.

A **mammary gland,** located within a breast, is composed of fatty tissue and a series of small milk-secreting sacs that drain into ducts that open at the nipple. The lack of estrogen in males prevents their mammary glands from enlarging and developing secretory functions.

Male Reproductive Anatomy   The external male reproductive organs include the scrotum and penis. Sperm are produced in the highly coiled **seminiferous tubules** of the **testes. Leydig cells** produce testosterone and other androgens. Sperm production requires a cooler temperature than the internal body temperature of most mammals, so the **scrotum** suspends the testes below the abdominal cavity.

Sperm pass from a testis into the coiled **epididymis,** in which they mature and gain motility. During **ejacu-**

lation, sperm are propelled through the **vas deferens,** into a short **ejaculatory duct,** and out through the **urethra,** which runs through the penis.

Three sets of accessory glands add secretions to the **semen.** The **seminal vesicles** contribute an alkaline fluid containing mucus, fructose (as an energy source for the sperm), a coagulating enzyme, ascorbic acid, and prostaglandins. The **prostate gland** produces a secretion that contains anticoagulant enzymes and citrate. Benign enlargement of the prostate and prostate cancer are common medical problems in older men. The **bulbourethral glands** produce a mucus that neutralizes urine remaining in the urethra.

The 2–5 ml of ejaculated semen may contain more than 100 to 650 million sperm. The slightly alkaline semen neutralizes the vagina. Semen initially coagulates, then liquifies as the sperm start swimming through the female tract.

The **penis** is composed of spongy tissue that engorges with blood during sexual arousal, producing an erection that facilitates insertion of the penis into the vagina. Some mammals have a **baculum,** a bone that helps stiffen the penis. The head of the penis, called the **glans penis,** is covered by a fold of skin called the **prepuce,** or foreskin.

■ **INTERACTIVE QUESTION   46.3**

Label the indicated structures in these diagrams of the human male and female reproductive systems.

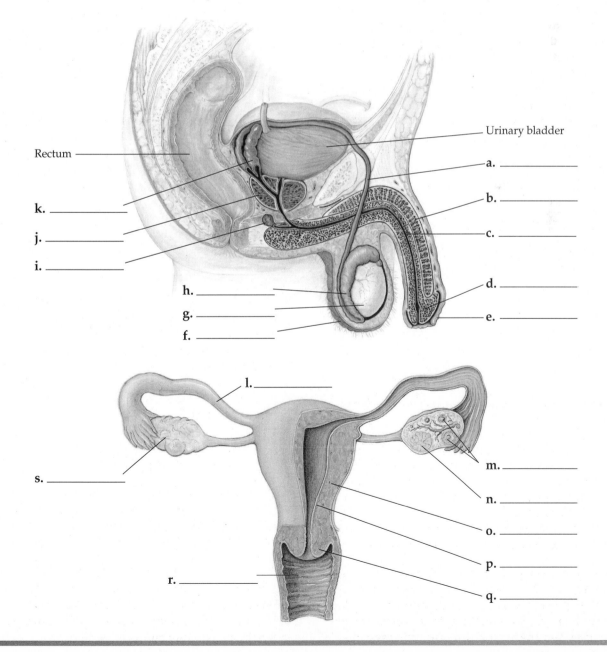

Human Sexual Response   The human sexual response cycle includes two types of physiological reactions: **vasocongestion,** increased blood flow to a tissue, and **myotonia,** increased muscle tension. The excitement phase involves vasocongestion and vaginal lubrication, preparing the vagina and penis for **coitus,** or sexual intercourse. The plateau phase continues vasocongestion and myotonia, and breathing rate and heart rate increase. In both sexes, **orgasm** is characterized by rhythmic, involuntary contractions of reproductive structures. In males, emission deposits semen in the urethra, and ejaculation occurs when the urethra contracts and semen is expelled. In the resolution phase, vasocongested organs return to normal size, and muscles relax.

46.4   In humans and other mammals, a complex interplay of hormones regulates gametogenesis

Gametogenesis, the production of gametes, differs in females and males. **Oogenesis** begins in the female embryo. **Oogonia,** the stem cells that give rise to ova, divide and differentiate into **primary oocytes,** which are arrested in prophase I of meiosis. Following puberty, FSH periodically stimulates a primary oocyte to finish meiosis I and develop into a **secondary oocyte,** which is arrested in metaphase II. In humans, meiosis is completed if a sperm penetrates the oocyte. Meiotic cytokinesis is unequal, producing one large ovum and up to three small haploid polar bodies that disintegrate. Researchers recently reported finding multiplying oogonia in the ovaries of adult mice.

Spermatogenesis, the production of mature sperm cells, occurs continuously in the seminiferous tubules of the testes as **spermatogonia,** the stem cells that give rise to sperm, differentiate into spermatocytes that undergo meiosis to produce spermatids. The haploid nucleus of a sperm is contained in a head, tipped with an **acrosome,** which contains enzymes that help the sperm penetrate the egg. Mitochondria provide ATP for movement of the flagellum, or tail.

■ INTERACTIVE QUESTION   46.4

List the three important ways in which oogenesis differs from spermatogenesis.

a.

b.

c.

The Reproductive Cycles of Females   Female humans and many primates have **menstrual cycles,** during which the endometrium lining thickens to prepare for the implantation of the embryo and then is shed if fertilization does not occur. This bleeding, called **menstruation,** occurs on a cycle of approximately 28 days in humans. Other mammals have **estrous cycles,** during which the endometrium thickens, but if fertilization does not occur, it is reabsorbed. Estrous cycles are often coordinated with season, and females are receptive to sexual activity only during **estrus,** or heat, the period surrounding ovulation.

The human female reproductive cycle involves the integration of the **uterine cycle** (menstrual cycle) and the **ovarian cycle.** Five hormones coordinate the ovarian and uterine cycles using positive and negative feedback.

The ovarian cycle begins with the release from the hypothalamus of GnRH (gonadotropin-releasing hormone), which stimulates the anterior pituitary to secrete small amounts of FSH (follicle-stimulating hormone) and LH (luteinizing hormone). In this **follicular phase,** FSH stimulates follicular growth, and the cells of the follicle secrete estrogen. The slow rise in estrogen inhibits the release of pituitary gonadotropins (FSH and LH).

When the secretion of estrogen by the growing follicle rises sharply, the hypothalamus is stimulated to increase GnRH output, which results in a rise in LH and FSH release. The increase in LH induces maturation of the follicle, which enlarges and forms a bulge near the ovary surface. **Ovulation** occurs about a day after the LH surge, with the rupture of the follicle and adjacent ovary wall.

In the **luteal phase,** LH stimulates the transformation of the ruptured follicle to form the corpus luteum, which secretes estrogen and progesterone. The rising level of these hormones exerts negative feedback on the hypothalamus and pituitary, inhibiting secretion of LH and FSH. Without LH to maintain it, the corpus luteum disintegrates, thereby dropping the levels of estrogens and progesterone. This drop releases the inhibition of the hypothalamus and pituitary, and FSH and LH secretion begins again, stimulating growth of new follicles and the start of the next ovarian cycle.

The uterine (menstrual) cycle is controlled by the hormones secreted by the ovaries. Estrogen secreted by the growing follicles causes the endometrium to begin to thicken in the **proliferative phase.** After ovulation, estrogen and progesterone stimulate increased vascularization of the endometrium and the development of glands that secrete a nutrient fluid. Thus, the luteal phase of the ovarian cycle corresponds with the **secretory phase** of the uterine cycle.

The rapid drop of ovarian hormones caused by the disintegration of the corpus luteum reduces blood supply to the endometrium and begins its disintegration, leading to the **menstrual flow phase** of the uterine cycle. By convention, the uterine and ovarian cycles begin with the first day of menstruation.

Estrogens are also responsible for female secondary sex characteristics and influence sexual behavior.

Menopause, the cessation of ovulation and menstruation, results from a decline in the production of estrogens as the ovaries become less responsive to FSH and LH.

■ INTERACTIVE QUESTION   46.5

In the diagram of the human female reproductive cycle, label the lines indicating the levels of the gonadotropic and the ovarian hormones in the blood and the phases of the ovarian and uterine cycles.

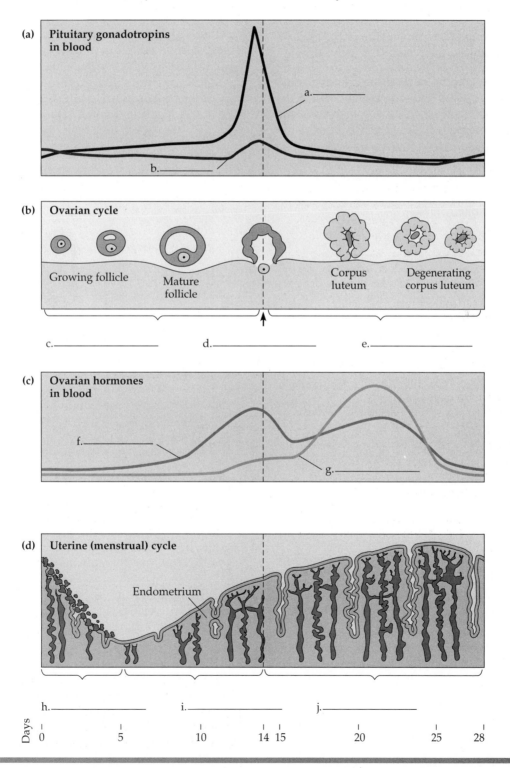

(a) **Pituitary gonadotropins in blood**

a._____

b._____

(b) **Ovarian cycle**

Growing follicle   Mature follicle   Corpus luteum   Degenerating corpus luteum

c._____   d._____   e._____

(c) **Ovarian hormones in blood**

f._____

g._____

(d) **Uterine (menstrual) cycle**

Endometrium

h._____   i._____   j._____

Days   0   5   10   14 15   20   25   28

Hormonal Control of the Male Reproductive System

Androgens are responsible for the male primary (associated with reproduction) and secondary (associated with voice deepening, hair distribution, and muscle growth) sex characteristics. The most important androgen is testosterone, produced mainly by Leydig cells of the testes. Androgens are also determinants of sexual and other behaviors.

■ **INTERACTIVE QUESTION 46.6**

Fill in the blanks in the following description of the control of male reproductive hormones.

a. _____, produced by the hypothalamus, regulates the release of gonadotropic hormones from the anterior pituitary. b. _____ stimulates Sertoli cells of the seminiferous tubules, which nourish sperm and promote spermatogenesis. c. _____ stimulates the production of d. _____ by Leydig cells. Testosterone contributes to spermatogenesis and development of e. _____. The production of GnRH, LH, and FSH is regulated by f. _____.

46.5 In humans and other placental mammals, an embryo grows into a newborn in the mother's uterus

Conception, Embryonic Development, and Birth The development of one or more embryos in the uterus is called **pregnancy** or **gestation.** The gestation period correlates with body size and the degree of development of the young at birth. Human pregnancy averages 266 days (38 weeks).

Fertilization, or **conception,** occurs in the oviduct. **Cleavage** produces a ball of cells by the time the embryo reaches the uterus in about 3 to 4 days. The **blastocyst,** a sphere of cells around a cavity and inner cell mass, implants into the endometrium about 7 days after conception.

The embryo secretes **human chorionic gonadotropin (HCG),** which maintains the corpus luteum's secretion of progesterone and estrogens through the first few months.

Human gestation can be divided into three **trimesters.** After implantation, the outer layer of the blastocyst, the **trophoblast,** grows and mingles with the endometrium and form the **placenta,** a disk-shaped organ in which gas and nutrient exchange and waste removal take place between maternal blood and embryonic capillaries. Development proceeds to **organogenesis,** and by the eighth week, the embryo has all the rudimentary structures of the adult and is called a **fetus.** High progesterone levels stimulate growth of

the uterus and maternal part of the placenta, cessation of menstrual cycling, and breast enlargement.

During the second trimester, HCG declines, the corpus luteum degenerates, and the placenta secretes its own progesterone, which maintains the pregnancy. The fetus grows rapidly and is quite active. The third trimester is a period of rapid fetal growth.

Through a positive feedback system, hormones and local regulators induce **labor.** High levels of estrogen stimulate the development of oxytocin receptors on the uterus. Oxytocin produced by the fetus and the mother's posterior pituitary stimulates contractions and placental secretion of prostaglandins, which further enhance uterine contractions.

A series of strong contractions of the uterus results in birth, or **parturition.** During the first stage of labor, the cervix dilates. The second stage consists of the contractions that force the fetus out of the uterus and through the vagina. The placenta is delivered in the final stage of labor.

Lactation is unique to mammals. Prolactin secretion by the anterior pituitary stimulates milk production, and oxytocin controls the release of milk during nursing.

■ **INTERACTIVE QUESTION 46.7**

List the functions of the following hormones and the locations that secrete them:

a. human chorionic gonadotropin

b. progesterone

c. oxytocin

The Mother's Immune Tolerance of the Embryo and Fetus Several hypotheses attempt to explain why a mother does not reject an embryo, which has paternal as well as maternal chemical markers. There is evidence that the trophoblast, a protective layer that develops from the blastocyst and surrounds the embryo, releases signals, such as HCG, a variety of protein "factors," interleukins, and other substances, that suppress the mother's immune response. Another hypothesis is that the trophoblast and the placenta secrete an enzyme that breaks down tryptophan, an amino acid needed for T cell function. Another possible mechanism involves secretion of a hormone that induces the synthesis of a "death activator" membrane protein (FasL) on placental cells that causes T cells to self-destruct.

Contraception and Abortion **Contraception,** the deliberate prevention of pregnancy, can be accomplished by several methods: preventing release of egg or sperm, preventing fertilization, or preventing implantation. Complete abstinence from sexual intercourse is the most effective means of birth control. The **rhythm method,** or **natural family planning,** is based on refraining from intercourse during the period in which conception is most likely, the few days before and after ovulation.

Barrier methods of contraception include **condoms** and **diaphragms,** which, when used in conjunction with spermicidal foam or jelly, present a physical and chemical barrier to fertilization. Condoms are the only form of birth control that offers some protection against sexually transmitted diseases.

The release of gametes may be prevented by chemical contraception, as in **birth control pills,** and by sterilization. Most birth control pills are combinations of synthetic estrogen and progestin, which act by negative feedback to stop the release of GnRH by the hypothalamus and FSH and LH by the pituitary, resulting in a cessation of ovulation and follicle development. High doses of combinational birth control pills can be taken as morning–after pills (MAPS) following unprotected intercourse. A progestin-only minipill alters the cervical mucus so that it blocks sperm entry to the uterus. Longer-lasting progestin treatments include capsules implanted under the skin and injections. Cardiovascular problems are a potential effect of estrogen-containing birth control pills.

Tubal ligation in women and **vasectomy** in men are sterilization procedures that permanently prevent gamete release.

Abortion is the termination of a pregnancy. Spontaneous abortion, or miscarriage, occurs in as many as one-third of all pregnancies. The drug mifepristone (RU-486) is an analog of progesterone that blocks progesterone receptors in the uterus and can be used to terminate a pregnancy within the first 7 weeks.

■ INTERACTIVE QUESTION 46.8

List the three general types, along with examples, of birth control methods. Which examples are most likely to prevent pregnancy? Which are least likely to do so?

a.

b.

c.

Modern Reproductive Technology Some genetic diseases and congenital defects can be detected while the fetus is in the uterus, using such techniques as amniocentesis, chorionic villus sampling, ultrasound imaging, and a new technique that identifies fetal cells in the mother's blood that can be tested.

Reproductive technology can help couples unable to conceive. Fertilization procedures called **assisted reproductive technology (ART)** include *in vitro* **fertilization (IVF),** zygote intrafallopian transfer (ZIFT), and gamete intrafallopian transfer (GIFT). IVF, the most common ART procedure, involves the removal of ova from a woman, fertilization within a culture dish, and implantation of the developing embryo in the uterus.

■ Word Roots

a- = not, without (*asexual reproduction:* a type of reproduction involving only one parent that produces genetically identical offspring)

acro- = tip; **-soma** = body (*acrosome:* an organelle at the tip of a sperm cell that helps the sperm penetrate the egg)

bacul- = a rod (*baculum:* a bone that is contained in, and helps stiffen, the penis of rodents, raccoons, walruses, and several other mammals)

blasto- = produce; **-cyst** = sac, bladder (*blastocyst:* a hollow ball of cells produced one week after fertilization in humans)

coit- = a coming together (*coitus:* the insertion of a penis into a vagina, also called sexual intercourse)

contra- = against (*contraception:* the prevention of pregnancy)

-ectomy = cut out (*vasectomy:* the cutting of each vas deferens to prevent sperm from entering the urethra)

endo- = inside (*endometrium:* the inner lining of the uterus, which is richly supplied with blood vessels)

epi- = above, over (*epididymis:* a coiled tubule located adjacent to the testes where sperm are stored)

labi- = lip; **major-** = larger (*labia majora:* a pair of thick, fatty ridges that enclose and protect the labia minora and vestibule)

lact- = milk (*lactation:* the production of milk)

menstru- = month (*menstruation:* the shedding of portions of the endometrium during a menstrual cycle)

minor- = smaller (*labia minora:* a pair of slender skin folds that enclose and protect the vestibule)

myo- = muscle (*myotonia:* increased muscle tension)

oo- = egg; **-genesis** = producing (*oogenesis:* the process in the ovary that results in the production of female gametes)

partheno- = a virgin (*parthenogenesis:* a type of reproduction in which females produce offspring from unfertilized eggs)

partur- = giving birth (*parturition:* the expulsion of a baby from the mother, also called birth)
-theca = a cup, case (*spermatheca:* a sac in the female reproductive system where sperm are stored)
tri- = three (*trimester:* a three-month period)
vasa- = a vessel (*vasocongestion:* the filling of a tissue with blood caused by increased blood flow through the arteries of that tissue)

Structure Your Knowledge

1. Trace the path of a human sperm from the point of production to the point of fertilization, briefly commenting on the functions of both the structures it passes through and the associated glands.

2. Answer the following questions concerning the human menstrual cycle.
   a. What does GnRH do?
   b. What does FSH stimulate?
   c. What causes the spike in LH level?
   d. What does this LH surge induce?
   e. What does LH maintain during the luteal phase?
   f. What inhibits secretion of LH and FSH?
   g. What allows LH and FSH secretion to begin again?

3. Describe how birth control pills work. How does the French drug RU-486 (mifepristone) function?

Test Your Knowledge

FILL IN THE BLANKS

_____ 1. type of asexual reproduction in which a new individual grows while attached to the parent's body

_____ 2. small, volatile chemicals that may act as mate attractants

_____ 3. development of egg without fertilization

_____ 4. individual with functioning male and female reproductive systems

_____ 5. common opening of digestive, excretory, and reproductive systems in nonmammalian vertebrates

_____ 6. type of reproductive cycle in which thickened endometrium is reabsorbed

_____ 7. hormone that maintains the uterine lining during pregnancy

_____ 8. common duct for urine and semen in mammalian males

_____ 9. filling of a tissue with blood due to increased blood flow

_____ 10. period when ovulation and menstruation cease in humans

MULTIPLE CHOICE: *Choose the one best answer.*

1. Which of the following is an explanation for the periodicity of reproductive cycles in animals?
   a. Reproduction may correspond with periods of increased food supply, during which energy can be invested in gamete formation.
   b. Seasonal cycles may allow offspring to be produced during favorable environmental conditions when chances of survival are highest.
   c. Hormonal control of reproduction may be tied to biological clocks and seasonal cues.
   d. Synchronicity in release of gametes increases probability of fertilization.
   e. All of these may contribute to periodic reproductive activity.

2. Which of the following is *least* likely to be hermaphroditic?
   a. earthworm
   b. barnacle (sessile crustacean)
   c. tapeworm
   d. grasshopper
   e. liver fluke

3. Which of the following is *incorrectly* paired with its function?
   a. seminiferous tubules—add fluid containing mucus, fructose, and prostaglandins to semen
   b. scrotum—encases testes and suspends them below abdominal cavity
   c. epididymis—tubules in which sperm gain motility
   d. prostate gland—adds fluid to semen
   e. vas deferens—transports sperm from epididymis to ejaculatory duct

4. The function of the corpus luteum is to
   a. nourish and protect the egg cell.
   b. produce prolactin in the milk sacs of the mammary gland.
   c. produce progesterone and estrogen.
   d. produce estrogen and disintegrate following ovulation.
   e. maintain pregnancy by production of human chorionic gonadotropin.

5. Which of the following hormones is *incorrectly* paired with its function?
   a. androgens—responsible for primary and secondary male sex characteristics
   b. oxytocin—stimulates uterine contractions during parturition
   c. estrogens—responsible for primary and secondary female sex characteristics
   d. FSH—acts on Sertoli cells that nourish sperm, promoting spermatogenesis
   e. prolactin—stimulates breast development at puberty

6. Myotonia is
   a. a congenital birth defect.
   b. the hormone responsible for breast development.
   c. muscle tension.
   d. the filling of a tissue with blood.
   e. responsible for delivery of the placenta.

7. The secretory phase of the uterine cycle
   a. begins with dropping levels of estrogens and progesterone.
   b. is when the endometrium begins to degenerate and menstrual flow occurs.
   c. involves the initial proliferation of the endometrium.
   d. corresponds with the follicular phase of the ovarian cycle.
   e. corresponds with the luteal phase of the ovarian cycle.

8. Examples of birth control methods that prevent the production or release of gametes are
   a. sterilization and chemical contraception.
   b. birth control pills and IUDs.
   c. condoms and diaphragms.
   d. abstinence and chemical contraception.
   e. the progestin minipill and RU-486.

9. The ability of a pregnant woman not to reject her "foreign" fetus may be due to the
   a. separation of fetal and maternal blood.
   b. the suppression of her immune response by several possible mechanisms.
   c. the protection of the fetus within the trophoblast that is made of maternal tissue.
   d. the production of human chorionic gonadotropin by the fetus.
   e. the masking of paternal markers on fetal cells by specialized white blood cells.

10. Certain maternal diseases, drugs, alcohol, and radiation are most dangerous to embryonic development
    a. during the first 2 weeks when the embryo has not yet implanted and spontaneous abortion may occur.
    b. during the first 2 months when organogenesis is occurring.
    c. during the first and second trimesters when the embryonic liver is not yet filtering toxins.
    d. during the second trimester when the corpus luteum no longer secretes progesterone.
    e. during the third trimester when the most rapid growth is occurring.

Use the following to answer questions 11–15.
   a. estrogen
   b. progesterone
   c. LH (luteinizing hormone)
   d. FSH (follicle-stimulating hormone)
   e. HCG (human chorionic gonadotropin)

11. Which hormone stimulates ovulation and the development of the corpus luteum?

12. Which hormone is produced by the developing follicle and initiates thickening of the endometrium?

13. Which hormone is produced by the embryo during the first trimester and is necessary for maintaining a pregnancy?

14. Which hormone stimulates Leydig cells to make testosterone, which in turn stimulates sperm production?

15. Which hormone is produced by the corpus luteum and later by the placenta and is responsible for maintaining a pregnancy?

16. In the birth process, which of the following is/are involved in triggering and maintaining labor?
    a. HCG produced by the fetus
    b. oxytocin produced by fetus and mother, and prostaglandins produced by the placenta
    c. a drop in progesterone caused by the disintegration of the corpus luteum
    d. prolactin produced by the fetus and mother
    e. a surge in the production of LH

17. In which location does fertilization usually take place in a human female?
    a. ovary
    b. oviduct
    c. uterus
    d. cervix
    e. vagina

18. The major advantage of sexual reproduction is that
    a. it is easier than asexual reproduction.
    b. offspring have a better start when produced from a union of egg and sperm than when simply budded off a parent.
    c. it produces diploid offspring, whereas asexually produced offspring are haploid.
    d. it produces variation in the offspring, and new genetic combinations may be better adapted to a changing environment.
    e. it requires both male and female members of a species to be present in a population and have behavioral interactions.

19. Which of the following are the stem cells that give rise to sperm?
    a. spermatogonia
    b. spermatozoa
    c. Sertoli cells
    d. spermatocytes
    e. spermatids

20. How does meiosis differ in the production of human sperm and ova?
    a. Meiosis occurs while a female is an embryo, but does not begin until puberty in males.
    b. Each meiotic division produces four sperm but only two ova.
    c. Meiosis occurs in the testes of males but in the oviducts of females.
    d. Primary oocytes stop dividing by mitosis before birth, whereas male stem cells continue to divide throughout life.
    e. Meiosis is an uninterrupted process in males, whereas it begins again in females when a follicle matures and is only completed when a sperm penetrates the egg cell.

Chapter 47
Animal Development

Framework

Fertilization initiates physical and molecular changes in the egg cell. Early embryonic development includes cleavage of the fertilized egg to form a blastula, gastrulation, and organogenesis. The amount of yolk in the egg and the evolutionary history of the animal determine how these processes occur in any particular animal group.

Development and the differentiation of cells result from the control of gene expression by both cytoplasmic determinants and cell-cell induction. Pattern formation depends on the positional information a cell receives within a developing structure. Developmental biologists, using transplant experiments and techniques from molecular biology, are gradually unraveling some of the mechanisms underlying the complex processes of animal development.

Chapter Review

Animal development charts the course from a single fertilized egg to an organism made of many differentiated cells organized into specialized tissues and organs.

Preformation, or the notion that the egg or sperm contains a miniature embryo—a "homunculus"—was once the favored explanation of animal development. *Epigenesis* is the theory that the form of an embryo gradually develops from an egg. The developmental plan of an embryo is predetermined by the zygote genome and by the distribution of maternal substances called **cytoplasmic determinants** in the egg. As cell division separates cytoplasmic components, nuclei are exposed to different environments that affect which genes are expressed by different cells. The location of cells within embryonic regions also affects their early development.

Cell signaling conveys developmental messages between cells that influence gene expression and lead to **cell differentiation.** **Morphogenesis** produces the body shape. The combination of molecular genetics, classical studies, and the use of well-known research organisms is helping developmental biologists to study the physical, cellular, and molecular events of embryonic development.

47.1 After fertilization, embryonic development proceeds through cleavage, gastrulation, and organogenesis

Fertilization Fertilization, the union of egg and sperm, combines the haploid sets of chromosomes of these specialized cells and activates the egg by initiating metabolic reactions that trigger embryonic development.

In sea urchin fertilization, the **acrosome** at the tip of the sperm discharges hydrolytic enzymes when it comes in contact with the jelly coat of an egg. This **acrosomal reaction** allows the *acrosomal process* to elongate through the jelly coat. Fertilization within the same species is assured when proteins on the surface of the acrosomal process attach to specific receptor molecules extending through the *vitelline layer* of the egg, which is external to the plasma membrane.

In response to the fusion of sperm and egg plasma membranes, ion channels open and sodium ions flow into the egg. The resulting depolarization of the membrane prevents other sperm from fusing with the egg, providing a **fast block to polyspermy.**

Membrane fusion also initiates the **cortical reaction,** involving a signal-transduction pathway that triggers release of calcium ions from the egg's endoplasmic reticulum. In response to the Ca^{2+} increase in the cytoplasm, **cortical granules** located in the outer cortex of the egg release their contents into the *perivitelline space*. The released enzymes and mucopolysaccharides cause the vitelline layer to elevate and harden to form the **fertilization envelope,** which functions as a **slow block to polyspermy.**

The rise in Ca^{2+} concentration activates the egg by increasing cellular respiration and protein synthesis. Parthenogenetic development can be initiated by injecting calcium into an egg. Even an enucleated egg can be activated to begin protein synthesis, showing that inactive mRNA had been stockpiled in the egg.

After the sperm nucleus fuses with the egg nucleus, DNA replication begins in preparation for the cell division that begins the development of the embryo.

■ INTERACTIVE QUESTION   47.1

a. What assures that only sperm of the right species fertilize sea urchin eggs?

b. What assures that only one sperm will fertilize an egg?

Following internal fertilization in mammals, secretions of the female reproductive tract enhance motility and alter surface molecules of the sperm. A sperm migrates through the follicle cells released with the egg to the **zona pellucida,** the extracellular matrix of the egg. Complementary binding of a molecule on the sperm head to a receptor in the zona pellucida induces an acrosomal reaction. The hydrolytic enzymes released from the acrosome enable the sperm cell to penetrate the zona pellucida.

Binding of a sperm membrane protein with the egg membrane leads to the release of enzymes from cortical granules. This cortical reaction causes alterations of the zona pellucida and a slow block to polyspermy.

The entire sperm enters the egg. A centrosome forms around the basal body of the sperm flagellum and divides to form two centrosomes (with centrioles) that function in cell division. Nuclear envelopes disperse and the first mitotic division occurs.

Cleavage   Cleavage is a succession of rapid cell divisions during which the embryo becomes partitioned into many small cells, called **blastomeres.** The first divisions produce a cluster of cells called the **morula.** A fluid-filled cavity called the **blastocoel** develops, creating the stage called the **blastula.** Cleavage parcels different regions of the cytoplasm, which contain different cytoplasmic determinants, into cells and sets the stage for later development.

The polarity of the egg of many frogs and other animals is defined by the **vegetal pole,** where the stored nutrients in **yolk** are most concentrated, and the opposite end—called the **animal pole**—where the polar bodies budded from the egg during meiosis and the anterior part of the embryo often forms.

In the eggs of many frogs, the animal hemisphere has melanin granules in the outer cytoplasm (the cortex), whereas the vegetal hemisphere contains the yellow yolk. Fertilization results in a rotation of the animal pole cortex toward the point of sperm entry. This cortical rotation sets up the dorsal-ventral axis. An opposite narrow **gray crescent,** which the first cleavage division will bisect, marks the future dorsal side of the embryo.

The first two cleavage divisions are vertical, or meridional; the third division is equatorial. Yolk impedes cell division, and smaller cells are produced in the animal hemisphere in the frog. The frog blastula has its blastocoel restricted to the animal hemisphere due to unequal cell division.

In species in which eggs have little yolk, blastomeres are of equal size, although an animal-vegetal axis results from unequal distribution of other substances.

Deuterostomes—echinoderms and chordates—share similarities in embryonic development.

The large amount of yolk in the eggs of birds results in **meroblastic cleavage** restricted to the small disk of cytoplasm on top of the yolk. **Holoblastic cleavage** is the complete division of eggs with small or moderate amounts of yolk.

Meroblastic cleavage of the fertilized bird egg produces a cap of cells called the **blastoderm** on top of the large, undivided yolk. The blastomeres separate into an upper *epiblast* and a lower *hypoblast* layer, forming a cavity between them comparable to a blastocoel.

In the yolk-rich eggs of insects, the zygote nucleus undergoes repeated divisions and the nuclei migrate to the egg surface. Plasma membranes eventually form around the nuclei, creating a blastula made of cells surrounding a mass of yolk.

■ **INTERACTIVE QUESTION 47.2**

Identify the early embryonic stages diagrammed below, and label the indicated regions or structures.

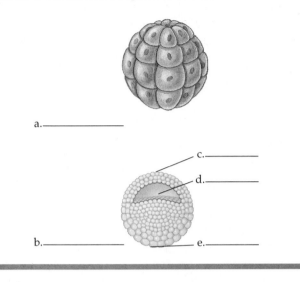

a._____

c._____

d._____

b._____ e._____

Gastrulation Changes in cell motility, shape, and adhesion are part of the morphogenetic rearrangements of cells in **gastrulation** that result in a three-layered embryo called the **gastrula.** The gastrula consists of the embryonic **germ layers**—the outer **ectoderm,** the middle **mesoderm,** and the inner **endoderm** that lines the embryonic digestive tract. This three-layered body plan is characteristic of most animal phyla.

In a sea urchin, gastrulation occurs as cells from the vegetal pole detach and move into the blastocoel as migratory *mesenchyme cells.* The remaining cells at the vegetal pole flatten into a *vegetal plate* that then buckles inward by a process known as **invagination.** The invagination deepens to form a narrow pouch called the **archenteron,** a cavity that will become the digestive tract. The archenteron opening is the **blastopore,** which develops into the anus. Cytoplasmic extensions (filopodia) from mesenchyme cells at the tip of the archenteron pull the archenteron across the blastocoel, where the endoderm fuses with the ectoderm to form the mouth.

Gastrulation in frog development is more complicated because of the multilayered blastula wall and the large, yolk-filled cells of the vegetal hemisphere. A group of invaginating cells begins gastrulation in the region of the gray crescent, forming a tuck that will become the **dorsal lip.** In a process called **involution,** surface cells roll over the dorsal lip and move inside, becoming organized into the endoderm and mesoderm. The blastopore eventually forms a circle surrounding a **yolk plug** of large, yolk-filled cells. Gastrulation produces an endoderm-lined archenteron surrounded by mesoderm, with the outer layer of the gastrula composed of ectoderm.

In the chick, a linear invagination, called the **primitive streak,** marks the anterior-posterior axis of the

embryo and is the site of gastrulation. Cells of the epiblast move through the primitive streak, forming the mesoderm and the endoderm. Cells remaining in the epiblast become ectoderm. The hypoblast cells form part of the yolk sac and connecting stalk.

■ **INTERACTIVE QUESTION 47.3**

Label the parts in these diagrams of gastrulation in a sea urchin and a frog gastrula.

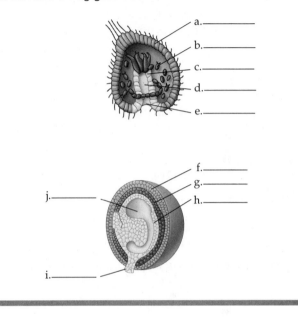

a._____

b._____

c._____

d._____

e._____

f._____

g._____

j._____ h._____

i._____

Organogenesis Organs begin to develop from the embryonic germ layers in a process known as **organogenesis.** Morphogenetic changes include folds, splits, and dense clustering (condensation) of cells. In a frog embryo, the **notochord,** the skeletal rod characteristic of chordates, forms from condensation of dorsal mesoderm just above the archenteron. Ectoderm above the developing notochord forms a neural plate, which folds inward to form a hollow **neural tube,** from which will develop the central nervous system.

In vertebrates, a band of cells called the **neural crest** separates from the side of the neural tube when it forms. These cells migrate to form bones of the skull, teeth, components of the peripheral nervous system, and many other cell types.

Mesoderm along the sides of the notochord condenses and separates into blocks, called **somites,** which give rise to vertebrae and muscles associated with the axial skeleton, arranged in segmented fashion. Lateral to the somites, the mesoderm splits to form the lining of the coelom.

Organogenesis in the chick proceeds in a fashion similar to that of the frog embryo. The borders of the blastoderm fold down and join, forming a three-layered tube attached by a stalk to the yolk.

■ INTERACTIVE QUESTION   47.4

Label the indicated structures in this diagram of organo-genesis in a frog embryo.

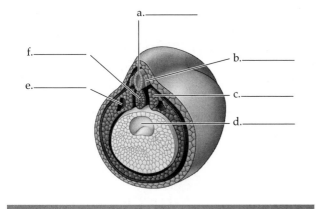

Developmental Adaptations of Amniotes   Reptiles (including birds) and mammals are **amniotes,** meaning that they create the aqueous environment necessary for embryonic development with a fluid-filled sac surrounded by a membrane called the amnion. The primary germ layers also give rise to four **extraembryonic membranes,** which are essential to development within the eggs of birds and other reptiles and the uterus of mammals.

■ INTERACTIVE QUESTION   47.5

List the four extraembryonic membranes found in birds and other reptiles, and briefly describe their functions.

a.

b.

c.

d.

Mammalian Development   Fertilization occurs in the oviduct, and embryonic development begins on the journey to the uterus. The mammalian egg and zygote have not been shown to have an obvious polarity. There is little yolk; thus cleavage is holoblastic. Gastru-

lation and early organogenesis, however, are similar in pattern to those of birds and other reptiles.

The **blastocyst** consists of an outer epithelium, the **trophoblast,** surrounding a cavity into which protrudes a cluster of cells, called the **inner cell mass.** These inner cells develop into the embryo and some of the extraembryonic membranes. The trophoblast secretes enzymes that enable the blastocyst to embed in the uterine lining (endometrium) and extends projections that will develop, along with mesodermal tissue, into the fetal portion of the placenta.

The inner cell mass forms a flat embryonic disk with an upper *epiblast* and lower *hypoblast,* homologous to the two layers in birds. In gastrulation, cells from the epiblast move through a primitive streak to form the mesoderm and endoderm.

The four extraembryonic membranes are homologous to those of birds and other reptiles. The **chorion** surrounds the embryo and the other membranes and functions in gas exchange. The **amnion** encloses the embryo in a fluid-filled amniotic cavity. The **yolk sac,** enclosing a small fluid-filled cavity, is the site of early formation of blood cells. The **allantois** forms blood vessels that connect the embryo with the placenta through the umbilical cord.

Organogenesis begins with the formation of the notochord, neural tube, and somites. In the human embryo, all major organs have begun development by the end of the first trimester.

■ INTERACTIVE QUESTION   47.6

Label the indicated cell layers in this diagram of a human blastocyst implanting into the endometrium. List the embryonic tissues and membranes that will develop from each cell layer.

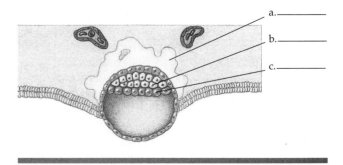

47.2   Morphogenesis in animals involves specific changes in cell shape, position, and adhesion

The Cytoskeleton, Cell Motility, and Convergent Extension   Changes in cell shape usually involve reorganization of the cytoskeleton. Extension of microtubules to elongate a cell and then contraction of actin filaments at one end create a wedge shape, initiating invaginations and evaginations.

Migrating embryonic cells use their cytoskeleton to extend and retract cellular protrusions. Cells at the leading edge of migrating tissue may extend filopodia and then drag along sheets of cells. By **convergent extension,** cells of a tissue layer crawl between each other, causing the sheet of cells to elongate and narrow.

Roles of the Extracellular Matrix and Cell Adhesion Molecules Glycoproteins such as fibronectins in the extracellular matrix (ECM) provide anchorage for crawling cells and tracts that direct the movement of migrating cells. Cells may also secrete other ECM substances that inhibit migration. Glycoproteins called **cell adhesion molecules (CAMs)** help regulate movement and tissue building by holding specific cells together. **Cadherins** are important adhesion molecules whose genes are differentially expressed in specific locations during development.

47.3 The developmental fate of cells depends on their history and on inductive signals

The mechanisms underlying the differentiation of cells during embryonic development are organized into two basic concepts. First, heterogeneous cytoplasm in the egg is partitioned through cleavage into different blastomeres, or local environmental differences establish early differences between embryonic cells. Second, the interactions between cells, called **induction,** affect a cell's developmental fate. Both of these phenomena influence gene expression, which leads to the differentiation of cells.

Fate Mapping In the 1920s, W. Vogt developed a **fate map** for amphibian embryos by labeling regions of the blastula with dye to determine where specific cells showed up in later developmental stages. Cell lineage analysis follows the mitotic descendants of individual blastomeres. The developmental history of every cell has been determined for *Caenorhabditis elegans.*

In most animals, early "founder cells" can be identified that will generate specific tissues. Experiments have also shown that a cell's *developmental potential* becomes restricted as development proceeds.

Establishing Cellular Asymmetries In nonamniotic vertebrates, the major axes are established in the egg or at fertilization. In amniotes, polarity does not seem to be established until later. In chicks, gravity may be involved in establishing the anterior-posterior axis while the egg moves down the oviduct, and pH differences may determine the dorsal-ventral axis of the blastoderm.

In mammals, polarity becomes obvious after cleavage, but the location of sperm entry may play a role.

The zygote is the only **totipotent** cell in many species. Even though cytoplasmic determinants are asymmetrically distributed in the frog egg, however, the first cleavage division bisects the gray crescent and equally separates these determinants.

Mammalian blastomeres remain totipotent until the 16-cell stage. The progressive restriction of a cell's developmental potency is characteristic of animal development. Experimental manipulation shows that the tissue-specific fates of cells of late gastrulae are usually fixed.

■ INTERACTIVE QUESTION 47.7

If an eight-cell stage of a sea urchin embryo is split vertically into two four-cell groups, both develop into normal larvae. If the division is made horizontally, abnormal larvae develop. Explain this experimental result.

Cell Fate Determination and Pattern Formation by Inductive Signals Cells influence the development of other cells through signals that switch on or off sets of genes.

Using transplant experiments with amphibian embryos, H. Mangold and H. Spemann established that the dorsal lip is a primary organizer of the embryo because of its influence in early organogenesis. The molecular basis of induction by *Spemann's organizer* appears to be tied to the inactivation of bone morphogenic protein 4 (BMP-4) on the dorsal side of the embryo. Proteins similar to BMP-4 and its inhibitors appear to play a role in the development of many different organisms.

Development of the vertebrate eye involves a series of inductions from cells in all three germ layers, progressively determining the fate of the ectodermal cells that become lens cells.

Pattern formation is the ordering of cells and tissues into the characteristic structures in the proper locations in the animal. Cells and their offspring differentiate based on **positional information,** molecular cues that indicate their position with respect to body axes.

A limb bud of a chick consists of a core of mesoderm and a layer of ectoderm. Researchers have identified two organizer regions that secrete proteins that provide positional information to cells.

Cells of the **apical ectodermal ridge (AER)** at the tip of the limb bud secrete proteins of the fibroblast growth factor (FGF) family. These growth signals appear to promote limb bud outgrowth and patterning along the proximal-distal axis. The AER and other ectoderm also appear to signal the limb's dorsal-ventral axis.

The **zone of polarizing activity (ZPA),** mesodermal tissue located at the posterior attachment of the bud, appears to communicate position along the anterior-posterior axis in a developing limb. Cells of the ZPA secrete a protein growth factor called Sonic hedgehog. Similar proteins have been identified as important positional cues in a number of development processes in other organisms, such as fruit flies and mice.

Cells respond to positional information relating to their location in a developing organ according to their developmental histories. Thus, patterns of gene expression first determine whether cells will become part of a wing or of a leg; then positional cues indicate what part of the wing or leg the cells will help form. A hierarchy of gene activations leads to the expression of genes that are involved in specifying various regions of the developing limb.

■ INTERACTIVE QUESTION   47.8

What would be the effect of implanting cells genetically engineered to produce and secrete Sonic hedgehog to the anterior attachment of a limb bud?

▶Word Roots

acro- = the tip (*acrosomal reaction:* the discharge of a sperm's acrosome when the sperm approaches an egg)

arch- = ancient, beginning (*archenteron:* the endoderm-lined cavity, formed during the gastrulation process, that develops into the digestive tract of an animal)

blast- = bud, sprout; **-pore** = a passage (*blastopore:* the opening of the archenteron in the gastrula that develops into the mouth in protostomes and the anus in deuterostomes)

blasto- = produce; **-mere** = a part (*blastomere:* small cells of an early embryo)

cortex- = shell (*cortical reaction:* a series of changes in the cortex of the egg cytoplasm during fertilization)

ecto- = outside; **-derm** = skin (*ectoderm:* the outermost of the three primary germ layers in animal embryos)

endo- = within (*endoderm:* the innermost of the three primary germ layers in animal embryos)

epi- = above; **-genesis** = origin, birth (*epigenesis:* the progressive development of form in an embryo)

extra- = beyond (*extraembryonic membrane:* four membranes that support the developing embryo in reptiles, birds, and mammals)

fertil- = fruitful (*fertilization:* the union of haploid gametes to produce a diploid zygote)

gastro- = stomach, belly (*gastrulation:* the formation of a gastrula from a blastula)

holo- = whole (*holoblastic cleavage:* a type of cleavage in which there is complete division of the egg)

in- = into; **vagin-** = a sheath (*invagination:* the infolding of cells)

involut- = wrapped up (*involution:* cells rolling over the edge of a lip into the interior)

mero- = a part (*meroblastic cleavage:* a type of cleavage in which there is incomplete division of yolk-rich egg, characteristic of avian development)

meso- = middle (*mesoderm:* the middle primary germ layer of an early embryo)

morul- = a little mulberry (*morula:* a solid ball of blastomeres formed by early cleavage)

noto- = the back; **-chord** = a string (*notochord:* a flexible rod that runs along the dorsal axis of the body in the future position of the vertebral column)

poly- = many (*polyspermy:* fertilization by more than one sperm)

soma- = a body (*somites:* paired blocks of mesoderm just lateral to the notochord of a vertebrate embryo)

tropho- = nourish (*trophoblast:* the outer epithelium of the blastocyst, which forms the fetal part of the placenta)

zona = a belt; **pellucid-** = transparent (*zona pellucida:* the extracellular matrix of a mammalian egg)

▶Structure Your Knowledge

1. Create a flowchart that shows the sequence of key events in the fertilization of a sea urchin egg, and indicate the functions of these events.

2. Fill in the table on page 399, briefly describing the early stages of development for sea urchin, frog, bird, and mammalian embryos.

Animal	Cleavage	Blastula	Gastrula
Sea urchin			
Frog			
Bird			
Mammal			

Test Your Knowledge

MULTIPLE CHOICE: *Choose the one best answer.*

1. The vitelline layer of a sea urchin egg
   a. is inside the fertilization envelope.
   b. releases Ca^{2+}, initiating the cortical reaction.
   c. has receptor molecules that are specific for proteins on the acrosomal process of sperm.
   d. is a thick jelly coat that is hydrolyzed by the release of enzymes from the acrosome.
   e. fuses with the sperm plasma membrane, initiating the depolarization of the membrane.

2. The slow block to polyspermy
   a. prevents sperm from other species from fertilizing the egg.
   b. is directly produced by the depolarization of the membrane.
   c. is a result of the formation of the fertilization envelope.
   d. is caused by the expulsion of calcium ions.
   e. involves all of the above.

3. The blastocoel
   a. develops into the archenteron or embryonic gut.
   b. is a fluid-filled cavity in the blastula.
   c. opens to the exterior through a blastopore.
   d. forms a hollow chamber during gastrulation.
   e. is lined with mesoderm.

4. Which of the following groups does *not* have eggs with an obvious polarity?
   a. sea urchin
   b. frog
   c. fruit fly
   d. bird
   e. mammal

5. Cytoplasmic determinants
   a. are unevenly distributed cytoplasmic components that influence the developmental fates of cells.
   b. are involved in regulation of gene expression.
   c. usually include maternal mRNA.
   d. are often separated in the first few cleavage divisions.
   e. are all of the above.

6. Which of the following is *incorrectly* paired with its embryonic germ layer?
   a. muscles—mesoderm
   b. central nervous system—ectoderm
   c. lens of the eye—ectoderm
   d. heart—endoderm
   e. notochord—mesoderm

7. In a frog embryo, gastrulation
   a. is impossible because of the large amount of yolk.
   b. proceeds by involution as cells roll over the dorsal lip of the blastopore.
   c. produces a blastocoel displaced into the animal hemisphere.
   d. occurs along the primitive streak.
   e. involves the formation of the notochord and neural tube.

8. The primitive streak of mammalian embryos is analogous to
   a. the dorsal lip of the frog embryo.
   b. the blastoderm of a bird embryo.
   c. the multinucleated stage of a fruit fly embryo.
   d. the archenteron of a sea urchin embryo.
   e. none of the above.

9. A function of the allantois in birds is to
   a. provide for nutrient exchange between the embryo and yolk sac.
   b. store metabolic wastes.
   c. form a respiratory organ in conjunction with the chorion.
   d. provide an aqueous environment for the developing embryo.
   e. both b and c.

10. Somites are
    a. blocks of mesoderm circling the archenteron.
    b. condensations of cells from which the notochord arises.
    c. serially arranged mesoderm blocks that develop into vertebrae and skeletal muscles.
    d. mesodermal derivatives of the notochord that line the coelom.
    e. produced during organogenesis in sea urchins, frogs, birds, and mammals, but not in fruit flies.

11. What forms the fetal portion of the placenta?
    a. the epiblast
    b. the trophoblast and some mesoderm
    c. the allantois and yolk sac
    d. the endometrium
    e. the amnion

12. In her transplant experiments with frog embryos, Mangold found that a part of the dorsal lip moved to another location would result in a second gastrulation and even the development of a doubled, face-to-face tadpole. Which of the following is the best explanation for these results?
    a. morphogenetic movements
    b. separation of cytoplasmic determinants
    c. totipotency of cells
    d. fate map determination
    e. induction by dorsal lip organizer

13. Pattern formation appears to be determined by
    a. positional information a cell receives from gradients of chemical signals called morphogens.
    b. differentiation of cells, which then migrate into developing organs.
    c. the movement of cells along fibronectin fibers.
    d. the induction of cells by mesoderm cells found in the center of organs.
    e. gastrulation and the formation of the three tissue layers.

Use this chick embryo diagram for questions 14–16.

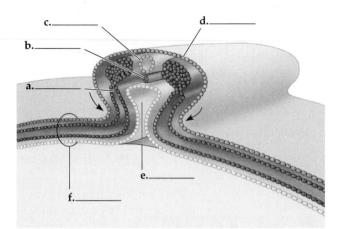

14. Which line points to the notochord?

15. Which line points to the archenteron?

16. To what structure does line f point?
    a. the primitive streak through which gastrulation occurs
    b. the hypoblast
    c. the mesodermal lining of the coelom
    d. the blastocoel between the epiblast and the hypoblast
    e. the developing extraembryonic membranes

17. The early cleavage divisions of a human zygote most closely resemble
    a. those of a sea urchin because neither egg has a large store of yolk and cleavage is holoblastic.
    b. those of a chick because both birds and mammals share a common reptilian ancestor.
    c. those of a frog because the cells of the animal hemisphere are smaller than those of the vegetal hemisphere in both types of embryos.
    d. those of a *Drosophila* because both pass through an early multinucleated stage before cell membranes form.
    e. all of the above animals because as deuterostomes, they have similar embryonic patterns.

18. Gastrulation and organogenesis in mammals most closely resemble
    a. those of a sea urchin because both types of eggs have little yolk and cleavage is holoblastic.
    b. those of a chick because both birds and mammals share a common reptilian ancestor.
    c. those of a frog because gastrulation takes place through the dorsal lip of the blastopore in both types of eggs.
    d. neither a chick nor a frog because mammals have some embryonic cells that develop into extraembryonic membranes.
    e. those of a *Drosophila* because homeotic genes control pattern development in both organisms.

19. Which of the following groups of structures is formed from endoderm?
    a. notochord, nerve cord, medulla of adrenal gland
    b. skeletal muscles, vertebrae, circulatory system
    c. lining of coelom and archenteron
    d. lining of digestive tract, liver, pancreas
    e. lens of eye, teeth, reproductive system

20. What is secreted by cells of the apical ectodermal ridge?
    a. Sonic hedgehog that communicates location along the anterior-posterior axis
    b. cadherins that help to hold migrating cells together as they form tissues of the leg
    c. ZPA, which forms fibronectin fibers that guide migrating cells
    d. fibroblast growth factors that promote limb bud outgrowth
    e. CAMs that cause the limb bud to elongate by convergent extension

Chapter 48

Nervous Systems

Framework

This chapter describes the structural components of nervous systems and how they functionally integrate, coordinate, and transmit information from the internal and external environment.

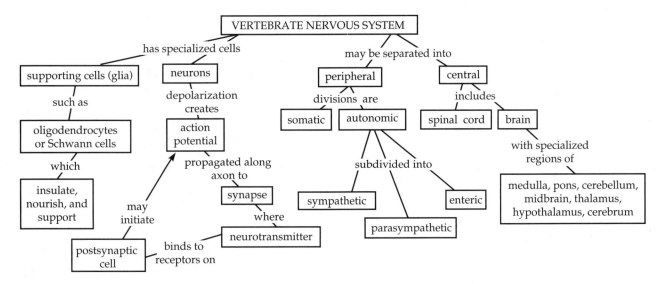

Chapter Review

The estimated 10^{11} nerve cells, or **neurons,** of the human brain communicate in complex circuits that are involved in different tasks. Mechanisms to sense and respond to changes in the environment evolved in prokaryotes. Modification of this ability to respond to environmental changes allowed for the development of communication between the cells of multicellular organisms.

48.1 Nervous systems consist of circuits of neurons and supporting cells

Organization of Nervous Systems A simple diffuse **nerve net** controls the radially symmetrical body cavity

403

in cnidarians such as hydras. Sea stars have a central nerve ring with radial **nerves,** bundles of projections from neurons, connected to a nerve net in each arm.

Cephalization, the concentration of neurons in a brain in the head, evolved in bilaterally symmetrical animals. Flatworms have a **central nervous system (CNS)** consisting of a simple brain and two longitudinal nerve cords. Annelids and arthropods have a more complicated brain and a ventral nerve cord with segmentally arranged clusters of neurons called **ganglia.** The **peripheral nervous system (PNS)** consists of nerves connecting the CNS with the rest of the body. Vertebrates have a brain and dorsal spinal cord; nerves and ganglia make up the PNS.

■ **INTERACTIVE QUESTION   48.1**

The organization of an organism's nervous system correlates with how the animal interacts with the environment. How do the different nervous systems of clams and squid illustrate this generalization?

Information Processing   The three stages of information processing are sensory input, integration, and motor output. Sensors detect external stimuli or internal conditions. **Sensory neurons** transmit this information to the CNS, where **interneurons** integrate it and send output through **motor neurons** to **effector cells** such as muscle or endocrine cells. The simplest type of nerve circuit regulates an automatic response to stimuli, called a **reflex.**

Neuron Structure   A neuron consists of a **cell body,** which contains the nucleus and organelles, and the extensions, **dendrites** and **axons.** The highly branched, short dendrites receive signals from other neurons; the longer axon transmits signals to other cells. Axons originate from a region of the cell body called the **axon hillock.** Many axons are wrapped in an insulating **myelin sheath.** Terminal branches of axons end in **synaptic terminals,** which usually release **neurotransmitters** that relay signals across the **synapse** to a **postsynaptic cell,** another neuron or effector cell. The transmitting cell is called the **presynaptic cell.**

Supporting Cells (Glia)   The very numerous supporting cells called **glia** give structural integrity and physiological support to the nervous system. **Astrocytes** are supporting glia in the CNS. Some facilitate information transfer at synapses, perhaps as part of the mechanism of learning and memory. During development, astrocytes induce the formation of the **blood-brain barrier,** which restricts the passage of most substances into the brain. **Radial glia** guide the embryonic growth of neurons. Radial glia and astrocytes can act as stem cells. Both **oligodendrocytes** (in the CNS) and **Schwann cells** (in the PNS) insulate axons in a myelin sheath by wrapping around them and forming concentric membrane layers. Gaps between adjacent Schwann cells are called nodes of Ranvier.

■ **INTERACTIVE QUESTION   48.2**

Label the indicated structures on this diagram of a vertebrate neuron. Indicate the direction of impulse transmission.

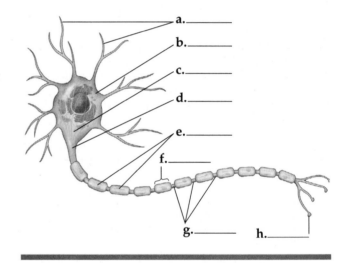

a._____
b._____
c._____
d._____
e._____
f._____
g._____   h._____

48.2   Ion pumps and ion channels maintain the resting potential of a neuron

An electrical potential difference, or **membrane potential,** exists across the plasma membrane of all cells. Electrophysiologists measure membrane potential by placing microelectrodes connected to a voltage recorder inside and outside a cell.

The Resting Potential The membrane potential of a typical nontransmitting neuron, called the **resting potential,** is between −60 and −80 mV. A membrane potential depends on differences in the concentrations of ions on either side of the membrane. The principal cation outside cells is sodium (Na^+) and the principal anion is chloride (Cl^-). Inside cells, potassium (K^+) is the principal cation, and negatively charged amino acids and other molecules are anions, A^-. The concentration gradient, or ratio of outside concentration/inside concentration, of Na^+ is 150/15, or 10; the K^+ gradient is 5/150, or 1/30. The sodium-potassium pump maintains these gradients.

Ions diffuse through channels across a membrane down their concentration gradient until balanced by the electrical gradient across the membrane. The membrane voltage at this equilibrium, or **equilibrium potential (E_{ion}),** is determined for an ion with a +1 charge by the Nernst equation: $E_{ion} = 62$ mV log ([ion]outside/[ion]inside). This equation applies to a membrane permeable to only a single type of ion. The equilibrium potential for K^+ (E_K) is −92 mV (at 37°C). Using the concentration gradients for Na^+ in a neuron, E_{Na} is +62 mV (the inside of the membrane is more positive than outside).

Neurons at rest have more K^+ channels open than Na^+ channels, and the resting potential is closer to E_K than E_{Na}. When the permeability to a particular ion changes, the membrane potential will move away from resting potential and toward the equilibrium potential for that ion.

■ **INTERACTIVE QUESTION 48.3**

a. What is the principle cation inside the cell? _____ outside the cell? _____

b. Which side of the membrane has a negative charge?

c. If a cell's membrane potential shifts from −70 mV to −90 mV, how has the permeability of the cell's membrane to K^+ and/or Na^+ changed?

Gated Ion Channels In addition to the ungated K^+ and Na^+ ion channels that create the resting potential, neurons also have **gated ion channels** that open or close in response to stimuli. **Stretch-gated ion channels,** found in stretch sensors, open in response to mechanical stimuli. **Ligand-gated ion channels,** found in synapses, respond to a chemical stimulus, such as a neurotransmitter. **Voltage-gated ion channels,** found in axons, respond to a change in membrane potential.

48.3 Action potentials are the signals conducted by axons

Stimuli that open or close gated ion channels may increase or decrease the membrane potential. A stimulus that opens potassium channels will result in **hyperpolarization,** as K^+ flows out and the membrane potential moves toward E_K (−92 mV). When sodium channels open and Na^+ flows in, a **depolarization** occurs as the inside of the cell becomes less negative. With this type of **graded potential,** the magnitude of a voltage change is proportional to the strength of the stimulus: the stronger the stimulus, the more gated ion channels that open.

Production of Action Potentials Once depolarization of a typical neuron reaches a certain membrane voltage called the **threshold,** an **action potential** is triggered. The action potential is an all-or-none event, always creating the same voltage spike once the threshold is reached. The frequency of action potentials, however, increases with the intensity of a stimulus.

Both Na^+ and K^+ voltage-gated ion channels are involved in an action potential. Sodium activation gates open rapidly in response to depolarization. The sodium inactivation gates, which are open in the resting state, close slowly in response to depolarization. Voltage-gated potassium channels open slowly in response to depolarization.

As a stimulus depolarizes the membrane to threshold, Na^+ activation gates open; the influx of Na^+ causes further depolarization, which opens more activation gates. During the *rising phase*, the membrane potential approaches E_{Na}. The closing of Na^+ inactivation gates and opening of K^+ activation gates rapidly brings the membrane potential back toward E_K during the *falling phase*.

During the *undershoot*, the membrane's permeability to K^+ is higher than at rest, and the continued outflow of K^+ temporarily hyperpolarizes the membrane. During the **refractory period,** which occurs before the sodium inactivation gates have reopened, the neuron cannot respond to another stimulus.

■ INTERACTIVE QUESTION   48.4

This diagram shows the changes in voltage-gated ion channels during an action potential. Label the channels and gates, ions, and five phases of the action potential. Label the axes of the graph and show where each phase occurs. Describe the ion movements associated with each phase.

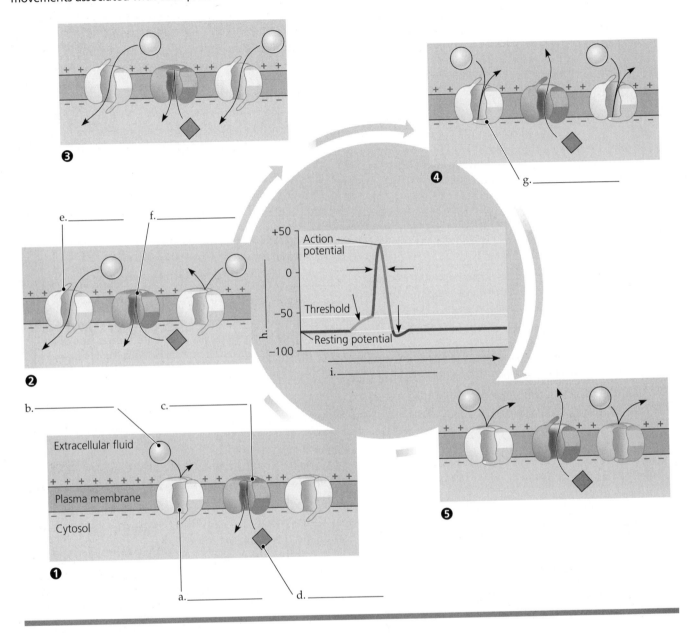

Conduction of Action Potentials Na^+ influx in the rising phase depolarizes adjacent sections of the membrane, bringing them to the threshold. Local depolarizations and action potentials across the membrane result in the propagation of serial action potentials along the length of the neuron. Because of the brief re-fractory period, the action potential is propagated in only one direction.

Resistance to current flow is inversely proportional to the cross-sectional area of the conducting "wire." The greater the axon diameter, the faster action potentials are propagated. Some invertebrates, such as squid and lobsters, have giant axons that conduct impulses very rapidly.

In vertebrates, voltage-gated ion channels are concentrated in the nodes of Ranvier, small gaps between successive Schwann cells. Action potentials can be generated only at these nodes, and a nerve impulse "jumps" from node to node, resulting in a faster mode of transmission known as **saltatory conduction.**

48.4 Neurons communicate with other cells at synapses

Electrical synapses allow electrical current to flow directly from cell to cell via gap junctions. Electrical synapses are found in the giant axons of some crustaceans but are less common than chemical synapses in vertebrates and most invertebrates.

A *chemical synapse* involves the release of neurotransmitters. A synaptic terminal contains **synaptic vesicles,** in which thousands of molecules of neurotransmitter are stored. The depolarization of the presynaptic membrane opens voltage-gated calcium channels in the membrane. The influx of Ca^{2+} causes the synaptic vesicles to fuse with the presynaptic membrane and release neurotransmitter into the **synaptic cleft.**

■ **INTERACTIVE QUESTION 48.5**

Identify the components of this chemical synapse following the depolarization of the synaptic terminal.

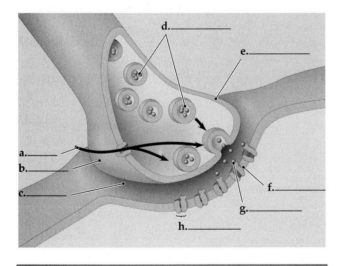

Direct Synaptic Transmission In *direct synaptic transmission,* binding of neurotransmitter to a receptor on ligand-gated channels allows ions to cross the membrane, creating a *postsynaptic potential.* If the channel allows both Na^+ to flow into and K^+ to flow out of the cell, the net inflow of positive charge depolarizes the membrane, creating an **excitatory postsynaptic potential (EPSP)** that brings the membrane potential closer to threshold. If binding of neurotransmitter opens K^+ channels, the membrane hyperpolarizes, producing an **inhibitory postsynaptic potential (IPSP).**

The effect of neurotransmitters is terminated when the molecules diffuse away, are taken up and repackaged into synaptic vesicles, or are broken down by enzymes.

Postsynaptic potentials are graded potentials. Their magnitude depends on the number of neurotransmitter molecules that bind to receptors, and they become smaller as they spread. The membrane potential of the axon hillock at any given time is determined by the sum of all EPSPs and IPSPs.

■ **INTERACTIVE QUESTION 48.6**

a. _____ **summation** occurs with repeated release of neurotransmitter from one or more synaptic terminals before the postsynaptic potential returns to its resting potential.

b. _____ **summation** occurs when several different presynaptic terminals, usually from different neurons, release neurotransmitter simultaneously.

Indirect Synaptic Transmission Some neurotransmitters bind to receptors that trigger a signal transduction pathway in the postsynaptic cells. This *indirect synaptic transmission* begins more slowly but lasts longer. Binding of receptor often leads to the production of cAMP as a second messenger and the phosphorylation of channel proteins, opening or closing ion channels.

Neurotransmitters Neurotransmitters may have many different receptors and may produce very different effects in postsynaptic cells.

Acetylcholine is a common neurotransmitter in invertebrates and vertebrates. Depending on the type of receptor, it can be inhibitory or excitatory in the vertebrate CNS. In neuromuscular junctions, acetylcholine released from a motor axon produces an EPSP in a muscle cell.

Biogenic amines, neurotransmitters derived from amino acids, are often involved in indirect synaptic transmission in the CNS. **Epinephrine, norepinephrine,** and **dopamine** are derived from tyrosine. **Serotonin** (synthesized from tryptophan) and dopamine affect sleep, mood, attention, and learning. Imbalances of these neurotransmitters have been associated with several disorders.

The amino acids **gamma aminobutyric acid (GABA), glycine, glutamate,** and **aspartate** function as neurotransmitters in the CNS. GABA is the most common inhibitory transmitter in the brain.

Neuropeptides are short chains of amino acids that are released by most neurons in addition to their nonpeptide neurotransmitter and often activate signal-transduction pathways. **Substance P** is an excitatory neurotransmitter that functions in pain perception. **Endorphins** are neuropeptides produced in the brain during physical or emotional stress that have pain-killing and other functions. Opiates bind to endorphin receptors in the brain.

Neurons of the vertebrate CNS and PNS use nitric oxide (NO) and carbon monoxide as local regulators. Some neurons signal endothelial cells in blood vessels to synthesize and release NO, triggering relaxation of smooth muscle cells and vessel dilation. CO produced by neurons in the brain affects the release of hypothalamic hormones. In the PNS, CO hyperpolarizes intestinal smooth muscle cells.

■ INTERACTIVE QUESTION 48.7

Acetylcholine stimulates skeletal muscle contraction but inhibits or slows cardiac muscle contraction. How can this neurotransmitter have such opposite effects?

48.5 The vertebrate nervous system is regionally specialized

All vertebrate nervous systems have a high degree of cephalization. The spinal cord integrates simple responses to stimuli and transports information to and from the brain. The CNS develops from the embryonic dorsal hollow nerve cord. Spaces in the brain called **ventricles** are continuous with the narrow **central canal** of the spinal cord, and all are filled with **cerebrospinal fluid.** This fluid, formed by filtration of the blood, cushions the brain and carries out circulatory functions. Axons are in bundles or tracts; **white matter** is named for the white color of their myelin sheaths. Neuron cell bodies, dendrites, and unmyelinated axons make up the **gray matter.**

The Peripheral Nervous System   The PNS carries information to and from the CNS, regulating movement and homeostasis. Paired **cranial nerves** and **spinal nerves,** with associated ganglia, make up the vertebrate PNS. Most cranial and all spinal nerves contain both sensory and motor neurons.

Functionally, the peripheral nervous system consists of the **somatic nervous system,** which carries signals to and from skeletal muscles, the movement of which is largely under conscious control. The **autonomic nervous system** maintains the internal environment by its control over smooth and cardiac muscles and various organs.

The autonomic nervous system is separated into three divisions. The **sympathetic division** accelerates the heart and metabolic rate, arousing an organism for action and generating energy. The **parasympathetic division** carries signals that enhance self-maintenance activities that conserve energy, such as digestion and slowing the heart rate. The **enteric division** includes complex networks of neurons that control the secretions of the digestive tract, pancreas, and gallbladder, and the contractions of smooth muscles that produce peristalsis. The enteric division is normally regulated by the sympathetic and parasympathetic divisions.

■ **INTERACTIVE QUESTION 48.8**

Complete this concept map to help you learn the functional organization of the vertebrate nervous system.

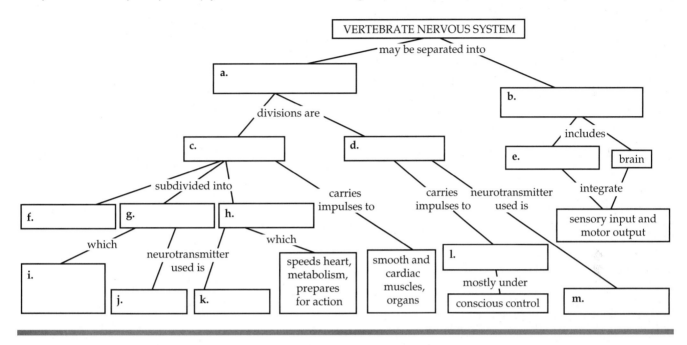

Embryonic Development of the Brain As a vertebrate embryo develops, the **forebrain, midbrain,** and **hindbrain** become evident as three bilaterally symmetrical, anterior bulges of the neural tube. In birds and mammals, the forebrain becomes quite large. In humans, five brain regions develop from these initial bulges: the *telencephalon* and *diencephalon* from the forebrain, the *mesencephalon* from the midbrain, and the *metencephalon* and *myelencephalon* from the hindbrain.

During the second and third months of human development, the **cerebrum** expands greatly and its outer portion, called the **cerebral cortex,** extends over many of the other brain regions. The thalamus, hypothalamus, and epithalamus develop from the diencephalon, an evolutionarily older portion of the forebrain.

The adult **brainstem** includes the midbrain (from the mesencephalon), the pons (from the metencephalon), and the medulla oblongata (from the myelencephalon). The cerebellum, which is not part of the brainstem, develops from the metencephalon.

The Brainstem The brainstem, a more ancient brain structure, is a stalk with caplike swellings at the anterior end of the spinal cord. Neuron cell bodies in several centers in the brainstem send axons to the cerebral cortex and cerebellum, causing changes in attention, alertness, appetite, and motivation. The **medulla oblongata,** or **medulla,** contains control centers for such homeostatic functions as breathing, swallowing, heart and blood vessel actions, and digestion. The **pons** functions with the medulla in some of these activities and in conducting information between the rest of the brain and the spinal cord. The tracts of motor neurons from the mid- and forebrain cross in the medulla, so that the right side of the brain controls much of the movement of the left side of the body, and vice versa.

The midbrain receives and integrates sensory information and sends this information to specific regions of the forebrain. Sensory axons involved in hearing pass through or terminate in the inferior colliculi. The superior colliculi form prominent optic lobes in nonmammalian vertebrates but only coordinate visual reflexes in mammals, in which vision is integrated in the cerebrum.

Arousal is a state in which an individual is aware of the external world. The **reticular formation** is a network of more than 90 clusters of cell bodies in the brain stem. The reticular activating system (RAS) regulates sleep and arousal. The level of arousal relates to the amount of filtered input the RAS sends to the cerebral cortex. Sleep-producing nuclei are located in the pons and medulla, and serotonin may be the neurotransmitter involved in these areas. A center that causes arousal is found in the midbrain.

All birds and mammals have a characteristic sleepwake cycle. The hormone melatonin, produced by the pineal gland appears to play an important part in this cycle. Sleep may be important in learning and memory.

The Cerebellum   The **cerebellum** is involved in learning and remembering motor skills, as well as in coordination and error-checking during motor, perceptual, and cognitive functions. The cerebellum integrates information from the auditory and visual systems with sensory input from the joints and muscles as well as motor pathways from the cerebrum to provide automatic coordination of movements and balance.

The Diencephalon   The **epithalamus** includes the pineal gland and the choroid plexus, one of several clusters of capillaries producing cerebrospinal fluid. The **thalamus** is a major input and sorting center for sensory information going to the cerebrum and an output center for motor information from the cerebrum. It also receives input from the cerebrum and other parts of the brain regulating emotion and arousal.

The **hypothalamus** is the major brain region for homeostatic regulation. It produces the posterior pituitary hormones and the releasing hormones that control the anterior pituitary. The hypothalamus contains the regulating centers for many survival functions and also plays a role in sexual and mating behaviors, the alarm response, and pleasure.

Studies of daily, regular behaviors show that an internal timekeeper, called the **biological clock,** is important for maintaining these circadian rhythms. In mammals, the **suprachiasmatic nuclei (SCN),** located in the hypothalamus, function as the biological clock. Sensory neurons provide visual information to the SCN to keep the mammalian clock synchronized with natural cycles of light and dark. Humans kept in free-running conditions have recently been found to have about a 24-hour, 11-minute cycle.

The Cerebrum   The cerebrum is divided into right and left **cerebral hemispheres,** each with an outer covering of gray matter, called the cerebral cortex, inner white matter, and the **basal nuclei.** These groups of neurons, located deep in the white matter, are important in planning and learning movements.

As the largest and most complex part of the mammalian brain, the cerebral cortex has changed the most during vertebrate evolution. The **neocortex,** the outer six layers of neurons running along the cerebrum, is unique to mammals. Convolutions greatly increase its surface area in primates and cetaceans.

The right side of the cortex receives information from and controls the movement of the left side of the body, and vice versa. Communication between the two hemispheres travels through the **corpus callosum,** a thick band of axons.

■ INTERACTIVE QUESTION   48.9

Identify the structures of the human brain. Match the functions listed below to these structures.

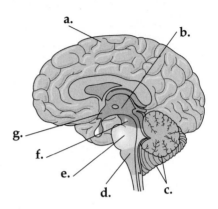

Structure		Function
a. _____		_____
b. _____		_____
c. _____		_____
d. _____		_____
e. _____		_____
f. _____		_____
g. _____		_____

Functions

1. Coordination of balance, movement, perceptual, and cognitive functions

2. Aids medulla in some functions, conducts information between brain and spinal cord

3. Sorts and relays information to cerebrum

4. Regulates breathing, heart rate, digestion

5. Integrates sensory and motor information, thinking

6. Produces hormones, homeostatic regulation

7. Sends sensory information to forebrain, contains nuclei involved in hearing and vision

---

48.6   The cerebral cortex controls voluntary movement and cognitive functions

Each side of the cerebral cortex is divided into four lobes with specialized functions localized in each lobe. Functional areas include *primary sensory areas,* which receive and process specific types of sensory information, and *association areas,* which integrate information from parts of the brain.

Information Processing in the Cerebral Cortex Sensory information (visual, auditory, somatosensory) is directed to primary sensory areas located in different lobes of the cortex. Olfactory information is routed to "primitive" cortical regions, then, via the thalamus, to the interior frontal lobe. Association areas near the primary sensory areas integrate ("associate") the different aspects of sensory inputs. The cerebral cortex generates motor response plans, which the primary motor cortex directs to the skeletal muscles. The proportion of somatosensory and motor cortex devoted to controlling each part of the body is correlated with the importance of that area.

Lateralization of Cortical Function As the brain of a child develops, functions become segregated in different cerebral hemispheres. In this **lateralization** of function, the left hemisphere usually becomes specialized for language, math, and logic, whereas the right hemisphere is involved in pattern recognition, spatial perception, and emotional processing. The left hemisphere is better at processing fine visual and auditory data and directing detailed movement. The right hemisphere seems to specialize in perceiving images within the whole context.

Language and Speech Studies of brain injuries and imaging of brain activity using fMRI and PET show that several association areas in different lobes are involved in understanding and generating language. *Broca's area* in the frontal lobe is involved in speech generation; *Wernicke's area* in the posterior temporal lobe is involved in speech comprehension. Other areas are involved in reading, speaking, and attaching meaning to words.

Emotions The amygdala, hippocampus, and olfactory bulb, along with sections of the thalamus, hypothalamus, and inner portions of the cortex, form a ring around the brain stem called the **limbic system.** This system interacts with sensory and other areas of the neocortex in generating emotions, both primary emotions, such as laughing and crying, and emotional feelings associated with the survival-related functions of the brain stem, such as aggression, feeding, and sexuality.

Emotional brain circuits of the limbic system form early; thus infants can bond to a caregiver; recognize elements of a human face; and express fear, distress, and anger. The amygdala, located in the temporal lobe, functions in laying down emotional memories. This emotional memory system appears to develop earlier than the system that allows recall of events, which involves the hippocampus. The association of primary emotions with different situations involves portions of the neocortex, particularly the prefrontal cortex.

Memory and Learning Human memory consists of **short-term memory**, information held briefly, and **long-term memory,** the retention of this information for later recall. Transferring information from short-term to long-term memory involves the hippocampus. Various association areas of the cortex are involved with storing and retrieving words and images.

Remembering facts and numbers may involve rapid changes in the strength of existing nerve connections. Learning and remembering skills and procedures seems to involve nerve cells making new connections. Skill memories are formed by repetition of motor activities and do not require conscious recall of specific steps.

Studies of simple forms of learning in a mollusc have shown the effect of interneurons in changing ion channels between sensory and motor neurons.

In **long-term potentiation (LTP),** the strength of synaptic transmission is increased following a brief burst of action potentials from a presynaptic cell. The presynaptic cell releases glutamate, which binds with AMPA receptors on ligand-gated Ca^{2+} channels and NMDA receptors on channels that are both ligand and voltage gated. The influx of Ca^{2+} triggers signal transduction pathways that increase both the number and responsiveness of the ligand-gated channels and that cause the production and release of NO into the synapse. NO causes the presynaptic neuron to release more glutamate. LTP may be involved in learning and memory storage.

Consciousness The study of human consciousness, the awareness of self and one's own experiences, has been facilitated by brain-imaging techniques but remains a challenging endeavor. Consciousness appears to be an emergent property of the brain that perhaps involves a repetitive scanning mechanism of multiple brain areas to produce a unified conscious moment.

■ **INTERACTIVE QUESTION 48.10**

What three factors facilitate the transfer of information from short-term to long-term memory?

48.7 CNS injuries and diseases are the focus of much research

The mammalian central nervous system cannot fully repair itself when brain cells are destroyed by injury or disease, although surviving brain cells can make new connections and sometimes compensate for damage.

Nerve Cell Development New research in neurobiology centers on the differentiation and migration of neurons, the growth of axons, and the formation of synapses with the proper target cells. The **growth cone** at the leading edge of an axon appears to respond to signal molecules released from cells along the growth path. These signals may act as attractants or repellents to direct axonal growth. Cell adhesion molecules (CAMs) on the growth cone also interact with molecules on surrounding cells that provide tracks for the axon to follow. Growth is stimulated by nerve growth factor released by astrocytes and by growth-promoting proteins produced by the axon itself. The genes and basic mechanism of this process are quite similar in nematodes, insects, and vertebrates and have apparently been conserved during evolution.

Neural Stem Cells   In 1998, F. Gage and P. Eriksson reported finding newly formed nerve cells in the hippocampus of cancer victims who had been treated with a chemical marker that labels the DNA of dividing cells. In 2001, Gage and his colleagues reported that they had cultured neural progenitor cells removed from brain tissue samples. These cells were not as developmentally flexible as embryonic stem cells, but were able to differentiate into neurons and astrocytes. A goal of future research is to find ways to induce neural stem cells to differentiate into glia or neurons when needed, and to culture neural stem cells for transplantation into damaged neurological tissue.

Diseases and Disorders of the Nervous System
Schizophrenia is characterized by psychotic episodes involving hallucinations and delusions. The disease has both genetic and environmental components. Treatments have focused on drugs that block dopamine receptors, although evidence indicates that the neurotransmitters serotonin, norepinephrine, and glutamate are also involved. Drugs that reduce the major symptoms often have negative side affects.

Depressive illnesses include **bipolar disorder**, which involves swings of mood from high to low, and **major depression**, which involves a persistent low mood. Both illnesses have a genetic and environmental component. Drug treatments against depression often increase the activity of biogenic amines.

Alzheimer's disease (AD) is dementia characterized by confusion, memory loss, and personality changes. This age-related, progressive disease involves the death of neurons in large areas of the brain. Postmortem diagnosis involves finding neurofibrillary tangles (bundles of degenerated processes) and senile plaques (aggregates of beta-amyloid). Membrane secretases cleave a protein into this insoluble peptide that accumulates outside neurons and appears to trigger death of neighboring neurons. PIB is a chemical that accumulates in regions of amyloid deposits and may help in early detection and identifying treatments for AD.

Parkinson's disease is a progressive, age-related motor disorder characterized by difficulty in movements, rigidity, and muscle tremors. The death of neurons in the substantia nigra, which normally release dopamine in the basal nuclei, leads to the motor symptoms of this disease. Accumulation of the protein α-synuclein is associated with neuron degeneration. The disease appears to result from both genetic and environmental factors. Brain surgery, deep-brain stimulation, and drugs such as L-dopa, a dopamine precursor, are used to manage the symptoms. A potential cure may be to implant dopamine-secreting neurons in the substantia nigra or basal ganglia.

Word Roots

astro- = a star; **-cyte** = cell (*astrocytes:* glial cells that provide structural and metabolic support for neurons)

auto- = self (*autonomic nervous system:* the branch of the peripheral nervous system of vertebrates that regulates the internal environment)

bio- = life; **-genic** = producing (*biogenic amines:* neurotransmitters derived from amino acids)

cephalo- = head (*cephalization:* the clustering of sensory neurons and other nerve cells to form a brain near the anterior end and mouth of animals with elongated, bilaterally symmetrical bodies)

dendro- = tree (*dendrite:* one of usually numerous, short, highly branched processes of a neuron that receive signals from other neurons)

de- = down, out (*depolarization:* an electrical state in an excitable cell whereby the inside of the cell is made less negative relative to the outside)

endo- = within (*endorphin:* a hormone produced in the brain and anterior pituitary that inhibits pain perception)

epi- = above, over (*epithalamus:* a brain region, derived from the diencephalon, that contains several clusters of capillaries that produce cerebrospinal fluid; it is located above the thalamus)

glia = glue (*glia:* supporting cells that are essential for the structural integrity of the nervous system and for the normal functioning of neurons)

hyper- = over, above, excessive (*hyperpolarization:* an electrical state whereby the inside of the cell is made more negative relative to the outside than at the resting membrane potential)

hypo- = below (*hypothalamus:* the ventral part of the vertebrate forebrain that functions in maintaining homeostasis, especially in coordinating the endocrine and nervous systems; it is located below the thalamus)

inter- = between (*interneurons:* an association neuron; a nerve cell within the central nervous system that forms synapses with sensory and motor neurons and integrates sensory input and motor output)

neuro- = nerve; **trans-** = across (*neurotransmitter:* a chemical messenger released from the synaptic terminal of a neuron at a chemical synapse that diffuses across the synaptic cleft and binds to and stimulates the postsynaptic cell)

oligo- = few, small (*oligodendrocytes:* glial cells that form insulating myelin sheaths around the axons of neurons in the central nervous system)

para- = near (*parasympathetic division:* one of three divisions of the autonomic nervous system)

post- = after (*postsynaptic cell:* the target cell at a synapse)

pre- = before (*presynaptic cell:* the transmitting cell at a synapse)

salta- = leap (*saltatory conduction:* rapid transmission of a nerve impulse along an axon resulting from the action potential jumping from one node of Ranvier to another, skipping the myelin-sheathed regions of membrane)

soma- = body (*somatic nervous system:* the branch of the vertebrate peripheral nervous system that carries signals to and from skeletal muscles in response to external stimuli)

supra- = above, over (*suprachiasmatic nuclei:* a pair of structures in the hypothalamus of mammals that functions as a biological clock)

syn- = together (*synapse:* the locus where a neuron communicates with a postsynaptic cell in a neural pathway)

Structure Your Knowledge

1. Develop a flow chart, diagram, or description of the sequence of events in the creation and propagation of an action potential and in the transmission of this potential across a chemical synapse.

2. List the location and functions of these important brain nuclei or functional systems.
   a. reticular formation
   b. suprachiasmatic nucleus
   c. basal nuclei
   d. limbic system
   e. amygdala and hippocampus

Test Your Knowledge

MULTIPLE CHOICE: *Choose the one best answer.*

1. Which of the following animals is mismatched with its nervous system?
   a. sea star (echinoderm)—modified nerve net, central nerve ring with radial nerves
   b. hydra (cnidarian)—ring of ganglia, paired ventral nerve cords
   c. annelid worm—brain, ventral nerve cord with segmental ganglia
   d. vertebrate—central nervous system of brain and spinal cord and peripheral nervous system
   e. squid (mollusc)—large complex brain, ventral nerve cord, some giant axons

2. Interneurons
   a. may connect sensory and motor neurons.
   b. are confined to the PNS.
   c. are confined to the spinal cord.
   d. have electrical synapses between neurons.
   e. do not have cell bodies.

3. Nodes of Ranvier are
   a. gaps where Schwann cells abut and at which action potentials are generated.
   b. neurotransmitter-containing vesicles located in the synaptic terminals.
   c. the major components of the blood-brain barrier that restrict the passage of substances into the brain.
   d. clusters of receptor proteins located on the postsynaptic membrane.
   e. ganglia adjacent to the spinal cord.

4. Which of the following is *not* true of the resting potential of a typical neuron?
   a. The inside of the cell is more negative than is the outside.
   b. There are concentration gradients with more sodium outside the cell and a higher potassium concentration inside the cell.
   c. It is about $-70\,mV$ and can be measured by using microelectrodes placed inside and outside the cell.
   d. It is formed by the sequential opening of voltage-gated channels.
   e. It results from the combined equilibrium potentials of potassium and sodium.

5. After the depolarization of an action potential, the fall in the membrane potential occurs due to the
   a. closing of sodium activation and inactivation gates.
   b. opening of sodium activation gates.
   c. refractory period in which the membrane is hyperpolarized.
   d. delay in the action of the sodium-potassium pump.
   e. opening of voltage-gated potassium channels and the closing of sodium inactivation gates.

6. The threshold of a membrane
   a. is exactly midway between E_K and E_{Na}.
   b. opens voltage-gated channels and permits the rapid outflow of sodium ions.
   c. is the depolarization that is needed to generate an action potential.
   d. is a graded potential that is proportional to the strength of a stimulus.
   e. is an all-or-none event.

7. Which of the following is *not* true of chemical synapses?
    a. Synaptic terminals at the ends of branching axons contain synaptic vesicles, which enclose the neurotransmitter.
    b. The influx of sodium when an action potential reaches the presynaptic membrane causes synaptic vesicles to release their neurotransmitter into the cleft.
    c. The binding of neurotransmitter to receptors on the postsynaptic membrane usually opens or closes ion channels.
    d. An excitatory postsynaptic potential forms when sodium channels open and the membrane potential moves closer to an action potential threshold.
    e. Neurotransmitter is often rapidly degraded in the synaptic cleft.

8. An inhibitory postsynaptic potential occurs when
    a. sodium flows into the postsynaptic cell.
    b. enzymes do not break down the neurotransmitter in the synaptic cleft.
    c. binding of neurotransmitter opens ion gates that result in the membrane becoming hyperpolarized.
    d. acetylcholine is the neurotransmitter.
    e. norepinephrine is the neurotransmitter.

9. Which of the following do (does) *not* function as a neurotransmitter or local regulator released by neurons?
    a. amino acids such as glutamate
    b. neuropeptides such as endorphin
    c. biogenic amines such as dopamine
    d. steroids
    e. NO

10. Which of the following pairs is mismatched?
    a. Ca^+ channels in presynaptic membrane—voltage-gated ion channel
    b. NMDA receptors for glutamate on postsynaptic neuron—ligand- and voltage-gated Ca^+ channels
    c. Na^+ or K^+ channels in postsynaptic membrane—voltage-gated ion channels
    d. receptor in indirect synaptic transmission—no ion channel; activates signal transduction pathway
    e. Na^+ channels in axon hillock—voltage-gated ion channel

11. Which of the following is *not* true of the autonomic nervous system?
    a. It is a subdivision of the somatic nervous system.
    b. It consists of the sympathetic, parasympathetic, and enteric divisions.
    c. It is part of the peripheral nervous system.
    d. It controls smooth and cardiac muscles.
    e. Control is generally involuntary.

12. If you needed to obtain nuclei from the cell bodies of neurons for an experiment, you would want to make preparations of
    a. the corpus callosum.
    b. the gray matter of the brain.
    c. basal nuclei of the brain.
    d. sympathetic ganglia near the spinal cord.
    e. either b, c, or d.

13. The superior and inferior colliculi
    a. control biorhythms and are found in the thalamus.
    b. are the part of the limbic system found in the midbrain.
    c. are located in the parietal and frontal lobes, respectively, and are involved in language and speech.
    d. are nuclei in the midbrain involved in hearing and vision.
    e. are found in the inner cortex and are involved in memory storage.

14. Which of the following structures is *incorrectly* paired with its function?
    a. pons—conducts information between spinal cord and brain
    b. cerebellum—contains the tracts that cross motor neurons from one side of the brain to the other side of the body
    c. thalamus—sorts and relays incoming impulses to the cerebrum
    d. corpus callosum—band of axons connecting left and right hemispheres
    e. hypothalamus—homeostatic regulation, pleasure centers

15. Which of the following is evidence indicating that schizophrenia, bipolar disorder, and major depression have a genetic as well as an environmental component?
    a. All three seem to involve an excess of dopamine.
    b. All can be treated with the same drugs.
    c. If one identical twin has the illness, there is a 50% chance that the other will, indicating that genes and the environment contribute equally to these disorders.
    d. The genes have been identified for these illnesses, and they share a common sequence.
    e. Identical twins raised separately do not share these illnesses; those raised in the same home are likely to both be affected by the disorder.

16. Long-term potentiation seems to be involved in
    a. memory storage and learning.
    b. enhanced response of a presynaptic cell resulting from strong depolarization from bursts of action potentials.
    c. the attachment of emotion to information to be learned.
    d. the release of the excitatory neurotransmitter glutamate that opens sodium-gated channels.
    e. all of the above.

17. What is the main effect of the neurotransmitter GABA in the CNS?
    a. stimulate instinctual behavior
    b. create inhibitory postsynaptic potentials
    c. create excitatory postsynaptic potentials
    d. induce sleep
    e. decrease pain and induce euphoria

18. The function of motor neurons is to carry impulses
    a. from the spinal cord to the brain.
    b. from the cerebellum to the cerebrum.
    c. from sensory receptors to the spinal cord.
    d. from the CNS to effectors (muscles and organs).
    e. between interneurons and the PNS.

19. Which of the following is *incorrectly* paired with its function?
    a. axon hillock—region of neuron where action potential originates
    b. Schwann cells—create myelin sheath around axon in PNS
    c. synapse—space between presynaptic and postsynaptic cell into which neurotransmitter is released
    d. synaptic terminal—receptor that is part of ion channel that is keyed to specific neurotransmitter
    e. dendrite—receive signals from other neurons

20. If the binding of a neurotransmitter to its receptor opens Cl^- channels, what would be the effect on the postsynaptic cell? (Cl^- is in higher concentration outside the cell.)
    a. It would hyperpolarize, receiving an IPSP.
    b. It would reach threshold, and an action potential would fire.
    c. It would depolarize and form an EPSP but probably not fire an action potential.
    d. It would initiate a signal-transduction pathway.
    e. Its membrane potential would not change because neither sodium nor potassium channels opened.

21. Movement of an action potential in only one direction along a neuron is a function of
    a. saltatory conduction.
    b. the pathway from dendrite to axon.
    c. the refractory period when sodium inactivation gates are still closed.
    d. the localized depolarization of the surrounding membrane.
    e. the reaching of threshold, which creates an all-or-none firing.

22. What makes up the white matter of the spinal cord?
    a. the ventricles
    b. myelinated sheaths of axons
    c. cerebrospinal fluid
    d. motor and interneuron cell bodies
    e. nodes of Ranvier

23. Why is signal transmission faster in myelinated axons?
    a. These axons are thinner, and there is less resistance to the voltage flow.
    b. These axons use electrical synapses rather than chemical synapses.
    c. The action potential can jump from node to node along the insulating myelin sheath.
    d. These axons are thicker and provide less resistance to voltage flow.
    e. These axons have higher depolarization values than do unmyelinated axons.

24. How is an increase in the strength of a stimulus communicated by a neuron?
    a. The spike of the action potential reaches a higher voltage.
    b. The frequency of action potentials generated along the neuron increases.
    c. The length of an action potential (the duration of the rising phase) increases.
    d. The action potential travels along the neuron faster.
    e. All action potentials are the same; the nervous system cannot discriminate between different strengths of stimuli.

25. Lateralization of the cerebrum refers to the fact that
    a. the left hemisphere usually controls verbal and analytic ability, while the right hemisphere is specialized for emotional processing and spatial relations.
    b. the right side of the brain controls the left side of the body, and vice versa.
    c. the primary somatosensory cortex receives sensory information that is mapped onto one region of the cerebral cortex, while the primary motor cortex localized in another region controls motor output to the body.
    d. visceral functions are controlled by the hindbrain, whereas thinking and language are centered in association areas of the cortex.
    e. memory formation involves circuits that include the limbic system, hippocampus, and association areas of the cortex.

26. Which of the following neurotransmitters is released by neurons from the sympathetic division of the autonomic nervous system at synapses with smooth muscle in the stomach?
    a. GABA
    b. Ca^{2+}
    c. acetylcholine
    d. substance P
    e. norepinephrine

27. The release of the neurotransmitter glutamate opens Ca^{2+} channels in neurons, activating signal transduction pathways that make the postsynaptic cell more sensitive to stimulation and increase the release of glutamate from the presynaptic cell. What phenomenon does this describe?
    a. an emotion
    b. a spinal reflex
    c. short-term memory
    d. sleep
    e. long-term potentiation

28. A function of the reticular formation is to
    a. keep you awake during a lecture.
    b. coordinate movement and balance.
    c. produce emotions.
    d. control visceral functions such as breathing and heart rate.
    e. create consciousness.

29. When did the ability of cells to sense and respond to chemical signals evolve?
    a. during the Cambrian explosion within the various animal phyla
    b. with the development of chemical signals that control the path of axon growth
    c. in prokaryotes who were able to sense and respond to environmental chemicals
    d. in the ancestor of the first animal
    e. in the first multicellular organism

30. Which of these structures is found only in the mammalian brain?
    a. telencephalon and diencephalon
    b. optic lobes
    c. olfactory cortex
    d. neocortex
    e. hippocampus

31. Which of the following is part of the mechanism by which axons grow along a path to their proper target cell?
    a. production of nerve growth factor by astrocytes
    b. reception of signal molecules by the growth cone that either attract or repel the growing axon
    c. interaction of cell adhesion molecules on the growth cone with molecules of cells along the growth path
    d. activation of genes in the axon that produce growth-promoting proteins
    e. All of the above are involved in this process.

32. Neural progenitor cells
    a. were first cultured in 1998 from the hippocampus of some cancer victims.
    b. retain the ability to divide and differentiate into both neurons and astrocytes.
    c. are similar to embryonic stem cells in that they can differentiate into all kinds of body cells.
    d. have been transplanted into individuals with spinal cord injuries and were able to grow and make connections with the proper target cells.
    e. are the first cells formed as the neural tube rolls up, and they give rise to both glia and neurons.

Chapter 49

Sensory and Motor Mechanisms

Framework

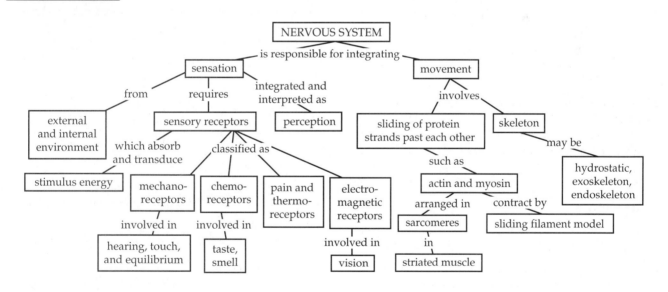

Chapter Review

On a cellular level, all animal movement depends on energy expended to move protein strands past each other, either in microtubules, which are responsible for beating of cilia and flagella, or in microfilaments, which are involved in amoeboid movement and muscle contraction. In a continuous cycle, animals are constantly moving; exploring and sensing their environment; and processing that information in the brain, which then directs their next movements.

49.1 Sensory receptors transduce stimulus energy and transmit signals to the central nervous system

Information is transmitted in the nervous system as action potentials. **Sensations,** or impulses arriving from sensory neurons, are routed to different parts of the brain that interpret them, producing **perceptions** of various stimuli.

Sensory reception is the detection of energy from a particular stimulus by **sensory receptors,** which are usually modified neurons or epithelial cells that occur singly or in groups within sensory organs. Sensory receptors may be **exteroreceptors** or **interoreceptors,** which detect stimuli from the external or internal environment, respectively.

Functions Performed by Sensory Receptors **Sensory transduction** is the conversion of stimulus energy into a **receptor potential,** a graded change in the membrane potential of a receptor cell that is proportional to the strength of the stimulus. Receptor potentials result from the opening or closing of ion channels in response to a stimulus.

The strengthening of stimulus energy, either by accessory structures of sense organs or as part of signal transduction pathways, is called **amplification.**

Transmission of the receptor potential to the central nervous system may occur either as a result of the generation of an action potential when the sensory receptor is a sensory neuron, or by the release of neurotransmitter from a receptor into a synapse with a sensory neuron, which then may transmit an action potential. The magnitude of the receptor potential correlates with the frequency of action potentials or the quantity of neurotransmitter released, which then translates into the frequency of action potentials in the sensory neuron.

Hair cells detect motion and are found in the vertebrate ear and lateral line organs of fishes. When motion produces bending in the cilia or microvilli projecting from a hair cell, the cell depolarizes, releases more neurotransmitter, and the rate of action potential firing increases. Bending in the other direction reduces neurotransmitter release and slows the formation of action potentials in the sensory neuron.

Integration, or processing, of sensory information begins with the summation of graded potentials from receptors. Ongoing stimulation may lead to **sensory adaptation** in which continuous stimulation results in a decline in sensitivity of the receptor cell. Sensory information is further integrated within complex receptors and the CNS.

■ **INTERACTIVE QUESTION   49.1**

List the four functions common to all sensory receptors as they absorb stimulus energy and transmit information about the stimulus to the nervous system.

a.

b.

c.

d.

Types of Sensory Receptors Sensory receptors can be grouped into five categories. **Mechanoreceptors** respond to the mechanical energy of pressure, touch, stretch, motion, and sound. Bending or stretching of the mechanoreceptor cell membrane increases its permeability to sodium and potassium ions, leading to depolarization. Crayfish stretch receptors and vertebrate hair cells are mechanoreceptors.

Vertebrate stretch receptors monitor the length of skeletal muscles. **Muscle spindles,** containing dendrites of sensory neurons wound around a few specialized skeletal muscle fibers, are stimulated by stretching of the muscles. In humans, dendrites of sensory neurons function as touch receptors; the type of stimulus they respond best to depends on the structure of their connective tissue covering.

Chemoreceptors include both general receptors that monitor the total solute concentration and specific receptors that respond to individual kinds of important molecules. Membrane permeability of the receptor cells changes in response to binding of a stimulus molecule.

Electromagnetic receptors respond to various forms of electromagnetic energy. **Photoreceptors** detect visible light and are often organized into eyes. Some animals have receptors that detect infrared rays, electric currents, and magnetic fields.

Thermoreceptors in the skin and hypothalamus respond to heat or cold and help to regulate body temperature in mammals. The exact identity of thermoreceptors in the skin is still uncertain.

Pain receptors (nociceptors) in humans are naked dendrites in the epidermis. Different groups of receptors respond to excess heat, pressure, or chemicals released by injured cells. Histamines and acids stimulate pain receptors, and prostaglandins increase pain by sensitizing receptors.

■ **INTERACTIVE QUESTION   49.2**

a. Where are pain receptors located in humans?

b. How do aspirin and ibuprofen reduce pain?

49.2   The mechanoreceptors involved with hearing and equilibrium detect settling particles or moving fluid

Sensing Gravity and Sound in Invertebrates   Most invertebrates have **statocysts,** which often consist of a layer of ciliated receptor cells around a chamber containing **statoliths,** dense granules or grains of sand. The stimulation of the receptors under the statoliths provides positional information to the animal.

Many insects sense sounds with body hairs that vibrate in response to sound waves of specific frequencies. Localized "ears" are found on many insects, consisting of a tympanic membrane with attached receptor cells stretched over an internal air chamber.

Hearing and Equilibrium in Mammals   In mammals and most terrestrial vertebrates, the sensory organs for hearing and balance are in the ear.

The ear consists of three regions: The external pinna and the auditory canal make up the **outer ear.** The **tympanic membrane** (eardrum), separating the outer ear from the **middle ear,** transmits sound waves to three small bones—the **malleus** (hammer), **incus** (anvil), and **stapes** (stirrup)—which conduct the waves to the inner ear by way of a membrane called the **oval window.** The **Eustachian tube,** connecting the pharynx and the middle ear, equalizes pressure within the middle ear. **The inner ear** is a fluid-filled labyrinth of channels within the temporal bone. The **semicircular canals** function in equilibrium, and the **cochlea** functions in hearing.

The coiled cochlea has two large, connected fluid-filled canals—an upper vestibular canal and a lower tympanic canal—separated by the smaller cochlear duct. The **organ of Corti,** located on the basilar membrane that forms the floor of the cochlear duct, contains hair cells. Their hairs extend into the cochlear duct, and some attach to the tectorial membrane, which overhangs the organ of Corti.

Pressure waves in air—transmitted by the tympanic membrane, the three bones of the middle ear, and the oval window—produce pressure waves in the cochlear fluid, which travel from the vestibular canal through the tympanic canal and dissipate when they strike the **round window.** These pressure waves vibrate the basilar membrane, bending the hairs of its receptor cells against the tectorial membrane. The hair cells depolarize, increasing neurotransmitter release and the frequency of action potentials generated in associated sensory neurons that travel in the auditory nerve to the brain.

Volume is a result of the amplitude, or height, of the sound wave; a stronger wave bends the hair cells more and results in more action potentials. **Pitch** is related to the frequency of sound waves, usually expressed in hertz (Hz). Different regions of the basilar membrane vibrate in response to different frequencies, and neurons associated with the vibrating region transmit action potentials to specific auditory regions of the cortex where the sensation is perceived as a particular pitch.

■ **INTERACTIVE QUESTION   49.3**

Label the parts in this diagram of the human ear.

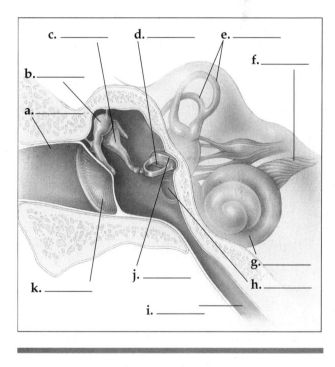

Within the inner ear are two chambers, the **utricle** and **saccule,** and three semicircular canals, responsible for detection of body position and balance. Hair cells in the utricle and saccule project into a gelatinous material containing calcium carbonate particles called otoliths. These heavy particles exert a pull on the hairs in the direction of gravity; different head angles stimulate different hair cells, thereby altering the output of neurotransmitters and the resulting action potentials of sensory neurons of the vestibular nerve. Other hair cells in the swellings at the base of the semicircular canals (the ampullae) project into a gelatinous mass called the cupula and are able to sense changes in rotational movements by the pressure of the endolymph in the semicircular canals against the cupula.

Hearing and Equilibrium in Other Vertebrates   Fishes and aquatic amphibians have inner ears, consisting of a saccule, utricle, and semicircular canals, within which sensory hairs are stimulated by the movement of otoliths. Sound-wave vibrations pass through the skeleton of the head to the inner ear. Some fishes have a Weberian apparatus, a series of bones that transmits vibrations from the swim bladder to the inner ear.

Mechanoreceptor units called neuromasts—clusters of hair cells with sensory hairs embedded in a gelatinous cap or cupula—are contained in the **lateral line system** of fishes and aquatic amphibians. Water moving through a tube along both sides of the body bends the cupula and stimulates the hair cells, enabling the fish to perceive its own movement, water currents, and the pressure waves generated by other moving objects.

The inner ear is the organ of hearing and equilibrium in terrestrial vertebrates. In amphibians, sound vibrations are conducted by a tympanic membrane on the body surface and a single middle ear bone to the inner ear. A cochlea has evolved in birds and mammals.

■ **INTERACTIVE QUESTION   49.4**

What organs in mammals, fishes, and invertebrates are used to sense movement or equilibrium? How are these organs similar?

a. Mammals:

b. Fishes:

c. Invertebrates:

d. Similarities:

49.3   The senses of taste and smell are closely related in most animals

For terrestrial animals, the sense of **gustation** (taste) detects chemicals in solution, whereas **olfaction** (smell) detects airborne chemicals. Both rely on chemoreceptors.

Sensillae, or sensory hairs containing taste receptors, are found on the feet and mouthparts of insects. Insects also have olfactory hairs, usually located on their antennae.

Taste in Humans   **Taste buds,** which contain groups of modified epithelial cells, are scattered on the tongue and mouth. The primary taste perceptions are sweet, sour, salty, bitter, and umami. The brain integrates the input from the taste receptors as they differentially respond to various chemicals, and it creates the perception of a complex flavor. Depolarization of chemoreceptor in response to their particular stimuli occurs by several mechanisms that may involve ion influxes or closing of ion channels. In all taste receptors, depolarization leads to neurotransmitter release, and a sensory neuron transmits action potentials to the brain.

Smell in Humans   Olfactory receptor cells are neurons that line the upper part of the nasal cavity. Binding of an odorous substance, or odorant, to specific proteins, called odorant receptors, on their cilia triggers a signal-transduction pathway, which opens ion channels. The membrane depolarizes and sends action potentials to the olfactory bulb of the brain. Approximately 3% of the genes in the human genome are odorant receptor genes.

■ **INTERACTIVE QUESTION   49.5**

For what purposes do animals use their senses of gustation and olfaction?

49.4   Similar mechanisms underlie vision throughout the animal kingdom

Despite the diversity of photoreceptors that have evolved in the animal kingdom, all contain similar light-absorbing pigment molecules and may be homologous with a common genetic basis.

Vision in Invertebrates   The ocellus or eye cup of planarians, which detects light intensity and direction, consists of a layer of cells with a screening pigment that partially encloses photoreceptors. The brain compares impulses from the left and right ocelli to help the animal navigate a direct path away from a light source.

The **compound eye** of insects and crustaceans contains up to thousands of light detectors called **ommatidia,** each with its own lens. The differing intensities of light entering the many ommatidia produce a mosaic image. The compound eyes of insects are adept at detecting movement and color, and some detect ultraviolet radiation.

Some jellies, polychaetes, spiders, and many molluscs have a **single-lens eye,** in which light is focused through the single lens onto a layer of photoreceptors.

The Vertebrate Visual System   The eyeball consists of a tough, outer connective tissue layer called the **sclera** and a thin, pigmented inner layer called the **choroid.** The **conjunctiva** is a mucous membrane covering the sclera. At the front of the eye, the sclera becomes the transparent **cornea,** which functions as a fixed lens, and the choroid forms the colored **iris,** which regulates the amount of light entering through the **pupil.** The **retina,** the innermost layer of the eyeball, contains the photoreceptors. The optic nerve attaches to the eye at the optic disk, forming a blind spot on the retina.

The transparent **lens** focuses an image onto the retina. The **ciliary body** produces the **aqueous humor** that fills the anterior eye cavity; jellylike **vitreous humor** fills the posterior cavity. Many fishes focus by moving the lens backward or forward. Mammals focus on close objects by **accommodation,** in which ciliary muscles contract, causing suspensory ligaments to slacken and the elastic lens to become rounder.

Rods and **cones** are the photoreceptors in the retina. The relative proportion of each of these receptors correlates with the activity pattern of the animal: Rods are more light sensitive and enable night vision, whereas cones distinguish colors. In the human eye, rods are most concentrated toward the edge of the retina, whereas the center of the visual field, the **fovea,** is filled with cones.

■ **INTERACTIVE QUESTION   49.6**

Label the parts of the vertebrate eye.

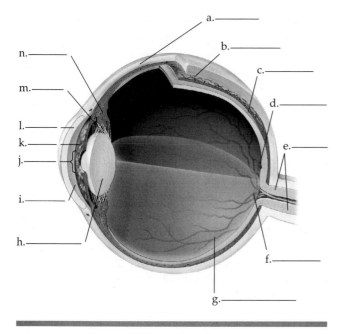

Both rods and cones have visual pigments embedded in a stack of disks in the outer segment of each cell. **Retinal** is the light-absorbing molecule and is bonded to a membrane protein called an **opsin.** Rods contain the visual pigment **rhodopsin.** When retinal absorbs light, it changes shape and separates from its opsin. In the dark, retinal is converted back to its original shape, and it recombines with its opsin. In bright light, the rhodopsin remains "bleached," and cones are responsible for vision.

The three subclasses of red, green, and blue cones, each with its own type of opsin that binds with retinal to form a **photopsin,** are named for the color of light they are best at absorbing.

Visual processing begins in the retina, where rods and cones synapse with neurons called **bipolar cells.** In the dark, rods and cones are depolarized and continually release glutamate, which, depending on the postsynaptic receptor, either depolarizes or hyperpolarizes bipolar cells. Following the light-induced dissociation of retinal, opsin activates a membrane-bound G protein called transducin, which then activates an enzyme that detaches cGMP bound to sodium ion channels in the plasma membrane. Sodium channels close, hyperpolarizing the membrane and decreasing its release of glutamate. In response, the bipolar cells reverse their polarization, either becoming hyperpolarized or depolarized, depending on the receptor.

Bipolar cells synapse with **ganglion cells,** whose axons form the optic nerve and transmit action potentials to the brain. **Horizontal cells** and **amacrine cells** in the retina help to integrate visual information.

Signals from rods and cones may follow the vertical pathway directly from photoreceptors to bipolar cells to ganglion cells. A lateral pathway involves lateral integration by horizontal cells, which carry signals from one receptor to others and to several bipolar cells, and by amacrine cells, which relay information from one bipolar cell to several ganglion cells. **Lateral inhibition** occurs when a horizontal cell inhibits more distant cells and is a form of integration that enhances contrast. It also occurs in interactions of amacrine cells and also at higher levels of processing.

The left and right optic nerves meet at the **optic chiasma** at the base of the cerebral cortex. Information from the left visual field of both eyes travels to the right side of the brain, whereas what is sensed in the right field of view goes to the left side. Most axons of the ganglion cells go to the **lateral geniculate nuclei** of the thalamus, where neurons lead to the **primary visual cortex** in the occipital lobe of the cerebrum. Interneurons also carry information to other visual processing and integrating centers in the cortex. At least 30% of the cerebral cortex may be devoted to creating the three-dimensional perception of what we see.

■ INTERACTIVE QUESTION   49.7

The receptive field of a ganglion cell includes the rods or cones that supply information to the bipolar cells connected to it.

a. Would a large or small receptive field produce the sharpest image?

b. Where would the smallest receptive fields be found?

49.5   Animal skeletons function in support, protection, and movement

Types of Skeletons   Fluid under pressure in a closed body compartment creates a **hydrostatic skeleton.** As muscles change the shape of the fluid-filled compartment, the animal elongates or moves. Earthworms and other annelids move by **peristalsis,** using rhythmic waves of contractions of circular and longitudinal muscles.

Typical of molluscs and arthropods, **exoskeletons** are hard coverings deposited on the surface of animals. The cuticle of an arthropod contains fibrils of the polysaccharide **chitin** embedded in a protein matrix. Where protection is most needed, this flexible support is hardened by cross-linking proteins and addition of calcium salts.

The supporting elements of an **endoskeleton** are embedded in the soft tissues of the animal. Sponges have endoskeletons composed of spicules or fibers, and echinoderms have hard plates beneath the skin. The endoskeletons of chordates are composed of cartilage and/or bone. The vertebrate axial skeleton consists of skull, vertebral column, and rib cage; the appendicular skeleton contains the pectoral and pelvic girdles and the limb bones. Joints allow for flexibility in body movement.

■ INTERACTIVE QUESTION   49.8

List an advantage and a disadvantage for each of the three types of skeletons and indicate in which animal groups each are found.

a. hydrostatic skeleton

b. exoskeleton

c. endoskeleton

Physical Support on Land   Body proportions of small and large animals vary, partly as a result of the physical relationship between the diameter of a support (strength increases with the square of its diameter) and the strain produced by increasing weight (weight increases with the cube of the height). Posture is an important structural factor in supporting body weight. Most of the weight of large land mammals is supported by muscles and tendons, which hold the legs relatively straight and directly under the body.

49.6   Muscles move skeletal parts by contracting

Because muscles can only contract, antagonistic muscle pairs are required to move body parts in opposite directions.

Vertebrate Skeletal Muscle   Vertebrate **skeletal muscle** consists of a bundle of multinucleated muscle cells, called fibers, running the length of the muscle. Each fiber is a bundle of **myofibrils,** each composed of two kinds of **myofilaments: Thin filaments** consist of two strands of actin coiled with one strand of regulatory protein, and **thick filaments** are arrays of myosin molecules.

The regular arrangement of myofilaments produces repeating light and dark bands; thus skeletal muscle is also called striated muscle. The repeating units are called **sarcomeres** and are delineated by **Z lines.** Thin filaments attach to the Z line and project toward the center of the sarcomere. Thick filaments lie in the center, forming the **A band,** with an H zone in the center where the thin filaments do not reach. At the edges of the sarcomere, where thick filaments do not extend when the muscle is at rest, is the **I band** of only thin filaments.

According to the **sliding-filament model** of muscle contraction, the thick and thin filaments do not change length but simply slide past each other, increasing their area of overlap.

The mechanism for the sliding of filaments is the hydrolyzing of ATP by the globular head of a myosin molecule, which then changes to a high-energy configuration that binds to actin and forms a cross-bridge. When relaxing to its low-energy configuration, the myosin head bends and pulls the attached thin filament toward the center of the sarcomere. When a new molecule of ATP binds to the myosin head, it breaks its bond to actin, and the cycle repeats with the high-energy configuration attaching to an actin molecule farther along on the thin filament. Energy for muscle contraction comes from creatine phosphate, which regenerates ATP, and from glycogen, which is broken down to glucose. Glucose can generate ATP by glycolysis or aerobic respiration.

■ **INTERACTIVE QUESTION   49.9**

Identify the components in this diagram of a section of a skeletal muscle fiber.

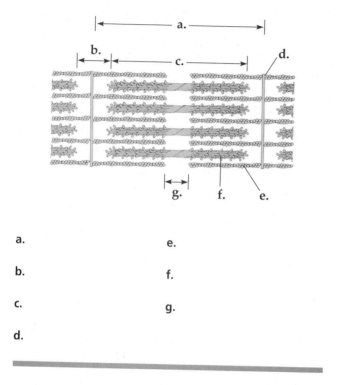

a.

b.   e.

c.   f.

d.   g.

When the muscle is at rest, the myosin-binding sites of the actin molecules are blocked by the regulatory protein **tropomyosin,** whose position is controlled by another set of regulatory proteins, the **troponin complex.** When troponin binds to calcium ions, the tropomyosin-troponin complex changes shape, and myosin-binding sites on the thin filament are exposed.

Calcium ions are actively transported into the **sarcoplasmic reticulum (SR)** of a muscle cell. When an action potential of a motor neuron causes the release of acetylcholine into the synapse with a muscle fiber, the fiber depolarizes and an action potential spreads into the **T (transverse) tubules,** infoldings of the muscle cell plasma membrane. The action potential opens calcium channels in the sarcoplasmic reticulum, releasing calcium into the cytosol. Calcium binding with troponin exposes myosin-binding sites, and contraction begins. Contraction ends when the sarcoplasmic reticulum pumps calcium ions back out of the cytosol, and the tropomyosin-troponin complex again blocks the binding sites.

The response of a single muscle fiber to an action potential of a motor neuron is an all-or-none contraction, producing a twitch. Graded contraction of whole muscles may result from involving additional fibers. A single motor neuron and all the muscle fibers it innervates make up a **motor unit.** The strength of a muscle contraction depends on the size and number of motor

units involved. Muscle force (tension) can be increased by the activation of additional motor units, called **recruitment.** Prolonged contraction can cause muscle fatigue by depletion of ATP, loss of ion gradients required for depolarization, and accumulation of lactate. Some postural support muscles are always in a state of partial contraction, in which different motor units take turns contracting.

Repeated, rapid-fire action potentials produce a state of smooth and sustained contraction called **tetanus.**

Slow muscle fibers, common in muscles that must sustain long contractions, have less sarcoplasmic reticulum and slower calcium pumps; thus calcium remains in the cytosol longer and twitches last longer. Fast muscle fibers produce rapid and powerful contractions. Oxidative fibers, which rely mostly on aerobic respiration, have many mitochondria, a good blood supply, and **myoglobin,** a pigment that extracts oxygen from the blood and stores it. Glycolytic fibers primarily use glycolysis to produce ATP.

■ **INTERACTIVE QUESTION   49.10**

a. Identify the three main types of skeletal muscle fibers.

b. Which type of fiber would have few mitochondria; little myoglobin; and produce brief, powerful contractions?

c. Which type of fiber would you expect to be in the highest proportion in postural muscles?

Other Types of Muscle   Vertebrate **cardiac muscle** cells are connected by **intercalated disks** through which action potentials spread to all the cells of the heart. Ion channels in the plasma membrane of a cardiac muscle cell produce rhythmic depolarizations that generate action potentials. The action potentials of cardiac muscle cells last quite long.

The scattered arrangement of actin and myosin filaments in **smooth muscle** accounts for its nonstriated appearance. Smooth muscle lacks T tubules, a troponin complex, and a well-developed sarcoplasmic reticulum. Calcium ions enter the cell in response to an action potential and bind to calmodulin, which activates an enzyme that phosphorylates myosin. Its relatively slow contractions can occur over a greater range of lengths than striated muscle.

Invertebrates have muscle cells similar to the skeletal and smooth muscle cells of vertebrates. The flight muscles of insects are capable of independent and rapid contraction.

49.7 Locomotion requires energy to overcome friction and gravity

Locomotion is the active movement from place to place. Animals move to obtain food, escape from danger, and find mates.

Swimming Overcoming gravity is less of a problem for swimming animals due to the buoyancy of water. The density of water, however, makes drag (friction) a major problem. Most fast swimmers have a fusiform shape that reduces friction. Animals may swim by using their legs as oars, using water to jet-propel themselves, or by undulating their bodies and tails from side to side or up and down.

Locomotion on Land Walking, running, hopping, or crawling on land requires the animal to support itself and move against gravity; but air offers little resistance to movement. Powerful leg muscles and strong skeletal support are important for propelling an animal forward and supporting it. A running or hopping animal may store some energy in its tendons to facilitate the next step. Maintaining balance is another requirement for walking, running, or hopping. A crawling animal must overcome the friction produced by its contact with the ground.

Flying A flying animal uses its wings to provide enough lift to overcome the force of gravity.

Comparing Costs of Locomotion The cost of locomotion depends on the mode of locomotion and the environment. Adaptations that maximize efficiency of locomotion increase fitness in that more of an animal's energy is available for growth and reproduction.

■ INTERACTIVE QUESTION 49.11

a. Which mode of locomotion is the most energy efficient?

b. Which mode is the most energetically expensive per distance traveled?

c. Which mode consumes the most energy per unit of time?

d. In animals specialized for the same mode of transport, does a larger animal or a smaller one expend more energy per kilogram of body mass?

Word Roots

ama- = together (*amacrine cell:* neurons of the retina that help integrate information before it is sent to the brain)

aqua- = water (*aqueous humor:* the clear, watery solution that fills the anterior cavity of the eye)

bi- = two (*bipolar cell:* neurons that synapse with the axons of rods and cones in the retina of the eye)

chemo- = chemical (*chemoreceptor:* a receptor that transmits information about the total solute concentration in a solution or about individual kinds of molecules)

coch- = a snail (*cochlea:* the complex, coiled organ of hearing that contains the organ of Corti)

electro- = electricity (*electromagnetic receptor:* receptors of electromagnetic energy, such as visible light, electricity, and magnetism)

endo- = within (*endoskeleton:* a hard skeleton buried within the soft tissues of an animal)

exo- = outside (*exoskeleton:* a hard encasement on the surface of an animal that provides protection and points of attachment for muscles)

externo- = outside (*exteroreceptor:* sensory receptors that detect stimuli outside the body, such as heat, light, pressure, and chemicals)

fovea- = a pit (*fovea:* center of the visual field of the eye)

gusta- = taste (*gustatory receptors:* taste receptors)

hydro- = water (*hydrostatic skeleton:* a skeletal system composed of fluid held under pressure in a closed body compartment; the main skeleton of most cnidarians, flatworms, nematodes, and annelids)

inter- = between; **-cala** = insert (*intercalated disks:* specialized junctions between cardiac muscle cells, which provide direct electrical coupling)

interno- = inside (*interoreceptor:* sensory receptors that detect stimuli within the body, such as blood pressure and body position)

mechano- = an instrument (*mechanoreceptor:* a sensory receptor that detects physical deformations in the body's environment associated with pressure, touch, stretch, motion, and sound)

myo- = muscle; **-fibro** = fiber (*myofibril:* a fibril collectively arranged in longitudinal bundles in muscle cells; composed of thin filaments of actin and a regulatory protein and thick filaments of myosin)

noci- = harm (*nociceptor:* pain receptors in the epidermis of the skin)

olfact- = smell (*olfactory receptor:* smell receptors)

omma- = the eye (*ommatidia:* the facets of the compound eye of arthropods and some polychaete worms)

peri- = around; **-stalsis** = a constriction (*peristalsis:* rhythmic waves of contraction of smooth muscle that push food along the digestive tract)

photo- = light (*photoreceptor:* receptors of light)

rhodo- = red (*rhodopsin:* a visual pigment consisting of retinal and opsin)

sacc- = a sack (*saccule:* a chamber in the vestibule behind the oval window that participates in the sense of balance)

sarco- = flesh; **-mere** = a part (*sarcomere:* the fundamental, repeating unit of striated muscle, delimited by the Z lines)

sclero- = hard (*sclera:* a tough, white outer layer of connective tissue that forms the globe of the vertebrate eye)

semi- = half (*semicircular canals:* a three-part chamber of the inner ear that functions in maintaining equilibrium)

stato- = standing; **-lith** = a stone (*statolith:* sensory organs that contain mechanoreceptors and function in the sense of equilibrium)

tetan- = rigid, tense (*tetanus:* the maximal, sustained contraction of a skeletal muscle caused by a very fast frequency of action potentials elicited by continual stimulation)

thermo- = heat (*thermoreceptor:* an interoreceptor stimulated by either heat or cold)

trans- = across; **-missi** = send (*transmission:* the conduction of impulses to the central nervous system)

tropo- = turn, change (*tropomyosin:* the regulatory protein that blocks the myosin-binding sites on the actin molecules)

tympan- = a drum (*tympanic membrane:* another name for the eardrum)

utric- = a leather bag (*utricle:* a chamber behind the oval window that opens into the three semicircular canals)

vitre- = glass (*vitreous humor:* the jellylike material that fills the posterior cavity of the vertebrate eye)

Structure Your Knowledge

1. Describe the path of a light stimulus from where it enters the vertebrate eye to its transmission as an action potential through the optic nerve.

2. Arrange the following structures in an order that enables you to describe the passage of sound waves from the external environment to the perception of sound in the brain.

round window	basilar membrane
oval window	cerebral cortex
pinna	vestibular and tympanic canals
auditory nerve	tectorial membrane
cochlea	tympanic membrane
auditory canal	malleus, incus, stapes
cochlear duct	organ of Corti with hair cells

3. Trace the sequence of events in muscle contraction from an action potential in a motor neuron to the relaxation of the muscle.

Test Your Knowledge

MATCHING: *Match the term with its description.*

_____ 1. oxygen-storing compound in muscles

_____ 2. stretch receptor in muscle, monitors position

_____ 3. stimulus energy converted into receptor potential

_____ 4. light-absorbing pigment

_____ 5. naked dendrites that detect pain

_____ 6. bones that transmit vibrations to inner ear of some fish

_____ 7. invertebrate structure for sensing position

_____ 8. energy-storing compound in muscles

_____ 9. changing shape of lens to focus

_____ 10. decrease in sensitivity of receptor during constant stimulation

A. accommodation	J. nociceptors
B. adaptation	K. ommatidia
C. ampulla	L. retinal
D. creatine phosphate	M. statocyst
E. cupula	N. transduction
F. ganglion cells	O. transmission
G. muscle spindle	P. tropomyosin
H. myofibrils	Q. Weberian apparatus
I. myoglobin	

MULTIPLE CHOICE: *Choose the one best answer.*

1. What is a sensory perception?
   a. the opening of ion channels in response to mechanical stimulation
   b. the conversion of stimulus energy into a receptor potential by a sensory receptor or sensory neuron
   c. the interpretation of sensation according to the region of the brain that receives the nerve impulse
   d. the amplification and transmission of sensory data to the CNS
   e. the emotional response to a sensation

2. Which of the following is an interoreceptor?
   a. temperature receptor in hypothalamus
   b. pain receptor in fingertip
   c. taste receptor on tongue
   d. chemoreceptor in nasal passages
   e. rod or cone cell in eye

3. The compound eye found in insects and crustaceans
   a. is composed of thousands of prisms that focus light onto a few photoreceptor cells.
   b. cannot sense colors.
   c. contains many ommatidia, each of which contains a lens that focuses light from a tiny portion of the field of view.
   d. may detect infrared and ultraviolet radiation.
   e. does all of the above.

4. Which of the following statements is *not* true?
   a. The iris regulates the amount of light entering the pupil.
   b. The choroid is the thin, pigmented inner layer of the eye that contains the photoreceptor cells.
   c. The ciliary muscle functions in changing the shape of the lens.
   d. Thin aqueous humor fills the anterior eye cavity.
   e. The conjunctiva is a mucous membrane surrounding the sclera.

5. Which of the following is *incorrectly* paired with its function?
   a. cones—respond to different light wavelengths, producing color vision
   b. horizontal cells—produce lateral inhibition of receptor and bipolar cells
   c. bipolar cells—relay impulses between rods or cones and ganglion cells
   d. ganglion cells—synapse with the optic nerve at the optic disk or blind spot
   e. fovea—center of the visual field, containing only cones in humans

6. The molecule rhodopsin
   a. dissociates when struck by light and is functional only in complete dark.
   b. is recombined by enzymes from retinal and opsin in the light.
   c. is more sensitive to light than are photopsins and is involved in black-and-white vision in dim light.
   d. activates a signal-transduction pathway that increases a rod cell's permeability to sodium.
   e. is or does all of the above.

7. Rotation of the head of a vertebrate is sensed by
   a. the organ of Corti located in the utricle and saccule.
   b. changes in the action potentials of hair cells in response to the pull on them by calcium carbonate particles.
   c. bending of hair cells in the ampulla of a semicircular canal in response to the inertia of the endolymph.
   d. hair cells in response to movement of fluid in the coiled cochlea.
   e. statocysts when statoliths stimulate hair cells.

8. Fish can perceive pressure waves from moving objects with their
   a. lateral line system containing mechanoreceptor units called neuromasts.
   b. tectorial membrane as it is stimulated by the bending of sensory hairs embedded in a gelatinous cap.
   c. inner ears, within which sensory hairs are stimulated by the movement of otoliths.
   d. Weberian apparatus, a series of bones that transmits vibrations from the swim bladder to the inner ear.
   e. sensillae collected into ampullae in the lateral line system.

9. The volume of a sound is determined by
   a. the area in the brain that receives the signal.
   b. the section of the basilar membrane that vibrates.
   c. the number of hair cells that are stimulated.
   d. the frequency of the sound wave.
   e. the degree to which hair cells are bent and the resulting increase in the number of action potentials produced in the sensory neuron.

10. Hydrostatic skeletons are used for movement by all of the following *except*
    a. cnidarians.
    b. arthropods.
    c. nematodes.
    d. annelids.
    e. flatworms.

11. When striated muscle fibers contract,
    a. the Z lines are pulled closer together.
    b. the I band remains the same.
    c. the A band becomes shorter.
    d. the H zone widens slightly.
    e. all of the above occur.

12. Tetanus
    a. is muscle fatigue resulting from a buildup of lactate during anaerobic respiration.
    b. is the all-or-none contraction of a single muscle fiber.
    c. is the result of stimulation of additional motor neurons.
    d. is the result of a volley of action potentials, whose summation produces an increased and sustained contraction.
    e. is the rigidity of muscles following death.

13. What is the role of ATP in muscle contraction?
    a. to form cross-bridges between thick and thin filaments
    b. to release myosin head from actin when it binds to myosin and to provide energy to form myosin's high-energy configuration
    c. to remove the tropomyosin-troponin complex from blocking the binding sites on actin
    d. to bend the cross-bridge and pull the thin filaments toward the center of the sarcomere
    e. to replace the supply of creatine phosphate required for movement of actin past myosin

14. How does calcium affect muscle contraction?
    a. It is released from the T tubules in response to an action potential to initiate contraction.
    b. The binding of acetylcholine opens calcium channels in the plasma membrane, creating an action potential that travels down the T tubules.
    c. It binds to tropomyosin and helps to stabilize cross-bridge formation.
    d. Its binding to troponin causes tropomyosin to move away from the myosin-binding sites on the actin filament.
    e. Its release from the sarcoplasmic reticulum changes the membrane potential of the muscle cell so that contraction can occur.

15. A motor unit is
    a. a sarcomere that extends from one Z line to the next Z line.
    b. a motor neuron and the muscle fibers it innervates.
    c. the myofibrils of a fiber and the transverse tubules that connect them.
    d. the muscle fibers that make up a muscle.
    e. the antagonistic set of muscles that flex and extend a body part.

16. Which of the following is *not* a characteristic of cardiac muscle?
    a. scattered arrangement of actin and myosin filaments
    b. intercalated disks that spread action potentials between cells
    c. action potentials that last a long time
    d. ability to generate action potentials without nervous input
    e. striations

17. Smooth muscle contracts relatively slowly because
    a. the only ATP available is supplied by fermentation.
    b. its contraction is stimulated by hormones, not motor neurons.
    c. it does not have a well-developed sarcoplasmic reticulum, and calcium enters the cell through the plasma membrane during an action potential.
    d. it is not striated.
    e. it is composed exclusively of slow muscle fibers.

18. The absorption of light by rhodopsin in a rod cell leads to
    a. an increase in the release of neurotransmitter.
    b. the opening of sodium channels caused by the binding of cGMP.
    c. a hyperpolarized membrane and a decrease in the release of neurotransmitter.
    d. the bleaching of rhodopsin, which increases the sensitivity of rod cells to light.
    e. the closing of sodium channels and the release of neurotransmitter.

19. What is the function of the cornea?
    a. regulate the amount of light that enters the eye
    b. accommodation to focus on close objects
    c. tough, protective connective tissue layer surrounding the eye
    d. transparent covering that acts as a fixed lens to bend light rays
    e. produce aqueous humor that helps to focus images on the optic disk

20. A receptive field in the retina would involve a *single*
    a. bipolar cell.
    b. photoreceptor (rod or cone) cell.
    c. amacrine cell.
    d. horizontal cell.
    e. ganglion cell.

21. What is the function of the three bones of the middle ear?
    a. respond to different frequencies of sound waves
    b. transmit pressure waves to the oval window
    c. equalize pressure within the inner ear by communicating through the Eustachian tube
    d. sense rotation in three different planes (x, y, and z axes)
    e. support the tympanic membrane (eardrum)

22. The sensory taste receptors on the feet of flies are
    a. mechanoreceptors.
    b. photoreceptors.
    c. pheromone receptors.
    d. chemoreceptors.
    e. naked dendrites.

23. What do the lateral line of fish, invertebrate statocysts, and the cochlea of your ear all have in common?
    a. They are used to sense sound or pressure waves.
    b. They are organs of equilibrium.
    c. They use hair cells as mechanoreceptors.
    d. They use a second-messenger pathway of signal transduction.
    e. They use granules to stimulate their receptor cells.

24. In a contracting muscle fiber, creatine phosphate is used to
    a. supply extra O_2 for aerobic respiration to generate ATP.
    b. phosphorylate ADP to replenish ATP supplies.
    c. provide the energy for the sliding of the myosin filament toward the center of the sarcomere.
    d. recycle lactate to glucose.
    e. transduce the action potential of the T tubules into stimulation of the sarcoplasmic reticulum.

25. When you pull up your lower arm and "make a muscle" in your biceps, you are
    a. contracting a flexor.
    b. relaxing a flexor.
    c. contracting an extensor.
    d. contracting a tendon.
    e. producing a simple muscle twitch.

26. Different regions of the basilar membrane vibrate in response to different
    a. volumes of sounds.
    b. amplitudes of sound waves.
    c. pitches.
    d. action potential frequencies.
    e. directions of rotation of the head.

27. How does an animal with a jointed exoskeleton move?
    a. alternate contractions of circular and longitudinal muscles
    b. expansion of the exoskeleton at joints, followed by contraction of the skeleton
    c. active contraction and extension of muscles that attach to opposite sides of a joint
    d. contraction of muscles that cause the exoskeleton to work against the hydrostatic skeleton, creating movement of body parts
    e. alternate contraction of antagonistic muscles (extensors and flexors) that span a joint

28. Which of the following modes of locomotion is the most energy efficient for an animal that is specialized for moving in that fashion?
    a. swimming
    b. running
    c. flying
    d. hopping
    e. crawling

Chapter 50

An Introduction to Ecology and the Biosphere

▶ Key Concepts

50.1 Ecology is the study of interactions between organisms and the environment

50.2 Interactions between organisms and the environment limit the distribution of species

50.3 Abiotic and biotic factors influence the structure and dynamics of aquatic biomes

50.4 Climate largely determines the distribution and structure of terrestrial biomes

▶ Framework

This chapter describes the organizational levels at which ecological questions are asked; the abiotic factors to which organisms have adapted in both an ecological and an evolutionary time frame; and the major world communities, or biomes, in which adaptations to climate and abiotic factors have produced similar and characteristic life forms.

▶ Chapter Review

Ecology studies the distribution and abundance of organisms and their interactions with the environment.

50.1 Ecology is the study of interactions between organisms and the environment

Ecology had its foundation in and continues to make use of the descriptive science of natural history. Rigorous experimental designs, however, are commonly used to investigate complex ecological questions.

Ecology and Evolutionary Biology Interactions between organisms and their environments occur within ecological time. The cumulative effects of these interactions are realized on the scale of evolutionary time.

Organisms and the Environment Organisms are affected by and, in turn, affect both the **abiotic** and **biotic** components of the environment. All the organisms in the environment are called the **biota.** Ecologists try to determine what environmental factors, either directly or indirectly, limit the geographic range (distribution) and determine the abundance of a species.

Subfields of Ecology   Levels of ecological study range from individual organisms to ecosystems, although several areas are often combined in a single study.

Organismal ecology, which may include the disciplines of behavioral, physiological, and evolutionary ecology, considers the responses and adaptations of an organism to its environment. **Population ecology** is concerned with the factors that control the size of **populations,** which are groups of individuals of a species occupying a particular area. The **community** includes all the populations of organisms in an area; **community ecology** looks at interactions such as predation and competition.

An **ecosystem** includes abiotic factors as well as the biological community, and **ecosystem ecology** addresses such topics as the flow of energy and chemical cycling. A landscape or seascape is the arrangement of several ecosystems in a geographic region; **landscape ecology** looks at the flow of energy, materials, and organisms among different ecosystems. A characteristic of landscapes is *patchiness*, a mosaic of different environments or ecosystems. The **biosphere** includes all Earth ecosystems.

Ecology and Environmental Issues   The science of ecology is not synonymous with the growing awareness of environmental issues, but it helps us to understand these complicated problems and their possible solutions. Environmental issues are scientifically complex, and the **precautionary principle** should guide decisions in which not all answers are known.

■ **INTERACTIVE QUESTION   50.1**

List the five areas of ecological study and describe the focus of inquiry at each level.

a. Ecology & Evolutionary Biology: the interactions bet. organisms and their environments occur within ecological time.

b. Organisms and their environment.—Organisms are affected by and, in turn, affect both biotic (living) and abiotic (nonliving). Biota means all organisms in the environment.

c. Organismal Ecology—Population ecology: concerned w/ the factors that control size

d. Ecosystem Ecology—addresses such topics as the flow of energy and chemical cycling

e. Landscape ecology—the flow of energy, materials and other organisms among different ecosystem.

50.2 Interactions between organisms and the environment limit the distribution of species

Biogeography studies the past and present distribution of species. Ecologists often work through a series of logical questions to determine what limits geographic distributions of species.

Dispersal and Distribution   The cattle egret is an example of a species that has naturally extended its range through **dispersal.** Transplants of a species can indicate whether dispersal limits its distribution. A successful transplant shows that the potential range of a species is larger than its actual range. Introduced species, which are introduced to new areas either purposely or accidentally, often disrupt their new ecosystem.

Behavior and Habitat Selection   Sometimes the behavior of organisms in habitat selection keeps them from occupying all their potential range. Habitat selection by ovipositing insects, which often choose only certain host plants, may limit their distribution.

Biotic Factors   The inability of transplanted organisms to survive and reproduce may be due to predation, disease, parasitism, competition, or lack of mutual symbiosis. "Removal and addition" experiments test whether predators limit the distribution of prey species. Sea urchins were shown to limit the abundance and distribution of seaweeds.

Abiotic Factors   Global patterns of geographic distributions are influenced by abiotic factors such as regional differences in temperature, rainfall, and light. An environment may have both *spatial* and *temporal heterogeneity* (differ in both space and time) for abiotic factors.

Temperature is an important environmental factor because of its effects on metabolism and enzyme activity. Most organisms cannot maintain body temperatures that differ much from ambient temperature. Even endotherms function best within a narrow range of environmental temperatures.

The availability of water in different habitats can vary greatly. Organisms must maintain their water balance, compensating for different osmolarities in aquatic environments and avoiding desiccation in terrestrial habitats.

Light energy drives almost all ecosystems. The intensity and quality of light are limiting factors in aquatic environments. Many plants and animals are sensitive to photoperiod, which serves as a reliable indicator of seasonal changes.

Wind increases the rate of heat and water loss in organisms and can affect plant morphology.

Soils; which vary in their physical structure, pH, and mineral composition; affect the distribution of plants and, in turn, the distribution of animals. Substrate composition in aquatic environments influences water chemistry and the types of organisms that can inhabit those areas.

Climate The **climate,** or prevailing weather conditions of a locality, is determined by temperature, water, sunlight, and wind. **Macroclimate** is the climatic pattern on a local to global level; **microclimate** is the fine variations within a habitat patch.

The absorption of solar radiation heats the atmosphere, land, and water, setting patterns for temperature variations, air circulation, and water evaporation that cause latitudinal variations in climate. The shape of the Earth and the tilting of its axis create seasonal variations in day length and temperature that increase with latitude. The **tropics** receive the greatest amount of and least variation in solar radiation.

The global circulation of air begins as intense solar radiation near the equator, causes warm, moist air to rise, producing the characteristic wet tropical climate, the arid conditions around 30° north and south as dry air descends, the fairly wet though cool climate about 60° latitude as air rises again, and the cold and rainless climates of the arctic and antarctic regions. Air flowing close to the Earth's surface produces predictable global wind patterns, such as the cooling trade winds in the tropics and subtropics and the prevailing westerlies in temperate zones.

Regional climatic patchiness is influenced by proximity to water and topographical features. Coastal areas are generally more moist, and large bodies of water moderate the climate. A *Mediterranean climate*, however, is created when cool, dry ocean breezes warm as they cross land, absorbing moisture and creating a hot, rainless climate.

■ **INTERACTIVE QUESTION 50.2**

Mountains affect local climate. Describe their influence in the following three areas:

a. solar radiation: It heats the land and water & atmosphere

b. temperature: When warm, moist air approaches a mountain and the air rises & cools, releasing moisture on the air

c. rainfall: In windward are rain but in leeward side of the mountain, cooler, dry air

Seasonal changes affect local climate. Seasonal changes in wind patterns can produce wet and dry seasons and affect ocean currents, sometimes causing upwellings of cold, nutrient-rich water. Seasonal temperature changes produce the semiannual **turnover** of waters in lakes that brings oxygenated water to the bottom and nutrient-rich water to the surface.

Climate also varies on a very small scale. Microclimates within an area have differences in abiotic features that affect the local distributions of organisms.

Global warming will have a great effect on the distributions of plants and animals. Fossil pollen deposits have documented the rates of northward expansion of various tree species following the last continental glacier. As the geographic climatic limits change with global warming, seed dispersal may not be rapid enough for some species of plants to migrate into new ranges.

■ **INTERACTIVE QUESTION 50.3**

List and give examples of the four factors that may limit the geographic distribution of a species.

a.

b.

c.

d.

50.3 Abiotic and biotic factors influence the structure and dynamics of aquatic biomes

Biomes are major types of ecological groupings that are found in broad geographic regions of land or water.

One of the chemical differences in aquatic biomes is salt concentration—less than 1% for freshwater biomes versus an average of 3% for marine biomes. Three-fourths of Earth is covered by oceans, which influence global rainfall, climate, and wind patterns. Marine algae and photosynthetic bacteria produce a large portion of the world's oxygen and consume enormous amounts of carbon dioxide. Freshwater biomes are closely linked with and shaped by the surrounding terrestrial biomes.

Many aquatic biomes are stratified in availability of light and temperature. The **photic zone** receives sufficient light for photosynthesis, whereas little light penetrates into the lower **aphotic zone**. The bottom substrate, called the **benthic zone,** is home to organisms collectively called **benthos.** Settling **detritus** (dead organic material) provides food for the benthos. In both lakes and marine environments, open water is called the pelagic zone. Aquatic environments are also classified based on distance from shore and water depth: **littoral** and **limnetic zones** in lakes, and intertidal, neritic, and oceanic zones in oceans. The abyssal zone is the deepest region of the ocean floor.

In the ocean and in many lakes, a narrow **thermocline** separates warmer surface waters from the cold bottom layer. In temperate lakes and oceans, seasonal turnover mixes surface and bottom layers.

Oligotrophic lakes are often deep, nutrient poor, fairly nonproductive, and generally oxygen rich. The shallower, nutrient-rich waters of **eutrophic lakes** support large, productive phytoplankton communities. Runoff carrying nutrients and sediment may gradually convert oligotrophic lakes into eutrophic lakes. Dumping of municipal wastes and runoff from fertilized lands can cause algal blooms and fish kills.

Streams and **rivers** are flowing habitats, whose physical and chemical characteristics vary from the headwaters to the mouth or point of entry into oceans or lakes. Overhanging vegetation contributes to nutrient content. Oxygen levels are high in turbulently flowing water and low in murky, warm waters. Human impact on streams and rivers includes pollution and damming.

Defined as areas covered with water often enough to support aquatic plants, **wetlands** are among the most productive biomes. Topography creates basin, riverine, and fringe wetlands. Much of these richly diverse biomes has been lost to draining and filling.

Where a freshwater river meets the ocean, an **estuary** is formed. Salinity varies both spatially and daily with the rise and fall of tides. Salt marsh grasses and algae, including phytoplankton, are the major producers. Estuaries serve as feeding and breeding areas for marine invertebrates, fish, and waterfowl. Pollution, filling, and dredging have extensively disrupted these highly productive areas.

■ INTERACTIVE QUESTION   50.4

Indicate with a + or − whether the following are relatively high or low in oxygen level, nutrient content, and productivity.

Biome	Oxygen Level	Nutrient Content	Productivity
Oligotrophic lake	a.		
Eutrophic lake	b.		
Headwater of stream	c.		
Turbid river	d.		
Estuary	e.		

In **intertidal zones** the daily cycle of tides exposes the shoreline to variations in water, nutrients, and temperature, and to the mechanical force of wave action. Rocky intertidal communities are vertically stratified, with organisms adapted to firmly attach to the hard substrate. Sandy intertidal zones are home to burrowing worms, clams, and crustaceans. Recreational use and oil pollutants have severely reduced numbers of beach-nesting birds and sea turtles.

Found in the photic zone of clear tropical waters, **coral reefs** are highly diverse and productive biomes. The structure of the reef is produced by the calcium carbonate skeletons of the coral (various cnidarians) and serves as a substrate for red and green algae. The coral animals are nourished by symbiotic unicellular algae. Overfishing, coral collecting, pollution, and global warming are destroying coral reefs and fishes.

The water of the **oceanic pelagic biome** is typically nutrient poor but oxygen rich. Seasonal mixing of temperate oceans stimulates phytoplankton growth. Phytoplankton flourish in the photic region and are grazed on by numerous types of zooplankton and larvae of invertebrates and fish. Overfishing, waste dumping, and oil spills have all damaged the Earth's oceans.

Nutrients reach the marine **benthic zone** as detritus falling from the waters above. **Neritic** benthic communities receive sunlight and are very diverse and productive. Various invertebrates and fishes inhabit the **abyssal** zone, the deep benthic region, and are adapted to cold and high water pressure. Chemoautotrophic prokaryotes form the basis of a collection of organisms adapted to the hot, low-oxygen environment surrounding **deep-sea hydrothermal vents.** Overfishing has eliminated many benthic fish populations.

■ **INTERACTIVE QUESTION   50.5**

Different marine environments can be classified on the basis of light penetration, distance from shore, and open water or bottom. Match the following zones to their corresponding numbers on the diagram below:

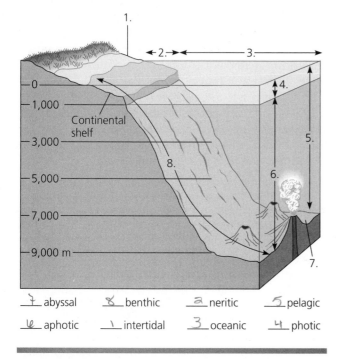

_7_ abyssal	_8_ benthic	_2_ neritic	_5_ pelagic
_6_ aphotic	_1_ intertidal	_3_ oceanic	_4_ photic

50.4   Climate largely determines the distribution and structure of terrestrial biomes

Climate and Terrestrial Biomes   The geographic distribution of the world's major terrestrial biomes is related to abiotic factors—in particular, the prevailing climate. A **climograph** plots annual mean temperature and rainfall for a region; generally these values correlate with the distribution of various biomes. Overlaps of biomes on a climograph indicate the importance of seasonal patterns of variation in rainfall and temperatures.

General Features of Terrestrial Biomes   Biomes are usually named for their predominant vegetation and major climatic features. Each biome also has characteristic microorganisms, fungi, and animals. Terrestrial biomes may have vertical stratification, such as the layers in a forest from **canopy,** low-tree stratum, shrub understory, herbaceous plant ground layer, and forest floor (litter) to root layer. Vertical stratification of vegetation provides diverse habitats for animals. The area where biomes grade into each other is called an **ecotone.**

Species composition of any one biome varies locally. Often, convergent evolution has produced a superficial resemblance of unrelated "ecological equivalents."

The extensive patchiness typical of most biomes results from natural or human disturbances. Grasslands, savannas, chaparral, and many coniferous forests are maintained by the periodic disturbance of fire. Urban and agricultural biomes now cover a large portion of Earth's land mass.

Tropical forests occur in equatorial and subequatorial regions. Variations in rainfall result in tropical dry forests, where rainfall is seasonal, and **tropical rain forests,** where rainfall is more abundant. Temperature is uniformly warm. The tropical rain forest has pronounced vertical stratification, and animal diversity is higher than in any other terrestrial biome. Agriculture and development are destroying large amounts of tropical forests.

Savannas are equatorial and subequatorial grasslands with scattered trees and rainy and dry seasons. Fires restrict vegetation to grasses and forbs, small broad-leaved plants. Large grazing mammals and their predators are common, although insects are the dominant herbivores. Ranching and overhunting have reduced large mammal populations.

Characterized by low and unpredictable precipitation, **deserts** may be hot or cold, depending on location. Desert animals have physiological and behavioral adaptations to dry conditions. Plants may use C_4 and CAM photosynthesis, and have reduced leaf surface area, water storage adaptations, and protective spines and toxins. Irrigated agriculture and urbanization are now common in deserts, reducing natural biodiversity.

Chaparral, common along coastlines in midlatitudes, has cool, rainy winters and hot, dry summers. The dominant vegetation—evergreen shrubs and small trees—is maintained by and adapted to periodic fires. Browsing and small mammals are common. Urbanization and agriculture have reduced areas of chaparral.

Temperate grasslands are maintained by fire, seasonal drought, and grazing by large mammals. Winters are generally cold and dry; summers are hot and wet. Soils are deep and fertile, and most North American grasslands have been converted to farmland.

Characterized by broad-leaved deciduous trees, **temperate broadleaf forests** grow in midlatitude regions that have adequate moisture to support the growth of large trees. Winters are cold and summers hot and humid. Humans have heavily logged broadleaf forests, clearing land for agriculture and development.

The largest terrestrial biome, the **coniferous forest,** or *taiga,* is found in northern latitudes and characterized by harsh winters with heavy snowfall, periodic droughts, and hot summers. Birds and large mammals are common animals. Coastal coniferous forests of the U.S. Pacific Northwest are temperate rain forests. Old-growth stands of trees in coniferous forests are rapidly being logged.

Tundra, covering large areas of the Arctic, is characterized by long, cold winters; short, mild summers; and dwarfed or matlike vegetation. A layer of frozen soil called **permafrost** prevents water penetration. Migratory large mammals and birds are common. The *alpine tundra,* found at all latitudes on very high mountains, has similar flora and fauna. Mineral and oil extraction may damage areas of arctic tundra.

■ INTERACTIVE QUESTION   50.6

Temperature and precipitation are two of the key factors that influence the vegetation found in a biome. On the climograph shown below, label the North American biomes (arctic and alpine tundra, coniferous forest, desert, temperate grassland, temperate broadleaf forest, and tropical forest) represented by each plotted area of temperature and precipitation.

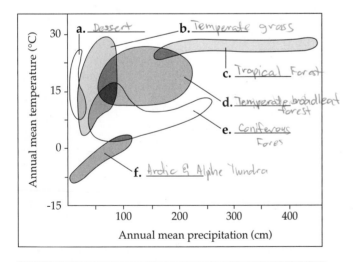

Word Roots

a- = without; **bio-** = life (*abiotic components:* nonliving chemical and physical factors in the environment)

abyss- = deep, bottomless (*abyssal zone:* the very deep benthic communities near the bottom of the ocean; this region is characterized by continuous cold, extremely high water pressure, low nutrients, and near or total absence of light)

bentho- = the depths of the sea (*benthic zone:* the bottom surfaces of aquatic environments)

estuar- = the sea (*estuary:* the area where a freshwater stream or river merges with the ocean)

eu- = good, well; **troph-** = food, nourishment (*eutrophic:* shallow lakes with high nutrient content in the water)

geo- = the Earth (*biogeography:* the study of the past and present distribution of species)

hydro- = water; **therm-** = heat (*deep-sea hydrothermal vents:* a dark, hot, oxygen-deficient environment associated with volcanic activity; the food producers are chemoautotrophic prokaryotes)

inter- = between (*intertidal zone:* the shallow zone of the ocean where land meets water)

limn- = a lake (*limnetic zone:* the well-lit, open surface waters of a lake farther from shore)

littor- = the seashore (*littoral zone:* the shallow, well-lit waters of a lake close to shore)

oligo- = small, scant (*oligotrophic lake:* a nutrient-poor, clear, deep lake with minimum phytoplankton)

micro- = small (*microclimate:* very fine scale variations of climate, such as the specific climatic conditions underneath a log)

pelag- = the sea (*oceanic pelagic biome:* most of the ocean's waters far from shore, constantly mixed by ocean currents)

perman- = remaining (*permafrost:* a permanently frozen stratum below the arctic tundra)

-photo = light (*aphotic zone:* the part of the ocean beneath the photic zone, where light does not penetrate sufficiently for photosynthesis to occur)

profund- = deep (*profundal zone:* the deep aphotic region of a lake)

thermo- = heat; **-clin** = slope (*thermocline:* a narrow stratum of rapid temperature change in the ocean and in many temperate-zone lakes)

Structure Your Knowledge

1. **a.** Define *ecology*. - Study off the house
   b. How does ecology relate to evolutionary biology?
2. **a.** What are biomes? major terrestrial or aquatic life zones
   b. What accounts for the similarities in life forms found in the same type of biome in geographically separated areas?

Test Your Knowledge

MULTIPLE CHOICE: *Choose the one best answer.*

1. Which level of ecology considers energy flow and chemical cycling?
   a. abiotic
   b. community
   c. ecosystem
   d. organismal
   e. population

2. Which level of ecological research would consider how a community is affected by neighboring ecosystems?
   a. ecosystem ecology
   b. landscape ecology
   c. community ecology
   d. population ecology
   e. biosphere ecology

3. Ecologists often use mathematical models and computer simulations because
   a. ecological experiments are always too broad in scope to be performed.
   b. most of them are mathematicians.
   c. ecology is becoming a more descriptive science.
   d. these approaches allow them to study the interactions of multiple variables and simulate large-scale experiments.
   e. variables can be manipulated with computers but cannot be manipulated in field experiments.

4. Which of the following would affect the distribution of a species?
   a. dispersal ability
   b. interactions with mutualistic symbionts
   c. climate and physical factors of the environment
   d. predators, parasites, and competitors
   e. All of the above influence where species are found.

5. According to the precautionary principle,
   a. ecological research should provide all the answers before policy decisions are made.
   b. environmental decisions must ignore political, economic, and social concerns and be based strictly on science.
   c. ecological research should not try to manipulate variables in natural settings but only in the laboratory or computer simulations.
   d. environmental decisions should be made carefully, taking into account the complexity of ecosystems and the potential effects of such decisions.
   e. ecologists, not legislators, should make environmental policy and funding decisions.

6. Which of the following is a concern about the effects of global warming on tree species?
   a. The increased ozone may damage leaf cells, reducing photosynthetic rates.
   b. Trees may not be able to disperse fast enough to reach new habitats that meet their climatic requirements.
   c. Warmer temperatures may speed tree growth, producing trees that are too tall and spindly.
   d. The additional CO_2 in the atmosphere may actually increase photosynthetic rates and prove beneficial to tree growth.
   e. All of the above are correct.

7. In which of the following biomes is light most likely to be a limiting factor?
   a. desert
   b. estuary
   c. coral reef
   d. grassland
   e. ocean pelagic zone

8. Which of the following is *incorrectly* paired with its description?
   a. neritic zone—shallow area over continental shelf
   b. abyssal zone—benthic region where light does not penetrate
   c. littoral zone—area of open water
   d. intertidal zone—shallow area at edge of water
   e. profundal zone—deep, aphotic region of lakes

9. Phytoplankton are the basis of the food chain in
   a. streams.
   b. wetlands.
   c. the oceanic photic zone.
   d. rocky intertidal zones.
   e. deep-sea hydrothermal vents.

10. The ample rainfall of the tropics and the arid areas around 30° north and south latitudes are caused by
    a. ocean currents that flow clockwise in the northern hemisphere and counterclockwise in the southern hemisphere.
    b. the global circulation of air initiated by intense solar radiation near the equator producing wet and warm air.
    c. the tilting of the earth on its axis and the resulting seasonal changes in climate.
    d. the heavier rain on the windward side of mountain ranges and the drier climate on the leeward side.
    e. the location of tropical rain forests and deserts.

11. The permafrost of the arctic tundra
    a. prevents plants from getting established and growing.
    b. protects small animals during the long winters.
    c. anchors plant roots in the frozen soil, helping them withstand the area's high winds.
    d. keeps the surface soil wet because water cannot penetrate through.
    e. Both c and d are correct.

12. Many plant species have adaptations for dealing with the periodic fires typical of a
    a. savanna.
    b. chaparral.
    c. temperate grassland.
    d. temperate broadleaf forest.
    e. a, b, or c.

13. Two communities have the same annual mean temperature and rainfall but very different biota and characteristics. The best explanation for this phenomenon is that the two
    a. are found at different altitudes.
    b. are composed of species that have very low dispersal rates.
    c. are found on different continents.
    d. receive different amounts of sunlight.
    e. have different seasonal temperatures and patterns of rainfall throughout the year.

14. Upwellings in oceans
    a. support reef communities.
    b. occur over deep-sea hydrothermal vents.
    c. are responsible for ocean currents.
    d. bring nutrient-rich water to the surface.
    e. are most common in tropical waters, where they bring oxygen-rich water to the surface.

15. Why do the tropics and the windward side of mountains receive more rainfall than areas around 30° latitude or the leeward side of mountains?
    a. Rising air expands, cools, and drops its moisture.
    b. Descending air condenses and drops its moisture.
    c. The tropics and the windward side of mountains are closer to the ocean.
    d. There is more solar radiation in the tropics and on the windward side of mountains.
    e. The rotation of the earth determines global wind patterns.

MATCHING: *Match the biotic description with its biome.*

Biome	Biotic Description
_____ 1. chaparral	A. broad-leaved deciduous trees
_____ 2. desert	B. lush growth, vertical layers
_____ 3. savanna	C. evergreen shrubs, fire-adapted vegetation
_____ 4. coniferous forest	D. scattered thorny trees, grasses, and forbs
_____ 5. temperate forest	E. tall stands of cone-bearing trees
_____ 6. temperate grassland	F. low shrubby or matlike vegetation
_____ 7. tropical rain forest	G. grasses adapted to fire and drought
_____ 8. tundra	H. widely scattered shrubs, cacti, succulents

Chapter 51
Behavioral Ecology

Framework

This chapter introduces the complex and fascinating subject of animal behavior.

Behaviors can range from simple fixed-action patterns in response to specific stimuli to problem solving in novel situations. Behaviors result from interactions among environmental stimuli, experience, and individual genetic makeup. The parameters of behavior are controlled by genetics and thus are acted upon by natural selection. Behavioral ecology focuses on the ultimate cause of reproductive fitness, which can be used to interpret foraging behavior, mating patterns, and altruistic behavior. Sociobiology extends evolutionary interpretations to social behavior.

Chapter Review

51.1 Behavioral ecologists distinguish between proximate and ultimate causes of behavior

Behavior ecology studies the control, development, and evolution of animal behavior.

What Is Behavior? **Behavior** is what an animal does and how it does it. Animal behaviors are observable movements or actions as well as nonmuscular physiological or neural changes, such as those involved in learning.

Proximate and Ultimate Questions **Proximate questions** explore the immediate cause of a behavior in terms of the cues or stimuli that trigger it and the mechanisms that produce it. **Ultimate questions** concern the evolutionary basis of the behavior—why it has been favored by natural selection.

■ INTERACTIVE QUESTION 51.1

Many animals breed in the spring or early summer.

a. What is a probable proximate cause of this behavior?

b. What is the probable ultimate cause of this behavior?

Ethology   The scientific study of how animals behave, particularly in their natural environment, is called **ethology.** The work of K. Lorenz, N. Tinbergen, and K. von Frisch provided ethology's conceptual foundation. Tinbergen's set of questions to guide behavioral studies emphasizes the importance of both proximate and ultimate bases of behavior.

A **fixed-action pattern (FAP)** is a highly stereotyped sequence of behaviors that, once begun, is usually carried through to completion. A FAP is triggered by a **sign stimulus**—some external stimulus that is often a limited subset of available sensory information.

Learning that occurs during a specific time and is generally irreversible is called **imprinting.** Imprinting is characterized by a limited **sensitive period** during which learning may occur. The ability to respond is innate; the environment provides the *imprinting stimulus.*

■ INTERACTIVE QUESTION   51.2

a. What are the proximate and ultimate causes for attack behavior in male stickleback fish?

b. What are the proximate and ultimate causes for geese imprinting on their mother?

c. How might imprinting interfere with captive rearing conservation efforts?

51.2   Many behaviors have a strong genetic component

All of an animal's anatomical, physiological, and behavioral traits have genetic and environmental components, and exhibit a norm of reaction (a range of variation) that relates to the environment in which traits develop and are expressed. Behavior that is performed virtually the same by all individuals, regardless of environmental differences, is *developmentally fixed* and called **innate.**

Directed Movements   A **kinesis** is a simple change in activity or turning rate in response to a stimulus. Although kinetic movements are randomly directed, they tend to maintain organisms in favorable environments. A **taxis** is an oriented movement toward or away from a stimulus, usually performed automatically.

■ INTERACTIVE QUESTION   51.3

Sow bugs are placed in experimental chambers that are either humid or dry and have both light and dark areas. In the humid chamber, the sow bugs move into the dark area and stop moving. In the dry chamber, they move into the dark area and continue to move about in that area. Explain these experimental results.

Even more complex behaviors, such as aspects of migratory behavior in birds, can be under strong genetic control. The genetic basis of "migratory restlessness" in captive birds has been documented by breeding migratory and nonmigratory blackcaps.

Animal Signals and Communication   **Communication** between animals is often under strong genetic control and involves the transmission of, reception of, and response to special behaviors called **signals.** Communication may involve visual, auditory, chemical, tactile, and electrical signals, depending on the lifestyle and sensory specializations of a species.

Pheromones are chemical signals commonly used by mammals and insects in reproductive behavior to attract mates and to trigger specific courtship behaviors. Pheromones can also function as alarm signals.

Auditory communication in many insects is involved in mating rituals and under direct genetic control. Morphologically identical green lacewings have been separated into different species based on courtship songs. Hybrid songs are sung by hybrid offspring bred in laboratory conditions, indicating genetic control.

■ INTERACTIVE QUESTION   51.4

Why is most communication among mammals olfactory and auditory, whereas communication among birds is visual and auditory?

Genetic Influences on Mating and Parental Behavior   A number of mammalian behaviors are under strong genetic control. Male prairie voles are unusual mammals because they are monogamous, help their mates care for the young, and are aggressive to all other voles when they are mated. Transgenic laboratory mice were created with a distribution of receptors for a neuropeptide (AVP) in their brains similar to that of monogamous prairie voles. These transgenic mice showed many mating behaviors similar to male prairie vole and unlike those of wild type mice.

51.3 Environment, interacting with an animal's genetic makeup, influences the development of behaviors

Dietary Influence on Mate Choice Behavior Experiments have demonstrated that the type of food eaten during larval development influences later mate selection by female *Drosophila majavensis,* possibly as a result of differences in the composition of exoskeletons of the flies resulting from different diets. *Drosophila* "taste" the exoskeletons of potential mates.

Social Environment and Aggressive Behavior California male mice, also monogamous, provide extensive parental care and are always highly aggressive toward other mice. When cross-fostered in nests of white-footed mice, California mice showed reduced aggression and reduced parental behavior. The behavior of white-footed male mice was also altered by cross-fostering in California mice nests.

Learning **Learning** is the modification of behavior as a result of experience and enables animals to change behavior in response to environmental changes.

A simple type of learning called **habituation** is the loss of sensitivity to unimportant stimuli or to stimuli not associated with appropriate feedback. Habituation may increase fitness by allowing an animal's nervous system to focus on important stimuli.

The experience of the spatial structure of its environment may modify an organism's behavior in **spatial learning.** Animals may learn and use a particular set of **landmarks,** or location indicators, to find their way within their area. The use and kind of landmarks may vary with the stability of the environment.

More complicated than a set of learned landmarks, **cognitive maps** are internal representations of the spatial relationships of objects in an animal's surroundings. Evidence for cognitive maps comes from research with jays and nutcrackers that are able to retrieve stored food from thousands of caches.

In **associative learning,** animals learn to associate one stimulus with another. In an example of associative learning called **classical conditioning,** *Drosophila* learned to avoid particular scents that had been coupled with an electric shock.

Associative learning may help animals avoid predators. In one experimental design, zebrafish exposed to water with their alarm substance and then water with pike odor reduced their activity when exposed only to pike odor a few days later.

Operant conditioning refers to trial-and-error learning through which an animal associates a behavior with a reward or punishment.

Cognition refers to an animal's ability to perceive, store, process, and use information from its sensory receptors. **Cognitive ethology** considers the connection between nervous system function and animal behavior. Research with insects indicate that they are capable of categorizing environmental objects as "same" or "different."

Problem-solving behavior is most often observed in mammals, especially in primates and dolphins, although such behavior has also been documented in some bird species. Many animals use the behavior of others as information used in problem solving.

Studies of bird songs show varying degrees of genetic and environmental influence in the learning of these songs. White-crowned sparrows appear to have a 50-day sensitive period in which they memorize the song of their species. If raised in isolation during this sensitive period, a bird does not develop a typical adult song. In a second learning phase, a juvenile bird sings a subsong, which gradually improves as the bird practices, apparently comparing its own singing with the memorized song. The adult sparrow then sings this crystallized song for the rest of its life.

Canaries have no sensitive period for song learning and do not crystallize an adult song. They learn a new, more elaborate song each breeding season.

■ INTERACTIVE QUESTION 51.5

Indicate the type of learning illustrated by the following examples:

a. Ewes will adopt and nurse a lamb shortly after they give birth to their own lamb but will butt and reject a lamb introduced a day or two later.

b. A dog, whose early "accidents" were cleaned up with paper towels accompanied with harsh discipline, hides any time a paper towel is used in the household.

c. Ducklings eventually ignore a cardboard silhouette of a hawk that is repeatedly flown over them.

d. Honeybees appear to use landmarks to locate their hive.

e. In Pavlov's experiments, the ringing of a bell caused a dog to salivate.

51.4   Behavioral traits can evolve by natural selection

Behavioral Variation in Natural Populations   Closely related species often have behavioral differences. Variations in behavior between populations within a species may correlate with variations in the environment. Most laboratory-born garter snakes from coastal areas where banana slugs are an abundant source of prey ate slugs when they were offered, whereas few laboratory-born snakes from inland populations ate the slugs. The researcher proposed that snakes in coastal areas with the ability to recognize slugs by chemoreception had higher fitness, leading to the evolution of this difference in prey selection behavior.

Foraging is behavior involved with searching for, recognizing, catching, and consuming food. Laboratory experiments showed that differences in foraging and territorial behavior between funnel-web spiders in riparian zones, where food is more plentiful but bird predation high, and funnel-web spiders in arid habitats, where food is less available, are genetically based.

Experimental Evidence for Behavior Evolution   Laboratory studies of *Drosophila* have documented an evolutionary change in foraging path length in populations of high or low density. The populations originally had equal frequencies of *for^R* (for rover) and *for^s* (for sitter) alleles.

A combination of field and laboratory studies have documented changes in migratory behavior in European blackcaps over the past few decades. Instead of migrating southwest to the Mediterranean for the winter, some German blackcaps are migrating west to Britain. Offspring of these new migrants had similar orientations of their migratory restlessness as documented in funnel cages.

■ INTERACTIVE QUESTION   51.6

Why do experiments looking at behavioral variations among and within natural populations often raise and test experimental animals in the laboratory?

51.5   Natural selection favors behaviors that increase survival and reproductive success

Foraging Behavior   According to the **optimal foraging theory,** feeding behaviors will maximize energy intake over expenditure and risk of being eaten while foraging. Some behavioral ecologists use cost-benefit analysis to study the proximate and ultimate causes of foraging.

Studies such as those on prey selection by bluegill sunfish have indicated that animals can modify their foraging behavior in ways that tend to maximize overall energy intake. This ability appears to be innate, although experience and physical maturation are thought to increase foraging efficiency.

Optimal foraging must also account for the risk of predation while foraging. Mule deer appear to forage more in open areas, where they are less likely to fall prey to mountain lions.

■ INTERACTIVE QUESTION   51.7

Explain how Zack's study of whelk-eating crows supports the optimal foraging theory.

Mating Behavior and Mate Choice   Many species have **promiscuous** mating, with no strong pair bonds forming. Longer-lasting relationships may be **monogamous** or **polygamous.** Polygamous relationships are most often **polygynous** (one male and many females), although a few are **polyandrous.** Monogamous species are less likely to be sexually dimorphic.

The needs of offspring are an ultimate factor in the reproductive pattern of the parents. If young require more food than one parent can supply, a male may increase his reproductive fitness by helping to care for offspring rather than going off in search of more mates. With mammals, the female often provides all the food, and males are often polygynous.

Certainty of paternity also influences mating systems and parental care. With internal fertilization, the acts of mating and egg laying or birth are separated, and paternity is less certain than when eggs are fertilized externally.

■ INTERACTIVE QUESTION   51.8

Exclusive male parental care is observed much more frequently in species with external fertilization, where the male's genetic contribution to the offspring is more certain. Explain how such behavior could evolve.

Sexual selection may be intrasexual, involving competition among members of one sex for mates, or intersexual, in which mates are chosen by one sex on the basis of particular characteristics.

Experiments involving feather ornaments added to zebra finch parents suggest that females imprint on their fathers, and that mate choices may play a role in the evolution of ornamentation in male zebra finches. Female choice in stalk-eyed fruit flies has been a selection factor in the evolution of long eyestalks, which correlate with male quality.

As with female choice, male competition for females can also reduce variation among males. **Agonistic behavior** involves a contest to determine which competitor gains access to a resource, such as food or a mate. The encounter may include a test of strength or, more commonly, symbolic behavior or ritual.

Intrasexual selection can also lead to the evolution of alternative male mating behavior and morphology. In a study of marine isopods, three genetically and morphologically distinctive male types obtain access to females in different ways—large males defend harems within intertidal sponges; female-mimicking males gain access to guarded harems, and tiny males invade and live within large harems. Variation among males is maintained because each type of male has high mating success at different densities of females.

Applying Game Theory Behavioral ecologists apply **game theory,** which evaluates strategies in situations where outcome depends on both an individual's strategy and the strategies of others, to explain how alternative morphologies and mating behaviors may be maintained in a population. Researchers studying the coexistence of three male phenotypes of the side-blotched lizard found that each one's reproductive success depended on the frequency of the other morphs.

■ INTERACTIVE QUESTION 51.9

In populations of the side-blotched lizard, aggressive orange throat males defend large territories with many females; blue throat males defend smaller territories and fewer females; and yellow throats mimic females and use "sneaky" tactics to obtain matings. Starting with a high abundance of orange throats in the population, which type of male will tend to increase its mating success and increase in frequency next, and which type of male will then replace this second type after it increases in numbers? Explain your answer.

51.6 The concept of inclusive fitness can account for most altruistic social behavior

Many social behaviors are selfish, benefiting one individual's reproductive success at the expense of others.

Altruism Selflessness, or **altruism,** is behavior that reduces an individual's fitness while increasing the fitness of other individuals.

Inclusive Fitness Natural selection favors traits that increase reproductive success, thus propagating the genes for those traits. W. Hamilton was the first to explain altruistic behavior in terms of **inclusive fitness,** the ability of an individual to pass on its genes either by producing its own offspring or by helping close relatives produce their offspring.

Hamilton developed a quantitative measure, called **Hamilton's rule,** that predicts that natural selection would favor altruistic acts among related individuals if $rB > C$. B and C are the benefit to the recipient and the cost to the altruist, measured by the change in the average number of offspring produced as a result of the altruistic act. The term r refers to the **coefficient of relatedness,** a measure of the probability of a gene being inherited by two individuals from a common ancestor. **Kin selection** is the term for the natural selection of altruistic behavior that enhances the reproductive success of related individuals.

Studies show that most cases of altruistic behavior involve close relatives, such as females in Belding's squirrel populations and worker bees in a hive, and thus improve the individual's inclusive fitness.

When altruistic behavior involves nonrelated animals, the explanation offered is **reciprocal altruism;** there is no immediate benefit for the altruistic individual, but some future benefit may occur when the helped animal may "return the favor." Cheaters obtain a large benefit, but behavioral ecologists, using game theory, propose that a *tit for tat* behavioral strategy, in which cheating is immediately retaliated against, may enable reciprocal altruism to evolve and persist in a population. Reciprocal altruism often is used to explain altruism in humans.

■ INTERACTIVE QUESTION 51.10

a. According to kin selection, would an individual be more likely to exhibit altruistic behavior toward a parent, a sibling, or a first (full) cousin?

b. Explain your answer in terms of the coefficient of relatedness and Hamilton's rule.

Social Learning   Learning can have a social as well as genetic and environmental components. **Social learning,** learning through the observation of others, forms the basis of **culture**—a system of information transfer that involves teaching and/or social learning and influences behavior in a population.

In **mate choice copying,** individuals may copy the mate choice of others in a population. Experiments with guppies have shown that females will preferentially mate with males who they have observed engaged in courtship with other females. Such mate choice copying was shown to mask a genetically controlled preference for a particular male coloration.

Vervet monkeys have an innate ability to give alarm calls in response to threatening objects. They learn to discriminate in their calls by observing other members of the group and receiving social confirmation.

Evolution and Human Culture   **Sociobiology** relates evolutionary theory to social behavior and to human culture. In his 1975 book, *Sociobiology,* E. O. Wilson speculated on the evolutionary basis of certain social behaviors of humans.

The parameters of human social behavior may be set by genetics, but the environment undoubtedly shapes behavioral traits just as it influences the expression of physical traits. Due to our capacity for learning, human behavior appears to be quite plastic. Our structured societies, with their definitions of acceptable behaviors that exclude some behaviors that might otherwise enhance an individual's fitness, may be the one unique characteristic separating humans and other animals.

Word Roots

agon- = a contest (*agonistic behavior:* a type of behavior involving a contest of some kind that determines which competitor gains access to some resource, such as food or mates)

andro- = a man (*polyandry:* a polygamous mating system involving one female and many males)

etho- = custom, habit (*ethology:* the study of animal behavior in natural conditions)

gyno- = a woman (*polygyny:* a polygamous mating system involving one male and many females)

kine- = move (*kinesis:* a change in activity rate in response to a stimulus)

mono- = one; **-gamy** = reproduction (*monogamous:* a type of relationship in which one male mates with just one female)

poly- = many (*polygamous:* a type of relationship in which an individual of one sex mates with several of the other sex)

socio- = a companion (*sociobiology:* the study of social behavior based on evolutionary theory)

Structure Your Knowledge

1. How does the nature-versus-nurture controversy apply to behavior ecology?
2. How does the concept of fitness in an evolutionary sense apply to all aspects of behavior?

Test Your Knowledge

MULTIPLE CHOICE: *Choose the one best answer.*

1. Behavioral ecology is the
    a. mechanistic study of the behavior of animals, focusing on stimulus and response.
    b. application of human emotions and thoughts to other animals.
    c. study of animal cognition.
    d. study of animal behavior from an evolutionary perspective of fitness.
    e. study of the ecological basis of behavior.

2. Proximate causes
    a. explain the evolutionary significance of a behavior.
    b. are immediate causes of behavior such as environmental stimuli.
    c. are environmental, whereas ultimate causes are genetic.
    d. are endogenous, although they may be set by exogenous cues.
    e. show that nature is more important than nurture.

3. What is the behavior called that maximizes an animal's energy intake-to-expenditure ratio?
    a. optimal foraging
    b. Hamilton's rule
    c. a fixed-action pattern
    d. cognition
    e. learning

4. Which of the following is an example of a fixed-action pattern?
   a. a bluegill sunfish feeding on larger *Daphnia* when prey are abundant
   b. a chick pecking at the red spot on a parent's moving beak
   c. a whale migrating long distances to its feeding territory
   d. a songbird learning its song after listening to a taped song of its species
   e. a bird learning to avoid monarch butterflies

5. Which of these terms includes all the others?
   a. habituation   d. learning
   b. imprinting   e. cognitive map
   c. problem-solving

6. A sensitive period
   a. is the time right after birth.
   b. usually follows the receiving of a sign stimulus.
   c. is a limited time in which imprinting can occur.
   d. is the period during which birds can learn to fly.
   e. is the time during which mate selection occurs.

7. In operant conditioning,
   a. an animal improves its performance of a fixed-action pattern (FAP).
   b. an animal learns as a result of trial and error.
   c. sensitivity to unimportant or repetitive stimuli decreases.
   d. a bird can learn the song of a related species if it hears only that song.
   e. an irrelevant stimulus can elicit a response because of its association with a normal stimulus.

8. Which modality of intraspecies communication signal would be best suited to a nocturnal species such as an owl?
   a. auditory   d. tactile
   b. visual   e. electrical
   c. chemical

9. A kinesis
   a. is a randomly directed movement that is not caused by external stimuli.
   b. is a movement that is directed toward or away from a stimulus.
   c. is a change in activity in response to a stimulus.
   d. is illustrated by trout swimming upstream.
   e. may involve landmarks but not cognitive maps.

10. Which of the following examples of behavior provide evidence of animal cognition?
    a. chimpanzee stacking up boxes to reach a banana
    b. ravens pulling up string to obtain attached food item
    c. trained honeybees that can match colors or patterns, indicating their concept of "same" and "different"
    d. the cognitive map of a jay that enables it to retrieve food from its many caches
    e. All of the above show evidence of information processing and animal cognition.

11. In a species in which females provide all the needed food and protection for the young,
    a. males are likely to be promiscuous.
    b. mating systems are likely to be monogamous.
    c. mating systems are likely to be polyandrous.
    d. males most likely will show sexual selection.
    e. females will have a higher Darwinian fitness than males.

12. A crow that aids its parents in raising siblings is increasing its
    a. reproductive success.
    b. altruistic behavior.
    c. inclusive fitness.
    d. coefficient of relatedness.
    e. certainty of paternity.

13. Sociobiology
    a. explores the evolutionary basis of behavioral characteristics within animal societies.
    b. applies evolutionary explanations to human social behaviors.
    c. studies the roles of culture and genetics in human social behavior.
    d. considers communication, mating systems, and altruism from the viewpoint of fitness.
    e. does all of the above.

14. According to the concept of kin selection,
    a. an animal would be more likely to aid a stranger if the "kindness" could be reciprocated.
    b. an animal would aid its parent before it would help its sibling.
    c. animals are more likely to choose close relatives as mates.
    d. examples of altruism usually involve close relatives and increase an animal's inclusive fitness.
    e. evolution is the proximate cause of animal behavior.

15. A female bird would most likely increase her fitness by
    a. mating with as many males as possible.
    b. choosing a mate based on evidence that he has "good genes."
    c. reproducing only once in her lifetime.
    d. being polyandrous.
    e. always foraging in a large flock.

16. When a white-crowned sparrow sings a subsong,
    a. it is practicing the songs of other bird species in its vicinity.
    b. it has developed the crystallized song that it will sing for the rest of its life.
    c. it is in the sensitive period during which it memorizes its song.
    d. it apparently compares the song it memorized during its sensitive period to its own singing.
    e. it is in the plastic phase, creating a new, more elaborate song for the next breeding season.

17. According to Hamilton's rule, natural selection would favor altruistic acts when
    a. the probability that the altruist will lose its life is less than 0.5 and the coefficient of relatedness is greater than 0.25 $(rC=B)$.
    b. the cost to the altruist times the coefficient of relatedness is less than the benefit to the receiver $(rC < B)$.
    c. the benefit to the receiver times the coefficient of relatedness is greater than the cost to the altruist $(rB > C)$.
    d. the cost to the receiver times the coefficient of relatedness is greater than the cost to the altruist $(rC > B)$.
    e. the benefit to the altruist times the coefficient of relatedness is less than the cost to the receiver $(rB < C)$.

18. The cross-fostering of California voles in white-footed mice nests provide evidence for
    a. almost total genetic control of aggression and parenting behavior.
    b. almost total environmental control of aggression and parenting behavior.
    c. an imprinting period during which behaviors related to aggression and parenting are set.
    d. the influence of the early social environment on the expression of aggressive and parental behaviors.
    e. the cognitive ability of voles to change their behavior in different environments.

19. Which of the following provides a way of analyzing situations in which the fitness of one behavioral phenotype is influenced by other phenotypes in the populations?
    a. zero sum game
    b. inclusive fitness
    c. social learning
    d. cognitive ethology
    e. game theory

20. Which of the following is *not* an example of social learning?
    a. garter snakes from coastal areas eating slugs
    b. mate choice copying in guppies
    c. alarm calls of adult vervet monkeys
    d. human culture
    e. chimpanzees using stones to crack nuts

Chapter 52

Population Ecology

Framework

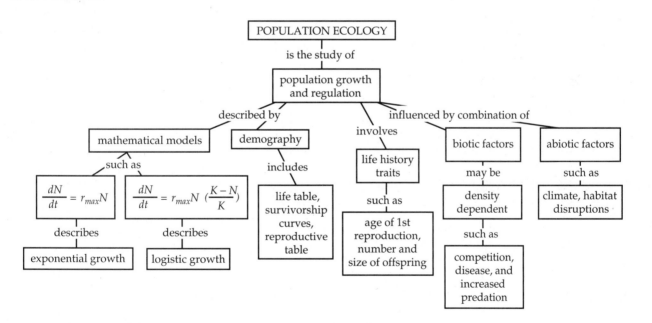

Chapter Review

The continuing growth of the human population in the face of limited resources is a critical biological phenomenon. **Population ecology** is the study of the influence of the environment on fluctuations in population size and composition.

52.1 Dynamic biological processes influence population density, dispersion, and demography

A **population** is a group of individuals of the same species that occupy the same area, use the same resources, and have a high probability of interacting and breeding with each other.

Density and Dispersion   Every population has geographic boundaries; ecologists define boundaries based upon the type of organism and the research question being asked. The number of individuals per unit area or volume is a population's **density;** the pattern of spacing of those individuals within the population is referred to as **dispersion.**

Population density is often measured by using one of a variety of sampling techniques to count and estimate population size. The **mark-recapture method** is a common sampling technique. Indirect indicators, such as burrows or nests, also may be used. Changes in population density reflect additions of members through birth (including all forms of reproduction) and **immigration,** and removal through death (mortality) and **emigration.**

Individuals may be dispersed in the population's geographic range in several patterns. *Clumping* may indicate a heterogeneous environment, with organisms congregating in suitable microenvironments. Clumping also may be related to social interactions between individuals.

Uniform distribution may be related to competition for resources and result from interactions between individuals. **Territoriality,** the defense of a physical space, can lead to uniform dispersion. *Random* spacing, indicating the absence of strong interactions between individuals or a fairly consistent habitat, is not that common.

■ **INTERACTIVE QUESTION   52.1**

What is the likely dispersion pattern of fish that swim in schools, seabirds nesting on a small island, and thistles growing in a fairly uniform field?

Demography   The study of the vital statistics of a population, such as birth and death rates, is called **demography.**

A **life table** presents age-specific survival data for a population. It can be constructed by following a **cohort** of organisms from birth to death, calculating the proportion of the cohort surviving at each age group.

A **survivorship curve** shows the number or proportion of members of a cohort still alive at each age. Survivorship curves are often based on a beginning cohort of 1000 individuals, with the y axis logarithmic and the x axis on a relative scale, so species with different life spans can be compared on the same graph. There are three general types of survivorship curves. Type I, with low mortality during early and middle age and a rapid increase with old age, is typical of populations that produce relatively few offspring and provide

parental care. In a Type II curve, death rate is relatively constant throughout the life span. A Type III curve is typical of populations that produce many offspring, most of which die off rapidly. The few that survive are likely to reach adulthood. Many species show intermediate or more complex survivorship patterns.

■ **INTERACTIVE QUESTION   52.2**

Identify the types of survivorship curves shown below and give examples of species that exhibit each curve.

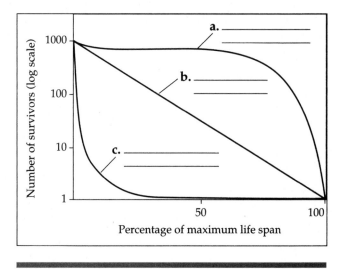

In sexually reproducing species, demographers usually follow only female reproduction, often only of female offspring. A **reproductive table** gives the age-specific reproductive rates in a population. Such a fertility schedule can be constructed by following the reproductive output of a cohort, measuring the number of female offspring by age group. The reproductive output of a sexual species—the average number of daughters per female in each age class—is the product of the proportion of females breeding and the number of female offspring they produce.

52.2   Life history traits are products of natural selection

The **life history** of an organism from birth through reproduction to death reflects evolutionary trade-offs between survival and reproduction. Life history traits include the age at first reproduction, how often an organism breeds, and the number of offspring produced during each reproductive episode.

Life History Diversity   Some species put all their reproductive resources into a single reproductive effort, often called **big-bang reproduction,** or **semelparity.** Other species follow the strategy of **iteroparity,** making repeated reproductive efforts over a span of time. A key factor in the selection for and evolution of big-bang versus **repeated reproduction** is the survival rate of young offspring. Big-bang reproduction may be favored when chances of offspring survival are low in unpredictable habitats. In more predictable environments with high competition for resources, fewer, better-provisioned offspring may have a better chance of surviving to reproduce.

"Tradeoffs" and Life Histories   Because organisms have a finite energy budget, they cannot maximize all life history traits simultaneously. Reproductive costs often include a reduction in survival. The production of large numbers of offspring is related to the selective pressures of high mortality rates of offspring in uncertain environments or from intense predation. Parental investments in the size of offspring, incubation or gestation, and parental care increase survival chances of offspring.

■ INTERACTIVE QUESTION   52.3

a. Explain why the life history of an organism can't be reproduce early, often, have large numbers of offspring, and live long.

b. In what way might high competition for limited resources in a predictable environment influence life history traits?

52.3   The exponential model describes population growth in an idealized, unlimited environment

Per Capita Rate of Increase   Growth of a small population in a very favorable environment will be restricted only by the biological limitations of their life history traits. Ignoring immigration and emigration, the change in population size during a specific time period is equal to the number of births minus deaths.

Births and deaths (mortality) can be expressed in terms of the average number per individual during a time period, or a *per capita birth rate (b)* and *per capita death rate (m)*. These rates can be calculated from estimates of population size *(N)* and data in life tables and reproductive tables. The population growth equation using per capita birth and death rates becomes $\Delta N/\Delta t = bN - mN$. The difference between per capita birth rate and death rate is the *per capita rate of increase,* symbolized by r: $r = b - m$. **Zero population growth (ZPG)** occurs when $r = 0$. The formula describing change in the population at any one instant uses differential calculus and is written as $dN/dt = rN$.

Exponential Growth   Under ideal conditions, a population may exhibit **exponential population growth,** or geometric population growth. The *intrinsic rate of increase* (r_{max}) is the fastest per capita rate of increase possible for a species. This exponential population growth, expressed as $dN/dt = r_{max}N$, produces a J-shaped growth curve when graphed. The larger the population *(N)* becomes, the faster the population grows. Periods of exponential growth may occur in some populations that exploit an unfilled environment or rebound from a catastrophic event.

52.4   The logistic growth model includes the concept of carrying capacity

A population may grow exponentially for only a short time before its increased density limits the resources available for its members. The **carrying capacity** *(K)* is the maximum sustainable population size that a particular environment can support at a particular time. Crowding and resource limitation may lead to decreased per capita birth rates and increased per capita death rates.

The Logistic Growth Model   The per capita rate of increase decreases from its maximum at low population size to zero as carrying capacity is reached. The mathematical model of **logistic population growth** $(dN/dt = r_{max}N(K - N)/K)$ includes the expression $(K - N)/K$ to reflect the impact of the increasing N on the per capita rate of increase as the population approaches the carrying capacity.

When N is small, $(K - N)/K$ is close to 1, and growth is approximately exponential $(r_{max}N)$. As population size approaches the carrying capacity, the $(K - N)/K$ term becomes a small fraction, and per capita rate of increase is small. When N reaches K, the term $(K - N)/K$ is 0, and the population stops growing. The logistic model produces an S-shaped growth curve, and maximum increase in population numbers occurs when N is intermediate.

■ INTERACTIVE QUESTION   52.4

Label the exponential and logistic growth curves, and show the equation associated with each curve. What is K for the population shown with curve b?

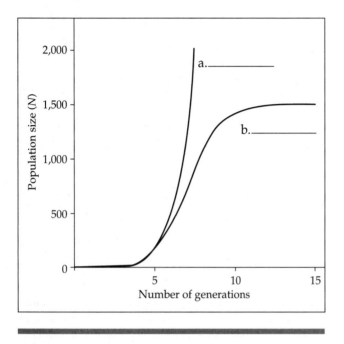

The Logistic Model and Real Populations   Some laboratory populations of small animals and microorganisms show logistic growth. Natural populations may grow logistically but few reach a stable carrying capacity.

The logistic model makes the assumption that any increase in population numbers will have a negative effect on population growth. The *Allee effect* is seen in some populations, however, when individuals benefit as the population grows, either from physical support, as in plants, or from social interactions important to reproduction.

The logistic model also assumes that populations approach their carrying capacity smoothly, but many populations overshoot and then oscillate above and below a general carrying capacity. There are populations for which population density is not an important factor. These populations are often reduced by environmental conditions before resources have a chance to become limiting.

The Logistic Model and Life Histories   Natural selection will favor different life history traits depending on population densities and environmental conditions. Populations at high density, close to their carrying capacity, may experience **K-selection** or density-dependent selection for traits such as competitive abil-

ity and efficient resource utilization. In environments in which population density fluctuates or where population density is low, **r-selection** or density-independent selection would favor traits that maximize population growth, such as production of numerous, small offspring. Laboratory studies that have varied population densities and conditions have produced different proportions of K-selected and r-selected traits in populations of the same species. Many ecologists criticize the concepts of r- and K-selection as oversimplified theories of life history evolution.

■ INTERACTIVE QUESTION   52.5

Indicate whether the following would be considered to be r-selected or K-selected life history traits.

a. early age at first reproduction; many small offspring produced

b. few, relatively large offspring produced every year

52.5   Populations are regulated by a complex interaction of biotic and abiotic influences

The ecological questions of what eventually stops population growth and what determines population fluctuations or stability have practical applications for conservation and agriculture.

Population Change and Population Density   When the birth rate or death rate does not change as population density changes, it is said to be **density independent.** Rates are said to be **density dependent** if death rate rises and birth rate falls with increasing population density. An equilibrium density may be reached in a population as long as birth rate or death rate or both are density dependent.

Density-Dependent Population Regulation   Density-dependent decreases in birth rate and increases in death rate may regulate populations through negative feedback.

Limited food supply often limits reproductive output. The availability of territorial space may be the limiting resource for some animals. Increased population densities may affect health and the transmission rate of disease in both plant and animal populations. The accumulation of toxic metabolic wastes may also be a limiting factor. Predation may be a density-dependent factor when a predator feeds preferentially on a prey population that has reached a high density.

Intrinsic factors may also regulate population size. Studies of mice have shown that even when food or shelter is not limiting, population size stabilizes when high densities induce a stress syndrome of hormonal changes that inhibit reproduction and increase mortality.

Population Dynamics All populations show some fluctuations in numbers. **Population dynamics** studies these variations in population size and the factors that cause them.

Fluctuations in populations of large grazers and browsers may be linked to the severity of winter. The erratic fluctuations of some populations have been linked to a combination of biotic and abiotic factors—in the case of the Dungeness crab, to density-dependent cannibalism and changes in temperature and ocean currents.

A group of populations may form a **metapopulation** in which immigration and emigration may significantly influence individual population sizes.

Population Cycles The ten-year cycles in density of snowshoe hares and lynx in boreal forests have been studied to determine whether food shortages, predator overexploitation, or a combination of both causes the cyclic collapses in hare populations. Experimentally increasing the food supply raised the carrying capacity for hares, but the density cycles continued to occur in both experimental and control areas. Field ecologists determined that 90% of hare deaths are due to predation. Experiments that excluded predators from one area, and both excluded predators and added food to another area, support the hypothesis that excessive predation mainly drives hare cycles, but available winter food supply is a contributing influence.

Predator cycles most likely follow the population cycles of their prey, and these may be accentuated when predators turn on one another as prey become scarce.

■ INTERACTIVE QUESTION 52.6

a. List some density-dependent factors that may limit population growth.

b. List some abiotic factors that may cause population fluctuations.

52.6 Human population growth has slowed after centuries of exponential increase

The Global Human Population From 1650, it took 200 years for the global population to double to 1 billion. In the next 80 years, it doubled to 2 billion; it doubled again in the next 45 years and is projected to reach 7.3 to 8.4 billion by the year 2025. The rate of growth has begun to slow, partly due to diseases such as AIDS and voluntary population control.

Population stability can be reached in one of two ways: Zero population growth = high birth rates − high death rates; or zero population growth = low birth rates − low death rates. The movement from the first configuration to the second is called the **demographic transition.** Death rates declined rapidly in most developing countries after 1950, but birth rate decline has been more uneven—from rapid in China to just beginning in much of Africa. The world's annual rate population growth rate is regional; it is near equilibrium (0.1%) in developed nations and 1.4% in developing countries, where 80% of the world's population lives.

Human population growth is unique in that it can be consciously controlled by voluntary contraception and family-planning programs. The key to the demographic transition is reduced family size. In many cultures, women are delaying marriage and reproduction, thus slowing population growth.

The **age structure** of a population influences present and future growth. A large proportion of individuals of reproductive age or younger results in more rapid growth now or in the near future. Age structure also predicts future social conditions and needs.

Infant mortality and **life expectancy at birth** vary among human populations, with mortality being much higher and life expectancy lower in developing countries.

Global Carrying Capacity Estimates of Earth's carrying capacity have varied greatly and average around 10–15 billion. These estimates may use different assumptions, such as the logistic equation, the amount of inhabitable land, or food as the limiting factor.

The concept of an **ecological footprint** takes into account multiple human needs in estimating carrying capacity. Six types of ecologically productive areas are used to calculate each country's ecological footprint: arable land, pasture, forest, ocean, built-up land, and fossil energy land (vegetative area required to absorb CO_2 from burning fossil fuels). Each type of ecologically productive area is converted to land area per person and totaled for the planet, adding up to about 2 hectares (ha) per person. Land for parks and conservation reduces this estimate to 1.7 ha per person.

Ecological footprints vary greatly by country, as do the available **ecological capacities** (resource base) of each nation. In a 1997 study, the United States had an ecological footprint of 8.4 ha per person and an available ecological capacity of only 6.2 ha. Thus, the United States has already exceeded its carrying capacity. This study also suggests that the world human population as a whole is already slightly above its carrying capacity.

The ultimate carrying capacity of Earth may be determined by food supplies, space, nonrenewable resources, degradation of the environment, or several interacting factors. When and how we reach zero population growth is an issue of great social and ecological consequence.

Word Roots

co- = together (*cohort:* a group of individuals of the same age, from birth until all are dead)

demo- = people; **-graphy** = writing (*demography:* the study of statistics relating to births and deaths in populations)

itero- = to repeat (*iteroparity:* a life history in which adults produce large numbers of offspring over many years; also known as repeated reproduction)

semel- = once; **-parity** = to beget (*semelparity:* a life history in which adults have but a single reproductive opportunity to produce large numbers of offspring, such as the life history of the Pacific salmon; also known as "big-bang reproduction")

Structure Your Knowledge

1. Create a concept map to organize your understanding of the exponential and logistic equations—the mathematical models of population growth.

2. What is the best collection of life history traits that would maximize reproductive success?

Test Your Knowledge

MULTIPLE CHOICE: *Choose the one best answer.*

1. In a range with a heterogeneous distribution of suitable habitats, the dispersion pattern of a population probably would be
   a. clumped.
   b. uniform.
   c. random.
   d. unpredictable.
   e. dense.

2. Which of the following is *not* true of life tables?
   a. They were first used by life insurance companies to estimate survival patterns.
   b. They show the age-specific mortality or death rate for a population.
   c. Ecologists have collected them for a large number of natural populations.
   d. They can be used to construct survivorship curves.
   e. They are often constructed by following a cohort from birth to death.

3. In a population in which offspring survival is quite low and the environment is inconsistent, one might expect
   a. the production of a small number of large offspring.
   b. the production of a large number of large offspring.
   c. iteroparity or repeated reproduction with a small number of offspring.
   d. semelparity or big-bang reproduction.
   e. more *K*-selected traits.

4. A Type I survivorship curve is level at first, with a rapid increase in mortality in old age. This type of curve is
   a. typical of many invertebrates that produce large numbers of offspring.
   b. typical of humans and other large mammals.
   c. found most often in *r*-selected populations.
   d. almost never found in nature.
   e. typical of all species of birds.

5. The middle of the S-shaped growth curve in the logistic growth model
   a. shows that at middle densities, individuals of a population do not affect each other.
   b. is best described by the term *rN*.
   c. shows that reproduction will occur only until the population size reaches *K* and *dN/dt* becomes 0.
   d. is the period when competition for resources is highest.
   e. is the period when the population growth rate is the highest.

6. A few members of a population have reached a favorable habitat with few predators and unlimited resources, but their population growth rate is slower than that of the parent population. What is a possible explanation for this situation?
   a. The genetic makeup of these founders may be less favorable than that of the parent population.
   b. The parent population may still be in the exponential part of its growth curve and not yet limited by density-dependent factors.
   c. The Allee effect may be operating; there are not enough population members present for successful reproduction.
   d. a, b, and c may apply.
   e. This scenario would not happen.

7. The term $(K - N)/K$
   a. is the carrying capacity for a population.
   b. is greatest when K is very large.
   c. is zero when population size equals carrying capacity.
   d. increases in value as N approaches K.
   e. accounts for the overshoot of carrying capacity.

8. Which of the following would *not* be a density-dependent factor limiting a population's growth?
   a. increased predation by a predator
   b. a limited number of available nesting sites
   c. a stress syndrome that alters hormone levels
   d. a very early fall frost
   e. intraspecific competition

9. The carrying capacity for a population is estimated at 500; the population size is currently 400; and r_{max} is 0.1. What is dN/dt?
   a. 0.01
   b. 0.8
   c. 8
   d. 40
   e. 50

10. In order to maintain the largest sustainable fish harvest, fishing efforts should
    a. take only postreproductive fish.
    b. maintain the population close to its carrying capacity.
    c. reduce the population to a very low number to take advantage of exponential growth.
    d. maintain the population density close to ½ K.
    e. be prohibited.

11. The human population is growing at such an alarmingly fast rate because
    a. technology has increased our carrying capacity.
    b. the death rate has greatly decreased since the Industrial Revolution.
    c. the age structure of many countries is highly skewed toward younger ages.
    d. fertility rates in many developing countries are above the 2.1 children per female replacement level.
    e. all of the above are true.

12. In which of the following would immigration and emigration likely play a role in population dynamics?
    a. metapopulations
    b. exponential growth
    c. demographic transition
    d. big-bang reproduction
    e. territoriality

Questions 13–15. Use the following choices to indicate how these life history traits would be affected by the described changes.
    a. increase
    b. decrease
    c. stay the same
    d. no relationship or unable to predict

13. For a population regulated by density-dependent factors, how might clutch or seed crop size change with increased population density?

14. In the study of European kestrels described in the text, how did reducing the brood size by transferring chicks to other nests affect the survivorship of those parents in the following winter?

15. In a population showing exponential growth, how would dN/dt be expected to change with an increase in N?

16. Experimental studies of the population cycles of the snowshoe hare and the lynx have shown that

   a. the hare population is regulated by its food resources because adding food increased the carrying capacity of experimental areas.

   b. hares are as likely to die of starvation as of predation.

   c. lynx are the only predators of hares and increases in the lynx population cause the cycles in the hare populations.

   d. the stress of overcrowding causes the population cycles in both hare and lynx.

   e. the hare population is regulated by a combination of food and predators (not just the lynx); the lynx population appears to cycle in response to its prey availability.

17. The demographic transition is the gradual shift from

   a. a Type I survivorship curve to a Type II curve.

   b. semelparity to iteroparity.

   c. an age structure skewed toward the younger ages to an even age distribution.

   d. high birth rates and high death rates to low birth rates and low death rates.

   e. exponential growth to logistic growth.

18. An ecological footprint is an estimate of

   a. the carrying capacity of each nation.

   b. the available ecological capacity of each nation.

   c. the amount of land needed per person to meet the current demand on resources.

   d. the size of a population in relationship to the resources it uses.

   e. how much land is needed to produce food for a vegetarian versus a meateater.

Chapter 53

Community Ecology

Framework

Communities are composed of populations of various species that may interact through competition, predation, herbivory, parasitism, or mutualism. The structure of a community—its species composition and relative abundance—is determined by these interactions and the trophic structure of the community. Disturbances keep most communities in a state of nonequilibrium. Species diversity relates to a community's size and geographic location.

Chapter Review

The collection of different species living close enough to allow for potential interaction is called a biological **community.** Community ecology studies the factors involved in determining a community's structure—its species composition and the relative abundance of species.

53.1 A community's interactions include competition, predation, herbivory, symbiosis, and disease

Interspecific interactions occur between the different species living in a community. The effect of these interactions on the survival and reproduction of a population can be signified by $+$ and $-$ signs. For example, in a $+/-$ interaction such as predation, the interaction is beneficial to the predator species and detrimental to the prey population.

Competition If populations of two species use the same limited resource, **interspecific competition** may affect the survival and reproduction of both populations.

Gause's laboratory experiments with *Paramecium* showed that two species of protists that rely on the same limited resource could not coexist in the same community. This **competitive exclusion** principle predicts that the less efficient competitor will be locally eliminated.

An organism's **ecological niche** is described as its role in an ecosystem—its habitat and use of biotic and abiotic resources. The competitive exclusion principle holds that two species with identical niches cannot coexist in a community. As a result of competition, a species' *realized niche* might be smaller than its *fundamental niche*.

Resource partitioning, slight variations in niche that allow ecologically similar species to coexist, provides circumstantial evidence that competition was a selection factor in evolution. **Character displacement** of some morphological trait or resource use allows closely related sympatric species to avoid competition. When these species are allopatric (geographically separate), their differences may be much less.

■ INTERACTIVE QUESTION   53.1

Two species of *Anolis* lizards are often found perched and feeding in the same trees, with species I in the upper and outer branches, and species II occupying shady inner branches. After removing one or the other species in test trees, an ecologist observes the following results: Species I is found throughout the branches of trees in which it is now the sole occupant. Species II is still found only in the shady interior when it is the sole occupant. What do these results indicate about the niches of these two species?

The realized niche is smaller than fundamental niche

Predation   **Predation** involves a predator killing and eating prey. Adaptations to increase success in predation may include acute senses, speed and agility, camouflage coloration, and physical structures such as claws, fangs, teeth, and stingers.

Animals can defend against predation by hiding, fleeing, or defending. Potential prey may use camouflage in the form of **cryptic coloration** to blend in with the background. Mechanical and chemical defenses discourage predation. Some animals passively accumulate compounds from the food they eat that are toxic to their predators; others may synthesize their own toxins. Bright, conspicuous, **aposematic coloration** warns predators not to eat animals with chemical defenses.

Mimicry may be used by prey to exploit the warning coloration of other species. Predators may use mimicry to "bait" their prey.

■ INTERACTIVE QUESTION   53.2

Name the following two types of mimicry:

a. harmless species resembling a poisonous or distasteful species *Batesian mimicry*

b. mutual imitation by two or more distasteful species

Müllerian mimicry

Herbivory   In **herbivory,** an herbivore eats parts of a plant or alga. Most herbivores are small invertebrates such as insects, which may have chemical sensors that recognize their food plants. Herbivores may have teeth or digestive systems adapted for processing vegetation.

Plants may defend themselves with mechanical devices, such as thorns, or chemical compounds. Distasteful or toxic chemicals include such well-known compounds as strychnine, morphine, nicotine, tannins, and various spices.

Parasitism   Symbiosis may be defined as an interaction between organisms of two species that live in direct contact. In **parasitism,** a **parasite** obtains its nourishment from its **host.** Parasites that live within a host are called **endoparasites;** those that feed on the surface of a host are called **ectoparasites.** In **parasitoidism,** insects lay eggs on or in hosts, on which their larvae then feed. Parasites may have complex life cycles with a number of hosts. Parasites can have substantial influence on their host population.

Disease   **Pathogens,** such as bacteria, viruses, protists, or even fungi or prions, are like microscopic parasites. Pathogens may kill their hosts; parasites do not usually cause lethal harm to the host on which they feed.

Mutualism   In **mutualism,** interactions between species benefit both participants. Mutualistic interactions may involve the evolution of related adaptations in both species.

Commensalism   In **commensalism,** only one member appears to benefit from the interaction. Examples include "hitchhiking" species and species that feed on food incidentally exposed by another.

Interspecific Interactions and Adaptations   **Coevolution** may be defined as reciprocal adaptations of two species that involve genetic changes in both interacting populations. While there is little evidence of true coevolution in most interspecific interactions, these interactions may lead to generalized adaptations of organisms to the presence of other species in a community.

■ INTERACTIVE QUESTION   53.3

Name and give examples of the interspecific interactions symbolized in the table.

	Interaction	Examples
+ / +	**a.** *Mutualism*	*only benefit on both*
+ / 0	**b.** *Commensalism*	*" " one participants*
− / −	**c.** *competition*	
+ / −	**d.** *predation*	*involves killing & eating prey*
+ / −	**e.** *Herbivory*	*Eats plants & alga*
+ / −	**f.** *Parasitism*	*endo- within ecto- on the surface*
+ / −	**g.** *Disease*	*Pathogens are microscopic parasites*

53.2 Dominant and keystone species exert strong controls on community structure

Species Diversity The **species diversity** of a community is determined both by **species richness,** the number of different species present, and by **relative abundance,** the relative numbers of individuals in each species. Estimating these two aspects of a community requires various sampling techniques and may be difficult due to the rarity of most species in a community.

Trophic Structure The **trophic structure** of a community is its feeding relationships. A **food chain** shows the transfer of food energy from one trophic level to the next: from producers to herbivores (primary consumers) to carnivores (secondary, tertiary, or quaternary consumers) and eventually to decomposers.

A **food web** diagrams the complex trophic relationships within a community. The complicated connections of a food web arise because many consumers feed at various trophic levels. Food webs can be simplified by grouping species into functional groups such as primary consumers, or by isolating partial food webs that interact little with the more complex web.

Within a food web, each food chain usually consists of five or fewer links. According to the **energetic hypothesis,** food chains are limited by the inefficiency of energy transfer (only about 10%) from one trophic level to the next. The **dynamic stability hypothesis** suggests that short food chains are more stable than long ones. An environmental disruption that reduces production at lower levels will be magnified at higher trophic levels as food supply is reduced all the way up the chain. The increasing size of animals at successive trophic levels may also limit food chain length, both due to the difficulty of eating large animals and the quantity of food required to support large animals.

■ INTERACTIVE QUESTION 53.4

Experimental data from tree hole communities showed that food chains were longest when food supply (leaf litter) was greatest. Which hypothesis about what limits food chain length do these results support?

Species with a Large Impact Species in a community that have the highest abundance or largest **biomass** are a major influence on the occurrence and distribution of other species. A species may become a **dominant species** due to its competitive use of resources or success at avoiding predation or disease. **Invasive species** may reach high biomass due to the lack of natural predators and pathogens. The removal of a dominant species from a community may adversely affect any species that relied exclusively on that species, but its role may quickly be filled by other species.

A **keystone species** has a large impact on community structure as a result of its ecological role. Paine's study of a predatory sea star demonstrated its role in maintaining species richness in an intertidal community by reducing the density of mussels, a highly competitive prey species.

Ecosystem "engineers" or foundation species influence community structure by changing the physical environment. Such **facilitators** may positively affect other species by modifying the environment.

Bottom-Up and Top-Down Controls Arrows can be used to indicate the effect of an increase in the biomass of one trophic level on another trophic level. $V \to H$ indicates that an increase in vegetation (V) would increase the number of herbivores (H); $V \gets H$ indicates that an increase in herbivores would decrease vegetation biomass. $V \leftrightarrow H$ means that interactions are reciprocal with each trophic level affected by changes in the other. According to the **bottom-up model** of community organization, $N \to V \to H \to P$, an increase in mineral nutrients (N) yields an increase in biomass at each succeeding trophic level: vegetation, herbivores, and predators (P). The **top-down model,** $N \gets V \gets H \gets P$, assumes that predation controls community organization, with a series of $+/-$ effects cascading down the trophic levels. According to this model, also called the *trophic cascade model,* increasing predators will decrease herbivores, which results in increased vegetation and then lowered levels of nutrients.

■ INTERACTIVE QUESTION 53.5

Many freshwater lake communities appear to be organized along the top-down model. What actions might ecologists take if they wanted to use **biomanipulation** to control excessive algal blooms in a lake with four trophic levels (algae, zooplankton, primary predator fish, and top predator fish)?

53.3 Disturbance influences species diversity and composition

Traditionally, biological communities were viewed as existing in a state of equilibrium, held there by interspecific interactions. The ability of a community to reach and maintain this relatively constant species composition and to return to this steady state following a disturbance is known as *stability*. The **nonequilibrium model,** in contrast, emphasizes that communities are constantly changing as a result of disturbances.

What Is Disturbance?   **Disturbances** such as fire, drought, storms, overgrazing, or human activities change resource availability, reduce or eliminate some populations, and may create opportunities for new species. According to the **intermediate disturbance hypothesis,** small-scale disturbances may enhance environmental patchiness and help maintain species diversity. Human prevention of some natural disturbances such as small fires, may lead to large-scale disturbances, such as large, destructive fires.

Human Disturbance   Human activities have altered the structure of communities all over the world through such activities as conversion of land for agriculture, logging, and clearing for urban development. A common result of human disturbance is a reduction in species diversity.

Ecological Succession   The sequential transitions in species composition in a community, usually following some disturbance, are known as **ecological succession.** If no soil was originally present, as on a new volcanic island or on the moraine left by a retreating glacier, the process is called **primary succession.** A series of colonizers usually begins with autotrophic prokaryotes and moves through lichens, mosses, grasses, shrubs, and trees until the community reaches its prevalent form of vegetation. **Secondary succession** occurs when an existing community is disrupted by fire, logging, or farming, but the soil remains intact. Herbaceous species may colonize first, followed by woody shrubs and eventually forest trees.

Early colonizers may *facilitate* the arrival of other species by improving the environment. Or the actions of early species may *inhibit* the establishment of later species. Species may be independent in their colonization and *tolerate* the arrival of later species.

Ecologists have studied moraine succession over the 250-year retreat of glaciers at Glacier Bay in Alaska. The first pioneering plant species include mosses and fireweed; *Dryas*, a mat-forming shrub, dominates after about 30 years. Alder is the dominant plant a few decades later, followed by Sitka spruce. The community becomes a spruce-hemlock forest by the third century after deglaciation, except in flat, poorly drained areas, where *Sphagnum* invades. These mosses make the soil waterlogged and acidic, killing the trees and creating *Sphagnum* bogs.

■ **INTERACTIVE QUESTION   53.6**

a. During the succession following glacial retreat, describe the effects of the alder stage on soil fertility.

b. What is the effect of the spruce forest on soil pH?

The soil pH changes about from 7.0 to 4.0

53.4   Biogeographic factors affect community biodiversity

Equatorial-Polar Gradients   Surveys of plant and animal species have documented much greater numbers of species in tropical habitats than in temperate and polar regions. Tropical communities are older, partly because of their longer growing season and partly because they have not had to "start over" after glaciation, as has been the case several times for many polar and temperate communities.

Solar energy input and water availability are important climatic explanations for the latitudinal gradient in biodiversity. **Evapotranspiration** is the amount of water evaporated from soil and transpired by plants, and is determined by solar energy, temperature, and water availability (*actual evapotranspiration*) or just solar radiation and temperature (*potential evapotranspiration*). Evapotranspiration rates have been shown to correlate with species richness of trees and vertebrates in North America.

■ **INTERACTIVE QUESTION   53.7**

Why would the fact that tropical communities are "older" than temperate or polar communities contribute to greater species diversity?

Area Effects   A **species-area curve** illustrates the correlation between the size of a community and the number of species found there. In general, the larger the area, the greater diversity of habitats and the greater the species richness. Use of such curves in conservation biology can allow predictions on how a loss of habitat may affect biodiversity.

Island Equilibrium Model Any habitat surrounded by a significantly different habitat is considered an island and allows ecologists to study factors that affect species diversity. In the 1960s R. MacArthur and E. O. Wilson developed a general hypothesis of island biogeography, stating that the size of the island and its closeness to the mainland (or source of dispersing species) are important variables directly correlated with species diversity. Larger islands closer to the mainland will have a higher species diversity than smaller or more distant islands. The eventual number of species on the island depends on the immigration rate of new species and the extinction rate of island species. These rates change as the number of species on the island increases, and when the rates become equal, an equilibrium in species diversity develops, although species composition may continue to change.

While the island biogeography hypothesis may apply over relatively short time periods in cases where colonization determines species composition, over long periods, adaptive evolutionary changes and abiotic disturbances on the island are probably more important in determining community structure and composition.

■ INTERACTIVE QUESTION 53.8

Many biogeographic studies have found that large islands have greater species richness than small islands. Label the lines on the following graph that show how immigration rate and extinction rate vary with the number of species on large and small islands. Indicate the location of the equilibrium number on the x axis for a small and a large island.

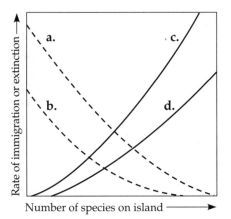

53.5 Contrasting views of community structure are the subject of continuing debate

Integrated and Individualistic Hypotheses Ecologists in the early 1900s developed two differing views on community composition based upon plant distributions. F. E. Clements advanced the **integrated hypothesis** of a community functioning as an integrated unit, with specific species linked together by their interrelationships. H. A. Gleason advocated the **individualistic hypothesis** that saw communities as chance groupings of species found in the same area because of similar abiotic requirements.

The integrated hypothesis predicts that species are clustered into discrete communities. The individualistic hypothesis, in contrast, predicts that species have independent distributions along environmental gradients and that boundaries between communities are indistinct.

■ INTERACTIVE QUESTION 53.9

In tests of the integrated versus individualistic hypotheses, which hypothesis is most often supported by the observed distribution of species in plant communities?

Rivet and Redundancy Models The **rivet model** of communities, first proposed by P. and A. Ehrlich in 1981, suggests that changing the composition or abundance of species in a community would affect many of the members of that interwoven community. The **redundancy model,** proposed by B. Walker in 1992, views communities as loosely connected assemblages, in which species are redundant—a disappearing species will be replaced by another species that fills the same role in the community. The relationships among members of most communities probably fall between these two polar models. These hypotheses are important to current environmental issues concerning how extinctions may affect community structure and ecosystem functioning.

▶ Word Roots

crypto- = hidden, concealed (*cryptic coloration:* a type of camouflage that makes potential prey difficult to spot against its background)

ecto- = outer (*ectoparasites:* parasites that feed on the external surface of a host)

a) Immigration - large island

b) Immigration - large island

c) Extinction - small island

d)   "   -   "

endo- = inner (*endoparasites:* parasites that live within a host)

herb- = grass; **-vora** = eat (*herbivory:* the consumption of plant material by an herbivore)

hetero- = other, different (*heterogeneity:* a measurement of biological diversity considering richness and relative abundance)

inter- = between (*interspecific competition:* competition for resources between plants, between animals, or between decomposers when resources are in short supply)

mutu- = reciprocal (*mutualism:* a symbiotic relationship in which both the host and the symbiont benefit)

Structure Your Knowledge

1. Complete this concept map to organize your understanding of the important factors that structure a community.

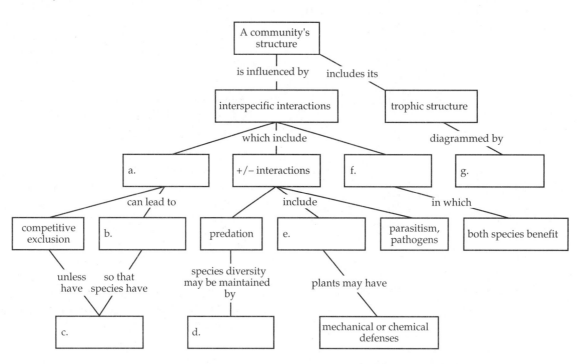

2. Community ecologists develop models or hypotheses to describe community structure and the factors that contribute to such structure. Briefly explain the following models that were described in this chapter.
   a. Competitive exclusion principle
   b. Energetic hypothesis
   c. Dynamic stability hypothesis
   d. Bottom-up model
   e. Top-down (trophic cascade) model
   f. Nonequilibrium model
   g. Individualistic hypothesis
   h. Integrated hypothesis
   i. Rivet model
   j. Redundancy model

Test Your Knowledge

MULTIPLE CHOICE: *Choose the one best answer.*

1. Which of the following is *not* descriptive of Gleason's individualistic concept of communities?
   a. Communities are chance collections of species that are in the same area because of similar environmental requirements.
   b. There should be no distinct boundaries between communities.
   c. The consistent composition of a community is based on interactions that cause it to function as an integrated unit.
   d. Species are distributed independently along environmental gradients.
   e. Most plant communities studied meet the predictions made by this concept.

2. Two species, A and B, occupy adjoining environmental patches that differ in several abiotic factors. When species A is experimentally removed from a portion of its patch, species B colonizes the vacated area and thrives. When species B is experimentally removed from a portion of its patch, species A does not successfully colonize the area. What might you conclude from these results?
   a. Both species A and species B are limited to their range by abiotic factors.
   b. Species A is limited to its range by competition, and species B is limited by abiotic factors.
   c. Both species are limited to their range by competition.
   d. Species A is limited to its range by abiotic factors, and species B is limited to its range because it cannot compete with species A.
   e. Species A is a predator of species B.

3. The species richness of a community refers to
   a. the relative numbers of individuals in each species.
   b. the number of different species found in a community.
   c. the feeding relationships or trophic structure within the community.
   d. the species diversity of that community.
   e. its stability or ability to persist through disturbances.

4. The rivet model of communities is most similar to
   a. the dynamic stability hypothesis.
   b. the top-down model.
   c. the redundancy model.
   d. the individualistic hypothesis.
   e. the integrated hypothesis.

5. Through resource partitioning,
   a. two species can compete for the same prey item.
   b. slight variations in niche allow closely related species to coexist in the same habitat.
   c. two species can share identical niches in a habitat.
   d. competitive exclusion results in the success of the superior species.
   e. two species with identical niches do not share the same habitat and thus avoid competition.

6. Which of the following organisms and trophic level is mismatched?
   a. algae—producer
   b. phytoplankton—primary consumer
   c. fungi—decomposer
   d. carnivorous fish larvae—secondary consumer
   e. eagle—tertiary or quaternary consumer

7. Aposematic coloring is most commonly found in
   a. prey whose body morphology is cryptic.
   b. predators who are able to sequester toxic plant compounds in their bodies.
   c. prey species that have chemical defenses.
   d. good-tasting prey that evolve to look like each other.
   e. prey species that are camouflaged to match their environment.

8. A palatable (good-tasting) prey species may defend against predation by
   a. Müllerian mimicry.
   b. Batesian mimicry.
   c. secondary compounds.
   d. aposematic coloration.
   e. either a or b.

9. When one species was removed from a tidepool, the species richness became significantly reduced. The removed species was probably
   a. a strong competitor.
   b. a potent parasite.
   c. a resource partitioner.
   d. a keystone species.
   e. the species with the highest relative abundance.

10. A highly successful parasite
    a. will not harm its host.
    b. may benefit its host.
    c. will be able to feed without killing its host.
    d. will kill its host fairly rapidly.
    e. will have coevolved into a commensalistic interaction with its host.

11. Why do most food chains consist of only three to five links?
    a. There are only five trophic levels: producers; primary, secondary, and tertiary consumers; and decomposers.
    b. Most communities are controlled bottom-up by mineral nutrient supply, and few communities have enough nutrients to support more links.
    c. The dominant species in most communities consumes the majority of prey; thus, not enough food is left to support higher predators.
    d. According to the energetic hypothesis, the inefficiency of energy transfer from one trophic level to the next limits the number of links that can exist.
    e. According to the trophic cascade model, increasing the biomass of top trophic levels causes a decrease in the biomass of lower levels, so that the top levels can no longer be supported.

12. During succession, inhibition by early species
    a. may prevent the achievement of a stable community.
    b. may slow down both the rate of colonization and the rate of extinction, depending on the size of the area and distance from the source of dispersing species.
    c. results from the frequent disturbances that often eliminate early colonizers.
    d. may slow down the successful colonization by other species.
    e. may involve changes in soil pH or accelerated accumulation of humus.

13. According to the nonequilibrium model,
    a. chance events such as disturbances play major roles in the structure and composition of communities.
    b. species composition in a community is always in flux as a result of human interventions.
    c. food chains are limited to a few links because long chains are more unstable in the face of environmental disturbances.
    d. the communities with the most diversity have the least stability or resistance to change.
    e. early colonizers inhibit other species, whereas later colonizers facilitate the arrival of new species.

14. An island that is small and far from the mainland, in contrast to a large island close to the mainland, would be expected to
    a. have lower species diversity.
    b. be in an earlier successional stage.
    c. have higher species diversity but a much lower abundance of organisms.
    c. have a higher rate of colonization but a higher rate of extinction.
    d. have a lower rate of colonization and a lower rate of extinction.

15. According to the top-down (trophic cascade) model of community control, which trophic level would you *decrease* if you wanted to *increase* the vegetation level in a community?
    a. nutrients
    b. vegetation
    c. secondary consumers (carnivores)
    d. tertiary consumers
    e. omnivores

16. A major explanation for the decline in species richness along an equatorial-polar gradient is the correlation of high levels of solar radiation and water availability with biodiversity. Which of the following is also suggested as a factor in the high species richness of tropical communities?
    a. the inverse relationship between biodiversity and evapotranspiration
    b. the greater age of these communities (longer growing season and fewer climatic setbacks), providing more time for speciation events
    c. the larger area of the tropics and corresponding richness predicted by the species-area curve
    d. the lack of disturbances in tropical areas
    e. the greater immigration rate and lower extinction rate found on large tropical islands

17. Ecologists survey the tree species in two forest plots of different ages. Plot 1 has six different species and 95% of all trees belong to just one species. Plot 2 has five different species, each of which is represented by approximately 20% of the trees. How would you describe plot 2 as compared with plot 1?
    a. higher species richness
    b. greater species diversity
    c. lower relative abundance
    d. lower species richness
    e. Both b and d are correct.

18. Which of the following interspecific interactions is *not* an example of a $+/-$ interaction?
    a. ectoparasite and host
    b. herbivore and plant
    c. honeybee and flower
    d. pathogen and host
    e. carnivore and prey

19. Which of the following organisms is mismatched with its community role?
    a. beaver—community "engineer"
    b. black rush *Juncus* in salt marsh—facilitator
    c. sea otter in North Pacific—keystone predator
    d. trees in spruce-hemlock forest—dominant species
    e. alder and *Dryas* (a mat-forming shrub)—inhibitor

20. Two allopatric species of Galapagos finches have beaks of similar size. There is a significant difference in beak size when the two species occur on the same island. What is this an example of?
    a. competitive exclusion
    b. coevolution
    c. commensalism
    d. character displacement
    e. trophic cascade

Chapter 54
Ecosystems

Framework

This chapter describes energy flow and chemical cycling through ecosystems. Producers convert light energy into chemical energy, which is then passed, with a loss of energy at each level, through the food web and ultimately to detritivores. Energy makes a one-way trip through ecosystems. Chemical elements are cycled in the ecosystem from abiotic reservoirs through producers, consumers, and detritivores, and back to the reservoirs.

An ecosystem's primary production may be limited by nutrients, temperature, or moisture. The low trophic efficiency in the transfer of energy from one level to the next is reflected in pyramids of production, biomass, and numbers.

Human activities are altering chemical cycles, causing climate change, depleting atmospheric ozone, and changing ecosystems.

Chapter Review

An **ecosystem** is a community and its physical or abiotic environment. Most ecosystems are powered by energy from sunlight, which is transformed to chemical energy by autotrophs, passed to a series of heterotrophs in the organic compounds of food, and continually dissipated in the form of heat. Chemical elements are cycled between the abiotic and biotic components of the ecosystem as autotrophs incorporate them into organic compounds and the processes of metabolism and decomposition return them to the soil, air, and water.

54.1 Ecosystem ecology emphasizes energy flow and chemical cycling

Ecosystem ecologists group species into trophic levels to follow energy transformations and chemical movements in an ecosystem.

Ecosystems and Physical Laws According to the first law of thermodynamics, energy cannot be created or destroyed, only transformed. Energy flows through ecosystems from its input as solar radiation to its conversion into chemical energy to its dissipation as heat. The second law of thermodynamics states that in each energy conversion, some energy is converted to heat. Ecologists trace the energy flow in ecosystems and the efficiency of ecological energy conversions.

■ INTERACTIVE QUESTION   54.1

Compare the movement of energy and chemicals in ecosystems.

Trophic Relationships Most **primary producers**, or autotrophs, use light energy to photosynthesize sugars for use as fuel in respiration and as building materials for other organic compounds. Heterotrophs depend on autotrophs for their organic compounds. The **primary consumers** are herbivores; **secondary consumers** are carnivores. **Tertiary consumers** eat other carnivores. **Detritivores**, or **decomposers**, consume **detritus**, which is organic wastes, fallen leaves, and dead organisms.

Decomposition   Detritivores that feed on plant remains often form a link between producers and consumers. Fungi and prokaryotes are the most important decomposers in most ecosystems, converting organic materials from all trophic levels to inorganic compounds that can be recycled by autotrophs.

54.2   Physical and chemical factors limit primary production in ecosystems

Primary production is the amount of light energy converted to chemical energy during a period of time—the photosynthetic output of an ecosystem's autotrophs.

Ecosystem Energy Budgets   The intensity of solar energy striking the Earth varies by latitude and, depending on cloud cover and dust in the air, by region. Only a small portion of incoming solar radiation strikes photosynthetic organisms, and, of that, only about 1% is converted to chemical energy. Nevertheless, worldwide photosynthetic production is about 170 billion tons of organic material per year.

Net primary production (NPP) is the ecosystem's **gross primary production (GPP)** minus the energy used by plants in their own cellular respiration (R). NPP = GPP − R.

Net primary production can be expressed as energy per unit area per unit time ($J/m^2/yr$) or as biomass measured in terms of dry weight of vegetation added ($g/m^2/yr$). *Standing crop* is the total biomass of photosynthetic organisms in an ecosystem. Primary production and the contribution to Earth's total production vary by ecosystem.

Primary Production in Marine and Freshwater Ecosystems   The depth to which light penetrates affects primary production in oceans. Nutrients, however, limit marine production more than light. Nitrogen and phosphorus levels are very low in the photic zone of the open ocean, limiting the growth of phytoplankton. Nutrient-addition experiments in polluted coastal waters indicate that nitrogen limits algal growth more than does phosphorus. Studies of the Sargasso Sea have shown that the micronutrient iron is the **limiting nutrient** in these unproductive waters. The addition of iron to test regions in tropical oceans resulted in an increase in cyanobacteria that fix nitrogen, which then stimulated the growth of eukaryotic phytoplankton.

In freshwater ecosystems, nutrient limitation also affects production. **Eutrophication,** the shift in composition of phytoplankton communities from domination by green algae and diatoms to blooms of cyanobacteria, has been linked to phosphorus pollution from sewage and fertilizer runoff.

■ INTERACTIVE QUESTION   54.2

a. List some ecosystems with high rates of production.

b. List some ecosystems with low rates of production.

c. The open ocean has low primary production yet contributes the greatest percentage of Earth's primary production. Explain.

d. Antarctic seas are often more productive than most tropical seas, even though they are colder and receive lower light intensity. Explain.

Primary Production in Terrestrial and Wetland Ecosystems   Production in terrestrial and wetland ecosystems is related primarily to moisture and temperature. The variation in precipitation and temperature in terrestrial ecosystems is reflected in a measure called **actual evapotranspiration**—the annual amount of water evaporated from a landscape and transpired by plants. Ecosystems usually show a positive correlation between actual evapotranspiration and primary production.

Nutrients may limit production on a local scale. Nitrogen or phosphorus is the limiting nutrient in many terrestrial and wetland ecosystems.

54.3   Energy transfer between trophic levels is usually less than 20% efficient

The rate at which consumers in an ecosystem produce new biomass from their food is called **secondary production.**

Production Efficiency   Herbivores consume only a fraction of the plant material produced; they cannot digest all they eat; and much of the energy they do absorb is used for cellular respiration. Only the chemical energy stored as growth or in offspring is available as food to higher trophic levels. The proportion of assimilated food energy that is used for net secondary production (growth and reproduction) is a measure of the efficiency of energy transformation: **production efficiency** = net secondary production/assimilation of primary production. Production efficiencies vary from 1–3% for "warm-blooded" birds and mammals, to 10% for fishes, to 40% for insects.

Trophic efficiency is the percentage of the energy of one trophic level that makes it to the next level, usually ranging from 5–20%. Trophic efficiencies take into account the loss of energy through respiration, feces, and the organic material not consumed by the next trophic level. A *pyramid of net production* shows this multiplicative loss of energy.

A *biomass pyramid* illustrates the standing crop biomass of organisms at each trophic level. This pyramid usually narrows rapidly from producers to the top trophic level. Some aquatic ecosystems have inverted biomass pyramids in which zooplankton (consumers) outlive and outweigh the highly productive, but heavily consumed, phytoplankton. Phytoplankton have a short **turnover time,** determined by dividing standing crop biomass by production. The production pyramid for this ecosystem, however, is normal in shape.

The *pyramid of numbers* illustrates that higher trophic levels contain small numbers of individuals, resulting from the larger size of these animals and the greatly decreased energy availability illustrated by the pyramid of production.

■ INTERACTIVE QUESTION 54.3

a. Why is production efficiency higher for fishes than for birds and mammals?

b. Assuming a 10% trophic efficiency (transfer of energy to the next trophic level), approximately what proportion of the chemical energy produced in photosynthesis makes it to a tertiary consumer?

The Green World Hypothesis Most terrestrial ecosystems are green. The **green world hypothesis** proposes five factors that keep herbivore populations from stripping Earth's vegetation: plant defenses; limited essential nutrients that restrict herbivore growth and reproduction; abiotic fluctuations; intraspecific competition; and interspecific interactions such as predation, parasitism, and disease. This final, "top-down" explanation is proposed as the most important factor limiting herbivore populations.

54.4 Biological and geochemical processes move nutrients between organic and inorganic parts of the ecosystem

Chemical elements are passed between abiotic and biotic components of ecosystems through **biogeochemical cycles.** Plants and other autotrophs use inorganic nutrients to build organic matter, which is passed through the food chain. Chemicals are returned to the atmosphere, water, or soil through respiration and the action of decomposers.

A General Model of Chemical Cycling The route of a biogeochemical cycle depends on the element and the trophic structure of an ecosystem. Gaseous forms of carbon, oxygen, sulfur, and nitrogen have global cycles involving atmospheric reservoirs. Less mobile elements, such as phosphorus, potassium, calcium, and the trace elements, have a more localized cycle in which soil is the main abiotic reservoir.

Most nutrients are found in four types of reservoirs or compartments: organic material in living organisms or detritus, available to other organisms; unavailable organic material in "fossilized" deposits; available inorganic elements and compounds in water, soil, or air; and unavailable elements in rocks. Nutrients may leave the unavailable reservoirs through weathering of rock, erosion, or burning of fossil fuels.

The movement of elements through biogeochemical cycles is quite complex, with influx and loss of nutrients from ecosystems occurring in many ways. Ecologists study this movement by adding radioactive tracers to chemical elements or by following naturally occurring nonradioactive isotopes through ecosystems.

Biogeochemical Cycles Ecologists studying biogeochemical cycles consider each chemical's biological importance, the forms in which it is used, the major reservoirs, and the processes that drive each cycle.

The water cycle involves evaporation, precipitation, and transpiration, with a net flow of water evaporating by solar energy from the oceans, moving as water vapor to the land where it precipitates, and returning to the oceans through runoff and groundwater. The oceans contain 97% of the water in the biosphere.

In the carbon cycle, plants take CO_2 from the atmosphere for photosynthesis and organisms release it in cellular respiration. Fossil fuel combustion is increasing atmospheric CO_2. Reservoirs include fossil fuels, dissolved carbon in the oceans, plant and animal biomass, CO_2 in the atmosphere, and sedimentary rock.

Plants require nitrogen in the form of NH_4^+ or NO_3^-. Animals obtain nitrogen in organic form from plants or other animals. Most nitrogen enters ecosystems through *nitrogen fixation:* Soil bacteria and symbiotic bacteria in root nodules fix nitrogen into ammonium. In *nitrification,* nitrifying bacteria convert NH_4^+ to NO_3^-. In *denitrification,* the metabolism of other bacteria returns N_2 to the atmosphere. Fertilizers add a significant amount of nitrogen to ecosystems. The major nitrogen reservoir is the atmosphere.

Bacterial and fungal decomposers, in a process called *ammonification,* break down organic compounds and return ammonium to the soil. Most local nitrogen cycling involves decomposition and reassimilation.

■ INTERACTIVE QUESTION   54.5

Label the organisms and compounds that are illustrated in this nitrogen cycle.

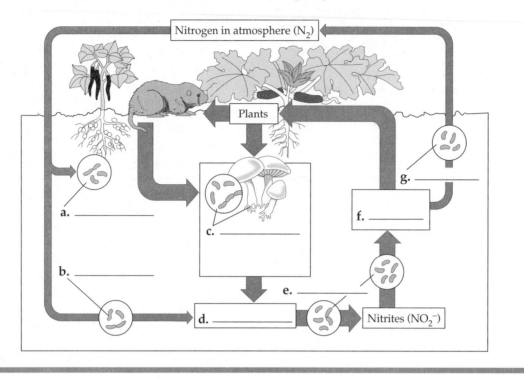

Weathering of rock adds phosphorus to the soil in the form of PO_4^{3-}, which is absorbed by plants. Organic phosphate is transferred from plants to consumers and returned to the soil through the action of decomposers or by animal excretion. Humus and soil particles usually bind phosphate, keeping it available locally for recycling. Sedimentary rocks of marine origin are the largest reservoirs.

■ INTERACTIVE QUESTION   54.4

What is the biological importance of water, carbon, nitrogen, and phosphorus?

Decomposition and Nutrient Cycling Rates   Temperature and the availability of water and O_2 influence decomposition rates; thus, nutrient cycling times vary in different ecosystems. Nutrients cycle rapidly in a tropical rain forest; the soil contains only about 10% of the ecosystem's nutrients. In temperate forests where decomposition is slower, 50% of organic nutrients are stored in the detritus and soil. Decomposition rates are slow in aquatic ecosystems, and sediments constitute a nutrient sink.

Vegetation and Nutrient Cycling: The Hubbard Brook Experimental Forest   A team of scientists has looked at nutrient cycling in the Hubbard Brook forest ecosystem since 1963, an example of long-term ecological research (LTER). The mineral budget for each of six valleys was determined by measuring the input of key nutrients in rainfall and their outflow through the creek that drained each watershed. About 60% of the precipitation exited through the stream; the rest was lost by transpiration and evaporation. Most minerals were recycled within the forest ecosystem.

The effect of deforestation on nutrient cycling was measured for three years in a valley that was completely logged and sprayed with herbicides. Compared with a control, water runoff from the deforested valley increased 30–40%; net loss of minerals such as Ca^{2+} and K^+ was large. Nitrate increased in concentration in the creek 60-fold, removing this critical soil nutrient and contaminating drinking water.

Long-term data from Hubbard Brook indicate that acid precipitation has removed most of the Ca^{2+} from the soils. This lack of Ca^{2+} appears to have halted forest growth in the past decade. The effects of the addition of Ca^{2+} to experimental watersheds are being monitored.

a. In which natural ecosystem do nutrients cycle the fastest? Why?

b. In which natural ecosystem do nutrients cycle the slowest? Why?

c. What is the effect of loss of vegetation on nutrient cycling?

54.5 The human population is disrupting chemical cycles throughout the biosphere

Nutrient Enrichment The harvesting of crops removes nutrients that would otherwise recycle in the soil. After depleting the organic and inorganic reserves of nutrients, crops require the addition of synthetic fertilizers. The addition of nitrogen fertilizers, increased legume cultivation, and burning have doubled Earth's supply of fixed nitrogen. Excess soil nitrogen can be released as nitrogen oxides by denitrifying bacteria and contributes to global warming, ozone thinning, and acid precipitation.

Nitrogen that exceeds the **critical load,** the amount of added nutrient that can be absorbed by plants without damaging the ecosystem, can contaminate groundwater, degrade lakes and rivers, and drain into the ocean.

The lack of mineral nutrients in an oligotrophic lake keeps primary production low. The additional nutrients in eutrophic lakes increase primary and overall production. Sewage, factory wastes, and runoff of animal wastes and fertilizers from agricultural lands have led to *cultural eutrophication*. The rapid increase in nutrients can cause an explosive increase in algae and cyanobacteria. Oxygen shortages, due to respiration at night and the metabolism of decomposers that work on the accumulating organic material, kill off many fish and other lake organisms.

Acid Precipitation The burning of coal and other fossil fuels as well as wood releases oxides of sulfur and nitrogen, which form sulfuric and nitric acid in the atmosphere. These acids return to the earth as acid precipitation, defined as rain, snow, or fog with a pH less than 5.6. Emissions from the tall exhaust stacks of ore smelters and electrical plants drift downwind and create acid precipitation over vast and distant areas. Nutrients leach from soils as acid precipitation changes soil chemistry, and forests have been damaged. Fish populations in lakes across North America and Europe

have declined and community compositions have changed. Regulations and new technologies have reduced sulfur dioxide emissions, and the streams and lakes of New England are gradually recovering.

Toxins in the Environment Humans release a huge variety of toxic chemicals into the environment. Organisms absorb these toxins from food and water and may retain them within their tissues. In a process known as **biological magnification,** the concentration of such compounds increases in each successive link of the food chain. Chlorinated hydrocarbons, such as DDT, and polychlorinated biphenyls, or PCBs, have been implicated in endocrine system problems in many animal species. Many toxic chemicals dumped into ecosystems are nonbiodegradable; others, such as mercury, may become more harmful as they react with other environmental factors.

Atmospheric Carbon Dioxide The concentration of CO_2 in the atmosphere has been increasing since the Industrial Revolution as a result of the combustion of fossil fuel and the burning of wood removed by deforestation. If C_3 plants become able to outcompete C_4 plants with the increase in CO_2, species composition in natural and agricultural communities may be significantly altered.

In the Forest-Atmosphere Carbon Transfer and Storage (FACTS-I) experiment, scientists are monitoring the effects of elevated CO_2 levels on sample plots in a forest ecosystem over ten years.

Through a phenomenon known as the **greenhouse effect,** CO_2 and water vapor in the atmosphere absorb infrared radiation reflected from Earth and rereflect it back to Earth, causing an increase in temperature.

Scientists use various models to try to estimate the extent and consequences of increasing CO_2 levels. A number of studies predict a doubling of CO_2 levels and a temperature rise of 2°C by the end of the 21st century. Ecologists study the effects on vegetation of previous global warming trends to try to predict the impact of increasing temperatures. Controlling the level of CO_2 emissions in increasingly industrialized societies is a huge international challenge.

List some of the potential consequences of global warming.

Depletion of Atmospheric Ozone A layer of ozone molecules (O_3) in the lower stratosphere absorbs damaging ultraviolet radiation. This layer has been gradually thinning since 1975, largely as a result of the accumulation of breakdown products of chlorofluorocarbons in the atmosphere. The dangers of ozone depletion may include increased incidence of skin cancer and cataracts and unpredictable effects on phytoplankton, crops, and natural ecosystems.

Word Roots

auto- = self; **troph-** = food, nourishment (*autotroph:* an organism that obtains organic food molecules without eating other organisms)

bio- = life; **geo-** = the Earth (*biogeochemical cycles:* the various nutrient circuits that involve both biotic and abiotic components of ecosystems)

de- = from, down, out (*denitrification:* the process of converting nitrate back to nitrogen)

detrit- = wear off; **-vora** = eat (*detritivore:* a consumer that derives its energy from nonliving organic material)

hetero- = other, different (*heterotroph:* an organism that obtains organic food molecules by eating other organisms or their by-products)

Structure Your Knowledge

1. Two processes that emerge at the ecosystem level of organization are energy flow and chemical cycling. Develop a concept map that explains, compares, and contrasts these two processes.

2. Describe four or five human intrusions in ecosystem dynamics that have detrimental effects.

Test Your Knowledge

MULTIPLE CHOICE: *Choose the one best answer.*

1. Which of the following groups is absolutely essential to the functioning of an ecosystem?
   a. producers
   b. producers and herbivores
   c. producers, herbivores, and carnivores
   d. detritivores
   e. producers and detritivores

2. Primary production
   a. is equal to the standing crop of an ecosystem.
   b. is greatest in freshwater lakes and streams.
   c. is the rate of conversion of light to chemical energy in an ecosystem.
   d. is inverted in some aquatic ecosystems.
   e. is all of the above.

3. Which of the following is an accurate statement about ecosystems?
   a. Energy is recycled through the trophic structure.
   b. Energy is usually captured from sunlight by primary producers, passed to secondary producers in the form of organic compounds, and lost to detritivores in the form of heat.
   c. Chemicals are recycled between the biotic and abiotic sectors, whereas energy makes a one-way trip through the food web and is eventually dissipated as heat.
   d. There is a continuous process by which energy is lost as heat, and chemical elements leave the ecosystem through runoff.
   e. A food web shows that all trophic levels may feed off each other.

4. In the experiment in which iron was added to the Sargasso Sea, the growth of eukaryotic phytoplankton was stimulated because
   a. iron was the limiting nutrient for eukaryotic phytoplankton growth.
   b. the iron interacted with bottom sediments, releasing nitrogen and phosphorus into the water.
   c. iron interacted with phosphorus, making that nutrient available to the phytoplankton.
   d. the iron reached the critical load necessary to promote photosynthesis.
   e. iron stimulated the growth of nitrogen-fixing cyanobacteria, which then made nitrogen available for phytoplankton growth.

5. The open ocean and tropical rain forest are the two largest contributors to Earth's net primary production because
   a. both have high rates of net primary production.
   b. both cover huge surface areas of Earth.
   c. nutrients cycle fastest in these two ecosystems.
   d. the ocean covers a huge surface area and the tropical rain forest has a high rate of production.
   e. both a and b are correct.

6. Production in terrestrial ecosystems is affected by
   a. temperature.
   b. light intensity.
   c. availability of nutrients.
   d. availability of water.
   e. all of the above.

7. Secondary production
   a. is measured by the standing crop.
   b. is the rate of biomass production in consumers.
   c. is greater than primary production.
   d. is 10% less than primary production.
   e. is the gross primary production minus the energy used for respiration.

8. Which of the following is *not* true of a pyramid of production?
   a. Only about 10% of the energy in one trophic level is passed into the next level.
   b. Because of the loss of energy at each trophic level, most food chains are limited to three to five links.
   c. The pyramid of production of some aquatic ecosystems is inverted because of the large zooplankton primary consumer level.
   d. Eating grain-fed beef is an inefficient means of obtaining the energy trapped by photosynthesis.
   e. A pyramid of numbers is usually the same shape as a pyramid of production.

9. In which of the following would you expect production efficiency to be the greatest?
   a. plants
   b. mammals
   c. fish
   d. insects
   e. birds

10. Biogeochemical cycles are global for elements
    a. that are found in the atmosphere.
    b. that are found mainly in the soil.
    c. such as carbon, nitrogen, and phosphorus.
    d. that are dissolved in water.
    e. in the nonavailable reservoirs.

11. Which of these processes is *incorrectly* paired with its description?
    a. nitrification—oxidation of ammonium in the soil to nitrite and nitrate
    b. nitrogen fixation—reduction of atmospheric nitrogen into ammonia
    c. denitrification—return of N_2 to air, produced by denitrifying bacteria metabolizing nitrate
    d. ammonification—decomposition of organic compounds into ammonium
    e. industrial fixation—nitrogen added to soil in rain or dust particles

12. Clear-cutting tropical forests yields agricultural land with limited productivity because
    a. it is too hot in the tropics for most food crops.
    b. the tropical forest regrows rapidly and chokes out agricultural crops.
    c. few of the ecosystem's nutrients are stored in the soil; most are in the forest trees.
    d. phosphorus, not nitrogen, is the limiting nutrient in those soils.
    e. decomposition rates are high but primary production is low in the tropics.

13. Which of the following was *not* shown by the Hubbard Brook Experimental Forest study?
    a. Most minerals recycle within a forest ecosystem.
    b. Deforestation results in a large increase in water runoff.
    c. Mineral losses from a valley were great following deforestation.
    d. Nitrate was the mineral that showed the greatest loss.
    e. Acid rain increased as a result of deforestation.

14. The finding of harmful levels of DDT in grebes (fish-eating birds) following years of trying to eliminate bothersome gnat populations in a lakeshore town is an example of
    a. eutrophication.
    b. biological magnification.
    c. the biomass pyramid.
    d. chemical cycling.
    e. increasing resistance to pesticides.

15. Increasing atmospheric CO_2 levels
    a. could change global climate and lead to the flooding of coastal areas.
    b. could result in more C_4 plants in plant communities that were previously dominated by C_3 plants.
    c. causes an increase in temperature when CO_2 absorbs more sunlight entering the atmosphere.
    d. could increase precipitation in central continental areas.
    e. could do all of the above.

16. According to the green world hypothesis, herbivores eat only a small portion of an ecosystem's vegetation because
    a. primary production is much greater than secondary production.
    b. plants have a very short turnover time.
    c. herbivores cannot digest most of what they eat.
    d. predators, parasites, and disease keep herbivore populations in check.
    e. the production efficiency of herbivores is very low.

17. Which of the following trophic levels would most likely have the largest numbers of individuals?
    a. primary producers
    b. omnivores
    c. primary consumers
    d. herbivores
    e. tertiary consumers

18. Which of the following is a direct effect of the thinning of the ozone layer?
    a. a reduction in species diversity
    b. global warming
    c. acid precipitation
    d. an increase in harmful UV radiation reaching Earth
    e. cultural eutrophication

C_3 - photosynthesize normally

C_4 -

Chapter 55

Conservation Biology and Restoration Ecology

Framework

Biodiversity at the genetic, species, and ecosystem levels is crucial to human welfare. This chapter explores the threats to biodiversity and several of the approaches to preserving the diversity of species on Earth. Conservation biologists focus on determining the habitat needs of endangered species, establishing and managing nature reserves that are often in human-dominated landscapes, and restoring degraded areas. Ecological research and our biophilia may help to achieve the goal of sustainable development—the long-term perpetuation of human societies *and* the ecosystems that support them.

Chapter Review

Conservation biology integrates all areas of biology in the effort to sustain ecosystem processes and biodiversity. **Restoration ecology** uses ecological principles to return degraded ecosystems to a more natural state. About 1.8 million species have been formally identified; estimates of total numbers of species on Earth range from 10 to 200 million. Human activities are altering all ecosystem processes, and extinction rates resulting from these disruptions may be 1,000 times higher than at any time in the past 100,000 years.

55.1 Human activities threaten Earth's biodiversity

The Three Levels of Biodiversity   Loss of the genetic diversity within and between populations lessens a species' adaptive potential.

A second level of biodiversity is species diversity, the species richness of an ecosystem. An **endangered species,** according to the U.S. Endangered Species Act (ESA), is one that is "in danger of extinction throughout all or a significant portion of its range." A **threatened species** is defined as one that is likely to become endangered. There are many well-documented examples of recent extinctions and endangered species of most taxonomic groups. Because millions of the world's species are unknown, it is difficult to assess local and global extinctions and their effects on the structure and function of ecosystems.

The third level of biodiversity is ecosystem diversity; loss of an ecosystem may affect the whole biosphere.

Biodiversity and Human Welfare   There are both ethical and practical reasons for preserving biodiversity. A loss of biodiversity is a loss of the genetic potential held in the genomes of species. Biodiversity is a natural resource that can provide medicines, fibers, industrial chemicals, and food.

Humans depend on Earth's ecosystems. **Ecosystem services** include such things as purification of air and water, detoxification and decomposition of wastes, nutrient cycling, flood control, pollination of crops, and access to beauty and recreation. Some ecologists value these ecosystem services at twice the gross national product of all countries combined.

Four Major Threats to Biodiversity   The greatest threat to biodiversity is habitat destruction, caused by agriculture, urban development, forestry, mining, and pollution. Habitat destruction is cited for 73% of the species listed as extinct, endangered, vulnerable, or rare by the International Union for Conservation of Nature and Natural Resources (IUCN). Fragmentation of natural habitats is a common occurrence and almost always leads to species loss. Both terrestrial ecosystems and marine habitats have been damaged.

Introduced species, sometimes called invasive, nonnative, or exotic species, compete with or prey upon native species and have probably been responsible for about 40% of extinctions in the past 250 years. Humans have transplanted thousands of species, intentionally and unintentionally, with huge economic costs in damage and control efforts.

Overexploitation involves harvesting plants or animals at rates higher than the populations' abilities to reproduce. Species of large animals with low intrinsic reproductive rates and species on small islands are particularly vulnerable to extinction. Overfishing, particularly using new harvesting techniques, has drastically reduced populations of many commercially important fish species.

The extinction of a keystone species, an ecosystem engineer, or a species with a very specialized relationship to others may disrupt interaction networks and threaten other species.

■ INTERACTIVE QUESTION   55.1

Give an example of how each of the following threats to biodiversity has reduced population numbers or caused extinctions.

a. habitat destruction

b. introduced species

c. overexploitation

d. disruptions of interaction networks

55.2   Population conservation focuses on population size, genetic diversity, and critical habitat

Small-Population Approach   Very small populations are considered endangered. According to the small-population approach, the inbreeding and genetic drift characteristic of a small population may draw it into an **extinction vortex,** in which the loss of genetic variation leads, by positive feedback loops, to smaller and smaller numbers until the population becomes extinct.

As agriculture fragmented their habitat, the number of prairie chickens in Illinois declined from millions in the 19th century to 50 in 1993. A comparison with DNA from museum specimens indicated decreased genetic variation in the threatened population. This decline was associated with a decrease in fertility. After transplanting birds from larger populations in other states, researchers noted an increase in egg viability and the Illinois population rebounded.

Computer models that integrate many factors are used to estimate **minimum viable population (MVP),** the minimum population size necessary to sustain a population. **Population viability analysis (PVA)** uses MVP to predict long-term viability of a population, the probability of survival over a particular time.

The **effective population size** (N_e) is based on a population's breeding potential and is determined by a formula that includes data on the number of individuals that breed and the sex ratio of the population: $N_e = (4N_f N_m)/(N_f + N_m)$. Other formulas take into account other life history or genetic factors. Conservation efforts should be based on maintaining the minimum number of reproductively active individuals needed to prevent extinction.

■ INTERACTIVE QUESTION   55.2

Is the effective population size usually larger or smaller than the actual number of individuals in the population? Explain.

M. Shaffer performed a PVA as part of a long-term study of grizzly bears in Yellowstone National Park. He estimated the viable population size for threatened grizzly bear populations, given a suitable habitat, of

100 bears. Current estimates of the grizzly population in the greater Yellowstone ecosystem indicate a population that has grown to 400. The effective population size, N_e, is only 25% of the total population size, or 100 bears. Genetic analyses indicate that the Yellowstone grizzly population has less genetic variability than other populations in North America. Migration between isolated populations would increase both the effective size and genetic variation of the Yellowstone grizzly bear population.

■ **INTERACTIVE QUESTION 55.3**

Explain the basic premise of the small population approach. What conservation strategy is recommended for preserving small populations?

Declining-Population Approach The emphasis of the declining-population approach is to identify populations that may be declining, identify the environmental factors that caused that decline, and then recommend corrective measures.

The following logical steps are part of the declining-population approach: assess population trends and distribution to establish that a species is in decline; determine its environmental requirements; list all possible causes of the decline and the predictions that arise from each of these hypotheses; test the most likely hypothesis to see if the population rebounds if this suspected factor is altered; apply the results to the management of the threatened species.

Logging and agriculture have fragmented the mature pine forest habitats of the red-cockaded woodpecker, driving this species into decline. Historically, periodic fires kept the understory around the pines low, another habitat requirement. Recognition of the social organization of this species and the factors that slow its dispersal to new territories has aided in its recovery. Management strategies now include protection of some longleaf pine forests, controlled fires, and the excavation of breeding cavities in unoccupied habitat to encourage establishment of new breeding groups.

■ **INTERACTIVE QUESTION 55.4**

Describe the declining-population approach to the conservation of endangered species.

Weighing Conflicting Demands Preserving habitat for endangered species often conflicts with human economic and recreational desires. Keystone species exert more influence on community structure and ecosystem processes, and prioritizing the species to be saved on the basis of their ecological role may be key to the survival of whole communities.

55.3 Landscape and regional conservation aim to sustain entire biotas

Conservation efforts increasingly are directed at sustaining the biodiversity of whole communities and ecosystems. A goal of **landscape ecology** is to make biodiversity conservation a part of ecosystem management and landscape use.

Landscape Structure and Biodiversity Landscape dynamics are important to conservation efforts because species often use more than one ecosystem or live on borders between ecosystems. Landscapes include ecosystems separated by boundaries or *edges*, which have their own sets of physical conditions and communities of organisms. Edge communities may be important sites of speciation, but their proliferation due to human fragmentation of habitats may serve to reduce biodiversity as edge species become predominant. The long-term Biological Dynamics of Forest Fragments Project has identified groups of species that live in forest edges and those that live in forest interior. Landscapes with habitats that support both these groups will have the greatest biodiversity.

Movement corridors are narrow strips or clumps of habitat that connect isolated patches. Artificial corridors are sometimes constructed when habitat patches have been separated by major human disruptions.

■ **INTERACTIVE QUESTION 55.5**

What are some potential benefits of corridors? How may they be harmful?

Establishing Protected Areas Currently, about 7% of Earth's land has been set aside as reserves. Conservation biologists apply the study of community, ecosystem, and landscape dynamics to the designation and management of these protected areas. **Biodiversity hot spots,** small areas with very high concentrations of endemic, threatened, and endangered species, are good choices for nature reserves. Hot spots vary, however, by taxonomic group, and their protection would in no way conserve all of biodiversity.

Even though nature reserves provide islands of protected habitat, the concept of nonequilibrium ecology with natural disturbances applies to them as well as to their surrounding landscapes. Patch dynamics, edges, and corridor effects must be considered in the design and management of preserves.

New information on the requirements for minimum viable population sizes indicates that most national parks and reserves are much too small—the *biotic boundary* needed to sustain a population is usually much larger than the *legal boundary* set in a reserve.

■ INTERACTIVE QUESTION   55.6

What factors would favor the creation of larger, extensive preserves? What factors favor smaller, unconnected preserves?

Zoned reserves have protected core areas surrounded with buffer zones in which the human social and economic climate is stable and activities are regulated to promote the long-term viability of the protected zones. Costa Rica has established eight zoned reserves, but deforestation has continued in some buffer zones.

It is likely that less than 10% of the biosphere will ever be protected in nature reserves, so the preservation of biodiversity involves working to create reserves in landscapes that are human dominated.

55.4   Restoration ecology attempts to restore degraded ecosystems to a more natural state

Areas degraded by farming, mining, or environmental pollution are often abandoned. The natural time frame for recovery relates to the size of the area disturbed. Restoration ecologists attempt to identify and manipulate the factors that most limit recovery time in order to speed the successional processes involved in a community's recovery from human disturbances.

Bioremediation   **Bioremediation** uses prokaryotes, fungi, or plants to detoxify polluted ecosystems. Some plants may be able not only to extract metals from contaminated soils, but also to concentrate them for commercial use. Prokaryotes are being used to metabolize toxins in dump sites and clean up oil spills.

Biological Augmentation   An example of **biological augmentation** is using plant species that thrive in nutrient-poor soils to facilitate recolonization of native species.

Exploring Restoration   Restoration ecologists often apply adaptive management, in which they experiment with promising management approaches and learn as they work in each unique and complex disturbed ecosystem.

55.5   Sustainable development seeks to improve the human condition while conserving biodiversity

Sustainable Biosphere Initiative   **Sustainable development** emphasizes the long-term prosperity of human societies and the ecosystems that support them. The Ecological Society of America endorses a research agenda, the Sustainable Biosphere Initiative, to encourage studies of global change, biodiversity, and maintenance of the productivity of natural and artificial ecosystems. An important goal is developing the ecological knowledge necessary to make intelligent and responsible decisions concerning Earth's resources.

Case Study: Sustainable Development in Costa Rica   Partnerships between the government, nongovernment organizations (NGOs), and citizens have contributed to the success of conservation in Costa Rica. Living conditions in the country have improved, as evidenced by a decrease in infant mortality, increase in life expectancy, and high literacy rate. A projected increase in population from 4 to 6 million in the next 50 years, however, will present challenges to the goal of sustainable development.

Biophilia and the Future of the Biosphere   E. O. Wilson calls our attraction to Earth's diversity of life and our affinity for natural environments *biophilia*. Perhaps this connection is innate and will provide the ethical resolve to protect species from extinction and ecosystems from destruction. By coming to know and understand nature through the study of biology, we may be more able to appreciate and preserve the processes and diversity of the biosphere.

▶ Word Roots

bio- = life (*biodiversity hot spot:* a relatively small area with an exceptional concentration of species)

Structure Your Knowledge

1. What are the major threats to biodiversity, listed in order of importance?

2. How does the loss of biodiversity threaten human welfare?

3. What do edges and movement corridors have to do with habitat fragmentation?

Test Your Knowledge

MULTIPLE CHOICE: *Choose the one best answer.*

1. According to the Endangered Species Act, what is the definition of a threatened species?
   a. an exotic species that cannot successfully compete with indigenous organisms
   b. an endemic species that is found nowhere else in the world
   c. a species that is found in disturbed habitats
   d. a species that is in danger of extinction in all or a large part of its range
   e. a species that is likely to become endangered

2. Ecosystem services include all of the following *except*
   a. pollination of crops.
   b. production of antibiotics and drugs.
   c. access to aesthetic beauty.
   d. decomposition of wastes.
   e. moderation of weather extremes.

3. Which of the following is the most serious threat to biodiversity?
   a. competition from introduced species
   b. commercial harvesting
   c. habitat destruction
   d. overexploitation
   e. disruptions of interaction networks

4. Some grassland and conifer forest preserves have effective fire prevention programs. What is the most likely result of such programs?
   a. an increase in species diversity because fires are prevented
   b. a change in community composition because fires are natural disturbances that maintain the community structure
   c. the preservation of endangered species in the area
   d. no change in the species composition of the preserved community
   e. succession to a deciduous forest

5. According to the small population approach, what is the most important remedy for preserving an endangered species?
   a. establish a large nature reserve around its habitat
   b. control the populations of its natural predators
   c. determine the reason for its decline
   d. encourage dispersal and increase in genetic variability
   e. set up artificial breeding programs

6. Which of the following is typical of biodiversity hot spots?
   a. a large number of endemic species
   b. a high rate of habitat degradation
   c. little species diversity
   d. a large land or aquatic area
   e. very large populations of migratory birds

7. Which of the following may occur when a population drops below its minimum viable population size?
   a. genetic drift
   b. a further reduction in population size
   c. inbreeding
   d. a loss of genetic variability
   e. All of the above are characteristics of an extinction vortex that the population may enter.

8. What are movement corridors?
   a. strips or clumps of habitat that connect isolated habitats
   b. the routes taken by migratory animals
   c. a landscape that includes several different ecosystems
   d. the areas forming the boundary or edge between two ecosystems
   e. buffer zones that promote the long-term viability of protected areas

9. What does it mean if a population's effective population size (N_e) is the same as its actual population size?
   a. The population is not in danger of becoming extinct.
   b. The population has high genetic variability.
   c. All the members of the population breed.
   d. The population's minimum viable population will not sustain the population.
   e. The population is being drawn into an extinction vortex.

10. The focus of the declining-population approach to conservation is to

a. predict a species' minimum viable population size.

b. transplant members from other populations to increase genetic variation.

c. perform a population viability analysis to predict the long-term viability of a population in a particular habitat.

d. determine the cause of a species' decline and take remedial action.

e. establish zoned reserves that ensure that human landscapes surrounding reserves support the protected habitats.

11. With limited resources, conservation biologists need to prioritize their efforts. Of the following choices, which should receive the greatest conservation attention in order to preserve biodiversity?

a. the northern spotted owl

b. declining keystone species in a community

c. a commercially important species

d. endangered and threatened vertebrate species

e. all declining species

12. Restoration ecology

a. uses the zoned reserve system to buffer nature reserves.

b. identifies biodiversity hot spots for protection.

c. may use bioremediation and biological augmentation to return degraded areas to their natural state.

d. uses the research agenda of the Sustainable Biosphere Initiative to study biodiversity and preserve Earth's ecosystems.

e. uses adaptive management to restore and maintain the productivity of artificial ecosystems.

Answer Section

CHAPTER 1: EXPLORING LIFE

1.1 **a.** The biosphere includes all of Earth's ecosystems.
b. An ecosystem includes all the living organisms in an area, along with the nonliving components of the environment.
c. A community consists of all the organisms inhabiting a particular area.
d. A population includes the individuals of a single species in an area.
e. An organism is an individual living entity.
f. Organs and organ systems, found in more complex organisms, perform life's functions.
g. Tissues are collections of similar cells. Several tissues make up an organ.
h. Cells are the fundamental unit of life.
i. Organelles are the functional components that make up cells.
j. Molecules are composed of atoms. They are the chemical units of life.

1.2 The order of nucleotides in a DNA molecule that makes up a gene "spells" the instructions for making a protein with a specific shape and function.

1.3 **a.** biologists from all fields, engineers, medical scientists, physicists, chemists, mathematicians, and computer scientists
b. in predicting how a particular drug or treatment may affect various aspects of the body; in predicting effects on various parts of the ecosystem as CO_2 concentration climbs

1.4 They are characterized to a large extent by their mode of nutrition. Plants are photosynthetic, fungi absorb their nutrients from decomposing organic material, and animals ingest other organisms.

1.5 Most species tend to produce more offspring than can survive. Organisms with heritable traits best suited to the environment will tend to reproduce more successfully and leave more offspring. Over time, favorable adaptations will accumulate in a population. New species may arise as small populations are exposed to different environments and natural selection favors different traits.

1.6 **a.** Since most encounters with coral snakes are fatal, predators didn't "learn" to avoid them. But predators with genes that somehow made them instinctively avoid coral snakes would have been more likely to pass on those genes to offspring, gradually adapting the predator population to the presence of poisonous snakes in their area.
b. This experiment could not control for the number of predators in each area and thus the total number of attacks on snakes. By presenting data as the percent of total attacks, the effect of this variable was eliminated.

STRUCTURE YOUR KNOWLEDGE

1. **a.** Cells are the basic units of life. They come in two distinct forms: prokaryotic and eukaryotic.
b. DNA is the molecule of inheritance, coding information for proteins and the functions of a cell, and passed on from one generation to the next.
c. With every increase in biological level, the organization and interactions of component parts lead to the properties of the dynamic system.
d. Biological systems are regulated. Organisms regulate their internal environment through positive and negative feedback systems that either speed up or slow down body processes.
e. Organisms exchange materials and energy with the living and nonliving components of their environment.
f. All organisms require energy. Energy flows through ecosystems from sunlight to chemical

energy in producers and consumers to escape as heat.

g. Life is united by descent from a common ancestor, as evidenced by the universal genetic code. The diversity of life is grouped into three domains.

h. Evolution explains the unity and diversity of life. Darwin's theory of natural selection leading to differential reproductive success accounts for the adaptation of populations to Earth's varying environments.

i. At each level of biological organization, structure and function are correlated.

j. Science includes observation-based discovery and hypothesis-based inquiry. Hypotheses must be testable and falsifiable.

k. Technologies are goal-oriented applications of science. Science and technology provide many benefits yet pose many ethical dilemmas for society.

TEST YOUR KNOWLEDGE

1. d	**3.** b	**5.** a	**7.** e
2. d	**4.** b	**6.** d	**8.** c

CHAPTER 2: THE CHEMICAL CONTEXT OF LIFE

■ INTERACTIVE QUESTIONS

2.1 calcium, phosphorus, potassium, sulfur, sodium, chlorine, magnesium

2.2 neutrons; 15, 15, 16, 31

2.3 absorb; released

2.4 **a.** Nitrogen $_7$N   **c.** Magnesium $_{12}$Mg

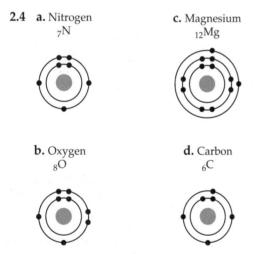

b. Oxygen $_8$O   **d.** Carbon $_6$C

2.5 Although each orbital can hold a pair of electrons, each electron fills a separate orbital until no empty orbitals remain.

2.6 **a.** protons
b. atomic number
c. element
d. neutrons
e. mass number or atomic mass
f. isotopes
g. electrons
h. electron shells or energy levels
i. valence shell

2.7 **a.** 1   **b.** 2   **c.** 3   **d.** 4

2.8 **a.** Nonpolar; even though N has a high electronegativity, the three pairs of electrons are shared equally between the two N atoms because each atom has an equally strong attraction for the electrons.

b. Polar; N is more electronegative than H and pulls the shared electrons in each covalent bond closer to itself.

c. Nonpolar; C and H have similar electronegativities and equally share electrons between them.

d. The C=O bond is polar because O is more electronegative than C; the C-H bonds are relatively nonpolar.

2.9 **a.** $CaCl_2$ **b.** Ca^{2+} is the cation.

2.10

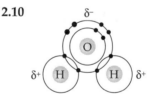

2.11 Water's two covalent bonds with H are spread apart at a 104.5° angle due to the hybridization of the s and 3 p orbitals.

2.12 6, 6, 6

SUGGESTED ANSWERS TO STRUCTURE YOUR KNOWLEDGE

1.

Particle	Charge	Mass	Location
Proton	+1	1 dalton	nucleus
Neutron	0	1 dalton	nucleus
Electron	−1	negligible	orbitals in electron shells

2. a. The atoms of each element have a characteristic number of protons in their nuclei, referred to as the atomic number. In a neutral atom, the atomic number also indicates the number of electrons.

The mass number is an indication of the approximate mass of an atom and is equal to the number of protons and neutrons in the nucleus.

The atomic mass is equal to the mass number and is measured in the atomic mass unit of daltons. Protons and neutrons both have a mass of approximately 1 dalton.

mass number and atomic weight → $^{12}_{6}$C

atomic number

b. The valence, an indication of bonding capacity, is the number of unpaired electrons that an atom has in its valence shell.

c. The valence of an atom is most related to the chemical behavior of an atom because it is an indication of the number of bonds the atom will make, or the number of electrons the atom must share in order to reach a filled valence shell.

3. Ionic and nonpolar covalent bonds represent the two ends of a continuum of electron sharing between atoms in a molecule. In ionic bonds the electrons are completely pulled away from one atom by the other, creating negatively and positively charged ions (anions and cations). In nonpolar covalent bonds the electrons are equally shared between two atoms. Polar covalent bonds form when a more electronegative atom pulls the shared electrons closer to it, producing a partial negative charge associated with that portion of the molecule and a partial positive charge associated with the atom from which the electrons are pulled.

ANSWERS TO TEST YOUR KNOWLEDGE

Multiple Choice:

1. b	6. e	11. b	16. c	21. b
2. d	7. d	12. c	17. d	22. c
3. a	8. b	13. c	18. b	23. d
4. e	9. e	14. b	19. a	24. e
5. a	10. c	15. e	20. c	25. e

CHAPTER 3: WATER AND THE FITNESS OF THE ENVIRONMENT

■ INTERACTIVE QUESTIONS

3.1

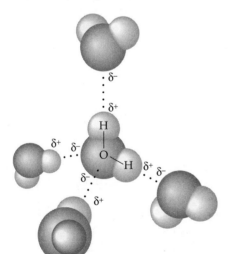

3.2 a. polar water molecules
b. absorbed
c. released
d. specific heat
e. heat of vaporization
f. evaporative cooling
g. solar heat
h. rain
i. ice forms

3.3 a. olive oil: hydrophobic nonpolar
b. sugar: hydrophilic polar
c. salt: hydrophilic ionic
d. candle wax: hydrophobic nonpolar

3.4 a. The molecular mass of $C_3H_6O_3$ is 90 d, the combined atomic masses of its atoms. A mole of lactic acid 90 g. A 0.5 M solution would require ½ mol, or 45 g.
b. 228 grams of NaCl. (57 g × 2 mol × 2 to make 2 liters)

3.5

[H⁺]	[OH⁻]	pH	Acidic, Basic, or Neutral?
10^{-3}	10^{-11}	3	acidic
10^{-8}	10^{-6}	8	basic
10^{-7}	10^{-7}	7	neutral
10^{-1}	10^{-13}	1	acidic

3.6 carbonic acid   bicarbonate   hydrogen ion

$$H_2CO_3 \rightleftharpoons HCO_3^- + H^+$$

    H^+donor H^+ acceptor

a. Bicarbonate acts as a base to accept excess H^+ ions when the pH starts to fall; the reaction moves to the left.

b. When the pH rises, H^+ ions are donated by carbonic acid, and the reaction shifts to the right.

SUGGESTED ANSWERS TO STRUCTURE YOUR KNOWLEDGE

1. **a.** Cohesion, adhesion
   b. A water column is pulled up through plant vessels.

c. Heat is absorbed or released when hydrogen bonds break or form. Water absorbs or releases a large quantity of heat for each degree of temperature change.

d. High heat of vaporization

e. Solar heat is dissipated from tropical seas.

f. Evaporative cooling

g. Evaporation of water cools surfaces of plants and animals.

h. Hydrogen bonds in ice space water molecules apart, making ice less dense.

i. Floating ice insulates bodies of water so they don't freeze solid.

j. Versatile solvent

k. Polar water molecules surround and dissolve ionic and polar solutes.

2.   See concept map below.

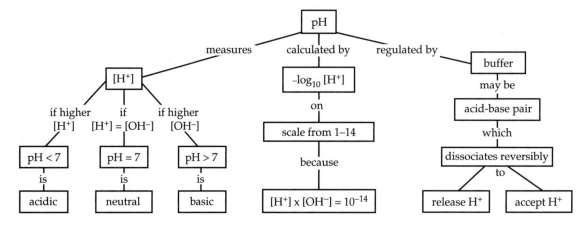

ANSWERS TO TEST YOUR KNOWLEDGE

1. c	**6.** d	**11.** b	**16.** a	**21.** b
2. a	**7.** e	**12.** d	**17.** b	**22.** e
3. e	**8.** b	**13.** c	**18.** d	
4. e	**9.** e	**14.** c	**19.** d	
5. b	**10.** d	**15.** e	**20.** e	

CHAPTER 4: CARBON AND THE MOLECULAR DIVERSITY OF LIFE

■ INTERACTIVE QUESTIONS

4.1 Ethanol and dimethyl ether, structural isomers, have the same number and kinds of atoms but a different bonding sequence and very different properties. Maleic acid and fumaric acid are geometric isomers whose double bonds fix the spatial arrangement of the molecule. Maleic acid is the *cis* isomer; both non-hydrogen groups are on the same side of the double bond. Although the enantiomers L- and D-lactic acid look similar in a flat representation of their structures, they are not superimposable.

SUGGESTED ANSWERS TO STRUCTURE YOUR KNOWLEDGE

1.

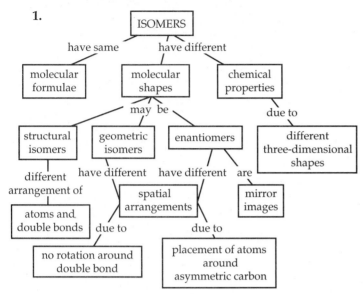

2.

Functional Group	Molecular Formula	Names and Characteristics of Organic Compounds Containing Functional Group
Hydroxyl	—OH	Alcohols; polar group
Carbonyl	$>C\!=\!O$	Aldehyde or ketone; polar group
Carboxyl	—COOH	Carboxylic acid; release H^+
Amino	—NH$_2$	Amines; basic, accept H^+
Sulfhydryl	—SH	Thiols; cross-links stabilize protein structure
Phosphate	—OPO$_3^{2-}$	Organic phosphates; used in energy transfers, acidic

ANSWERS TO TEST YOUR KNOWLEDGE

Multiple Choice:

1. c	**4.** a	**7.** d
2. b	**5.** d	**8.** d
3. e	**6.** a	**9.** c

Matching:

1. a, c	**4.** e	**7.** b, f	**10.** a, d
2. b, f	**5.** e	**8.** e	**11.** e
3. a, d, e	**6.** a, c, d, e	**9.** a, c	**12.** c

CHAPTER 5: THE STRUCTURE AND FUNCTION OF MACROMOLECULES

■ INTERACTIVE QUESTIONS

5.1 **a.** hydroxyl
b. carbonyl
c. aldose
d. ketose
e. rings

5.2

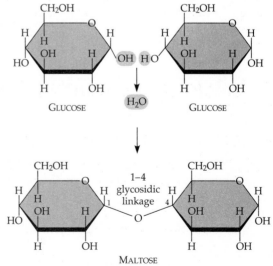

5.3 **a.** monosaccharides
   b. $(CH_2O)_n$
   c. energy compounds
   d. carbon skeletons, monomers
   e. glycosidic linkages
   f. disaccharides
   g. polysaccharides
   h. glycogen
   i. animals
   j. starch
   k. cellulose
   l. chitin

5.4

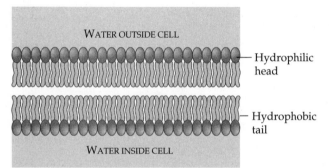

5.5 **a.** fats, triacylglycerides
   b. phospholipids
   c. glycerol
   d. fatty acids
   e. unsaturated: $C=C$ bonds
   f. saturated: no $C=C$, all possible $C-H$ bonds
   g. phosphate group
   h. cell membranes
   i. steroids
   j. cell membrane component (cholesterol), hormones

5.6

a.
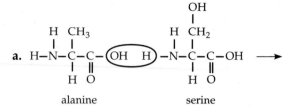
alanine   serine

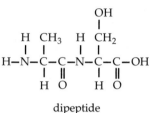

dipeptide

b. serine's R group is polar; alanine's R group is nonpolar
c. a polypeptide backbone

5.7 **a.** hydrogen bond
   b. hydrophobic and van der Waals interactions
   c. disulfide bridge
   d. ionic bond
   These interactions between R groups produce tertiary structure.

5.8 **a.** A change in pH alters the availability of H^+, OH^-, or other ions, thereby disrupting the hydrogen bonding and ionic bonds that maintain protein shape.
   b. A protein in an organic solvent would turn inside out as the hydrophilic regions became clustered on the inside of the molecule and the hydrophobic regions interacted with the nonpolar solvent.
   c. The return to its functional shape indicates that a protein's conformation is intrinsically determined by its primary structure—the sequence of its amino acids.

5.9

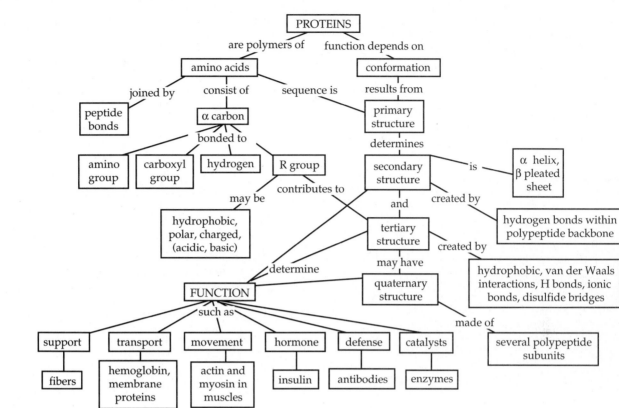

5.10   DNA → RNA → Protein

5.11   a.

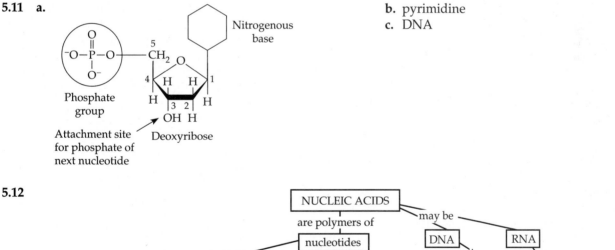

b. pyrimidine
c. DNA

5.12

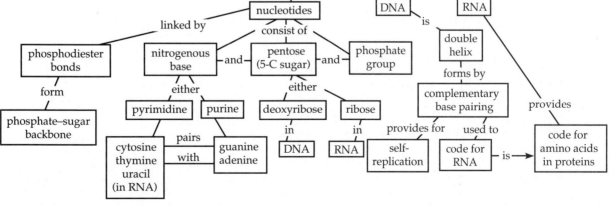

SUGGESTED ANSWERS TO STRUCTURE YOUR KNOWLEDGE

1. The primary structure of a protein is the specific, genetically coded sequence of amino acids in a polypeptide chain. The secondary structure involves the coiling (α helix) or folding (β pleated sheet) of the protein, stabilized by hydrogen bonds along the polypeptide backbone. The tertiary structure involves interactions between the side chains of amino acids and produces a characteristic three-dimensional shape for a protein. Quaternary structure occurs in proteins composed of more than one polypeptide chain.

2. **a.** amino acid (glycine)
   b. fatty acid
   c. nitrogenous base, purine (adenine)
   d. glycerol
   e. phosphate group
   f. pentose (ribose)
   g. sugar (triose)

1. b, d   **3.** c, e, f   **5.** c   **7.** e, f
2. a   **4.** f, g   **6.** a   **8.** b

ANSWERS TO TEST YOUR KNOWLEDGE

Matching:

1. A   **4.** C   **7.** B   **10.** A
2. B   **5.** C   **8.** C
3. D   **6.** D   **9.** A

Multiple Choice:

1. e   **6.** b   **11.** c   **16.** a   **21.** d
2. c   **7.** c   **12.** d   **17.** c   **22.** a
3. a   **8.** a   **13.** b   **18.** b   **23.** d
4. e   **9.** d   **14.** c   **19.** c   **24.** e
5. c   **10.** c   **15.** d   **20.** b

Fill in the Blanks:

1. F. Sanger   **6.** quaternary
2. guanine   **7.** glycogen
3. purines   **8.** phospholipids
4. nucleotide   **9.** cellulose
5. primary structure **10.** chaperonins (amino acid sequence)

CHAPTER 6: A TOUR OF THE CELL

■ INTERACTIVE QUESTIONS

6.1 **a.** the study of cell structure
b. the internal ultrastructure of cells
c. the three-dimensional surface topography of a specimen
d. Light microscopy enables study of living cells and may introduce fewer artifacts than do TEM and SEM.

6.2 **a.** 10^2, or 100 times the surface area
b. 10^3, or 1000 times the volume

6.3 The genetic instructions for specific proteins are transcribed from DNA into messenger RNA (mRNA), which then passes into the cytoplasm to complex with ribosomes where it is translated into the primary structure of proteins.

6.4 **a.** smooth ER—in different cells may house enzymes that synthesize lipids; metabolize carbohydrates; detoxify drugs and alcohol; store and release calcium ions in muscle cells
b. nuclear envelope—double membrane that encloses nucleus; pores regulate passage of materials
c. rough ER—attached ribosomes produce proteins that enter cisternae; produces secretory proteins and membranes
d. transport vesicle—carries products of ER and Golgi apparatus to various locations

e. Golgi apparatus—processes products of ER; makes polysaccharides, packages products in vesicles targeted to specific locations
f. plasma membrane—selective barrier that regulates passage of materials into and out of cell
g. lysosome—houses hydrolytic enzymes to digest macromolecules

6.5

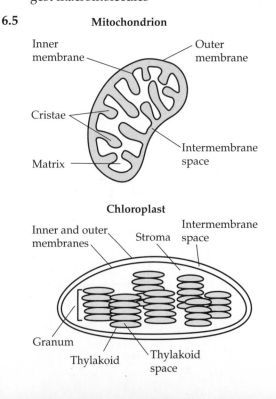

Mitochondrion

Inner membrane
Outer membrane
Cristae
Intermembrane space
Matrix

Chloroplast

Inner and outer membranes
Stroma
Intermembrane space
Granum
Thylakoid
Thylakoid space

6.6 Peroxisomes do not bud from the endomembrane system, but grow by incorporating proteins and lipids from the cytosol; they increase in number by dividing.

6.7 **a.** hollow tube, formed from columns of tubulin dimers; 25-nm diameter
b. cell shape and support (compression resistant), tracks for moving organelles, chromosome movement, beating of cilia and flagella
c. two twisted chains of actin molecules; 7-nm diameter
d. muscle contraction, maintain (tension bearing) and change cell shape, pseudopod movement, cytoplasmic streaming, sol-gel transformations
e. supercoiled fibrous proteins of keratin family; 8–12 nm
f. reinforce cell shape, anchor nucleus; nuclear lamina

6.8 **Two adjacent plant cells**

Middle lamella
Primary cell wall
Secondary cell wall
Plasma membrane
Cytoplasm
Vacuole   Plasmodesma

SUGGESTED ANSWERS TO STRUCTURE YOUR KNOWLEDGE

1. **a.** nucleus, chromosomes, centrioles, microtubules (spindle), microfilaments (actin-myosin aggregates pinch apart cell)
   b. nucleus, chromosomes, DNA → mRNA → ribosomes → enzymes and other proteins
   c. mitochondria
   d. ribosomes, rough and smooth ER, Golgi apparatus, vesicles
   e. smooth ER (peroxisomes also detoxify substances)
   f. lysosomes, food vacuoles
   g. peroxisomes
   h. cytoskeleton: microtubules, microfilaments, intermediate filaments; extracellular matrix
   i. cilia and flagella (microtubules), microfilaments (actin) in muscles and pseudopodia
   j. plasma membrane, vesicles
   k. desmosomes, tight and gap junctions, ECM

2. **a.** structural support, middle lamella glues cells together
   b. storage, waste disposal, protection, growth
   c. photosynthesis, production of carbohydrates
   d. starch storage
   e. cytoplasmic connections between cells

3. **a.** rough endoplasmic reticulum
   b. smooth endoplasmic reticulum
   c. chromatin
   d. nucleolus
   e. nuclear envelope
   f. nucleus
   g. ribosomes
   h. Golgi apparatus
   i. plasma membrane
   j. mitochondrion
   k. lysosome
   l. cytoskeleton
   m. microtubules
   n. intermediate filaments
   o. microfilaments
   p. microvilli
   q. peroxisome
   r. centrosome (contains a pair of centrioles)
   s. flagellum

4.

DNA ──transcription of gene──→ mRNA ──moves into cytosol and complexes with──→ RIBOSOME

mRNA translated into polypeptide; polypeptide moves into cisternal space; may be bonded to carbohydrate to form glycoprotein

becomes attached to

TRANSPORT VESICLES pinch off and join *cis* face of ←── ROUGH ER ←──

GOLGI APPARATUS polypeptide may be modified ──TRANSPORT VESICLES from *trans* face leave and fuse with──→ PLASMA MEMBRANE

ANSWERS TO TEST YOUR KNOWLEDGE

Multiple Choice:

1. c	**5.** a	**9.** b	**13.** b	**17.** c
2. b	**6.** d	**10.** d	**14.** c	**18.** e
3. d	**7.** a	**11.** e	**15.** e	**19.** b
4. e	**8.** c	**12.** b	**16.** d	**20.** a

Fill in the Blanks:

1. transport vesicle		**6.** basal body
2. cristae		**7.** cytosol
3. extracellular matrix		**8.** cytoskeleton
4. peroxisomes		**9.** tight junction
5. grana		**10.** tonoplast

CHAPTER 7: MEMBRANE STRUCTURE AND FUNCTION

■ INTERACTIVE QUESTIONS

7.1 **a.** phospholipid bilayer
b. hydrocarbon tail—hydrophobic
c. phosphate head—hydrophilic
d. hydrophobic region of protein
e. hydrophilic region of protein

7.2 **a.** In hybrid human/mouse cells, membrane proteins rapidly intermingle.
b. The cell may increase the proportion of unsaturated phospholipids in its membrane.

7.3 transport, enzymatic activity, signal transduction, intercellular attachment, cell-cell recognition, attachment to cytoskeleton and ECM

7.4 Ions and larger polar molecules, such as glucose, are impeded by the hydrophobic center of the plasma membrane's lipid bilayer. Passage through the center of a lipid bilayer is not easy even for small, polar water molecules.

7.5 Side A initially has fewer free water molecules; side B initially has more; water will move from B to A.

7.6 **a.** The protists will gain water from their hypotonic environment.
b. They may have membranes that are less permeable to water and contractile vacuoles that expel excess water.

7.7 **a.** isotonic   **b.** hypotonic

7.8 Although it may speed diffusion, facilitated diffusion is still passive transport because the solute is moving down its concentration gradient; the process is driven by the concentration gradient and not energy expended by the cell.

7.9 Three sodium ions are pumped out of the cell for every two potassium ions pumped in, resulting in a net movement of positive charge from the cytoplasm to the extracellular fluid.

7.10 **a.** Human cells use receptor-mediated endocytosis to take in cholesterol.
b. LDL receptor proteins in the plasma membrane are defective and low-density lipoproteins cannot bind and be transported into the cell.

SUGGESTED ANSWERS TO STRUCTURE YOUR KNOWLEDGE

1.

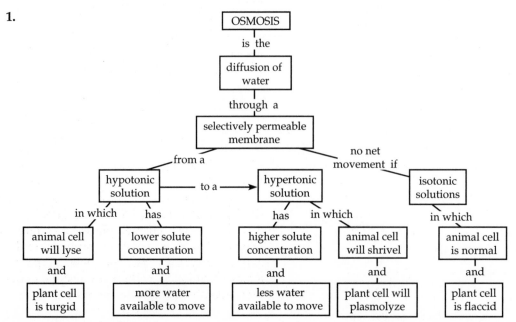

2. **a.** II represents facilitated diffusion. The solute is moving through a transport protein and down its concentration gradient. The cell does not expend energy in this transport. Polar molecules and ions may move by facilitated diffusion.
    b. III represents active transport because the solute is clearly moving against its concentration gradient and the cell is expending ATP to drive this transport against the gradient.
    c. I illustrates diffusion through the lipid bilayer. The solute molecules must be nonpolar or very small polar molecules.
    d. I and II. Both diffusion and facilitated diffusion are considered passive transport because the solute moves down its concentration gradient and the cell does not expend energy in the transport.

ANSWERS TO TEST YOUR KNOWLEDGE

Multiple Choice:

1. c	**6.** a	**11.** e	**16.** b	**21.** d
2. b	**7.** e	**12.** a	**17.** e	**22.** b
3. a	**8.** c	**13.** b	**18** c	**23.** d
4. e	**9.** b	**14.** c	**19.** e	**24.** c
5. d	**10.** a	**15.** b	**20.** a*	**25.** d

*Explanation for answer to question 20: This problem involves both osmosis and diffusion. Although the solution is initially isotonic, glucose will diffuse down its concentration gradient until it reaches dynamic equilibrium with a 1.5 *M* concentration on both sides. The increasing solute concentration on side A will cause water to move into this hypertonic side, and the water level will rise.

CHAPTER 8: AN INTRODUCTION TO METABOLISM

■ INTERACTIVE QUESTIONS

8.1 **a.** capacity to cause change
  b. kinetic
  c. motion
  d. potential
  e. position
  f. conserved
  g. created nor destroyed
  h. first
  i. transformed or transferred
  j. entropy
  k. second

8.2

	System with high free energy	System with low free energy
Stability	low	high
Spontaneous	change will be	change will not be
Equilibrium	moves toward	is at
Work capacity	high	low

8.3

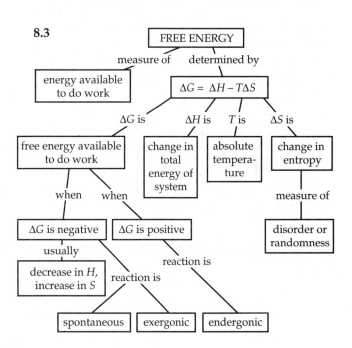

8.4 **a.** adenine
  b. ribose
  c. three phosphate groups
  d. A hydrolysis reaction breaks the terminal phosphate bond and releases a molecule of inorganic phosphate: $ATP + H_2O \rightarrow ADP + P_i$
  e. The negatively charged phosphate groups are crowded together, and their mutual repulsion makes this area instable. The chemical change to a more stable state of lower free energy accounts for the relatively high release of energy.

8.5 **a.** free energy
    b. transition state
    c. E_A (free energy of activation) without enzyme
    d. E_A with enzyme
    e. ΔG of reaction

8.6 Also see text figure 8.17, page 153

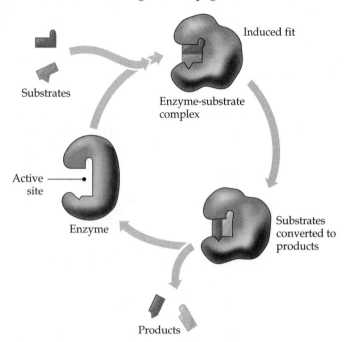

8.7 A competitive inhibitor would mimic the shape of the substrates and compete with them for the active site. A noncompetitive inhibitor would be a shape that could bind to another site on the enzyme molecule and would change the conformation of the active site such that the substrates could no longer fit.

8.8 ATP would act as an inhibitor to catabolic pathways, slowing the breakdown of fuel molecules if sufficient energy is available in the cell. ATP may act as an activator of anabolic pathways that store resources in more complex molecules.

SUGGESTED ANSWERS TO STRUCTURE YOUR KNOWLEDGE

1. Metabolism is the totality of chemical reactions that take place in living organisms. To create and maintain the structural order required for life requires an input of free energy—from sunlight for photosynthetic organisms and from energy-rich food molecules for other organisms. A cell couples catabolic, exergonic reactions ($-\Delta G$) with anabolic, endergonic reactions ($+\Delta G$), using ATP as the primary energy shuttle between the two.

2. Enzymes are essential for metabolism because they lower the activation energy of the specific reactions they catalyze and allow those reactions to occur extremely rapidly at a temperature conducive to life. By regulating the enzymes it produces, a cell can regulate which of the myriad of possible chemical reactions take place at any given time. Metabolic control also occurs through allosteric regulation and feedback inhibition. The compartmental organization of a cell facilitates a cell's metabolism.

ANSWERS TO TEST YOUR KNOWLEDGE

Multiple Choice:

1. c	6. c	11. b	16. d	21. e
2. b	7. a	12. c	17. b	22. a
3. a	8. e	13. b	18. d	
4. e	9. e	14. b	19. c	
5. c	10. c	15. e	20. b	

Fill in the Blanks:

1. metabolism
2. anabolic
3. kinetic
4. allosteric
5. entropy
6. free energy of activation
7. competitive inhibitors
8. coenzymes
9. feedback inhibition
10. phosphorylated intermediates

CHAPTER 9: CELLULAR RESPIRATION: HARVESTING CHEMICAL ENERGY

■ INTERACTIVE QUESTIONS

9.1 $C_6H_{12}O_6$, 6 CO_2, energy (ATP + heat)

9.2 **a.** oxidized
    b. oxidizing agent
    c. reduced

9.3 **a.** oxygen
    b. glucose
    c. Some is stored in ATP and some is released as heat.

9.4 **a.** electron acceptor or oxidizing agent
    b. NADH

9.5 **a.** glycolysis: glucose $\rightarrow$ pyruvate (not technically considered part of cellular respiration because it does not require O_2)
    b. citric acid cycle
    c. oxidative phosphorylation: electron transport and chemiosmosis
    d. substrate-level phosphorylation
    e. substrate-level phosphorylation
    f. oxidative phosphorylation

The top two arrows show electrons carried by NADH (and FADH$_2$) to the electron transport chain.

9.6 **a.** 2ATP

b. 2 three-carbon sugars (glyceraldehyde-3-phosphate)

c. 2 NAD$^+$

d. 2 NADH + 2 H$^+$

e. 4 ATP

f. 2 pyruvate

9.7 **a.** pyruvate

b. CO$_2$

c. NADH + H$^+$

d. Coenzyme A

e. acetyl CoA

f. oxaloacetate

g. citrate

h. CO$_2$

i. NADH + H$^+$

j. CO$_2$

k. NADH + H$^+$

l. ATP

m. FADH$_2$

n. NADH + H$^+$

9.8 **a.** intermembrane space

b. inner mitochondrial membrane

c. mitochondrial matrix

d. electron transport chain

e. NADH + H$^+$

f. NAD$^+$

g. FADH$_2$

h. chemiosmosis

i. 2 H$^+$ + ½ O$_2$

j. H$_2$O

k. ATP synthase

l. ADP + Ⓟ$_i$

m. ATP

9.9 **a.** −2   **f.** 2

b. 4   **g.** 6

c. citric acid cycle   **h.** 2

d. 32 or 34   **i.** 2

e. 38   **j.** 2

9.10 Respiration yields up to 19 times more ATP than does fermentation. By oxidizing pyruvate to CO$_2$ and passing electrons from NADH (and FADH$_2$) through the electron transport chain, respiration can produce a maximum of 38 ATP compared to the 2 net ATP that are produced by fermentation.

SUGGESTED ANSWERS TO STRUCTURE YOUR KNOWLEDGE

1. See Interactive Questions 9.5, 9.6, 9.7, and 9.8.

2.

Process	Main Function	Inputs	Output
Glycolysis	Oxidation of glucose to 2 pyruvate, 2 ATP net	glucose 2 ATP 2 NAD$^+$ 4 ADP + Ⓟ$_i$	2 pyruvate 4 ATP (2 net) 2 NADH + 2H$^+$ 2 H$_2$O
Pyruvate to acetyl CoA	Oxidation of pyruvate to acetyl CoA, which then enters citric acid cycle	2 pyruvate 2 CoA 2 NAD$^+$	2 acetyl CoA 2 CO$_2$ 2 NADH + 2H$^+$
Citric acid cycle	Acetyl CoA is combined with oxaloacetate to produce citrate, which is cycled back to oxaloacetate as redox reactions produce NADH and FADH$_2$, ATP is formed by substrate-level phosphorylation, and CO$_2$ is released.	2 acetyl CoA 2 ADP + Ⓟ$_i$ 6 NAD$^+$ 2 FAD	2 CoA 4 CO$_2$ 2 ATP 6 NADH + 6H$^+$ 2 FADH$_2$
Oxidative phosphorylation (Electron transport and chemiosmosis)	NADH (from glycolysis and citric acid) and FADH$_2$ transfer electrons to electron transport chain. In a series of redox reactions, H$^+$ is pumped into intermembrane space, and electrons are delivered to ½ O$_2$. Proton-motive force drives H$^+$ through ATP synthase to make ATP.	10 NADH + H$^+$ 2 FADH$_2$ H$^+$ + O$_2$ 34 ADP + Ⓟ$_i$	10 NAD$^+$ 2 FAD H$_2$O 34 ATP
Fermentation	Anaerobic catabolism: glycolysis followed by oxidation of NADH to NAD$^+$ so glycolysis can continue. Pyruvate is either reduced to ethyl alcohol and CO$_2$ or to lactate.	See glycolysis above 2 pyruvate 2 NADH	2 ATP 2 NAD$^+$ 2 ethanol and 2 CO$_2$ or 2 lactate

3.

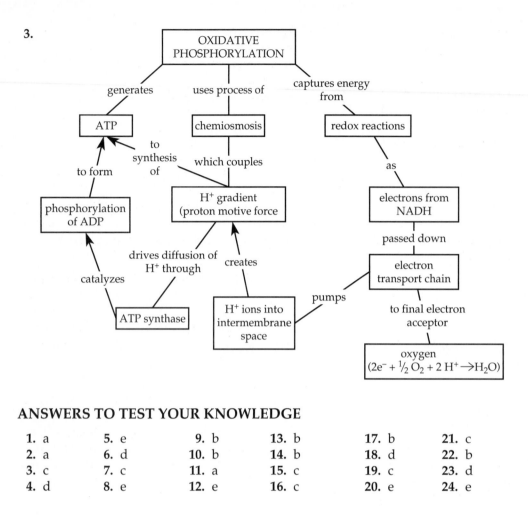

ANSWERS TO TEST YOUR KNOWLEDGE

1. a	**5.** e	**9.** b	**13.** b	**17.** b	**21.** c	**25.** a
2. a	**6.** d	**10.** b	**14.** b	**18.** d	**22.** b	**26.** e
3. c	**7.** c	**11.** a	**15.** c	**19.** c	**23.** d	**27.** c
4. d	**8.** e	**12.** e	**16.** c	**20.** e	**24.** e	**28.** d

CHAPTER 10: PHOTOSYNTHESIS

■ INTERACTIVE QUESTIONS

10.1   **a.** outer membrane
   b. granum
   c. inner membrane
   d. thylakoid space
   e. thylakoid
   f. stroma

10.2   **a.** light
   b. H_2O
   c. light reactions in thylakoid membranes
   d. O_2
   e. ATP
   f. NADPH
   g. CO_2
   h. Calvin cycle in stroma
   i. CH_2O (sugar)

10.3   blue light; the shorter the wavelength, the greater the energy

10.4   The solid line is the absorption spectrum; the dotted line is the action spectrum. Some wavelengths of light, particularly in the blue and the yellow-orange range, result in a higher rate of photosynthesis than would be indicated by the absorption of those wavelengths by chlorophyll *a*. These differences are partially accounted for by accessory pigments, such as chlorophyll *b* and the carotenoids, which absorb light energy from different wavelengths and make that energy available to drive photosynthesis.

10.5   A photosystem contains light-harvesting complexes of pigment molecules (chlorophyll *a*, chlorophyll *b*, and carotenoids) bound to particular proteins and a reaction center, which includes two chlorophyll *a* molecules (P700 or P680) and a primary electron acceptor.

10.6 **a.** photosystem II
b. photosystem I
c. water (H_2O)
d. oxygen ($\frac{1}{2} O_2$)
e. P680, reaction-center chlorophyll *a*
f. primary electron acceptor
g. electron transport chain
h. photophosphorylation by chemiosmosis
i. ATP
j. P700, reaction-center chlorophyll *a*
k. primary electron acceptor
l. $NADP^+$ reductase
m. NADPH
ATP and NADPH provide the chemical energy and reducing power for the Calvin cycle.

10.7 **a.** Ferredoxin (Fd) passes the electrons to the cytochrome complex in the electron transport chain, from which they return to P700.
b. Electrons from P680 are not passed to P700. Without the oxidizing agent P680, water is not split. Fd does not pass electrons to $NADP^+$ reductase to form NADPH.
c. Electrons do pass down the electron transport chain, and the energy released by their "fall" drives photophosphorylation.

10.8 **a.** in the thylakoid space (pH of about 5)
b. (1) transport of protons into the thylakoid space as Pq transfers electrons to the cytochrome complex; (2) protons from the splitting of water remain in the thylakoid space; (3) removal of H^+ in the stroma during the reduction of $NADP^+$.

10.9 **a.** carbon fixation
b. reduction
c. regeneration of CO_2 acceptor (RuBP)
d. $3 CO_2$
e. ribulose bisphosphate (RuBP)
f. rubisco
g. 3-phosphoglycerate
h. $6 ATP \rightarrow 6 ADP$
i. 1,3-bisphosphoglycerate
j. $6 NADPH \rightarrow 6 NADP^+$
k. $6 \, \textcircled{P}_i$
l. glyceraldehyde-3-phosphate (G3P)
m. G3P
n. glucose and other organic compounds
o. $3 ATP \rightarrow 3 ADP$

10.10 Photorespiration may be an evolutionary relic from the time when there was little O_2 in the atmosphere and the ability of rubisco to distinguish between O_2 and CO_2 was not critical. Now, in our oxygen-rich atmosphere, photorespiration seems to be an agricultural liability.

10.11 **a.** in the bundle-sheath cells
b. Carbon is initially fixed into a four-carbon compound in the mesophyll cells by PEP carboxylase. When this compound is broken down in the bundle-sheath cells, CO_2 is maintained at a high enough concentration that rubisco does not accept O_2 and cause photorespiration.

SUGGESTED ANSWERS TO STRUCTURE YOUR KNOWLEDGE

1.

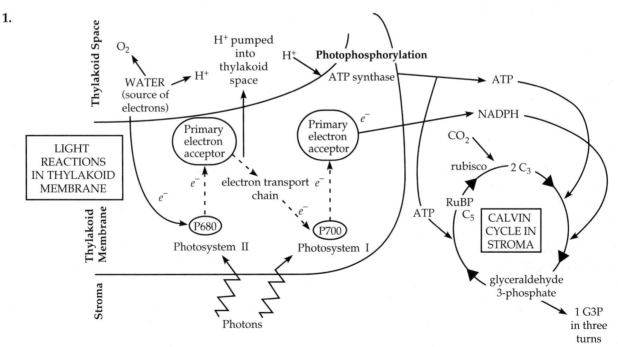

2.

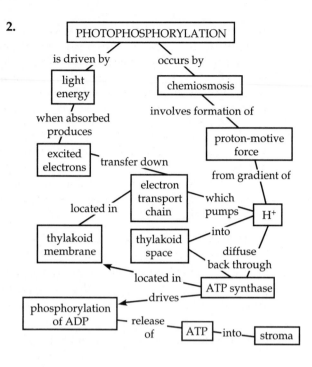

Multiple Choice:

1. d	7. e	13. d	19. e	25. b
2. a	8. c	14. d	20. d	26. a
3. b	9. a	15. d	21. d	27. d
4. a	10. e	16. b	22. e	28. a
5. b	11. a	17. c	23. c	
6. b	12. b	18. d	24. a	

CHAPTER 11: CELL COMMUNICATION

■ INTERACTIVE QUESTIONS

11.1 **a.** Yes

b. In plants, hormones (often called *growth regulators*) may reach their target cells by traveling through vessels, through cells via plasmodesmata, or even through the air as a gas.

11.2 The G protein is also a GTPase enzyme that hydrolyzes its bound GTP to GDP and inactivates itself. It then dissociates from the enzyme it had activated, and that enzyme returns to its original state.

11.3 **a.** signal molecules
b. α helix in the membrane
c. phosphates
d. ATP → ADP
e. tyrosines
f. activated relay proteins
g. cellular responses

11.4 **a.** A target cell has a protein receptor that is specific for that signal.
b. Signals that bind to surface receptors are large, polar, and/or ionic. These hydrophilic molecules cannot pass through the membrane. Intracellular signals are small or hydrophobic and can move through the lipid membrane.

11.5 **a.** A protein kinase transfers a phosphate group from ATP to a protein; adding a charged phosphate group causes a conformation change that usually activates the protein.
b. A protein phosphatase removes a phosphate group from a protein, usually inactivating the protein.
c. A phosphorylation cascade is a series of protein kinase relay molecules that are sequentially phosphorylated.

11.6 **a.** signal molecule (first messenger)
b. G-protein-linked receptor
c. activated G protein (GTP bound)
d. adenylyl cyclase
e. ATP
f. cAMP (second messenger)
g. protein kinase A
h. phosphorylation cascade to cellular response

11.7 **a.** signal molecule
b. G protein
c. membrane phospholipid
d. IP$_3$
e. Ca^{2+}
f. endoplasmic reticulum

11.8 **a.** Signal molecules reversibly bind to receptors, and when they leave a receptor, the receptor reverts to its inactive form. The concentration of signal molecules influences how many are bound at any time.

b. Activated G proteins are inactivated when the GTPase portion of the protein converts GTP to GDP.

c. Phosphodiesterase converts cAMP to AMP, thus damping out this second messenger.

d. Protein phosphatases remove phosphate groups from activated proteins. The balance of active protein kinases and active phosphatases regulates the activity of many proteins.

SUGGESTED ANSWERS TO STRUCTURE YOUR KNOWLEDGE

1. Cell signaling is essential for communication between cells. Unicellular organisms use signals to relay environmental or reproductive information. Communication in multicellular organisms allows for the development and coordination of specialized cells. Extracellular signals control the crucial activities of cells, such as cell division, differentiation, metabolism, and gene expression.

2. Cell signaling occurs through signal transduction pathways that include reception, transduction, and response. First a signal molecule binds to a specific receptor. The message is transduced as the activated receptor activates a protein that may relay the message through a sequence of activations, finally leading to the activation of proteins that produce the specific cellular response.

3. The sequence described in the previous answer results in the activation of cellular proteins. When the signal is transduced to activate a transcription factor, however, the cellular response is a change in gene expression and the production of new proteins.

4. In an enzyme cascade, each step in the pathway activates multiple substrates of the next step, thus amplifying the original message to produce potentially millions of activated proteins and thus a large cellular response to a few signals.

ANSWERS TO TEST YOUR KNOWLEDGE

1. c	**4.** b	**7.** c	**10.** a	**13.** b
2. d	**5.** e	**8.** e	**11.** c	**14.** d
3. a	**6.** d	**9.** c	**12.** e	

CHAPTER 12: THE CELL CYCLE

■ INTERACTIVE QUESTIONS

12.1 **a.** 46
  b. 23
  c. 92

12.2 **a.** Growth—most organelles and cell components are produced continuously throughout these subphases.
  b. DNA synthesis

12.3 Refer to main text Fig. 12.6 for chromosome diagrams.
  a. G$_2$ of interphase
  b. prophase
  c. prometaphase
  d. metaphase
  e. anaphase
  f. telophase and cytokinesis
  g. centrosomes (with centrioles)
  h. chromatin (duplicated)
  i. nuclear envelope
  j. nucleolus
  k. early mitotic spindle
  l. aster
  m. nonkinetochore microtubules
  n. kinetochore microtubules
  o. metaphase plate
  p. spindle
  q. cleavage furrow
  r. nuclear envelope forming

12.4 **a.** MPF is a complex of cyclin and Cdk that initiates mitosis by phosphorylating proteins and other kinases.
  b. MPF concentration is high as it triggers the onset of mitosis but is reduced at the end of mitosis because it depends on the concentration of cyclin in the cell. The Cdk level is constant through the cell cycle, but the level of cyclin varies because active MPF starts a process that degrades cyclin. Thus MPF regulates its own level and can only become active when sufficient cyclin accumulates again during the next interphase.

SUGGESTED ANSWERS TO STRUCTURE YOUR KNOWLEDGE

1. Interphase: 90% of cell cycle; growth and DNA replication.
  • G$_1$ phase: The chromosome consists of long, thin chromatin fibers made of DNA and associated proteins. RNA molecules are being transcribed from genes that are switched on.

- S phase—synthesis of DNA: The chromosome is replicated; two exact copies, called sister chromatids, are produced and held together by proteins along their length. Growth continues.
- G_2 phase: Growth continues.

Mitosis phase: cell division
- Prophase: The sister chromatids become tightly coiled and folded.
- Prometaphase: Kinetochore fibers from opposite ends of the mitotic spindle attach to the kinetochores of the sister chromatids; the chromosome moves toward midline.
- Metaphase: The centromere of the chromosome is aligned at the metaphase plate along with the centromeres of the other chromosomes.
- Anaphase: The sister chromatids separate (now considered to be individual chromosomes) and move to opposite poles.
- Telophase: Chromatin fiber of chromosome uncoils and is surrounded by reforming nuclear membrane.

2.

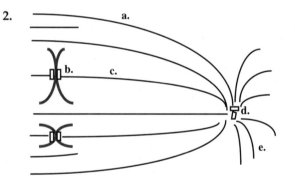

a. nonkinetochore microtubules: push poles apart by "walking" past microtubules from the opposite pole
b. kinetochore: protein and DNA structure in region of centromere where microtubules attach
c. kinetochore microtubules: move chromosomes to metaphase plate and separate chromosomes as motor proteins of kinetochores "walk" toward the pole and the microtubules disassemble
d. centrosome and centrioles: region of mitotic spindle formation; organizes microtubules
e. aster: radiating spindle fibers (in animal cells)

3.　a. anaphase
　　b. interphase
　　c. late telophase
　　d. metaphase

ANSWERS TO TEST YOUR KNOWLEDGE

Fill in the Blank:

1. G_0	5. S phase	9. prometaphase
2. anaphase	6. metaphase	10. G_1 phase
3. prophase	7. telophase	
4. cytokinesis	8. prophase	

Multiple Choice:

1. d	5. b	9. b	13. a
2. c	6. c	10. a	14. e
3. d	7. c	11. c	15. a
4. e	8. c	12. e	16. d

CHAPTER 13: MEIOSIS AND SEXUAL LIFE CYCLES

■ INTERACTIVE QUESTIONS

13.1　a. 14; 7
　　　b. 28; 1
　　　c. 56. Sister chromatids are produced when a chromosome duplicates and are joined at the centromere. Nonsister chromatids are chromatids on different homologous chromosomes.

13.2　a. meiosis
　　　b. fertilization
　　　c. zygote
　　　d. gametes
　　　e. fertilization
　　　f. 2n
　　　g. zygote
　　　h. meiosis

　　　i. meiosis
　　　j. spores
　　　k. mitosis
　　　l. gametes
　　　m. fertilization
　　　n. zygote

13.3　2^{23}, approximately 8 million

13.4　a. metaphase II
　　　b. prophase I
　　　c. anaphase I
　　　d. interphase
　　　e. metaphase I
　　　f. anaphase II
　　　Proper sequence: d.　b.　e.　c.　a.　f.

SUGGESTED ANSWERS TO STRUCTURE YOUR KNOWLEDGE

1. **a.** Chromosome replication, sister chromatids attached at centromere
   b. Synapsis of homologous pairs (tetrads), crossing over at chiasmata
   c. Homologous pairs line up independently at metaphase plate
   d. Homologous pairs of chromosomes separate and homologues move toward opposite poles, sister chromatids remain attached
   e. Haploid set of chromosomes, each consisting of two sister chromatids, aligns at metaphase plate; sister chromosomes not identical due to crossing over
   f. Centromeres separate and nonreplicated chromosomes move to opposite poles

2.

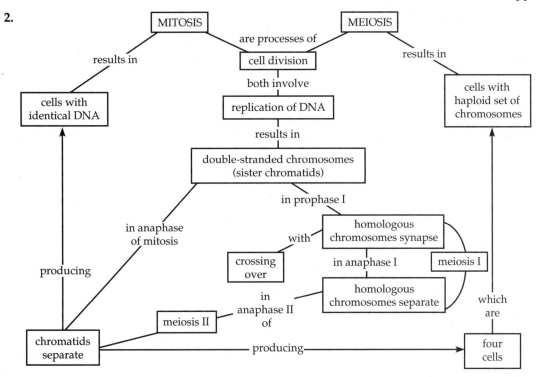

ANSWERS TO TEST YOUR KNOWLEDGE

Multiple Choice:

1. b	**4.** a	**7.** a	**10.** b	**13.** c	**16.** b	**19.** e
2. d	**5.** e	**8.** c	**11.** a	**14.** d	**17.** d	**20.** b
3. e	**6.** b	**9.** b	**12.** b	**15.** e	**18.** b	

CHAPTER 14: MENDEL AND THE GENE IDEA

■ INTERACTIVE QUESTIONS

14.1 **a.** *R*
  b. *r*
  c. F$_1$ Generation
  d. *Rr*
  e. F$_2$ Generation
  f. *R*
  g. *r*
  h. *Rr* ◯
  i. *Rr* ◯
  j. *rr* ⬭
  k. 3 round : 1 wrinkled
  l. 1 *RR* : 2 *Rr* : 1 *rr*

14.2 **a.** all tall (*Tt*) plants
  b. 1:1 tall (*Tt*) to dwarf (*tt*)

14.3 **a.** tall purple plants
b. *TtPp*
c. *TP, Tp, tP, tp*
d.

sperm

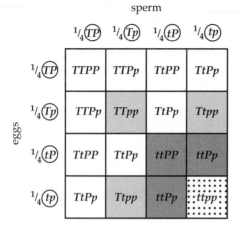

e. 9 tall purple:3 tall white:3 dwarf purple:
1 dwarf white
f. 12:4 or 3:1 tall to dwarf; 12:4 or 3:1 purple to white

14.4 You could determine this in two ways. The probability of getting both recessive alleles in a gamete is ¼, and for two such gametes to join is ¼ × ¼, or ¹⁄₁₆. An easier way is to treat each gene involved as a monohybrid cross. The probability of obtaining a homozygous recessive offspring in a monohybrid cross is ¼. The probability of the offspring being homozygous recessive for both genes is ¼ × ¼, or ¹⁄₁₆.

14.5 **a.** Consider the outcome for each gene as a monohybrid cross. The probability that a cross of *Aa* × *Aa* will produce an *A_* offspring is ¾. The probability that a cross of *Bb* × *bb* will produce a *B_* offspring is ½. The probability that a cross of *cc* × *CC* will produce a *C_* offspring is 1. To have all of these events occur simultaneously, multiply their probabilities: ¾ × ½ × 1 = ⅜.
b. Offspring could be *A_bbC_*, *aaB_C_*, or *A_B_C_*. The genotype *A_B_cc* is not possible.
Probability of *A_bbC_* = ¾ × ½ × 1 = ⅜
Probability of *aaB_C_* = ¼ × ½ × 1 = ⅛
Probability of *A_B_C_* = ¾ × ½ × 1 = ⅜
Probability of offspring showing at least two dominant traits is the total of these independent probabilities, or ⅞. The probability of only one dominant trait (*aabbCc*) is ¼ × ½ × 1 = ⅛.

14.6 **a.** A:*IᴬIᴬ* and *Iᴬi*
b. B:*IᴮIᴮ* and *Iᴮi*
c. AB:*IᴬIᴮ*
d. O:*ii*

14.7 The ratio of offspring from this *Mm Bb* × *Mm Bb* cross would be 9:3:4, a common ratio when one gene is epistatic to another.

Phenotype	Genotype	Ratio
Black	*M_B_*	¾ × ¾ = ⁹⁄₁₆
Brown	*M_bb*	¾ × ¼ = ³⁄₁₆
White	*mm__*	¼ × 1 = ¼ or ⁴⁄₁₆

14.8 **a.** The parental cross produced 25-cm tall F_1 plants, all *AaBbCc* plants with 3 units of 5 cm added to the base height of 10 cm.
b. As a general rule in the polygenic inheritance of a quantitative character, the number of phenotypic classes resulting from a cross of heterozygotes equals the number of alleles involved plus one. In this case, 6 alleles (*AaBbCc*) + 1 = 7. So, there will be 7 different phenotypic classes in the F_2 among the 64 possible combinations of the 8 types of F_1 gametes. These 7 classes will go from 6 dominant alleles (40 cm), 5 dominant (35 cm), 4 dominant (30 cm), and so on, to all 6 recessive alleles (10 cm).

14.9 **a.** This trait is recessive. If it were dominant, then albinism would be present in every generation, and it would be impossible to have albino children with nonalbino (homozygous recessive) parents.
b. father = *Aa*; mother = *Aa*, because neither parent is albino and they have albino offspring (*aa*)
c. mate 1 = *AA* (probably); mate 2 = *Aa*; grandson 4 = *Aa*
d. The genotype of son 3 could be *AA* or *Aa*. If his wife is *AA*, then he could be *Aa* (both his parents are carriers) and the recessive allele never would be expressed in his offspring. Even if he and his wife were both carriers (heterozygotes), there would be a ²⁴³⁄₁₀₂₄ (¾ × ¾ × ¾ × ¾ × ¾) or 24% chance that all five children would be normally pigmented.

14.10 **a.** ¼
b. ⅔. There is a probability that ¾ of the offspring will have a normal phenotype. Of these, ⅔ would be predicted to be heterozygotes and, thus, carriers of the recessive allele.

14.11 Both sets of prospective grandparents must have been carriers. The parents do not have the disorder so they are not homozygous recessive. Thus each has a ⅔ chance of being a heterozygote carrier. The probability that both parents are carriers is ⅔ × ⅔ = ⁴⁄₉; the chance that two heterozygotes will have a recessive homozygous child is ¼. The overall chance that a child will inherit the disease is 4/9 × 4/9 = 1/9. Should this couple have a baby that has the disease, this would establish that they are both carriers, and the chance that a subsequent child would have the disease is ¼.

SUGGESTED ANSWERS TO STRUCTURE YOUR KNOWLEDGE

1. Mendel's law of segregation occurs in anaphase I, when alleles segregate as homologous chromosomes move to opposite poles of the cell. The two cells formed from this division have one-half the number of chromosomes and one copy of each gene. Mendel's law of independent assortment relates to the lining up of synapsed chromosomes at the equatorial plate in a random fashion during metaphase I. Genes on different chromosomes will assort independently into gametes.

2. All alleles operate independently within a cell, coding for their gene products, usually enzymes, as specified by their particular sequence of DNA nucleotides. If both alleles code for functional enzymes or products, then the alleles may be codominant with both traits expressed in the heterozygote, as illustrated by MN or AB blood types. If the recessive allele codes for a nonfunctional enzyme, then the resulting recessive trait may be obvious only when homozygous (such as O blood type in which no carbohydrate molecules are attached to the cell membrane). If one copy of an allele produces sufficient product that the dominant trait is expressed, then we have complete dominance where the heterozygote phenotype is indistinguishable from the dominant homozygote (for example, smooth peas in which the enzyme converts sugar to starch). With incomplete dominance, the heterozygote is distinguishable from both parental types. Genotypes translate into phenotypes by way of biochemical pathways, which are controlled by enzymes coded for by genes. Thus molecular processes produce physical, physiological, and behavioral results.

ANSWERS TO GENETICS PROBLEMS

1. White alleles are dominant to yellow alleles. If yellow were dominant, then you should be able to get white squash from a cross of two yellow heterozygotes.

2. a. ¼   [½ (to get *AA*) × ½ (*bb*)]
   b. ⅛   [¼ (*aa*) × ½ (*BB*)]
   c. ½   [1 (*Aa*) × ½ (*Bb*) × 1 (*Cc*)]
   d. 1/32   [¼ (*aa*) × ¼ (*bb*) × ½ (*cc*)]

3. Since flower color shows incomplete dominance, use symbols such as C^R and C^W for those alleles. There will be six phenotypic classes in the F_2 instead of the normal four classes found in a 9:3:3:1 ratio. You could find the answer with a Punnett square, but multiplying the probabilities of the monohybrid crosses is more efficient.

Tall red	$T\_ C^R C^R$	= ¾ × ¼	3/16
Tall pink	$T\_ C^R C^W$	= ¾ × ½	⅜ or 6/16
Tall white	$T\_ C^W C^W$	= ¾ × ¼	3/16
Dwarf red	$tt\ C^R C^R$	= ¼ × ¼	1/16
Dwarf pink	$tt\ C^R C^W$	= ¼ × ½	⅛ or 2/16
Dwarf white	$tt\ C^W C^W$	= ¼ × ¼	1/16

4. Determine the possible genotypes of the mother and child. Then find the blood groups for the father that could not have resulted in a child with the indicated blood group.
   a. no groups exonerated   d. AB only
   b. A or O   e. B or O
   c. A or O

5. The parents are *CcBb* and *Ccbb*. Right away you know that all *cc* offspring will die and no *BB* black offspring are possible because one parent is *bb*. Only four phenotypic classes are possible. Determine the proportion of each type by applying the law of multiplication.
   Lethal (*cc_ _*) = 1/4 of embryos do not develop
   Normal brown (*CCBb*) = ¼ × ½ = ⅛
   Normal white (*CCbb*) = ¼ × ½ = ⅛
   Deformed brown (*CcBb*) = ½ × ½ = ¼
   Deformed white (*Ccbb*) = ½ × ½ = ¼
   Ratio of viable offspring: 1:1:2:2

6. Father's genotype must be *Pp* since polydactyly is dominant and he has had one normal child. Mother's genotype is *pp*. The chance of the next child having normal digits is ½ or 50% because the mother can only donate a *p* allele and there is a 50% chance that the father will donate a *p* allele.

7. The genotypes of the puppies were ⅜ *B_S_*, ⅜ *B_ss*, ⅛ *bbS_*, and ⅛ *bbss*. Because recessive traits show up in the offspring, both parents had to have had at least one recessive allele for both genes. Black:chestnut occurs in a 6:2 or 3:1 ratio, indicating a heterozygous cross. Solid:spotted occurs in a 4:4 or 1:1 ratio, indicating a cross between a heterozygote and a homozygous recessive. Parental genotypes were *BbSs* × *Bbss*.

8. The 1:1 ratio of the first cross indicates a cross between a heterozygote and a homozygote. The second cross indicates that the hairless hamsters cannot have a homozygous genotype, because then only hairless hamsters would be produced. Let us say that hairless hamsters are *Hh* and

normal-haired are *HH*. The second cross of heterozygotes should yield a 1:2:1 genotype ratio.
The 1:2 ratio indicates that the homozygous recessive genotype may be lethal, with embryos
that are *hh* never developing.

9. **a.** Tail length appears to be a quantitative character. A general rule for the number of alleles in
a cross involving a quantitative character is one
less than the number of phenotypic classes. Here
there are five phenotypic classes, thus four alleles or two gene pairs. In this cross, the phenotypic ratio is approximately 1:4:6:4:1. The ratio
base of 16 also indicates a dihybrid cross. Each
dominant gene appears to add 5 cm in tail
length. The 15-cm pigs are double recessives; the
35-cm pigs are double dominants.
b. A cross of a 15-cm and a 30-cm pig would be
equivalent to *aabb* × *AABb*. Offspring would
have either one or two dominant alleles and
would have tail lengths of 20 or 25 cm.

10. **a.** The black parent would be CC^{ch}. The Himalayan parent could be either $C^h c^h$ or $C^h c$. First
list the alleles in order of dominance: C, C^{ch}, C^h,
c. To approach this problem you should write
down as much of the genotype as you can be
sure about for each phenotype involved. The
parents were black (C_) and Himalayan (C^h_).
The offspring genotypes would be black (C_)
and Chinchilla (C^{ch}_). Right away you can tell
that the black parent must have been CC^{ch} because half of the offspring are Chinchilla; C^{ch} is
dominant over C^h and c and could not have been
present in the Himalayan parent. The genotype
of this parent could be either $C^h C^h$ or $C^h c$. A testcross would be necessary to determine this
genotype.
b. You cannot definitely determine the genotype
of the parents in the second cross. You can, however, eliminate some genotypes. The black parent
could not be *CC* or CC^{ch}, because both the C and
C^{ch} alleles are dominant to C^h, and Himalayan offspring were produced. The Chinchilla parent
could not be $C^{ch}C^{ch}$ for the same reason. One or
both parents had to have C^h as their second allele;
one, but not both, might have c as their second allele.

11. **a.** First figure out possible genotypes: _ _E_ =
golden (any *B* combination with at least 1 *E*),
B_ee = black, *bbee* = brown. All you know of the
first parents at the start is _ _E_ × _ _E_. Since
you see black and brown offspring, you know
that both had to be heterozygous for *Ee*, and at
least one was heterozygous for *Bb* (to get a black
dog). Since black and brown are in a 1:1 ratio

(like a testcross ratio), then one dog was *Bb* and
the other was *bb*. You need to consider the results
from the second cross to know which dog was *Bb*.
b. For the second cross, you now know that Dog 2
is *?bEe* and Dog 3 is *B?ee*. A ratio of approximately 3:1 black to brown looks like a cross of
heterozygotes, so both parents must be *Bb*. So
Dog 3 is *Bbee*, and Dog 2 must be *BbEe*. That
means that Dog 1 must be *bbEe*.

12.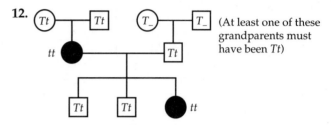

13. Since the parents were true-breeding for two
characters, the F_1 would be dihybrids. Since all
F_1 had red, terminal flowers, those two traits
must be dominant and their genotypes could be
represented as *RrTt*. One would predict an F_2
phenotypic ratio of 9 red, terminal:3 red, axial:3
white, terminal:1 white, axial. If 100 offspring
were counted, one would expect to get 19 (3/16 ×
100) plants with red, axial flowers.

14. Because the F_2 generation is a ratio based on 16,
one can conclude that two genes are involved in
the inheritance of this flower color character. The
parent plants were probably double homozygotes (*AABB* for red and *aabb* for white). The all-
red F_1 must be *AaBb*, and the red color must require at least one dominant allele of both genes.
The 9:6:1 ratio of the F_2 indicates that 9 plants
had at least one dominant allele of both genes to
produce the red color; 1 plant was doubly homozygous recessive and thus white; and the 6
pale purple were caused by having at least one
dominant allele of only one of the two genes involved in flower color (*A_bb* or *aaB_*).

ANSWERS TO TEST YOUR KNOWLEDGE

Matching:

1. I		3. F		5. L		7. H		9. C
2. A		4. E		6. J		8. D		10. K

Multiple Choice:

1. c	5. b	9. b	13. d
2. a	6. d	10. c	14. d
3. e*	7. c	11. d	15. c
4. c	8. d	12. a	16. b

* There are three different ways to get this outcome:
HHT, HTH, THH. Each outcome has a probability of 1/8.

CHAPTER 15: THE CHROMOSOMAL BASIS OF INHERITANCE

 INTERACTIVE QUESTIONS

15.1

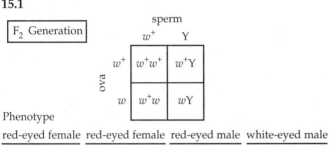

F$_2$ Generation		sperm	
		w^+	Y
ova	w^+	w^+w^+	w^+Y
	w	w^+w	wY

Phenotype

red-eyed female	red-eyed female	red-eyed male	white-eyed male

Genotype

w^+w^+	w^+w	w^+Y	wY

15.2 **a.** tall, purple-flowered and dwarf, white-flowered
b. tall, white-flowered and dwarf, purple-flowered

15.3 If linked genes have their loci close together on the same chromosome, they travel together during meiosis and more parental offspring are produced. Recombinants are the result of crossing over between nonsister chromatids of homologous chromosomes.

15.4 Solving a linkage problem is often a matter of trial and error. Sometimes it helps to lay out the loci with the greatest distance between them and fit the other genes between or on either side by adding or substracting map unit distances.

15.5 The gene is sex-linked, so a good notation is X^C, X^c, and Y so that you will remember that the Y does not carry the gene. Capital C indicates normal sight. Genotypes are:

1. X^{C}Y
2. X^{C}X^c
3. X^{C}X^C or X^{C}X^c (probably X^{C}X^C since 4 sons are X^{C}Y)
4. X^{C}X^c
5. X^{C}Y
6. X^{c}Y
7. X^{C}X^c

15.6 **a.** A trisomic organism ($2n + 1$) has an extra copy of one chromosome, usually caused by a nondisjunction during meiosis. A triploid organism ($3n$) has an extra set of chromosomes, possibly caused by a total nondisjunction in gamete formation.

b. The genetic balance of a trisomic organism would be more disrupted than that of an organism with a complete extra set of chromosomes.

15.7 translocation, deletion, and inversion

15.8 Aneuploidies of sex chromosomes appear to upset genetic balance less, perhaps because relatively few genes are located on the Y chromosome and extra X chromosomes are inactivated as Barr bodies.

SUGGESTED ANSWERS TO STRUCTURE YOUR KNOWLEDGE

1. Genes that are not linked assort independently, and the ratio of offspring from a testcross with a dihybrid heterozygote should be 1:1:1:1 (*AaBb* × *aabb* gives *AaBb, Aabb, aaBb, aabb* offspring and a frequency of recombination frequency of 50%). Genes that are linked and do not cross over should produce a 1:1 ratio in this testcross (*AaBb* and *aabb* because a heterozygote derived from a parental cross of *AABB* × *aabb* produces only *AB* and *ab* gametes). If crossovers between distant genes always occur, the heterozygote will produce equal quantities of *AB, Ab, aB,* and *ab* gametes, and the genotype ratio of offspring will be 1:1:1:1, the same as it is for unlinked genes. Because each 1% recombination frequency is equal to 1 map unit, this measurement ceases to be meaningful at relative distances of 50 or more map units. However, crosses with intermediate genes on the chromosome could establish both that the genes *A* and *B* are on the same chromosome and that they are a certain map unit distance apart.

2. A cross between a mutant female fly and a normal male should produce all normal females (who get a wild-type allele from their father) and all mutant male flies (who get the mutant allele on the X chromosome from their mother). X^{m}X^m × X$^+$Y produces X$^+$X^m and X^{m}Y offspring (normal females and mutant males).

3. The serious phenotypic effects that are associated with these chromosomal alterations indicate that normal development and functioning is dependent on genetic balance. Most genes appear to be vital to an organismís existence, and extra copies of genes upset genetic balance. Inversions and translocations, which do not disrupt the balance of genes, can alter phenotype because of effects on gene functioning from neighboring genes.

ANSWERS TO GENETICS PROBLEMS

1. **a.** The trait is recessive and probably sex-linked. Two sets of unaffected parents in the second generation have offspring with the trait, indicating that it must be recessive. More males than females display the trait. Females #3 and #5 must be carriers since their father has the trait, and they each pass it on to a son.
   b. Using the symbols X^T for the dominant allele and X^t for the recessive:
   1. X^tY **5.** X^TX^t
   2. X^TX^t **6.** X^TX^t or X^TX^T
   3. X^TX^t **7.** X^TX^t
   4. X^TY
   c. Individual #6 has a brother who has the trait and her mother's father had the trait, so her mother must be a carrier of the trait, meaning there is a ½ probability that #6 is a carrier. Since she is mated to a phenotypically normal male, none of her daughters will show the trait (0 probability). Her sons have a ½ chance of having the trait if she is a carrier. ½ × ½ = ¼ probability that her sons will have the trait. There is a ½ chance that a child would be male, so the probability of an affected child is ½ × ¼, or ⅛.

2. c, a, b, d

3. The genes appear to be linked because the parental types appear most frequently in the offspring. Recombinant offspring represent 10 out of 40 total offspring for a recombination frequency of 25%, indicating that the genes are 25 map units apart.

4. One of the mother's X chromosomes carries the recessive lethal allele. One-half of male fetuses would be expected to inherit that chromosome and spontaneously abort. Assuming an equal sex ratio at conception, the ratio of girl to boy children would be 2:1, or 6 girls and 3 boys.

5. **a.** For female chicks to be black, they must have received a recessive allele from the male parent. If all female chicks are black, the male parent must have been Z^bZ^b. If the male parent was homozygous recessive, then all male offspring will receive a recessive allele, and the female parent would have to be Z^BW to produce all barred males.
   b. For female chicks to be both black and barred, the male parent must have been Z^BZ^b. If the female parent were Z^BW, only barred male chicks would be produced. To get an equal number of black and barred male chicks, the female parent must have been Z^bW.

ANSWERS TO TEST YOUR KNOWLEDGE

Multiple Choice:

1. e	**5.** a	**9.** a	**13.** c	**17.** e
2. a	**6.** e	**10.** e	**14.** a	**18.** d
3. b	**7.** b	**11.** d	**15.** e	**19.** c
4. d	**8.** c	**12.** b	**16.** d	**20.** d

CHAPTER 16: THE MOLECULAR BASIS OF INHERITANCE

■ INTERACTIVE QUESTIONS

16.1 **a.** They grew T2 with *E. coli* with radioactive sulfur to tag phage proteins.
b. T2 was grown with *E. coli* in the presence of radioactive phosphorus to tag phage DNA.
c. Radioactivity was found in the supernatant, indicating that the phage protein did not enter the bacterial cells.
d. In the samples with the labeled DNA, most of the radioactivity was found in the bacterial cell pellet.
e. They concluded that viral DNA is injected into the bacterial cells and serves as the hereditary material for viruses.

16.2 **a.** sugar-phosphate backbone
b. 3′ end of chain
c. pyrimidine base
d. purine base
e. hydrogen bonds
f. cytosine
g. guanine
h. adenine
i. thymine
j. 5′ end of chain
k. nucleotide
l. deoxyribose
m. phosphate group
n. 3.4 nm
o. 0.34 nm
p. 2 nm

16.3

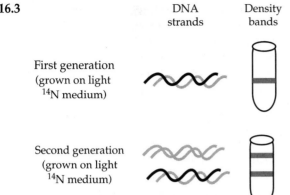

DNA strands   Density bands

First generation (grown on light ^{14}N medium)

Second generation (grown on light ^{14}N medium)

16.4 The phosphate end of each strand is the 5′ end, and the hydroxyl group extending from the 3′ carbon of the sugar marks the 3′ end. See Figure 16.7 in the text.

16.5
a. helicase
b. single-strand binding protein
c. DNA pol III
d. leading strand
e. lagging strand
f. DNA ligase
g. DNA pol I (replacing primer)
h. Okazaki fragment
i. RNA primer
j. replication fork
k. primase
l. 3′ end of parental strand
m. 5′ end

SUGGESTED ANSWERS TO STRUCTURE YOUR KNOWLEDGE

1. Watson and Crick used the X-ray diffraction photo of Franklin to deduce that DNA was a helix 2 nm wide, with nitrogenous bases stacked 0.34 nm apart, and making a full turn every 3.4 nm. Franklin had concluded that the sugar-phosphate backbones were on the outside of the helix with the bases extending inside. Using molecular models of wire, Watson and Crick experimented with various arrangements and finally paired of a purine base with a pyrimidine base, which produced the proper diameter. Specificity of base pairing (A with T and C with G) is assured by hydrogen bonds.

2. Replication bubbles form where proteins recognize specific base sequences and open up the two strands. *Helicase,* an enzyme that works at the replication fork, untwists the helix and separates the strands. *Topoisomerase* eases the twisting ahead of the replication fork. *Single-strand binding proteins* support the separated strands while replication takes place. *Primase* lays down about 10 RNA bases to start the new strand. After a proper base pairs up on the exposed template, *DNA polymerase III* joins the nucleotide to the 3′ end of the new strand. On the lagging strand, short Okazaki fragments are formed by primase and polymerase (again moving 5′ → 3′). *DNA polymerase I* replaces the primer. *Ligase* joins the 3′ end of one fragment to the 5′ end of its neighbor. Proofreading enzymes check for mispaired bases, and nucleases, DNA polymerase, and other enzymes repair damage or mismatches.

ANSWERS TO TEST YOUR KNOWLEDGE

Multiple Choice:

1. c	**6.** a	**10.** a	**14.** a	**18.** d
2. a	**7.** b	**11.** e	**15.** d	**19.** e
3. b	**8.** e	**12.** d	**16.** b	**20.** c
4. d	**9.** c	**13.** e	**17.** c	**21.** a
5. b				

CHAPTER 17: FROM GENE TO PROTEIN

■ INTERACTIVE QUESTIONS

17.1 DNA $\xrightarrow{\text{transcription}}$ RNA $\xrightarrow{\text{translation}}$ protein

17.2 Met Pro Asp Phe Lys stop

17.3
a. Initiation: Transcription factors bind to promoter and facilitate the binding of RNA polymerase II, forming a transcription initiation complex; RNA polymerase II separates DNA strands at initiation site.

b. Elongation: RNA polymerase II moves along DNA strand, connecting RNA nucleotides that have paired to the DNA template to the 3′ end of the growing RNA strand.

c. Termination: After polymerase transcribes past a termination and a polyadenylation signal sequence, the pre-mRNA is cut and released.

17.4 A 5′ cap consisting of a modified guanosine triphosphate is added to the 5′ UTR. A poly-A tail consisting of up to 250 adenine nucleotides is attached to the 3′ UTR. Spliceosomes have cut out the introns and spliced the exons together.

17.5

DNA Triplet $3' \rightarrow 5'$	mRNA Codon $5' \rightarrow 3'$	Anticodon $3' \rightarrow 5'$	Amino Acid
TAC	AUG	UAC	methionine
GGA	CCU	GGA	proline
TTC	AAG	UUC	lysine
ATC	UAG	AUC	stop

17.6 **1.** Codon recognition: An elongation factor (not shown) helps an aminoacyl tRNA into the A site where its anticodon base-pairs to the mRNA codon; two GTP increase accuracy and efficiency.
2. Peptide bond formation: Ribosome catalyzes peptide bond formation between new amino acid and polypeptide held in P site.
3. Translocation: The tRNA in the P site is moved to the E site and released; the tRNA now holding the polypeptide is moved from the A to the P site, taking the mRNA with it; one GTP is required.
4. Termination: Release factor binds to stop codon in the A site. Free polypeptide is released from the P site. Ribosomal subunits and other assembly components separate.

a. amino end of growing polypeptide
b. aminoacyl tRNA
c. large subunit of ribosome
d. A site
e. small subunit of ribosome
f. 5′ end of mRNA
g. peptide bond formation
h. E site
i. release factor
j. termination (stop) codon
k. P site
l. free polypeptide

17.7   A ribosome that is translating an mRNA that codes for a secretory or membrane protein will become bound to the ER when an SRP binds to the initial signal peptide and then to an ER receptor protein.

17.8   **a.** Messenger RNA carries the code from DNA that specifies the amino acid sequence of proteins to ribosomes.
b. Transfer RNA carries a specific amino acid to its position in a polypeptide based on matching its anticodon to an mRNA codon.
c. Ribosomal RNA makes up 60% of ribosomes and has structural and catalytic functions.
d. Small nuclear RNA is part of spliceosomes and plays a catalytic role in splicing pre-mRNA.
e. SRP RNA is part of signal-recognition particle that binds to signal peptides of polypeptides targeted to the ER.
f. Small nucleolar RNA aids in processing pre-rRNA transcripts in the nucleolus.
g. Small interfering RNA and microRNA are small single- and double-stranded RNA molecules that play role in regulating gene expression.

17.9   **a.** Silent: a base-pair substitution producing a codon that still codes for the same amino acid.
b. Missense: a base-pair substitution or frameshift mutation that results in a codon for a different amino acid.
c. Nonsense: a base-pair substitution or frameshift mutation that creates a stop codon and prematurely terminates translation.
d. Frameshift: an insertion or deletion of one, two, or more than three nucleotides that disrupts the reading frame and creates extensive missense and nonsense mutations.

SUGGESTED ANSWERS TO STRUCTURE YOUR KNOWLEDGE

1.

	Transcription	Translation
Template	DNA	RNA
Location	nucleus (cytoplasm in prokaryotes)	cytoplasm; ribosomes can be free or attached to ER
Molecules involved	RNA nucleotides, DNA template strand, RNA polymerase, transcription factors	amino acids, tRNA, mRNA, ribosomes, ATP, GTP, enzymes, initiation, elongation, and release factors
Enzymes involved	RNA polymerases, RNA processing enzymes, ribozymes	aminoacyl-tRNA synthetase, ribosomal enzymes
Control— start and stop	transcription factors locate promoter region with TATA box, polyadenylation signal sequence	initiation factors, initiation sequence (AUG), stop codons, release factor
Product	primary transcript (pre-mRNA)	protein
Product processing	RNA processing: 5′ cap and poly-A tail, splicing of pre-mRNA—introns removed by snRNPs in spliceosomes	spontaneous folding, disulfide bridges, signal peptide removed, cleaving, quaternary structure, modification with sugars, etc.
Energy source	ribonucleoside triphosphate	ATP and GTP

2. The genetic code is the RNA triplets that code for amino acids. The order of these codons is specified by the sequence of nucleotides on DNA, which is transcribed into the codons found on mRNA and translated into their corresponding amino acids. There are 64 possible mRNA codons created from the four nucleotides used in the triplet code (4^3).

Redundancy of the code refers to the fact that several triplets may code for the same amino acid. Often these triplets differ only in the third nucleotide. The wobble phenomenon explains the fact that there are only about 45 different tRNA molecules that pair with the 61 possible codons (three codons are stop codons). The third nucleotide of many tRNAs can pair with more than one base. Because of the redundancy of the genetic code, these wobble tRNAs still place the correct amino acid in position.

The genetic code is nearly universal; each codon codes for exactly the same amino acid in almost all organisms. (Some exceptions have been found.) This universality points to an early evolution of the code in the history of life and the evolutionary relationship of all life on Earth.

3.

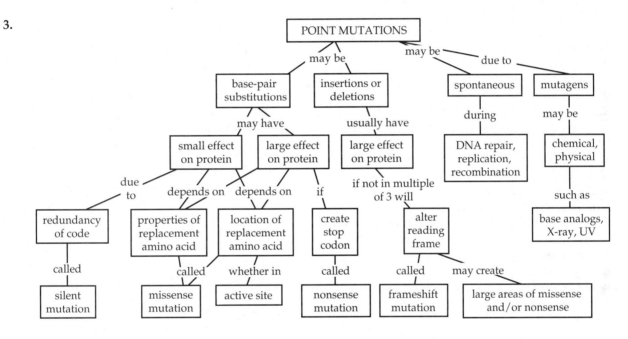

ANSWERS TO TEST YOUR KNOWLEDGE

Multiple Choice:

1. d	4. e	7. c	10. e	13. d	16. e	19. b	22. d
2. a	5. c	8. c	11. d	14. c	17. e	20. a	23. e
3. b	6. b	9. b	12. d	15. b	18. b	21. a	24. b

CHAPTER 18: THE GENETICS OF VIRUSES AND BACTERIA

■ INTERACTIVE QUESTIONS

18.1 1. Phage attaches to host cell and injects DNA.
2. Phage DNA forms circle. Certain factors determine which cycle is entered.
3. New phage DNA and proteins are synthesized and self-assemble into phages.
4. Bacterium lyses, releasing phages.
5. Phage DNA integrates into bacterial chromosome.
6. Bacterium reproduces, passing prophage to daughter cells.
7. Colony of infected bacteria forms.

8. Occasionally, prophage exits bacterial chromosome and begins lytic cycle.
a. phage DNA
b. bacterial chromosome
c. new phages
d. prophage
e. replicated bacterial chromosome with prophage

18.2 RNA → DNA → RNA; viral reverse transcriptase, host RNA polymerase

18.3
a. DNA
b. RNA
c. protein capsid
d. host
e. bacterium
f. lytic or lysogenic cycles
g. animal
h. viral envelope
i. reverse transcriptase
j. plant
k. viroids
l. plasmids
m. transposons

18.4 A temperate phage that leaves the lysogenic cycle and enters the lytic cycle may carry bacterial genes that were adjacent to the insertion site of the prophage.

18.5
a. circular chromosome
b. F plasmid, R plasmid
c. mutation
d. transformation
e. naked DNA
f. transduction
g. phage
h. conjugation
i. F$^+$ or Hfr, and F$^-$ cells
j. transposons
k. antibiotic-resistant genes
l. adaptation to environment/evolution

18.6
a. regulatory gene
b. promoter
c. operator
d. genes
e. operon
f. RNA polymerase
g. active repressor
h. inducer (allolactose)
i. mRNA for enzymes of pathway

18.7
a. anabolic; corepressor; on; inactive
b. catabolic; inducers; off; active

SUGGESTED ANSWERS TO STRUCTURE YOUR KNOWLEDGE

1.

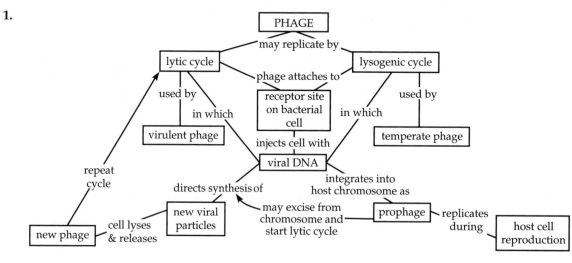

2.

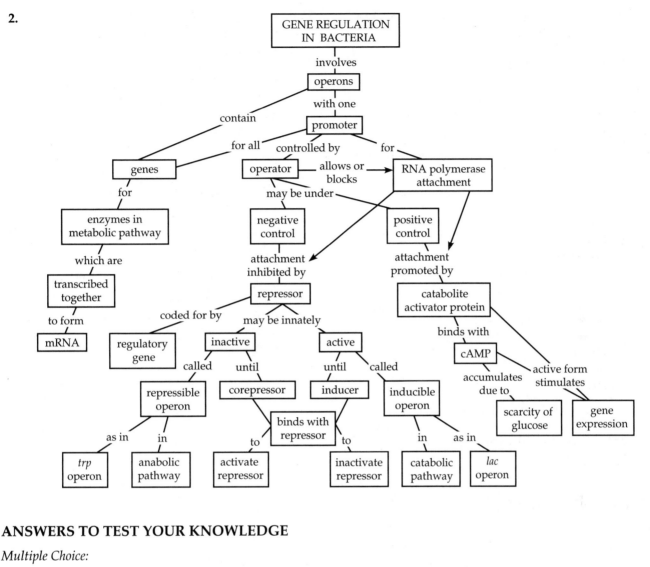

ANSWERS TO TEST YOUR KNOWLEDGE

Multiple Choice:

1. b	**4.** e	**7.** b	**10.** d	**13.** a	**16.** d	**19.** b	**22.** e	**25.** a
2. d	**5.** a	**8.** b	**11.** a	**14.** e	**17.** c	**20.** a or d	**23.** b	**26.** a
3. c	**6.** d	**9.** c	**12.** b	**15.** c	**18.** e	**21.** d	**24.** c	

Matching:

1. C	**3.** G	**5.** F	**7.** A
2. D	**4.** E	**6.** B	**8.** H

CHAPTER 19 EUKARYOTIC GENOMES: ORGANIZATION. REGULATION, AND EVOLUTION

■ INTERACTIVE QUESTIONS

19.1   nucleosomes (10-nm fiber); 30-nm fiber; looped domains (300-nm fiber); coiling and folding of looped domains into highly condensed metaphase chromosome

19.2   **a.** Barr body—compacted X chromosome in cells of female

b. Histone tail deacetylation would decrease transcription because it would make genes in the nucleosome less accessible.

19.3   **a.** distal control elements in enhancer
b. activators
c. DNA-bending protein
d. promoter

e. mediator proteins
f. transcription factors
g. TATA box
h. RNA polymerase

19.4 **a.** Regulatory proteins may bind to sequences in the 5′ UTR and block attachment of ribosomes, thus decreasing gene expression. Sequences in the 3′ UTR may affect the length of time an mRNA remains intact, thus either increasing or decreasing gene expression.
b. These small, single-stranded RNA molecules (miRNAs) join with a complex of proteins and act to decrease gene expression by base-pairing with target mRNA and either blocking translation or degrading the mRNA.

19.5 Cell-cycle simulating pathway involving Ras: (1) and (2) growth factor binds with receptor. (3) G protein Ras is activated, and (4) sets off protein kinase cascade. (5) Transcription factor is activated that turns on gene. (6) Protein produced that stimulates cell cycle. Mutation to the *Ras* gene may create a hyperactive Ras protein that signals without binding of growth factor.

Cell-cycle inhibiting pathway involving p53: (1) Damage to DNA signals (2) protein kinase cascade that (3) activates p53. This transcription factor turns on gene for (4) protein that inhibits cell cycle so that damaged DNA does not replicate. A mutation may result in a missing or defective p53 transcription factor. The protein that inhibits the cell cycle would not be produced.

19.6
1.	D	1.5%
2.	F	24
3.	B	44
4.	A	3
5.	E	5
6.	C	15

19.7 The lysozyme gene, which codes for a bacterial infection-fighting enzyme, was present in the last common ancestor of birds and mammals. After their lineages split, the gene underwent a duplication event in the mammalian lineage, and a copy of the lysozyme gene evolved into a gene coding for a protein involved in milk production.

SUGGESTED ANSWERS TO STRUCTURE YOUR KNOWLEDGE

1. **a.** DNA packing into nucleosomes; histone tail acetylation increases, whereas deacetylation and methylation of tails decreases transcription; methylation of DNA may be involved in long-term inactivation of genes

b. Transcription factors (activators) bind with enhancers, then interact with mediator proteins and promoter regions to form transcription initiation complex; repressors can inhibit transcription; steroid hormones or other chemical messages may bind with receptor proteins, producing transcription factors
c. RNA processing (alternative splicing, 5′ cap and poly-A tail added); mRNA degradation by shortening of poly-A tail, removal of 5′cap, and miRNA targeting
d. Repressor proteins may prevent ribosome binding so mRNA can be stockpiled (awaiting fertilization in ovum); activation of initiation factors
e. Protein processing by cleavage or modification; transport to target location; selective degradation by proteosomes of proteins marked with ubiquitin

2. **a.** Proto-oncogenes are key genes that control cellular growth and division. When such genes mutate to form a more active product, become amplified (multiple copies), or have changes in their normal control mechanisms, they may become oncogenes and produce the uncontrolled cell division that leads to formation of a tumor.
b. Tumor-suppressor genes code for proteins that regulate cell division. Loss or mutation of both alleles of a tumor-suppressor gene may allow tumors to develop. Usually mutations or other changes must occur both in oncogenes and in several suppressor genes for cancer to develop.

3. **a.** Retrotransposons are transposable elements that move about a genome as an RNA intermediate, which is converted back into DNA by reverse transcriptase, coded for by the retrotransposon.
b. By moving copies of themselves around the genome, retrotransposons provide locations for recombination between different chromosomes. They may also transport genes or exons to new locations and interrupt coding or regulatory sequences.

ANSWERS TO TEST YOUR KNOWLEDGE

Multiple Choice:

1. b	**6.** e	**11.** d	**16.** d
2. a	**7.** c	**12.** e	**17.** c
3. e	**8.** d	**13.** a	**18.** e
4. b	**9.** c	**14.** b	**19.** c
5. d	**10.** e	**15.** b	**20.** e

CHAPTER 20: DNA TECHNOLOGY AND GENOMICS

■ INTERACTIVE QUESTIONS

20.1 The third sequence, because it has the same sequence running in opposite directions. The enzyme would probably cut between G and A, producing AATT and TTAA sticky ends.

20.2
a. bacterial plasmid
b. *lacZ* gene
c. restriction site
d. *amp^R* (ampicillin resistance) gene
e. gene of interest
f. complementary sticky ends
g. human DNA fragments
h. recombinant plasmid
i. recombinant bacteria
j. plate with ampicillin and X-gal

1. Plasmid and human DNA are isolated.
2. Both DNAs are cut with same restriction enzyme. Single cut in plasmid disrupts *lacZ* gene; multiple fragments of human DNA formed.
3. Fragments are mixed and some foreign fragments base-pair with plasmid. DNA ligase seals ends.
4. Recombinant plasmids transform bacteria with mutation in their *lacZ* gene.
5. Plate bacteria on agar containing ampicillin and X-gal.
   Cells containing recombinant plasmid are identified by their ability to grow in presence of the antibiotic and by their white color. Blue colonies contain plasmids that resealed and thus have a functioning *lacZ* gene. Identify clones carrying gene of interest with nucleic acid probe or by protein product.

20.3 A genomic library contains copies of DNA segments from the entire genome. Thus, all genes should be represented, along with the regulatory sequences and introns. A cDNA library allows you to sequence only the exons of a gene, and also indicates which genes are expressed either in different cell types or at different stages of development in the same cell type.

20.4
a. By amplifying the gene prior to cloning, the later task of identifying clones carrying the desired gene is simplified.
b. There is a limit to the number of accurate copies that can be made due to the accumulation of relatively rare copying errors. Large quantities of a gene are better prepared by DNA cloning in cells, and the gene may be expressed in the cells.

20.5 Southern blotting procedure:
a. restriction enzyme treatment of samples
b. gel electrophoresis
c. blotting
d. hybridization with labeled probe
e. autoradiography
   Crime samples contain blood from the victim and, presumably, the perpetrator. After identifying the bands from the victim, the remaining fragments match those of Suspect 2's fragment profile.

20.6 The public consortium followed a hierarchy of three stages: (1) genetic (linkage) mapping that established about 200 markers/chromosome; (2) physical mapping that cloned and ordered smaller and smaller overlapping fragments (using YAC or BAC vectors for cloning the large fragments); and (3) DNA sequencing of each small fragment, followed by assembly of the overall sequence. The Celera whole-genome shotgun approach omitted the first two stages. Each chromosome was cut into small fragments, which were cloned in plasmid or phage vectors. The sequence of each fragment was determined, and powerful computers assembled the overlapping fragments to determine the overall sequence.

20.7
a. Comparing noncoding sequences in the human genome with those in the much smaller yeast genome has revealed highly conserved sequences that have turned out to be important regulatory control sequences in both. The functions of some human disease genes have been identified through the study of similar yeast genes.
b. Once the sequence and organization of the smaller genome is determined, it can be used as a framework for organizing the sequences from the larger genome.

20.8
a. Before, members of the extended family had to be tested to determine the variant of the RFLP marker to which the disease-causing allele was linked. Then RFLP analysis was done with the patient's blood to determine which marker, and thus which allele, he or she had inherited. Now the cloned gene can be used as a probe, and hybridization analysis can detect the abnormal allele.
b. The major difficulty is to assure that proper control mechanisms are present so that the gene is expressed at the proper time, in the proper place, and to the proper degree. Insertion of the therapeutic gene must not harm other cell functions. Ethical considerations include whether

genetic engineering should be done on germ cells, thus influencing the genetic makeup of future generations.

SUGGESTED ANSWERS TO STRUCTURE YOUR KNOWLEDGE

1. **a.** Bacterial enzymes that cut DNA at restriction sites, creating "sticky ends" that can base-pair with other fragments. Use: make recombinant DNA, form restriction fragments used for many other techniques

    b. Mixture of molecules applied to gel in electric field; molecules separate, moving at different rates due to charge and size. Use: separate restriction fragments into pattern of distinct bands, fragments can be removed from gel and retain activity or can be identified with probes

    c. mRNA isolated from cell is treated with reverse transcriptase to produce a complementary DNA strand, and then a double-stranded DNA gene, minus introns and control regions. Use: creates genes that are easier to clone in bacteria, produces library of genes that are active in cell

    d. Radioactively or fluorescently labeled single-stranded DNA or mRNA used to base-pair with complementary sequence of DNA or RNA. Use: locate gene in clone of bacteria, identify bands on gels, diagnose infectious diseases

    e. DNA fragments separated by gel electrophoresis, transferred by blotting onto paper, labeled probe added, rinsed, autoradiography. Use: analyze DNA for homologous sequences, DNA fingerprinting

    f. Single-stranded DNA fragments are incubated with four nucleotides, DNA polymerase, and four labeled dideoxy nucleotides that interrupt synthesis, samples separated by size, sequence of nucleotides read from sequence of fluorescent tags. Use: determine nucleotide sequence

    g. Polymerase chain reaction: DNA is mixed with heat-resistant DNA polymerase, nucleotides, and primers having complementary sequences for targeted DNA section and repeatedly heated to separate, cooled to pair with primers and replicate. Use: rapidly produce multiple copies of a gene or section of DNA *in vitro*

    h. Restriction fragments analyzed by Southern blotting to compare different band patterns caused by DNA differences in restriction sites. Use: DNA fingerprints for forensic use, map chromosomes using RFLP markers, diagnose genetic diseases

    i. Fluorescently labeled cDNA is made from a cell's mRNA and applied to a DNA microarray of single-stranded DNA from many different genes attached to a glass grid. The intensity and location of fluorescence indicates gene expression in the cell. Use: test thousands of genes simultaneously to compare gene expression in different tissues or at different developmental stages or conditions.

2. Agricultural applications: (1) production of vaccines and hormones, which will improve the health or productivity of livestock, (2) improvement of the genomes of agricultural plants and animals, (3) "pharm" animals or plants to produce human proteins, (4) development of plant varieties that have genes for resistance to diseases, herbicides, and insects

    Medical applications: (1) development of vaccines, (2) diagnosis of genetic diseases, (3) treatment of genetic disorders, (4) production of insulin, human growth factor, TPA, and other useful products

ANSWERS TO TEST YOUR KNOWLEDGE

Multiple Choice:

1. c	**6.** c	**11.** c	**16.** d	**21.** a
2. b	**7.** c	**12.** b	**17.** b	**22.** a
3. e	**8.** a	**13.** d	**18.** d	**23.** e
4. e	**9.** a	**14.** a	**19.** c	**24.** d
5. c	**10.** b	**15.** c	**20.** b	**25.** d

CHAPTER 21: THE GENETIC BASIS OF DEVELOPMENT

■ INTERACTIVE QUESTIONS

21.1 Model organisms ideally should have easily observable embryos, short generation times, high reproductive rates, be easy to grow in the lab, a sequenced genome, and there should be a preexisting knowledge base of their genetics and biology.

21.2 DNA methylation helps to regulate gene expression. An adult cell must have these epigenetic changes in its chromatin reprogrammed in order to support normal gene expression during development. The DNA of many cloned embryos has been found to be improperly methylated.

21.3 The action of MyoD must depend on a combination of regulatory proteins, some of which may be lacking in the cells that it is not able to transform into muscle cells.

21.4 a. a mutation in both *bicoid* alleles of the mother fly so that no *bicoid* mRNA was secreted into the anterior of the egg and thus no Bicoid protein gradient was available to establish the anterior end
b. a mutation in a segmentation gene
c. a mutation in a homeotic gene so that a head segment was identified as a thoracic segment and genes for leg development were activated

21.5 a. sequential inductions
b. concentration
c. signal transduction pathways; transcriptional regulation

21.6 a. The basic mechanism for programmed cell death evolved early in animal evolution.
b. Programmed cell death in humans is important in the normal development of the nervous system, fingers, and toes; for normal functioning of the immune system; and to prevent the development of cancerous cells (by the internal triggering of apoptosis in cells with DNA damage).

21.7 All of these genes are master regulatory genes that control the expression of batteries of other genes. The homeobox codes for the DNA-binding homeodomain of a transcription factor.

SUGGESTED ANSWERS TO STRUCTURE YOUR KNOWLEDGE

1. Proteins translated from maternal mRNA deposited in the egg cell (transcripts of egg-polarity genes) diffuse through the embryo's multinucleated cytoplasm and regulate the activity of zygotic genes. These morphogen gradients activate gap genes, whose products activate pair-rule genes, whose products activate the segment polarity genes. Most of the products of these segmentation genes (as well as the maternal effect genes) are transcription factors that control this hierarchy of gene activation responsible for the segmentation pattern of the embryo. The appropriate homeotic genes are activated in each segment, and they regulate the transcription of the genes that control development of segmental structures.

2. Most cytoplasmic determinants are mRNA for transcription factors that are divided by the first few mitotic divisions. They are present in the cells and can enter the nucleus and regulate transcription. Inducers must communicate between cells. They are often proteins that bind to cell surface receptors and initiate a signal transduction pathway involving a cascade of enzyme activations, usually leading to the activation of transcription factors within the target cell.

ANSWERS TO TEST YOUR KNOWLEDGE

Multiple Choice:

1. e	**5.** d	**9.** c	**13.** a
2. b	**6.** c	**10.** c	**14.** c
3. c	**7.** e	**11.** a	**15.** d
4. e	**8.** d	**12.** b	**16.** e

CHAPTER 22: DESCENT WITH MODIFICATION: A DARWINIAN VIEW OF LIFE

■ INTERACTIVE QUESTIONS

22.1 a. 1. E f **3.** B e **5.** F a **7.** G g
2. A b **4.** C c, d, g **6.** D c
b. 5, 1, 2, 4, 3, 7, 6

22.2 The excessive production of offspring sets up the struggle for existence; only a small proportion can live to leave offspring of their own. Natural selection is the differential reproductive success of individuals within a population that are best suited to the environment, which leads over generations to greater adaptation of populations of organisms to their environment.

22.3 Treatment with the drug 3TC prevents most HIV from reproducing when their enzyme reverse transcriptase inserts this cytosine mimic into its DNA, halting replication and production of new HIV. The 3TC-resistant HIV have a version of reverse transcriptase that discriminates between cytosine and 3TC, and they are still able to reproduce. With HIV's short generation time, it takes very little time for 3TC-resistant HIV to be strongly selected for by this drug and 100% of a patient's HIV population to be 3TC resistant.

22.4 **a.** biogeography
b. fossil record
c. homologies
d. island species and mainland species or neighboring species in different habitats
e. ancestral and transitional forms
f. homologous structures
g. embryological development
h. molecular comparisons
i. DNA and proteins
j. descent from a common ancestor

SUGGESTED ANSWERS TO STRUCTURE YOUR KNOWLEDGE

1. The two major components of Darwin's evolutionary theory are that all life has descended from a common ancestral form and that this evolution has been the result of natural selection. The theory of natural selection is based on several key observations and inferences. The overproduction of offspring in conjunction with limited resources leads to a struggle for existence and the differential reproductive success of those organisms best suited to the local environment. The unequal survival and reproduction of the most fit individuals in a population leads to the gradual accumulation of adaptive characteristics in a population.

ANSWERS TO TEST YOUR KNOWLEDGE

Multiple Choice:

1. b	**5.** c	**9.** d	**13.** b
2. c	**6.** a	**10.** d	**14.** d
3. e	**7.** e	**11.** c	**15.** c
4. a	**8.** d	**12.** e	**16.** c

CHAPTER 23: THE EVOLUTION OF POPULATIONS

■ INTERACTIVE QUESTIONS

23.1 **a.** The 98 *BB* mice contribute 196 *B* alleles, and the 84 *Bb* mice contribute 84 *B* alleles to the gene pool. These 84 *Bb* mice also contribute 84 *b* alleles, and the 18 *bb* mice contribute 36 *b* alleles. Of a total of 400 alleles, 280 are *B* and 120 are *b*. Allele frequencies are 0.7 *B* and 0.3 *b*.
b. The frequencies of genotypes are 0.49 *BB*, 0.42 *Bb*, and 0.09 *bb* ($^{98}/_{200}$, $^{84}/_{200}$, $^{18}/_{200}$).

23.2 0.7; 0.3; 0.49; 0.42; 0.09

23.3 **a.** 0.36; 0.48; 0.16. Plug *p* (0.6) and *q* (0.4) into the expanded binomial: $p^2 + 2pq + q^2$.
b. 0.6; 0.4. Add the frequency of the homozygous dominant genotype to ½ the frequency of the heterozygote for the frequency of *p*. For *q*, add the homozygous recessive frequency and ½ the heterozygote frequency. Alternatively, to determine *q*, take the square root of the homozygous recessive frequency if you are sure the population is in Hardy-Weinberg equilibrium. The frequency of *p* is then $1 - q$.

23.4 **a.** mutation, with some recombination
b. sexual recombination
c. Bacteria and viruses have very short generation times and a new beneficial mutation can increase in frequency rapidly in an asexually reproducing bacterial population. Although mutations are the source of new alleles, they are so infrequent that their contribution to genetic variation in a large, diploid population is minimal. However, sexual recombination in the production and union of gametes produces zygotes with fresh combinations of alleles each generation.

23.5 **a.** natural selection
b. genetic drift
c. gene flow
d. better reproductive success
e. small population
f. bottleneck effect
g. founder effect
h. genetic variation between populations

23.6 **a.** less; we have about a tenth of the variability found in fruit fly populations.
b. Humans have about a 0.1% nucleotide diversity.

23.7 **a.** The two given fitness values are less than 1, so *Bb* must produce the most offspring and have a relative fitness of 1.
b. 0

23.8 **a.** Diploidy—The sickle-cell allele is hidden from selection in heterozygotes.
b. Heterozygote advantage—Heterozygotes are resistant to malaria and have a selective advantage in areas where malaria is a major cause of death.

SUGGESTED ANSWERS TO STRUCTURE YOUR KNOWLEDGE

1. **a.** The Hardy-Weinberg theorem states that allele frequencies within a population will remain constant from one generation to the next as long as only Mendelian segregation and sexual recombination of alleles are involved. An equilibrium in which neither allele nor genotype frequencies will change from generation to generation requires five conditions: The population is large, mutation and migration are negligible, mating is random, and no selective pressure operates. This Hardy-Weinberg equilibrium provides a null hypothesis to allow population geneticists to test for evolution.
    b. $p^2 + 2pq + q^2 = 1$. In the Hardy-Weinberg equation, p and q refer to the frequencies of two alleles in the gene pool. The frequency of homozygous offspring is ($p \times p$) or p^2 and ($q \times q$) or q^2. Heterozygous individuals can be formed in two ways, depending on whether the ovum or sperm carries the p or q allele, and their frequency is equal to $2pq$.

2. Genetic variation is retained within a population by diploidy and balancing selection. Diploidy masks recessive alleles from selection when they occur in the heterozygote. Thus, less adaptive or even harmful alleles are maintained in the gene pool and are available should selection pressures change. Balanced selection maintains several alleles at a gene locus in a population and leads to balanced polymorphism. In situations in which there is heterozygote advantage, the two alleles will be retained in stable frequencies within the gene pool. Frequency-dependent selection, in which morphs present in higher numbers are selected against by predators or other factors, is another cause of balanced polymorphism.

ANSWERS TO TEST YOUR KNOWLEDGE

1. e	6. d	11. e	16. b	21. a
2. a	7. e	12. c	17. c	22. e
3. c	8. d	13. d	18. b	23. d
4. c	9. b	14. a	19. b	24. c
5. d	10. e	15. b	20. c	25. e

CHAPTER 24: THE ORIGIN OF SPECIES

■ INTERACTIVE QUESTIONS

24.1 **a.** reduced hybrid fertility   **b.** post-
    c. gametic isolation   **d.** pre-
    e. mechanical isolation   **f.** pre-
    g. temporal isolation   **h.** pre-
    i. hybrid breakdown   **j.** post-
    k. behavioral isolation   **l.** pre-
    m. reduced hybrid viability   **n.** post-
    o. habitat isolation   **p.** pre-

24.2 **a.** reproductive isolation
    b. morphological
    c. ecological
    d. paleontological
    e. evolutionary lineage leading to unique genetic history

24.3 **a.** 20 chromosomes
    b. 24 chromosomes

24.4 **a.** In allopatric speciation, a new species forms while geographically isolated from the parent population. In sympatric speciation, some reproductive barrier isolates the gene pool of a subgroup of a population within the same geographic range as the parent population.
    b. Reproductive barriers may evolve as a byproduct of the genetic change associated with the isolated population's adaptation to a new environment, genetic drift, or sexual selection. In sympatric speciation in plants, hybridization between closely related species followed by mitotic or meiotic errors that produce fertile polyploids is a common mechanism. A change in resource use or sexual selection may reproductively isolate a subset of an animal population.

24.5 The Hawaiian Archipelago is a series of relatively young, isolated, and physically diverse islands whose thousands of endemic species are examples of adaptive radiation resulting from multiple colonizations and allopatric and sympatric speciations.

24.6 In the gradualism model of evolution, small changes accumulate within populations as a result of chance events and natural selection, leading to the gradual divergence of species. The punctuated equilibrium model holds that evolution occurs in spurts of relatively rapid change interspersed with long periods of stasis. A species changes most as it arises from an ancestral species and then undergoes little change for the rest of its existence.

24.7 Feathers and wings may have first developed as structures for social displays, thermoregulation,

and camouflage; light, honeycombed bones may have increased the agility of bipedal dinosaurs.

24.8 **a.** allometric growth
       b. paedomorphosis
       c. heterochrony

SUGGESTED ANSWERS TO STRUCTURE YOUR KNOWLEDGE

1. Speciation, by which a new species evolves from a parent species, is part of macroevolution and the increase in biological diversity. Microevolution is the process by which changes occur within the gene pool of a population as a result of either chance events or natural selection. If the makeup of the gene pool changes enough, microevolution may lead to speciation.

2. Modification of existing structures: either the gradual development of a more complex structure from one that served the same function, or exaptation, in which a structure that evolved in one context is gradually changed to serve a novel function. "Evo-devo": changes in genes that control aspects of development, such as growth rates, timing, and spatial arrangements of body parts, may result in novel designs.

ANSWERS TO TEST YOUR KNOWLEDGE

1. b	**5.** c	**9.** b	**13.** b	**17.** c
2. c	**6.** a	**10.** d	**14.** c	**18.** b
3. d	**7.** b	**11.** b	**15.** c	**19.** e
4. e	**8.** c	**12.** a	**16.** e	**20.** d

CHAPTER 25: PHYLOGENY AND SYSTEMATICS

■ INTERACTIVE QUESTIONS

25.1 Organisms that were abundant, widespread, existed over a long period of time, had hard shells or skeletons, and lived in regions in which the geology favored fossil formation are most likely to be represented in the fossil record.

25.2 Adaptations to different environments may lead to large morphologic differences within related groups, and convergent evolution can produce similar structures in unrelated organisms in response to similar selective pressures. Examples are the Hawaiian silversword plants and the Australian and North American burrowing moles.

25.3 A C G T G C A C G
      A   G T G   A G G
      The second DNA segment shows two deletions and one base substitution.

25.4

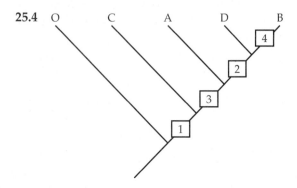

25.5 The four-chambered heart of birds and mammals is analogous, not homologous. It evolved independently in the two groups. An abundance of evidence supports the hypothesis that birds and reptiles are closer relatives, and that birds and mammals evolved from different reptilian ancestors.

25.6 **a.** rRNA genes
       b. mtDNA (mitochondrial DNA)
       c. genes for olfactory receptors

25.7 Assuming that genes have a constant rate of mutation, in a protein that has a crucial function, mutations would be more likely to be harmful and quickly removed from the population. Therefore, fewer mutations would be available to become fixed. If a protein has less than a critical function, more mutations would be neutral and remain, and the gene will change more quickly.

SUGGESTED ANSWERS TO STRUCTURE YOUR KNOWLEDGE

1. **a.** biological diversity
    b. phylogeny
    c. species
    d. cladistics analysis
    e. shared primitive characters
    f. clades
    g. homologies
    h. molecular comparisons

ANSWERS TO TEST YOUR KNOWLEDGE

Multiple Choice:

1. c	5. d	9. e	13. a
2. a	6. b	10. b	14. e
3. c	7. d	11. a	15. d
4. b	8. e	12. d	16. b

True or False:

1. True
2. False—change the first *more* to the *less* they vary.
3. True
4. True
5. True
6. False—change *phlogram* to *ultrametric trees*

CHAPTER 26: THE TREE OF LIFE: AN INTRODUCTION TO BIOLOGICAL DIVERSITY

■ INTERACTIVE QUESTIONS

26.1 A cell needs metabolic machinery to provide the enzymes, energy, and building blocks to replicate its genetic information. This genetic information codes for that machinery. Thus both replication and metabolism are required for life.

26.2 17,190 years. Carbon-14 has a half-life of 5,730 years. In 5,730 years the ratio of C-14 to C-12 would be reduced by ½; in 11,460 years it would be reduced by ¼; and in 17,190 years it would be reduced by ⅛.

26.3 Mass extinctions empty many biological niches, which may then be exploited by species that survived extinction.

26.4 Some organisms had enzymes that were less sensitive to oxygen or could convert hydrogen peroxide and free oxygen to less damaging substances. Others were able to adapt electron transport to aerobic respiration, using oxygen as an electron acceptor and producing more ATP than anaerobic mechanisms.

26.5 The inner membranes of both mitochondria and chloroplasts have enzymes and transport system homologous to those of modern prokaryotes. They replicate by a splitting process similar to binary fission. They contain a single, circular DNA molecule that is not associated with histones. They contain the machinery needed to make their own proteins. Their ribosomes are similar to prokaryotic ribosomes.

26.6 Marsupials probably migrated to Australia through South America and Antarctica while the continents were still joined. When Pangaea broke up, marsupials adaptively radiated on isolated Australia. Eutherians (placental mammals) diversified on the other continents.

26.7 A huge ice age from 750 to 570 million years ago restricted living organisms to deep sea vents or hot springs and restricted the size, diversity, and distribution of life until the Earth thawed.

SUGGESTED ANSWERS TO STRUCTURE YOUR KNOWLEDGE

1. The primitive Earth is believed to have had seas, volcanoes, and large amounts of ultraviolet radiation passing through a thin atmosphere, which probably consisted of H_2O, CO, CO_2, N_2, CH_4, and NH_3. Life could arise in this environment because the reducing (electron-adding) atmosphere and the availability of energy from lightning and UV radiation may have facilitated the abiotic synthesis of organic molecules, and hot lava rocks or clay may have aided the polymerization of these monomers.

   An alternative hypothesis is that deep-sea vents may have provided the energy and chemical precursors needed for the origin of life. Organic molecules may also have arrived on meteorites.

2.

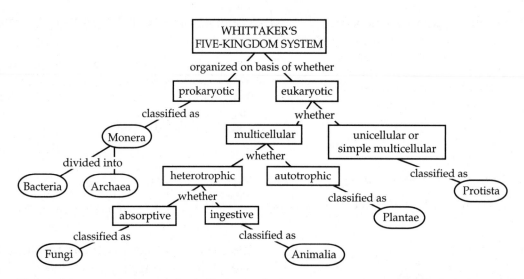

3. Molecular systematics and cladistic analysis show that prokaryotes diverged early into two distinct lineages, each of which can be assigned to a separate domain. Molecular systematics also shows that the traditional kingdom Protista is not monophyletic, but can be divided into five or more kingdoms. The goal of a classification system is to reconstruct phylogeny. New data allow systematists to revise and refine their hypotheses of the evolutionary history of the diversity of life.

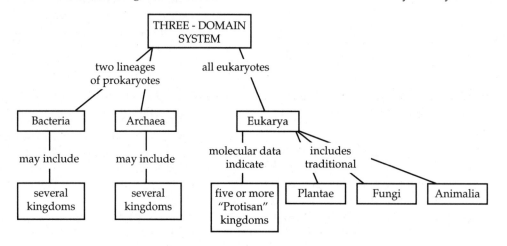

ANSWERS TO TEST YOUR KNOWLEDGE

Multiple Choice:

1. b	3. a	5. d	7. a	9. c	11. a	13. c	15. a
2. e	4. b	6. e	8. c	10. c	12. e	14. e	16. b

CHAPTER 27: PROKARYOTES

■ INTERACTIVE QUESTIONS

27.1 **a.** spheres (cocci), rods (bacilli), or spirals
b. 1–5 μm in diameter
c. cell wall made of peptidoglycan; gram-negative bacteria also have outer lipopolysaccharide membrane; archaea lack peptidoglycan; sticky capsule for adherence; fimbriae and pili for attachment or conjugation
d. flagella; may show taxis to stimuli
e. infoldings of plasma membrane may be used in metabolic functions

f. circular DNA molecule with little associated protein found in nucleoid region; may have plasmids with antibiotic resistance and other genes
g. binary fission; rapid population growth; some horizontal gene transfer; rapid adaptive evolution

27.2 **a.** light
b. phototroph
c. chemicals
d. CO_2
e. organic nutrients
f. heterotroph
g. photoautotroph
h. photoheterotroph
i. chemoheterotroph

27.3 **a.** endospore; thick-walled, resistant cell that can withstand harsh conditions for long periods of time, breaking dormancy when conditions are favorable
b. cyanobacteria; the cell is a heterocyte, specialized for nitrogen fixation

27.4 Genetic prospecting is a technique of sampling genetic material directly from the environment. Before its development, researchers could study only those species that could be cultured in the lab. This technique has lead to the discovery of new species and even new branches in prokaryotic phylogeny.

27.5 The photosynthesis of cyanobacteria brings carbon from CO_2 into the food chain and releases O_2. Cyanobacteria also fix atmospheric nitrogen, supplying other organisms with the nitrogen they need to make proteins.

SUGGESTED ANSWERS TO STRUCTURE YOUR KNOWLEDGE

1. Molecular evidence, especially comparison of rRNA signature sequences, indicates that bacteria and archaea diverged very early in evolutionary history and that archaea may share a more recent common ancestor with eukaryotes.

2. Decomposers—recycle nutrients; bioremediation: sewage treatment, clean up oil spills.

   Nitrogen fixers—provide nitrogen to the soil by bacteria in nodules on legume roots or by cyanobacteria.

   Mutualism—enteric bacteria, some digest food and produce vitamins.

   Biotechnology—production of antibiotics, hormones, and many useful products.

ANSWERS TO TEST YOUR KNOWLEDGE

Multiple Choice:

1. c	**4.** c	**7.** a	**10.** d
2. e	**5.** e	**8.** c	**11.** e
3. b	**6.** d	**9.** b	**12.** e

Fill in the Blanks:

1. cocci	**6.** endospore
2. nucleoid region	**7.** biofilms
3. Gram stain	**8.** exotoxins
4. fimbriae and pili	**9.** bioremediation
5. taxis	**10.** anaerobic respiration

CHAPTER 28: PROTISTS

■ INTERACTIVE QUESTIONS

28.1 The four membranes indicate secondary endosymbiosis of a green alga (whose chloroplasts originated from the primary endosymbiont, a cyanobacterium). The inner two membranes belonged to the ancient cyanobacterium. The third membrane is from the engulfed green alga's plasma membrane and the outer membrane from the heterotrophic eukaryote's food vacuole. The vestigial nucleus indicates that this process occurred relatively recently. In more ancient secondary endosymbioses, many components of the engulfed alga have been lost.

28.2 Frequent global changes in the surface proteins of these parasites enables them to evade the host's immune system.

28.3 *Paramecium* is a ciliate in the clade Alveolata.
a. cilia
b. oral groove
c. food vacuoles, combine with lysosomes
d. contractile vacuole
e. anal pore
f. macronucleus
g. micronucleus

28.4 These seaweeds may have a holdfast that maintains a strong hold on intertidal rocks.

Biochemical adaptations include cell walls containing gel-forming polysaccharides that protect the algae from the abrasive action of waves and reduce drying when exposed at low tides.

28.5 The calcareous tests or forams are long-lasting fossils in marine sediments and sedimentary rocks. Foram fossils are useful as index fossils for dating sedimentary rocks.

28.6
a. diploid
b. multinucleate mass, plasmodium
c. amoeboid or flagellated cells
d. haploid
e. solitary amoeboid cells until aggregate to form fruiting body
f. two amoebas fuse to form resistant zygote

SUGGESTED ANSWERS TO STRUCTURE YOUR KNOWLEDGE

1.
a. cyanobacterium
b. primary endosymbiosis
c. red algae

d. green algae
e. secondary endosymbiosis
f. dinoflagellates
g. apicomplexans
h. stramenopiles
i. euglenids

ANSWERS TO TEST YOUR KNOWLEDGE

Matching:

1. E	**4.** H	**7.** A
2. D	**5.** I	**8.** F
3. B	**6.** C	**9.** G

Multiple Choice:

1. d	**4.** d	**7.** b	**10.** c
2. a	**5.** c	**8.** e	
3. c	**6.** d	**9.** b	

CHAPTER 29: PLANT DIVERSITY I: HOW PLANTS COLONIZED LAND

■ INTERACTIVE QUESTIONS

29.1 homologies in nuclear and chloroplast genes, rosette cellulose-synthesizing complexes, peroxisome enzymes, phragmoplast formation, and similarities in sperm

29.2
a. gametes
b. mitosis
c. zygote
d. mitosis
e. sporophyte
f. spores
g. meiosis
h. gametophyte

29.3
a. bryophytes (nonvascular plants)
b. seedless vascular plants
c. gymnosperms
d. angiosperms

29.4
a. gametophyte
b. archegonia
c. antheridia
d. swim
e. sporophyte
f. meiosis
g. capsule (sporangium)
h. gametophytes (germinate to form protonema)

29.5
a. meiosis
b. spore
c. gametophyte
d. antheridium
e. sperm
f. archegonium
g. egg
h. fertilization
i. zygote
j. gametophyte
k. young sporophyte
l. mature sporophyte
m. sporangia (in sori)

The gametophyte (upper portion of the diagram) is haploid; the sporophyte (lower portion) is diploid.

SUGGESTED ANSWERS TO STRUCTURE YOUR KNOWLEDGE

1. protected and nourished embryo, sporopollenin-walled spores, apical meristems, alternation of generations, multicellular gametangia, vascular tissue strengthened with lignin, cuticle, secondary compounds, stomata

2. **a.** land plants
    b. bryophytes
    c. vascular plants
    d. seedless vascular plants
    e. seed plants
    f. charophyceans
    g. liverworts
    h. hornworts
    i. mosses
    j. pterophytes (ferns, horsetails, whisk ferns)

ANSWERS TO TEST YOUR KNOWLEDGE

Multiple Choice:

1. c	**5.** d	**9.** d	**13.** b
2. e	**6.** b	**10.** e	**14.** d
3. d	**7.** d	**11.** d	**15.** c
4. a	**8.** a	**12.** e	

CHAPTER 30: PLANT DIVERSITY II: THE EVOLUTION OF SEED PLANTS

■ INTERACTIVE QUESTIONS

30.1 a. integuments (2*n*)
   b. megasporangium (2*n*)
   c. megaspore (*n*)
   d. seed coat (2*n*)
   e. food supply (from female gametophyte tissue) (*n*)
   f. embryo (2*n*)

30.2 Resistant, airborne or animal-carried pollen grains can transport sperm, even over long distances, to the female gametophyte, eliminating the need for a moist environment for sperm to swim to reach eggs.

30.3 a. The smaller pollen cones have many sporophylls holding microsporangia. Microsporocytes (microspore mother cells) undergo meiosis to produce microspores that develop into pollen grains, the male gametophyte.
   b. Ovulate cones have scales the hold two ovules, each containing a megasporangium. A megasporocyte (megaspore mother cell) undergoes meiosis to form megaspores. The surviving megaspore develops into a multicellular female gametophyte containing a few archegonia, each with an egg.

30.4 a. sepals, petals, stamens, carpels
   b. an embryo surrounded by food enclosed in a seed coat
   c. mature ovary that may be modified to help disperse seeds

30.5 a. sepal
   b. petal
   c. anther (top of stamen)
   d. stigma (top of carpel)
   e. receptacle
   f. ovary
   g. ovule with megasporocyte
   h. microsporocytes
   i. meiosis
   j. microspore
   k. pollen grain (male gametophyte)
   l. sperm
   m. surviving megaspore
   n. pollen tube
   o. style
   p. polar nuclei
   q. embryo sac (female gametophyte)
   r. egg
   s. fertilization
   t. zygote
   u. endosperm nucleus
   v. seed coat
   w. endosperm
   x. embryo
   y. germinating seed

30.6

Characteristic	Monocot	Eudicot
Number of cotyledons	one	two
Leaf venation	parallel	netlike
Vascular bundles in stems	scattered	in ring
Root system	fibrous	taproot
Flower parts	multiples of three	multiples of four or five
Openings in pollen grain	one	three

SUGGESTED ANSWERS TO STRUCTURE YOUR KNOWLEDGE

1. **a.** Phyla Ginkgophyta, Cycadophyta, Gnetophyta, and Coniferophyta
    b. In addition to the monocots, phylum Anthophyta now includes the clades *Amborella*, water

lilies, star anise and relatives, magnoliids, and eudicots.

2. gametophytes develop within the resistant walls of spores; the protected female gametophyte retained on sporophyte plant; nonswimming pollen; seed for protection, dispersal, and nourishment of embryo

3. flower as enclosed reproductive structure; animal pollinators; fruit for protection and dispersal of seeds

ANSWERS TO TEST YOUR KNOWLEDGE

Multiple Choice:

1. e	**3.** a	**5.** c	**7.** b	**9.** c
2. b	**4.** d	**6.** d	**8.** b	**10** e

True or False:

1. False—add *or algae.*
2. True
3. False—change *gymnosperms* to *angiosperms.*
4. True
5. True
6. False—change *fruit* to *seed,* or change *an embryo . . .* to *a mature ovary.*
7. True
8. True
9. False—change *angiosperms* to *gymnosperms,* or change *haploid cells . . .* to *an embryo sac with a few haploid cells.*
10. True

CHAPTER 31: FUNGI

■ INTERACTIVE QUESTIONS

31.1 The fungus provides phosphate ions and other minerals to the plants; the plants supply the fungi with organic nutrients.

31.2 **a.** network of hyphae, typical body form
b. cross-walls between cells, with pores
c. multinucleated hyphae, aseptate
d. two nuclei from different sources exist together in hyphae
e. fusion of two parent hyphae
f. hyphae with two different nuclei paired up in cells
g. fusion of nuclei to form diploid state

31.3 **(1)** ascomycete
(2) zygomycete
(3) basidiomycete
a. conidia (asexual spores)
b. ascocarp
c. ascospores
d. asci
e. sporangia
f. zygosporangium
g. basidiocarp
h. basidium
i. basidiospores

SUGGESTED ANSWERS TO STRUCTURE YOUR KNOWLEDGE

1. **a.** chytrids, aquatic and soil saprobes or parasites
b. flagellated zoospores

c. zygomycetes, molds and parasites; black bread mold; microsporidia
d. resistant zygosporangia in sexual stage; asexual spores in sporangium on aerial hyphae; microsporidia are highly derived parasites
e. glomeromycetes; all form mycorrhizae
f. endomycorrhizae that form arbuscles; 90% of all plants have arbuscular mycorrhizae
g. ascomycetes; sac fungi, yeasts, molds, morels, mycorrhizae
h. dikaryotic hyphae in ascocarp, karyogamy in asci, ascospores; asexual spores called conidia at tips of hyphae
i. basidiomycetes; club fungi, mushrooms, shelf fungi, rusts
j. fruiting body called basidiocarp, basidiospores produced in basidia

2. The role of fungi as saprobes and decomposers of organic material is central to the recycling of chemicals between living organisms and their physical environment. Parasitic fungi have economic and health effects on humans. Plant pathogens cause extensive loss of food crops and trees. Mycorrhizae are found on the roots of almost all vascular plants. These mutualistic associations greatly benefit the plant. Lichens colonize barren habitats and prepare the way for plant succession.

3. Molecular systematics indicates that both fungi and animals shared a common flagellated protistan ancestor.

ANSWERS TO TEST YOUR KNOWLEDGE

Fill in the Blanks:

1. septum
2. heterokaryon
3. chitin
4. exoenzymes
5. basidium
6. asci
7. mycorrhizae
8. zygosporangia
9. coenocytic
10. chytrids
11. arbuscular mycorrhizae
12. Opisthokonta

Multiple Choice:

1. a	3. c	5. e	7. a	9. c
2. c	4. d	6. b	8. a	10. e

CHAPTER 32: INTRODUCTION TO ANIMAL DIVERSITY

■ INTERACTIVE QUESTIONS

32.1 (1) hollow sphere of unspecialized flagellated cells; (2) beginning of cell specialization; (3) infolding of one side of sphere; (4) forming of digestive cavity, producing a gastrula-like "protoanimal"

32.2 Changes is ecological relationship, such as predator-prey interactions; increased oxygen levels that could support more active animals; evolution of *Hox* complex of genes that regulate development

32.3
a. spiral and determinate
b. radial and indeterminate
c. schizocoelous, split in solid mass of mesoderm
d. enterocoelous, buds from archenteron
e. mouth, second opening for anus
f. anus, second opening for mouth

SUGGESTED ANSWERS TO STRUCTURE YOUR KNOWLEDGE

1.
a. metazoa: multicellular heterotrophic eukaryotes that feed by ingestion
b. Parazoa: Porifera (sponges); assymetrical; no true tissues

c. Eumetazoa: "true" animals; have true tissues
d. Radiata: radial symmetry; diploblastic
e. Bilateria: bilateral symmetry; triploblastic
f. Deuterostomes: radial, indeterminate cleavage; enterocoelous; blastopore forms anus
g. Protostomes: spiral, determinate cleavage; schizocoelous; blastopore forms mouth
h. Lophotrochozoa: molecular similarities; named for lophophore feeding structure and trochophore larva of some phyla
i. Ecdysozoa: molecular similarities; named for molting of exoskeleton typical of some phyla

ANSWERS TO TEST YOUR KNOWLEDGE

Multiple Choice:

1. c	4. e	7. d	10. e
2. d	5. c	8. d	
3. e	6. e	9. b	

CHAPTER 33: INVERTEBRATES

■ INTERACTIVE QUESTIONS

33.1
a. line body cavity (spongocoel); create water current, trap and phagocytose food particles, may produce gametes
b. wander through mesohyl; digest food and distribute nutrients, produce skeletal fibers, may produce gametes

33.2
a. polyp
b. medusa
c. mouth/anus
d. tentacle
e. mesoglea
f. gastrovascular cavity

33.3 **a.** Planarians have an extensively branched gastrovascular cavity; food is drawn in and undigested wastes are expelled through the mouth at the tip of a muscular pharynx.

b. Tapeworms are bathed in predigested food in their host's intestine.

33.4 **a.** **1.** muscular foot for movement
2. visceral mass containing internal organs
3. mantle that may secrete shell and enclose visceral mass to form mantle cavity that houses gills

b. **1.** Snails creep slowly, using their radula to rasp up algae.
2. Clams are sedentary suspension feeders that use siphons to draw water over the gills, where food is trapped in mucus.
3. Squid are active predators that capture and crush prey with jaws.

33.5 **a.** septum
b. intestine
c. metanephridium
d. chaetae
e. ventral nerve cord
f. dorsal and ventral blood vessels
g. longitudinal and circular muscles

33.6 Earthworms have longitudinal and circular muscle layers and a segmented body cavity that allows them to move by peristalsis, alternately elongating and contracting body sections and anchoring to the substrate with chaetae. Nematodes are not segmented and have only longitudinal muscles. These pull against the hydrostatic skeleton of the roundworm's tapered body, producing thrashing movements.

33.7 **a.** anterior feeding appendages modified as pincers or fangs
b. Many spiders trap their prey in webs they have spun from abdominal gland secretions. Spiders kill prey with poison-equipped, fanglike chelicerae. After masticating and adding digestive juices, they suck up partially digested food.

33.8 **a.** mostly terrestrial
b. mostly aquatic
c. fly, walk, burrow
d. swim, float (planktonic)
e. tracheal system
f. gills or body surface
g. Malpighian tubules
h. diffusion across cuticle, glands regulate salt balance

i. one pair
j. two pairs
k. unbranched; mandibles and modified mouthparts for chewing, piercing, or sucking; thorax has three pairs of walking legs and may have wings (extensions of cuticle)
l. branched; mandibles and multiple mouthparts, walking legs on thorax, abdominal appendages

33.9 **a.** radial anatomy as adults, usually five spokes; sessile or slow moving
b. endoskeleton of calcareous plates
c. water vascular system with tube feet for locomotion, feeding, respiration

SUGGESTED ANSWERS TO STRUCTURE YOUR KNOWLEDGE

1. **a.** Eumetazoa
   b. Bilateria
   c. Deuterostomia
   d. Porifera (sponges)
   e. Cnidaria (hydras, jellies, sea anemones, corals)
   f. Platyhelminthes (flatworms: planaria, flukes, tapeworms)
   g. Rotifera (rotifers)
   h. Lophophorates (bryozoans, phoronids, brachiopods)
   i. Nemertea (proboscis worms)
   j. Mollusca (clams, snails, squids)
   k. Annelida (segmented worms: earthworms, polychaetes, leeches)
   l. Nematoda (roundworms)
   m. Arthropoda (crustaceans, insects, spiders, millipedes)
   n. Echinodermata (sea stars, sea urchins)
   o. Chordata (vertebrates, two invertebrate subphyla)

ANSWERS TO TEST YOUR KNOWLEDGE

Matching:

1. C l	**4.** A h	**7.** E b	**10.** C i
2. B e	**5.** H d	**8.** B a	**11.** H m
3. E g	**6.** B j	**9.** D f	**12.** E c

Multiple Choice:

1. b	**4.** e	**7.** c	**10.** b	**13.** d
2. c	**5.** a	**8.** b	**11.** a	**14.** a
3. c	**6.** c	**9.** a	**12.** b	**15.** e

CHAPTER 34: VERTEBRATES

■ INTERACTIVE QUESTIONS

34.1 lancelet, subphylum Cephalochordata
   a. mouth
   b. *pharyngeal slits
   c. intestine
   d. segmental muscles
   e. anus
   f. *post-anal tail
   g. *dorsal, hollow nerve cord
   h. *notochord
   * Chordate characteristics

34.2 two clusters of *Hox* genes, other duplicated families of genes, neural crest, pharyngeal clefts that develop into gill slits, higher metabolism, more muscles, chambered heart, red blood cells with hemoglobin, and kidneys that remove wastes

34.3 more extensive skull, vertebrae (backbone) that encloses spinal cord, and, in aquatic vertebrates, fin rays.

34.4 **a.** Chordates
   b. tunicates (urochordates)
   c. lancelet (cephalochordates)
   d. Craniates
   e. hagfishes
   f. Vertebrates
   g. lampreys
   h. Gnathostomes
   i. chondrichthyans: sharks and rays
   j. Osteichthyans
   k. ray-finned fishes
   l. Lobe-fins
   m. coelacanths
   n. lungfishes
   o. tetrapods
   p. amphibians (frogs, salamanders)
   q. amniotes
   r. reptiles (turtles, snakes, crocodiles, and birds)
   s. mammals

34.5 **a.** jaws, lateral line system, mineralized endoskeleton, paired appendages, enlarged forebrain with better olfaction and vision, duplication of *Hox* genes.
   b. increase the area for absorption and slow the passage of food through the short shark intestine
   c. common opening for the reproductive tract, excretory system, and digestive tract

34.6 See answer to Interactive Question 34.4.

34.7 **a.** Coelacanths, lungfishes, and terrestrial tetrapods
   b. It had the derived appendages of tetrapods with fully formed legs, ankles, and digits. But it retained primitive adaptations for living in water, such as gills and a finned tail. Its pectoral and pelvic girdles were too weak to support its body on land.

34.8 **a.** chorion—functions in gas exchange
   b. amnion—encases embryo in fluid, prevents dehydration, and cushions shocks
   c. allantois—stores metabolic wastes, functions with chorion in gas exchange
   d. yolk sac—expands over yolk; blood vessels transport stored nutrients into embryo

34.9 Archosaurs include the extinct dinosaurs and pterosaurs as well as crocodilians and birds. Lepidosaurs include extinct huge marine reptiles, tuatara, lizards, and snakes.

34.10 strong skeleton; no teeth and absence of some organs to reduce weight; feathers that shape wings into airfoil; strong keel on sternum to which flight muscles attach; high metabolic rate supported by efficient circulatory system, respiratory system, and endothermy; well-developed visual and motor areas of brain

34.11 **a.** hair; milk produced by mammary glands; endothermic with active metabolism; diaphragm to ventilate lungs; four-chambered heart; larger brains, increased learning capacity; differentiated teeth and remodeled jaw
   b. See answer to Interactive Question 34.4.

34.12 increased brain size; change in jaw shape and dentition; bipedal posture; reduced size difference between the sexes; capacity for language and symbolic thought; manufacture and use of complex tools

SUGGESTED ANSWERS TO STRUCTURE YOUR KNOWLEDGE

1. See Interactive Question 34.4 on page 269.

2. (1) Pharyngeal slits used for suspension feeding became adapted for gas exchange; (2) skeletal supports for gills became adapted for use as hinged jaws; (3) lungs of bony fish were transformed to

air bladders, aiding in buoyancy; (4) fleshy fins supported by skeletal elements developed into limbs for terrestrial animals; (5) feathers, probably originally used for insulation, came to be used for flight; (6) dexterous hands important for an arboreal life evolved to manipulate tools.

3. Amphibians remain in damp habitats, burrowing in mud during droughts; some secrete foamy protection for eggs laid on land. Reptiles have scaly, waterproof skin, an amniote egg that provides an aquatic environment for the developing embryo, and behavioral adaptations to modulate changing temperatures. All terrestrial groups are tetrapods, using limbs for locomotion on land.

ANSWERS TO TEST YOUR KNOWLEDGE

Fill in the Blanks:

1. Urochordata	6. amniotic egg
2. Cephalochordata	7. monotremes
3. somites	8. diaphragm
4. gnathostomes	9. anthropoids
5. operculum	10. Australopithecus

Multiple Choice:

1. a	5. a	9. b
2. c	6. d	10. e
3. e	7. a	11. d
4. e	8. c	12. b

CHAPTER 35: PLANT STRUCTURE, GROWTH, AND DEVELOPMENT

■ INTERACTIVE QUESTIONS

35.1 a. reproductive shoot (flower)
b. terminal bud
c. node
d. internode
e. vegetative shoot
f. leaf
g. blade
h. petiole
i. axillary bud
j. stem
k. taproot
l. lateral roots
m. shoot system
n. root system

35.2 a. sclerenchyma (fibers and sclereids), tracheids, and vessel elements
b. the cells listed in a. and sieve-tube members

35.3 Primary growth continues to add to the tips of roots and shoots, while secondary growth thickens and strengthens older regions of a woody plant.

35.4 a. epidermis (dermal)—root hairs for absorption
b. cortex (ground)—uptake of minerals and water, food storage
c. vascular cylinder (vascular)—transport
d. endodermis—regulates water and mineral movement into vascular cylinder
e. pericycle—origin of lateral roots
f. xylem—water and mineral transport
g. phloem—nutrient transport

35.5 a. cuticle
b. upper epidermis
c. palisade mesophyll
d. spongy mesophyll
e. guard cells
f. xylem
g. phloem
h. stomata

35.6 Primary growth is produced by the apical meristems in the terminal buds at the tips of branches. Once these cells mature, secondary growth may begin as vascular cambium produces secondary xylem (wood) and secondary phloem, and cork cambium produces the tough periderm.

35.7 H, F, A, B, G, D, C, E

35.8 Microtubules of the preprophase band leave behind ordered actin microfilaments that determine the orientation of cell division. Microtubules may orient the cellulose microfibrils in cell walls that will determine the direction in which cells can expand.

35.9 a.

Whorl	Genes Active	Organs in Normal	Genes Active Mutant A	Organs in Mutant A
1	*A*	sepals	*C*	carpels
2	*AB*	petals	*BC*	stamens
3	*BC*	stamens	*BC*	stamens
4	*C*	carpels	*C*	carpels

b. Gene *A* would be expressed alone in each whorl, producing a flower with four whorls of sepals.

SUGGESTED ANSWERS TO STRUCTURE YOUR KNOWLEDGE

1. Except for leaves and flowers, which stop growing when they reach maturity (determinate growth), plant roots and shoots continue to grow throughout the life of a plant. Although certain characteristic forms have been favored by natural selection in species adapted to particular environments, the specific environment in which a plant grows greatly influences its individual body form. For example, a plant may grow taller or produce different shapes of leaves depending on its light exposure. This developmental plasticity may help compensate for a plant's immobility.

2. This drawing is a eudicot stem, indicated by the ring of vascular bundles and the central pith.
   a. sclerenchyma
   b. phloem
   c. xylem
   d. cortex
   e. pith
   f. vascular bundle
   g. epidermis

ANSWERS TO TEST YOUR KNOWLEDGE

Matching:

1. H	**5.** I	**9.** J
2. F	**6.** A	**10.** G
3. B	**7.** C	
4. E	**8.** D	

Multiple Choice:

1. a	**5.** d	**9.** e	**13.** c
2. e	**6.** a	**10.** d	**14.** c
3. b	**7.** e	**11.** e	**15.** e
4. c	**8.** c	**12.** b	**16.** d

CHAPTER 36: TRANSPORT IN PLANTS

■ NTERACTIVE QUESTIONS

36.1 **a.** $\Psi_P = 0$, $\Psi_S = -0.6$, $\Psi = -0.6$
b. $\Psi_P = 0.6$, $\Psi_S = -0.6$, $\Psi = 0$. The turgor pressure of the cell that develops with movement of water into the cell finally offsets the solute potential of the cell. No net osmosis occurs, and the water potentials of both cell and solution are 0.
c. $\Psi_P = 0$, $\Psi_S = -0.8$, $\Psi = -0.8$. The cell would plasmolyze, losing water until the solute potential of the cell would equal that of the bathing solution. There would be no turgor pressure.

36.2 **a.** symplastic
b. apoplastic
c. Casparian strip
d. xylem vessels
e. vascular cylinder
f. endodermis
g. cortex
h. epidermis
i. root hair

36.3 **a.** The loss of water vapor from the air spaces through the stomata causes evaporation from the water film coating the mesophyll cells, which creates a tension on the remaining water film.
b. Water molecules hold together due to hydrogen bonding, and a pull on one water molecule is transmitted throughout the column of water.
c. Water molecules adhere to the hydrophilic walls of narrow xylem cells, which helps to support the column of water against gravity.
d. The transpiration of water from the leaf creates a tension, which produces a gradient of water potentials that extends from the leaf to the root.

36.4 In order to provide sufficient CO_2 for photosynthesis, stomata must be open and large amounts of water will be lost to transpiration. A sunny, windy, dry day will increase the rate of transpiration, perhaps to the point that insufficient water is available in the soil. Due to a lack of water, guard cells may lose tugor (and also be signaled by abscisic acid) and stomata will close, slowing transpiration. In reducing excessive water loss, however, the plant mesophyll cells now receive insufficient CO_2, and the photosynthetic rate will decline.

36.5 A storage organ, such as a root or tuber, stores carbohydrates during the summer and is a sugar sink, but when its starch is broken down to sugar in the spring to supply growing buds, it becomes a sugar source.

36.6 **a.** cotransport protein
b. sucrose
c. H^+
d. proton pump
e. ATP

SUGGESTED ANSWERS TO STRUCTURE YOUR KNOWLEDGE

1. Solutes may move across membranes by passive transport when they move down their concentration gradient. Transport proteins, which speed this passive transport, may be specific carrier

proteins or selective channels. Active transport usually involves a proton pump that creates a membrane potential as positively charged hydrogen ions are moved out of the cell. Cations may now move down their electrochemical gradient. In cotransport, a solute is moved by a transport protein driven by the inward diffusion of H^+.

2. The transpiration-cohesion-tension mechanism explains the ascent of xylem sap. The cohesion of water molecules transmits the pull resulting from transpiration throughout the column, and the adhesion of water to the hydrophilic walls of the xylem vessels aids the flow. Tension created by the evaporation of water from the menisci that form in the leaf's air spaces lowers the water potential in the leaf and sets up this bulk flow of water.

A pressure flow mechanism also explains the movement of phloem sap, but the plant must expend energy to create the differences in water potential that drive translocation. The active accumulation of sugars in sieve-tube members lowers water potential at the source end, resulting in an inflow of water. The removal of sugar from the sink end of a phloem tube is followed by the osmotic loss of water. The resulting difference in pressure between the source and sink end of the phloem tube causes the bulk flow of phloem sap.

ANSWERS TO TEST YOUR KNOWLEDGE

1. b	6. b	11. a	16. d
2. e	7. d	12. c	17. d
3. d	8. c	13. e	18. b
4. a	9. a	14. b	
5. c	10. c	15. c	

CHAPTER 37: PLANT NUTRITION

■ INTERACTIVE QUESTIONS

37.1 **a.** component of chlorophyll, activates enzymes
b. component of nucleic acids, phospholipids, ATP, some coenzymes
c. component of cytochromes, cofactor for chlorophyll synthesis, activates enzymes

37.2 in the younger, growing parts of a plant

37.3 A loam, which is a mixture of sand, silt, and clay, has enough large particles to provide air spaces and enough small particles to retain water and minerals. Cations bind to clay particles and are made available to roots by cation exchange. Humus in a soil helps to retain water and provide a steady supply of mineral nutrients, such as nitrogen, phosphorous, and potassium.

37.4 **a.** By linking a promoter that responds to a decrease in a nutrient to a reported gene, a grower can tell when a deficiency is imminent, thus only supplying fertilizer in response to need.
b. variety of conservation-minded, environmentally safe, and profitable farming methods that permit the use of soil as a renewable resource

37.5 **a.** nitrogen-fixing bacteria
b. ammonifying bacteria
c. nitrifying bacteria
d. denitrifying bacteria
e. nitrate and nitrogenous organic compounds

37.6 **a.** Legume crops, which increase nitrogen in the soil, are often followed the following growing season with a nonlegume crop.

b. "Green manure" is a legume crop that is plowed under to improve soil fertility by adding both humus and nitrogen.

37.7 The fossil record shows that some of the earliest land plants had mycorrhizae, and these symbiotic associations may have been an important adaptation helping plants to colonize nutrient-poor soils.

SUGGESTED ANSWERS TO STRUCTURE YOUR KNOWLEDGE

1.

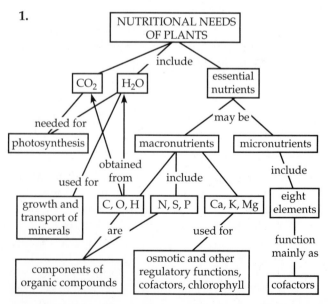

2. Root nodules are mutualistic associations between legumes and nitrogen-fixing bacteria of the genus *Rhizobium*. These bacteria receive nourishment from the plant and provide the plant with ammonium. Mycorrhizae are mutualistic associations between the roots of most plants and various species of fungi. The fungi stimulate root growth, increase surface area for absorption, and secrete antibiotics that may help protect the plant. In return, the fungi receive nourishment from the plant.

3. Both these types of symbiotic relationships involve chemical recognition between specific plant species and bacterial or fungal species.

ANSWERS TO TEST YOUR KNOWLEDGE

1. b	**4.** b	**7.** e	**10.** b	**13.** c
2. a	**5.** a	**8.** d	**11.** c	**14.** e
3. c	**6.** e	**9.** c	**12.** d	**15.** b

CHAPTER 38: ANGIOSPERM REPRODUCTION AND BIOTECHNOLOGY

■ INTERACTIVE QUESTIONS

38.1
  a. stamen
  b. anther
  c. filament
  d. petal
  e. carpel
  f. stigma
  g. style
  h. ovary
  i. sepal
  j. receptacle
  k. ovule

Pollen is formed in the anther. Pollination occurs when pollen lands on the stigma. Fertilization occurs within the embryo sac in the ovule.

38.2
  a. a pollen grain with a patterned spore wall surrounding a generative cell (which will divide to form two sperm) and a tube cell
  b. the embryo sac, often containing seven cells and eight nuclei

38.3 Double fertilization conserves resources by allowing for nutrient development only when fertilization has occurred.

38.4
  a. seed coat
  b. epicotyl
  c. hypocotyl
  d. cotyledons
  e. radicle
  f. cotyledon (scutellum)
  g. endosperm
  h. epicotyl
  i. hypocotyl
  j. radicle
  k. coleorhiza
  l. coleoptile

38.5 Hormonal interactions induce softening of the pulp, a change in color, and an increase in sugar content.

38.6 The coleoptile pushes through the soil, and the shoot tip is protected as it grows up through the tubular sheath.

38.7 Apomixis provides for the dispersal of seeds which are clones of the parent plant.

SUGGESTED ANSWERS TO STRUCTURE YOUR KNOWLEDGE

1. One version of the life cycle of an angiosperm. See also textbook Figure 38.2.

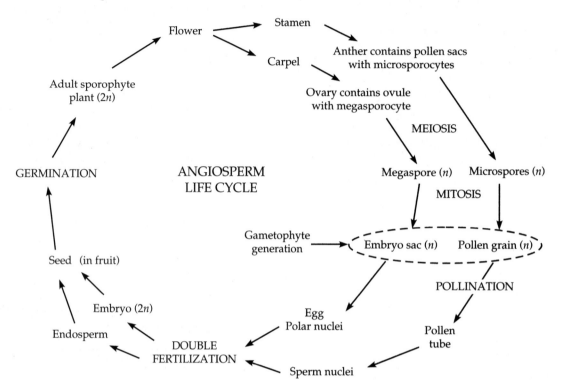

2. *Sexual:* Advantages include increased genetic variability, which provides the potential to adapt to changing conditions, and the dispersal and dormancy capabilities provided by seeds. Disadvantages are that the seedling stage is very vulnerable, sexual reproduction is very energy intensive (since most seeds and seedlings don't survive), and genetic recombination may separate adaptive traits.

   Asexual: Advantages include the hardiness of vegetative propagation and the maintenance of genetically well-adapted plants in a given environment. Disadvantages relate to the advantages of sexual reproduction: There is no genetic variability from which to choose should conditions change, and there are no seeds for dormancy or dispersal.

3. *Benefits:* reduce use of chemical pesticides; weed crops with herbicides instead of erosion-producing tillage; increase nutritional value of crops; produce disease-resistant crops

   Dangers: introduce allergens into human foods; harm nontarget species with pesticides; create hybrid "superweeds" due to crop-to-weed transgene escape; produce other unanticipated harmful results that cannot be stopped once GMOs are released into the environment

ANSWERS TO TEST YOUR KNOWLEDGE

Fill in the Blanks:

1. ovary
2. sporophyte
3. monoecious
4. embryo sac
5. radicle
6. epicotyl
7. coleoptile
8. scion
9. protoplast
10. callus

Multiple Choice:

1. e	5. e	9. c	13. e
2. a	6. a	10. d	14. a
3. d	7. a	11. a	15. c
4. d	8. e	12. d	16. e

CHAPTER 39: PLANT RESPONSES TO INTERNAL AND EXTERNAL SIGNALS

■ INTERACTIVE QUESTIONS

39.1 **a.** The signal is light; the receptor is a phytochrome located in the cytosol.

b. Light-activated phytochrome activates at least two pathways, one producing the second messenger cGMP and one leading to an increase in cytosolic Ca^{2+} levels. Both affect ion channels and activate specific kinases. Some kinase cascades may activate various transcription factors that change gene expression.

c. The plant response involves growth changes (slowing of stem elongation, expansion of leaves, elongation of roots). The de-etiolation process involves decreases in levels of growth-regulating hormones and the activation or new production of enzymes involved in producing chlorophyll or in photosynthesis.

39.2 **a.** ethylene
b. cytokinins
c. abscisic acid
d. auxin
e. brassinosteroids
f. gibberellins

39.3 The relative amount of red and far-red light is communicated to a plant by the ratio of the two forms of phytochrome. Canopy trees absorb red light, and a shaded tree will have a higher ratio of P_r to P_{fr}, inducing the tree to grow taller.

39.4 Free-running periods are circadian rhythms that vary from exactly 24 hours when an organism is kept in a constant environment without environmental cues.

39.5 **a.** not flower
b. flower
c. not flower
d. flower
e. not flower
f. flower
g. not flower
h. flower
i. not flower
j. flower

In the next to last light regimen, a flash of far-red light cancels the effect of the red flash, and the red flash in the last one cancels the effect of the far-red flash.

39.6 **a.** positive phototropism, negative gravitropism
b. thigmotropism; mechanical stimulation causing unequal growth rates of cells on opposite sides of the tendril

39.7 **a.** reduce transpirational water loss by stomatal closing and wilting; deep roots continue to grow in moister soil; heat-shock proteins produced to stabilize plant proteins
b. change membrane lipid composition by increasing unsaturated fatty acids; air tubes may develop in cortex of root

SUGGESTED ANSWERS TO STRUCTURE YOUR KNOWLEDGE

1. cytokinin spray used to keep flowers fresh; ethylene used to ripen stored fruit; synthetic auxins to induce seedless fruit set; gibberellins sprayed on Thompson seedless grapes; 2,4-D (auxin) used as herbicide

2.

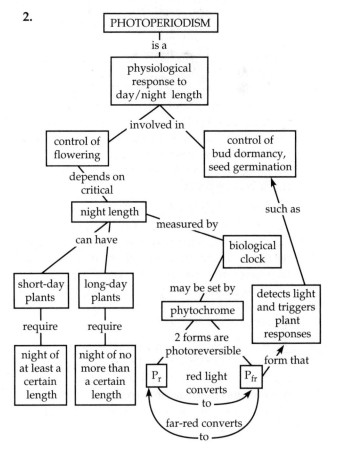

3. Specific resistance begins with an *R-Avr* match. A pathogen molecule binds to a specific plant receptor and triggers a signal-transduction pathway that initiates a hypersensitive response and production of alarm signal. The HR response includes production of antimicrobial molecules, sealing off of infected areas, and destruction of infected cells. The distribution of the signal hormone, probably salicylic acid, triggers signal-transduction pathways in distant cells that activate a systemic acquired resistance. Antimicrobial molecules are produced that protect the cell against pathogens for several days.

ANSWERS TO TEST YOUR KNOWLEDGE

True or False:

1. False—change *abscisic acid* to *gibberellin*, or say *The removal of abscisic acid*.
2. True
3. False—change *gibberellins* to *cytokinins*.
4. False—change *thigmomorphogenesis* to *triple response*; or say *Thigmomorphogenesis is a change in growth form in response to mechanical stress*.
5. True
6. True
7. False—change *circadian rhythm* to *photoperiodism*; or change *day or night length* to *on a 24-hour cycle*.
8. True
9. False—change *A virulent* to *An avirulent*
10. False—change *cryptochrome* to *chromophore*; or say *Cryptochrome is a blue-light photoreceptor*.

Multiple Choice:

1. b	4. b	7. b	10. b	13. d
2. c	5. a	8. e	11. b	14. d
3. e	6. b	9. c	12. e	15. c

CHAPTER 40: BASIC PRINCIPLES OF ANIMAL FORM AND FUNCTION

■ INTERACTIVE QUESTIONS

40.1 The animal's organ systems can control the composition of the fluid bathing its cells. Specialized organs can efficiently meet the animal's needs. A complex body is especially advantageous in the highly variable environments on land.

40.2 a. Stratified squamous; lining of esophagus; thick layer is protective, and new cells produced near basement membrane replace those abraded off.
b. Simple columnar; lining of digestive tract; cells with large cytoplasmic volumes are specialized for secretion or absorption.

40.3 a. osteon (Haversian system)
b. central canal
c. bone
d. collagenous fiber
e. elastic fiber
f. loose connective tissue
g. red blood cells
h. plasma
i. blood

40.4 a. Cardiac; dark bands are intercalated discs that relay electrical signals from cell to cell during heartbeat.
b. smooth muscle

40.5 digestive, circulatory, respiratory, immune and lymphatic, excretory, endocrine, reproductive, nervous, integumentary, skeletal, muscular

40.6 Short bursts of activity rely on ATP present in muscle cells and generated anaerobically by glycolysis. Sustained activity depends on the generation of ATP by aerobic respiration, and an endotherm's cellular respiration rate is about 10 times greater than an ectotherm's rate. An endotherm's respiratory and circulatory systems support their higher metabolic rate.

40.7 Because the environmental temperature determines its body temperature, an ectotherm that lives where temperatures fluctuate very little or where temperatures are very high can have a body temperature more stable than or higher than that of an endotherm.

40.8 Countercurrent heat exchangers warm the cold arterial blood coming from the gills as it moves in smaller arteries inward to the body core, retaining the heat generated by their strong swimming muscles.

40.9 a. vasodilation of surface arterioles, activation of sweat glands, panting
b. vasoconstriction of surface vessels, erection of fur, shivering by contraction of skeletal muscles, nonshivering thermogenesis

SUGGESTED ANSWERS TO STRUCTURE YOUR KNOWLEDGE

1.

Tissue	Structural Characteristics	General Functions	Specific Examples
Epithelial	Tightly packed cells; basement membrane; cuboidal, columnar, squamous shapes; simple or stratified	Protection, absorption, secretion; lines body surfaces	Mucous membrane, may be ciliated; glandular epithelium; lining of blood vessels
Connective	Few cells that secrete extracellular matrix of protein fibers in liquid, gel, or solid substance	Connect and support other tissues	Loose connective; adipose; fibrous connective; bone; cartilage; blood
Muscle	Long cells (fibers) with myofibrils of actin and myosin	Contraction, movement	Skeletal (voluntary); smooth (involuntary); cardiac
Nervous	Neurons with cell bodies, axons, and dendrites	Sense stimuli, conduct impulses	Nerves, brain

2. All organs are covered with epithelial tissue, and internal lumens, ducts, and blood vessels are lined with epithelium. Most organs contain some types of connective tissue. Muscle tissue would be present if the organ must contract. Nervous tissue innervates most organs. The stomach is an example of an organ composed of many tissues. An epithelial mucosa lines the lumen; the submucosa contains blood vessels and loose connective tissue; the muscularis, a layer of circular and longitudinal smooth muscle, comes next; and the outer serosa is an epithelium.

3. Endotherms have thermoregulatory mechanisms that allow them to maintain a constant internal temperature across a wider range of environmental temperatures. While ectotherms may use behavioral mechanisms to reduce temperature fluctuations, they may still be exposed to greater internal fluctuations as environmental temperatures change. Ectotherms may adjust their cellular physiology in response to gradual changes in environmental (and thus body) temperature.

ANSWERS TO TEST YOUR KNOWLEDGE

Multiple Choice:

1. b	**6.** d	**11.** c	**16.** b
2. a	**7.** b	**12.** d	**17.** c
3. d	**8.** a	**13.** d	**18.** a
4. e	**9.** b	**14.** a	**19.** d
5. b	**10.** e	**15.** e	**20.** d

CHAPTER 41: ANIMAL NUTRITION

■ INTERACTIVE QUESTIONS

41.1 Human craving for fatty foods may have evolved from the feast-and-famine existence of our ancestors. Natural selection may have favored individuals who gorged on and stored high-energy molecules, as they were more likely to survive famines.

41.2 Vegetarians must eat a combination of plant foods, complementary in amino acids, to prevent protein deficiencies.

41.3 Recommended daily allowances – the nutrient intakes that nutritionists propose for maintaining health

41.4 The proteins, fats, and carbohydrates that make up food are too large to pass through cell membranes. Monomers also must be reorganized to form the macromolecules specific to each animal.

41.5 **a.** pharynx—sucks food through mouth
b. esophagus—passes food to crop
c. crop—stores and moistens food
d. gizzard—grinds food
e. intestine—digests and absorbs
The typhlosole increases the surface area for absorption.

41.6 **a.** Polysaccharides (starch and glycogen) broken down in the mouth by salivary amylase to smaller polysaccharides and maltose; hydrolysis continues until salivary amylase is inactivated by the low pH in the stomach.
b. Proteins in the stomach are broken down by pepsin to smaller polypeptides.

41.7 **a.** pancreatic amylases, disaccharidases
b. trypsin, chymotrypsin, aminopeptidase, carboxypeptidase, and dipeptidases
c. nucleases, nucleotidases, nucleosidases, phosphatases
d. lipase (aided by bile salts)

41.8 Carnivores are characterized by sharp incisors, fanglike canines, and jagged premolars and molars; herbivores have broad molars for grinding plant material. The intestine of an herbivore is longer to facilitate digestion of plant material, and, unlike carnivores, its cecum may contain cellulose-digesting microorganisms.

SUGGESTED ANSWERS TO STRUCTURE YOUR KNOWLEDGE

1. **a.** energy/fuel
    b. carbon skeletons/organic raw materials
    c. essential nutrients
    d. undernourished
    e. biosynthesis
    f. malnourishment
    g. essential amino acids
    h. vitamins
    i. coenzymes or parts of coenzymes
    j. minerals

2. **a.** salivary glands
    b. oral cavity
    c. pharynx
    d. esophagus
    e. stomach
    f. small intestine
    g. large intestine (colon)
    h. anus
    i. rectum
    j. pancreas
    k. pyloric sphincter
    l. gallbladder
    m. liver

ANSWERS TO TEST YOUR KNOWLEDGE

Matching:

1. M	**4.** B	**7.** J	**10.** E
2. I	**5.** F	**8.** H	
3. D	**6.** G	**9.** N	

Multiple Choice:

1. d	**6.** d	**11.** b	**16.** e	**21.** b
2. c	**7.** a	**12.** a	**17.** a	**22.** e
3. b	**8.** c	**13.** c	**18.** d	
4. b	**9.** b	**14.** c	**19.** c	
5. d	**10.** e	**15.** b	**20.** d	

CHAPTER 42: CIRCULATION AND GAS EXCHANGE

■ INTERACTIVE QUESTIONS

42.1 **a.** The heart pumps hemolymph through vessels into sinuses, and body movements squeeze the sinuses, forcing hemolymph through the body. When the heart relaxes, hemolymph is drawn back in through pores.
b. Open circulatory systems take less energy to build, maintain, and operate. It may serve as a hydrostatic skeleton in molluscs.

42.2 **a.** capillaries of head and forelimbs
b. left pulmonary artery
c. aorta
d. capillaries of left lung
e. left pulmonary vein
f. left atrium
g. left ventricle
h. aorta
i. capillaries of abdomen and hind limbs
j. posterior vena cava
k. right ventricle
l. right atrium

m. right pulmonary vein
n. capillaries of right lung
o. right pulmonary artery
p. anterior vena cava

The pulmonary veins, left side of the heart, and aorta and arteries leading to the systemic capillary beds carry oxygen-rich blood. See text figure 42.5 for numbers. Note that oxygen-rich blood in the aorta splits and travels either to the head region or lower body, returning to the heart through the anterior or posterior vena cava, respectively. Thus, numbers 7–10 do not represent the sequential flow of blood.

42.3 **a.** atrioventricular
b. semilunar
c. SA (sinoatrial) node
d. AV (atrioventricular) node

42.4 **a.** artery walls
b. contraction of ventricle
c. diastole
d. systole

e. diastolic pressure (lower number)
f. systolic pressure (higher number)
g. peripheral resistance
h. contract
i. relax
j. nerves, hormones

42.5 **a.** plasma
b. electrolytes
c. osmotic balance, buffering, muscle and nerve functioning
d. proteins
e. buffers, osmotic factors, lipid escorts, antibodies, clotting factors
f. nutrients, wastes, respiratory gases, hormones
g. erythrocytes (red blood cells)
h. transport O_2 and NO; help transport CO_2
i. leukocytes (white blood cells)
j. phagocytes
k. defense and immunity
l. platelets
m. blood clotting

42.6 smoking, not exercising, obesity, and eating a fat-rich diet

42.7 **a.** water
b. gill
c. pump water through mouth and out operculum
d. air
e. tracheoles of tracheal system
f. body movements compress and expand air tubes
g. air (and water)
h. lungs and moist skin
i. positive pressure breathing: fill mouth, close nostrils, raise floor of mouth, and blow up lungs
j. air
k. alveoli in lungs
l. negative pressure breathing: lower diaphragm, raise ribs, thus increasing volume and decreasing pressure of lungs; air flows in

42.8 This graph shows the Bohr shift. The dissociation curve for hemoglobin is shifted to the right at a lower pH, meaning that at any given partial pressure of O_2, hemoglobin is less saturated with oxygen. A rapidly metabolizing tissue produces more CO_2, which lowers pH, and hemoglobin will unload more of its O_2 to that tissue.

SUGGESTED ANSWERS TO STRUCTURE YOUR KNOWLEDGE

1. **a.** aorta
b. pulmonary artery
c. pulmonary veins
d. left atrium
e. left ventricle
f. right ventricle
g. posterior vena cava
h. atrioventricular valve
i. semilunar valve
j. right atrium
k. anterior vena cava
Blood flow from venae cavae → right atrium → right ventricle → pulmonary artery → lung → pulmonary vein → left atrium → left ventricle → aorta

2. Nasal cavity → pharynx → through glottis to larynx → trachea → bronchus → bronchiole → alveoli → interstitial fluid → alveolar capillary → hemoglobin molecule → venule → pulmonary vein → left atrium → left ventricle → aorta → renal artery → arteriole → capillary in kidney

ANSWERS TO TEST YOUR KNOWLEDGE

Multiple Choice:

1. c	8. b	15. a	22. e	29. e
2. c	9. a	16. d	23. d	30. c
3. d	10. e	17. b	24. e	31. c
4. e	11. d	18. c	25. b	32. a
5. a	12. d	19. c	26. d	33. b
6. b	13. c	20. b	27. d	34. d
7. a	14. a	21. b	28. d	

CHAPTER 43: THE IMMUNE SYSTEM

■ INTERACTIVE QUESTIONS

43.1 Innate immunity includes nonspecific defenses that are present at birth and before exposure to pathogens. Acquired immunity includes specific defenses that develop only after exposure to microbes, abnormal cells, or foreign substances.

43.2 **a.** short-lived cells that phagocytose microbes; about 70% of white blood cells
b. leukocytes that migrate from blood to tissues, where they develop into macrophages
c. long-lived, large phagocytes that engulf microbes and dead tissue cells

d. leukocytes that attack large parasites with enzymes

e. leukocytes that attack membranes of the body's infected or abnormal cells, stimulate apoptosis

f. release histamine to initiate inflammatory response

g. causes vasodilation and increased permeability of blood vessels

h. proteins released by virus-infected cells that stimulate neighboring cells to produce substances that inhibit viral replication

i. set of serum proteins that cause lysis of microbes and are involved with innate and acquired defenses

j. enzyme in tears, saliva, and mucus that attacks bacterial cell walls

k. released by activated macrophages; promote blood flow

l. chemical attractants that guide phagocyte migration

43.3 B cell receptors and antibodies recognize intact antigens that are either molecules on the surfaces of infectious agents or molecules free in the body. T cell receptors recognize pieces of antigens that have complexed with an MHC molecule inside a cell and are then presented on the cell surface. Cytotoxic T cells recognize foreign antigens that have been manufactured by a body cell and displayed with class I MHC molecules. Helper T cells recognize antigen fragments from microbes that have been engulfed by antigen-presenting cells (dendritic cells, macrophages, and B cells) and displayed with class II MHC molecules.

43.4 **a.** During development, the *V* and *J* gene segments of the genes coding for the polypeptide chains of the receptors are randomly combined, producing a large number of different antigen-binding specificities.

b. As lymphocytes mature in the thymus or bone marrow, they are tested for self-reactivity, and those that do react against self components are inactivated or destroyed.

c. Once a lymphocyte recognizes its antigen, the cell divides into a clone of effector and memory cells specific for that antigen.

43.5 **a.** antigen-presenting cell (dendritic cell)

b. bacterium

c. peptide antigen

d. class II MHC molecule

e. T cell receptor

f. CD4

g. helper T cell

h. cytokines stimulate helper T, cytotoxic T, and B cells

i. cytotoxic T cell

j. B cell

k. cell-mediated immunity (attack on infected cells)

l. humoral immunity (secretion of antibodies by plasma cells)

43.6 **a.** CD4

b. CD8

c. cytokines that stimulate cytotoxic T cells, B cells, and themselves

d. perforin and enzymes that kill the target cell

43.7 **a.** viral neutralization—blocks viral binding sites; coats microbes in opsonization

b. agglutination of antigen-bearing microbes

c. precipitation of soluble antigens

d. complement activation and formation of a membrane attack complex

The first three mechanisms tag antigens for phagocytosis.

43.8

Blood Type	Antigens on RBCs	Antibodies in Plasma	Can Receive Blood from	Can Donate Blood to
A	A	anti-B	A, O	A, AB
B	B	anti-A	B, O	B, AB
AB	AB	none	A, B, AB, O	AB
O	none	anti-A + B	O	A, B, AB, O

43.9 **a.** HIV destroys helper T cells, thus crippling the humoral and cell-mediated immune systems and leaving the body unable to fight the HIV virus and other opportunistic diseases such as Kaposi's sarcoma and *Pneumocystis* pneumonia.

b. HIV is readily passed through unprotected sex and needle sharing. The antigenic changes of HIV during replication complicate the development of a vaccine. Mutational changes also lead to drug resistant strains of HIV.

SUGGESTED ANSWERS TO STRUCTURE YOUR KNOWLEDGE

1.

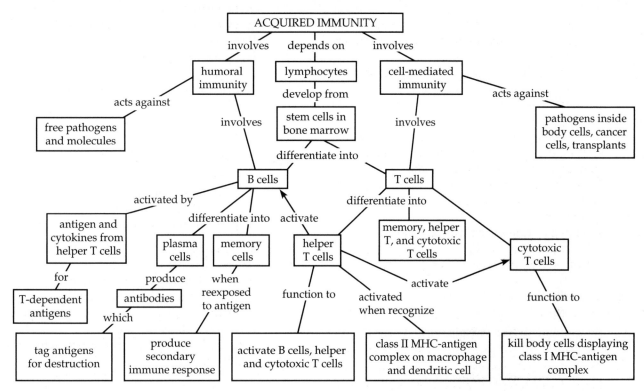

2. An antibody is a protein that consists of two identical light polypeptide chains and two identical heavy chains, held together by disulfide bonds in a Y-shaped molecule. The amino acid sequences in the variable sections of the light and heavy chains in the arms of the Y account for the specificity in binding between antibodies and epitopes of an antigen. The two arms of an antibody allow it to cross-link antigens, contributing to agglutination and precipitation, which enhance phagocytosis. The constant region of the antibody tail determines its effector function: five types of these constant regions correspond to five classes of antibodies. As an example, IgE immunoglobulins are attached to mast cells, which create allergic responses, and IgA is found in body secretions and is the major antibody in colostrum, the first breast milk produced.

3. Macrophages and dendritic cells of innate immunity phagocytose microbes in a nonspecific manner, but present peptide antigens in their class II MHC molecules to helper T cells. This interaction, along with synthesis of cytokines by these phagocytes, helps to activate helper T cells. The complement system functions in both innate and acquired immunity. Some activated complement proteins promote inflammation or stimulate phagocytosis. The complement system may lyse microbes after activation by substances on many microbes or by antigen-antibody complexes on specific microbes.

ANSWERS TO TEST YOUR KNOWLEDGE

Multiple Choice:

1. d	7. a	13. d	19. e	25. c
2. c	8. c	14. c	20. c	26. e
3. e	9. a	15. d	21. e	27. a
4. e	10. b	16. d	22. b	28. c
5. b	11. a	17. a	23. b	29. b
6. e	12. c	18. e	24. d	30. c

CHAPTER 44: OSMOREGULATION AND EXCRETION

■ INTERACTIVE QUESTIONS

44.1 **a.** isoosmotic
b. conformer but may regulate specific solutes
c. slightly hyperosmotic
d. maintain urea and TMAO at high concentrations, rectal gland and kidneys remove salt
e. hypoosmotic
f. drink water to compensate for loss to hyperosmotic seawater, gills pump out salt, little urine excreted.
g. hyperosmotic
h. copious dilute urine to compensate for osmotic gain, may pump salts in through gills
i. hyperosmotic
j. body coverings reduce evaporation, behavior adaptations, drink water

44.2 **a.** An endotherm because it must eat more food to maintain its higher metabolic rate. A carnivore because it eats more protein and thus produces more nitrogenous wastes.
b. *Uric acid* can be stored safely within the egg while the embryo develops. *Urea* is soluble and less toxic than ammonia and can be removed by the mother's blood. Urea is less expensive to produce than uric acid.

44.3 **a.** osmoregulation; remove excess water
b. both; remove excess water
c. both; conserve water
d. both; usually conserve water

44.4 **a.** water, salts, nitrogenous wastes, glucose, vitamins, and other small molecules
b. blood cells and large molecules such as plasma proteins

44.5 **a.** Bowman's capsule
b. glomerulus
c. proximal tubule
d. descending limb, loop of Henle
e. ascending limb, loop of Henle
f. distal tubule
g. collecting duct
(1) HCO_3^-, NaCl*, H_2O, nutrients*, and K^+ are reabsorbed from the proximal tubule. H^{+*} and NH_3 are secreted into the proximal tubule.
(2) Water is reabsorbed from the descending limb of the loop of Henle.
(3) NaCl diffuses from the thin segment of the ascending limb and is pumped out* of the thick segment.

(4) NaCl*, H_2O, and HCO_3^{-*} are reabsorbed from the distal tubule. K^{+*} and H^{+*} are secreted into the tubule.
(5) NaCl* is reabsorbed; urea and H_2O diffuse out of the collecting tubule.

* indicates active transport.

44.6 **a.** high blood osmolarity
b. ADH, antidiuretic hormone
c. distal tubules and collecting ducts
d. JGA, juxtaglomerular apparatus
e. renin
f. NaCl and water reabsorption
g. adrenal glands
h. aldosterone
i. Na^+ and water reabsorption
j. increase in blood pressure or volume
k. ANP, atrial natriuretic protein

44.7 **a.** Juxtamedullary nephrons; the longer the loop of Henle, the more hyperosmotic the urine.
b. Mammals and birds; loops of Henle are especially long in those taxa inhabiting dry habitats.

SUGGESTED ANSWERS TO STRUCTURE YOUR KNOWLEDGE

1. **a.** water, salts, glucose, amino acids, vitamins, urea, other small molecules
   b. ammonia, drugs and poisons, H^+, K^+
   c. water, glucose, amino acids, vitamins, K^+, NaCl, bicarbonate

2. Two solutes, NaCl and urea, contribute to the high osmolarity of the medulla. As the filtrate moves up the ascending limb, NaCl first diffuses and is then pumped out, contributing to the gradient of NaCl in the medulla. As the filtrate flows through the collecting duct, some urea diffuses out and also contributes to the osmotic gradient. As the filtrate in the collecting duct moves through the high osmolarity of the medulla, water flows out of the collecting duct by osmosis, creating a more concentrated urine.

ANSWERS TO TEST YOUR KNOWLEDGE

Multiple Choice:

1. a	**5.** d	**9.** d	**13.** a	**17.** d
2. c	**6.** a	**10.** b	**14.** c	**18.** e
3. c	**7.** c	**11.** e	**15.** a	
4. b	**8.** e	**12.** c	**16.** c	

CHAPTER 45: HORMONES AND THE ENDOCRINE SYSTEM

■ INTERACTIVE QUESTIONS

45.1 In negative feedback, the response reduces the initial stimulus, thus turning off the response. In positive feedback, such as in the release of milk during nursing, the stimulus is reinforced and the response increases.

45.2 **a.** chemical messages that usually travel through the bloodstream to target cells
b. chemical messages, such as cytokines, neurotransmitters, NO, growth factors, and prostaglandins, that communicate locally between cells
c. cell that produces hormones
d. modified nerve cell that produces hormones (neurohormones)

45.3 Oxytocin and ADH are synthesized by neurosecretory cells and transported down their processes to the posterior pituitary. Releasing and inhibiting tropic hormones are secreted by neurosecretory cells into capillaries that drain through portal vessels to a capillary bed in the anterior pituitary.

45.4 **a.** stimulates contraction of uterus and milk release
b. increases water reabsorption in kidney
c. stimulates thyroid gland
d. stimulate activity of gonads
e. stimulates adrenal cortex to produce and secrete hormones
f. various effects, depending on species, including stimulation of milk production
g. regulates skin pigment cells in amphibians
h. reduce perception of pain
i. promotes growth and the release of insulin-like growth factors

45.5 Insufficient iodine results in reduced production of T_3 and T_4. Thus, there is no negative feedback to turn off production of TRH by the hypothalamus and TSH by the anterior pituitary. TSH continues to stimulate the thyroid, which enlarges.

45.6 **a.** insulin
b. glucagon
c. uptake of glucose
d. glycogen breakdown in liver
e. islets of Langerhans in pancreas
f. glycogen hydrolysis in liver and glucose release
g. blood glucose level

45.7 **a.** Nervous stimulation from the hypothalamus to the adrenal medulla causes the secretion of epinephrine, which increases blood pressure, rate and volume of heart beat, breathing rate, and metabolic rate, stimulates glycogen hydrolysis and fatty acid release, and changes blood flow patterns. Norepinephrine release maintains blood pressure.
b. A releasing hormone from the hypothalamus stimulates ACTH release from the pituitary, which stimulates the adrenal cortex to release corticosteroids. Glucocorticoids increase blood glucose through conversion of proteins and fats. Mineralocorticoids increase blood volume and pressure by stimulating the kidney to reabsorb sodium ions and water.

SUGGESTED ANSWERS TO STRUCTURE YOUR KNOWLEDGE

1. Chemical signals that bind to plasma membrane receptors initiate signal-transduction pathways that may activate cellular enzymes or affect gene expression. Steroid hormones bind with protein receptors inside a cell, and the hormone-receptor complex acts as a transcription factor to turn on (or off) specific genes.

2. Calcitonin is secreted by the thyroid in response to a rise in Ca^{2+} levels above a set point. Its effects are to stimulate Ca^{2+} deposit in bones and reduce Ca^{2+} uptake in the intestines and kidneys. As Ca^{2+} levels in the blood fall, PTH (parathyroid hormone) is secreted by the parathyroid glands and reverses the effects of calcitonin. PTH also activates vitamin D, which increases Ca^{2+} uptake from food in the intestines.

ANSWERS TO TEST YOUR KNOWLEDGE

Matching:

1. I f	**5.** A g	**9.** C c, released from g
2. G a	**6.** E b	**10.** J c, released from g
3. F d	**7.** L j	
4. B h	**8.** D j	

Multiple Choice:

1. e	**5.** d	**9.** e	**13.** a	**17.** a
2. d	**6.** a	**10.** e	**14.** d	**18.** b
3. d	**7.** c	**11.** d	**15.** c	
4. c	**8.** b	**12.** c	**16.** d	

CHAPTER 46: ANIMAL REPRODUCTION

■ INTERACTIVE QUESTIONS

46.1 a. Rapid colonization of new areas and perpetuation of successful genotypes in stable habitats; eliminates need to locate mate.
b. Varying genotypes and phenotypes of offspring may enhance reproductive success of parents in fluctuating environments.

46.2 a. courtship behaviors that trigger the release of gametes and may provide a means for mate selection
b. chemical signals (pheromones) from individual releasing gametes
c. environmental signals (temperature, day length)

46.3 a. vas deferens
b. erectile tissue of penis
c. urethra
d. glans penis
e. prepuce
f. scrotum
g. testis
h. epididymis
i. bulbourethral gland
j. prostate gland
k. seminal vesicle
l. oviduct
m. follicles
n. corpus luteum
o. uterine wall (muscular layer)
p. endometrium (lining of uterus)
q. cervix
r. vagina
s. ovary

46.4 a. Spermatogonia continue to divide by mitosis and spermatogenesis occurs continuously, whereas a female's supply of primary oocytes may be present at birth.
b. Each meiotic division produces four sperm, but only one ovum (and polar bodies that disintegrate).
c. Spermatogenesis is an uninterrupted process. Oogenesis occurs in stages: prophase I before birth; after puberty, meiosis I and II up to metaphase II in a maturing follicle just before ovulation; and completion of meiosis II (in humans) when a sperm cell penetrates the oocyte.

46.5 a. LH
b. FSH
c. follicular phase
d. ovulation
e. luteal phase
f. estrogen
g. progesterone
h. menstrual flow phase
i. proliferative phase
j. secretory phase

46.6 a. GnRH (gonadotropin-releasing hormone)
b. FSH (follicle-stimulating hormone)
c. LH (luteinizing hormone)
d. androgens
e. primary and secondary sex characteristics
f. negative feedback loops involving testosterone

46.7 a. secreted by embryo, maintains corpus luteum (and thus progesterone) in first trimester
b. secreted by corpus luteum (estrogen and progesterone secretion) and later by placenta, regulates growth and maintenance of placenta, formation of mucus plug in cervix, cessation of ovulation, growth of uterus
c. secreted by fetus and posterior pituitary; stimulates uterine contractions and uterine secretion of prostaglandins during birth; controls release of milk during nursing

46.8 a. prevent release of gametes—sterilization, combination birth control pills, (or injection, patch, or vaginal ring)
b. prevent fertilization—abstinence, rhythm method, condom, diaphragm, progestin minipill, coitus interruptus, cervical cap, spermicides
c. prevent implantation of embryo—morning after pills (MAP)

The most effective methods are abstinence, sterilization, and chemical contraception. Least effective are the rhythm method and coitus interruptus.

SUGGESTED ANSWERS TO STRUCTURE YOUR KNOWLEDGE

1. Path of sperm from formation to fertilization

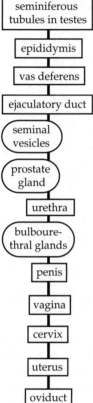

seminiferous tubules in testes — site of spermatogenesis, Leydig cells produce testosterone

epididymis — tubules in which sperm mature and become motile

vas deferens — duct running from epididymis

ejaculatory duct — short duct joining vas deferens from both testes

seminal vesicles — contribute alkaline fluid containing mucus fructose, coagulant, prostaglandins

prostate gland — contributes fluid containing anticoagulants and citrate (a nutrient) to semen

urethra — carries urine or semen through penis

bulboure-thral glands — secretes small amount of mucus prior to ejaculation

penis — erection allows for insertion into vagina

vagina — receptacle for sperm

cervix — narrow opening into uterus

uterus — organ in which fetus develops

oviduct — tube leading to ovary, sperm fertilizes ovum here

2. **a.** GnRH is a releasing hormone of the hypothalamus that stimulates the anterior pituitary to secrete FSH and LH.
   b. FSH stimulates growth of follicles.
   c. The LH surge is caused by an increase in GnRH production that resulted from increasing levels of estrogens produced by the developing follicle.
   d. The LH surge induces the maturation of the follicle and ovulation, and it transforms the ruptured follicle to the corpus luteum.
   e. LH maintains the corpus luteum, which secretes estrogens and progesterone.
   f. High levels of estrogens and progesterone act on the hypothalamus and pituitary to inhibit the secretion of LH and FSH.
   g. When the corpus luteum disintegrates, the production of estrogens and progesterone ceases, which allows LH and FSH secretion to begin again.

3. Birth control pills, which contain a combination of synthetic estrogens and progestin, act by negative feedback to prevent the release of GnRH by the hypothalamus and FSH and LH by the pituitary, thus preventing the development of follicles and ovulation. RU-486 is an analog of progesterone that blocks progesterone receptors in the uterus, thus preventing progesterone from maintaining the endometrium.

ANSWERS TO TEST YOUR KNOWLEDGE

Fill in the Blanks:

1. budding
2. pheromomes
3. parthenogenesis
4. hermaphrodite
5. cloaca
6. estrous cycle
7. progesterone
8. urethra
9. vasocongestion
10. menopause

Multiple Choice:

1. e	5. e	9. b	13. e	17. b
2. d	6. c	10. b	14. c	18. d
3. a	7. e	11. c	15. b	19. a
4. c	8. a	12. a	16. b	20. e

CHAPTER 47: ANIMAL DEVELOPMENT

■ INTERACTIVE QUESTIONS

47.1 a. specific binding between receptors on the plasma membrane (extending through the vitelline layer) and proteins on acrosomal process
b. fast block to polyspermy caused by membrane depolarization during acrosomal reaction and slow block to polyspermy caused by formation of the fertilization envelope during cortical reaction

47.2 a. morula
b. blastula of frog embryo
c. animal pole
d. blastocoel
e. vegetal pole

47.3 a. ectoderm
b. mesenchyme cells
c. endoderm

d. archenteron
e. blastopore
f. ectoderm
g. mesoderm
h. endoderm
i. yolk plug in blastopore
j. archenteron

47.4 **a.** neural tube
b. neural crest
c. somite
d. archenteron
e. coelom
f. notochord

47.5 **a.** The yolk sac grows to enclose the yolk and develops blood vessels to carry nutrients to the embryo.
b. The amnion encloses a fluid-filled sac, which provides an aqueous environment for development and acts as a shock absorber.

c. The allantois serves as a receptacle for certain metabolic wastes.
d. The chorion (pressed against the inside of the egg shell), in conjunction with the allantois, provides gas exchange for the developing embryo.

47.6 **a.** expanding region of trophoblast—chorion, fetal portion of placenta
b. epiblast—three primary germ layers, amnion, some mesoderm becomes part of placenta, allantois
c. hypoblast—yolk sac

47.7 Cytoplasmic determinants in the sea urchin egg must have an unequal polar distribution. The developmental fates of cells in the animal and vegetal halves of the embryo are determined by the first horizontal division.

47.8 The developing limb would be a mirror image with posterior digits facing both forward and backward. This is the same effect as transplanting a donor ZPA to that location.

SUGGESTED ANSWERS TO STRUCTURE YOUR KNOWLEDGE

1.

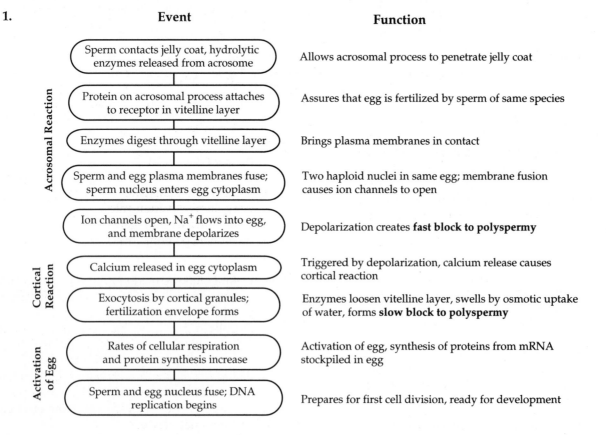

Event	Function
Sperm contacts jelly coat, hydrolytic enzymes released from acrosome	Allows acrosomal process to penetrate jelly coat
Protein on acrosomal process attaches to receptor in vitelline layer	Assures that egg is fertilized by sperm of same species
Enzymes digest through vitelline layer	Brings plasma membranes in contact
Sperm and egg plasma membranes fuse; sperm nucleus enters egg cytoplasm	Two haploid nuclei in same egg; membrane fusion causes ion channels to open
Ion channels open, Na⁺ flows into egg, and membrane depolarizes	Depolarization creates **fast block to polyspermy**
Calcium released in egg cytoplasm	Triggered by depolarization, calcium release causes cortical reaction
Exocytosis by cortical granules; fertilization envelope forms	Enzymes loosen vitelline layer, swells by osmotic uptake of water, forms **slow block to polyspermy**
Rates of cellular respiration and protein synthesis increase	Activation of egg, synthesis of proteins from mRNA stockpiled in egg
Sperm and egg nucleus fuse; DNA replication begins	Prepares for first cell division, ready for development

Acrosomal Reaction

Cortical Reaction

Activation of Egg

2.

	Cleavage	Blastula	Gastrula
Sea urchin	Holoblastic; equal-sized blastomeres	Hollow ball of cells, one cell thick, surrounding blastocoel	Invagination through blastopore to form endoderm of archenteron, migrating mesenchyme forms mesoderm
Frog	Holoblastic, but yolk impedes divisions; unequal-sized blastomeres	Blastocoel in animal hemisphere; walls several cells thick	Involution of cells at dorsal lip of blastopore, migrating cells form mesoderm and endoderm around archenteron
Bird	Meroblastic, only in cytoplasmic disk on top of yolk	Blastoderm; blastocoel cavity between epiblast and hypoblast	Migration of epiblast cells through primitive streak forms mesoderm and endoderm, lateral folds join to form archenteron
Mammal	Holoblastic, no apparent polarity to egg, blastomeres equal	Embryonic disc forms from innercell mass with epiblast and hypoblast; trophoblast surrounds embryo	Migration of epiblast cells through primitive streak forms mesoderm and endoderm

ANSWERS TO TEST YOUR KNOWLEDGE

Multiple Choice:

1. c	**4.** e	**7.** b	**10.** c	**13.** a	**16.** e	**19.** d
2. c	**5.** e	**8.** a	**11.** b	**14.** b	**17.** a	**20.** d
3. b	**6.** d	**9.** e	**12.** e	**15.** e	**18.** b	

CHAPTER 48: NERVOUS SYSTEMS

■ INTERACTIVE QUESTIONS

48.1 Sessile molluscs such as clams have little cephalization and only simple sense organs. Cephalopod molluscs such as squid have large brains, image-forming eyes, and giant axons, all of which contribute to their active predatory life and their ability to learn.

48.2
a. dendrites
b. cell body
c. axon hillock
d. axon
e. myelin sheath
f. Schwann cell
g. nodes of Ranvier
h. synaptic terminal
The impulse moves from the cell body along the axon to the synaptic terminals.

48.3.
a. K^+ inside the cell; Na^+ outside the cell
b. The inside of the cell is more negative compared to outside the cell.
c. The membrane potential moved closer to E_K; more potassium channels must have opened, and/or sodium channels must have closed.

48.4
a. sodium channel
b. Na^+
c. potassium channel
d. K^+
e. sodium activation gate
f. potassium activation gate
g. sodium inactivation gate
h. membrane potential (mV)
i. time

1. Resting state: activation gates on sodium and potassium channels are closed, sodium inactivation gates are open.
2. Depolarization: some Na^+ channels open; if Na^+ influx reaches threshold, additional Na^+ gates open and action potential is triggered.
3. Rising phase: sodium activation gates open and Na^+ enters cell, inside of cell beomes positive with respect to outside; activation gates on K^+ channels not yet open.
4. Falling phase: sodium inactivation gates close, potassium gates open, K^+ leaves cell and inside of cell becomes negative.
5. Undershoot: both gates of the sodium channels are closed, potassium channels still open, and membrane becomes hyperpolarized.

48.5 **a.** Ca^{2+} flowing into presynaptic cell
     b. synaptic terminal
     c. synaptic cleft
     d. synaptic vesicles containing neurotransmitter
     e. presynaptic membrane
     f. postsynaptic membrane
     g. receptor with bound neurotransmitter
     h. closed ligand-gated ion channel

48.6 **a. Temporal**
     b. Spatial

48.7 The type of postsynaptic receptor and its mode of action determine neurotransmitter function. Binding of acetylcholine to receptors in heart muscle activates a signal transduction pathway that makes it harder to generate an action potential, thus reducing the strength and rate of contraction. The acetylcholine receptor in skeletal muscle cells opens ion channels and has a stimulatory effect, depolarizing the muscle cell membrane.

48.8 **a.** peripheral nervous system
     b. central nervous system
     c. autonomic

     d. somatic
     e. spinal cord
     f. enteric
     g. parasympathetic
     h. sympathetic
     i. slows heart, conserves energy
     j. acetylcholine
     k. norepinephrine at target organ (acetylcholine by preganglionic neuron)
     l. skeletal muscles
     m. acetylcholine

48.9 **a.** cerebral cortex 5
     b. thalamus   3
     c. cerebellum   1
     d. medulla   4
     e. pons   2
     f. midbrain   7
     g. hypothalamus 6

48.10 This transfer is facilitated by rehearsal, a positive or negative emotional state, and associations with previously learned and stored information.

SUGGESTED ANSWERS TO STRUCTURE YOUR KNOWLEDGE

1.

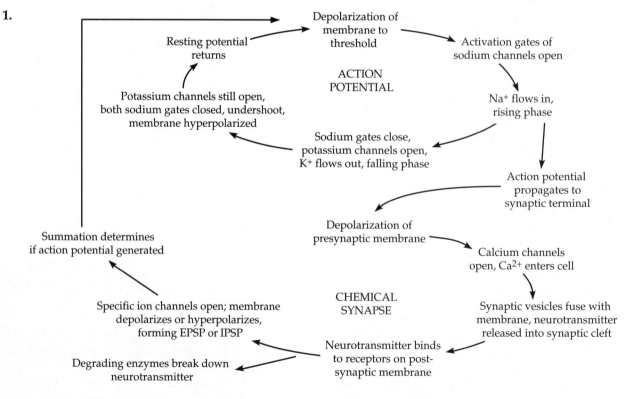

2.   **a.** system of over 90 clusters of neurons extending through the brainstem; regulates state of arousal; filters all sensory information going to cortex
b. in hypothalamus; functions as biological clock for circadian rhythms
c. clusters of neurons deep in white matter of cerebrum; centers for motor coordination and planning and learning movements
d. functional center in parts of thalamus, hypothalamus, and inner cerebral cortex (olfactory bulb, hippocampus, and amygdala); centers of human emotions and memories
e. two structures in inner cerebral cortex; involved in limbic system and memory; amygdala is involved in forming emotional memories; hippocampus functions in learning and memory storage

ANSWERS TO TEST YOUR KNOWLEDGE

Multiple Choice:

1. b	**8.** c	**15.** c	**22.** b	**29.** c
2. a	**9.** d	**16.** a	**23.** c	**30.** d
3. a	**10.** c	**17.** b	**24.** b	**31.** e
4. d	**11.** a	**18.** d	**25.** a	**32.** b
5. e	**12.** e	**19.** d	**26.** e	
6. c	**13.** d	**20.** a	**27.** e	
7. b	**14.** b	**21.** c	**28.** a	

CHAPTER 49: SENSORY AND MOTOR MECHANISMS

■ INTERACTIVE QUESTIONS

49.1 **a.** sensory transduction
b. amplification
c. transmission
d. integration

49.2 **a.** Most are concentrated in the skin; some are associated with other organs.
b. They inhibit prostaglandin synthesis.

49.3 **a.** auditory canal
b. malleus (hammer)
c. incus (anvil)
d. stapes (stirrup)
e. semicircular canals
f. auditory nerve
g. cochlea
h. round window
i. Eustachian tube
j. oval window
k. tympanic membrane (ear drum)

49.4 **a.** utricle and saccule for position, semicircular canals for movement
b. lateral line system with neuromasts
c. statocysts containing statoliths
d. The balance organs of mammals have otoliths and those of invertebrates contain statoliths. These granules settle by gravity and stimulate hair cells to signal position and movement. Sensory hairs on the hair cells in the neuromasts of fishes are embedded in a gelatinous cap, which is bent by moving water and stimulates the hair cells. Hair cells in the ampullae of the semicircular canals project into a gelatinous cap, which is bent by moving endolymph. All of these are mechanoreceptors.

49.5 find mates, recognize territory, navigate, communicate, and locate and taste food

49.6 **a.** sclera
b. choroid
c. retina
d. fovea (center of visual field)
e. optic nerve
f. optic disk (blind spot)
g. vitreous humor
h. lens
i. aqueous humor
j. pupil
k. iris
l. cornea
m. suspensory ligament
n. ciliary body

49.7 **a.** small
b. fovea

49.8 **a.** Hydrostatic: no extra materials needed, good shock absorber; not much protection or support for lifting animal off the ground; cnidarians, flatworms, nematodes, and annelids
b. Exoskeleton: quite protective; in arthropods, must be molted in order to grow, restricts size; most molluscs and arthropods
c. Endoskeleton: various types of joints allow for flexible movement in vertebrates, good structural support; not very protective; sponges, echinoderms, chordates

49.9 **a.** sarcomere
b. I band
c. A band
d. Z line
e. thin filament (actin)
f. thick filament (myosin)
g. H zone

49.10 **a.** fast glycolytic, fast oxidative, slow oxidative
b. fast glycolytic
c. slow oxidative

49.11 **a.** swimming
b. running
c. flying
d. smaller

SUGGESTED ANSWERS TO STRUCTURE YOUR KNOWLEDGE

1. Light enters the eye through the pupil and is focused by the lens onto the retina. When rhodopsin (visual pigment in rods) absorbs light energy, it sets off a signal-transduction pathway that decreases the receptor cell's permeability to sodium and hyperpolarizes the membrane. The reduction in the release of neurotransmitter by a rod cell releases some connected bipolar cells from inhibition, which in turn generates action potentials in ganglion cells—the sensory neurons that form the optic nerve. Horizontal and amacrine cells provide lateral integration of visual information.

2. Sound waves are collected by the *pinna* and travel through the *auditory canal* to the *tympanic membrane,* where they are transmitted by the *malleus, incus,* and *stapes.* Vibration of the stapes against the *oval window* sets up pressure waves in the fluid in the *vestibular canal* within the *cochlea,* from which they are transmitted to the *tympanic canal* and then dissipated when they strike the *round window.* The pressure waves vibrate the *basilar membrane,* on which the *organ of Corti* is located within the *cochlear duct.* Tips of the hair cells arising from the organ of Corti are embedded in the *tectorial membrane.* When vibrated, they bend, triggering a depolarization, release of neurotransmitter, and initiation of action potentials in the sensory neurons of the *auditory nerve,* which leads to the *cerebral cortex.*

3. A motor neuron releases acetylcholine into the synapse with a muscle, initiating an action potential within the muscle fiber, which spreads into the center of the cell along the transverse tubules. The action potential changes the permeability of the sarcoplasmic reticulum membrane, which releases calcium ions into the cytosol. The calcium binds with troponin, changes the shape of the tropomyosin-troponin complex, and exposes the myosin-binding sites of the actin molecules. Heads of myosin molecules in their high-energy configuration bind to these sites, forming cross-bridges. Myosin relaxes to its low-energy state and its head bends, pulling the thin filament toward the center of the sarcomere. When myosin binds to an ATP, the cross-bridge breaks. Hydrolysis of the ATP returns the myosin to its high-energy configuration, and it binds further along the actin molecule. This sequence continues as long as there is ATP and until calcium is pumped back into the sarcoplasmic reticulum, when the tropomyosin-troponin complex again blocks the actin sites.

ANSWERS TO TEST YOUR KNOWLEDGE

Matching:

1. I	**4.** L	**7.** M	**10.** B
2. G	**5.** J	**8.** D	
3. N	**6.** Q	**9.** A	

Multiple Choice:

1. c	**8.** a	**15.** b	**22.** d
2. a	**9.** e	**16.** a	**23.** c
3. c	**10.** b	**17.** c	**24.** b
4. b	**11.** a	**18.** c	**25.** a
5. d	**12.** d	**19.** d	**26.** c
6. c	**13.** b	**20.** e	**27.** e
7. c	**14.** d	**21.** b	**28.** a

CHAPTER 50: AN INTRODUCTION TO ECOLOGY AND THE BIOSPHERE

■ INTERACTIVE QUESTIONS

50.1 **a.** organismal: physiological, behavioral, and morphological adaptations of organisms to the biotic and abiotic environment

b. population: growth and regulation of population size

c. community: interactions among populations that affect community structure

d. ecosystem: energy flow and chemical cycling between biotic and abiotic components

e. landscape: exchanges among patches or ecosystems of a landscape or seascape

50.2 **a.** South-facing slopes in the northern hemisphere receive more sunlight and are warmer and drier than north-facing slopes.

b. Air temperature drops with an increase in elevation, and high-altitude communities may be similar to communities in higher latitudes.

c. The windward side of a mountain range receives much more rainfall than the leeward side. The warm, moist air rising over the mountain releases moisture, and the drier, cooler air absorbs moisture as it descends the other side.

50.3 **a.** dispersal: an area may be beyond the dispersal ability of a species

b. behavior and habitat selection: insect larvae may be able to feed on more plants, but females oviposit on a single type of plant

c. biotic factors: the presence of predators or competitors may restrict a species' range

d. abiotic factors: climate (sunlight, water, temperature, wind) may determine whether a species can inhabit an area

50.4 **a.** + − −
b. − + +
c. + − −
d. − + −
e. + + +

50.5

_7_ abyssal	_2_ neritic		
_6_ aphotic	_3_ oceanic		
_8_ benthic	_5_ pelagic		
_1_ intertidal	_4_ photic		

50.6 **a.** desert
b. temperate grassland
c. tropical forest
d. temperate broadleaf forest
e. coniferous forest
f. arctic and alpine tundra

SUGGESTED ANSWERS TO STRUCTURE YOUR KNOWLEDGE

1. **a.** Ecology is the study of the distribution and abundance of organisms and their relationship to the abiotic and biotic environments.
b. An important cause of evolutionary change is the interactions of organisms with their environment. Thus, interactions occurring within an ecological time frame translate into adaptations that are evident on the scale of evolutionary time.

2. **a.** Biomes are characteristic communities, usually identified by the predominant vegetation and climate, that range over broad geographic areas.
b. Convergent evolution, common adaptations of different organisms to similar environments, accounts for similarities in life forms within geographically separated biomes.

ANSWERS TO TEST YOUR KNOWLEDGE

Multiple Choice:

1. c	**5.** d	**9.** c	**13.** e
2. b	**6.** b	**10.** b	**14.** d
3. d	**7.** e	**11.** d	**15.** a
4. e	**8.** c	**12.** e	

Matching:

1. C	**3.** D	**5.** A	**7.** B
2. H	**4.** E	**6.** G	**8.** F

CHAPTER 51: BEHAVIORAL ECOLOGY

■ INTERACTIVE QUESTIONS

51.1 **a.** The stimulus of an increase in day length may result in reproductive behavior.

b. Production of viable offspring is most successful at this time due to warm temperature and abundant food supply; thus, this behavior has been selected for by natural selection.

51.2 **a.** The red belly of an intruder into its territory is the sign stimulus that releases attack behavior in the fish. By chasing away competitors, the male decreases the chance that another male will fertilize eggs laid in his nesting territory.

b. The imprinting stimulus for newly hatched geese is the movement of an object, whether it be their mother or Konrad Lorenz, away from them. Mother-offspring bonding is critical for survival of offspring and for the reproductive fitness of the parent.

c. Should captive birds imprint on other species or on humans, they will not form pair bonds with their own species.

51.3 The sow bugs were showing negative phototaxis and a kinesis for relative humidity.

51.4 Mammals are mostly nocturnal; birds are diurnal.

51.5 **a.** imprinting
b. operant conditioning (a type of associative)
c. habituation
d. spatial learning
e. classical conditioning (associative)

51.6 Raising animals in the lab allows researchers to search for evidence of genetic bases of behavior by controlling for environmental influences. In some cases, crossbreeding experiments provide evidence of genetic control of behavior.

51.7 Zack measured the average number of drops required to break whelk shells at various heights and calculated the average total effort required by computing the total flight height (number of drops × height per drop). He measured the average flight height for crows in their whelk-breaking behavior (5.23m) and found it was very close to the 5-meter height he predicted based on an optimal compromise between energy gained versus energy expended.

51.8 If a male's parental care resulted in his greater reproductive success, then the genes for this behavior would increase in frequency in a population. But if the male were parenting offspring that were not his own, then he would not pass these "parental care" genes on to the next generation in a higher proportion because he was also helping to perpetuate the genes of other males.

51.9 Orange throats cannot successful defend all their females from sneaky yellow throats, so yellow throats mate more often and their frequency increases in the population. But blue throats restrict the access of yellow throats to their small number of females, and blue throat numbers will increase. The aggressive orange throats can take over blue throat territories, and once again, orange throat numbers will increase.

51.10 **a.** a parent or a sibling
b. The individual shares more genes in common with a parent or a sibling. The coefficient of relatedness is 0.5, whereas it is only 0.125 for a cousin. Thus, with Hamilton's rule $rB > C$, the benefit to the recipient is discounted less by a larger value of r.

SUGGESTED ANSWERS TO STRUCTURE YOUR KNOWLEDGE

1. Behavior ecology seeks to identify the aspects of animal behavior that are innate and genetically programmed as well as those that are a product of experience and learning. Fixed-action patterns are clearly developmentally fixed, although experience may improve the performance of such behaviors. In other cases, genetics may set the parameters for an organism's behavior; however, experience can modify behavior, and learning is clearly evident.

2. According to the concept of fitness, an animal's behavior should help to increase its chance of survival and production of viable offspring. Survival behaviors that are genetically programmed would be most likely to be passed on, and species would evolve many innate behaviors for foraging, migrating, mating, and care of offspring. The evolution of cognitive ability would help individuals deal with novel situations. Social behaviors and communication may help individuals succeed. Parental investment and certainty of paternity may result in differences in mating systems and parental care that influence an individual's reproductive success. Altruistic behavior may be explained on the basis of kin selection; the inclusive fitness of an animal increases if its altruistic behavior benefits related animals who may be carrying a high proportion of the same genes.

ANSWERS TO TEST YOUR KNOWLEDGE

Multiple Choice:

1. d	**6.** c	**11.** a	**16.** d
2. b	**7.** b	**12.** c	**17.** c
3. a	**8.** a	**13.** e	**18.** d
4. b	**9.** c	**14.** d	**19.** e
5. d	**10.** e	**15.** b	**20.** a

CHAPTER 52: POPULATION ECOLOGY

■ INTERACTIVE QUESTIONS

52.1 clumped; uniform; random

52.2 **a.** Type I, humans and many large mammals
b. Type II, Belding's ground squirrels and some other rodents, some annual plants, various invertebrates, and some lizards
c. Type III, many fishes and marine invertebrates such as oysters, long-lived plants

52.3 **a.** An organism has limited resources to divide between growth, survival and reproduction.
b. Selection would most likely favor iteroparity, with fewer, larger, better-provisioned or cared-for offspring.

52.4 **a.** exponential growth; $dN/dt = r_{max}N$
b. logistic growth; $dN/dt = r_{max}N(K-N)/K$; K is 1500

52.5 a. *r*-selected; **b.** *K*-selected

52.6 **a.** nutrients, space for nests, accumulation of toxic wastes, predation, intrinsic limiting factors
b. extremes in weather, natural disasters, fires

2. Reproductive success is measured in the number of offspring that survive and live to reproduce. Many "choices" are available in life history traits: big-bang versus repeated reproduction, age at first reproduction, number of reproductive episodes, number of offspring, parental investment in size of offspring or care. There are always trade-offs between reproduction and survival due to limited energy budgets. Considering that diverse environments (with both biotic and abiotic factors) and different population densities (*K*- or *r*-selection) create different selection pressures on a population, there is no one "best" reproductive strategy.

SUGGESTED ANSWERS TO STRUCTURE YOUR KNOWLEDGE

1.

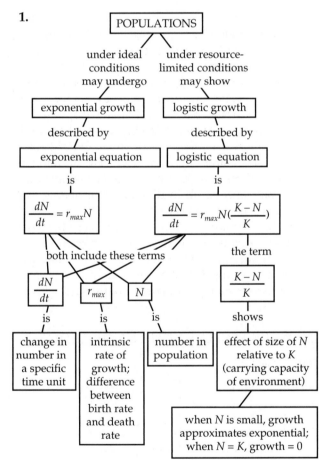

ANSWERS TO TEST YOUR KNOWLEDGE

Multiple Choice:

1. a	**6.** d	**11.** e	**16.** e
2. c	**7.** c	**12.** a	**17.** d
3. d	**8.** d	**13.** b	**18.** c
4. b	**9.** c	**14.** a	
5. e	**10.** d	**15.** a	

CHAPTER 53: COMMUNITY ECOLOGY

■ INTERACTIVE QUESTIONS

53.1 The realized niche of species I is smaller than its fundamental niche when it is in competition with species II. In these trees, species II's fundamental and realized niches are the same.

53.2 **a. Batesian mimicry**
b. Müllerian mimicry

53.3 **a.** mutualism: mycorrhizae, flowering plants and pollinators, ants on acacia trees, cellulose-digesting microorganisms in termites and ruminants
b. commensalism: cattle egrets and cattle that flush insects (although cattle may benefit when ectoparasites are eaten or be harmed if the birds make them more obvious to predators)
c. competition: competition of barnacle species
d. predation: animal predators killing prey
e. herbivory: herbivores eating parts of plants
f. parasitism: endo- and ectoparasites feed in or on host
g. disease: pathogens may kill host

53.4 energetic hypothesis

53.5 The $+/-$ cascade that would be needed to end with a decrease in algae would require an increase in zooplankton, which would require a decrease in primary predators caused by an increase in top predators. More top predators could be added to the lake or primary predators could be removed.

53.6 **a.** Soil nitrogen levels begin quite low, but rise due to the symbiotic nitrogen-fixing bacteria of alder.
b. The soil pH changes from about 7.0 to 4.0 as the acid spruce leaves decompose.

53.7 More interspecific interactions would have had time to develop, and this longer span of evolutionary time would allow for more speciation events to have occurred.

53.8 **a.** immigration—large island
b. immigration—small island
c. extinction—small island
d. extinction—large island

The small island equilibrium number is projected down from the intersection of lines b and c. The large island number is projected down from the intersection of a and d.

53.9 the individualistic hypothesis: each species appears to have an independent distribution along environmental gradients

SUGGESTED ANSWERS TO STRUCTURE YOUR KNOWLEDGE

1. **a.** competition
   b. resource partitioning and character displacement
   c. slightly different niches
   d. keystone predator
   e. herbivory
   f. mutualism
   g. food web

2. **a.** No two species with the same niche can coexist in a habitat; the more competitive species will cause the local elimination of the other.
   b. Food chains are limited to a few links because of the inefficiency of energy transfer (only about 10%) from one trophic level to the next.
   c. Food chains are more stable with fewer links because the effect of environmental disruptions becomes magnified in higher trophic levels.
   d. Each trophic level controls the next higher level; adding nutrients will increase the biomass of all other trophic levels.
   e. Community organization is controlled from the top by predation, with a cascade of $+/-$ effects down the trophic levels with changes in predator numbers.
   f. Most communities are constantly changing in composition due to the effects of disturbances.
   g. Plant communities are chance groupings of species that share similar abiotic requirements.
   h. Plant communities are interrelated groups of species that function as integrated units.
   i. Most species in a community are interconnected and necessary for community function.
   j. Species in a community are only loosely connected, and individual species can easily be replaced by other community members.

ANSWERS TO TEST YOUR KNOWLEDGE

Multiple Choice:

1. c	5. b	9. d	13. a	17. e
2. d	6. b	10. c	14. a	18. c
3. b	7. c	11. d	15. d	19. e
4. e	8. b	12. d	16. b	20. d

CHAPTER 54: ECOSYSTEMS

■ INTERACTIVE QUESTIONS

54.1 Whereas energy makes a one-way trip through ecosystems, chemical elements move through the trophic levels and are then recycled to abiotic reservoirs and back into producers.

54.2 **a.** tropical rain forest, coral reef, swamp and marsh, estuary
b. desert, tundra, lake and stream, open ocean
c. The open ocean covers 65% of Earth's surface area.
d. Upwellings in these cold seas bring nitrogen and phosphorus to the surface. The lack of these nutrients limits production in many tropical waters.

54.3 **a.** Much of a bird or a mammal's assimilated energy is used to maintain a warm body temperature and is not available for net secondary production (growth and reproduction).
b. $\frac{1}{1000}$ (10% of 10% of 10%)

54.4 Water is essential to all organisms and its availability influences primary production and decomposition. Carbon forms the backbone for all organic molecules. Nitrogen is a component of amino acids and nucleic acids. Phosphorus is found in nucleic acids, phospholipids, and ATP.

54.5 **a.** nitrogen-fixing bacteria in root nodules
b. nitrogen-fixing soil bacteria
c. decomposers
d. ammonium (NH_4^+)
e. nitrifying bacteria
f. nitrates (NO_3^-)
g. denitrifying bacteria

54.6 **a.** tropical rain forests, because decomposition proceeds rapidly in the warm, wet climate, and nutrients are rapidly assimilated by new growth
b. lakes and oceans; lack of oxygen slows decomposition and, unless there are upwellings, nutrients are not available to producers
c. Nutrients are not recycled and may leave the ecosystem through runoff.

54.7 melting of ice caps and flooding of coastlines; change in precipitation patterns such that many currently important agricultural areas may become arid; changes in species composition or increased extinction rate as temperatures rise faster than communities can adjust and species can migrate

SUGGESTED ANSWERS TO STRUCTURE YOUR KNOWLEDGE

1.

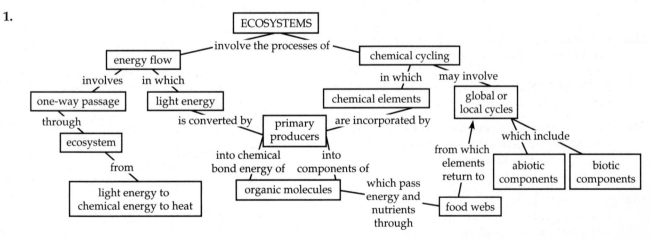

2. **a.** Deforestation can cause an increase in water runoff with an accompanying loss of soil and minerals. Clear-cutting of tropical forests results in loss of species diversity and a reduction in productivity of the area.
b. Dumping of wastes and runoff from agricultural lands has resulted in eutrophication of many lakes, killing fish and other organisms.

c. Toxic chemicals introduced into the environment have been incorporated into the food chain. As a result of biological magnification, these substances pose threats to top-level consumers.
d. An increase in atmospheric CO_2 from fossil fuel and wood combustion adds to the greenhouse effect. The resulting increase in temperature may have far-reaching effects on climate, sea level, and species composition.

e. Agriculture interrupts nutrient cycling and requires the addition of synthetic fertilizers, which upset the nitrogen cycle, can cause eutrophication, and add to the greenhouse effect, ozone depletion, and acid precipitation.
f. Acid precipitation, caused by fossil fuel combustion, is harming forests and lakes.
g. Chlorine-containing pollutants are thinning the ozone layer, allowing more harmful UV radiation to reach Earth.

ANSWERS TO TEST YOUR KNOWLEDGE

Multiple Choice:

1. e	5. d	9. d	13. e	17. a
2. c	6. e	10. a	14. b	18. d
3. c	7. b	11. e	15. a	
4. e	8. c	12. c	16. d	

CHAPTER 55: CONSERVATION BIOLOGY AND RESTORATION ECOLOGY

■ INTERACTIVE QUESTIONS

55.1 Many examples are provided in the text.
a. Coral reefs have been damaged by human activities; 40–50% of the reefs could be lost in the next few decades. About a third of all marine fish species utilize these reefs.
b. The introduction of the Nile perch to Lake Victoria has caused the loss of a huge number of fish species.
c. Commercial harvest or illegal hunting have reduced populations of whales, the African elephant, and numerous fishes.
d. The hunting of "flying fox" bats threatens the pollination of plants.

55.2 Smaller, because usually not all individuals in a population successfully breed. Thus, the MVP must refer to the number of reproductively active individuals in a population.

55.3 The loss of genetic variation within a small population due to inbreeding and genetic drift can force it into an extinction vortex in which the population grows smaller and smaller. Promoting migration between small populations to increase genetic variation is proposed as an urgent conservation need.

55.4 This proactive approach relies on early detection of population decline, identification of the species' habitat needs, testing to determine which factor is contributing to the decline, recommending corrective measures, and monitoring results.

55.5 Corridors promote dispersal between populations and may be essential to species that migrate between different habitats. However, they may also contribute to the spread of diseases.

55.6 Large preserves are required for large, far-ranging animals that require extensive habitats. They also have proportionally less border area and thus have fewer edge effects. An advantage of smaller preserves that collectively have the same area as a large one is the slower spread of disease within a population.

SUGGESTED ANSWERS TO STRUCTURE YOUR KNOWLEDGE

1. habitat destruction, introduced species, overexploitation, disruption of interaction networks

2. Biodiversity is a natural resource from which we obtain medicines, crops, fibers, and other products; many potentially valuable species will become extinct before they are known to scientists. A loss of biodiversity may disrupt ecosystem processes in harmful ways. Humans have evolved within the context of living communities and may be affected in unknown ways by changes in our ecosystem.

3. Fragmentation of habitats produces more interfaces or *edges* between different ecosystems (between forests and cleared areas, between deserts and housing developments). The species that inhabit edges are able to use both types of ecosystems. As edges proliferate, edge-adapted species may become more dominant than the species in the adjoining habitats. Strips of quality habitat that connect fragmented habitat patches may serve as *movement corridors* that promote dispersal between isolated populations and help maintain genetic variation.

ANSWERS TO TEST YOUR KNOWLEDGE

Multiple Choice:

1. e	4. b	7. e	10. d
2. b	5. d	8. a	11. b
3. c	6. a	9. c	12. c